OXFORD STUDIES IN NUCLEAR PHYSICS

General Editor: P. E. Hodgson

1. J. McL. Emmerson: *Symmetry principles in particle physics* (1972)
2. J. M. Irvine: *Heavy nuclei, superheavy nuclei, and neutron stars* (1975)
3. I. S. Towner: *A shell-model description of light nuclei* (1977)
4. P. E. Hodgson: *Nuclear heavy-ion reactions* (1978)
5. R. D. Lawson: *Theory of the nuclear shell model* (1980)
6. W. E. Frahn: *Diffractive processes in nuclear physics* (1985)
7. S. S. M. Wong: *Nuclear statistical spectroscopy* (1986)
8. N. Ullah: *Matrix ensembles in the many-nucleon problem* (1987)
9. A. N. Antonov, P. E. Hodgson, and I. Zh. Petkov: *Nucleon momentum and density distributions in nuclei* (1988)
10. D. Bonatsos: *Interacting boson models of nuclear structure* (1988)
11. H. Ejiri and M. J. A. de Voigt: *Gamma-ray and electron spectroscopy in nuclear physics* (1989)
12. B. Castel and I. S. Towner: *Modern theories of nuclear moments* (1990)
13. R. F. Casten: *Nuclear structure* (1990)
14. I. Zh. Petkov and M. V. Stoitsov: *Nuclear density function theory* (1991)
15. E. Gadioli and P. E. Hodgson: *Pre-equilibrium nuclear reactions* (1992)
16. J. D. Walecka: *Theoretical nuclear and subnuclear physics* (1995)

THEORETICAL NUCLEAR AND SUBNUCLEAR PHYSICS

THEORETICAL NUCLEAR AND SUBNUCLEAR PHYSICS

John Dirk Walecka
CEBAF
College of William and Mary

New York Oxford
OXFORD UNIVERSITY PRESS
1995

Oxford University Press

Oxford New York
Athens Auckland Bangkok Bombay
Calcutta Cape Town Dar es Salaam Delhi
Florence Hong Kong Istanbul Karachi
Kuala Lumpur Madras Madrid Melbourne
Mexico City Nairobi Paris Singapore
Taipei Tokyo Toronto

and associated companies in
Berlin Ibadan

Published by Oxford University Press, Inc.,
200 Madison Avenue, New York, New York 10016

Library of Congress Cataloging-in-Publication Data
Walecka, John Dirk, 1932–
Theoretical nuclear and subnuclear physics /
John Dirk Walecka.
p. cm. — (Oxford studies in nuclear physics; 16)
Includes bibliographical references and index.
ISBN 0-19-507214-6
1. Nuclear structure. 2. Nuclear reactions.
3. Electroweak interactions.
4. Continuous Electron Beam Accelerator Facility.
I. Title. II. Series.
QC793.3.S8W35 1995 539.7—dc20 94-13930

1 3 5 7 9 8 6 4 2
Printed in the United States of America
on acid-free paper

Dedicated to the memory of Jim Walecka
1966–1993

PREFACE

In the summer of 1986 I left Stanford University, after 26 years on the faculty, to assume the job of Scientific Director at the Continuous Electron Beam Accelerator Facility (CEBAF). Each year I gave a physics lecture series at the site. The first year was on electron scattering, then on graduate-level quantum mechanics, then advanced quantum mechanics and field theory, and finally on special topics in nuclear physics. The last three lecture series were repeated in 1990–1993. The goal of these lectures was to provide an informed base from which research contributions can be made at the forefront of the field. This book is based on the lecture series on special topics in nuclear physics.

CEBAF is the top priority new construction project for the field of nuclear physics in the United States. It is a high-energy, high-intensity, high-duty factor electron accelerator that will provide a powerful microscope for studying the spatial distribution of the charges and currents in the nucleus with finer and finer resolution. My version of CEBAF's scientific goal is the following:

> *CEBAF's scientific goal is to study the structure of the nuclear many-body system ($B \geq 1$), its quark substructure, and the nature of the strong and electroweak interactions governing the behavior of this fundamental form of matter.*

Why do we do nuclear physics? Why is nuclear physics interesting? The nucleus is a unique form of matter, consisting of many baryons (protons and neutrons) in close proximity. All of the forces of nature are present in the nucleus — strong, electromagnetic, and weak. Even gravity is important if we include neutron stars, which are nothing more than enormous nuclei held together by the gravitational attraction. The nucleus provides a unique microscopic laboratory to test the structure of the fundamental interactions. The nucleus manifests remarkable properties as a strongly interacting quantum mechanical many-body system. Furthermore, most of the mass and energy in the visible universe comes from nuclei and nuclear reactions. In addition, we now know that there are new underlying degrees of freedom in the nucleus, quarks and gluons, interacting through remarkable new forces described by *quantum chromodynamics* (QCD). The baryon itself is now a complicated nuclear many-body system. Nuclear physics is also crucial for understanding the universe, for example, the early universe, formation of the elements, supernovas, and neutron stars. Finally, to me, nuclear physics is really the study of the structure of matter.

How do we do nuclear physics? The *traditional* approach to nuclear physics starts from static two-body potentials fit to two-body scattering and bound state data. These two-body potentials are then inserted in the nonrelativistic many-body Schrödinger equation and that equation is solved in some approximation — it can be solved exactly for few-body systems using modern computing techniques. Electromagnetic and weak currents are then constructed from the properties of free nucleons and used to probe the structure of the nuclear system.

Although this traditional approach to nuclear physics has had a great many successes, it is clearly inadequate for a more detailed understanding of the nuclear system. A more appropriate set of degrees of freedom consists of the *hadrons*, the strongly interacting mesons and baryons. There are many arguments that one can give for this. For example, the long-range part of all modern two-nucleon potentials consists of the exchange of mesons including π with $(J^\pi, T) = (0^-, 1), \sigma(0^+, 0), \omega(1^-, 0)$, and $\rho(1^-, 1)$. We know that at long range the force between two nucleons comes from meson exchange. As another example, one of the significant achievements in the field of electromagnetic nuclear physics in recent years has been the unambiguous identification of exchange currents, additional currents present in the nuclear system arising from the flow of charged mesons between the nucleons in the nucleus.

One of the current goals of nuclear physics is to describe nuclear matter under extreme conditions—at high density and high pressure as is achieved in supernovas or in relativistic heavy-ion collisions, or at very high momentum transfer q^2 as will be achieved at CEBAF. It is this high momentum transfer that probes short distance scales. In any theoretical extrapolation away from the gross properties of observed terrestrial nuclei, it is important to incorporate general principles of physics such as quantum mechanics, special relativity, and microscopic causality. The only consistent theoretical framework we have for describing such a relativistic, interacting, many-body system is relativistic quantum field theory based on a local lagrangian density. It is convenient to refer to renormalizable relativistic quantum field theories of the nuclear system based on hadronic degrees of freedom as *quantum hadrodynamics* (QHD).

In recent years our colleagues in high energy particle physics have demonstrated that the hadrons are themselves composite objects made up of quarks held together by the exchange of gluons. We also now have a theory of the strong interactions binding quarks and gluons into the observed hadrons. This theory is based on an internal color symmetry $SU(3)_C$ and is known as *quantum chromodynamics* (QCD). The theory QCD has two absolutely remarkable properties. The first is *asymptotic freedom*, which roughly states that at very high momenta, or very short distances, the renormalized coupling constant for the basic processes in the theory goes to zero; as a consequence, one can do perturbation theory in this regime. The second property is *confinement*. The basic underlying degrees of freedom in the theory, quarks and gluons, do not exist as asymptotic, free, scattering states in the laboratory. They exist and interact only inside hadrons. You cannot hold a single quark, or single gluon in your hand. There are strong indications from lattice gauge theory, where

QCD is solved at a finite number of space-time points, that confinement is indeed a dynamic property of QCD arising from the nonlinear gluon couplings. Ultimately, nuclear physics is the study of strong-coupling QCD.

As for the other basic forces in nature, surely one of the great intellectual achievements of our era is the unification of the theories of electromagnetism and of the weak interactions. It is essential to continue to put this theory of the electroweak interactions to rigorous tests and fully explore its consequences. Nuclei provide unique laboratories in which to conduct such tests and explorations.

Thus there now exists a *standard model* of the strong and electroweak interactions; it is a relativistic quantum field theory based on the underlying local gauge symmetry structure $SU(3)_C \bigotimes SU(2)_W \bigotimes U(1)_W$.

The current picture of the nucleus in the standard model is that of a bound system of baryons and mesons, which are in turn confined triplets of quarks and of quark-antiquark pairs, respectively. The exact nature of confinement is still an open problem. It involves solving a strong-coupling, nonlinear, relativistic quantum field theory. One gets experimental insight into the nature of confinement by varying the environment in which the hadrons find themselves, and this is done by putting the hadrons in a nucleus. The electroweak interactions of leptons (electrons and neutrinos) with the nucleus are mediated by the photon and the heavy weak vector bosons, the Z^0 and $W^{\pm}$. The electroweak interactions couple directly to the quarks; the gluons are absolutely neutral to the electroweak interactions. Thus every time one studies a nuclear gamma decay, for example, one is directly probing the quark structure of the nucleus. Once the quark is struck, it is not a quark which is emitted from the target, but a hadron. Nuclei are the ideal laboratories for studying this process of *hadronization*.

Another truly remarkable property of QCD is that the effective degrees of freedom at low energy and long wavelengths *are* the hadrons, the baryons and mesons. Now it is self-evident that *the appropriate set of degrees of freedom depends on the distance scale at which we probe the system.* It was an observation of the Vogt Committee, the last national committee to examine the top priority given to the construction of CEBAF by the nuclear physics community, that

> *The search for new nuclear degrees of freedom and the relationship of nucleon-meson degrees of freedom to quark-gluon degrees of freedom in nuclei is one of the most challenging and fundamental questions of physics.* [Report of Vogt Subcommittee of NSAC (1985)]

With this motivation, these lectures concentrate on the following four special topics in nuclear physics:

Basic Nuclear Structure. Here the nature of the nuclear force, as manifested by the two-nucleon potential, is reviewed. The properties of nuclear matter are summarized and the independent-particle and independent-pair descriptions of nuclear matter developed. The nuclear shell model is introduced via a canonical transformation to particles and holes. The theory of electromagnetic interactions with nuclei is presented. The description of collective particle-hole excitations of nuclei is then introduced with examples.

The Relativistic Nuclear Many-Body Problem. A simple QHD model with (σ, ω) fields is first developed and solved in the mean-field approximation (MFT). Several applications are discussed. The Feynman rules are written for the full (σ, ω) theory and various classes of corrections to the MFT examined. The inclusion of pions in a QHD description of nuclei is discussed in detail. The concept of chiral invariance is developed and the σ-model, which provides a specific realization of this symmetry, and of spontaneous symmetry breaking, is analyzed. The chiral transformation and the nonlinear sigma model are also discussed, as is the dynamic generation of a low-mass σ and the $\Delta(1232)$. A QHD model with $(\pi, \sigma, \omega, \rho)$ is then presented. QCD is introduced and the Feynman rules given for QCD. The relation between QHD and QCD is discussed, and a simple model calculation of the phase diagram of nuclear matter is carried out exhibiting the transition to the quark-gluon plasma.

Strong-Coupling QCD. The nature of Yang-Mills nonabelian gauge theories is reviewed, as is the path-integral formulation of quantum mechanics and field theory. Lattice gauge theory is then developed in enough detail so that students can perform relevant numerical calculations. First, as illustration, QED in 1-space and 1-time dimension is formulated as a lattice gauge theory and solved in the MFT. The nonabelian $SU(2)$ theory is formulated on the lattice and solved analytically in both the MFT [which is actually done for $SU(n)$] and strong-coupling limits. A discussion of observables in lattice gauge theory is presented, and the basic concepts of Monte Carlo evaluation of multidimensional sums and integrals using the Metropolis algorithm developed. Finally, fermions are included. There is then a discussion of QCD-inspired models of hadrons, including the M.I.T. bag model, Skyrme's model, and other extensions. The topic of deep-inelastic scattering, an important limit of QCD for nuclear physics, is discussed, and the evolution equations for the structure functions, which lead to the renormalization-group improved perturbation theory results, are presented. Finally, the EMC effect is described.

Electroweak Interactions with Nuclei. The basic phenomenology of the weak interactions is reviewed and the standard model of the electroweak interactions developed. Semileptonic weak interactions with nuclei, β-decay, μ-capture, ν-reactions, ν-scattering, are discussed in enough detail for the reader, in principle, to proceed to calculate any semileptonic weak-interaction process involving the nucleus. Various applications of electroweak interactions with nuclei are summarized. The Feynman rules for radiative corrections within the standard model of electroweak interactions are presented, and an application discussed, as is the mixing matrix in the quark sector with three generations of quarks. Finally a general expression for parity violation in $A(\vec{e}, e')A^*$ is developed as an application.

CEBAF's role in all four topics is discussed.

In my opinion, CEBAF's nuclear physics goals are to explore the following:

1. Limits of a single-baryon description;
2. Role of subnucleonic, hadronic degrees of freedom;
3. Limits of a hadronic description;
4. Transition from a baryon-meson to a quark-gluon description;
5. Limits of quark models;
6. Experimental implications of QCD;
7. Experimental implications of the standard model;
8. Weak neutral currents in nuclei;
9. At all levels, the *phenomena* manifest by this remarkable, strongly coupled, quantum-mechanical, many-body system.

The solution to problems forms an integral part of learning, and the author has tried to include interesting exercises of varying degrees of difficulty. There are over 260 problems, some at the end of each chapter. Answers are given in most of the problems so that they, too, can serve as a source of reference. There are, in addition, 14 appendixes containing many details inappropriate for direct inclusion in the text.

A comprehensive bibliography has not been attempted since it is impossible to include all developments in nuclear physics within a single text. While for the most part the references quoted are only those directly relevant to the discussion, there has been some attempt to steer the reader to recent comprehensive monographs or texts which provide a basis for further study.

It is expected that the reader has a working knowledge of graduate-level quantum mechanics, classical continuum mechanics, nonrelativistic many-body theory, and advanced quantum mechanics, although every attempt is made to be complete at the more advanced level. A basic introductory course in the qualitative phenomena of nuclear and particle physics is also assumed, as is a good general physics background.

The author is extremely grateful to his colleagues Siu Chin, John Dubach, Dick Furnstahl, and Brian Serot for their valuable comments on the manuscript. He would also like to thank the students in Physics 641-42 at the College of William and Mary, in particular Sergei Ananyan, for their assistance in transferring this material to the classroom.

This manuscript was typed by the author in LaTeX, from which the book is printed. Figures are reproduced by permission.

Newport News, Virginia
July 21, 1994

John Dirk Walecka
Scientific Director CEBAF, 1986-1992
Professor of Physics
College of William and Mary
Senior Fellow CEBAF, 1992-present

CONTENTS

Part IV: ELECTROWEAK INTERACTIONS WITH NUCLEI

Part I
BASIC NUCLEAR STRUCTURE

1

NUCLEAR FORCES: A REVIEW

The motivation and goals for this book have been discussed in detail in the preface. Part I of the book is on *Basic Nuclear Structure*, where Refs. [N1, N2, N3, N4, N5, N6, N7] provide good backgound texts.[1] This first section is concerned with the essential properties of the nuclear force as described by phenomenological two-nucleon potentials. The discussion summarizes many years of extensive experimental and theoretical effort; it is meant to be a brief *review* and *summary*. It is assumed that the concepts, symbols, and manipulations in this first section are familiar to the reader.

Attractive. That the strong nuclear force is basically attractive is demonstrated in many ways: a bound state of two nucleons, the deuteron, exists in the spin triplet state with $(J^\pi, T) = (1^+, 0)$; interference with the known Coulomb interaction in p-p scattering demonstrates that the force is also attractive in the spin singlet ${}^1S_0$ state; and, after all, atomic nuclei are self-bound systems.

Short-Range. Nucleon-nucleon scattering is observed to be isotropic, or s-wave with $l = 0$, up to $\approx$ 10 MeV in the center-of-mass (C-M) system. The reduced mass is $1/\mu_{\rm red} = 1/m + 1/m = 2/m$. This allows one to make a simple estimate of the range of the nuclear force through the relations

$$\begin{aligned} \hbar l_{\rm max} &= rp \\ l_{\rm max} &= r\sqrt{\frac{2\mu_{\rm red}E}{\hbar^2}} \\ l_{\rm max} &\approx r({\rm Fermis})\sqrt{\frac{E}{40}\,{\rm MeV}} \end{aligned} \tag{1.1}$$

Here we have used the numerical relations (worth remembering)

$$\begin{aligned} 1\,{\rm Fermi} &\equiv 1\,{\rm fm} \\ &\equiv 10^{-13}{\rm cm} \\ \frac{\hbar^2}{2m_p} &\approx 20.7\,{\rm MeV\ fm}^2 \end{aligned} \tag{1.2}$$

[1] These books, in particular Ref. [N1], provide an extensive set of references to the original literature. It is impossible to include all the developments in nuclear structure in this part of the book. The references quoted in the text are only those directly relevant to the discussion.

A combination of these results indicates that the range of the nuclear force is

$$r \approx \text{few Fermis} \tag{1.3}$$

Spin-Dependent. The neutron-proton cross section σ_{np} is much too large at low energy to come from any reasonable potential fit to the properties of the deuteron alone

$$\begin{aligned} \sigma_{np} &= \frac{3}{4}(^3\sigma) + \frac{1}{4}(^1\sigma) \\ \sigma_{np} &= 20.4 \times 10^{-24}\text{cm}^2 \\ &\equiv 20.4 \text{ barns} \end{aligned} \tag{1.4}$$

At low energies, it is a result of effective range theory that the scattering measures only two parameters

$$k \cot \delta_0 = -\frac{1}{a} + \frac{1}{2} r_0 k^2 \tag{1.5}$$

where a is the scattering length and r_0 is the effective range. The best current values for these quantities for np in the spin singlet and triplet states are (Ref. [N1])

$$\begin{aligned} {}^1a &= -23.714 \pm 0.013 \text{ fm} & {}^3a &= 5.425 \pm 0.0014 \text{ fm} \\ {}^1r_0 &= 2.73 \pm 0.03 \text{ fm} & {}^3r_0 &= 1.749 \pm 0.008 \text{ fm} \end{aligned} \tag{1.6}$$

The singlet state just fails to have a bound state ($a = -\infty$), while the triplet state has just one, the deuteron, bound by 2.225 MeV.

Noncentral. The fact that the deuteron has a nonvanishing quadrupole moment indicates that there must be some $l = 2$ mixed into the $l = 0$ ground state. Therefore the two-nucleon potential cannot be invariant under spatial rotations alone. The most general velocity-independent potential that is invariant under overall rotations and reflections is

$$\begin{aligned} V &= V_0(r) + \boldsymbol{\sigma}_1 \cdot \boldsymbol{\sigma}_2 V_1(r) + S_{12} V_T(r) \\ S_{12} &\equiv \frac{3(\boldsymbol{\sigma}_1 \cdot \boldsymbol{r})(\boldsymbol{\sigma}_2 \cdot \boldsymbol{r})}{r^2} - \boldsymbol{\sigma}_1 \cdot \boldsymbol{\sigma}_2 \end{aligned} \tag{1.7}$$

The term $S_{12}V_T(r)$ gives rise to the tensor force. Several properties are of interest here:

1. Since

$$\begin{aligned} \mathbf{S} &= \frac{1}{2}(\boldsymbol{\sigma}_1 + \boldsymbol{\sigma}_2) \\ 4\mathbf{S}^2 &= 4S(S+1) = 6 + 2\boldsymbol{\sigma}_1 \cdot \boldsymbol{\sigma}_2 \end{aligned} \tag{1.8}$$

It follows that

Table 1.1: States of the two-nucleon system.

States	$^1S_0$	$^1P_1$	$^1D_2$	$^3S_1 + ^3D_1$	$^3P_0$	$^3P_1$	$^3P_2 + ^3F_2$	$^3D_2$
Parity	+	−	+	+	−	−	−	+
Particle exchange	−	+	−	+	−	−	−	+
Particles	nn np pp	np	nn np pp	np[a]	nn np pp	nn np pp	nn np pp	np

[a] The deuteron.

$$\begin{aligned} \boldsymbol{\sigma}_1 \cdot \boldsymbol{\sigma}_2 &= -3\,; && \text{singlet } (S = 0) \\ &= +1\,; && \text{triplet } (S = 1) \end{aligned} \tag{1.9}$$

2. The total spin S is a good quantum number for the two-nucleon system if the hamiltonian H is symmetric under interchange of particle spins [as in Eq. (1.7)], for then the wave function must be either symmetric ($S = 1$) or antisymmetric ($S = 0$) under this symmetry.[2]

3. Higher powers of the spin operators can be reduced to the form in Eq. (1.7) for spin-1/2 particles.

4. Since the total spin operator annihilates the singlet state, $(\boldsymbol{\sigma}_1 + \boldsymbol{\sigma}_2)^1\chi = 0$, so does the tensor operator S_{12}

$$S_{12}[^1\chi] = 0 \tag{1.10}$$

Charge Independent. Charge independence states that the force between any two nucleons is the same $V_{pp} = V_{pn} = V_{nn}$ in the same state. The Pauli principle limits the states that are available to two identical nucleons. For two spin-1/2 nucleons, a complete basis can be characterized by eight quantum numbers, for clearly the states $|\mathbf{p}_1, s_1; \mathbf{p}_2, s_2\rangle$ form such a basis. Alternatively, one can take as the good quantum numbers $|E, J, M_J, S, \pi, \mathbf{P}_{\text{CM}}\rangle$. Table 1.1 lists the first few states available to the two-nucleon system. The Pauli principle states that nn and pp must go into an overall antisymmetric state.[3] Charge independence states that the forces are equal in those states where one can have all three types of particles including np; the nuclear force is independent of the charge in these states. At low energy, the cross sections are given in terms of the singlet

[2] If P_σ is the spin exchange operator then $P_\sigma[^1\chi(1,2)] \equiv {}^1\chi(2,1) = -{}^1\chi(1,2)$ is odd and, similarly, $P_\sigma[^3\chi(1,2)] = +^3\chi(1,2)$ is even. Thus from Eqs. (1.9) $P_\sigma = (1 + \vec{\sigma}_1 \cdot \vec{\sigma}_2)/2$.

[3] In terms of isospin we assign $T = 0$ to the states that are even under particle interchange and $T = 1$ to those that are odd, so that the overall wave function is antisymmetric.

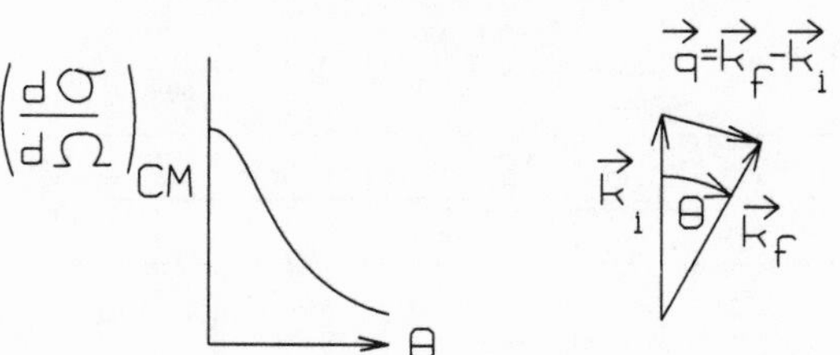

Figure 1.1: Sketch of cross section in Born approximation.

and triplet amplitudes by

$$\begin{aligned}\left(\frac{d\sigma}{d\Omega}\right)_{np} &= \frac{1}{4}|f(^1S_0)|^2 + \frac{3}{4}|f(^3S_1)|^2 \\ \left(\frac{d\sigma}{d\Omega}\right)_{nn} &= \frac{1}{4} \times 4|f(^1S_0)|^2 = |f(^1S_0)|^2 \end{aligned} \tag{1.11}$$

Exchange Character. At higher energies more partial waves contribute to the cross section. At high enough energies, one can use the Born approximation

$$\left(\frac{d\sigma}{d\Omega}\right)_{\text{CM}} = \left|\frac{2\mu_{\text{red}}}{4\pi\hbar^2}\int e^{-i\mathbf{q}\cdot\mathbf{r}}V(r)d^3r\right|^2 \tag{1.12}$$

where the momentum transfer $\mathbf{q}$ is defined in Fig. 1.1. For large $\mathbf{q}$ the integrand oscillates rapidly and the integral goes to zero as sketched in Fig. 1.1. The experimental results for np scattering are shown in Fig. 1.2. There is significant backscattering, in fact, the cross section is approximately symmetric about 90°. If $f(\pi-\theta) = f(\theta)$ then only even l partial waves contribute to the cross section; the odd l's will distort $d\sigma/d\Omega$.

To describe this situation one introduces the concept of an exchange force — a force that depends on the symmetry of the wave function. The interaction is written $V(r)P_{\text{M}}$ where the Majorana space exchange operator is defined by[4]

$$P_{\text{M}}\phi(\mathbf{r_2},\mathbf{r_1}) \equiv \phi(\mathbf{r_1},\mathbf{r_2}) \tag{1.13}$$

Hence since $\mathbf{r} = \mathbf{r}_2 - \mathbf{r}_1$

$$\begin{aligned} P_{\text{M}}\phi(\mathbf{r}) &= \phi(-\mathbf{r}) \\ P_{\text{M}}Y_{lm}\left(\frac{\mathbf{r}}{|\mathbf{r}|}\right) &= (-1)^l Y_{lm}\left(\frac{\mathbf{r}}{|\mathbf{r}|}\right) \end{aligned} \tag{1.14}$$

The odd l in the amplitude can evidently be eliminated with a Serber force defined by

$$V \equiv V(r)\frac{1}{2}(1+P_{\text{M}}) \tag{1.15}$$

[4]Since the overall wave function is antisymmetric $P_{\text{M}}P_\sigma P_\tau = -1$ (Note $P_\sigma^2 = P_\tau^2 = +1$). Thus $P_{\text{M}} = -P_\sigma P_\tau = -(1+\vec{\sigma}_1\cdot\vec{\sigma}_2)(1+\vec{\tau}_1\cdot\vec{\tau}_2)/4$ provides an alternate definition.

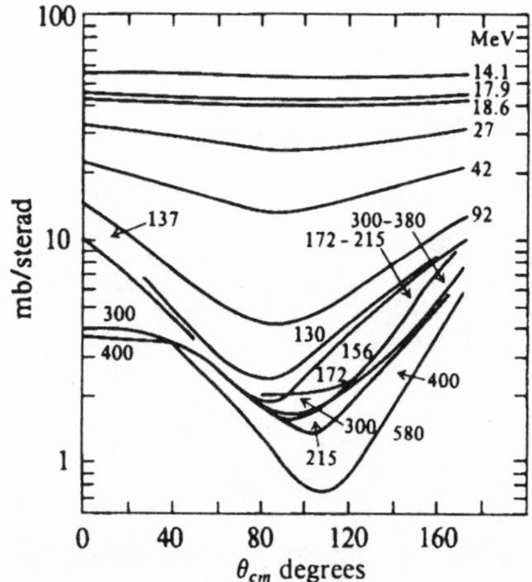

Figure 1.2: The n–p differential cross section in C-M system as a function of laboratory energy. From Ref. [N1].

The differential cross section in Born approximation with this interaction is

$$\begin{aligned} \left(\frac{d\sigma}{d\Omega}\right)_{\text{CM}} &= \left|\frac{2\mu_{\text{red}}}{4\pi\hbar^2}\int e^{-i\mathbf{k}_f\cdot\mathbf{r}}V(r)\frac{1}{2}(1+P_{\text{M}})e^{i\mathbf{k}_i\cdot\mathbf{r}}d^3r\right|^2 \\ &= \left|\frac{2\mu_{\text{red}}}{4\pi\hbar^2}\int e^{-i\mathbf{k}_f\cdot\mathbf{r}}V(r)\frac{1}{2}(e^{i\mathbf{k}_i\cdot\mathbf{r}}+e^{-i\mathbf{k}_i\cdot\mathbf{r}})d^3r\right|^2 \qquad (1.16) \end{aligned}$$

This result is sketched in Fig. 1.3. The nuclear force has roughly a Serber exchange nature; it is very weak in the odd-l states.

Hard Core. The pp cross section is illustrated in Fig. 1.4. Recall that since the particles are here identical, one necessarily has the relation $[d\sigma(\pi - \theta)/d\Omega]_{\text{CM}} = [d\sigma(\theta)/d\Omega]_{\text{CM}}$. Although the cross sections shown in Figs. 1.2 and 1.4 are very different, it is possible to make a charge-independent analysis of np and pp scattering as first shown in detail by Breit and coworkers (Ref. [N8]). The overall magnitude of the pp cross section indicates that more than s-wave

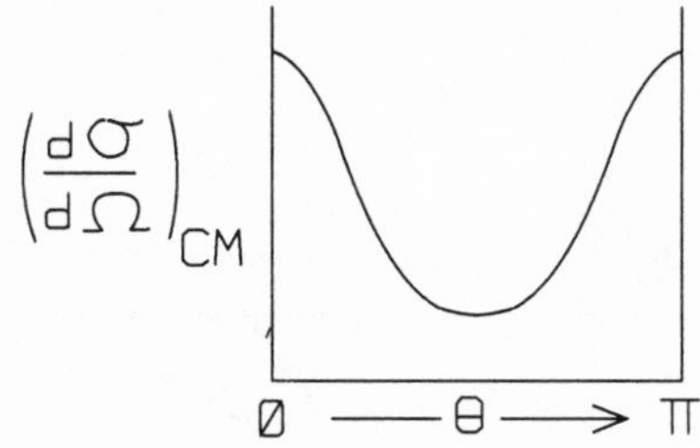

Figure 1.3: Sketch of cross section in Born approximation with a Serber force.

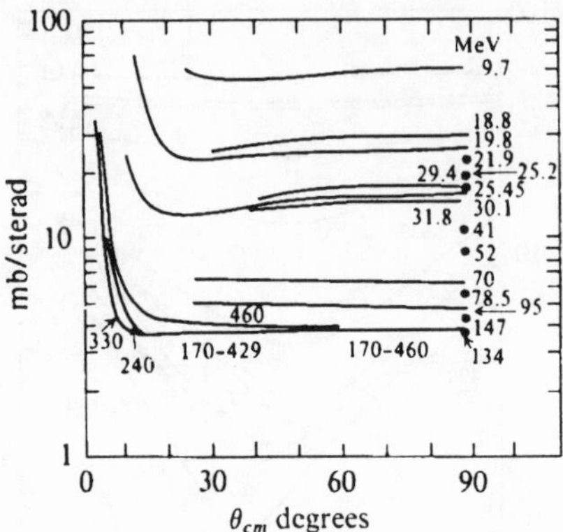

Figure 1.4: Same as Fig. 1.2 for p–p scattering. From Ref. [N1].

nuclear scattering must be included (recall the unitarity bound of π/k^2), and the higher partial waves must interfere so as to give the observed flat angular distribution beyond the Coulomb peak. A hard core will change the sign of the s-wave phase shifts at high energy and allow the ${}^1S - {}^1D$ interference term in pp scattering to yield a uniform angular distribution as first demonstrated by Jastrow (Ref. [N9]); with a Serber force, it is only the states (${}^1S_0$, ${}^1D_2$) in Table 1.1 that contribute to nuclear pp scattering. Recall that for a pure hard core potential the s-wave phase shift is negative $\delta_0 = -ka$ as illustrated in Fig. 1.5. With a finite attractive well outside of the hard core, one again expects to see the negative phase shift arising from the hard core at high enough energy. The experimental situation for the s-wave phase shifts in both pp and np scattering is sketched in Fig. 1.6. From an analysis of the data, one concludes that there is a hard core of radius

$$r_c \approx 0.4 \text{ to } 0.5 \text{ fm} \tag{1.17}$$

in the relative coordinate in the nucleon-nucleon interaction.

Spin-Orbit Force. It is difficult to explain the large nucleon polarizations observed perpendicular to the plane of scattering with just the central and

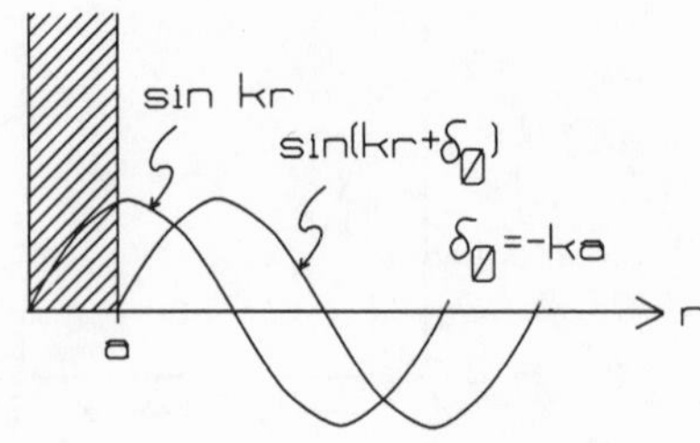

Figure 1.5: The s-wave phase shift for scattering from a hard-core potential.

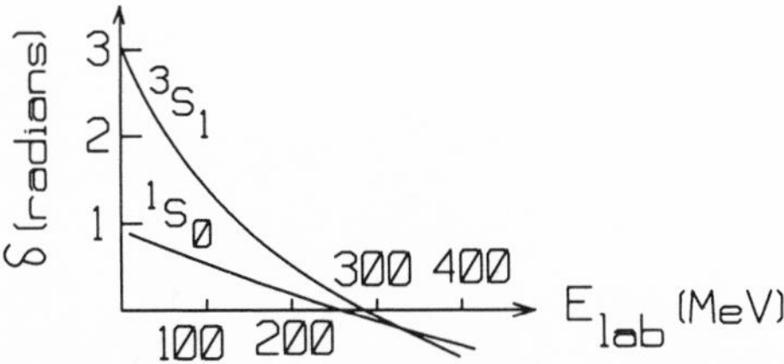

Figure 1.6: Sketch of s-wave nucleon-nucleon phase shifts. After Ref. [N1].

tensor forces discussed above. To explain the data one must also include a spin-orbit potential of the form

$$\begin{aligned} V &= -V_{SO}\mathbf{L}\cdot\mathbf{S} \\ \mathbf{L}\cdot\mathbf{S} &= \frac{1}{2}[J(J+1)-l(l+1)-S(S+1)] \end{aligned} \tag{1.18}$$

This last expression vanishes if either $S = 0$ $(l = J)$ or $l = 0$ $(S = J)$. The spin-orbit force vanishes in s-states and is empirically observed to have a short range; thus it is only effective at higher energies.

In *summary*, the present situation with respect to our phenomenological knowledge of the nucleon-nucleon force is the following:

- The experimental scattering data can be fit up to laboratory energies of $\approx$ 300 MeV with a set of potentials depending on spins and parities ${}^1V_C^+$, ${}^3V_C^+$, ${}^1V_C^-$, ${}^3V_C^-$, ${}^3V_T^+$, ${}^3V_T^-$, etc.
- The potentials contain a hard core[5] with $r_c \approx 0.4$ to 0.5 fm.
- The forces in the odd-l states are relatively weak at low energies, and on the average slightly repulsive.
- The tensor force is necessary to understand the quadrupole moment of the deuteron (and its binding).
- A strong, short-range, spin-orbit force is necessary to explain the polarization at high energy.

Commonly used nucleon-nucleon potentials include the "Bonn potential" in Ref. [N6], the "Paris potential" in Ref. [N10], and the "Reid potential" in Ref. [N11]. The first two contain the one-meson (boson) exchange potentials (OBEP) at large distances.

Meson Theory of Nuclear Forces. The exchange of a neutral scalar meson of Compton wavelength $1/m \equiv \hbar/mc$ (Fig. 1.7) in the limit of infinitely heavy sources gives rise to the celebrated Yukawa potential (Ref. [N12])

[5] Or, more generally, a strong, short-range repulsion.

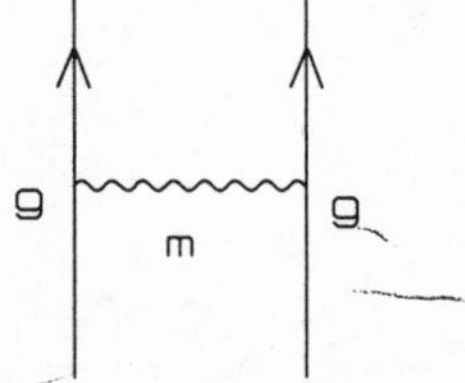

Figure 1.7: Contribution of neutral scalar meson exchange to the N–N interaction.

$$V(r) = -\frac{g^2}{4\pi c^2}\frac{e^{-mr}}{r} \tag{1.19}$$

A derivation of this result, as well as the potentials arising from other types of meson exchange, is given in Appendix A. In charge-independent pseudoscalar meson theory with a nonrelativistic coupling of $\boldsymbol{\tau}(\boldsymbol{\sigma}\cdot\nabla)$ at each vertex, one obtains a tensor force of the correct sign in the N–N interaction. In fact, for this reason, Pauli (Ref. [N13]) claimed there had to be a long-range pseudoscalar meson exchange before the π-meson was discovered. Since the π is the lightest known meson, the 1–π exchange potential is exact at large distances $r \to \infty$; mesons with higher mass $\bar{m}$ give a potential that goes as $e^{-\bar{m}r}/r$ by the uncertainty principle. The existence of this 1–π exchange tail in the N–N interaction has by now been verified experimentally in many ways.

The Paris and Bonn potentials in Refs. [N10, N6] include the exchange of $(\boldsymbol{\pi}, \sigma, \boldsymbol{\rho}, \omega)$ mesons with spin and isospin $(J^\pi, T) = (0^-, 1), (0^+, 0), (1^-, 1), (1^-, 0)$, respectively, in the long-range part of the N–N potential. The short-distance behavior of the interaction is then parameterized.

One can get a qualitative understanding of the short-range repulsion and spin-orbit force in the strong N–N interaction by considering meson exchange and using the analogy with quantum electrodynamics (QED). Suppose one couples a neutral vector meson field, the ω, to the conserved baryon current. Then just as with the Coulomb interaction in atomic physics, which is described by the coupling of a neutral vector meson field (the photon) to the conserved electromagnetic current:

1. Like baryonic charges repel;
2. Unlike baryonic charges (e.g., p–$\bar{p}$) attract;
3. There will be a spin-orbit force;
4. While the range of the Coulomb potential $1/r$ is infinite because the mass of the photon vanishes $m_\gamma = 0$, the range of the strong nuclear effects will be $\sim \hbar/m_\omega c$. Since the ω has a large mass, the force will be shortrange.

2

NUCLEAR MATTER

Nuclear Radii and Charge Distributions. The best information we have about nuclear charge distributions comes from electron scattering, where one uses short-wavelength electrons to explore the structure (Ref. [N14]). In the work of Hofstadter and colleagues at Stanford (Ref. [N15]) a phaseshift analysis was made of elastic electron scattering from an arbitrary charge distribution through the Coulomb interaction. The best fit to the data, on the average, was found with the following shape

$$\rho = \frac{\rho_0}{1 + e^{(r-R)/a}} \tag{2.1}$$

This is illustrated in Fig. 2.1. Several features of the empirical results are worthy of note:

1. $(A/Z)\rho_0$, the central nuclear density, is observed to be *constant* from nucleus to nucleus.

2. The radius to 1/2 the maximum ρ is observed to vary with nucleon number A according to[6]

$$R = r_0 A^{1/3} \tag{2.2}$$

where the half-density radius parameter r_0 is given by

$$r_0 \approx 1.07\,\text{fm} \tag{2.3}$$

We assume that the neutron density tracks the proton density and that the neutrons are confined to the same nuclear volume.[7] This means that the nuclear density A/V is given by

$$\begin{aligned} \frac{A}{V} &= \frac{3}{4\pi r_0^3} \\ &\approx 1.95 \times 10^{38}\ \text{particles/cm}^3 \end{aligned} \tag{2.4}$$

One thus concludes *nuclear matter has a constant density* from nucleus to nucleus.

[6]The nucleon number A is identical to the baryon number B, which, to the best of our current experimental knowledge, is an exactly conserved quantity; the notation will be used interchangeably throughout this book.

[7]This assumption is verified quite well in nucleon-nucleus scattering.

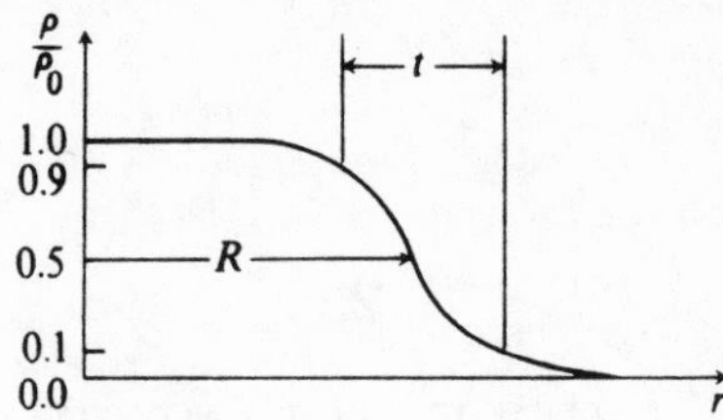

Figure 2.1: Two parameter fit to nuclear charge distribution in Eq. (2.1).

3. It is an experimental fact that for a proton (Ref. [N16])

$$\langle r^2 \rangle_p^{1/2} \approx 0.77 \text{ fm} \tag{2.5}$$

If the nucleus is divided into cubical boxes so that $A/V \equiv 1/l^3$ then Eq. (2.4) implies that $l = 1.72$ fm. Is $l \gg 2r_p$ so that the nucleus can, to a first approximation, be treated as a collection of undistorted nonrelativistic nucleons interacting through static two-body potentials? We will certainly start our description of nuclear physics working under this assumption! It is evident, however, that these dimensions are very close, and the internal structure of the nucleon will have to be taken into account as we progress.

4. The surface thickness of the nucleus, defined to be the distance over which the density in Eq. (2.1) falls from $0.9\,\rho_0$ to $0.1\,\rho_0$ is given by

$$t \approx 2.4 \text{ fm} \tag{2.6}$$

for nuclei from Mg to Pb. (Note $a \approx t/2\ln 9$.)

5. The mean square radius of a sphere of radius $R_{\rm C}$ of uniform charge density ρ (Fig. 2.2) is given by

$$\begin{aligned} \langle r^2 \rangle &= \frac{\int_0^{R_{\rm C}} 4\pi r^2 dr \cdot r^2}{\int_0^{R_{\rm C}} 4\pi r^2 dr} \\ &= \frac{3}{5} R_{\rm C}^2 \end{aligned} \tag{2.7}$$

One can thus also define an equivalent uniform density parameter $r_{0\rm C}$ through

$$R_{\rm C} = r_{0\rm C} A^{1/3} \tag{2.8}$$

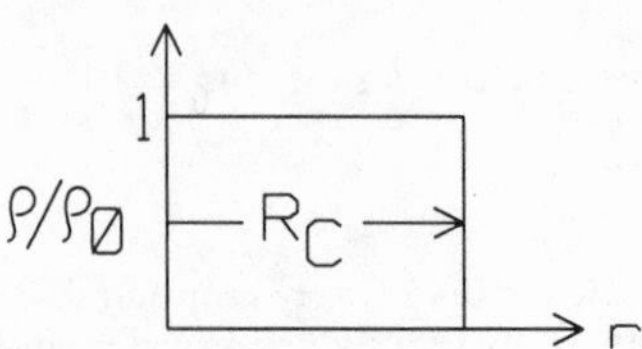

Figure 2.2: Equivalent uniform charge density.

The values of r_{0C} for two typical nuclei are

$$\begin{array}{cc} r_{0C} & \text{Nucleus} \\ 1.32\text{ fm} & {}^{40}_{20}\text{Ca} \\ 1.20\text{ fm} & {}^{209}_{83}\text{Bi} \end{array} \tag{2.9}$$

6. It is important to remember that it is the nuclear charge distribution that is measured in electron scattering;[8] the nuclear force range may extend beyond this.

The Semiempirical Mass Formula (Ref. [N17]). A useful expression for the average energy of nuclei in their ground states, or nuclear masses, can be obtained by picturing the nucleus as a liquid drop. With twice as much liquid, there will be twice the energy of condensation, or binding energy. A first term in the energy will thus represent this *bulk property of nuclear matter*

$$E_1 = -a_1 A \tag{2.10}$$

The nucleons at the surface are only attracted by the nucleons inside. This gives rise to a surface tension and *surface energy* which decreases the binding

$$\begin{aligned} E_2 &= \sigma(\text{surface tension}) \times (\text{area}) \\ &= 4\pi\sigma R^2 \\ &= (4\pi r_0^2\sigma)A^{2/3} \\ &\equiv a_2 A^{2/3} \end{aligned} \tag{2.11}$$

To this will be added the *Coulomb interaction* of Z protons, assumed uniformly distributed over the nucleus[9]

$$\begin{aligned} E_3 &= \frac{3}{5}\frac{Z(Z-1)}{4\pi R_C}e^2 \\ &= \frac{3}{5}\frac{e^2}{4\pi r_{0C}}\frac{Z(Z-1)}{A^{1/3}} \\ &\approx a_3\frac{Z^2}{A^{1/3}} \end{aligned} \tag{2.12}$$

To proceed further, some specifically nuclear effects must be included:

1. It is noted empirically that existing nuclei prefer to have $N = Z$. A *symmetry energy* will be added to take this into account. If one has twice as many particles with the same N/Z, one will have twice the symmetry energy as a consequence of the bulk property of nuclear matter. With a parabolic

[8] For more recent measurements of nuclear charge distributions see Ref. [N72].

[9] In this book we use rationalized c.g.s. (Heaviside-Lorentz) units such that the fine structure constant is $\alpha = e^2/4\pi\hbar c \approx (137.0)^{-1}$.

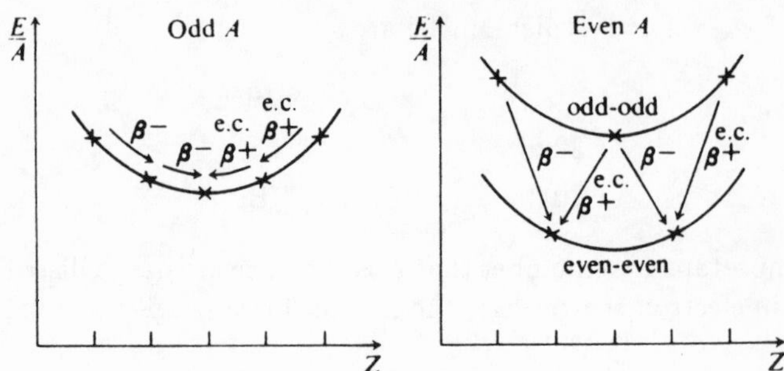

Figure 2.3: Nuclear energy surfaces for odd A, and even A nuclei.

approximation (C is just a constant), the symmetry energy takes the form

$$\begin{aligned} E_4 &\approx C\left(\frac{1}{2} - \frac{Z}{N+Z}\right)^2 A \\ &= \frac{C}{4A^2}(A-2Z)^2 A \\ &\equiv a_4 \frac{(A-2Z)^2}{A} \end{aligned} \tag{2.13}$$

2. It is also noted experimentally that nuclei prefer to have even numbers of the same kinds of particles, protons or neutrons. For example

- There are only 4 stable odd-odd nuclei: ${}^2_1\text{H}$, ${}^6_3\text{Li}$, ${}^{10}_5\text{B}$, ${}^{14}_7\text{N}$.
- There is only 1 stable odd A nuclear isobar.
- For even A there may be 2 or more stable nuclei with even Z and even N.

The schematic representation of the nuclear energy surfaces for these different cases is shown in Fig. 2.3, along with the possible β-decay transitions. It is a general rule, which follows from energetics, that of two nuclei with the same A, and with Z differing by 1, at least one is β-unstable. The bottom two even-even nuclei in Fig. 2.3 can only get to each other by double β-decay, which is extremely rare. To represent these observations, a *pairing energy* will be included in the nuclear energy

$$E_5 = \lambda \frac{a_5}{A^{3/4}} \tag{2.14}$$

Here λ=+1 for odd-odd, 0 for odd-even, and -1 for even-even nuclei, representing the contributions of 1 or 2 extra pairs of identical nucleons. The $A^{-3/4}$ dependence is empirical.

A combination of these terms leads to the Weizsäcker semiempirical mass formula (Ref. [N17])

$$E = -a_1 A + a_2 A^{2/3} + a_3 \frac{Z^2}{A^{1/3}} + a_4 \frac{(A-2Z)^2}{A} + \lambda \frac{a_5}{A^{3/4}} \tag{2.15}$$

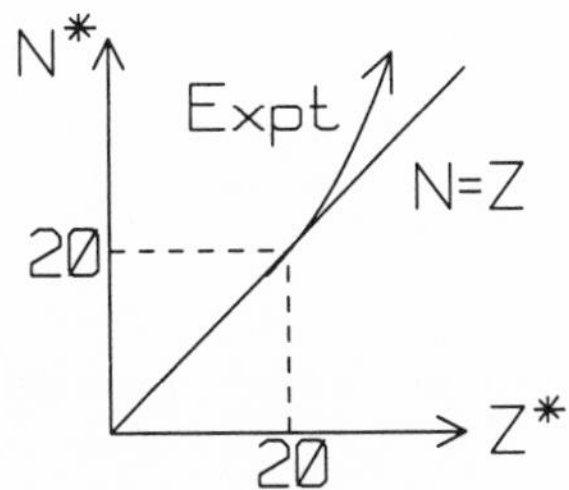

Figure 2.4: Equilibrium neutron number $N^* = A - Z^*$ vs Z^* compared with $N = Z = A/2$.

One empirical determination of the parameters appearing in this expression is given by (Ref. [N18])[10]

$$\begin{aligned} a_1 &= 15.75 \text{ MeV} & a_2 &= 17.8 \text{ MeV} \\ a_3 &= 0.710 \text{ MeV} & a_4 &= 23.7 \text{ MeV} \\ a_5 &= 34 \text{ MeV} \end{aligned} \tag{2.16}$$

The empirical value of a_3 implies $r_{0C} = 1.22$ fm, in remarkably good agreement with the values in Eq. (2.9).

There are only two terms in Eq. (2.15) that depend on Z. The stable value Z^* is found by minimization at fixed A; for odd A nuclei $\lambda = 0$, and there is a single Z^*. The condition $dE/dZ|_A = 0$ yields the value

$$Z^* = \frac{A}{2 + a_3 A^{2/3}/2a_4} \tag{2.17}$$

For light nuclei, say up to $A = 40$, this gives $Z^* \approx A/2$. The result in Eq. (2.17) is sketched in Fig. 2.4 in terms of the equilibrium neutron number $N^* = A - Z^*$ vs Z^* and compared with $N = Z = A/2$.

The mass formula and value of Z^* here are mean values, and the average fit to nuclear masses is excellent.[11] The semiempirical mass formula is of great utility, for example, in discussing nuclear fission, permitting one to tell when the energy of two nuclear fragments is less than the energy of an initial (perhaps excited) nucleus, and hence when there will be an energy release in the fission process. This semiempirical approach can also be extended to take into account fluctuations of masses about the mean values. Such fluctuations can arise, for example, from shell structure, which is discussed later in this part of the book.

Nuclear Matter. We are now in a position to define a substance called *nuclear matter.*

1. Let $A \to \infty$ so that surface properties are negligible with respect to bulk properties; set $N = Z$ so that the symmetry energy vanishes; and then turn

[10] A very useful number to remember is $\hbar c = 197.3$ MeV fm; also $c = 2.998 \times 10^{10}$ cm/sec.
[11] For a more recent fit see Ref. [N73].

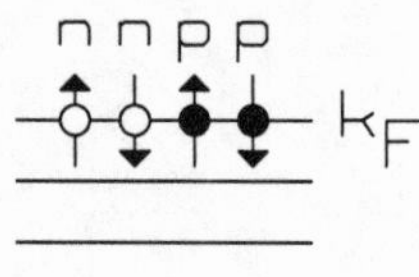

Figure 2.5: Nuclear matter as a degenerate Fermi gas

off the electric charge so that there is no Coulomb interaction. The resulting extended, uniform material is known as nuclear matter. It evidently has a *binding energy/nucleon* of

$$\frac{E}{A} \approx -15.7\,\mathrm{MeV} \tag{2.18}$$

That this expression is a constant independent of A is known as the *saturation of nuclear forces.*

2. Picture nuclear matter as a degenerate Fermi gas (Fig. 2.5). The degeneracy factor is 4 corresponding to neutrons and protons with spin up and spin down. The total number of occupied levels is A. Thus

$$A = \frac{4V}{(2\pi)^3} \int_0^{k_\mathrm{F}} d^3k \tag{2.19}$$

This yields

$$\frac{A}{V} = \frac{2}{3\pi^2} k_\mathrm{F}^3 \tag{2.20}$$

or, with the aid of Eq. (2.4)

$$k_\mathrm{F} r_0 = \left(\frac{9\pi}{8}\right)^{1/3} = 1.52 \tag{2.21}$$

The insertion of Eq. (2.3) then gives

$$k_\mathrm{F} \approx 1.42 \times 10^{13}\,\mathrm{cm}^{-1} = 1.42\,\mathrm{fm}^{-1} \tag{2.22}$$

This Fermi wave number provides a convenient parametrization of the *density* of nuclear matter.[12]

The theoretical challenge is to understand the bulk properties of nuclear matter in Eqs. (2.18) and (2.22) in terms of the strong nuclear force. In Part II of this book we discuss nuclear matter within the framework of relativistic hadronic field theories. In this first Part I, we work within the context of phenomenological two-body potentials and the nonrelativistic many-particle Schrödinger equation. We start the discussion with a simple independent-particle, Fermi-gas model of the extended uniform system of nuclear matter.

[12]Fits to interior densities of the heaviest known nuclei yield somewhat lower values $k_\mathrm{F} \approx 1.36 \pm 0.06$ fm (see Ref. [R1]).

3

THE INDEPENDENT-PARTICLE FERMI-GAS MODEL

The goal of this section is to develop an initial description of a large uniform sample of nuclear matter with A nucleons in a cubical volume V where one takes $A, V \to \infty$ at fixed baryon density $A/V = \rho_{\mathrm{B}} = 2k_{\mathrm{F}}^3/3\pi^2$. Periodic boundary conditions will be applied. The single-particle eigenfunctions are plane waves

$$\phi_{\mathbf{k}}(\mathbf{x}) = \frac{1}{\sqrt{V}} e^{i\mathbf{k}\cdot\mathbf{x}} \tag{3.1}$$

The form of these coordinate space wave functions follows from translation invariance. They are solutions to the Hartree-Fock equations, and since the Hartree-Fock equations follow from a variational principle, they form the *best* single-particle wave functions. This is the appeal of nuclear matter; the starting single-particle wave functions are known and simple.

Nuclear matter is composed of both protons and neutrons with spin up and spin down. We know from charge independence that protons and neutrons look like identical particles as far as the nuclear force is concerned. We will treat them as just two different charge states of the same particle, a *nucleon*.

Isotopic Spin. The nucleons will be given an additional internal degree of freedom that takes two values and distinguishes protons and neutrons. In strict analogy with ordinary spin 1/2 one introduces

$$\eta_p = \begin{pmatrix} 1 \\ 0 \end{pmatrix} \qquad \eta_n = \begin{pmatrix} 0 \\ 1 \end{pmatrix} \tag{3.2}$$

The operators in this simple two-dimensional space are $1, \boldsymbol{\tau}$, which form a complete set of 2×2 matrices. If the spin and isospin dependence is included, the single-particle wave functions then take the form

$$\psi_{\mathbf{k},\lambda,\rho} = \phi_{\mathbf{k}} \chi_\lambda \eta_\rho \tag{3.3}$$

Here χ_λ is the spin wave function with $\chi_\uparrow = \begin{pmatrix} 1 \\ 0 \end{pmatrix}, \chi_\downarrow = \begin{pmatrix} 0 \\ 1 \end{pmatrix}$ and η_ρ is the isospin wave function given above. The new isospin coordinate identifies the nature of the nucleon through its charge $q = \frac{1}{2}(1 + \tau_3)$.

Second Quantization. The hamiltonian for this system is given in second quantization by

$$\hat{H} = \sum_{\mathbf{k}\lambda\rho}\sum_{\mathbf{k}'\lambda'\rho'} a^\dagger_{\mathbf{k}\lambda\rho}\langle\mathbf{k}\lambda\rho|T|\mathbf{k}'\lambda'\rho'\rangle a_{\mathbf{k}'\lambda'\rho'} + \frac{1}{2}\sum_{\mathbf{k}_1\lambda_1\rho_1}\cdots\sum_{\mathbf{k}_4\lambda_4\rho_4}$$

$$\times\, a^\dagger_{\mathbf{k}_1\lambda_1\rho_1} a^\dagger_{\mathbf{k}_2\lambda_2\rho_2}\langle\mathbf{k}_1\lambda_1\rho_1,\mathbf{k}_2\lambda_2\rho_2|V|\mathbf{k}_3\lambda_3\rho_3,\mathbf{k}_4\lambda_4\rho_4\rangle a_{\mathbf{k}_4\lambda_4\rho_4} a_{\mathbf{k}_3\lambda_3\rho_3} \tag{3.4}$$

Here the fermion creation and destruction operators satisfy the canonical anticommutation relations[13]

$$\{a_{\mathbf{k}\lambda\rho}, a^\dagger_{\mathbf{k}'\lambda'\rho'}\} = \delta_{\mathbf{k}\mathbf{k}'}\delta_{\lambda\lambda'}\delta_{\rho\rho'} \tag{3.5}$$

It is assumed for the present purposes that the two-body potential is nonsingular and that all the matrix elements in Eq. (3.4) exist.

Let $|F\rangle$ be the normalized noninteracting Fermi gas ground state with neutron and proton levels with spin up and spin down filled equally to the Fermi wave number k_F (Fig. 2.5). First-order perturbation theory then gives the total energy of the system according to

$$E_0 + E_1 = \langle F|\hat{H}|F\rangle \tag{3.6}$$

Variational Estimate. In fact, this first-order calculation also provides a rigorous bound since the variational principle tells us that

$$E \le \langle F|\hat{H}|F\rangle \tag{3.7}$$

where E is the *exact* ground state energy of the fully interacting system. It follows from the expectation value that

$$E_0 + E_1 =$$

$$4\sum_{\mathbf{k}}^{k_\mathrm{F}}\frac{\hbar^2k^2}{2m} + \frac{1}{2}\sum_{\mathbf{k}_1\lambda_1\rho_1}\cdots\sum_{\mathbf{k}_4\lambda_4\rho_4}\langle\mathbf{k}_1\lambda_1\rho_1,\mathbf{k}_2\lambda_2\rho_2|V|\mathbf{k}_3\lambda_3\rho_3,\mathbf{k}_4\lambda_4\rho_4\rangle$$

$$\times\langle F|a^\dagger_{\mathbf{k}_1\lambda_1\rho_1} a^\dagger_{\mathbf{k}_2\lambda_2\rho_2} a_{\mathbf{k}_4\lambda_4\rho_4} a_{\mathbf{k}_3\lambda_3\rho_3}|F\rangle \tag{3.8}$$

All the operators in this last expression must refer to particles within the Fermi sea or the matrix element vanishes. The expectation value then gives $[\delta_{\mathbf{k}_1\mathbf{k}_3}\delta_{\lambda_1\lambda_3}\delta_{\rho_1\rho_3}]\ [\delta_{\mathbf{k}_2\mathbf{k}_4}\delta_{\lambda_2\lambda_4}\delta_{\rho_2\rho_4}] - [\delta_{\mathbf{k}_1\mathbf{k}_4}\delta_{\lambda_1\lambda_4}\delta_{\rho_1\rho_4}]\ [\delta_{\mathbf{k}_2\mathbf{k}_3}\delta_{\lambda_2\lambda_3}\delta_{\rho_2\rho_3}]$ from

[13] In writing these anticommutation relations, we have introduced a generalized Pauli principle. The state vector of a collection of nucleons is *antisymmetric* under the interchange of all coordinates including isotopic spin. There is, to this point in our development, no new physics implied by this assumption.

which the energy follows as

$$E_0 + E_1 = 4\sum_{\mathbf{k}}^{k_F} \frac{\hbar^2 k^2}{2m} + \frac{1}{2}\sum_{\mathbf{k}\lambda\rho}^{k_F}\sum_{\mathbf{k}'\lambda'\rho'}^{k_F} \times$$
$$\{\underbrace{\langle \mathbf{k}\lambda\rho, \mathbf{k}'\lambda'\rho'|V|\mathbf{k}\lambda\rho, \mathbf{k}'\lambda'\rho'\rangle}_{V_D} - \underbrace{\langle \mathbf{k}\lambda\rho, \mathbf{k}'\lambda'\rho'|V|\mathbf{k}'\lambda'\rho', \mathbf{k}\lambda\rho\rangle}_{V_E}\} \tag{3.9}$$

The first term in brackets is known as the *direct* interaction (V_D) and the second is the *exchange* interaction (V_E).

As an example, consider

$$V = V(r)(a_W + a_M P_M) \tag{3.10}$$

where a_M and a_W are positive constants and P_M is the Majorana space exchange operator (Section 1). Assume that $V(r)$ is nonsingular and that the volume integral of the potential $v = \int V(r) d^3r$ exists; take it to be a negative quantity representing the attractive nature of the nuclear force. Assume further that $V(r)$ has no spin dependence; in fact, the actual spin dependence of the nuclear force is quite weak — the $^1S_0$ state is just unbound, while the $^3S_1$ is just bound. We proceed to evaluate the required matrix elements of the potential in Eq. (3.9). The direct term is given by

$$\begin{aligned} V_D &= \frac{1}{V^2}\int\int d^3x d^3y\, e^{-i\mathbf{k}\cdot\mathbf{x}} e^{-i\mathbf{k}'\cdot\mathbf{y}} \chi_\lambda^\dagger(1)\chi_{\lambda'}^\dagger(2)\eta_\rho^\dagger(1)\eta_{\rho'}^\dagger(2) \\ &\quad \times V(\mathbf{x}-\mathbf{y}) e^{i\mathbf{k}\cdot\mathbf{x}} e^{i\mathbf{k}'\cdot\mathbf{y}} \chi_\lambda(1)\chi_{\lambda'}(2)\eta_\rho(1)\eta_{\rho'}(2) \\ &= \frac{1}{V}\left[a_W \int V(z) d^3z + a_M \int e^{-i(\mathbf{k}-\mathbf{k}')\cdot\mathbf{z}} V(z) d^3z\right] \end{aligned} \tag{3.11}$$

where we have defined $\mathbf{z} \equiv \mathbf{x} - \mathbf{y}$. The exchange term follows in exactly the same fashion as

$$V_E = \frac{1}{V}\delta_{\lambda\lambda'}\delta_{\rho\rho'}\left[a_W \int e^{-i(\mathbf{k}-\mathbf{k}')\cdot\mathbf{z}} V(z) d^3z + a_M \int V(z) d^3z\right] \tag{3.12}$$

The total energy in Eq. (3.9) thus becomes

$$\begin{aligned} E_0 + E_1 &= \frac{3}{5}\frac{\hbar^2 k_F^2}{2m} A + \frac{V}{2}\frac{1}{(2\pi)^6}\int\int^{k_F} d^3k d^3k' \\ &\quad \times \left\{16\left(a_W v + a_M \int e^{-i(\mathbf{k}-\mathbf{k}')\cdot\mathbf{z}} V(z) d^3z\right)\right. \\ &\quad \left. -4\left(a_M v + a_W \int e^{-i(\mathbf{k}-\mathbf{k}')\cdot\mathbf{z}} V(z) d^3z\right)\right\} \end{aligned} \tag{3.13}$$

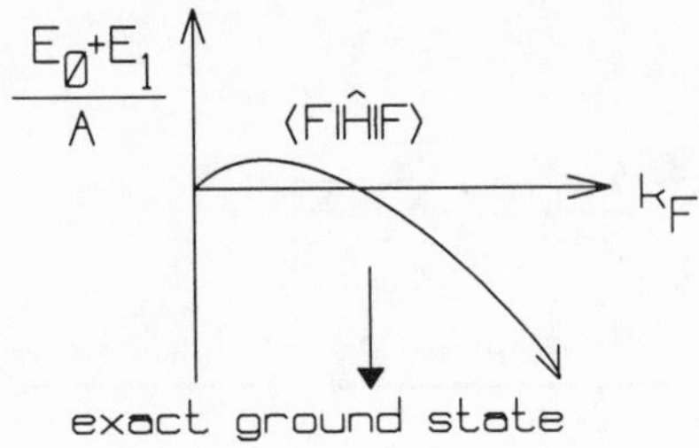

Figure 3.1: Variational estimate for energy of true ground state of nuclear matter with the nonsingular potential in Eq. (3.10).

The required momentum integrals are evaluated as

$$\begin{aligned}\int_0^{k_F} d^3k e^{i\mathbf{k}\cdot\mathbf{x}} &= 4\pi\int_0^{k_F} k^2 j_0(kx)dk = \frac{4\pi}{x^3}\int_0^{k_F x}\rho^2 j_0(\rho)d\rho \\ &= \frac{4\pi}{x^3}(k_F x)^2 j_1(k_F x) = \frac{4\pi k_F^3}{3}\left[\frac{3j_1(k_F x)}{k_F x}\right]\end{aligned} \tag{3.14}$$

The remaining volume is expressed in terms of the total number of nucleons as $V = 3\pi^2 A/2k_F^3$ and the energy becomes

$$\begin{aligned}\frac{E_0+E_1}{A} &= \frac{3}{5}\frac{\hbar^2 k_F^2}{2m} + \frac{k_F^3}{12\pi^2}\left\{(4a_W - a_M)\int V(z)d^3z\right. \\ &\left. +(4a_M - a_W)\int\left[\frac{3j_1(k_F z)}{k_F z}\right]^2 V(z)d^3z\right\}\end{aligned} \tag{3.15}$$

Since $[3j_1(k_F z)/k_F z]^2 \to 0$ as $k_F \to \infty$ for all finite z, the second term in brackets becomes negligible at high baryon density. Thus unless $(4a_W - a_M) < 0$ or $a_M > 4a_W$ the system will be *unstable against collapse.* This is a rigorous result since this lowest order calculation is variational; the true ground state energy must lie below this value. Thus, as illustrated in Fig. 3.1, the true system must become more bound as $k_F \to \infty$ and hence the system is unstable against collapse. Experimentally, the nuclear potential is approximately a Serber force with $a_M \approx a_W$. We have proven the following theorem for such an interaction:

> *A nonsingular Serber force with* $\int V(z)d^3z < 0$ *does not lead to nuclear saturation.*

Single-Particle Potential. The single-particle Hartree-Fock potential that a nucleon feels in nuclear matter at any density can readily be identified from this calculation as

$$\begin{aligned}U(\mathbf{k})_{\lambda\rho} &= \sum_{\mathbf{k}'\lambda'\rho'}^{k_F}\{\langle\mathbf{k}\lambda\rho,\mathbf{k}'\lambda'\rho'|V|\mathbf{k}\lambda\rho,\mathbf{k}'\lambda'\rho'\rangle \\ &-\langle\mathbf{k}\lambda\rho,\mathbf{k}'\lambda'\rho'|V|\mathbf{k}'\lambda'\rho',\mathbf{k}\lambda\rho\rangle\}\end{aligned} \tag{3.16}$$

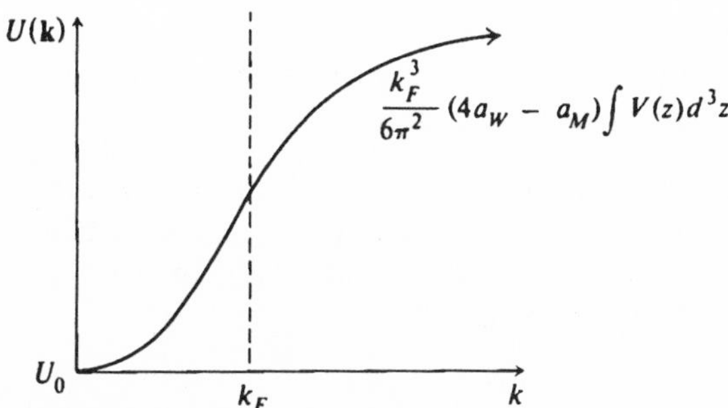

Figure 3.2: Sketch of single-nucleon potential in nuclear matter with a nonsingular potential; for a Serber force $a_{\mathrm{W}} = a_{\mathrm{M}}$.

The required integrals have all been evaluated above and one finds

$$U(k) = \frac{k_{\mathrm{F}}^3}{6\pi^2}\left\{(4a_{\mathrm{W}} - a_{\mathrm{M}})\int_0^\infty V(z)d^3z \right.$$
$$\left. +(4a_{\mathrm{M}} - a_{\mathrm{W}})\int_0^\infty j_0(kz)\left[\frac{3j_1(k_{\mathrm{F}}z)}{k_{\mathrm{F}}z}\right]V(z)d^3z\right\} \tag{3.17}$$

This result is independent of λ and ρ and is only a function of k^2. It is sketched in Fig. 3.2. The expansion $j_0(kz) \cong 1-(kz)^2/3!+(kz)^4/5!$ allows one to identify the first few terms in an expansion for small k. The momentum dependence, which gives rise to an effective mass for the nucleons, arises both from the exchange interaction through the potential $V_{\mathrm{W}} \equiv V(r)a_{\mathrm{W}}$ and from the direct interaction through the exchange potential $V_{\mathrm{M}} \equiv V(r)a_{\mathrm{M}}P_{\mathrm{M}}$. Since $j_0(kz)$ will oscillate rapidly for large k causing the second integral to vanish, the asymptotic form of the single-particle potential takes the form

$$U(k) \stackrel{k\to\infty}{\longrightarrow} \frac{k_{\mathrm{F}}^3}{6\pi^2}(4a_{\mathrm{W}} - a_{\mathrm{M}})\int_0^\infty V(z)d^3z \tag{3.18}$$

One is faced with two problems:

1. *How does one explain nuclear saturation in terms of the two-nucleon interaction?* It is an empirical fact that the static nucleon-nucleon potential is *singular* at short distances; there is empirical evidence for a strong short-range repulsion (Section 1). A method must be developed for incorporating this possiblity into our theoretical description.

2. *Why does the independent-particle model work at all if the forces are indeed so strong and singular?* It is an empirical fact that single-particle models do provide an amazingly successful first description of nuclei, for example, the shell model of nuclear properties and spectra, the optical

model for nucleon scattering, and the Fermi gas model for quasielastic nuclear response. How can this success be understood in the light of such strong, singular internucleon forces?

We proceed to develop the independent-pair approximation that allows one to incorporate the strong short-range part of the internucleon interaction into the theoretical description of nuclear matter within the framework of the non-relativistic many-body problem.

4

THE INDEPENDENT-PAIR APPROXIMATION

In the previous section the expectation value of the hamiltonian was computed using the wave functions of a noninteracting Fermi gas. If the two-body force is strong at short distances, then it is essential to include the effects of the interaction back on the wave function. This shall be carried out within the framework of the *independent-pair approximation*, which provides a simple summary of the theory of Brueckner, Bethe, and others (Refs. [N19, N20, N21, N22]). The present discussion uses the Bethe-Goldstone equation (Ref. [N23]) and is based on Ref. [N2] (see also Refs. [N1, N3]). In Ref. [N2] the relation to the full analysis of nuclear matter using Green's functions is developed in detail; here a more intuitive approach will be employed.

Bethe-Goldstone Equation. The basic idea is to write the Schrödinger equation for two interacting particles in the nuclear medium. The nature of the potential at short distances is then unimportant, for the differential equation is simply solved exactly in that region. This approach takes into account the effect of the two-body potential on the wave function to *all orders in V*. The effects of the surrounding nuclear medium are then taken into account in two ways:

1. First, the two particles interact in the presence of a degenerate Fermi gas — there is a certain set of levels already occupied by other nucleons and the Pauli principle prohibits the pair of interacting particles from making transitions into these already-occupied states.

2. Second, the interacting particles move in a self-consistent single-particle potential generated by the average interaction with all of the other particles in the nuclear medium.

We start from the Schrödinger equation for two free particles interacting through a potential V(1,2)

$$[T_1 + T_2 + V(1,2)]\Psi(1,2) = E\Psi(1,2)$$

$$\Psi(1,2) = \Phi_0(1,2) + \sum_{n\neq 0} \Phi_n(1,2)\frac{1}{E - E_n}\langle\Phi_n|V|\Psi\rangle \qquad (4.1)$$

Here E is the exact lowest-energy eigenvalue and

$$H_0\Phi_n = (T_1 + T_2)\Phi_n = E_n\Phi_n \qquad (4.2)$$

Figure 4.1: Quantization volume and levels available to two interacting particles in nuclear matter.

Equation (4.1) represents the expansion of the full two-particle wave function Ψ in the complete set Φ_n; it is an integral equation since the wave function Ψ appears in the matrix elements on the right-hand side. A particularly convenient *normalization* has been chosen for the wave function, $\langle\Phi_0|\Psi\rangle = 1$; this has been accomplished by setting the first coefficient on the right-hand side of Eq. (4.1) equal to unity, or

$$E - E_0 = \langle\Phi_0|V|\Psi\rangle \tag{4.3}$$

The proof of Eq. (4.1) follows by operating on it with $E - H_0$ and using the completeness of the unperturbed wave functions; these simply represent particles in a big box of volume V with periodic boundary conditions (Fig. 4.1)

$$\phi_{\mathbf{k}_1\mathbf{k}_2}(1,2) = \frac{1}{V}e^{i\mathbf{k}_1\cdot\mathbf{x}_1}e^{i\mathbf{k}_2\cdot\mathbf{x}_2} \tag{4.4}$$

Spin and isospin indices are here suppressed for clarity, and nonidentical nucleons are assumed; if the nucleons are identical, the wave function must be antisymmetrized.

The Pauli principle now restricts the possible intermediate states that can be admixed into the wave function of the interacting pair in Eq. (4.1) since some are already occupied by other nucleons (Fig. 4.1). This restricts the sum in Eq. (4.1) according to

$$\sum_n \to \sum_{|\mathbf{k}_1|,|\mathbf{k}_2|>k_F} \tag{4.5}$$

Next center-of-momentum (C-M) and relative coordinates may be introduced through the relations

$$\begin{aligned}
\mathbf{P} &= \mathbf{k}_1 + \mathbf{k}_2 & \mathbf{k} &= \frac{1}{2}(\mathbf{k}_1 - \mathbf{k}_2) \\
\mathbf{R} &= \frac{1}{2}(\mathbf{x}_1 + \mathbf{x}_2) & \mathbf{x} &= \mathbf{x}_1 - \mathbf{x}_2 \\
v &\equiv \frac{2\mu_{\text{red}}V(x)}{\hbar^2} & E &\equiv \frac{\hbar^2\kappa^2}{2\mu_{\text{red}}} + \frac{\hbar^2\mathbf{P}^2}{2M_{\text{tot}}}
\end{aligned} \tag{4.6}$$

Here $\mu_{\text{red}} = m/2$ and $M_{\text{tot}} = 2m$ are the reduced mass and total mass of the pair, respectively. The quantity κ^2 appearing in E parameterizes the exact eigenvalue.

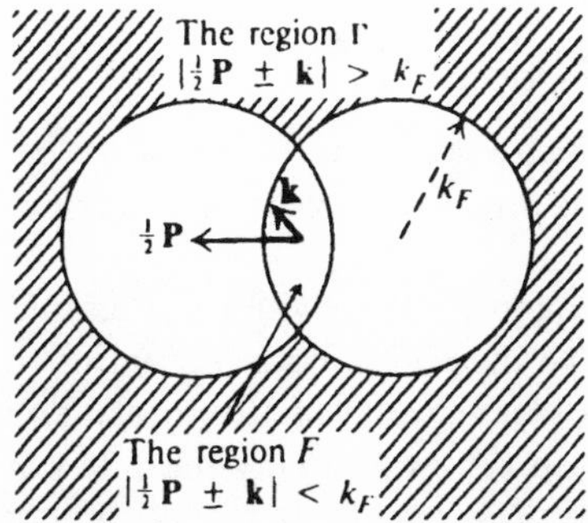

Figure 4.2: Momentum space integrations in the Bethe-Goldstone equation. Here F is the region $|\mathbf{P}/2 \pm \mathbf{k}| < k_{\rm F}$ where both the starting particles are inside the Fermi sphere and Γ is $|\mathbf{P}/2 \pm \mathbf{k}| > k_{\rm F}$ where both intermediate particles are outside the Fermi sphere.

The solution to the Schrödinger Eq. (4.1) takes the form

$$\Psi(1,2) = \frac{1}{\sqrt{V}} e^{i\mathbf{P}\cdot\mathbf{R}} \frac{1}{\sqrt{V}} \psi_{\mathbf{P},\mathbf{k}}(\mathbf{x}) \tag{4.7}$$

Substitution of this form and cancellation of common factors [assuming $V(1,2) = V(\mathbf{x_1} - \mathbf{x_2})$] gives[14]

$$\begin{aligned} \psi_{\mathbf{P},\mathbf{k}}(\mathbf{x}) &= e^{i\mathbf{k}\cdot\mathbf{x}} + \int_\Gamma \frac{d^3t}{(2\pi)^3} e^{i\mathbf{t}\cdot\mathbf{x}} \frac{1}{\kappa^2 - t^2} \int d^3y\, e^{-i\mathbf{t}\cdot\mathbf{y}} v(y) \psi_{\mathbf{P},\mathbf{k}}(\mathbf{y}) \\ \kappa^2 - k^2 &= \frac{1}{V} \int e^{-i\mathbf{k}\cdot\mathbf{x}} v(x) \psi_{\mathbf{P},\mathbf{k}}(\mathbf{x}) d^3x \end{aligned} \tag{4.8}$$

Here the initial momenta lie in F *inside* the Fermi sphere $|\mathbf{P}/2 \pm k| < k_{\rm F}$ while the admixed momenta lie in Γ *outside* the Fermi sphere $|\mathbf{P}/2 \pm \mathbf{t}| > k_{\rm F}$. The regions of momentum space integration are indicated pictorially in Fig. 4.2. Note that the C-M momentum **P** is a constant of the motion for the interacting pair.

Start with a pair of particles with $(\mathbf{P}, \mathbf{k})$; two-particle states above the Fermi sea are then admixed due to the two-particle potential. From this, the two-particle energy shift is determined. In analogy to Eq. (3.9), it is assumed that the total energy shift for the system is obtained by summing the energy shifts over the interacting pairs

$$\Delta E = \frac{1}{2} \sum_{\mathbf{k}_1\lambda_1\rho_1}^{k_{\rm F}} \sum_{\mathbf{k}_2\lambda_2\rho_2}^{k_{\rm F}} \Delta\epsilon_{\mathbf{k}\mathbf{P}}$$

[14]Use $\int d^3x_1 d^3x_2 = \int d^3R\, d^3x$ and $\int d^3R \exp\{i\mathbf{R}\cdot(\mathbf{P} - \mathbf{P}')\} = V\delta_{\mathbf{P},\mathbf{P}'}$.

$$\Delta\epsilon_{\mathbf{kP}} = \frac{1}{V}\int e^{-i\mathbf{k}\cdot\mathbf{x}}V(x)\psi_{\mathbf{Pk}}(\mathbf{x})d^3x \tag{4.9}$$

For identical particles (i.e., $p\uparrow p\uparrow$ etc.) the wave function must be antisymmetrized and there will be direct and exchange contributions to the two-particle energy shifts.

The set of Eqs. (4.8) and (4.9) provides the simplest way of including the effects of a singular short-range interaction potential on the wave function and approximating the expectation value of the total hamiltonian $H = \sum_i t(i) + \frac{1}{2}\sum_i\sum_j V(i,j)$ taken with the exact many-particle wave function.

As a generalization of these results, it will be assumed that each member of the interacting pair moves in a single-particle potential coming from the average interaction with the particles in the other occupied states. In analogy with Eq. (3.16) one writes

$$\begin{aligned}\varepsilon(\mathbf{k}_1) &= \frac{\hbar^2k_1^2}{2m} + U(\mathbf{k}_1)\\ U(\mathbf{k}_1\lambda_1\rho_1) &= \sum_{\mathbf{k}_2\lambda_2\rho_2}^{k_F}\Delta\epsilon_{\mathbf{kP}}\end{aligned} \tag{4.10}$$

Effective Mass Approximation. This fully coupled problem is still numerically complicated. It will here be simplified by making the *effective mass approximation.*[15] A Taylor series expansion is made about some momentum k_0, and only the first two terms retained

$$U(\mathbf{k}_1^2) \approx U(k_0^2) + \frac{\hbar^2}{2m}(k_1^2 - k_0^2)U_1 \equiv U_0 + \frac{\hbar^2}{2m}k_1^2U_1 \tag{4.11}$$

Thus

$$\varepsilon(\mathbf{k}_1) = U_0 + \frac{\hbar^2k_1^2}{2m^*} \qquad\qquad \frac{m^*}{m} = \frac{1}{1+U_1} \tag{4.12}$$

Since only energy differences enter in Eq. (4.8) for $\psi_{\mathbf{Pk}}$, U_0 cancels in that equation, and the only effect of this modification of the single-particle spectrum is then to replace $m \to m^*$ in the potential

$$v(x) = \frac{2\mu^*_{\text{red}}}{\hbar^2}V(|\mathbf{x}_1 - \mathbf{x}_2|) \tag{4.13}$$

where $\mu^*_{\text{red}} = m^*/2$ is the reduced effective mass.

Equations (4.8) are still involved. The wave function $\psi_{\mathbf{Pk}}$ satisfies an integral equation where the kernel depends on the eigenvalue κ^2, which itself can

[15] While exact over a small enough interval of the spectrum, Fig. 3.2 indicates that this can only be a crude approximation to the *entire* spectrum. (See, in this regard, Refs. [N74, N75].)

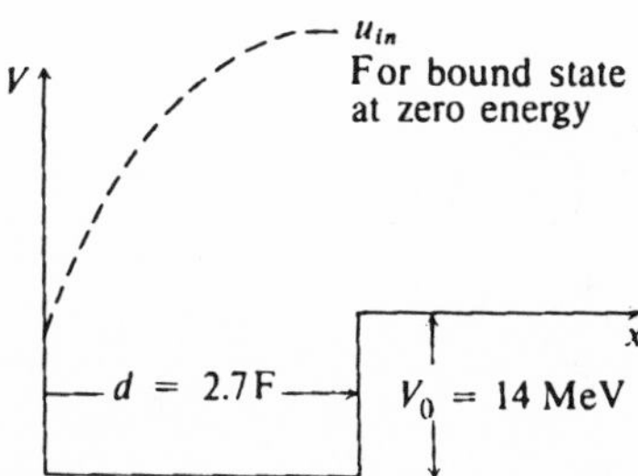

Figure 4.3: The s-wave wave function in a square well potential with a bound state at zero energy. Here $x \equiv |\mathbf{x}|$.

be determined only from $\psi_{\mathbf{Pk}}$; however, it is possible to make an important simplification if one is primarily interested only in the *bulk* properties of nuclear matter. The energy shift $\kappa^2 - k^2$ of a pair of particles in Eq. (4.8) goes as $1/V$ where V is the volume (Fig. 4.1). Since $\kappa^2 - t^2$ cannot vanish except close to the Fermi surface (Fig. 4.2), one can make the replacement $\kappa^2 - t^2 \rightarrow k^2 - t^2$ in the equation for $\psi_{\mathbf{Pk}}$ as $V \rightarrow \infty$.[16]

Solution for a Nonsingular Square Well Potential. Consider a square well potential fit to low-energy $^1S_0$ scattering. In this channel, the nucleon-nucleon force has approximately a bound state at zero energy (Section 1). The wave function inside such a potential in the free scattering problem contains 1/4 wavelength (Fig. 4.3) $u_{\text{in}} = N \sin(\pi x/2d) = N \sin[(2\mu_{\text{red}} V_0/\hbar^2)^{1/2} x]$. Thus the depth of such a potential is related to its range by

$$V_0 = \frac{\hbar^2 \pi^2}{8\mu_{\text{red}} d^2} \quad = \quad \frac{\hbar^2 \pi^2}{4md^2} \tag{4.14}$$

The effective range of a square well potential with a bound state at zero energy is given by $r_0 = d$ (Prob. 1.8). Use of the singlet effective range of $r_0 = 2.7$ fm [Eq. (1.6)] leads to $V_0 = 14$ MeV. This is a very *weak potential* in nuclear matter; for example, $V_0 \ll \varepsilon_{\text{F}}^0$ where the Fermi energy is given by $\varepsilon_{\text{F}}^0 = \hbar^2 k_{\text{F}}^2/2m = 42$ MeV with $k_{\text{F}} = 1.42\,\text{fm}^{-1}$.

Let us calculate the Bethe-Goldstone wave function in nuclear matter in the limiting case where $\mathbf{P} = \mathbf{k} = 0$. At zero energy, only the s-waves feel the effects of the potential. Assume that the modification of the wave function is small (an approximation readily verified at the end of the calculation) and replace $u(x)/x \approx j_0(kx) \rightarrow 1$ in the region of the potential. The modification of the

[16]This argument gets one into trouble with the *Cooper pairs* — the extraordinary eigenvalues that lead to superconductivity (see Ref. [N2] Chap.10). These eigenvalues are a problem only very close to the Fermi surface, and are unimportant when discussing the bulk properties of nuclear matter.

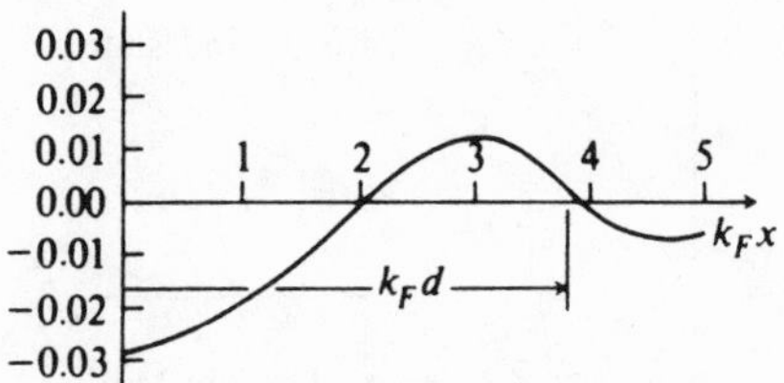

Figure 4.4: Difference $\Delta\psi_{\rm SW}$ between Bethe-Goldstone wave function and unperturbed plane wave for a zero energy pair in nuclear matter interacting through the ${}^1S_0$ square well potential in Fig. 4.3. From Ref. [N2].

wave function is then calculated from Eq. (4.8) to be

$$\Delta\psi_{\rm SW} \equiv \frac{u(x)}{x} - 1 \approx \frac{2v_0}{\pi}\int_{k_{\rm F}}^{\infty} dt j_0(tx)\int_0^d j_0(ty)y^2 dy \qquad (4.15)$$

The use of $\int_0^d y^2 j_0(ty)dy = d^3 j_1(td)/td$ allows this to be rewritten as

$$\Delta\psi_{\rm SW} = \frac{2v_0}{\pi k_{\rm F}^2}(k_{\rm F}d)^2\int_{k_{\rm F}d}^{\infty}\frac{d\rho}{\rho}j_1(\rho)j_0(\rho\frac{x}{d}) \qquad (4.16)$$

Here $\rho \equiv td$ and

$$\frac{v_0}{k_{\rm F}^2} = \frac{V_0}{\hbar^2 k_{\rm F}^2/2\mu_{\rm red}^*} = \frac{m^*}{m}\frac{\pi^2}{4(k_{\rm F}d)^2} = 0.17\frac{m^*}{m} \qquad (4.17)$$

It will be argued in the next section that $m^*/m \approx 0.6$ and thus the ratio in Eq. (4.17) is $v_0/k_{\rm F}^2 \approx 0.10$. The result in Eq. (4.16) is plotted in Fig. 4.4. We observe that $|\Delta\psi_{\rm SW}| \ll 1$ and this potential has almost *no effect on the wave function.* The wave function of this pair in nuclear matter is essentially the unperturbed value. Due to the relatively large k_F, this nucleon-nucleon potential cannot easily excite pairs out of the Fermi sea; the fact that $m^*/m < 1$ makes it even more difficult to do so.

Solution for a Pure Hard Core Potential (Ref. [N23]). Consider the Bethe-Goldstone (B-G) equation for a two-body potential that is an infinite barrier at $r = a$, and take $\mathbf{P} = 0$ as the simplest example. It is convenient in this case to first convert the B-G Eq. (4.8) with $\kappa^2 \approx k^2$ to differential-integral form by applying the operator $(\nabla^2 + k^2)$. This gives

$$\begin{aligned}(\nabla^2 + k^2)\psi(\mathbf{x}) &= \int_\Gamma \frac{d^3t}{(2\pi)^3}e^{i\mathbf{t}\cdot\mathbf{x}}\int d^3y\, e^{-i\mathbf{t}\cdot\mathbf{y}}v(y)\psi(\mathbf{y}) \\ &= v(x)\psi(\mathbf{x}) - \int_{\overline{\Gamma}} \frac{d^3t}{(2\pi)^3}e^{i\mathbf{t}\cdot\mathbf{x}}\int d^3y\, e^{-i\mathbf{t}\cdot\mathbf{y}}v(y)\psi(\mathbf{y}) \qquad (4.18)\end{aligned}$$

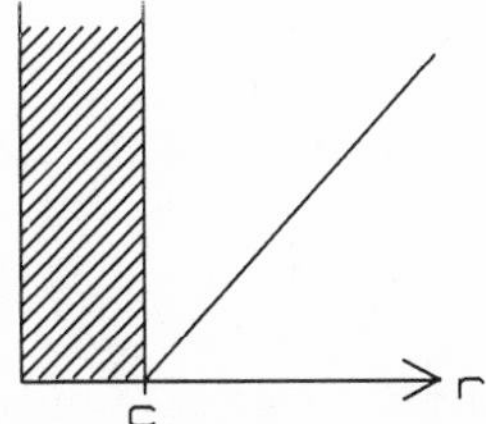

Figure 4.5: Sketch of s-wave wave function with a hard core.

Here we have used the relation $\int_\Gamma \equiv \int - \int_{\overline{\Gamma}}$ where $\overline{\Gamma}$ is the complement of Γ (see Fig. 4.2). Look for s-wave solutions to this equation of the form $\psi = u(x)/x$ where $x \equiv |\mathbf{x}|$; it is the s-waves that get in closest and will feel the singular potential

$$\begin{aligned}(\frac{d^2}{dx^2} + k^2)u(x) &= v(x)u(x) - \int_0^\infty \chi(x,y)v(y)u(y)dy \\ \chi(x,y) = \frac{2xy}{\pi}\int_0^{k_\mathrm{F}} j_0(tx)j_0(ty)t^2 dt &= \frac{1}{\pi}\left[\frac{\sin k_\mathrm{F}(x-y)}{x-y} - \frac{\sin k_\mathrm{F}(x+y)}{x+y}\right]\end{aligned} \tag{4.19}$$

Introduce dimensionless variables $r = k_\mathrm{F}x$, $r' = k_\mathrm{F}y$, $K = k/k_\mathrm{F}$, and

$$\nu(r) = \frac{V(x)}{\hbar^2 k_\mathrm{F}^2/2\mu^*_\mathrm{red}} \qquad\qquad u(r) = k_\mathrm{F}u(x) \tag{4.20}$$

Equation (4.19) then takes the form

$$\begin{aligned}(\frac{d^2}{dr^2} + K^2)u(r) &= \nu(r)u(r) - \int_0^\infty \chi(r,r')\nu(r')u(r')dr' \\ \chi(r,r') &= \frac{1}{\pi}\left[\frac{\sin(r-r')}{r-r'} - \frac{\sin(r+r')}{r+r'}\right]\end{aligned} \tag{4.21}$$

The wave function will in general have a discontinuity in slope at the core radius $c = k_\mathrm{F}a$ as illustrated in Fig. 4.5. This implies that the product of the potential and the wave function must contain a delta function.[17] Thus one must have

$$\nu(r)u(r) = \mathcal{A}\delta(r-c) + w(r) \tag{4.22}$$

Here $w(r) = 0$ outside the core $r > c$ since the potential vanishes there. Inside the core $r < c$ the wave function $u(r)$ must vanish, and $w(r)$ must be chosen to

[17] To see this, just integrate the radial differential equation over an infinitesimal region across the core boundary.

ensure that this happens. Thus

$$\begin{aligned}(\frac{d^2}{dr^2}+K^2)u(r) &= 0\,; \qquad\qquad r<c \\ w(r) &= \mathcal{A}\chi(r,c)+\int_0^c \chi(r,r')w(r')dr' \end{aligned} \tag{4.23}$$

For nuclear matter c is effectively a small parameter

$$c = k_{\rm F}r_c \approx (1.42\,{\rm fm}^{-1})(0.4\,{\rm fm}) \approx 0.57 \tag{4.24}$$

The kernel can then be expanded for small c

$$\begin{aligned}\chi(r,r') &= \frac{2rr'}{3\pi}\,; \qquad\qquad r,r'\to 0 \\ \frac{w(r)}{\mathcal{A}} &= \frac{2rc}{3\pi}\theta(c-r)+O(c^5)\end{aligned} \tag{4.25}$$

For small c, $w(r)$ may be neglected. Thus[18]

$$(\frac{d^2}{dr^2}+K^2)u(r) \approx \mathcal{A}[\delta(r-c)-\chi(r,c)] \equiv F(r) \tag{4.26}$$

The solution to this inhomogeneous differential equation can be written in the form

$$u(r)=\frac{1}{K}\int_0^r \sin\left[K(r-s)\right]F(s)ds \tag{4.27}$$

This is immediately checked by differentiation $u'=\int_0^r \cos\left[K(r-s)\right]F(s)ds$ and $u(r)''+K^2u(r)=F(r)$. A trigonometric expansion in Eq. (4.27) then gives

$$u(r)=\frac{\sin Kr}{K}\int_0^r \cos Ks\,F(s)ds-\frac{\cos Kr}{K}\int_0^r \sin Ks\,F(s)ds \tag{4.28}$$

Two features of this result are of particular interest:

1. The excluded region in momentum space in the B-G equation is just such as to guarantee that *there is no phase shift* in the wave function at large interparticle separation. All of the states with the same $|\mathbf{k}|$ are already occupied by other nucleons and it is therefore not possible to scatter into them. As a consequence, the potential can distort the B-G wave function only at short distances; there is no long-range effect. At large distances the B-G wave function goes over to an unperturbed plane wave. The proof that there is no phase shift follows by letting $r\to\infty$ in Eq. (4.28)

$$\begin{aligned}\int_0^\infty \sin Ks\,F(s)ds &\propto \int_0^\infty s^2ds\int_\Gamma t^2dt\; j_0(Ks)j_0(ts)\cdots \\ &= \int_\Gamma t^2dt\frac{\pi}{2Kt}\delta(K-t)\cdots=0\end{aligned} \tag{4.29}$$

[18]From Eq. (4.25) $w(r)$ is of order c^2 and any integral over $w(r)$ is at least of order c^3; its effects can readily be included in a power series expansion in c.

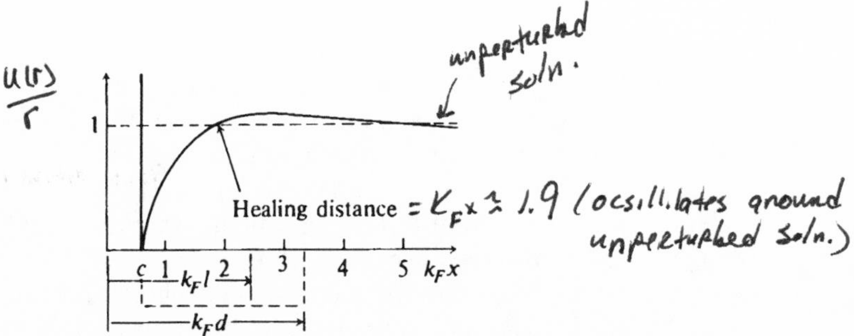

Figure 4.6: The Bethe-Goldstone s-wave wave function $u(r)/r$ for a pair with $\mathbf{K} = \mathbf{P} = 0$ interacting through a hard-core potential in nuclear matter. Recall that here $x = |\mathbf{x}|$ and $r = k_{\mathrm{F}}x$. Also shown are the average interparticle distance $k_{\mathrm{F}}l$ and range $k_{\mathrm{F}}d$ of the potential in Fig. 5.1. From Ref. [N2].

This last result follows since K is in F and t is in Γ (Fig. 4.2) and the argument of the delta function can never vanish.

2. The wave function must now be *normalized*, and to be normalized in the volume V one must demand $\psi(\mathbf{x}) \to e^{i\mathbf{k}\cdot\mathbf{x}}$ with unit amplitude as $x \to \infty$, or from Eq. (4.28)

$$\int_0^\infty \cos Ks\, F(s)ds = 1 \tag{4.30}$$

Insertion of the definition of $F(s)$ in Eq. (4.26) gives

$$\mathcal{A} = [\cos Kc - \int_0^\infty \cos Ks\ \chi(s,c)ds]^{-1} \tag{4.31}$$

These equations allow one to calculate the wave function $u(r)$ and the result is shown in Fig. 4.6.[19] The correlation function $\Delta\psi \equiv u(r)/r - j_0(Kr)$ oscillates to zero with large $r = k_{\mathrm{F}}x$. The *healing distance* will be defined as the point where $\Delta\psi$ first vanishes; from Fig. 4.6 we see[20]

$$k_{\mathrm{F}}x \approx 1.9 \qquad \text{healing distance} \tag{4.32}$$

Also plotted in Fig. 4.6 is the interparticle separation in nuclear matter $1/l^3 \equiv A/V = 2k_{\mathrm{F}}^3/3\pi^2$

$$k_{\mathrm{F}}l = \left(\frac{3\pi^2}{2}\right)^{1/3} = 2.46 \qquad \text{interparticle distance} \tag{4.33}$$

Justification of the Independent-Particle Model. We have seen that the attractive part of the nucleon-nucleon potential has only a relatively small effect

[19]The calculations presented are actually of $u(r) = \int_{c_-}^{r} \frac{1}{K}\sin[K(r-s)]\,F(s)ds$ so that $u(c) = 0$ as is guaranteed by the presence of the term $w(r)$.

[20]These results are insensitive to $\mathbf{k}$ and $\mathbf{P}$ (Ref. [N2]).

on the B-G wave function, leaving it essentially a plane wave. The hard core modifies the wave function at short distances, but the B-G wave function *heals* rapidly and oscillates with decaying amplitude about the plane wave value. Thus, except for a modification at small internucleon separation, a nucleon in nuclear matter moves as if it were in a plane wave state. The Pauli principle, acting through the already occupied orbitals, suppresses the role of correlations; there are no long-range correlations here, only short-range correlations.[21]

Justification of the Independent-Pair Approximation. The B-G wave function is modified only at short distance; at large distances it goes over to an unperturbed plane wave. We observe that the *healing distance is less than the interparticle separation.* It is therefore unlikely to find a third particle in the region of interaction of any pair. Thus it suffices to find the pair wave function at short distances to determine the properties of the system that depend on this short distance behavior, for example, the energy.

(1) Attractive well

$$E/A = -\frac{V_0 k_F^3}{6\pi}\left[(d^3 - b^3) + \frac{2r}{k_F^2}\int_b^d j_1^2(k_F r)\,dr\right]$$

(2) Fermi Energy

$$E_{(F)}/A = \frac{3}{5}\frac{\hbar^2 k_F^2}{2m}$$

(3) Hard Core Energy

$$\Delta E = \frac{1}{2}\sum_{k_1\lambda_1\rho_1}^{k_F}\sum_{k_2\lambda_2\rho_2}^{k_F}\Delta\varepsilon_{kP}$$

[21] See in this connection Ref. [N24].

5

NUCLEAR MATTER WITH A "REALISTIC" INTERACTION

The goal of this section is to obtain a qualitative, and even semiquantitative, description of nuclear matter with a static nucleon-nucleon potential determined from the behavior of two free nucleons. The analysis is based on the independent-pair approximation of Section 4; much of it is taken from Ref. [N2].[22] The discussion is a great oversimplification of many detailed calculations of the properties of nuclear matter starting from the free nucleon-nucleon interaction (see, e.g., Refs. [N27, N28, N22, N29, N30, N31, N32]). Furthermore, this pair contribution represents only the leading term in a nonrelativistic Green's function framework that systematically includes all many-body contributions (Ref. [N2]); detailed evaluation of many of these terms has also been carried out (Refs. [N29, N30, N31, N32]). We return to these points at the conclusion of this section. The goal of the present discussion is to obtain simple physical insight into the saturation properties of nuclear matter as determined in a nonrelativistic many-body approach using static two-nucleon potentials fit to free scattering data.

Assume a simple Serber square well potential fit to low-energy scattering data and containing a hard core (see Fig. 5.1)

$$\begin{aligned} V = & +\infty & |\mathbf{x}| < b \\ = & -V_0 \tfrac{1}{2}(1 + P_{\mathrm{M}}) & b < |\mathbf{x}| < b + b_{\mathrm{W}} \\ = & \; 0 & b + b_{\mathrm{W}} < |\mathbf{x}| \end{aligned} \tag{5.1}$$

The parameters of the potential will be fit to the free $^1S_0$ data. Recall that the observed spin dependence of the nucleon-nucleon force is not strong; furthermore, if the $^1S_0 - {}^3S_1$ difference is attributed entirely to the tensor interaction, the spin dependence is suppressed even further in nuclear matter since the expectation value of the tensor potential vanishes in a spin and isospin saturated Fermi gas (Prob. 3.1). The condition for a bound state at zero energy is that 1/4 wavelength be contained in the attractive potential, just as in Section 4

$$V_0 = \frac{\hbar^2 \pi^2}{4 m b_{\mathrm{W}}^2} \tag{5.2}$$

[22] See also Refs. [N25, N26].

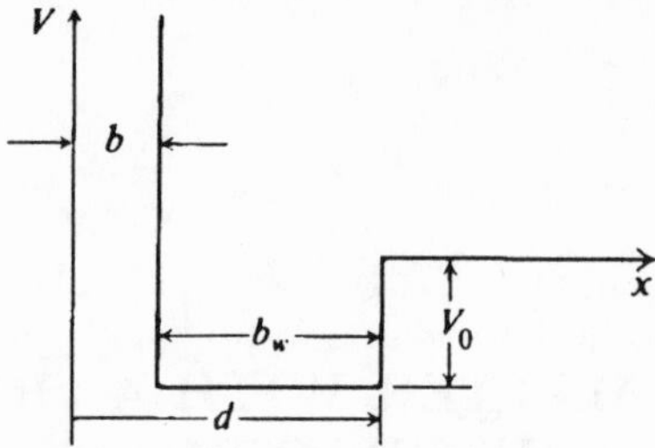

Figure 5.1: Square well Serber potential fit to low-energy nucleon-nucleon scattering and containing a hard core (see text). Here $x = |\mathbf{x}|$.

The effective range of such a potential is given by (Prob. 1.8)

$$2b + b_{\mathrm{W}} = r_0 \tag{5.3}$$

We take [1]$r_0 = 2.7$ fm and $b = 0.4$ fm, which implies $b_{\mathrm{W}} = 1.9$ fm. These results give an overall spatial extent for the potential in Fig. 5.1 of $k_{\mathrm{F}}d = k_{\mathrm{F}}(b + b_{\mathrm{W}}) = (1.42\ \mathrm{fm}^{-1})(2.3\ \mathrm{fm}) = 3.27$ in dimensionless form.

Again, just as in Section 4, this attractive well can be expected to have little effect on the B-G wave function (it pulls it in slightly), and the significant correlations come from the hard core alone. The B-G wave function for a hard core interaction was shown in Fig. 4.6, where the distance $k_{\mathrm{F}}d$ for this potential is also indicated. The wave function is pushed out to the core radius, then heals at a distance $k_{\mathrm{F}}x \approx 1.9$, and then bulges over the unperturbed plane wave result. Based on these observations, it is reasonable to make the following simplifying approximations in the expression for the energy shift $\Delta\varepsilon_{\mathbf{kP}}$ in Eq. (4.9):

1. $\psi \approx \psi_{\mathrm{c}}$ where ψ_{c} is the solution to the B-G equation with the hard core interaction alone. This assumption reflects the observation that this attractive potential has very little effect on the B-G wave function; it has the consequence that one can calculate the *wave function* by considering a gas of hard spheres alone, an appealing theoretical physics problem.

2. $\psi_{\mathrm{c}} \approx e^{i\mathbf{k}\cdot\mathbf{x}}$ *in the region of the attractive potential.* This approximation reflects the observation that *on the average* the unperturbed plane wave reproduces the matrix element taken with ψ_{c} over this attractive well; it has the significant consequence that one can use Born approximation for calculating the contribution of the attractive well to the total energy —a calculation carried out in Section 3.

The above assumptions imply that these calculations will provide only rough estimates, and they are carried out for illustrative purposes. The approximations hold only for the particular form of the potential illustrated in Fig. 5.1. For a strong attractive potential concentrated closer to the hard core, more work is required, for example, actual solutions to the B-G equation in the combined potential must be obtained.

With the above two approximations, the energy shift of an interacting pair takes the form

$$\Delta\epsilon_{\mathbf{Pk}} \approx \Delta\epsilon_{\mathbf{Pk}}^{\text{hard core}} + \Delta\epsilon_{\mathbf{Pk}}^{\text{attractive}}(\text{independent-particle model}) \tag{5.4}$$

Nuclear Binding Energies. There are three contributions to the total nuclear binding energy with these approximations:

Attractive Well Contribution. The results from the independent-particle model in Eq. (3.15) may now be used for this contribution to the energy, with the specific values $a_{\text{W}} = a_{\text{M}} = 1/2$ and with the integral running from $\int_b^d$ (see Fig. 5.1)

$$\frac{E^{(\text{a})}}{A} = -\frac{V_0 k_{\text{F}}^3}{6\pi}\left[(d^3 - b^3) + \frac{27}{k_{\text{F}}^2}\int_b^d j_1^2(k_{\text{F}}z)dz\right] \tag{5.5}$$

This integral can be done analytically (Ref. [N2]).

Fermi Energy. The energy shifts are all calculated relative to the kinetic energy of a noninteracting Fermi gas

$$\frac{E^{(\text{f})}}{A} = \frac{3}{5}\frac{\hbar^2 k_{\text{F}}^2}{2m} \tag{5.6}$$

Note that this equation contains m and not m^* — the effective mass m^* represents the effects of the single-particle potential $U(\mathbf{k})$, which arises from the interactions.

Hard Core Energy. The energy shift due to the presence of a hard core in the nucleon-nucleon potential is given in the independent-pair approximation by

$$\Delta E^{(\text{c})} = \frac{1}{2}\sum_{\mathbf{k}_1\lambda_1\rho_1}^{k_{\text{F}}}\sum_{\mathbf{k}_2\lambda_2\rho_2}^{k_{\text{F}}} \Delta\epsilon_{\mathbf{Pk}}^{\text{hard core}} \tag{5.7}$$

It is in a relative s-wave where the hard core first plays a role, and in this state the energy shift is given by

$$\Delta\epsilon_{\mathbf{Pk}}^{(l=0)} = \frac{1}{V}\int e^{-i\mathbf{k}\cdot\mathbf{x}} V_{\text{c}} \frac{u_{\mathbf{Pk}}(x)}{x} d^3x \tag{5.8}$$

For a hard core, the product of the potential times the wave function is given by Eqs. (4.20) and (4.22) as

$$\begin{aligned} v u_{\mathbf{Pk}}(r) \equiv \frac{2\mu_{\text{red}}^*}{\hbar^2 k_{\text{F}}^2} V_{\text{c}} u_{\mathbf{Pk}}(r) &= \mathcal{A}\delta(r-c) + w(r) \\ &\approx \mathcal{A}\delta(r-c) \end{aligned} \tag{5.9}$$

For the general case of a pair with C-M momentum $\mathbf{P} \neq 0$ the amplitude $\mathcal{A}$ is given by [cf. Eq. (4.31)]

$$\mathcal{A}(\mathbf{P}, \mathbf{K}) = [\cos Kc - \int_0^\infty \cos Ks\, \chi_P(s,c) ds]^{-1}$$

$$\chi_P(r,r') \quad = \quad \frac{2rr'}{\pi}\int_{\overline{\Gamma}} \frac{d\Omega_t}{4\pi} j_0(tr)j_0(tr')t^2\,dt \tag{5.10}$$

Insertion of this result in Eq. (5.8) yields

$$\Delta\epsilon_{\mathbf{Pk}}^{(l=0)} \approx \left(\frac{4\pi c}{k_{\mathrm{F}}^3 V}\right) j_0(Kc)\mathcal{A}(\mathbf{P},\mathbf{K}) \left(\frac{\hbar^2 k_{\mathrm{F}}^2}{2\mu_{\mathrm{red}}^*}\right)(1-\delta_{\lambda_1\lambda_2}\delta_{\rho_1\rho_2}) \tag{5.11}$$

The last factor comes from the antisymmetrization of the wave function for particles of the same spin and isotopic spin; they cannot be in relative s-states. A combination of numerical factors then gives the s-wave contribution to the ground state energy of a gas of hard spheres interacting in the presence of a single-particle potential characterized by the effective mass m^*

$$\frac{E_{l=0}^{(c)}}{A} \approx \frac{\hbar^2 k_{\mathrm{F}}^2}{2m^*}\frac{2c}{\pi}\int\int_{\mathrm{F}} \frac{d^3k_1}{4\pi/3}\frac{d^3k_2}{4\pi/3} j_0(Kc)\mathcal{A}(\mathbf{P},\mathbf{K}) \tag{5.12}$$

The integrals are in dimensionless units and run over the initially occupied states in the Fermi gas (the region F).

A power series expansion of this result provides a useful orientation. The first term is particularly simple, since to leading order in c one has $j_0\mathcal{A} = 1$. Evaluation of the required integrals gives (Ref. [N2])

$$\frac{E_{l=0}^{(c)}}{A} = \frac{\hbar^2 k_{\mathrm{F}}^2}{2m^*}\left[\frac{2c}{\pi} + \frac{12c^2}{35\pi^2}(11 - 2\ln 2) + 0.26c^3 + \cdots\right] \tag{5.13}$$

The s-wave interaction gives the exact answer for the ground state energy of a hard sphere gas through second order in the dimensionless hard core radius c (Refs. [N33, N34, N35, N36]). The discussion in Section 4 indicates that the s-wave leak term $w(r)$, representing the product of the wave function and the potential inside the hard core [Eq. (5.9)], first contributes to this power series in order c^4 (Prob. 5.2).

An argument exactly analogous to that given here for s-waves yields the leading contribution to the energy from a hard core in p-states (Prob. 5.3)

$$\frac{E_{l=1}^{(c)}}{A} = \frac{\hbar^2 k_{\mathrm{F}}^2}{2m^*}\frac{c^3}{\pi} \tag{5.14}$$

All of these results for the hard core energy have been generalized to include the possibility of an effective mass coming from the attractive part of the interaction.

Numerical integration of the full s-wave contribution to the hard-core energy in Eq. (5.12) gives the results shown in Fig. 5.2, where it is compared with the power series expansion in c over the region of interest in nuclear matter. Also shown is the p-wave contribution in Eq. (5.14).

Leak contribution $w(r)$ 1st contributes to $O(c^4)$

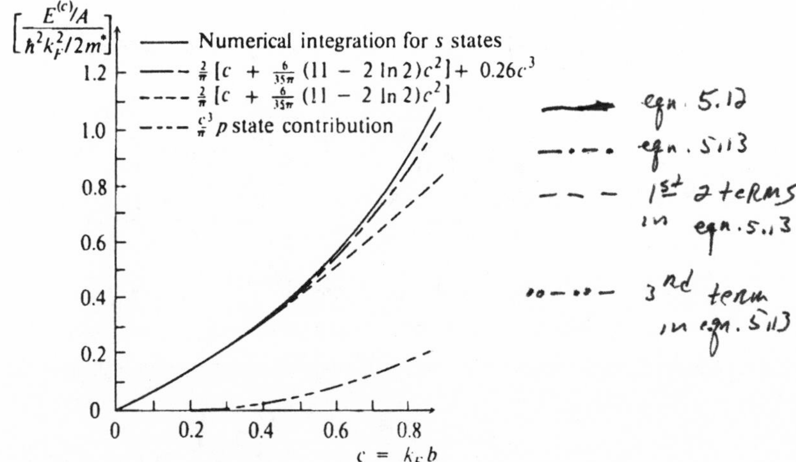

Figure 5.2: The ground state energy of a hard sphere gas as a function of the core size $c = k_\mathrm{F}b$; expressed per particle in units of $\hbar^2k_\mathrm{F}^2/2m^*$. From Ref. [N2].

Single-Particle Potential. The single-particle potential in the independent pair approximation is given by Eq. (4.10) as

$$U(\mathbf{k_1}\lambda_1\rho_1) = \sum_{\mathbf{k}_2\lambda_2\rho_2}^{k_\mathrm{F}} \Delta\epsilon_{\mathbf{Pk}} \tag{5.15}$$

Several features of this result are of interest:

Attractive Well Contribution. Use of the previous independent-particle approximation results in Eq. (3.17) with $a_\mathrm{W} = a_\mathrm{M} = 1/2$ and $\int_b^d$ gives

$$U^{(a)}(k^2) = -\frac{V_0 k_\mathrm{F}^3}{3\pi}\left[(d^3 - b^3) + \frac{9}{k_\mathrm{F}}\int_b^d j_0(kz)j_1(k_\mathrm{F}z)zdz\right] \tag{5.16}$$

Here $k \equiv k_1$. The momentum dependence of this single-particle potential enters through the factor $j_0(kz)$ in the integrand.

Effective Mass Approximation. The great advantage of the effective mass approximation in Section 4 is its transparency, leading to a simple modification of the differential form of the B-G equation and allowing one to have B-G wave functions in coordinate space that provide important physical insight into the behavior of nuclear matter.[23] One way to determine the effective mass is to match the single-particle spectrum at $k = k_\mathrm{F}$. This should provide an accurate

[23] An improved result is obtained by keeping the full expression $U(k^2)$ and solving the B-G equation in integral form.

Table 5.1: Effective mass determined from the attractive well at two different nuclear densities.

k_F	$U_1(k_F)$	$(m^*/m)_{k_F}$	$U_1(0)$	$(m^*/m)_{k=0}$
1.25 fm^{-1}	0.54	0.65	0.96	0.51
1.48 fm^{-1}	0.54	0.65	1.2	0.45

representation for transitions across the Fermi surface. In this case one has

$$\begin{aligned} U(k^2) &\approx U_0 + \frac{\hbar^2 k^2}{2m} U_1 \\ U_1 &= \frac{m}{\hbar^2 k_F} \left[\frac{dU}{dk}\right]_{k=k_F} \end{aligned} \tag{5.17}$$

Recall from Eq. (4.12) that

$$\frac{m^*}{m} = \frac{1}{1+U_1} \tag{5.18}$$

Another way to determine m^*/m is to expand about $k^2 = 0$; this gives the correct effective mass at the bottom of the Fermi sea. Both results can be determined analytically from Eq. (5.16). The numerical results for the effective mass arising from the attractive part of the nucleon-nucleon potential are shown in Table 5.1 for two different nuclear densities spanning the region of interest in nuclear matter. Note that m^*/m is insensitive to changes in the nuclear density in this region, and there is not a large difference in going from $k = 0$ to $k = k_F$.

Contribution of Hard Core to the Effective Mass. The energy shift due to the hard core is obtained to leading order in c from Eq. (5.11) (recall that to this order $j_0\mathcal{A} = 1$)

$$\Delta\epsilon^{(l=0)}_{\mathbf{Pk}} = \frac{4\pi c}{k_F^3 V}\left(\frac{\hbar^2 k_F^2}{2\mu^*_{\rm red}}\right)(1 - \delta_{\lambda_1\lambda_2}\delta_{\rho_1\rho_2}) \tag{5.19}$$

This result is *independent* of $\mathbf{k}_1$ and $\mathbf{k}_2$; hence, there is no contribution from this term to the effective mass. In fact, the contribution of the hard core interaction to the effective mass in nuclear matter is very small (Refs. [N2, N26, N36]) and shall be neglected here. As a consequence, the self-consistency problem is simplified enormously — m^*/m is determined by the contribution of the attractive well in Eq. (5.16), and this contribution is independent of m^*/m .

Results. The total nuclear energy/particle computed from these expressions is shown as a function of k_F (or density) in Fig. 5.3. Here a value of $m^*/m = 0.65$ from Table 5.1 was used in the calculation. Several features of these results are of interest:

1. Nuclear matter obviously *saturates.* The hard core energy becomes infinite at close packing due to the uncertainty principle; the hard cores put curvature

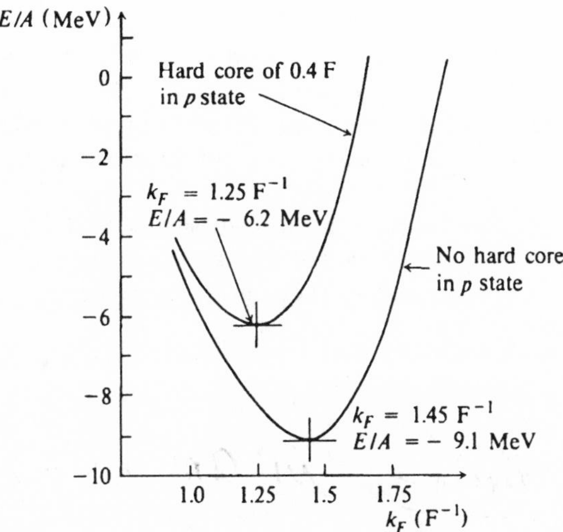

Figure 5.3: Energy/particle as a function of k_{F} for nuclear matter calculated as described in the text. Here $1\,\text{F} \equiv 1\,\text{fm}$. From Ref. [N2].

into the wave function, and this increases the kinetic energy of the system. It is evident that this positive quantum mechanical hard core energy is important well before the hard spheres actually touch.

2. Nuclear matter saturates at about the right density.

3. The binding energy at saturation is somewhat too small. If the ${}^1S_0 - {}^3S_1$ interaction difference is attributed entirely to the tensor force, then the spin and isospin average of the tensor force vanishes in the noninteracting Fermi gas (Prob. 3.1), and the second-order contribution will always *increase* the binding.

4. The net binding energy results from the *cancellation* of large contributions. For example, at $k_{\text{F}} = 1.50\ \text{fm}^{-1}$

$$\frac{E^{(a)}}{A} = -76.9\ \text{MeV} \qquad \frac{E^{(f)}}{A} = +28.0\ \text{MeV}$$
$$\frac{E^{(c)}_{l=0}}{A} = +39.9\ \text{MeV} \qquad \frac{E^{(c)}_{l=1}}{A} = +4.9\ \text{MeV} \tag{5.20}$$

5. The hard core is one element that keeps nuclear matter density relatively low; the Serber force, which weakens the force in the odd-l states, is another.

6. Nuclear matter has low enough density so that the pair wave function *heals* by the nearest neighbor distance. This serves as a justification for both the independent-pair approximation and the independent-particle model.

7. The vanishing of the relative wave function at the hard core radius implies that two nucleons never get to the region of short-distance overlap in nuclear matter; thus this calculation should not be sensitive to what actually goes on at these very short distances.

Many problems in the description of nuclear matter still remain within the framework of the present calculation based on static internucleon potentials and the nonrelativistic Schrödinger equation: the validity of the effective mass approximation, the role of higher clusters, uncertainty in our knowledge of V_{ij}, and the nature of many-body forces. In addition, the consequences of a fully relativistic theory remain to be investigated. In fact, in Part II we will see that a simple relativistic hadronic mean field theory leads to quite a different saturation mechanism — which viewed in the present light would consist of an infinite series of velocity-dependent (nonstatic) interactions.

Problems (NRQM & static V_{ij})

- Effective Mass Approx
- Higher clusters
- Many-Body forces (Tough ↓)
- Knowledge of V_{ij}
- Relativistic Theory
- Correct Degree of Freedom

6

THE SHELL MODEL

The previous discussion was concerned with the properties of infinite nuclear matter. We now turn our attention to finite nuclei. The starting point is the nuclear shell model (Ref. [N37]), which provides both a remarkably successful single-particle description of nuclei as well as a complete basis in which to describe more complicated nuclear states. The section starts with a discussion of the general canonical transformation to particles and holes in a finite nuclear many-body system (Ref. [N2]).

General Canonical Transformation to Particles and Holes. For clarity, assume first a spherical system and just one kind of fermion (i.e., p or n); the discussion will subsequently be generalized. The single-particle states will be characterized by the good quantum numbers in such a system

$$|\alpha\rangle = |nlsjm_j\rangle \quad \equiv \quad |a, m_\alpha\rangle \tag{6.1}$$

It is convenient to use the notation

$$|-\alpha\rangle = |nlsj - m_j\rangle \quad \equiv \quad |a, -m_\alpha\rangle \tag{6.2}$$

Assume that the ground state is well described in first approximation by a set of single-particle levels completely filled up to F (Fig. 6.1). Carry out the following canonical transformation on the creation and destruction operators $c_\alpha^\dagger, c_\alpha$

$$\begin{aligned} a_\alpha^\dagger &\equiv c_\alpha^\dagger\,; & \alpha > F \quad &\text{particles} \\ b_\alpha^\dagger &\equiv S_{-\alpha}c_{-\alpha}\,; & \alpha < F \quad &\text{holes} \end{aligned} \tag{6.3}$$

Here the phase S_α is defined by

$$S_\alpha \equiv (-1)^{j_\alpha - m_\alpha} \quad = \quad -S_{-\alpha} \tag{6.4}$$

where j_α is half-integral. Clearly

$$\{a_\alpha, a_{\alpha'}^\dagger\} = \{b_\alpha, b_{\alpha'}^\dagger\} \quad = \quad \delta_{\alpha,\alpha'} \tag{6.5}$$

and all other operators anticommute. Equations (6.3) therefore represent a *canonical transformation*. The reason for the phase S_α in Eq. (6.3) is that $b_\alpha^\dagger$ is now an *irreducible tensor operator* (ITO) of rank j. Thus it properly creates an eigenstate of angular momentum. The proof of this statement is

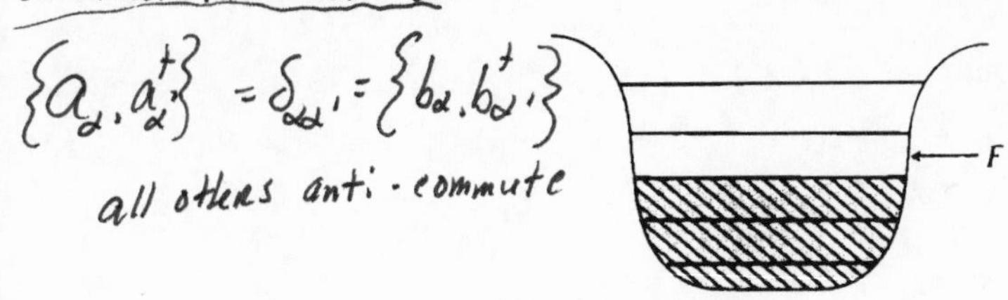

Figure 6.1: Single-particle levels filled up to F.

given in Appendix B. The complete canonical transformation in Eqs. (6.3) can be rewritten as

$$\begin{aligned} c_\gamma &= \theta(\gamma - F)a_\gamma + \theta(F-\gamma)S_\gamma b^\dagger_{-\gamma} \\ c^\dagger_\gamma &= \theta(\gamma - F)a^\dagger_\gamma + \theta(F-\gamma)S_\gamma b_{-\gamma} \end{aligned} \tag{6.6}$$

The goal is now to rewrite the nuclear hamiltonian

$$\hat{H} = \sum_{\alpha\beta} c^\dagger_\alpha \langle\alpha|T|\beta\rangle c_\beta + \frac{1}{2}\sum_{\alpha\beta\gamma\delta} c^\dagger_\alpha c^\dagger_\beta \langle\alpha\beta|V|\gamma\delta\rangle c_\delta c_\gamma \tag{6.7}$$

in normal-ordered form with respect to the new creation and destruction operators in Eq. (6.6). This hamiltonian governs the dynamics of the nucleus at the equilibrium nuclear density. Note that in writing Eq. (6.7) it has been assumed that the matrix elements of the potential V are finite.[24]

Wick's theorem (Ref. [N2]) may now be used to normal order the operators.[25] For example,

$$c^\dagger_\alpha c_\beta = N(c^\dagger_\alpha c_\beta) + c^{\dagger\bullet}_\alpha c^\bullet_\beta \tag{6.8}$$

Here N indicates a normal ordering with respect to the new particle and hole operators a and b [i.e., all destruction operators are placed to the right of the creation operators and a sign is affixed equal to (-1) raised to the number of interchanges of fermion operators needed to achieve this normal ordering]. The pair of dots in Eq. (6.8) indicates a contraction (i.e., the terms remaining after achieving the normal-ordered form). The new noninteracting "vacuum" is defined to contain neither particles nor holes (see Fig. 6.1)

$$a_\alpha|0\rangle = b_\alpha|0\rangle = 0 \tag{6.9}$$

[24] If the potential is singular, then the interaction must initially be treated in the independent pair approximation, as in Section 4, and the Bethe-Goldstone equation solved in the finite nuclear system (Ref. [N2]).

[25] To make the formal connection with Wick's theorem complete, one may consider the operators to be time dependent with the time of the operator on the left infinitesimally later than that on the right; however, a little reflection will convince the reader that this artifice is unnecessary; Eq. (6.12) is an algebraic identity.

It follows from these relations that

$$
\begin{aligned}
c_\alpha^{\dagger\bullet} c_\beta^{\bullet} &= \langle 0|c_\alpha^\dagger c_\beta|0\rangle = \delta_{\alpha\beta}\theta(F-\alpha) \\
c_\alpha^{\dagger\bullet} c_\beta^{\dagger\bullet} &= c_\alpha^{\bullet} c_\beta^{\bullet} = 0
\end{aligned}
\tag{6.10}
$$

Thus Wick's theorem applied to the bilinear kinetic energy term gives

$$
c_\alpha^\dagger c_\beta = \delta_{\alpha\beta}\theta(F-\alpha) + N(c_\alpha^\dagger c_\beta)
\tag{6.11}
$$

Wick's theorem applied to the quadralinear potential energy term gives

$$
\begin{aligned}
c_\alpha^\dagger c_\beta^\dagger c_\delta c_\gamma &= N(c_\alpha^\dagger c_\beta^\dagger c_\delta c_\gamma) + (\delta_{\alpha\gamma}\delta_{\beta\delta} - \delta_{\alpha\delta}\delta_{\beta\gamma})\theta(F-\alpha)\theta(F-\beta) \\
&\quad + \delta_{\alpha\gamma}\theta(F-\alpha)N(c_\beta^\dagger c_\delta) + \delta_{\beta\delta}\theta(F-\beta)N(c_\alpha^\dagger c_\gamma) \\
&\quad - \delta_{\alpha\delta}\theta(F-\alpha)N(c_\beta^\dagger c_\gamma) - \delta_{\beta\gamma}\theta(F-\beta)N(c_\alpha^\dagger c_\delta)
\end{aligned}
\tag{6.12}
$$

This result may now be substituted into the hamiltonian $\hat{H}$ in Eq. (6.7). Note that the potential is symmetric under particle interchange, which implies

$$
\langle\alpha\beta|V|\gamma\delta\rangle = \langle\beta\alpha|V|\delta\gamma\rangle
\tag{6.13}
$$

A change of dummy summation indices then indicates that the two terms in the second line of Eq. (6.12) make equal contributions to $\hat{H}$, as do the two terms in the third line.

Hartree-Fock Equations. A major simplification is obtained if the bilinear terms in $\hat{H}$ are *diagonal*, for then part of the finite nucleus problem has been solved exactly. The bilinear terms will be diagonal if the following equations are satisfied

$$
\langle\beta|T|\delta\rangle + \sum_{\alpha<F}[\langle\alpha\beta|V|\alpha\delta\rangle - \langle\alpha\beta|V|\delta\alpha\rangle] = \epsilon_\beta\delta_{\beta\delta}
\tag{6.14}
$$

The form of the single-particle wave functions, except for the good quantum numbers, has until now been unspecified. Choose them as solutions to the *Hartree-Fock* (H-F) equations

$$
\begin{aligned}
T\phi_\delta(2) + \sum_{\alpha<F}[\int \phi_\alpha(1)^\dagger V(1,2)\phi_\alpha(1)\phi_\delta(2)d^3x_1 & \\
- \int \phi_\alpha(1)^\dagger V(1,2)\phi_\delta(1)\phi_\alpha(2)d^3x_1] &= \epsilon_\delta\phi_\delta(2)
\end{aligned}
\tag{6.15}
$$

Equations (6.14) then follow immediately from Eqs. (6.15) for it is readily established that the H-F eigenvalues ϵ_α in these equations are real and that the solutions to these H-F equations are orthonormal (Prob. 6.1). Insertion of these results puts the hamiltonian $\hat{H}$ into the following form

$$\begin{aligned}
\hat{H} &= H_0 + \hat{H}_1 + \hat{H}_2 \\
H_0 &= \sum_{\alpha<F} (T_\alpha + \frac{1}{2}V_\alpha) \\
\hat{H}_1 &= \sum_{\alpha>F} \epsilon_\alpha a_\alpha^\dagger a_\alpha - \sum_{\alpha<F} \epsilon_\alpha b_\alpha^\dagger b_\alpha \\
\hat{H}_2 &= \frac{1}{2} \sum_{\alpha\beta\gamma\delta} \langle\alpha\beta|V|\gamma\delta\rangle N(c_\alpha^\dagger c_\beta^\dagger c_\delta c_\gamma)
\end{aligned} \tag{6.16}$$

This is an *exact* result. The H-F energies in these expressions are given by Eq. (6.14) as

$$\begin{aligned}
\epsilon_\alpha &\equiv T_\alpha + V_\alpha \\
T_\alpha &= \langle\alpha|T|\alpha\rangle \\
V_\alpha &= \sum_{\beta<F} [\langle\alpha\beta|V|\alpha\beta\rangle - \langle\alpha\beta|V|\beta\alpha\rangle]
\end{aligned} \tag{6.17}$$

Several comments are appropriate here:

1. These are completely general results. They are, in fact, not restricted to spherically symmetric systems (note the phase S_α never enters). The index α need denote only a complete set of solutions to the H-F equations.[26]

2. The H-F result $E_0 \equiv H_0$ also provides a *variational estimate* for the energy since

$$\langle 0|\hat{H}_2|0\rangle = \langle 0|\hat{H}_1|0\rangle = 0 \tag{6.18}$$

Therefore

$$\langle 0|\hat{H}|0\rangle = H_0 \tag{6.19}$$

The H-F wave functions are the best possible single-particle wave functions.

3. So far only one kind of fermion has been assumed. Thus the preceding results are directly applicable, for example, to atoms; however, the arguments are immediately generalized to the nuclear system with the inclusion of isospin

$$\begin{aligned}
|\alpha\rangle &= |nl\frac{1}{2}jm_j;\frac{1}{2}m_t\rangle \\
S_\alpha &= (-1)^{j_\alpha - m_{j\alpha}}(-1)^{\frac{1}{2}-m_{t\alpha}}
\end{aligned} \tag{6.20}$$

Then $b_\alpha^\dagger$ is an ITO with respect to both angular momentum and isospin (Appendix B).

[26] Strictly speaking $\hat{H}_1 = \sum_{\alpha>F} \epsilon_\alpha a_\alpha^\dagger a_\alpha - \sum_{\alpha<F} \epsilon_{-\alpha} b_\alpha^\dagger b_\alpha$. In the spherically symmetric case $\epsilon_\alpha = \epsilon_a = \epsilon_{-\alpha}$. It is assumed here that the pairs α and $-\alpha$ are filled below F.

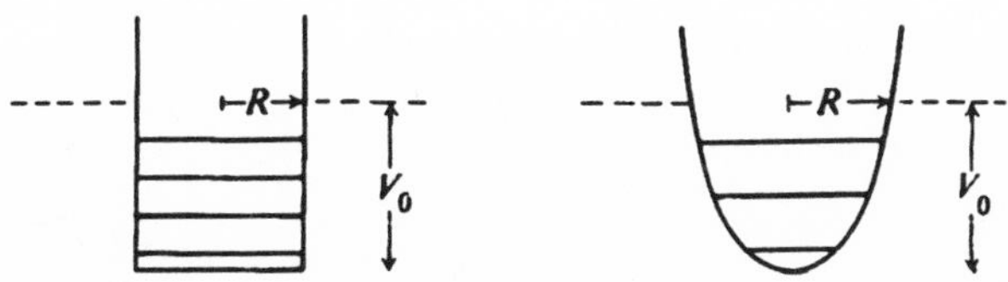

Figure 6.2: Solvable infinite square well and isotropic three-dimensional simple harmonic oscillator potentials.

4. One can now readily add particles and holes to the ground state $|0\rangle$ with the creation operators $a^\dagger$ and $b^\dagger$.

Single-Particle Shell Model. Although the H-F equations describe single-particle wave functions, they are still complicated coupled, nonlinear, integro-differential equations. Nonetheless, they have been solved numerically in many cases, and they provide remarkably accurate descriptions of finite nuclear systems (Refs. [N38, N39, N40, N41]). Typically the calculations use phenomenological, nonsingular, density-dependent interactions (Refs. [N39, N40, N41]), although interactions based on more fundamental nuclear matter studies with the free nucleon-nucleon force have also been employed (Ref. [N38]).

Many properties of finite nuclei can be understood with wave functions generated with approximate, solvable, single-particle potentials. Such wave functions always provide a first orientation and physical insight. We therefore consider two solvable single-particle models: first a spherical cavity with infinite potential walls, and second an isotropic three-dimensional simple harmonic oscillator (which also, of course, has infinite walls). Figure 6.2 illustrates these two cases; the bottom of the well at a radius $r = R$ has been adjusted to be at a potential $-V_0$ in both cases so a comparison can be made between them.

The wave functions for the infinite square well potential take the form

$$\begin{aligned} \Psi_{nlm} &= N_{nl} j_l(kr) Y_{lm}(\Omega_r) \\ \epsilon &= \frac{\hbar^2 k^2}{2m} - V_0 \end{aligned} \tag{6.21}$$

This solution is finite at the origin; it must also vanish at the boundary $j_l(kR) = 0$, which implies

$$k_{nl} R = X_{nl} \tag{6.22}$$

Here X_{nl} is the nth zero of the lth spherical Bessel function, excluding the origin and including the zero at the boundary. The level orderings are shown on the right-hand side (r.h.s.) of Fig. 6.3. The required integral over the spherical Bessel functions (see Ref. [N42]) gives the normalization constant

$$N_{nl}^2 = \frac{2}{R^3 j_{l+1}^2(X_{nl})} \tag{6.23}$$

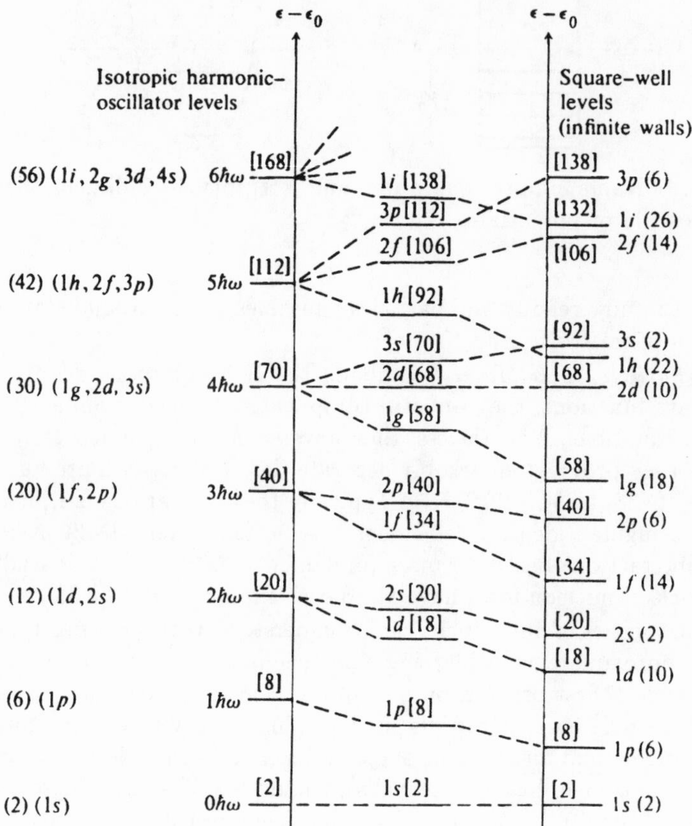

Figure 6.3: Level orderings in the two potentials shown in Fig. 6.2. The ground states have been arbitrarily normalized to the same energy (Refs. [N37, N2]).

The potential for the infinite oscillator also shown in Fig. 6.2 takes the form

$$\begin{aligned} V(r) &= -V_0\left[1-\left(\frac{r}{R}\right)^2\right] = -V_0 + \frac{1}{2}m\omega^2 r^2 \\ \frac{V_0}{R^2} &= \frac{1}{2}m\omega^2 \end{aligned} \tag{6.24}$$

One looks for solutions to the Schrödinger equation of the form

radial eqn.

$$\Psi_{nlm} = \frac{u_{nl}(r)}{r}Y_{lm}(\Omega_r) \tag{6.25}$$

$$\left[\frac{-\hbar^2}{2m}\frac{d^2}{dr^2} + \frac{1}{2}m\omega^2 r^2 + \frac{\hbar^2}{2m}\frac{l(l+1)}{r^2} - (\mathcal{E}_{nl}+V_0)\right]u_{nl}(r) = 0$$

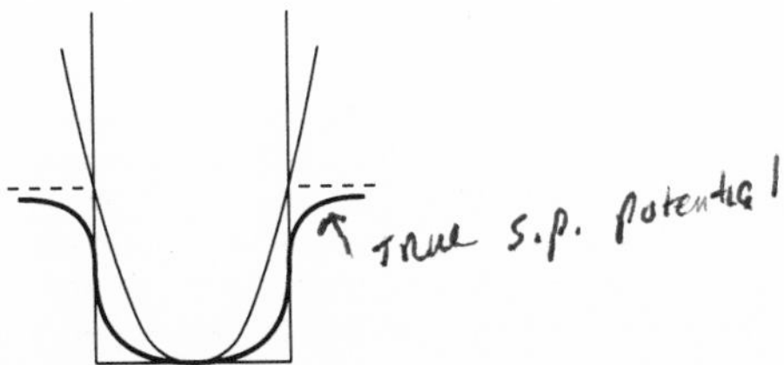

Figure 6.4: Expected shape of true single-particle potential compared with the two solvable models.

With the introduction of dimensionless variables $\hbar\omega \equiv \hbar^2/mb^2$ and $q \equiv r/b$ the radial equation takes the form

$$\left[-\frac{d^2}{dq^2} + q^2 + \frac{l(l+1)}{q^2} - \frac{2(\epsilon_{nl} + V_0)}{\hbar\omega}\right] u_{nl}(q) = 0 \tag{6.26}$$

The acceptable solutions are the *Laguerre polynomials* (Ref. [N43])

$$\begin{aligned} u_{nl}(q) &= N_{nl} q^{l+1} e^{-\frac{1}{2}q^2} L_{n-1}^{l+\frac{1}{2}}(q^2) \\ L_p^a(z) &\equiv \frac{\Gamma(a+p+1)}{\Gamma(p+1)} \frac{e^z}{z^a} \frac{d^p}{dz^p}(z^{a+p} e^{-z}) \end{aligned} \tag{6.27}$$

As in the previous case, $n = 1, 2, 3, \ldots, \infty$ is the number of nodes in the radial wave function, including the one at infinity. The eigenvalues are given by

$$\begin{aligned} \epsilon_{nl} &= \hbar\omega\left(N + \frac{3}{2}\right) - V_0 \\ N &= 2(n-1) + l = 0, 1, 2, \ldots, \infty \end{aligned} \tag{6.28}$$

The normalization follows from the integrals in Ref. [N43]

$$N_{nl}^2 = \frac{2(n-1)!}{b[\Gamma(n+l+\frac{1}{2})]^3} \tag{6.29}$$

The level orderings in this potential are shown on the left-hand side (l.h.s.) of Fig. 6.3.

The true nuclear single-particle potential will be finite; however, the low-lying levels in the potential will not be greatly affected by extending the walls of the potential to infinity. The shape of the true single-particle well over the nuclear volume can be expected to lie somewhere between the two extreme cases shown in Fig. 6.2 (see Fig. 6.4). It is possible to interpolate between the two solvable cases by flattening out the oscillator, or rounding off the square well. The highest-l states in each shell in the oscillator will be lowered the most, since they spend most of their time at the edge of the potential (see Fig. 6.4). Note

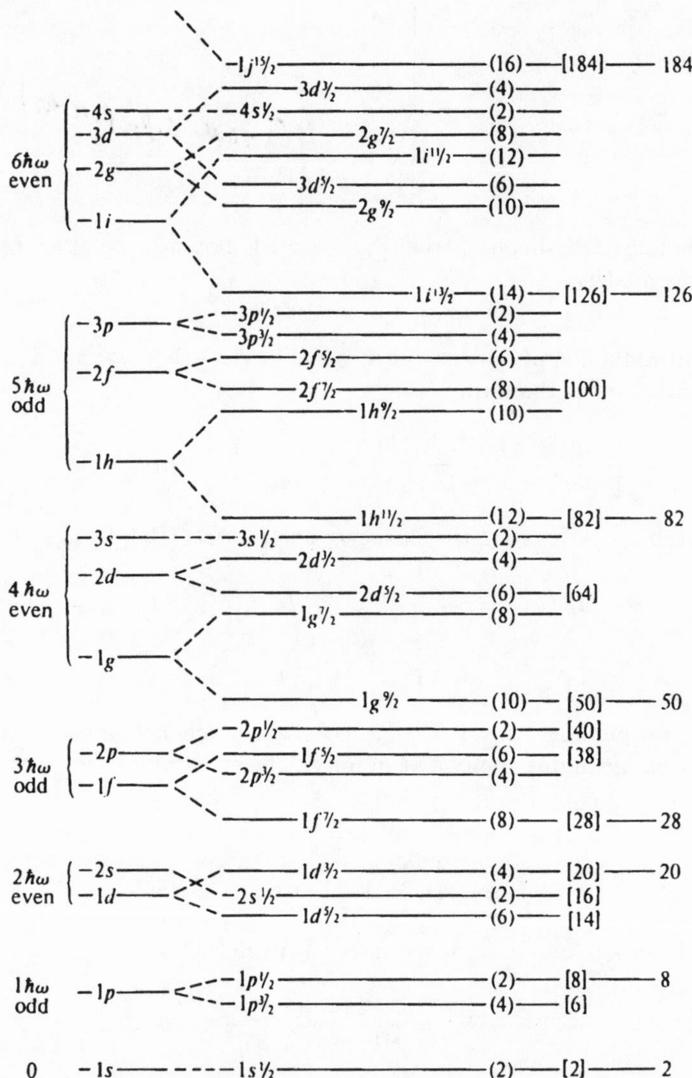

Figure 6.5: Level orderings in the nuclear single-particle potential with an additional atractive spin-orbit interaction (Refs. [N37, N2]).

^{40}Ca last stable one

both (nl) remain good quantum numbers in this interpolation, whose results are indicated in the center column of Fig. 6.3.

Spin-Orbit Splitting. It is known experimentally that certain groups of identical nucleons (p or n) possess special stability. These so-called *magic numbers* occur at $2, 8, 20, 28, 50, 82, 126, \ldots$. One does not obtain these numbers from the preceding analysis, where the total occupancies of the closed shells with the more realistic nuclear well (Fig. 6.4) are shown in the center column in Fig. 6.3. Mayer and Jensen (see Ref. [N37]) suggested that there is an additional *attractive single-particle spin-orbit term* in the H-F potential of the form[27]

$$H' = -\alpha(r)\,\mathbf{l}\cdot\mathbf{s} \tag{6.30}$$

The single-particle eigenstates are characterized by the eigenvalues $|nlsjm_j\rangle$. Note

$$\begin{aligned} 2\mathbf{l}\cdot\mathbf{s}|nlsjm_j\rangle &= (\mathbf{j}^2 - \mathbf{l}^2 - \mathbf{s}^2)\,|nlsjm_j\rangle \\ &= [j(j+1) - l(l+1) - s(s+1)]\,|nlsjm_j\rangle \end{aligned} \tag{6.31}$$

In the present case $s = 1/2$. Thus if $j = l + 1/2$ then $\mathbf{l}\cdot\mathbf{s} = l/2$, and if $j = l - 1/2$ then $\mathbf{l}\cdot\mathbf{s} = -(l+1)/2$. Therefore

$$\begin{aligned} \epsilon_{l-\frac{1}{2}} - \epsilon_{l+\frac{1}{2}} &= \alpha_{nl}\left(l + \frac{1}{2}\right) \\ \alpha_{nl} &\equiv \int u_{nl}^2(r)\alpha(r)dr \end{aligned} \tag{6.32}$$

This shell model of Mayer and Jensen now predicts the correct magic numbers if the spin-orbit interaction is strong enough to push the state of highest l and highest j in a major oscillator shell down into the next lower shell (Fig. 6.5). Upon filling the levels in Fig. 6.5 consecutively, one *predicts* the spins and parities of nuclei if all the properties of the nucleus are ascribed to the last nucleon. This extreme single-particle model of the nucleus is remarkably successful in this regard (Ref. [N37]). Also predicted are islands of isomerism where the only levels available to a particle in a low lying excited state differ greatly in angular momentum and have different parity; in such cases, the electromagnetic transition rates will be very small and nuclear metastability (isomerism) will occur.

It is to the nuclear electromagnetic properties and transition rates that we now turn our attention.

[27] We will find a natural explanation for the occurrence of this spin-orbit interaction within QHD in Part II of this book.

7

ELECTROMAGNETIC INTERACTIONS

In this section we discuss the interaction of nuclei, or any finite quantum mechanical system, with electromagnetic fields. Much of what we know about nuclei comes from such interactions. We start with the general multipole analysis of the interaction of a nucleus with the quantized radiation field (see Refs. [N4, N14, N45, N46, N47]). In the following $e_{\mathrm{p}} = |e|$ is the proton charge.

Multipole Analysis. The starting point in this analysis is the total hamiltonian for the nuclear system, the free photon field, and the electromagnetic interaction

$$\begin{aligned} H_{\text{total}} &= H_{\text{nuclear}} + \sum_{\mathbf{k}} \sum_{\rho=1,2} \hbar\omega_k a^\dagger_{\mathbf{k}_\rho} a_{\mathbf{k}_\rho} \\ &\quad - \frac{e_{\mathrm{p}}}{c} \int \mathbf{J}_N(\mathbf{x})\cdot\mathbf{A}(\mathbf{x}) d^3x + \frac{e_{\mathrm{p}}^2}{8\pi} \int\int \frac{\rho_N(\mathbf{x})\rho_N(\mathbf{x}')}{|\mathbf{x}-\mathbf{x}'|} d^3x d^3x' \end{aligned} \tag{7.1}$$

This is the hamiltonian of *quantum electrodynamics* (QED); it is written in the Coulomb gauge. $\mathbf{A}$ is the vector potential for the quantized radiation field

$$\mathbf{A}(\mathbf{x}) = \sum_{\mathbf{k}} \sum_{\rho=1,2} \frac{1}{(2\omega_k \Omega)^{1/2}} [\mathbf{e}_{\mathbf{k}_\rho} a_{\mathbf{k}_\rho} e^{i\mathbf{k}\cdot\mathbf{x}} + \text{h.c.}] \tag{7.2}$$

Here $\mathbf{e}_{\mathbf{k}_\rho}$ with $\rho = (1,2)$ represent two unit vectors transverse to $\mathbf{k}$ (see Fig. 7.1). The hermitian conjugate is denoted by h.c. We quantize with periodic boundary conditions (p.b.c.) in a large box of volume Ω, and in the end let $\Omega \to \infty$. Furthermore, here and henceforth we adopt a system of units where

$$\hbar = c = 1 \tag{7.3}$$

All final formulas will be written back in dimensionally correct units.

The only assumption made about the nucleus to this point is the existence of local current and charge density operators $\mathbf{J}(\mathbf{x}), \rho(\mathbf{x})$.[28] This must be the case for any true quantum mechanical system.[29]

[28] H_{nuclear} could be given in terms of potentials, or it could be for a coupled baryon and meson system, or it could be for a system of quarks and gluons; it does not matter at this point. $\mathbf{J}(\mathbf{x})$ and $\rho(\mathbf{x})$ could, for example, contain exchange currents.

[29] Although Eq. (7.1) is correct in QCD, some models may have an additional term of $O(e^2\mathbf{A}^2)$ in the hamiltonian; the arguments in this section are unaffected by such a term.

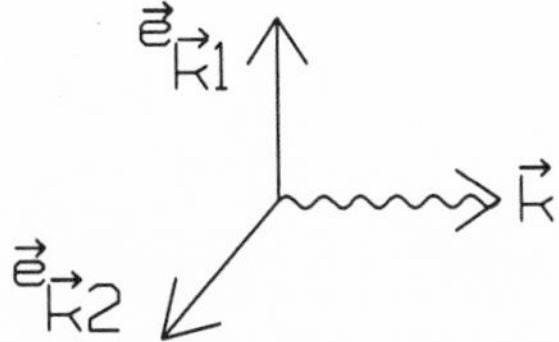

Figure 7.1: Transverse unit vectors.

It is convenient to first go from plane polarization to circular polarization with the transformation (cf. Fig. 7.1)

$$\mathbf{e}_{\pm 1} \equiv \mp\frac{1}{\sqrt{2}}(\mathbf{e}_1 \pm i\mathbf{e}_2) \qquad\qquad \mathbf{e}_0 \equiv \mathbf{e}_z \equiv \frac{\mathbf{k}}{|\mathbf{k}|} \tag{7.4}$$

These circular polarization vectors satisfy the relations

$$\mathbf{e}^{\dagger}_{\mathbf{k}\lambda} = (-1)^{\lambda}\mathbf{e}_{\mathbf{k}-\lambda} \qquad\qquad \mathbf{e}^{\dagger}_{\lambda}\cdot\mathbf{e}_{\lambda'} = \delta_{\lambda\lambda'} \tag{7.5}$$

If, at the same time, one defines

$$a_{\mathbf{k}\pm 1} \equiv \mp\frac{1}{\sqrt{2}}(a_{\mathbf{k}1} \mp i a_{\mathbf{k}2}) \tag{7.6}$$

then the transformation is *canonical*

$$[a_{\mathbf{k}\lambda}, a^{\dagger}_{\mathbf{k}'\lambda'}] = \delta_{\mathbf{k}\mathbf{k}'}\delta_{\lambda\lambda'} \tag{7.7}$$

Since $\mathbf{e}_1 a_1 + \mathbf{e}_2 a_2 = \mathbf{e}_{+1}a_{+1} + \mathbf{e}_{-1}a_{-1}$ the vector potential takes the form

$$\mathbf{A}(\mathbf{x}) = \sum_{\mathbf{k}}\sum_{\lambda=\pm 1}\frac{1}{(2\omega_k\Omega)^{1/2}}[\mathbf{e}_{\mathbf{k}\lambda}a_{\mathbf{k}\lambda}e^{i\mathbf{k}\cdot\mathbf{x}} + \text{h.c.}] \tag{7.8}$$

The index $\lambda = \pm 1$ is the circular polarization, as we shall see, and only $\lambda = \pm 1$ appears in the expansion so $\nabla\cdot\mathbf{A}(\mathbf{x}) = 0$, characterizing the Coulomb gauge.

Now proceed to calculate the transition probability for the nucleus to make a transition between two states and emit (or absorb) a photon. Work to lowest order in the electric charge e, use the *Golden Rule*, and compute the nuclear matrix element $\langle J_f M_f \mathbf{k}\lambda|H'|J_i M_i\rangle$ where H' is here the term linear in the vector potential in Eq. (7.1); it is this interaction term that can create (or destroy) a photon. All that will be specified about the nuclear state at this point is that it is an eigenstate of angular momentum. It will be assumed that the target is massive and its position will be taken to define the origin; transition current densities occur over the nuclear volume and hence *all transition current densities will be localized in space.* Since the photon matrix element is $\langle\mathbf{k}\lambda|a^{\dagger}_{\mathbf{k}'\lambda'}|0\rangle =$

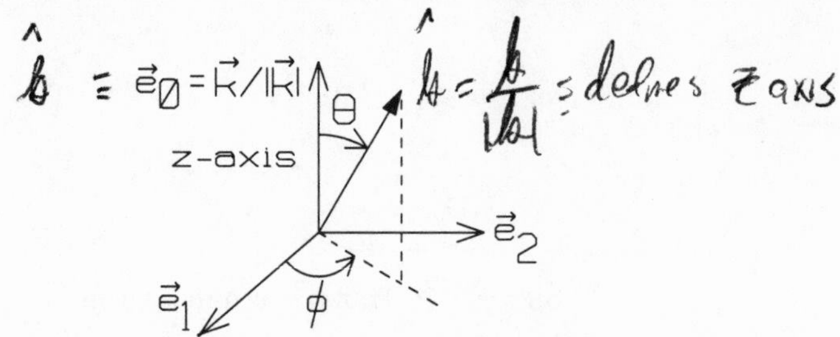

Figure 7.2: Coordinate system with z-axis defined by photon momentum.

$\delta_{kk'}\delta_{\lambda\lambda'}$, the required transition matrix element takes the form[30]

$$\langle J_f M_f \mathbf{k}\lambda|\hat{H}'|J_i M_i\rangle = \frac{-e_{\mathrm{p}}}{(2\omega_k\Omega)^{1/2}}\langle J_f M_f|\int e^{-i\mathbf{k}\cdot\mathbf{x}}\mathbf{e}^{\dagger}_{\mathbf{k}\lambda}\cdot\hat{\mathbf{J}}(\mathbf{x})d^3x|J_i M_i\rangle \tag{7.9}$$

Start by taking the photon momentum to define the z-axis (Fig. 7.2); the generalization follows below. In this case the plane wave can be expanded as

$$e^{i\mathbf{k}\cdot\mathbf{x}} = \sum_l i^l\sqrt{4\pi(2l+1)}j_l(kx)Y_{l0}(\Omega_x) \tag{7.10}$$

The vector spherical harmonics are defined by the relations (Ref. [N44])

$$\boldsymbol{\mathcal{Y}}^M_{Jl1} \equiv \sum_{m\lambda}\langle lm1\lambda|l1JM\rangle Y_{lm}(\Omega_x)\mathbf{e}_\lambda \tag{7.11}$$

Note this sum goes over all three spherical unit vectors, $\lambda = \pm 1, 0$. The definition in Eq. (7.11) can be inverted with the aid of the orthogonality properties of the Clebsch-Gordan (C-G) coefficients

$$Y_{lm}\mathbf{e}_\lambda = \sum_{JM}\langle lm1\lambda|l1JM\rangle\boldsymbol{\mathcal{Y}}^M_{Jl1} \tag{7.12}$$

The $\mathbf{e}_\lambda$ are now just fixed vectors; they form a complete orthonormal set. Therefore any vector can be expanded in spherical components as

$$\begin{aligned} \mathbf{v} &= \sum_\lambda(\mathbf{v}\cdot\mathbf{e}_\lambda)\,\mathbf{e}^\dagger_\lambda = \sum_\lambda v_\lambda\mathbf{e}^\dagger_\lambda \\ v_{\pm1} &= \mp\frac{1}{\sqrt{2}}(v_x \pm iv_y) \qquad v_0 = v_z \end{aligned} \tag{7.13}$$

As we shall see, the vector spherical harmonics project an irreducible tensor operator (ITO) of rank J from any vector density operator in the nuclear Hilbert

[30] We now revert to the notation where a caret over a symbol denotes an operator in the nuclear Hilbert space.

space. A combination of Eqs. (7.10) and (7.12) and use of the properties of the C-G coefficients yields[31]

$$\mathbf{e}_{\mathbf{k}\lambda}e^{i\mathbf{k}\cdot\mathbf{x}} = \sum_l \sum_J i^l \sqrt{4\pi(2l+1)}\, j_l(kx) \langle l01\lambda|l1J\lambda\rangle \boldsymbol{\mathcal{Y}}^\lambda_{Jl1}(\Omega_x) \tag{7.14}$$

The C-G coefficient limits the sum on l to three terms $l = J, J \pm 1$, and these C-G coefficients can be explicitly evaluated to give for $\lambda = \pm 1$

$$\mathbf{e}_{\mathbf{k}\lambda}e^{i\mathbf{k}\cdot\mathbf{x}} = \sum_{J\geq 1} i^J \sqrt{\frac{4\pi(2J+1)}{2}} \left\{ \mp j_J(kx)\boldsymbol{\mathcal{Y}}^\lambda_{JJ1} \right.$$
$$\left. -i\left[\sqrt{\frac{J+1}{2J+1}}\, j_{J-1}(kx)\boldsymbol{\mathcal{Y}}^\lambda_{J,J-1,1} - \sqrt{\frac{J}{2J+1}}\, j_{J+1}(kx)\boldsymbol{\mathcal{Y}}^\lambda_{J,J+1,1}\right]\right\} \tag{7.15}$$

From Ref. [N44]

$$\begin{aligned} \nabla \times j_J(kx)\boldsymbol{\mathcal{Y}}^\lambda_{JJ1} &= i\left[\left(\frac{d}{dx} - \frac{J}{x}\right) j_J(kx)\sqrt{\frac{J}{2J+1}}\boldsymbol{\mathcal{Y}}^\lambda_{J,J+1,1}\right. \\ &\left. + \left(\frac{d}{dx} + \frac{J+1}{x}\right) j_J(kx)\sqrt{\frac{J+1}{2J+1}}\boldsymbol{\mathcal{Y}}^\lambda_{J,J-1,1}\right] \end{aligned} \tag{7.16}$$

The differential operators just raise and lower the indices on the spherical Bessel functions, giving $-kj_{J+1}(kx)$ and $kj_{J-1}(kx)$, respectively. A combination of these results gives for $\lambda = \pm 1$

$$\begin{aligned} \mathbf{e}_{\mathbf{k}\lambda}e^{i\mathbf{k}\cdot\mathbf{x}} &= \sum_{J\geq 1} \sqrt{2\pi(2J+1)}\, i^J \left\{ \mp j_J(kx)\boldsymbol{\mathcal{Y}}^\lambda_{JJ1}(\Omega_x) \right. \\ &\left. -\frac{1}{k}\nabla \times [j_J(kx)\boldsymbol{\mathcal{Y}}^\lambda_{JJ1}(\Omega_x)] \right\} \; ; \qquad \lambda = \pm 1 \end{aligned} \tag{7.17}$$

Note the divergence of both sides of this equation vanishes (see Ref. [N44]).[32] Now use

$$\boldsymbol{\mathcal{Y}}^{\lambda\dagger}_{JJ1} = -(-1)^\lambda \boldsymbol{\mathcal{Y}}^{-\lambda}_{JJ1} \tag{7.18}$$

$\hat{e}^\dagger_\lambda = (-1)^\lambda \hat{e}_{-\lambda}$ $\quad \lambda = \pm 1, 0$

to arrive at the basic result for $\lambda = \pm 1$

$$\frac{-e_{\rm p}}{\sqrt{2\omega_k\Omega}} \int e^{-i\mathbf{k}\cdot\mathbf{x}} \mathbf{e}^\dagger_{\mathbf{k}\lambda} \cdot \hat{\mathbf{J}}(\mathbf{x}) d^3x = e_{\rm p} \sum_{J\geq 1} (-i)^J \sqrt{\frac{2\pi(2J+1)}{2\omega_k\Omega}} [\hat{T}^{\rm el}_{J-\lambda}(k) + \lambda \hat{T}^{\rm mag}_{J-\lambda}(k)] \tag{7.19}$$

used algebra + def of C-G coefficients

[31] Note this is the amplitude for photon *absorption*.

[32] The relation to be used is $\vec{\nabla}\cdot[j_J(kx)\vec{\mathcal{Y}}^M_{JJ1}] = 0$.

Here the *transverse electric and magnetic multipole operators* are defined by

$$\begin{aligned}\hat{T}^{\text{el}}_{JM}(k) &\equiv \frac{1}{k}\int d^3x\left[\nabla\times j_J(kx)\boldsymbol{\mathcal{Y}}^M_{JJ1}(\Omega_x)\right]\cdot\hat{\mathbf{J}}(\mathbf{x}) \\ \hat{T}^{\text{mag}}_{JM}(k) &\equiv \int d^3x\left[j_J(kx)\boldsymbol{\mathcal{Y}}^M_{JJ1}(\Omega_x)\right]\cdot\hat{\mathbf{J}}(\mathbf{x}) \end{aligned} \tag{7.20}$$

This important result merits several comments:

1. In a nucleus both the convection current density arising from the motion of charged particles (e.g., protons) and the intrinsic magnetization density coming from the intrinsic magnetic moments of the nucleons contribute to the electromagnetic interaction. The appropriate interaction hamiltonian should actually be written as

$$\begin{aligned}H' &= -e_{\text{p}}\int\hat{\mathbf{J}}_c(\mathbf{x})\cdot\mathbf{A}(\mathbf{x})d^3x - e_{\text{p}}\int\hat{\boldsymbol{\mu}}(\mathbf{x})\cdot[\nabla\times\mathbf{A}(\mathbf{x})]d^3x \\ &= -e_{\text{p}}\int\left[\hat{\mathbf{J}}_c(\mathbf{x})+\nabla\times\hat{\boldsymbol{\mu}}(\mathbf{x})\right]\cdot\mathbf{A}(\mathbf{x})d^3x \end{aligned}\tag{7.21}$$

To obtain the second line, a vector identity has been employed

$$\nabla\cdot(\mathbf{a}\times\mathbf{b}) = \mathbf{b}\cdot(\nabla\times\mathbf{a}) - \mathbf{a}\cdot(\nabla\times\mathbf{b}) \tag{7.22}$$

The total divergence has been converted to a surface integral far away from the nucleus using Gauss' theorem

$$\int_V \nabla\cdot\mathbf{v}\,d^3x = \int_S \mathbf{v}\cdot d\mathbf{S} \tag{7.23}$$

Finally, the integral over the far-away surface can be discarded for a *localized source*. A second application of this procedure yields the relation

$$\begin{aligned}\int d^3x\left[\nabla\times j_J(kx)\boldsymbol{\mathcal{Y}}^M_{JJ1}\right]\cdot\nabla\times\hat{\boldsymbol{\mu}}(\mathbf{x}) &= \int d^3x\,\hat{\boldsymbol{\mu}}(\mathbf{x})\cdot\nabla\times[\nabla\times j_J(kx)\boldsymbol{\mathcal{Y}}^M_{JJ1}] \\ &= k^2\int d^3x\,\hat{\boldsymbol{\mu}}(\mathbf{x})\cdot[j_J(kx)\boldsymbol{\mathcal{Y}}^M_{JJ1}] \end{aligned}\tag{7.24}$$

In arriving at the second equality the relation $\nabla\times(\nabla\times\mathbf{v}) = \nabla(\nabla\cdot\mathbf{v}) - \nabla^2\mathbf{v}$ has been employed; the term $\nabla\cdot\mathbf{v}$ vanishes here, and in this application the remaining term satisfies the Helmholtz equation $(\nabla^2+k^2)\mathbf{v}=0$, as the reader can readily verify. Thus the multipole operators can be rewritten to explicitly exhibit the individual contributions of the convection current and the intrinsic magnetization densities (Refs. [N4, N14])

$$\begin{aligned}\hat{T}^{\text{el}}_{JM}(k) &= \frac{1}{k}\int d^3x\left\{[\nabla\times j_J(kx)\boldsymbol{\mathcal{Y}}^M_{JJ1}]\cdot\hat{\mathbf{J}}_c(\mathbf{x}) + k^2 j_J(kx)\boldsymbol{\mathcal{Y}}^M_{JJ1}\cdot\hat{\boldsymbol{\mu}}(\mathbf{x})\right\} \\ \hat{T}^{\text{mag}}_{JM}(k) &= \int d^3x\left\{j_J(kx)\boldsymbol{\mathcal{Y}}^M_{JJ1}\cdot\hat{\mathbf{J}}_c(\mathbf{x}) + [\nabla\times j_J(kx)\boldsymbol{\mathcal{Y}}^M_{JJ1}]\cdot\hat{\boldsymbol{\mu}}(\mathbf{x})\right\} \end{aligned}\tag{7.25}$$

2. The $\hat{T}_{JM}$ are now *irreducible tensor operators of rank J in the nuclear Hilbert space.* This can be proven in general by utilizing the properties of the vector density operator $\hat{\mathbf{J}}(\mathbf{x})$ under rotations. It is easier to prove this property explicitly in any particular application. For example, consider the case where the nucleus is pictured as a collection of nonrelativistic nucleons, and the intrinsic magnetization density at the point $\mathbf{x}$ is constructed in first quantization by summing over the contribution of the individual nucleons

$$e_{\mathrm{p}}\hat{\boldsymbol{\mu}}(\mathbf{x}) = \mu_N \sum_{i=1}^{A} \lambda_i \boldsymbol{\sigma}(i)\delta^{(3)}(\mathbf{x} - \mathbf{x}_i) \tag{7.26}$$

Here λ_i is the intrinsic magnetic moment of the ith nucleon in nuclear magnetons (see below).[33] The contribution to $\hat{T}^{\mathrm{el}}_{JM}$, for example, then takes the form

$$e_{\mathrm{p}} \int j_J(kx)\boldsymbol{\mathcal{Y}}^M_{JJ1} \cdot \hat{\boldsymbol{\mu}}(\mathbf{x})d^3x =$$

$$\mu_N \sum_{i=1}^{A} \lambda_i j_J(kx_i) \sum_{mq} \langle Jm1q|J1JM\rangle Y_{Jm}(\Omega_i)\sigma_{1q}(i) \tag{7.27}$$

Here the definition of the vector spherical harmonics in Eq. (7.11) has been introduced. Each term in this sum is now recognized, with the aid of Ref. [N44], to be a tensor product of rank J formed from two ITO of rank J and 1, respectively.[34] Thus $\hat{T}^{\mathrm{el}}_{JM}$ is evidently an ITO of rank J under commutation with the total angular momentum operator, which in this case takes the form

$$\hat{\mathbf{J}} = \sum_{i=1}^{A} \mathbf{J}(i) = \sum_{i=1}^{A} [\mathbf{L}(i) + \mathbf{S}(i)] \; ; \qquad \text{angular momentum} \tag{7.28}$$

As another example, the convection current in this same picture of the nucleus is

$$\hat{\mathbf{J}}_c(\mathbf{x}) = \sum_{i=1}^{Z} \frac{1}{m}\{\delta^{(3)}(\mathbf{x} - \mathbf{x}_i), \mathbf{p}(i)\}_{\mathrm{sym}} \doteq \sum_{i=1}^{Z} \delta^{(3)}(\mathbf{x} - \mathbf{x}_i)\frac{\mathbf{p}(i)}{m} \tag{7.29}$$

The need for symmetrization[35] arises from the fact that $\mathbf{p}(i)$ and $\mathbf{x}_i$ do not commute; the current density arising from the matrix element of this expression takes the appropriate quantum mechanical form $(1/2im)[\psi^* \nabla\psi - (\nabla\psi)^*\psi]$. The last equality in Eq. (7.29) follows since one of the symmetrized terms can be partially integrated in the required matrix elements of the current, using the

[33] One could be dealing with a density operator in second quantization, or expressed in collective coordinates, etc; to test for an ITO, one first constructs the appropriate total angular momentum operator $\hat{\mathbf{J}}$, and then examines the commutation relations (see Ref. [N44]).

[34] Any spherically symmetric factor does not affect the behavior under rotations.

[35] $\{A, B\}_{\mathrm{sym}} \equiv (AB + BA)/2$.

hermiticity of $\mathbf{p}(i)$ and the observation that $\nabla \cdot \mathbf{A} = 0$ in the Coulomb gauge. Multipoles constructed from the convection current density in Eq. (7.29) are now shown to be ITO by arguments similar to the above.

3. The *parity* of the multipole operators is (Ref. [N4])

$$\hat{\Pi}\, \hat{T}^{\text{el}}_{JM}\, \hat{\Pi}^{-1} = (-1)^J \hat{T}^{\text{el}}_{JM} \qquad\qquad \hat{\Pi}\, \hat{T}^{\text{mag}}_{JM}\, \hat{\Pi}^{-1} = (-1)^{J+1} \hat{T}^{\text{mag}}_{JM} \tag{7.30}$$

Again the general proof follows from the behavior of the current density $\hat{\mathbf{J}}(\mathbf{x})$ as a polar vector under spatial reflections. It is easy to see this behavior in any particular application. For example, it follows from Eqs. (7.27) and (7.29) if one uses the properties of the individual quantities under spatial reflection: $\sigma_{1q} \rightarrow \sigma_{1q}$; $p_{1q} \rightarrow -p_{1q}$; and $Y_{lm}(-\mathbf{x}/|\mathbf{x}|) = (-1)^l Y_{lm}(\mathbf{x}/|\mathbf{x}|)$. Parity selection rules on the matrix elements of the transverse multipole operators now follow directly.

4. There is no $J = 0$ term in the sum in Eq. (7.19). This arises from the fact that the vector potential is transverse, and hence there are only transverse unit vectors, or equivalently unit helicities $\lambda = \pm 1$, arising in its expansion into normal modes [see Eqs. (7.8) and (7.17)]. This has the consequence, for example, that there can be no $J = 0 \rightarrow J = 0$ real photon transitions in nuclei.

5. The *Wigner-Eckart theorem* (Ref. [N44]) can now be employed to exhibit the angular momentum selection rules and M-dependence of the matrix element of an ITO between eigenstates of angular momentum

$$\langle J_f M_f | \hat{T}_{JM} | J_i M_i \rangle = \frac{(-1)^{J_i - M_i}}{(2J+1)^{1/2}} \langle J_f M_f J_i - M_i | J_f J_i J M \rangle \langle J_f || \hat{T}_J || J_i \rangle \tag{7.31}$$

Note that the required matrix elements of Eq. (7.19) imply $M_f = M_i - \lambda$. This means that the photon carries away the angular momentum λ along the z-axis, which is the direction of emission of the photon in the preceding analysis (Fig. 7.2); thus the *helicity* of the photon (its angular momentum along $\mathbf{k}$) is $\lambda = \pm 1$.

Photon in an Arbitrary Direction. Let us extend the previous analysis to describe the situation where the photon is emitted in an arbitrary direction relative to the coordinate axes picked to describe the quantization of the nuclear system. The situation is illustrated in Fig. 7.3. The unit vectors describing the photon are assumed to have Euler angles $\{\alpha, \beta, \gamma\}$ with respect to the nuclear quantization axes. The difficulty in achieving this configuration is that the photon axes here are the axes that are assumed to be *fixed in space*, having been determined, for example, by the detection of the photon, and the rotations are to be carried out with respect to these axes. Now one knows how to carry out a rotation of the nuclear state vector relative to a fixed set of axes. For example, the rotation operator that rotates a physical state vector through the angle β relative to a laboratory-fixed y-axis is $\hat{R}_{-\beta} \equiv e^{-i\beta \hat{J}_y}$; this fact follows entirely from the defining commutation relations for the angular momentum (see Prob. 7.1). The goal is to rotate the nuclear state vector $|J_i M_i\rangle$ quantized with respect to the photon axes into a nuclear state vector $|\Psi_i(J_i M_i)\rangle$ correctly quantized

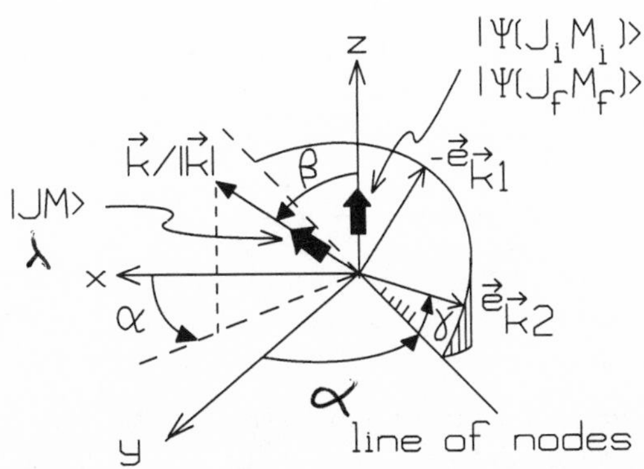

Figure 7.3: Photon emitted in arbitrary direction relative to quantization axes for nuclear system. Note $\{\alpha, \beta, \gamma\}$ are Euler angles.

with respect to the indicated $\{x, y, z\}$ coordinates. A concentrated effort, after staring at Fig. 7.3, will convince the reader that the following rotations, carried out with respect to the laboratory-fixed photon coordinate system in the indicated sequence, will achieve this end

1. $-\alpha$ about $\mathbf{k}/|\mathbf{k}|$
2. $-\beta$ about $\mathbf{e}_{\mathbf{k}2}$
3. $-\gamma$ about $\mathbf{k}/|\mathbf{k}|$

The rotation operator that accomplishes this rotation is

$$\hat{R}_{+\gamma,+\beta,+\alpha} = \exp\{i\gamma\hat{J}_3\}\exp\{i\beta\hat{J}_2\}\exp\{i\alpha\hat{J}_3\} \tag{7.32}$$

The $\{2, 3\}$ axes are now the laboratory-fixed $\{\mathbf{e}_{\mathbf{k}2}, \mathbf{k}/|\mathbf{k}|\}$ axes. Thus

$$|\Psi_i(J_iM_i)\rangle \;=\; \hat{R}_{\gamma\beta\alpha}|J_iM_i\rangle \;=\; \sum_{M_k}\mathcal{D}^{J_i}_{M_kM_i}(\gamma\beta\alpha)|J_iM_k\rangle \tag{7.33}$$

Here the rotation matrices have been introduced that characterize the behavior of the eigenstates of angular momentum under rotation (Ref. [N44]). It is clear from Fig. 7.3 that one can identify the usual polar and azimuthal angles that the photon makes with respect to the nuclear coordinate system according to $\beta \leftrightarrow \theta$ and $\alpha \leftrightarrow \phi$; the angle $\gamma \leftrightarrow -\phi$ of the orientation of the photon polarization vector around the photon momentum is a definition of the overall phase of the state vector, and, as such, merely involves a phase convention; the choice here is that of Jacob and Wick in Ref. [N48]. It will be apparent in the final result that this phase is irrelevant. Equation (7.33) expresses the required nuclear state vector as a linear combination of state vectors quantized along the photon axes. Since now only matrix elements between states quantized along $\mathbf{k}$ are required, *all the previous results can be utilized.* The required photon transition matrix element takes the form

$$\langle\Psi_f(J_fM_f)|\hat{H}_{J,-\lambda}|\Psi_i(J_iM_i)\rangle \;=\; \langle J_fM_f|\hat{R}^{-1}_{\gamma\beta\alpha}\hat{H}_{J,-\lambda}\hat{R}_{\gamma\beta\alpha}|J_iM_i\rangle \tag{7.34}$$

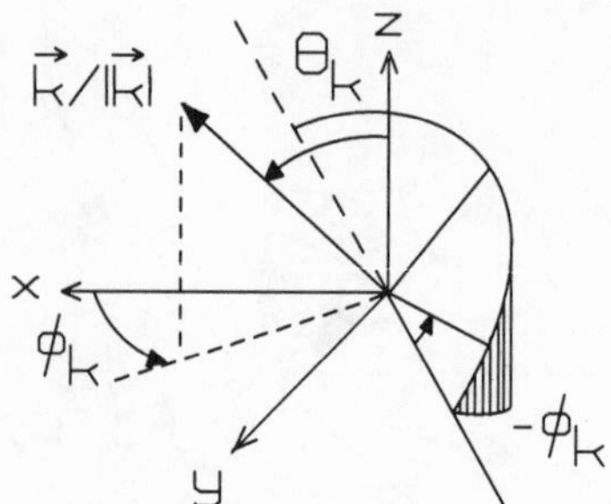

Figure 7.4: Configuration for transition matrix element describing photon emission and nuclear process $J_iM_i \rightarrow J_fM_f$ with nuclear quantization axis along the z-axis.

Here λ is the photon helicity, and $\hat{H}_{J,-\lambda}$ indicates one of the contributions to the operator in Eq. (7.19). Evidently

$$\hat{R}^{-1}_{\gamma\beta\alpha} = \hat{R}_{-\alpha-\beta-\gamma} \tag{7.35}$$

The definition of an ITO can now be used to simplify the calculation (Ref. [N44])

$$\hat{R}_{-\alpha-\beta-\gamma}\hat{H}_{J,-\lambda}\hat{R}^{-1}_{-\alpha-\beta-\gamma} = \sum_{M'} \mathcal{D}^J_{M'-\lambda}(-\alpha-\beta-\gamma)\hat{H}_{JM'} \tag{7.36}$$

The previous identification of the angles, and a combination of these results, permits one to write the transition matrix element describing the nuclear process $J_iM_i \rightarrow J_fM_f$ with the nuclear quantization axis along z and emission of a photon with helicity λ (Fig. 7.4) as

$$\langle\Psi_f(J_fM_f)|\hat{H}'(\mathbf{k}\lambda)|\Psi_i(J_iM_i)\rangle = \langle J_fM_f|\hat{H}^{\text{em}}_1(\mathbf{k}\lambda)|J_iM_i\rangle \tag{7.37}$$

where the appropriate transition operator is given by

$$\begin{aligned}\hat{H}^{\text{em}}_1(\mathbf{k}\lambda) &= e_p\sum_{JM}(-i)^J\sqrt{\frac{2\pi(2J+1)}{2\omega_k\Omega}} \\ &\quad\times[\hat{T}^{\text{el}}_{JM}(k)+\lambda\hat{T}^{\text{mag}}_{JM}(k)]\,\mathcal{D}^J_{M-\lambda}(-\phi_k,-\theta_k,\phi_k)\end{aligned} \tag{7.38}$$

The Wigner-Eckart theorem in Eq. (7.31) now permits one to extract all the angular momentum selection rules and M-dependence of the matrix element in Eq. (7.37). All M's now refer to a common set of coordinate axes.[36]

[36] These axes were originally the photon axes with the z-axis along $\mathbf{k}$, but they can now just as well be the nuclear $\{x, y, z\}$ axes in Fig. 7.3; the equivalence of these two interpretations is readily demonstrated by taking out the M-dependence in a C-G coefficient with the aid of the Wigner-Eckart theorem — it is the same in both cases. The two interpretations differ only by an *overall* rotation (with $\hat{R}^{-1}\hat{R}$ inserted everywhere), which leaves the physics unchanged.

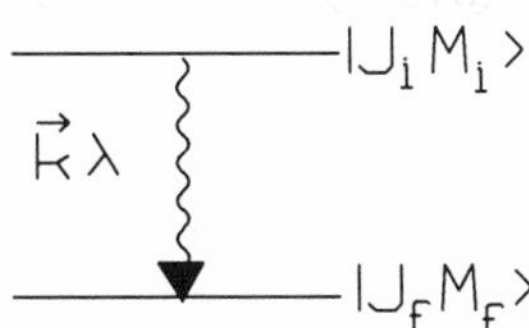

Figure 7.5: Nuclear transition with real photon emission.

The final $\mathcal{D}^J_{M-\lambda}$ in Eq. (7.38) plays the role of a "photon wave function," since the square of this quantity gives the intensity distribution in (θ_k, ϕ_k) of electromagnetic radiation carrying off $\{J, M, \lambda\}$ from the target.

Transition Probabilities and Lifetimes. We proceed to calculate the transition probability for the process indicated in Fig. 7.5. The total transition rate for an unoriented nucleus is given by the Golden Rule

$$\omega = 2\pi \sum_f \overline{\sum_i} |\langle J_f M_f \mathbf{k}\lambda | H' | J_i M_i \rangle|^2 \delta(E_f + \omega_k - E_i) \tag{7.39}$$

The appropriate sum over final states is given by

$$\sum_f = \frac{\Omega}{(2\pi)^3} \sum_\lambda \sum_{M_f} \int d^3k \tag{7.40}$$

The $\int dk$ allows one to integrate over the energy-conserving delta function $\int dk \delta(E_f + \omega_k - E_i) = 1$. The integral over final solid angles of the photon $\int d\Omega_k$ can be performed with the aid of the orthogonality properties of the rotation matrices (Ref. [N44])

$$\int_0^\pi \sin\theta d\theta \int_0^{2\pi} d\phi\, \mathcal{D}^J_{M-\lambda}(-\phi - \theta\phi)^* \mathcal{D}^{J'}_{M'-\lambda}(-\phi - \theta\phi) = \frac{4\pi}{2J+1} \delta_{JJ'} \delta_{MM'} \tag{7.41}$$

Note that since λ is the same in both functions, the dependence on the last ϕ (which was the phase convention adopted for the third Euler angle $-\gamma$ in Fig. 7.3) drops out of this expression, as advertised.

The average over initial nuclear states is performed according to $\overline{\sum_i} = (2J_i + 1)^{-1} \sum_{M_i}$. The use of the Wigner-Eckart theorem in Eq. (7.31) and the orthonormality of the C-G coefficients permits one to then perform the required sums over M_f and M_i

$$\sum_{M_f} \sum_{M_i} |\langle J_f M_f J_i - M_i | J_f J_i J M \rangle|^2 = 1 \tag{7.42}$$

The final sum on M gives $\sum_M = 2J + 1$.

Since the matrix element of one or the other multipoles must vanish by conservation of parity, assumed to hold for the strong interactions, it follows that

$$|\langle J_f||\hat{T}_J^{\text{el}} + \lambda \hat{T}_J^{\text{mag}}||J_i\rangle|^2 = |\langle J_f||\hat{T}_J^{\text{el}}||J_i\rangle|^2 + |\langle J_f||\hat{T}_J^{\text{mag}}||J_i\rangle|^2 \tag{7.43}$$

This expression is now independent of λ, and the sum over final photon polarizations gives $\sum_\lambda = 2$.

A combination of these results yields the total photon transition probability for the process illustrated in Fig. 7.5 [37]

$$\omega_{\text{f i}} = 8\pi\alpha\ kc \frac{1}{2J_i + 1} \sum_{J\geq 1} \left\{ |\langle J_f||\hat{T}_J^{\text{el}}(k)||J_i\rangle|^2 + |\langle J_f||\hat{T}_J^{\text{mag}}(k)||J_i\rangle|^2 \right\} \tag{7.44}$$

This is a very general result. For most nuclear transitions of interest involving real photons, the wavelength is large compared to the size of the nucleus. We thus consider next the long-wavelength reduction of the multipole operators, following closely the arguments developed in Ref. [N4].

Reduction of the Multipole Operators.[38] The use of the relations $1/\hbar c = 5.07 \times 10^{10}\ \text{cm}^{-1}/\text{MeV}$ and $R \approx 1.2\, A^{1/3} \times 10^{-13}$ cm allows one to write for real photons

$$kR \approx 6.1 \times 10^{-3} [E_\gamma(\text{MeV}) A^{1/3}] \tag{7.45}$$

Evidently $kR \ll 1$ for photons of a few MeV. In this case, the spherical Bessel functions can be expanded as[39]

$$j_J(kx) \to \frac{(kx)^J}{(2J+1)!!} \; ; \qquad kx \to 0 \tag{7.46}$$

One also needs from Ref. [N44]

$$\mathbf{L} Y_{lm} = \frac{1}{i}(\mathbf{r} \times \nabla) Y_{lm} = \sqrt{l(l+1)}\,\boldsymbol{\mathcal{Y}}_{ll1}^m \tag{7.47}$$

With this relation, the multipole operators in Eqs. (7.25) take the form

$$\begin{aligned}
\hat{T}_{JM}^{\text{el}} &= \frac{1}{k\sqrt{J(J+1)}} \int d^3x \left\{ [\nabla \times \mathbf{L} j_J(kx) Y_{JM}] \cdot \hat{\mathbf{J}}_c(\mathbf{x}) \right. \\
&\qquad \left. + k^2 [\mathbf{L} j_J(kx) Y_{JM}] \cdot \hat{\boldsymbol{\mu}}(\mathbf{x}) \right\} \\
\hat{T}_{JM}^{\text{mag}} &= \frac{1}{\sqrt{J(J+1)}} \int d^3x \, \{ [\nabla \times \mathbf{L} j_J(kx) Y_{JM}] \cdot \hat{\boldsymbol{\mu}}(\mathbf{x}) \\
&\qquad + [\mathbf{L} j_J(kx) Y_{JM}] \cdot \hat{\mathbf{J}}_c(\mathbf{x}) \}
\end{aligned} \tag{7.48}$$

[37] Recall $\alpha = e^2/4\pi\hbar c$ in the present units; the multipole operators are now dimensionless.
[38] Recall $\mathbf{x} \equiv \mathbf{r}$ and $x \equiv |\mathbf{x}| \equiv r$ in all these discussions.
[39] One has to get all the derivatives off the Bessel functions before they can be expanded — that is the point of the following exercise.

These expressions can now be manipulated in the following manner:

1. The differential orbital angular momentum operator $\mathbf{L}$ in Eq. (7.47) commutes with any function of the radial coordinate $[\mathbf{L}, f(r)] = 0$, and it is hermitian; thus it can be partially integrated in the last two terms on the r.h.s. in the above to get it over to the right [with a sign (-1)].

2. The divergence theorem in Eqs. (7.22) and (7.23) can be used on the first two terms on the r.h.s. of the above to get the curl to the right.

3. One can then get $\mathbf{L}$ to the right in these terms using the first argument [again with a (-1)]. This leads to two types of terms: first

$$\mathbf{L}\cdot\mathbf{v} = \frac{1}{i}(\mathbf{r}\times\nabla)\cdot\mathbf{v} = \frac{1}{i}(\nabla\times\mathbf{v})\cdot\mathbf{r} = -\frac{1}{i}\nabla\cdot(\mathbf{r}\times\mathbf{v}) \tag{7.49}$$

and second

$$\begin{aligned}\mathbf{L}\cdot(\nabla\times\mathbf{v}) &= \frac{1}{i}(\mathbf{r}\times\nabla)\cdot(\nabla\times\mathbf{v}) = \frac{1}{i}[\nabla\times(\nabla\times\mathbf{v})]\cdot\mathbf{r} \\ &= -\frac{1}{i}\nabla\cdot[\mathbf{r}\times(\nabla\times\mathbf{v})]\end{aligned} \tag{7.50}$$

Here the relation $\nabla\times\mathbf{r} = 0$ has been used in obtaining these equations.

4. Since all derivatives are now off the spherical Bessel functions and on the source terms, the Bessel functions may be expanded in the long-wavelength limit according to Eq. (7.46).

5. One next invokes the general vector identity

$$\int x^J Y_{JM}\nabla\cdot[\mathbf{r}\times(\nabla\times\mathbf{v})]\,d^3x = (J+1)\int x^J Y_{JM}\nabla\cdot\mathbf{v}\,d^3x \tag{7.51}$$

This identity holds as long as the source terms $\mathbf{v}(\mathbf{x})$ vanish outside the nucleus (Prob. 7.2).

With these steps the magnetic multipoles take the form

$$\hat{T}^{\text{mag}}_{JM} \approx \frac{1}{i}\frac{k^J}{(2J+1)!!}\sqrt{\frac{J+1}{J}}\int d^3x\, x^J Y_{JM}\left\{\nabla\cdot\hat{\boldsymbol{\mu}}(\mathbf{x}) + \frac{1}{J+1}\nabla\cdot[\mathbf{r}\times\hat{\mathbf{J}}_c(\mathbf{x})]\right\} \tag{7.52}$$

Partial integration of this result then gives for the long-wavelength limit of the transverse magnetic multipoles

$$\hat{T}^{\text{mag}}_{JM} \approx -\frac{1}{i}\frac{k^J}{(2J+1)!!}\sqrt{\frac{J+1}{J}}\int d^3x\,[\hat{\boldsymbol{\mu}}(\mathbf{x}) + \frac{1}{J+1}\mathbf{r}\times\hat{\mathbf{J}}_c(\mathbf{x})]\cdot\nabla x^J Y_{JM} \tag{7.53}$$

Similarly, the electric multipole operators take the form

$$\hat{T}^{\text{el}}_{JM} \approx \frac{1}{i}\frac{k^{J-1}}{(2J+1)!!}\sqrt{\frac{J+1}{J}}\int d^3x\left\{\nabla\cdot\hat{\mathbf{J}}_c(\mathbf{x}) + \frac{k^2}{J+1}\nabla\cdot[\mathbf{r}\times\hat{\boldsymbol{\mu}}(\mathbf{x})]\right\}x^J Y_{JM} \tag{7.54}$$

Now use the *continuity equation* on the first term

$$\nabla \cdot \hat{\mathbf{J}}_c(\mathbf{x}) = \nabla \cdot \hat{\mathbf{J}}(\mathbf{x}) \quad = \quad -\frac{\partial \hat{\rho}}{\partial t} = -i[\hat{H}, \hat{\rho}] \tag{7.55}$$

The matrix element of this relation yields

$$\langle f|[\hat{H}, \hat{\rho}]|i\rangle = (E_f - E_i)\langle f|\hat{\rho}|i\rangle \quad = \quad -k\langle f|\hat{\rho}|i\rangle \tag{7.56}$$

Thus, in the matrix element, one can replace[40] $\nabla \cdot \hat{\mathbf{J}}_c(\mathbf{x}) \to ik\hat{\rho}(\mathbf{x})$. Thus, for photon emission the long-wavelength limit of the transverse electric multipoles takes the form

$$\hat{T}^{\text{el}}_{JM} \approx \frac{k^J}{(2J+1)!!}\sqrt{\frac{J+1}{J}} \int d^3x \left\{ x^J Y_{JM}\hat{\rho}(\mathbf{x}) - \frac{ik}{J+1}\hat{\boldsymbol{\mu}}(\mathbf{x}) \cdot [\mathbf{r} \times \nabla x^J Y_{JM}] \right\} \tag{7.57}$$

Several features of these results are of interest:

1. The second term in Eq. (7.57) goes as $\hbar kc/mc^2 \ll 1$ and hence the contribution of this term is very small compared to that of the first term for real photons.[41]

2. The first term in Eq. (7.57) is just the JMth multipole of the charge density.

3. Make a model where the nucleus is composed of individual nucleons, and where only the leading terms to order $1/m$ are retained in the current, that is, the terms in $\mathbf{p}(i)$ and $\sigma(i)$ [see Eqs. (7.29) and (7.26)]. The $J = 1$ transverse magnetic dipole operator for $k \to 0$ then takes the form

$$\hat{T}^{\text{mag}}_{1M} \approx \frac{i\sqrt{2}}{3}\frac{\hbar k}{2mc}\sqrt{\frac{3}{4\pi}} \left\{ \sum_{i=1}^{Z} \mathbf{l}(i) + \sum_{i=1}^{A} \lambda_i \sigma(i) \right\}_{1M} \tag{7.58}$$

This is the familiar magnetic dipole operator to within a numerical factor and power of k. Here the nucleon magnetic moments in nuclear magnetons are given by $\lambda_p = 2.793$ for the proton and $\lambda_n = -1.913$ for the neutron (see Section 8).

It is useful to make the connection between these general results for the electromagnetic nuclear moments and the static nuclear moments measured in time-independent electric and magnetic fields; this connection is made as follows.

Static Moments. Consider first the static *electric* moments of the nucleus. Suppose one places a static charge distribution $\rho(\mathbf{r})$ in an *external* electrostatic potential $\Phi_{\text{el}}(\mathbf{r})$ where the external electric field is given by $\mathbf{E} = -\nabla\Phi_{\text{el}}(\mathbf{r})$ (see Fig. 7.6). A relevant example is a nucleus in the field of the atomic electrons. The interaction energy is given by

[40] Note this is for photon *emission*; for photon *absorption* one has the opposite sign for this term.

[41] This term can become large in electron scattering where, as we shall see, the appropriate ratio is $\hbar qc/mc^2$ with q the momentum transfer.

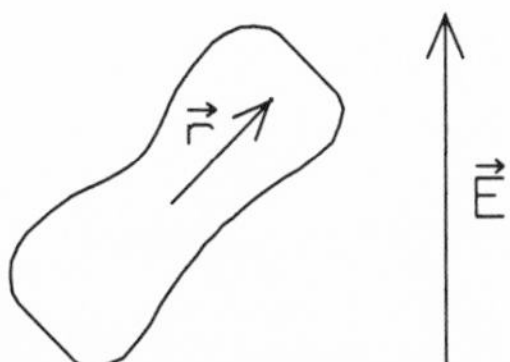

Figure 7.6: Static electric nuclear moments.

$$U = e_{\text{p}} \int \rho(\mathbf{r}) \Phi_{\text{el}}(\mathbf{r}) d^3 r \tag{7.59}$$

The external field satisfies Laplace's equation since it is source-free over the nucleus

$$\nabla^2 \Phi_{\text{el}}(\mathbf{r}) = 0 \tag{7.60}$$

It is also finite there. Thus the external field in the region of the nucleus can be expanded in terms of the acceptable solutions to Laplace's equation

$$\Phi_{\text{el}}(\mathbf{r}) = \sum_{lm} a_{lm} r^l Y_{lm}(\Omega_r) \tag{7.61}$$

The numerical coefficients a_{lm} can be related to various derivatives of the field at the origin. Substitution of Eq. (7.61) into Eq. (7.59) yields

$$U = e_{\text{p}} \sum a_{lm} \mathcal{M}_{lm}^{\text{el}} \tag{7.62}$$

Here the multipole moments of the charge density are defined by

$$\boxed{\mathcal{M}_{lm}^{\text{el}} = \int d^3x \, x^l Y_{lm}(\Omega_x) \rho(\mathbf{x})} \tag{7.63}$$

These are exactly the same expressions, to within a numerical factor and powers of k, as the first term in the transverse electric multipole operators in Eq. (7.57).[42] Note that the nuclear quadrupole moment is conventionally defined by

$$Q = \int (3z^2 - r^2) \rho(\mathbf{x}) \, d^3x \tag{7.64}$$

which differs by a numerical constant from $\mathcal{M}_{20}^{\text{el}}$.

Consider next the nuclear *magnetic* moments. Take the ground-state expectation value that gives $\langle \partial \hat{\rho}(\mathbf{x})/\partial t \rangle = i\langle [\hat{H}, \hat{\rho}] \rangle = 0$. This implies

$$\nabla \cdot \langle \hat{\mathbf{J}}(\mathbf{x}) \rangle \;=\; \nabla \cdot \langle \hat{\mathbf{J}}_c(\mathbf{x}) \rangle = 0 \tag{7.65}$$

Here the general decomposition of current has been invoked

$$\hat{\mathbf{J}} = \hat{\mathbf{J}}_c + \nabla \times \hat{\boldsymbol{\mu}} \tag{7.66}$$

[42]The charge multipole *operators* are defined in terms of the charge density *operator*.

Since the divergence of the last quantity in Eq. (7.65) vanishes everywhere, it can be expressed as the curl of another vector $\mathbf{M}(\mathbf{x})$

$$\langle \hat{\mathbf{J}}_c(\mathbf{x}) \rangle = \nabla \times \mathbf{M}(\mathbf{x}) \tag{7.67}$$

One can assume that the additional magnetization $\mathbf{M}(\mathbf{x})$ vanishes outside the nucleus, for suppose it does not. Then since its curl vanishes outside the nucleus by Eq. (7.67), it can be written as $\mathbf{M}(\mathbf{x}) = \nabla\chi(\mathbf{x})$ in this region. Now choose a new magnetization $\mathbf{M}'(\mathbf{x}) = \mathbf{M}(\mathbf{x}) - \nabla\chi(\mathbf{x})$. This new magnetization has the same curl everywhere, and now, indeed, vanishes outside the nucleus.

The expectation value of the interaction hamiltonian for the nucleus in an external magnetic field now takes the form

$$\langle \hat{H}_{\text{int}} \rangle = -e_{\text{p}} \int [\nabla \times \mathbf{M}(\mathbf{x})] \cdot \mathbf{A}^{\text{ext}}(\mathbf{x})\, d^3x - e_{\text{p}} \int \boldsymbol{\mu}(\mathbf{x}) \cdot \mathbf{B}^{\text{ext}}(\mathbf{x}) d^3x \tag{7.68}$$

Here $\boldsymbol{\mu} \equiv \langle \hat{\boldsymbol{\mu}} \rangle$. The use of Eqs. (7.22) and (7.23) permits this expression to be rewritten as

$$\langle \hat{H}_{\text{int}} \rangle = -e_{\text{p}} \int [\mathbf{M}(\mathbf{x}) + \boldsymbol{\mu}(\mathbf{x})] \cdot \mathbf{B}^{\text{ext}}(\mathbf{x})\, d^3x \tag{7.69}$$

Since $\mathbf{B}^{\text{ext}}(\mathbf{x})$ is an external magnetic field with no sources over the nucleus, it satisfies Maxwell's equations there

$$\nabla \cdot \mathbf{B}^{\text{ext}} \;=\; \nabla \times \mathbf{B}^{\text{ext}} = 0 \tag{7.70}$$

Thus one can write in the region of interest

$$\mathbf{B}^{\text{ext}} = -\nabla\Phi_{\text{mag}} \qquad\qquad \nabla^2\Phi_{\text{mag}} = 0 \tag{7.71}$$

One can now proceed with exactly the same arguments used on the electric moments. The energy of interaction is given by

$$\begin{aligned} \langle \hat{H}_{\text{int}} \rangle &= e_{\text{p}} \int [\mathbf{M}(\mathbf{x}) + \boldsymbol{\mu}(\mathbf{x})] \cdot \nabla\Phi_{\text{mag}}(\mathbf{x})\, d^3x \\ &= -e_{\text{p}} \int \Phi_{\text{mag}} \nabla \cdot (\mathbf{M} + \boldsymbol{\mu})\, d^3x \end{aligned} \tag{7.72}$$

The divergence in the last equation evidently plays the role of the "magnetic charge." Thus, just as before, when the general solution to Laplace's equation is substituted for the magnetic potential Φ_{mag}, all one needs are the magnetic charge multipoles given by

$$\begin{aligned} \mathcal{M}_{lm}^{\text{mag}} &= -\int x^l Y_{lm}(\Omega_x) \nabla \cdot (\mathbf{M} + \boldsymbol{\mu}) d^3x \\ &= -\int x^l Y_{lm}(\Omega_x) \nabla \cdot [\frac{1}{l+1} \mathbf{r} \times (\nabla \times \mathbf{M}) + \boldsymbol{\mu}] d^3x \end{aligned} \tag{7.73}$$

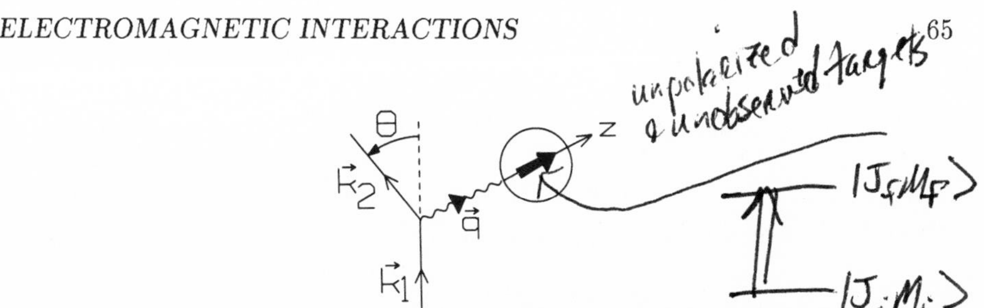

Figure 7.7: Electron scattering to discrete nuclear levels.

The second equality follows with the aid of the identity in Eq. (7.51). A partial integration, and the restoration to operator form yields the final result for the relevant static magnetic multipole operators

$$\hat{\mathcal{M}}_{lm}^{\text{mag}} = \int d^3x \left[\hat{\boldsymbol{\mu}}(\mathbf{x}) + \frac{1}{l+1}\mathbf{r} \times \hat{\mathbf{J}}_c(\mathbf{x})\right] \cdot \nabla x^l Y_{lm} \tag{7.74}$$

This is recognized to be, within a numerical factor and powers of k, the long wavelength limit of the transverse magnetic multipole operator in Eq. (7.53).

Electron Scattering to Discrete Levels. We conclude this part with a brief discussion of the results for the cross section for electron scattering with excitation of the nucleus to discrete nuclear levels; this can include both elastic and inelastic scattering. In first Born approximation, exactly the same electromagnetic multipoles as discussed here govern the cross section. Thus, while the detailed derivation of the cross section in this text is postponed to Part IV, it is useful to present the results here to tie into the present discussion. The scattering situation is illustrated in Fig. 7.7. One includes in the hamiltonian the interaction term between the nucleus and the electron as determined by QED

$$H' = -e\int \mathbf{j}_{\text{el}}(\mathbf{x})\cdot\mathbf{A}(\mathbf{x})d^3x + \frac{ee_{\text{p}}}{4\pi}\int\int \frac{\rho_{\text{el}}(\mathbf{x})\rho_N(\mathbf{x}')}{|\mathbf{x}-\mathbf{x}'|}d^3x d^3x' \tag{7.75}$$

To consistently calculate the scattering amplitude to order e^2, the Coulomb interaction is treated in first-order perturbation theory while the interaction with the transverse photon field [Eq. (7.8)] must be treated in second order. Plane waves and Dirac spinors are assumed for the initial and final electrons,[43] and one is evidently left with *nuclear* matrix elements of the form

$$\begin{aligned}
&\langle J_f M_f| \int e^{i\mathbf{q}\cdot\mathbf{x}}\hat{\rho}_N(\mathbf{x})d^3x|J_iM_i\rangle \\
&\langle J_f M_f| \int e^{i\mathbf{q}\cdot\mathbf{x}}\mathbf{e}_{\mathbf{q}\lambda}\cdot\hat{\mathbf{J}}_N(\mathbf{x})d^3x|J_iM_i\rangle\ ; \qquad \lambda = \pm 1
\end{aligned} \tag{7.76}$$

It simplifies the calculation to choose the z-axis for nuclear quantization along the direction of the three momentum transfer $\mathbf{q} = \mathbf{k_1} - \mathbf{k_2}$. In this case, from

[43] We work at energies where the electron mass can be neglected.

the previous analysis, transverse photon exchange has the nuclear selection rules $\Delta M = \pm 1$, and the Coulomb interaction has $\Delta M = 0$; this implies there will be no interference between these terms in the sum and average over final and initial nuclear orientations. The cross section takes the form (Ref. [N47])[44]

$$\frac{d\sigma}{d\Omega}(J_f \leftarrow J_i) = \frac{4\pi\sigma_{\rm M}}{2J_i+1}\left\{\frac{q_\mu^4}{\mathbf{q}^4}\sum_{J=0}^{\infty}|\langle J_f||\hat{M}_J^{\rm coul}(q)||J_i\rangle|^2 \right. \tag{7.77}$$
$$\left. + \left(\frac{q_\mu^2}{2\mathbf{q}^2} + \tan^2\frac{\theta}{2}\right)\sum_{J=1}^{\infty}(|\langle J_f||\hat{T}_J^{\rm el}(q)||J_i\rangle|^2 + |\langle J_f||\hat{T}_J^{\rm mag}(q)||J_i\rangle|^2)\right\}$$

Here $\sigma_{\rm M}$ is the Mott cross section for the scattering of a Dirac electron by a point charge

$$\sigma_{\rm M} \equiv \frac{\alpha^2\cos^2\theta/2}{4k_1^2\sin^4\theta/2} \tag{7.78}$$

Also $q_\mu^2 = \mathbf{q}^2 - (k_1-k_2)^2 = 4k_1k_2\sin^2\theta/2$ is the four-momentum transfer. The Coulomb multipoles appearing in this expression are defined by

$$\hat{M}_{LM}^{\rm coul} \equiv \int d^3x\, j_L(qx)Y_{LM}(\Omega_x)\hat{\rho}_N(\mathbf{x}) \tag{7.79}$$

In the long-wavelength limit, these become

$$\hat{M}_{LM}^{\rm coul}(q) \to \frac{q^L}{(2L+1)!!}\hat{\mathcal{M}}_{LM}^{\rm el} \tag{7.80}$$

where the static electric multipoles are defined in Eq. (7.63).

Several features of these results are worthy of note:

1. One can now use all the angular momentum and parity selection rules discussed previously.

2. These are the same transverse multipole operators that appear in real photon processes, only now the argument of the multipoles is $|\mathbf{q}|$, the three momentum transfer in the scattering rather than $|\mathbf{k}|$, the real photon momentum. In contrast to real photon transitions where the photon momentum is fixed by the energy it carries off, the quantity $|\mathbf{q}|$ can be varied up to arbitrarily large values in electron scattering.

3. For elastic scattering with $0^+ \to 0^+$ transitions, only the monopole moment of the charge density $\hat{M}_{00}^{\rm coul}(q)$ contributes to the cross section.

4. One can separate the contribution of the Coulomb multipoles from that of the transverse multipoles by varying θ at fixed $\{\mathbf{q}^2, q_\mu^2\}$.[45]

5. Equations (7.54), (7.55), and (7.80) imply there is a long-wavelength relation between the matrix elements of the Coulomb multipoles and those of

[44] Here $q = |\mathbf{q}|$. (Note we will later define $\mathbf{q}$ with opposite sign.)

[45] A "Rosenbluth plot."

the transverse electric multipoles

$$\langle f|\hat{T}^{\rm el}_{JM}(q)|i\rangle \approx -\frac{(E_f - E_i)}{\hbar c q}\sqrt{\frac{J+1}{J}}\langle f|\hat{M}^{\rm coul}_{JM}(q)|i\rangle \tag{7.81}$$

Thus one can get the rate for a real photon transition dominated by an electric multipole from Coulomb excitation of the level.

6. Nuclear recoil can be included in the density of final states. The effect is to multiply the r.h.s. of Eq. (7.77) by a factor r where $r^{-1} = 1+(2k_1/M_T)\sin^2\theta/2$ where M_T is the target inverse Compton wavelength. This is the leading correction in $1/M_T$ (see Prob. 7.4).

7. A much more detailed discussion and derivation of these results is presented in Refs. [N14, N47], where the evaluation of the required multipole operators in a coupled single-particle shell model basis is also developed. Extremely valuable tables of the required single-particle matrix elements of the multipole operators are available both with harmonic oscillator wave functions where the required angular momentum algebra and radial matrix elements can be evaluated analytically (Ref. [N49]), and with arbitrary radial wave functions where the radial matrix elements remain to be evaluated (Ref. [N50]).

It is to electromagnetic interactions in the nuclear shell model that we now turn our attention.

8

ELECTROMAGNETISM AND THE SHELL MODEL

We proceed to apply some of the general results on electromagnetic interactions with nuclei to the nuclear shell model where the basis for the description of the nucleus is a collection of nonrelativistic nucleons moving in a (complete) set of single-particle orbitals.

Extreme Single-Particle Model. It is an empirical result that the ground states of even-even nuclei all have $J^\pi = 0^+$. The simplest description of odd nuclei is to assign all of the ground-state nuclear properties to the last odd nucleon. This extreme single-particle shell model is remarkably successful in predicting ground state spins and parities as the levels in Fig. 6.5 are filled (Ref. [N37]). The model also allows a calculation of the electromagnetic properties of the ground state.

The magnetic dipole operator is given in units of the nuclear magneton $\mu_N = |e|\hbar/2mc$ by

$$\begin{aligned} \frac{\boldsymbol{\mu}_p}{\mu_N} &= \mathbf{l} + 2\lambda_p \mathbf{s} & \lambda_p &= +2.793 \\ \frac{\boldsymbol{\mu}_n}{\mu_N} &= 2\lambda_n \mathbf{s} & \lambda_n &= -1.913 \end{aligned} \tag{8.1}$$

The magnetic moment of the nucleus is defined as the expectation value of the magnetic dipole operator in the state where the nucleus is lined up as well as possible along the z axis

$$\mu \equiv \langle j, m = j|\hat{\mu}_{10}|j, m = j\rangle \quad = \quad \frac{\langle jj10|j1jj\rangle}{\sqrt{2j+1}}\langle j||\boldsymbol{\mu}||j\rangle \tag{8.2}$$

The second equality follows from the Wigner-Eckart theorem. To evaluate the remaining reduced matrix element one needs $\langle l\frac{1}{2}j||\mathbf{l}||l\frac{1}{2}j\rangle$ and $\langle l\frac{1}{2}j||\mathbf{s}||l\frac{1}{2}j\rangle$. These are the reduced matrix elements of an ITO acting on the first and second part of a coupled scheme respectively; they may be evaluated by using the

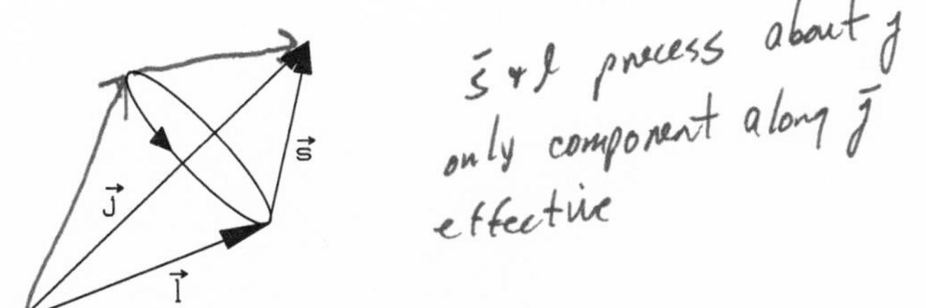

Figure 8.1: Vector model of angular momenta.

results in Ref. [N44][46]

$$\begin{aligned}\frac{\mu}{\mu_{\rm N}} &= \frac{1}{2(j+1)}\{[j(j+1)+l(l+1)-s(s+1)] \\ &\quad +2\lambda[j(j+1)+s(s+1)-l(l+1)]\}\ ; \qquad s=1/2 \end{aligned} \tag{8.3}$$

This result for the single-particle magnetic moment will be recognized as exactly the same expression obtained in the simple vector model of angular momenta. In this model the vectors $\mathbf{l}$ and $\mathbf{s}$ add to give the resultant $\mathbf{j}=\mathbf{l}+\mathbf{s}$ (Fig. 8.1). They then precess around this resultant so that the effective magnetic moment is only that component along $\mathbf{j}$

$$\boldsymbol{\mu} \longrightarrow \frac{(\boldsymbol{\mu}\cdot\mathbf{j})\,\mathbf{j}}{\mathbf{j}^2} \tag{8.4}$$

The insertion of the definition of the magnetic dipole operator in Eq. (8.1) gives

$$\frac{\mu}{\mu_{\rm N}} = \frac{j}{j(j+1)}[\mathbf{l}\cdot\mathbf{j}+2\lambda\mathbf{s}\cdot\mathbf{j}] \tag{8.5}$$

The square of the relations $\mathbf{j}-\mathbf{l}=\mathbf{s}$ and $\mathbf{j}-\mathbf{s}=\mathbf{l}$ gives

$$\begin{aligned} 2\mathbf{l}\cdot\mathbf{j}=\mathbf{j}^2+\mathbf{l}^2-\mathbf{s}^2 &= j(j+1)+l(l+1)-s(s+1) \\ 2\mathbf{s}\cdot\mathbf{j}=\mathbf{j}^2+\mathbf{s}^2-\mathbf{l}^2 &= j(j+1)+s(s+1)-l(l+1) \end{aligned} \tag{8.6}$$

A combination of these results indeed reproduces Eq. (8.3). Insertion of the allowed values $j=l\pm 1/2$ in Eq. (8.3) gives the final results

$$\begin{aligned} \frac{\mu}{\mu_{\rm N}} &= j-\frac{1}{2}+\lambda\ ; &\qquad j=l+\frac{1}{2} \\ \frac{\mu}{\mu_{\rm N}} &= j+\frac{j}{j+1}(\frac{1}{2}-\lambda)\ ; &\qquad j=l-\frac{1}{2} \end{aligned} \tag{8.7}$$

[46] A combination of the results in Ref. [N44] gives

$$\begin{aligned} \langle l\tfrac{1}{2}j||\mathbf{l}||l\tfrac{1}{2}j\rangle &= (-1)^{l+1/2+j+1}(2j+1)\left\{\begin{matrix} l & j & 1/2 \\ j & l & 1 \end{matrix}\right\}\langle l||\mathbf{l}||l\rangle \\ \langle l\tfrac{1}{2}j||\mathbf{s}||l\tfrac{1}{2}j\rangle &= (-1)^{l+1/2+j+1}(2j+1)\left\{\begin{matrix} 1/2 & j & l \\ j & 1/2 & 1 \end{matrix}\right\}\langle \tfrac{1}{2}||\mathbf{s}||\tfrac{1}{2}\rangle \end{aligned}$$

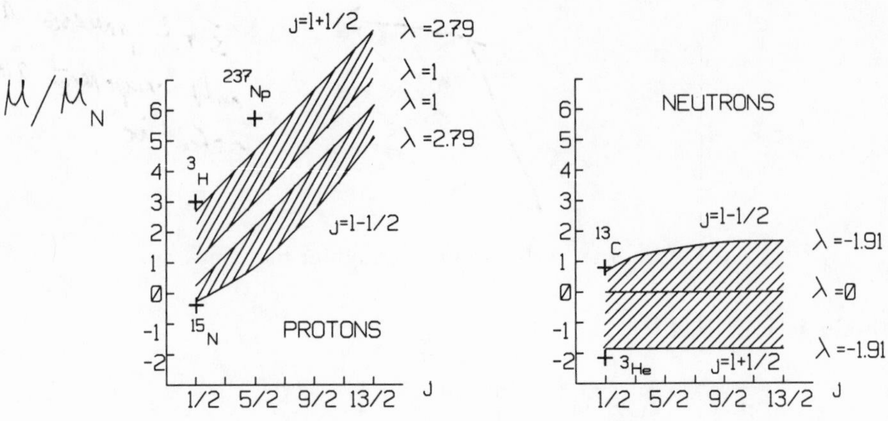

Figure 8.2: Schmidt lines for nuclear magnetic dipole moments (see Ref. [N1]).

Here the entire expression is applicable for protons, while only the term in λ is present for neutrons. When plotted against j, these results give rise to the celebrated *Schmidt lines* (Fig. 8.2). Several comments are relevant here (see Ref. [N1] for a summary of the data):

1. The results for the nuclear magnetic dipole moment are *independent of the radial wave functions.* They depend only on the angular momentum coupling scheme.

2. Of 137 odd-A nuclear magnetic dipole moments in Ref. [N1], only 5 lie outside the Schmidt lines; these five are indicated in Fig. 8.2.

3. Of 137 odd-A nuclear magnetic dipole moments in Ref. [N1], all but 10 lie between the Schmidt lines and the "fully quenched" moments obtained by setting $\lambda = 1$ for protons and $\lambda = 0$ for neutrons; these are the values obtained with the Dirac moment alone and vanishing anomalous magnetic moment.

4. There will be corrections to this extreme single-particle value of the magnetic moment coming from, among other things, configuration mixing, meson exchange currents, and a relativistic treatment of the nucleus.

The electric quadrupole operator for a single nucleon is defined by

$$Q_{20} \equiv 3z^2 - r^2 \;\; = \;\; 2r^2 P_2(\cos\theta) = 2r^2\sqrt{\frac{4\pi}{5}}Y_{20} \equiv 2r^2 C_{20} \qquad (8.8)$$

The quadrupole moment of the nucleus is the expectation value of this operator in the state where the nucleus is aligned as well as possible along the z axis[47]

$$Q \equiv \langle j, m = j | Q_{20} | j, m = j \rangle \;\; = \;\; \begin{pmatrix} j & 2 & j \\ -j & 0 & j \end{pmatrix} \langle j || Q_2 || j \rangle \qquad (8.9)$$

[47] We use indiscriminately the relation between C-G coefficients and 3-j symbols (Ref. [N44]).

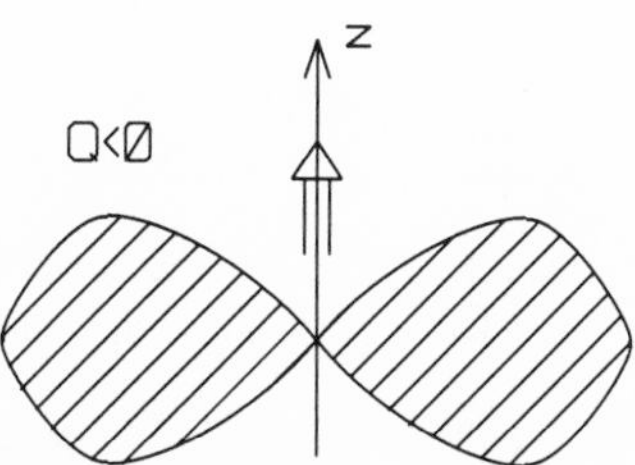

Figure 8.3: Charge distribution in single proton state with $m = j$ giving rise to a negative quadrupole moment.

In the single-particle shell model, one needs the reduced matrix element $\langle nl\frac{1}{2}j||C_2||nl\frac{1}{2}j\rangle$, which is again an ITO in a coupled scheme. The result from Ref. [N51] is[48]

$$Q = -\langle r^2\rangle_{nl}\frac{2j-1}{2j+2}\ ; \qquad j \geq 3/2 \tag{8.10}$$

That the quadrupole moment is indeed negative for a single odd proton is indicated pictorially in Fig. 8.3. Rough estimates of the required single-particle mean square radius can be obtained as follows;

$$\begin{aligned} \langle r^2\rangle_{nl} &= R^2 && \text{particle at surface} \\ &= \tfrac{3}{5}R^2 && \text{uniform distribution} \\ &= (\tfrac{2}{3}\text{ to }\tfrac{4}{5})R^2 && \text{for square well} \end{aligned} \tag{8.11}$$

It is thus evident that *in the single-particle shell model* $|Q|/R^2 < 1$.

Selected nuclear quadrupole moments appear in Table 8.1. Several comments are of interest here:

1. The single-particle shell model does all right for single protons outside closed shells.

2. If the shell is more than 1/2 filled, one gets the *wrong sign* for Q from this extreme single-particle model. A consistent many-body treatment of many particles in a shell is needed in this situation (see, e.g., Ref. [N2]).[49]

3. Odd neutrons outside of closed shells behave just as if they were odd protons, contributing as if they had an *effective charge* $+e_{\text{p}}$; in reality, a free neutron is uncharged.

[48] This is a marvelous reference and set of tables, which everyone should learn to use. In particular, it is shown in this book that

$$\langle l'\frac{1}{2}j'||C_L||l\frac{1}{2}j\rangle = (-1)^{j'+1/2}\sqrt{(2j'+1)(2j+1)}\begin{pmatrix} j' & L & j \\ 1/2 & 0 & -1/2 \end{pmatrix}\frac{1}{2}[1+(-1)^{l'+l+L}]$$

[49] A contemplation of Fig. 8.3 will convince the reader that a single hole in a closed shell has the opposite sign of the quadrupole moment from that of a single particle in the shell.

Table 8.1: Quadrupole moments of selected nuclei (Refs. [N1, N37, N52]).

Element	Spin	$Q \times 10^{26}\ \mathrm{cm}^2$	State	$\frac{Q}{(1.2A^{1/3}\,\mathrm{fm})^2}$	$\frac{Q}{\langle r^2\rangle_{nl}}\ \vert^a$
${}^{27}_{13}\mathrm{Al}$	5/2	+14.9	$(d_{5/2})^5 = d^{-1}_{5/2}$	+1.15	−0.57
${}^{35}_{17}\mathrm{Cl}$	3/2	−7.9	$d_{3/2}$	−0.51	−0.40
${}^{63}_{29}\mathrm{Cu}$	3/2	−16.0	$p_{3/2}$	−0.70	−0.40
${}^{175}_{71}\mathrm{Lu}$	3/2	+590.0	-	+13.10	mag< 1
${}^{17}\mathrm{O}_9$	5/2	−2.7	$d_{5/2}$	−0.28	−0.57
${}^{33}\mathrm{S}_{17}$	3/2	−6.0	$d_{3/2}$	−0.40	−0.40
${}^{83}\mathrm{Kr}_{47}$	9/2	+22.0	$(g_{9/2})^7$	+0.80	−0.73
${}^{167}\mathrm{Er}_{99}$	7/2	+ 1000.0	-	+22.90	mag< 1

[a] Single-particle shell model values. For the odd neutron nuclei, the single-particle values are obtained by assigning the odd neutron an *effective charge* $+e_{\mathrm{p}}$.

4. The inequality $|Q|/R^2 < 1$ derived above in the single-particle shell model is very badly violated in selected regions of the periodic table where one is far from a few particles or a few holes outside of major closed shells.

5. The explanation of the observed quadrupole moments is due to Rainwater (Ref. [N53]) and Bohr and Mottelson (Ref. [N5]). The core of the nucleus carries a large charge Z. A small deformation of this core can have a large effect on the nuclear quadrupole moment. In particular, if the attractive interaction between a valence neutron and the nucleons in the core drags the positively charged core along with the neutron as it moves, one has a simple qualitative understanding of the behavior of odd-neutron nuclei described above. Although we will not go into detail on the discussion of deformed nuclei, it is one of the most beautiful aspects of nuclear structure. Fortunately, Ref. [N5] provides a thorough treatise on the subject.

This discussion can be extended to higher nuclear moments, and to all momentum transfers, as indicated at the end of Section 7. First, however, it is necessary to construct the nuclear current operator in the nuclear shell model.

Nuclear Current Operator. If the nucleus is modeled as a quantum mechanical system of point nucleons with intrinsic magnetic moments, then we know how to construct the charge density, convection current density, and intrinsic magnetization density from basic quantum mechanics. In first quantization these quantities are given by

$$\hat{\rho}_{\mathrm{N}}(\mathbf{x}) = \sum_{j=1}^{A} e(j)\delta^{(3)}(\mathbf{x}-\mathbf{x}_j)$$

$$\hat{\mathbf{J}}_{\mathrm{c}}(\mathbf{x}) = \sum_{j=1}^{A} e(j)\{\frac{\mathbf{p}(j)}{m}, \delta^{(3)}(\mathbf{x}-\mathbf{x}_j)\}_{\mathrm{sym}}$$

$$\hat{\boldsymbol{\mu}}(\mathbf{x}) = \sum_{j=1}^{A} \mu(j) \frac{\boldsymbol{\sigma}(j)}{2m} \delta^{(3)}(\mathbf{x} - \mathbf{x}_j) \tag{8.12}$$

Here $\mathbf{p} \equiv (1/i)\nabla$. Thus one has for one particle, for example

$$\langle \hat{\rho}_{\mathrm{N}}(\mathbf{x}) \rangle = \int \psi^*(\mathbf{x}_p) \delta^{(3)}(\mathbf{x} - \mathbf{x}_p) \psi(\mathbf{x}_p) \, d^3x_p \quad = \quad |\psi(\mathbf{x})|^2 \tag{8.13}$$

and also

$$\begin{aligned} \langle \hat{\mathbf{J}}_c(\mathbf{x}) \rangle &= \int \psi^*(\mathbf{x}_p) \frac{1}{2im} [\nabla_p \delta^{(3)}(\mathbf{x} - \mathbf{x}_p) + \delta^{(3)}(\mathbf{x} - \mathbf{x}_p) \nabla_p] \psi(\mathbf{x}_p) \, d^3x_p \\ &= \frac{1}{2im} \{ \psi^*(\mathbf{x}) \nabla \psi(\mathbf{x}) - [\nabla \psi(\mathbf{x})]^* \psi(\mathbf{x}) \} \end{aligned} \tag{8.14}$$

These are the familiar results. A partial integration has been used in arriving at the second equality in Eq. (8.14). Here we have defined

$$\begin{aligned} e(j) &\equiv \frac{1}{2}[1 + \tau_3(j)] \\ \mu(j) &\equiv \lambda_p \frac{1}{2}[1 + \tau_3(j)] + \lambda_n \frac{1}{2}[1 - \tau_3(j)] \end{aligned} \tag{8.15}$$

In this section we shall write the *anomalous* magnetic moment $\lambda'(j)$ of the nucleon as

$$\begin{aligned} \lambda'(j) &= \lambda'_p \frac{1}{2}[1 + \tau_3(j)] + \lambda_n \frac{1}{2}[1 - \tau_3(j)] \\ \mu(j) &= e(j) + \lambda'(j) \end{aligned} \tag{8.16}$$

This discussion presents a consistent nonrelativistic treatment in a picture where the nucleus is made up of point nucleons with appropriate charges and intrinsic magnetic moments; however, a central future thrust of nuclear physics is the measurement and calculation of nuclear electromagnetic transition densities out to momentum transfers of the order of a GeV, or $q = O(m)$, and well beyond. It is essential to consider corrections to the nonrelativistic current operator as one moves into this regime. In order to do this, a fully relativistic treatment of the interacting many-body system is required, and Part II of this book is devoted to this topic. For the present, we simply consider the nuclear current density arising from the full relativistic electromagnetic vertex of a free nucleon.

Relativistic Corrections to the Current. The relativistic electromagnetic vertex of a free nucleon is illustrated in Fig. 8.4. The most general structure of the matrix element of the current for a free nucleon is given by (see, e.g., Ref. [R2])

$$\langle \mathbf{p}'\sigma'\rho' | J_\mu(0) | \mathbf{p}\sigma\rho \rangle = \frac{i}{\Omega} \bar{u}(\mathbf{p}', \sigma') \eta^\dagger_{\rho'} [F_1 \gamma_\mu + F_2 \sigma_{\mu\nu} q_\nu] \eta_\rho u(\mathbf{p}, \sigma) \tag{8.17}$$

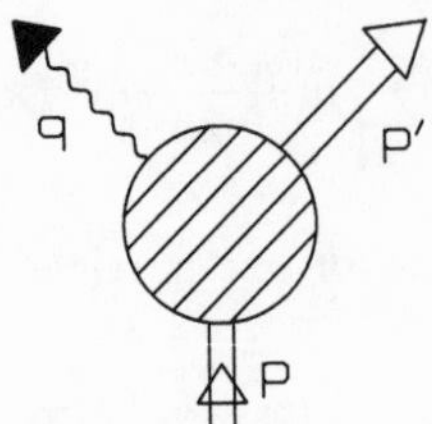

Figure 8.4: Electromagnetic vertex for a free nucleon.

Here the spin and isospin quantum numbers have been made explicit; $\bar{u}, u$ are Dirac spinors and η_p, η_n are two-component Pauli isospinors. The four-momentum transfer is defined by $q = p - p'$ (Fig. 8.4), and the form factors $F_1(q^2), F_2(q^2)$ are functions of q^2. The isospin structure of the form factors must be of the form

$$F_i = \frac{1}{2}(F_i^S + \tau_3 F_i^V)\,; \qquad i = 1, 2 \tag{8.18}$$

Relevant numerical values are

$$\begin{aligned} F_1^S(0) &= F_1^V(0) = 1 \\ 2mF_2^S(0) &= \lambda_p' + \lambda_n = -0.120 \\ 2mF_2^V(0) &= \lambda_p' - \lambda_n = +3.706 \end{aligned} \tag{8.19}$$

To construct the *nuclear* current density we now carry out the following series of steps (see Ref. [N14]):

1. Substitute the explicit form of the Dirac spinors for a free nucleon

$$u(\mathbf{p}, \sigma) = \left(\frac{E_p + m}{2E_p}\right)^{1/2} \begin{pmatrix} \chi_\sigma \\ \dfrac{\boldsymbol{\sigma}\cdot\mathbf{p}}{E_p + m}\chi_\sigma \end{pmatrix} \tag{8.20}$$

Here $\chi_\uparrow, \chi_\downarrow$ are two-component Pauli spinors for spin up and down along the z axis. Now expand the matrix element in Eq. (8.17) consistently to order $1/m^2$. The result is (see Prob. 8.5)

$$\begin{aligned} \langle \mathbf{p}'\sigma'\rho'|J_\mu(0)|\mathbf{p}\sigma\rho\rangle &= \frac{1}{\Omega}\eta_{\rho'}^\dagger \chi_{\sigma'}^\dagger \mathcal{M}_\mu \chi_\sigma \eta_\rho \\ \boldsymbol{\mathcal{M}} &= F_1 \frac{1}{2m}(\mathbf{p} + \mathbf{p}') + (F_1 + 2mF_2)\left[\frac{-i\boldsymbol{\sigma}\times\mathbf{q}}{2m}\right] + O(\frac{1}{m^3}) \\ \mathcal{M}_0 &= F_1 - (F_1 + 4mF_2)\left[\frac{\mathbf{q}^2}{8m^2} - \frac{i\mathbf{q}\cdot(\boldsymbol{\sigma}\times\mathbf{p})}{4m^2}\right] + O(\frac{1}{m^3}) \end{aligned} \tag{8.21}$$

Here $\mathcal{M}_\mu = (\boldsymbol{\mathcal{M}}, i\mathcal{M}_0)$.[50]

[50] It is assumed here that both q_0 and F_2 are $O(1/m)$.

2. Take as the prescription for constructing the nuclear current density operator at the origin, in second quantization, the following expression

$$\hat{J}_\mu(0) = \sum_{\mathbf{p}'\sigma'\rho'} \sum_{\mathbf{p}\sigma\rho} c^\dagger_{\mathbf{p}'\sigma'\rho'} \langle \mathbf{p}'\sigma'\rho' | J_\mu(0) | \mathbf{p}\sigma\rho \rangle c_{\mathbf{p}\sigma\rho} \tag{8.22}$$

Here the single-particle matrix element is precisely that of Eq. (8.17).

3. Use the general procedure for passing from first quantization to second quantization (Ref. [N2]). If, in first quantization the one-body nuclear density operator has the form

$$\hat{J}_\mu(\mathbf{x}) = \sum_{i=1}^{A} \{ J_\mu^{(1)}(i) \delta^{(3)}(\mathbf{x} - \mathbf{x}_i) \} \tag{8.23}$$

then in second quantization the operator density is

$$\hat{J}_\mu(\mathbf{x}) = \sum_{\mathbf{p}'\sigma'\rho'} \sum_{\mathbf{p}\sigma\rho} c^\dagger_{\mathbf{p}'\sigma'\rho'} \langle \mathbf{p}'\sigma'\rho' | J_\mu(\mathbf{x}) | \mathbf{p}\sigma\rho \rangle c_{\mathbf{p}\sigma\rho} \tag{8.24}$$

with

$$\langle \mathbf{p}'\sigma'\rho' | J_\mu(\mathbf{x}) | \mathbf{p}\sigma\rho \rangle = \int d^3y \, \phi^\dagger_{\mathbf{p}'\sigma'\rho'}(\mathbf{y}) \{ J_\mu^{(1)}(\mathbf{y}) \delta^{(3)}(\mathbf{x} - \mathbf{y}) \} \phi_{\mathbf{p}\sigma\rho}(\mathbf{y}) \tag{8.25}$$

4. The discussion in Section 7 shows that physical rates and cross sections are expressed in terms of the Fourier transform of the transition matrix element of the current

$$\int e^{-i\mathbf{q}\cdot\mathbf{x}} \langle f | \hat{J}_\mu(\mathbf{x}) | i \rangle \, d^3x \tag{8.26}$$

Here $q = p - p'$, and in electron scattering $q = k' - k$. We define

$$\langle f | \hat{J}_\mu(\mathbf{x}) | i \rangle \equiv J_\mu(\mathbf{x})_{\mathrm{fi}} \tag{8.27}$$

and observe that by partial integration in Eq. (8.26) with localized densities one can make the replacement

$$\nabla \leftrightarrow i\mathbf{q} \tag{8.28}$$

We then anticipate the presence of terms in $i\mathbf{q}$ in the elementary nucleon amplitudes by defining

$$\begin{aligned} \mathbf{J}(\mathbf{x})_{\mathrm{fi}} &\equiv \mathbf{J}_{\mathrm{c}}(\mathbf{x})_{\mathrm{fi}} + \nabla \times \boldsymbol{\mu}(\mathbf{x})_{\mathrm{fi}} \\ \rho(\mathbf{x})_{\mathrm{fi}} &\equiv \rho_{\mathrm{N}}(\mathbf{x})_{\mathrm{fi}} + \nabla \cdot \mathbf{s}(\mathbf{x})_{\mathrm{fi}} + \nabla^2 \phi(\mathbf{x})_{\mathrm{fi}} \end{aligned} \tag{8.29}$$

The use of Eq. (8.25) evaluated at $\mathbf{x} = 0$ now permits the identification of the nuclear density operators in first quantization, which give rise to the required result in second quantization of Eq. (8.22). The operators take the form

$$\begin{aligned} \hat{\mathbf{J}}(\mathbf{x}) &= \hat{\mathbf{J}}_{\mathrm{c}}(\mathbf{x}) + \nabla \times \hat{\boldsymbol{\mu}}(\mathbf{x}) \\ \hat{\rho}(\mathbf{x}) &= \hat{\rho}_{\mathrm{N}}(\mathbf{x}) + \nabla \cdot \hat{\mathbf{s}}(\mathbf{x}) + \nabla^2 \hat{\phi}(\mathbf{x}) \end{aligned} \tag{8.30}$$

Here the densities are defined by Eqs. (8.12), (8.15), (8.16), and

$$\begin{aligned} \hat{\phi}(\mathbf{x}) &= \sum_{j=1}^{A} s(j)\frac{1}{8m^2}\delta^{(3)}(\mathbf{x}-\mathbf{x}_j) \\ \hat{\mathbf{s}}(\mathbf{x}) &= \sum_{j=1}^{A} s(j)\frac{1}{4m^2}\boldsymbol{\sigma}(j)\times\{\mathbf{p}(j),\delta^{(3)}(\mathbf{x}-\mathbf{x}_j)\}_{\rm sym} \end{aligned} \tag{8.31}$$

with

$$s(j) \equiv e(j) + 2\lambda'(j) \tag{8.32}$$

5. It is an empirical result that in the nuclear domain[51]

$$\begin{aligned} \frac{F_1(q^2)}{F_1(0)} &\approx f_{\rm SN}(q^2) \approx \frac{F_2(q^2)}{F_2(0)} \\ f_{\rm SN}(q^2) &= \frac{1}{(1+q^2/0.71\,{\rm GeV}^2)^2} \end{aligned} \tag{8.33}$$

The (e, e') cross section is now determined by an *effective* Mott cross section

$$\bar{\sigma}_{\rm M} \equiv \sigma_{\rm M}|f_{\rm SN}(q^2)|^2 \tag{8.34}$$

The use of this effective Mott cross section represents an approximate way of taking into account in the nuclear domain the spatial extent of the *internal* charge and magnetization densities of a single constituent nucleon.

6. The present analysis gives the leading relativistic corrections to the nuclear current, assuming it is a one-body operator. It *neglects*, among other things: meson exchange currents, other multibody currents, relativistic terms in the *wave functions*, and off-shell corrections to the nucleon vertex in the nuclear medium.

7. The *goal* is to develop

- a consistent, relativistic, hadronic description of the nucleus
- a consistent, relativistic, quark description of the nucleus.

It is to this task that we turn our attention in Parts II and III of this book. First, however, as the conclusion of this review in Part I of basic nuclear structure, we discuss collective nuclear excitations built on the shell-model ground state.

[51] A more accurate representation of the experimental data for the proton and neutron out to very large q^2 is given by (see Ref. [N14])

$$\begin{aligned} G_{\rm M}(q^2) \equiv F_1 + 2mF_2 &= f_{\rm SN}(q^2)G_{\rm M}(0) \\ G_{\rm E}(q^2) \equiv F_1 - (q^2/2m)F_2 &= f_{\rm SN}(q^2)G_{\rm E}(0) \end{aligned}$$

although $G_{\rm E}^n(q^2)$ remains to be measured well.

9

EXCITED STATES—EQUATIONS OF MOTION

This section is based on Ref. [N2], which in addition contains a discussion of a systematic procedure using Green's functions to determine the nuclear excitation spectrum to all orders in the two-body interaction. References [N54, N55, N56, N57] provide important background material here.

Consider the general problem of describing the collective excitations built on the Hartree-Fock ground state of a finite nonrelativistic system. If the system is excited through a one-body operator of the form $\hat{\psi}^\dagger O \hat{\psi}$, then the resulting states must have a single particle promoted to a higher shell; in other words, as indicated in Fig. 9.1, the excited state must contain a *particle-hole pair*. Coherent superposition of particle-hole states can build up transition strength. One can describe the strongly excited collective excitations built on the Hartree-Fock ground state as linear combinations of particle-hole states. The goal is to develop a set of quantum mechanical equations of motion with which to describe the properties of these excitations. To that end, introduce the particle-hole pair creation operator

$$\hat{\zeta}^\dagger_{\alpha\beta} \equiv a^\dagger_\alpha b^\dagger_\beta \tag{9.1}$$

The notation is that of Section 6. Consider the matrix element

$$\langle \Psi_n | [\hat{H}, \hat{\zeta}^\dagger_{\alpha\beta}] | \Psi_0 \rangle = (E_n - E_0) \langle \Psi_n | \zeta^\dagger_{\alpha\beta} | \Psi_0 \rangle \tag{9.2}$$

Here $|\Psi_{0,n}\rangle$ are the exact ground and excited states. The hamiltonian in Eq. (6.16) is used to evaluate the commutator required on the l.h.s. of this relation

$$\begin{aligned}
[H_0, \hat{\zeta}^\dagger_{\alpha\beta}] &= 0 \\
[\hat{H}_1, \hat{\zeta}^\dagger_{\alpha\beta}] &= [(\sum_{\alpha > F} \epsilon_\alpha a^\dagger_\alpha a_\alpha - \sum_{\alpha < F} \epsilon_\alpha b^\dagger_\alpha b_\alpha), \hat{\zeta}^\dagger_{\alpha\beta}] \\
&= (\epsilon_\alpha - \epsilon_\beta) \hat{\zeta}^\dagger_{\alpha\beta} \\
[\hat{H}_2, \hat{\zeta}^\dagger_{\alpha\beta}] &= \frac{1}{2} \sum_{\rho\sigma\mu\nu} \langle \rho\sigma | V | \mu\nu \rangle [N(c^\dagger_\rho c^\dagger_\sigma c_\nu c_\mu), a^\dagger_\alpha b^\dagger_\beta]
\end{aligned} \tag{9.3}$$

Recall that

$$c_\alpha \equiv \theta(\alpha - F) a_\alpha + \theta(F - \alpha) S_\alpha b^\dagger_{-\alpha} \tag{9.4}$$

These are exact equations of motion of the system. Their solution is equivalent to that of the full Schrödinger equation for a finite many-body system. To

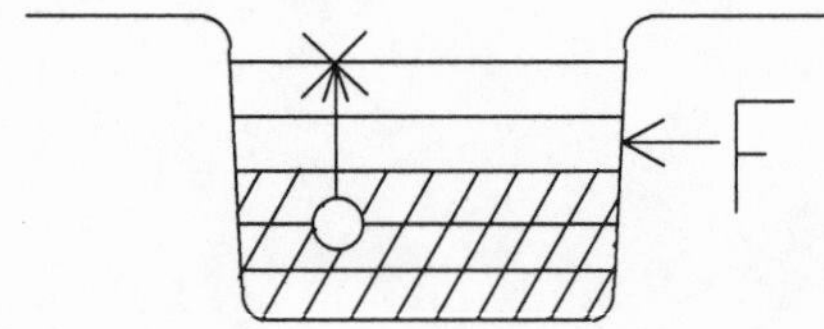

Figure 9.1: One-body excitations built on the Hartree-Fock ground state.

be tractable, some approximation is needed to deal with the last term in Eq. (9.3), which contains the entire remaining effects of the two-body interaction. This shall be done with the method of *linearization of the equations of motion.* We start with the simplest case, which provides a widely used approximation scheme.

Tamm-Dancoff Approximation (TDA). Make the following two approximations [see Fig. (9.1)]:

1. Assume the ground state corresponds to the filled core of Hartree-Fock states, that is

$$|\Psi_0\rangle \approx |0\rangle$$
$$a_\alpha|0\rangle = b_\alpha|0\rangle = 0 \tag{9.5}$$

2. Assume the excited state is some linear combination of particle-hole states

$$|\Psi_n\rangle \approx \sum_{\alpha\beta} \psi^{(n)*}_{\alpha\beta}\, \hat{\zeta}^\dagger_{\alpha\beta}|0\rangle \tag{9.6}$$

The sum here is assumed to go over some finite, albeit arbitrarily large, set of $\mathcal{N}$ particle-hole states.

With these two assumptions, only the terms proportional to $a^\dagger b^\dagger$ need be kept in the final commutator in Eq. (9.3); all other operators make a vanishing contribution to the matrix element in Eq. (9.2). Thus only the following terms in the commutator will contribute

$$\begin{aligned} &[N\{S_\mu S_\sigma a^\dagger_\rho b_{-\sigma} a_\nu b^\dagger_{-\mu} + S_\rho S_\nu b_{-\rho} a^\dagger_\sigma b^\dagger_{-\nu} a_\mu + S_\nu S_\sigma a^\dagger_\rho b_{-\sigma} b^\dagger_{-\nu} a_\mu \\ &\quad + S_\rho S_\mu b_{-\rho} a^\dagger_\sigma a_\nu b^\dagger_{-\mu}\}\ ,\ a^\dagger_\alpha b^\dagger_\beta] \end{aligned} \tag{9.7}$$

Use the symmetry property of matrix elements of the potential

$$\langle\rho\sigma|V|\mu\nu\rangle = \langle\sigma\rho|V|\nu\mu\rangle \tag{9.8}$$

This reduces the required commutator in Eq. (9.7) to the form[52]

$$2[(S_\mu S_\sigma a^\dagger_\rho b^\dagger_{-\mu} b_{-\sigma} a_\nu - S_\nu S_\sigma a^\dagger_\rho b^\dagger_{-\nu} b_{-\sigma} a_\mu)\ ,\ a^\dagger_\alpha b^\dagger_\beta] \tag{9.9}$$

[52] To evaluate this commutator, just move the destruction operators over to the right.

The required result is then

$$[\hat{H}_2, \hat{\zeta}^{\dagger}_{\alpha\beta}] \doteq \sum_{\lambda\mu} S_{-\beta}S_{-\mu}[\langle\lambda - \beta|V| - \mu\alpha\rangle - \langle\lambda - \beta|V|\alpha - \mu\rangle]\hat{\zeta}^{\dagger}_{\lambda\mu} \tag{9.10}$$

Here the symbol $\doteq$ indicates that only these terms contribute to the required matrix element in Eq. (9.2) in the TDA.

Define [see Eqs. (9.5) and (9.6)]

$$\langle\Psi_n|\hat{\zeta}^{\dagger}_{\alpha\beta}|\Psi_0\rangle \equiv \psi^{(n)}_{\alpha\beta} \tag{9.11}$$

A combination of the previous results then yields the resulting equations of motion

$$[(E_0 + \epsilon_\alpha - \epsilon_\beta) - E_n]\psi^{(n)}_{\alpha\beta} + \sum_{\lambda\mu} v_{\alpha\beta;\lambda\mu}\psi^{(n)}_{\lambda\mu} = 0 \tag{9.12}$$

Here $\epsilon_\alpha - \epsilon_\beta$ is the Hartree-Fock particle-hole *configuration energy* and the *particle-hole interaction* is defined in terms of matrix elements of the two-body potential according to

$$v_{\alpha\beta;\lambda\mu} \equiv S_{-\beta}S_{-\mu}[\langle\lambda - \beta|V| - \mu\alpha\rangle - \langle\lambda - \beta|V|\alpha - \mu\rangle] \tag{9.13}$$

Equations (9.12) provide a set of *linear, homogeneous, algebraic equations* for the numerical coefficients $\psi^{(n)}_{\lambda\mu}$. The vanishing of the determinant of the matrix of coefficients in these algebraic relations gives the set of energy *eigenvalues* E_n with $n = 1, \ldots, \mathcal{N}$. Substitution of these eigenvalues allows one to determine the *eigenvectors* $\psi^{(n)}_{\lambda\mu}$. In fact, since one of the equations is linearly dependent upon substitution of the eigenvalue, it is only $\mathcal{N} - 1$ ratios that are determined. The entire eigenvector can be obtained with the aid of a phase convention and the normalization condition. The latter is derived by noting that

$$\langle 0|\hat{\zeta}_{\lambda\mu}\hat{\zeta}^{\dagger}_{\alpha\beta}|0\rangle = \delta_{\alpha\lambda}\delta_{\mu\beta} \tag{9.14}$$

Thus from Eq. (9.6)

$$\langle\Psi_{n'}|\Psi_n\rangle = \delta_{nn'} \quad = \quad \sum_{\alpha\beta} \psi^{(n')}_{\alpha\beta}\psi^{(n)*}_{\alpha\beta} \tag{9.15}$$

The orthogonality of solutions for different n follows directly from the linear algebraic Eqs. (9.12) (see Prob. 9.2).

The transition matrix elements of any multipole operator can be calculated from

$$\hat{T} = \sum_{\alpha\beta} c^{\dagger}_{\alpha}\langle\alpha|T|\beta\rangle c_{\beta} \quad \doteq \quad \sum_{\alpha\beta}\langle\alpha|T| - \beta\rangle S_{-\beta}\hat{\zeta}^{\dagger}_{\alpha\beta} \tag{9.16}$$

The symbol $\doteq$ here has the same meaning as above; only these terms contribute to the transition matrix element in the TDA. Thus

$$\langle\Psi_n|\hat{T}|\Psi_0\rangle = \sum_{\alpha\beta}\langle\alpha|T|-\beta\rangle S_{-\beta}\psi_{\alpha\beta}^{(n)} \tag{9.17}$$

The result is a linear combination of single-particle transition matrix elements weighted with the previously determined numerical coefficients $\psi_{\alpha\beta}^{(n)}$.

The TDA provides an excellent framework for obtaining a qualitative, and semiquantitative, description of nuclear excitations. It assumes a closed-shell Hartree-Fock ground state and excited states that are linear combinations of only singly excited particle-hole pairs. This is clearly a simple model and often fails, for example, to exhibit the observed degree of coherence in nuclear transition strengths. It is of interest to ask how these results are modified if one allows for particle-hole pairs in the ground state, and additional particle-hole pairs in the excited state, while retaining the utility of linearized equations of motion. The random phase approximation provides such a framework.

Random Phase Approximation (RPA). The linearized equations of motion can be extended by allowing the ground state to have particle-hole pairs. This is accomplished by retaining *both the particle-hole pair creation and destruction operators* in the evaluation of the required commutators in Eqs. (9.3)

$$\hat{\zeta}_{\alpha\beta}^{\dagger} \equiv a_{\alpha}^{\dagger}b_{\beta}^{\dagger} \qquad\qquad \hat{\zeta}_{\alpha\beta} \equiv b_{\beta}a_{\alpha} \tag{9.18}$$

All matrix elements are retained of the form

$$\begin{aligned}\psi_{\alpha\beta}^{(n)} &\equiv \langle\Psi_n|\hat{\zeta}_{\alpha\beta}^{\dagger}|\Psi_0\rangle \\ \phi_{\alpha\beta}^{(n)} &\equiv \langle\Psi_n|\hat{\zeta}_{\alpha\beta}|\Psi_0\rangle\end{aligned} \tag{9.19}$$

Thus one can either create or annihilate a particle-hole pair in making a transition from the ground to the excited state. Consider now the following two matrix elements

$$\begin{aligned}\langle\Psi_n|[\hat{H},\hat{\zeta}_{\alpha\beta}^{\dagger}]|\Psi_0\rangle &= (E_n-E_0)\psi_{\alpha\beta}^{(n)} \\ \langle\Psi_n|[\hat{H},\hat{\zeta}_{\alpha\beta}]|\Psi_0\rangle &= (E_n-E_0)\phi_{\alpha\beta}^{(n)}\end{aligned} \tag{9.20}$$

The terms in $[\hat{H},\hat{\zeta}_{\alpha\beta}^{\dagger}]$ proportional to $\hat{\zeta}_{\alpha\beta}^{\dagger}$ were evaluated above. The additional terms proportional to $\hat{\zeta}_{\alpha\beta}$ are now required. An extension of the above analysis gives

$$\begin{aligned}&[N\{S_\rho S_\sigma b_{-\rho}b_{-\sigma}a_\nu a_\mu\}\ ,\ a_\alpha^{\dagger}b_\beta^{\dagger}] \\ &\quad\doteq 2[\delta_{\mu\alpha}\delta_{\beta-\rho}b_{-\sigma}a_\nu - \delta_{\nu\alpha}\delta_{\beta-\rho}b_{-\sigma}a_\mu]S_\rho S_\sigma\end{aligned} \tag{9.21}$$

Here Eq. (9.8) has again been used. One thus obtains

$$[\hat{H}_2,\hat{\zeta}_{\alpha\beta}^{\dagger}] \doteq \sum_{\lambda\mu}[v_{\alpha\beta;\lambda\mu}\hat{\zeta}_{\lambda\mu}^{\dagger} + u_{\alpha\beta;\lambda\mu}\hat{\zeta}_{\lambda\mu}] \tag{9.22}$$

The additional particle-hole interaction follows from the above as

$$u_{\alpha\beta;\lambda\mu} = S_{-\beta}S_{-\mu}[\langle -\beta - \mu|V|\alpha\lambda\rangle - \langle -\beta - \mu|V|\lambda\alpha\rangle] \tag{9.23}$$

The linearized equations of motion in the RPA thus take the form[53]

$$\begin{aligned} \{[E_0 + (\epsilon_\alpha - \epsilon_\beta)] - E_n\}\psi_{\alpha\beta}^{(n)} + \sum_{\lambda\mu}[v_{\alpha\beta;\lambda\mu}\psi_{\lambda\mu}^{(n)} + u_{\alpha\beta;\lambda\mu}\phi_{\lambda\mu}^{(n)}] = 0 \\ \{[E_0 - (\epsilon_\alpha - \epsilon_\beta)] - E_n\}\phi_{\alpha\beta}^{(n)} - \sum_{\lambda\mu}[v^*_{\alpha\beta;\lambda\mu}\phi_{\lambda\mu}^{(n)} + u^*_{\alpha\beta;\lambda\mu}\psi_{\lambda\mu}^{(n)}] = 0 \end{aligned} \tag{9.24}$$

These again form *linear, homogeneous, algebraic equations.* If the complete set of single-particle states is taken to include the bound states plus the continuum states with standing wave boundary conditions, then all the matrix elements of the potential are *real*

$$\begin{aligned} v_{\alpha\beta;\lambda\mu} = v^*_{\alpha\beta;\lambda\mu} &\equiv v_{\lambda\mu;\alpha\beta} \\ u_{\alpha\beta;\lambda\mu} = u^*_{\alpha\beta;\lambda\mu} &\equiv u_{\lambda\mu:\alpha\beta} \end{aligned} \tag{9.25}$$

With the aid of these conditions, one readily establishes the orthogonality condition on the eigenvectors (Prob. 9.2)

$$\sum_{\alpha\beta}(\psi_{\alpha\beta}^{(n')*}\psi_{\alpha\beta}^{(n)} - \phi_{\alpha\beta}^{(n')*}\phi_{\alpha\beta}^{(n)}) = \delta_{nn'} \tag{9.26}$$

The choice of normalization must be justified. So far only the matrix elements have been approximated through the linearization of the equations of motion in Eq. (9.20); the normalization of the eigenvectors is a bilinear relation that requires further restrictions (see Ref. [N55]). To this end, define the operator

$$\hat{Q}_n^\dagger \equiv \sum_{\alpha\beta}[\psi_{\alpha\beta}^{(n)}\hat{\zeta}_{\alpha\beta}^\dagger - \phi_{\alpha\beta}^{(n)}\hat{\zeta}_{\alpha\beta}] \tag{9.27}$$

It follows from the matrix element in Eqs. (9.20) that

$$[\hat{H}, \hat{Q}_n^\dagger] \doteq (E_n - E_0)\hat{Q}_n^\dagger \tag{9.28}$$

Assume this holds as an operator identity. Assume also that the interacting ground state can be defined by the relation

$$\hat{Q}_n|\Psi_0\rangle = 0 \tag{9.29}$$

for all n. The collective excitations can then be explicitly constructed by letting $\hat{Q}_n^\dagger$ act on the ground state

$$|\Psi_n\rangle = \hat{Q}_n^\dagger|\Psi_0\rangle \tag{9.30}$$

[53]Use $[\hat{H}, \hat{\zeta}_{\alpha\beta}]^\dagger = [\hat{\zeta}_{\alpha\beta}^\dagger, \hat{H}^\dagger] = -[\hat{H}, \hat{\zeta}_{\alpha\beta}^\dagger]$.

for Eq. (9.28) now implies

$$\hat{H}|\Psi_n\rangle \equiv [\hat{H}, \hat{Q}_n^\dagger]|\Psi_0\rangle + E_0\hat{Q}_n^\dagger|\Psi_0\rangle \quad \doteq \quad E_n|\Psi_n\rangle \tag{9.31}$$

Furthermore, the normalization condition can be derived from Eqs. (9.29) and (9.30) as

$$\langle\Psi_{n'}|\Psi_n\rangle = \delta_{nn'} \quad = \quad \langle\Psi_0|[\hat{Q}_{n'}, \hat{Q}_n^\dagger]|\Psi_0\rangle \tag{9.32}$$

The normalization condition in Eq. (9.26) now follows from this relation provided $[\hat{\zeta}_{\lambda\mu}, \hat{\zeta}_{\alpha\beta}^\dagger] \doteq \delta_{\lambda\alpha}\delta_{\beta\mu}$; however by explicit calculation

$$[\hat{\zeta}_{\lambda\mu}, \hat{\zeta}_{\alpha\beta}^\dagger] = \delta_{\mu\beta}\delta_{\lambda\alpha} - \delta_{\mu\beta}a_\alpha^\dagger a_\lambda - \delta_{\alpha\lambda}b_\beta^\dagger b_\mu \tag{9.33}$$

Necessary conditions for this "quasiboson" description of the collective excitations to hold are

$$\begin{aligned} \langle\Psi_0|a_\gamma^\dagger a_\gamma|\Psi_0\rangle &\ll 1 \\ \langle\Psi_0|b_\gamma^\dagger b_\gamma|\Psi_0\rangle &\ll 1 \end{aligned} \tag{9.34}$$

The first expression yields the probability that a *particle* is present in any level in the ground state, and the second is the probability that a *hole* is present; evidently these quantities must be small for the approximation to work. If these conditions hold, then we have, consistently,

$$\langle\Psi_0|[\hat{Q}_{n'}, \hat{Q}_n^\dagger]|\Psi_0\rangle = \sum_{\alpha\beta}[\psi_{\alpha\beta}^{(n')*}\psi_{\alpha\beta}^{(n)} - \phi_{\alpha\beta}^{(n')*}\phi_{\alpha\beta}^{(n)}] \quad = \quad \delta_{nn'} \tag{9.35}$$

The transition matrix elements of any single-particle operator can now be evaluated in the RPA as before by first retaining the appropriate terms in the operator

$$\hat{T} \doteq \sum_{\alpha\beta}[\langle\alpha|T|-\beta\rangle S_{-\beta}\hat{\zeta}_{\alpha\beta}^\dagger + \langle-\beta|T|\alpha\rangle S_{-\beta}\hat{\zeta}_{\alpha\beta}] \tag{9.36}$$

The matrix element of this relation then yields

$$\langle\Psi_n|\hat{T}|\Psi_0\rangle = \sum_{\alpha\beta} S_{-\beta}[\langle\alpha|T|-\beta\rangle\psi_{\alpha\beta}^{(n)} + \langle-\beta|T|\alpha\rangle\phi_{\alpha\beta}^{(n)}] \tag{9.37}$$

This is the desired result; it expresses the transition amplitude as a coherent superposition of the single-particle matrix elements, with the eigenvectors obtained by solution of the linear RPA equations as numerical coefficients.

Reduction of the Basis. The introduction of eigenstates of total angular momentum J and total isospin T will reduce the basis since these are good quantum numbers for the nucleus that cannot be mixed by the interaction. The analysis is carried out in detail in Section 59 of Ref. [N2], and the reader is urged to work through this material in detail. Here, instead, we analyze a simple model where the two-body interaction is assumed to be attractive and independent of spin and isospin.[54] This model illustrates many systematic features of more detailed particle-hole calculations. We here explicitly carry out the reduction of the basis in the TDA.

The situation is illustrated in Fig. 9.1. Consistent with this model it will be assumed that:

1. The ground state of even-even nuclei corresponds to closed shells with $S = L = T = 0$.

2. The single-particle levels can be characterized with the quantum numbers $\{\alpha\} = \{nl\frac{1}{2}\frac{1}{2}, m_l m_s m_t\} \equiv \{a, m_l m_s m_t\}$.

One can now explicitly transform from j-j to L-S coupling; however, it is simpler to start over with a slightly different canonical transformation to particles and holes[55]

$$\begin{aligned} c_\alpha &\equiv \theta(\alpha - F)a_\alpha + \theta(F-\alpha)\mathcal{S}_\alpha B^\dagger_{-\alpha} \\ \mathcal{S}_\alpha &\equiv (-1)^{l_\alpha - m_{l\alpha}}(-1)^{\frac{1}{2}-m_{s\alpha}}(-1)^{\frac{1}{2}-m_{t\alpha}} \end{aligned} \tag{9.38}$$

Start with Eqs. (9.12) and (9.13)

$$(\epsilon_a - \epsilon_b - \epsilon_n)\psi^{(n)}_{\alpha\beta} + \sum_{\lambda\mu}[\langle\lambda - \beta|V| - \mu\alpha\rangle - \langle\lambda - \beta|V|\alpha - \mu\rangle]\mathcal{S}_{-\beta}\mathcal{S}_{-\mu}\psi^{(n)}_{\lambda\mu} = 0 \tag{9.39}$$

To reduce the basis, sum this relation with the appropriate C-G coefficients. One can simply proceed with the angular momentum coupling since care has been taken to work with ITO at each step (Appendix B). Thus one defines

$$\begin{aligned} \psi^{(n)}_{LST}(ab) \equiv & \sum_{\{\text{all m's}\}} \langle l_a m_{l\alpha} l_b m_{l\beta}|l_a l_b L M_L\rangle\langle\frac{1}{2}m_{s\alpha}\frac{1}{2}m_{s\beta}|\frac{1}{2}\frac{1}{2}SM_S\rangle \\ & \times\langle\frac{1}{2}m_{t\alpha}\frac{1}{2}m_{t\beta}|\frac{1}{2}\frac{1}{2}TM_T\rangle\psi^{(n)}_{\alpha\beta} \end{aligned} \tag{9.40}$$

If V is independent of spin and isospin, then this dependence *factors* in the two-body matrix element. In the first term in the interaction potential above, one finds the following expression for the spin part of the matrix element

$$\begin{aligned} & \delta_{m_{s\lambda},-m_{s\mu}}\delta_{m_{s\beta},-m_{s\alpha}}(-1)^{\frac{1}{2}-m_{s\alpha}}(-1)^{\frac{1}{2}-m_{s\lambda}} = \\ & (\sqrt{2})^2\langle\frac{1}{2}m_{s\alpha}\frac{1}{2}m_{s\beta}|\frac{1}{2}\frac{1}{2}00\rangle\langle\frac{1}{2}m_{s\lambda}\frac{1}{2}m_{s\mu}|\frac{1}{2}\frac{1}{2}00\rangle \end{aligned} \tag{9.41}$$

[54]If V is independent of spin and isotopic spin, then the nuclear hamiltonian is invariant under the symmetry group $SU(4)$, which mixes the spin and isotopic spin of a nucleon in the fundamental representation (Ref. [N58]). All states belonging to an irreducible representation of $SU(4)$ are then degenerate, and with attractive interactions, the ground state belongs to the identity representation. For the particle-hole excitations we have $[4]\bigotimes\overline{[4]} = [1]\bigoplus[15]$ as explicitly illustrated in this model calculation. The situation is analogous to that in particle physics where the internal symmetry structure of the meson multiplets is that of $q\bar{q}$ pairs.

[55]Note $B^\dagger_\alpha = \pm b^\dagger_\alpha$ for $j_\alpha = l_\alpha \pm 1/2$. Thus there is simply an *a-dependent phase* relating the two canonical transformations.

with a similar expression for isospin. Thus when the first term gets summed with the C-G coefficients in Eq. (9.40) only $S = T = 0$ will contribute; in the second term in the interaction, these spin and isospin C-G coefficients go right through the potential and onto the eigenvectors.

For the l-dependence of the matrix elements, the use of the completeness of the eigenstates of angular momentum allows one to write

$$\langle l_a m_a l_b m_b|V|l_c m_c l_d m_d\rangle = \sum_{LM}\sum_{L'M'}\langle l_a m_a l_b m_b|l_a l_b LM\rangle \\ \times\langle l_a l_b LM|V|l_c l_d L'M'\rangle\langle l_c l_d L'M'|l_c m_c l_d m_d\rangle \tag{9.42}$$

Now use the fact that V is a scalar under rotations:

$$\langle l_a l_b LM|V|l_c l_d L'M'\rangle = \delta_{LL'}\delta_{MM'}\langle l_a l_b L|V|l_c l_d L\rangle \tag{9.43}$$

Thus

$$\langle l_a m_a l_b m_b|V|l_c m_c l_d m_d\rangle = \sum_{LM}\langle l_a m_a l_b m_b|l_a l_b LM\rangle \\ \times\langle l_c m_c l_d m_d|l_c l_d LM\rangle\langle l_a l_b L|V|l_c l_d L\rangle \tag{9.44}$$

Here the first two factors on the r.h.s. are C-G coefficients explicitly exhibiting the m-dependence, and the remaining matix element of the potential is diagonal in L and M and independent of M.

Substitution of Eq. (9.44) in Eq. (9.39) and projection with the C-G coefficient in Eq. (9.40) lead to the following sum over three C-G coefficients

$$\sum \equiv \sum_{m_\alpha m_\beta M'}\langle l_\alpha m_\alpha l_\beta m_\beta|l_\alpha l_\beta LM\rangle \\ \times\langle l_\lambda m_\lambda l_\beta - m_\beta|l_\lambda l_\beta L'M'\rangle\langle l_\mu - m_\mu l_\alpha m_\alpha|l_\mu l_\alpha L'M'\rangle\mathcal{S}_{-\beta} \tag{9.45}$$

A rearrangement of the order of the coupling in the C-G coefficients using formulas in Ref. [N44] puts this in standard form

$$\sum = (-1)^{L'-l_l-l_b}\sqrt{\frac{2L'+1}{2l_l+1}}\sum_{m_\alpha m_\beta M'}\langle l_\mu - m_\mu l_\alpha m_\alpha|l_\mu l_\alpha L'M'\rangle \\ \times\langle L'M' l_\beta m_\beta|L' l_\beta l_\lambda m_\lambda\rangle\langle l_\alpha m_\alpha l_\beta m_\beta|l_\alpha l_\beta LM\rangle \tag{9.46}$$

From Ref. [N44] this is a standard recoupling relation involving a 6-j symbol

$$\sum = (-1)^{l_m+l_a+L'}(2L'+1)\left\{\begin{matrix} l_m & l_a & L' \\ l_b & l_l & L\end{matrix}\right\}\mathcal{S}_{-\mu}\langle l_\lambda m_\lambda l_\mu m_\mu|l_\lambda l_\mu LM\rangle \tag{9.47}$$

The remaining C-G coefficient is now of the proper form to go right onto the eigenvector in Eq. (9.39) to give the proper coupling in Eq. (9.40). The result is that the TDA equations are reduced to the form

$$(\epsilon_a - \epsilon_b - \epsilon_n)\,\psi^{(n)}_{LST}(ab) + \sum_{lm} v^{LST}_{ab;lm}\,\psi^{(n)}_{LST}(lm) = 0 \tag{9.48}$$

Table 9.1: Quantum numbers of the degenerate states in the model of collective particle-hole interactions discussed in text.

T	S	L	J
1	0	L	L
0	1	L	$L-1, L, L+1$
1	1	L	$L-1, L, L+1$

The particle-hole interaction in this reduced basis is given by

$$v_{ab;lm}^{LST} = -\sum_{L'}(2L'+1)\left\{ \begin{array}{ccc} l_m & l_l & L \\ l_b & l_a & L' \end{array} \right\} \tag{9.49}$$
$$\times[\langle l_l l_b L'|V|l_a l_m L'\rangle - 4\delta_{S0}\delta_{T0}(-1)^{l_a+l_m+L'}\langle l_l l_b L'|V|l_m l_a L'\rangle]$$

Here the order of the first and second terms in the interaction has been interchanged. Several features of this result are of interest:

1. There is one set of linear equations for each $\{L, S, T\}$ independent of $\{M_L, M_S, M_T\}$; thus the *basis has been reduced.* The remaining sum in Eq. (9.48) goes over the set of particle-hole states characterized by the quantum numbers $\{ab\} \equiv \{n_a l_a \frac{1}{2}\frac{1}{2}; n_b l_b \frac{1}{2}\frac{1}{2}\}$; If there are $\mathcal{N}_L$ of these states contributing for a given L, then one is solving a set of $\mathcal{N}_L$-dimensional linear equations, and $n_L = 1, \ldots, \mathcal{N}_L$ labels the eigenvalues and eigenvectors.

2. $\epsilon_a - \epsilon_b$ are the Hartree-Fock single-particle energies of interaction with the *filled core.* The role of the particle-hole interaction v is to subtract off the interaction with the empty state.

3. The entire remaining dependence on S and T is contained in the coefficient of the last term in the interaction. If either $S \neq 0$ or $T \neq 0$ the last term in Eq. (9.49) vanishes. Thus the 15 spin and isospin states in Table 9.1 with $[2S+1, 2T+1] = [3,3]\bigoplus[3,1]\bigoplus[1,3]$ lie at the same energy or are *degenerate* for each of the $\mathcal{N}_L$ eigenvalues with given L. They can each be combined with these states of given L to produce states of good $\{L, S, J, T\}$.

4. The single state with $S = T = 0$ for each of the $\mathcal{N}_L$ eigenvalues with given L is split off from the 15 with $S \neq 0$ or $T \neq 0$ by the interaction in Eq. (9.49).

The transition matrix elements of the multipole operator in Eq. (9.17) are immediately expressed in terms of the reduced eigenvectors of Eq. (9.40) through the use of the Wigner-Eckart theorem on both the many-particle and single-particle matrix elements. Furthermore, this reduction of the basis is readily extended from the TDA to the RPA using analogous techniques. These results will be presented in the next section where the introduction of a simple contact two-body interaction will allow us to solve both the TDA and RPA equations analytically and investigate their consequences.

10

COLLECTIVE MODES — A SIMPLE MODEL WITH $-g\delta^{(3)}(\mathbf{r})$

In this section, which is based on Refs. [N2, N57], a simple attractive contact potential, independent of spin and isospin, will be assumed. This allows solution of the linearized equations of motion and investigation of the systematics of collective particle-hole excitations in nuclei. From $SU(4)$ invariance (Ref. [N58]) the degenerate particle-hole supermultiplets will belong to the $[4] \bigotimes [\bar{4}] = [15] \bigoplus [1]$ representations of $SU(4)$; as shown in Section 9, this structure is explicitly realized in this model dynamical calculation. Of course, there are significant spin-dependent effects present in nuclei, the most evident being the spin-orbit interaction, which gives rise to the nuclear shell model (Section 6). Thus the present model is applicable in detail, at most, to light nuclei.[56]

Take the two-body interaction to be a simple attractive short-range interaction of the form

$$V = -g\delta^{(3)}(\mathbf{x}_1 - \mathbf{x}_2) \tag{10.1}$$

In this case the matrix element of the two-body potential $\langle l_l l_b L'|V|l_a l_m L'\rangle$ required in Eq. (9.49) can be readily evaluated. First, the $\int d^3x_2$ is immediately performed. Then the use of Ref. [N44] gives

$$\sum_{m_1 m_2} \langle l_1 m_1 l_2 m_2 | l_1 l_2 LM\rangle Y_{l_1 m_1}(\Omega) Y_{l_2 m_2}(\Omega) = (-1)^{l_1 - l_2}$$
$$\times \frac{1}{\sqrt{4\pi}} \sqrt{(2l_1+1)(2l_2+1)} \begin{pmatrix} l_1 & l_2 & L \\ 0 & 0 & 0 \end{pmatrix} Y_{LM}(\Omega) \tag{10.2}$$

Thus

$$\begin{aligned} &\sum_{m_a m_b m_l m_m} \langle l_l m_l l_b m_b | l_l l_b L' M'\rangle \langle l_a m_a l_m m_m | l_a l_m L' M'\rangle \\ &\quad \times \int Y^*_{l_l m_l} Y^*_{l_b m_b} Y_{l_a m_a} Y_{l_m m_m} d\Omega \\ &= (-1)^{l_l - l_b + l_a - l_m} \frac{1}{4\pi} \sqrt{(2l_a+1)(2l_b+1)(2l_l+1)(2l_m+1)} \\ &\quad \times \begin{pmatrix} l_l & l_b & L' \\ 0 & 0 & 0 \end{pmatrix} \begin{pmatrix} l_a & l_m & L' \\ 0 & 0 & 0 \end{pmatrix} \end{aligned} \tag{10.3}$$

[56] It can be extended to j-j coupling and heavier nuclei (Refs. [N2, N57] and Prob. 9.3).

Further use of Ref. [N44] allows one to now perform the sum on L' required in Eq. (9.49)[57]

$$\sum_{L'}(2L'+1)\left\{\begin{array}{ccc} l_m & l_l & L \\ l_b & l_a & L' \end{array}\right\}\left(\begin{array}{ccc} l_l & l_b & L' \\ 0 & 0 & 0 \end{array}\right)\left(\begin{array}{ccc} l_a & l_m & L' \\ 0 & 0 & 0 \end{array}\right)$$

$$= (-1)^{l_b+l_l+L}\left(\begin{array}{ccc} l_a & L & l_b \\ 0 & 0 & 0 \end{array}\right)\left(\begin{array}{ccc} l_l & L & l_m \\ 0 & 0 & 0 \end{array}\right) \tag{10.4}$$

The [15] Supermultiplet in TDA. For the [15] supermultiplet the last term in Eq. (9.49) vanishes, and the particle-hole interaction takes the form

$$v_{ab;lm}^{[15]L} = \xi v_{ab}^{L} v_{lm}^{L} \tag{10.5}$$

Here the factors are defined by[58]

$$\begin{aligned} v_{ab}^{L} &\equiv (-1)^{l_a}\sqrt{(2l_a+1)(2l_b+1)}\left(\begin{array}{ccc} l_a & L & l_b \\ 0 & 0 & 0 \end{array}\right) \\ &\equiv \langle l_a||C_L||l_b\rangle \end{aligned} \tag{10.6}$$

The second equality follows from Ref. [N44]. The parameter ξ represents the remaining radial integral

$$\xi \equiv \frac{g}{4\pi}\int_0^\infty u_{n_l l_l} u_{n_b l_b} u_{n_a l_a} u_{n_m l_m} \frac{dr}{r^2} \tag{10.7}$$

Now the radial wave functions (assumed real) are peaked at the nuclear surface for particles in the first few unoccupied shells and holes in the last few filled shells, and the overlap integral ξ does not change much from one particle-hole pair to the next (Ref. [N59]). The assumption of constant ξ leads to a particle-hole interaction in Eq. (10.5) that is *separable*, allowing one to solve the linear Eqs. (9.48) *analytically*, for they now take the form[59]

$$(\epsilon_{ab}-\epsilon_n)\psi_{[15]L}^{(n)}(ab) + \xi v_{ab}^{L}\,[\sum_{lm} v_{lm}^{L}\,\psi_{[15]L}^{(n)}(lm)] = 0 \tag{10.8}$$

Multiplication by $v_{ab}^{L}/(\epsilon_{ab}-\epsilon_n)$ and $\sum_{ab}$ then leads to the *eigenvalue equation*

$$\frac{1}{\xi} = \sum_{ab}\frac{(v_{ab}^{L})^2}{\epsilon_n - \epsilon_{ab}} \tag{10.9}$$

This eigenvalue equation is solved graphically in Fig. 10.1. With $\mathcal{N}$ particle-

[57]Note that the 3-j symbol $\left(\begin{array}{ccc} l_1 & l_2 & L \\ 0 & 0 & 0 \end{array}\right)$ vanishes unless l_1+l_2+L is even.

[58]Recall $C_{LM} \equiv [4\pi/(2L+1)]^{1/2}Y_{LM}$.

[59]Note $\{ab\}$ and $\{lm\}$ are now simply sets of radial quantum numbers in this equation.

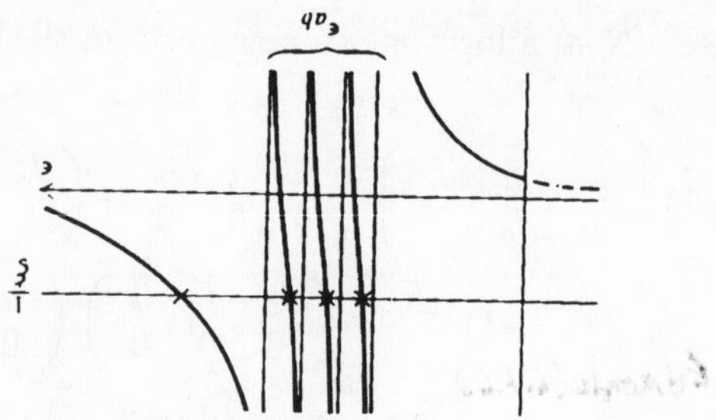

Figure 10.1: Graphic solution of TDA eigenvalue equation for simple model two-body interaction discussed in the text.

hole states, $\mathcal{N}-1$ eigenvalues lie between the configuration energies $\epsilon_{ab} \equiv \epsilon_a - \epsilon_b$. One eigenvalue ϵ_{top} is pushed up to arbitrarily high energy, depending on the value of $1/\xi$. If all the particle-hole states are degenerate (say $1\hbar\omega$ excitations in an oscillator) then

$$\epsilon_{ab} = \epsilon_a - \epsilon_b \quad \equiv \quad \epsilon_0 \tag{10.10}$$

The solution to Eq. (10.9) for the highest state therefore takes the form

$$\epsilon_{\text{top}} = \epsilon_0 + \xi \sum_{ab} (v_{ab}^L)^2 \tag{10.11}$$

The corresponding normalized eigenvector follows from Eq. (10.8) as

$$\psi_{[15]L}^{\text{top}}(ab) = \frac{v_{ab}^L}{\sqrt{\sum_{ab}(v_{ab}^L)^2}} \tag{10.12}$$

From Fig. 10.1 and Eq. (10.8) the other $\mathcal{N}-1$ eigenvalues and eigenvectors all satisfy the relations

$$\begin{aligned} \epsilon_n - \epsilon_0 &= 0 \\ \sum_{lm} v_{lm}^L \, \psi_{[15]L}^{(n)}(lm) &= 0 \end{aligned} \tag{10.13}$$

The general expression for the transition matrix element of an arbitrary multipole operator in this scheme[60] follows in direct analogy to Eq. (9.17)

$$\langle \Psi_n | \hat{T} | \Psi_0 \rangle = \sum_{\alpha\beta} \langle \alpha | T | -\beta \rangle \mathcal{S}_{-\beta} \psi_{\alpha\beta}^{(n)} \tag{10.14}$$

[60] Recall Eqs. (9.38).

Suppose $\hat{T}_{LM}$ is an ITO; use of the Wigner-Eckart theorem (Ref. [N44]) then gives for the l-dependence

$$
\begin{aligned}
\langle \Psi_L^n || \hat{T}_L || \Psi_0 \rangle &= \sum_{\alpha\beta} \langle l_a || T_L || l_b \rangle (\mathcal{S}_{-\beta})^2 \langle l_\alpha m_\alpha l_\beta m_\beta | l_\alpha l_\beta LM \rangle \psi_{\alpha\beta}^{(n)} \\
&= \sum_{ab} \langle l_a || T_L || l_b \rangle \psi_L^{(n)}(ab) \qquad (10.15)
\end{aligned}
$$

Exactly the same calculation can be repeated for isospin.

Suppose the multipole operator T_{LM} is independent of spin. In this case, the matrix element in Eq. (10.14) gives

$$
\delta_{m_{s\alpha},-m_{s\beta}}(-1)^{\frac{1}{2}-m_{s\alpha}} = \sqrt{2}\langle \frac{1}{2} m_{s\alpha} \frac{1}{2} m_{s\beta} | \frac{1}{2}\frac{1}{2} 00 \rangle \qquad (10.16)
$$

Hence only $S = 0$ will be connected to the ground state through this multipole.[61] Observe that if $S = 0$ within the [15] supermultiplet, then the isospin is $T = 1$. Apply these arguments to the transition multipoles of the charge density operator defined by

$$
\hat{Q}_{LM} \equiv \sum_{j=1}^{A} r^L(j) C_{LM}(\Omega_j) \frac{1}{2}[1 + \tau_3(j)] \qquad (10.17)
$$

Since the excited state has $T = 1$, only the τ_3 term can contibute to the transition matrix element, and recall that $\langle \frac{1}{2} || \frac{1}{2}\tau || \frac{1}{2} \rangle = \frac{1}{2}\sqrt{6}$ (Ref. [N44]). Thus

$$
\langle \Psi_{[15]L}^{(n)} \vdots\vdots\, \hat{Q}_L \,\vdots\vdots\, \Psi_0 \rangle = \sqrt{3}\delta_{S0} \sum_{ab} \langle l_a || C_L || l_b \rangle \langle r^L \rangle_{ab} \psi_{[15]L}^{(n)}(ab) \qquad (10.18)
$$

Here the notation $\langle \vdots\vdots \quad \vdots\vdots \rangle$ indicates a matrix element reduced with respect to both L and T.

Assume, as with the radial matrix elements, that $\langle r^L \rangle_{ab} \approx \langle r^L \rangle$ independent of $\{ab\}$, and make use of Eq. (10.6). Then

$$
\langle \Psi_{[15]L}^n \,\vdots\vdots\, \hat{Q}_L \,\vdots\vdots\, \Psi_0 \rangle \approx \sqrt{3}\delta_{S0} \langle r^L \rangle [\sum_{ab} v_{ab}^L \, \psi_{[15]L}^{(n)}(ab)] \qquad (10.19)
$$

Insertion of Eqs. (10.12) and (10.13) then gives

$$
\begin{aligned}
\langle \Psi_{[15]L}^{\text{top}} \,\vdots\vdots\, \hat{Q}_L \,\vdots\vdots\, \Psi_0 \rangle &= \sqrt{3}\delta_{S0} \langle r^L \rangle \sqrt{\sum_{ab} (v_{ab}^L)^2} \\
\langle \Psi_{[15]L}^n \,\vdots\vdots\, \hat{Q}_L \,\vdots\vdots\, \Psi_0 \rangle &= 0\,; \qquad \text{other } \mathcal{N}-1 \text{ supermultiplets} \qquad (10.20)
\end{aligned}
$$

[61] Recall the ground state here belongs to the identity, or [1], representation of $SU(4)$.

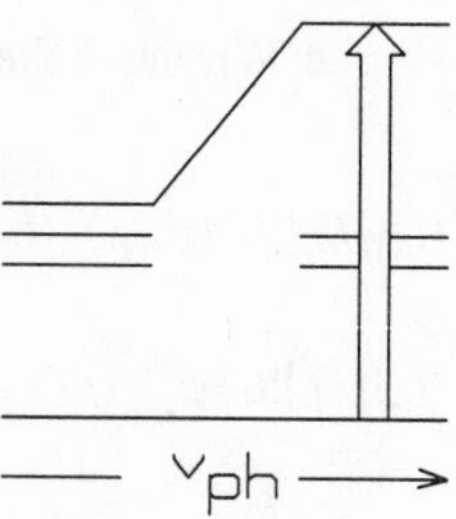

Figure 10.2: Schematic representation of results for [15] supermultiplets in the model discussed in the text.

Thus the top [15] supermultiplet for a given L gets *all* the strength of the transition matrix elements of the multipoles of the charge density; the other $\mathcal{N}-1$ are left with *no* transition strength whatsoever![62] The situation is illustrated schematically in Fig. 10.2.

Random Phase Approximation (RPA). We proceed to a calculation of nuclear excitations using the same model two-body interaction, but with the RPA equations of motion. The first term in the particle-hole interaction in Eqs. (9.24) is treated exactly as in the TDA; the reduced eigenvectors are defined by Eq. (9.40), and the interaction is given in this model by Eq. (10.5). The reduction of the new part of the eigenvector is accomplished by writing

$$\begin{aligned}\phi^{(n)}_{LST}(ab) &\equiv S_L S_S S_T \sum_{\{\text{all m's}\}} \langle l_a m_{l\alpha} l_b m_{l\beta} | l_a l_b L - M_L\rangle \\ &\times \langle \frac{1}{2} m_{s\alpha} \frac{1}{2} m_{s\beta} | \frac{1}{2}\frac{1}{2} S - M_S\rangle \langle \frac{1}{2} m_{t\alpha} \frac{1}{2} m_{t\beta} | \frac{1}{2}\frac{1}{2} T - M_T\rangle \phi^{(n)}_{\alpha\beta}\end{aligned} \tag{10.21}$$

Here $S_L \equiv (-1)^{L-M_L}$ with a similar definition for S_S and S_T. These phases are necessary, for while $\hat{\zeta}^\dagger_{\alpha\beta}$ is a product of ITO, only $S_\alpha S_\beta \hat{\zeta}_{-\alpha-\beta} = S_\alpha S_\beta b_{-\beta} a_{-\alpha}$ is such a product (Appendix B). Now one can proceed with the angular momentum couplings, and use the symmetry properties of the C-G coefficients (Ref. [N44]). A comparison of Eqs. (9.13) and (9.23) shows that a reduction of the basis analogous to that presented above for this model in the TDA leads to the expression (Prob. 10.1)

$$\begin{aligned}u^{[15]L}_{ab;lm} &= \xi v^L_{ab} v^L_{ml} (-1)^{l_m - l_l - L} (-1)^{S+T} \\ &= (-1)^{L+S+T} \xi v^L_{ab} v^L_{lm}\end{aligned} \tag{10.22}$$

[62]This model, for transitions to the giant dipole resonance with quantum numbers $J^\pi = 1^-, T = 1$ (in even nuclei with $N = Z$) was first examined in Ref. [N56].

see prob. 10.6

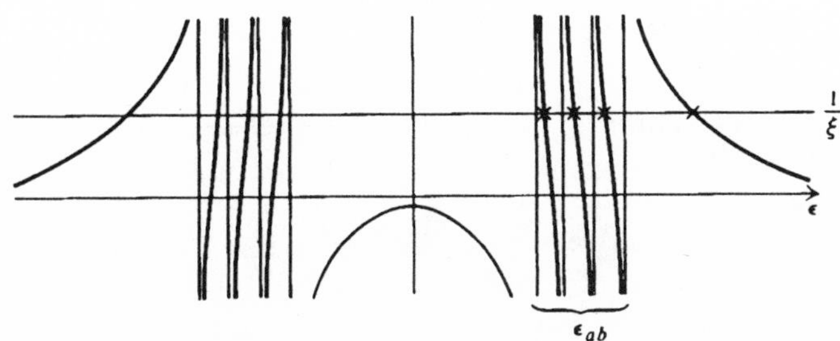

Figure 10.3: Graphic solution for eigenvalues with model problem in RPA.

It will again be assumed that ξ is a constant. The reduction of the RPA equations then takes the form

$$(\epsilon_{ab} - \epsilon_n)\psi^{(n)}_{[15]L}(ab) + \xi v^L_{ab} \left\{ \sum_{lm} [v^L_{lm}\psi^{(n)}_{[15]L}(lm) \right.$$
$$\left. +(-1)^{L+S+T} v^L_{lm}\phi^{(n)}_{[15]L}(lm)] \right\} = 0$$
$$(\epsilon_{ab} + \epsilon_n)\phi^{(n)}_{[15]L}(ab) + \xi v^L_{ab} \left\{ \sum_{lm} [v^L_{lm}\phi^{(n)}_{[15]L}(lm) \right.$$
$$\left. +(-1)^{L+S+T} v^L_{lm}\psi^{(n)}_{[15]L}(lm)] \right\} = 0 \quad (10.23)$$

To derive the eigenvalue equation from these relations divide by $(\epsilon_{ab} \mp \epsilon_n)$, respectively, then $\sum_{ab} v^L_{ab}$, multiply the second by $(-1)^{L+S+T}$, and add. The result is

$$\frac{1}{\xi} = \sum_{ab} (v^L_{ab})^2 \left[\frac{1}{\epsilon_n - \epsilon_{ab}} - \frac{1}{\epsilon_n + \epsilon_{ab}} \right] \quad (10.24)$$

This equation is symmetric under $\epsilon_n \leftrightarrow -\epsilon_n$; it is evident from Eq. (9.20) that the excitation energies $\epsilon_n \equiv E_n - E_0$ are to be interpreted as the solutions for positive ϵ_n. Note that the phase $(-1)^{L+S+T}$ cancels from this relation. This eigenvalue equation is solved graphically in Fig. 10.3. $\mathcal{N} - 1$ roots are again trapped between the configuration energies, and the top one is pushed up.

The eigenvalue equation again simplifies if the configuration energies are degenerate with $\epsilon_{ab} = \epsilon_0$, and the equations can be solved analytically just as before. The top eigenvalue is given by

$$\epsilon_{\text{top}} = \epsilon_0 \sqrt{1 + 2\frac{\xi}{\epsilon_0} \sum_{ab} (v^L_{ab})^2} \quad (10.25)$$

The normalized top eigenvector follows as

$$\begin{aligned}
\phi^{\rm top}_{[15]L}(ab) &= (-1)^{L+S+T}\frac{v^L_{ab}}{\sqrt{\sum_{ab}(v^L_{ab})^2}}\frac{\epsilon_0-\epsilon}{2\sqrt{\epsilon\epsilon_0}} \\
\psi^{\rm top}_{[15]L}(ab) &= \frac{v^L_{ab}}{\sqrt{\sum_{ab}(v^L_{ab})^2}}\frac{\epsilon_0+\epsilon}{2\sqrt{\epsilon\epsilon_0}}
\end{aligned} \tag{10.26}$$

Here $\epsilon\equiv\epsilon_{\rm top}$. The solution for the other $\mathcal{N}-1$ supermultiplets in this degenerate case also follows as before; for these $\epsilon_n=\epsilon_0$ and

$$\begin{aligned}
\phi^{(n)}_{[15]L}(ab) &= 0 \\
\sum_{ab} v^L_{ab}\psi^{(n)}_{[15]L}(ab) &= 0
\end{aligned} \tag{10.27}$$

The transition amplitude in RPA follows directly from Eq. (9.37)

$$\langle\Psi_n|\hat{T}|\Psi_0\rangle = \sum_{\alpha\beta} \mathcal{S}_{-\beta}[\langle\alpha|T|-\beta\rangle\psi^{(n)}_{\alpha\beta} + \langle-\beta|T|\alpha\rangle\phi^{(n)}_{\alpha\beta}] \tag{10.28}$$

The transition matrix elements of the charge density operator in Eq. (10.17) are calculated through the same analysis as described previously

$$\begin{aligned}
\langle\Psi^{(n)}_{[15]L} \,\vdots\vdots\, \hat{Q}_L \,\vdots\vdots\, \Psi_0\rangle &\approx \sqrt{3}\delta_{S0}\langle r^L\rangle \\
&\times\{\sum_{ab} v^L_{ab}[\psi^{(n)}_{[15]L}(ab) + (-1)^{L+S+T}\phi^{(n)}_{[15]L}(ab)]\}
\end{aligned} \tag{10.29}$$

Here equality has again been assumed for the radial matrix elements. Substitution of the eigenvectors in Eqs. (10.26) and (10.27) then gives the final result

$$\begin{aligned}
\langle\Psi^{\rm top}_{[15]L} \,\vdots\vdots\, \hat{Q}_L \,\vdots\vdots\, \Psi_0\rangle &= \sqrt{\frac{\epsilon_0}{\epsilon}}\sqrt{\sum_{ab}(v^L_{ab})^2}\sqrt{3}\delta_{S0}\langle r^L\rangle \\
\langle\Psi^{(n)}_{[15]L} \,\vdots\vdots\, \hat{Q}_L \,\vdots\vdots\, \Psi_0\rangle &= 0\,; \qquad \text{other}\,\mathcal{N}-1\,\text{supermultiplets}
\end{aligned} \tag{10.30}$$

Two features of these results are of particular interest:

1. A comparison of Eqs. (10.25) and (10.11) indicates that the top L state is not pushed up as far in the RPA as in the TDA. The rest of the states again remain degenerate at ϵ_0.

2. A comparison of Eqs. (10.30) and (10.20) shows that the transition strength to the top supermultiplet is similarly reduced in the RPA from that in the TDA. The transition strength to all the other degenerate supermultiplets again vanishes identically in the RPA.

The [1] Supermultiplet with $S = T = 0$. For the delta function potential in Eq. (10.1), the two-particle matrix elements satisfy the equality

$$(-1)^{l_a+l_m+L'}\langle l_l l_b L'|V|l_m l_a L'\rangle \equiv \langle l_l l_b L'|V|l_a l_m L'\rangle \tag{10.31}$$

It follows from Eq. (9.49) that if $S = T = 0$ then

$$v^{[1]}_{ab;lm} = -3v^{[15]}_{ab;lm} \tag{10.32}$$

An analysis analogous to that used to derive Eq. (10.22) gives exactly the same relation for the additional interaction term in the RPA (Prob. 10.2)

$$u^{[1]}_{ab;lm} = -3u^{[15]}_{ab;lm} \tag{10.33}$$

Thus the analysis for these states is exactly the same as that already carried out provided one makes the replacement

$$\xi \longrightarrow \xi' \equiv -3\xi \tag{10.34}$$

Contemplation of the previous results then immediately implies:

1. In contrast to the top state being pushed up in energy and gathering all the transition strength, the bottom state is now pushed down; it again contains all the transition strength for the charge density multipoles in this model.[63]
2. The bottom L state is pushed farther in the RPA than in the TDA.
3. The transition densities are more collective in the RPA than in the TDA.
4. In the TDA the bottom state will acquire a negative eigenvalue for sufficiently large ξ' (Fig. 10.1), while in the RPA it is possible that the lowest eigenvalue will in fact disappear under similar conditions (Fig. 10.3). Both of these results are indicative of an instability of the ground state with respect to these new modes in this strong-coupling limit.

Application to Nuclei. There is a rich variety of modes of motion of nuclei: single-particle excitations, collective shape oscillations, spin-isospin oscillations, rotations of deformed nuclei, superdeformed shape isomers, coupled combinations of these — the list can go on and on. It is not within the perspective of this text to go into each of these in detail. Many good books are available that do this, for example Refs. [N1, N5, N60]. Rather, the goal of the present development is to provide a theoretical basis for describing a wide variety of nuclear excitations. Reference [N2] shows how to consistently extend this description to arbitrary orders in the two-nucleon interaction.

[63]The isoscalar charge density operator for the dipole mode with $L = 1$ is given by Eq. (10.17) as $\hat{Q}_{1M} = \frac{1}{2}\sum_j \mathbf{r}(j)_{1M}$. This is proportional to the center-of-mass coordinate and cannot cause a true internal excitation of the nucleus. Thus there is no transition from the ground state to the [1] supermultiplet with $S = T = 0$ and $L = 1$ through the charge density operator. The present analysis is applicable to quadrupole $L = 2$ and higher charge oscillations of the nucleus.

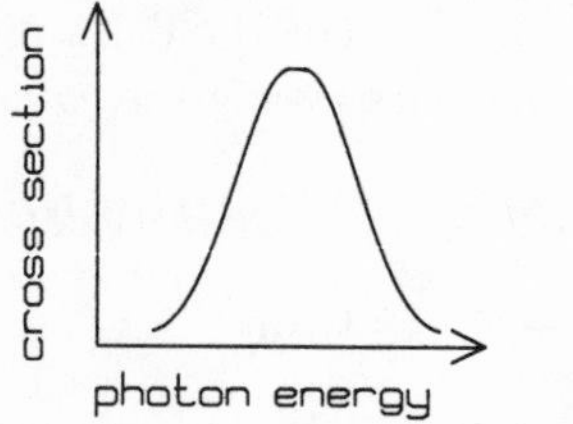

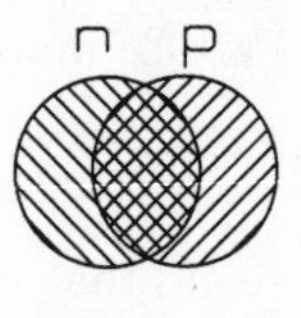

Figure 10.4: (a) Schematic representation of low-energy nuclear photoabsorption cross section and the giant dipole resonance. (b) Goldhaber-Teller model.

The present analysis does provide a framework for understanding a broad variety of nuclear phenomena. For example, it is an experimental fact that low-energy photoabsorption by nuclei is dominated by the giant dipole resonance (GDR) as illustrated schematically in Fig. 10.4. The GDR occurs at approximately 25 to 10 MeV in going from the lightest to the heaviest nuclei. It is a few MeV wide and the most important electromagnetic transition multipole in this energy regime. It systematically exhausts the $E1$ sum rule. Now the $E1$ operator has the form

$$\hat{\mathbf{Q}} = \sum_{j=1}^{A} \mathbf{r}(j)\frac{1}{2}[1+\tau_3(j)] \doteq \sum_{j=1}^{A} \mathbf{r}(j)\frac{1}{2}\tau_3(j) \qquad (10.35)$$

Therefore the GDR has quantum numbers $S = 0, T = 1, L^\pi = 1^-$ in nuclei whose ground states have quantum numbers $S = T = L = 0$.

The simplest picture of the giant dipole resonance is due to Goldhaber and Teller (Ref. [N61]); the protons oscillate as a unit against the neutrons as illustrated schematically in Fig. 10.4. The more sophisticated model for the giant dipole resonance presented here is due to Brown and Bolsterli (Ref. [N56]). The observed GDR does indeed lie at a higher energy than the configuration energies ϵ_{ab} determined from neighboring nuclei; it is also observed to carry all the dipole strength. The present model predicts that the GDR observed in photoabsorption comprises just three components with $(T = 1, S = 0)$ of a degenerate [15] dimensional spin-isospin supermultiplet of giant dipole resonances with $L = 1$. The simple picture of the other components is obtained in the framework of the Goldhaber-Teller model by considering the various oscillations of $\{p\uparrow, p\downarrow, n\uparrow, n\downarrow\}$ (Fig. 10.4). There is evidence from weak interactions and electron scattering that these other components are indeed present in light nuclei (see, e.g., Ref. [N62]).[64] The model calculation discussed here also predicts additional giant resonance [15] supermultiplets, in fact one for each L.

The collective states belonging to the identity, or [1], spin-isospin supermultiplet with $(S = 0, T = 0)$ correspond to pure charge oscillations of the nucleus.

[64] Reference [N76] clearly displays this supermultiplet in ^{4}He.

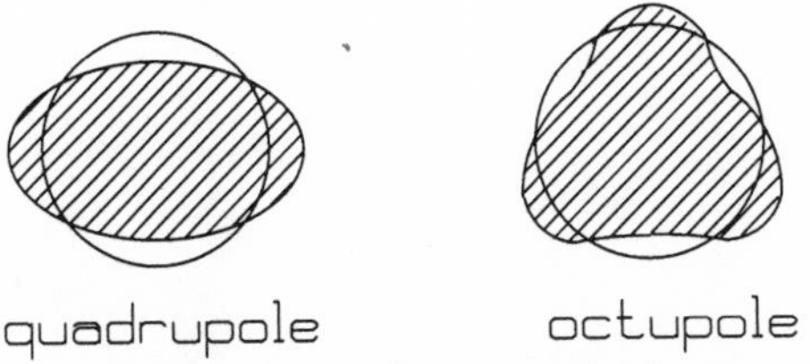

Figure 10.5: Schematic representation of collective shape oscillations of even-even nuclei.

For example, there can be quadrupole shape oscillations with $L = 2$, octupole shape oscillations with $L = 3$, and so on, as illustrated schematically in Fig. 10.5. These oscillations are seen systematically throughout the periodic table as low-lying collective excitations of even-even nuclei (Ref. [N1, N60]).

The RPA description of these phenomena tends to be in closer accord with the observations than that of the TDA (see, e.g., Ref. [N63]). We give one example of a more realistic calculation of the particle-hole excitations of a specific nucleus ^{16}O in the next section.

even-even nuclei
most common

picture

3^-
$0^+, 2^+, 4^+$ } octupole

2^+
0^+ } quadrupole

11

APPLICATION TO A REAL NUCLEUS - ^{16}O

Many calculations of nuclear spectra starting from realistic single-particle properties and two-nucleon interactions have been carried out (see, for example, Refs. [N64, N65, N66, N67, N68, N69]). It is impossible to summarize all these results here. Rather we present just one example of an attempt to calculate the excited states of a real nucleus. The calculation focuses on the negative-parity $T = 1$ states of ^{16}O; these are the states excited in inelastic electron scattering at large angles and high momentum transfer through the large isovector magnetic moment of the nucleon [Eqs. (7.77) and (8.30)].[65] The calculation is due to Donnelly and Walker in Ref. [N70] (see Ref. [N2]).

One starts with single-particle states of the form $|nljm_j; \frac{1}{2}m_t\rangle$, which diagonalize the strong spin-orbit force $H_{\text{so}} = V_{\text{so}}(r)\mathbf{l}\cdot\mathbf{s}$. The analysis of Sections 9 and 10 is readily generalized to this case (Prob. 9.3). The ground state of ^{16}O is assumed to form a closed p-shell. All particle-hole states corresponding to a hole in the p-shell and a particle in the next ($2s$-$1d$) oscillator shell are retained (Fig. 11.1).[66] The particle-hole configuration energies $\epsilon_a - \epsilon_b$ are taken from the neighboring oxygen isotopes as indicated in Fig. 11.2. They are shown in Table 11.1.

A nonsingular Serber-Yukawa potential fit to low-energy nucleon-nucleon scattering is used

$$V(1,2) = [{}^1V(r_{12}){}^1P + {}^3V(r_{12}){}^3P]\frac{1}{2}[1 + P_{\text{M}}(1,2)]$$

$${}^1P = \frac{1}{4}(1 - \boldsymbol{\sigma}_1\cdot\boldsymbol{\sigma}_2) \qquad {}^3P = \frac{1}{4}(3 + \boldsymbol{\sigma}_1\cdot\boldsymbol{\sigma}_2)$$

$$V(r_{12}) = V_0\frac{e^{-\mu r_{12}}}{\mu r_{12}}$$

$${}^1V_0 = -46.87\,\text{MeV} \qquad {}^1\mu = 0.8547\,\text{fm}^{-1}$$

$${}^3V_0 = -52.13\,\text{MeV} \qquad {}^3\mu = 0.7261\,\text{fm}^{-1} \tag{11.1}$$

The calculation employs harmonic oscillator single-particle solutions (Section 6) as approximate Hartree-Fock single-particle wave functions with an oscillator

[65] Note from Eq. (8.15) $\mu = \frac{1}{2}(\lambda_p + \lambda_n) + \frac{1}{2}\tau_3(\lambda_p - \lambda_n)$. Since $(\lambda_p - \lambda_n) \gg (\lambda_p + \lambda_n)$ it is the isovector transitions that dominate the transverse electron scattering cross section.

[66] The [15] supermultiplets here are obtained from the spatial states $(2s)(1p)^{-1}_{1^-}$ and $(1d)(1p)^{-1}_{1^-2^-3^-}$ where the total L is indicated with a subscript.

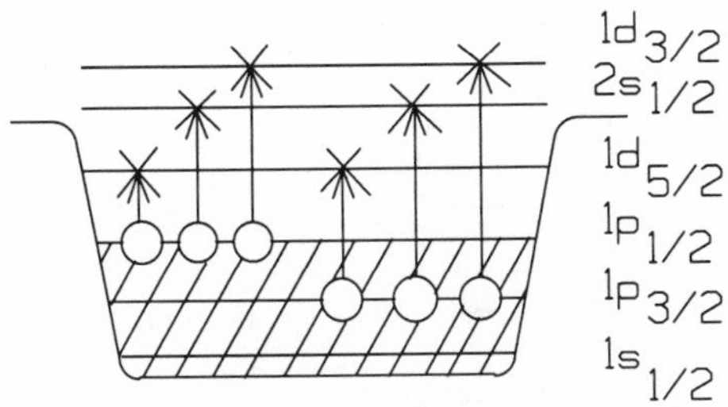

Figure 11.1: Particle-hole states retained in calculation of negative-parity $T = 1$ states of ^{16}O.

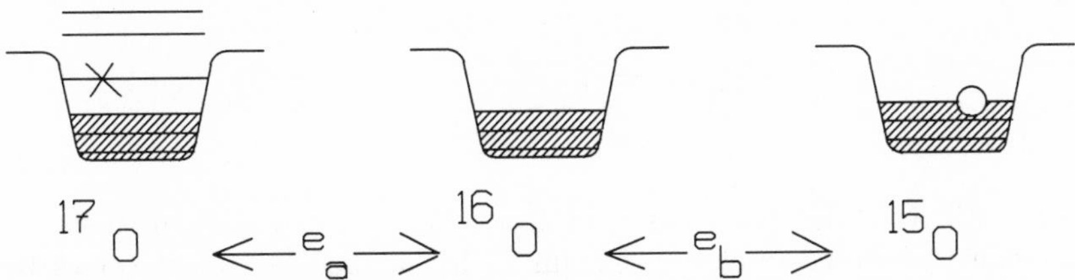

Figure 11.2: Particle-hole configuration energies for calculation in ^{16}O.

parameter $b = 1.77\,\text{fm}$ determined from a fit to elastic electron scattering. The calculated spectrum for ^{16}O is shown in Fig. 11.3. Also shown in this figure is the spectrum with the spin-dependent forces (H_{so} and $V_1\boldsymbol{\sigma}_1 \cdot \boldsymbol{\sigma}_2$) turned off. In this case one reproduces the previous model supermultiplet results of Section 10.

The cross section for photoabsorption involves the dipole states with (J^π, T) = $(1^-, 1)$; the comparison of the observed photoabsorption cross section for ^{16}O with the calculated values (arbitrary overall normalization) is indicated

Configurations	$\epsilon_a - \epsilon_b$ (MeV)	States
$(2s_{1/2})(1p_{3/2})^{-1}$	18.55	$1^-, 2^-$
$(1d_{5/2})(1p_{3/2})^{-1}$	17.68	$1^-, 2^-, 3^-, 4^-$
$(1d_{3/2})(1p_{3/2})^{-1}$	22.76	$0^-, 1^-, 2^-, 3^-$
$(2s_{1/2})(1p_{1/2})^{-1}$	12.39	$0^-, 1^-$
$(1d_{5/2})(1p_{1/2})^{-1}$	11.52	$2^-, 3^-$
$(1d_{3/2})(1p_{1/2})^{-1}$	16.60	$1^-, 2^-$

Table 11.1: Particle-hole configurations retained in calculation of negative-parity $T = 1$ states in ^{16}O and configuration energies obtained from neighboring nuclei.

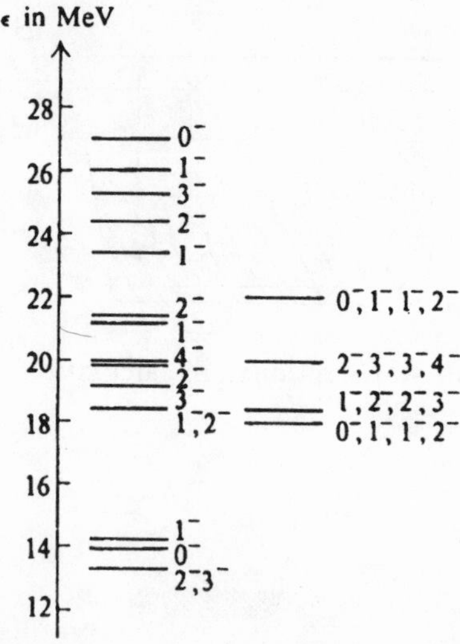

Figure 11.3: Calculated spectrum of $T = 1$ negative-parity excitations of ^{16}O. Also shown is the calculated spectrum with the spin-dependent forces turned off. From Refs. [N70, N2].

schematically in Fig. 11.4. The total calculated strength is too high by about a factor of 2.

The use of the current and magnetization operators in Eqs. (8.12) allows one to compute the electron scattering cross section [Eq. (7.77)] to the discrete levels in Fig. 11.3. The results are compared with the experimentally observed (e, e') spectrum at $\theta = 135^o$ and $\epsilon_i = 224\,\text{MeV}$ in Fig. 11.5. The solid curve is an estimate at this momentum transfer of the nonresonant background above the threshold for nucleon emission.

The *form factors* for the various complexes observed in Fig. 11.5 are compared with the experimental data (area under the resonance peaks) in Fig. 11.6.

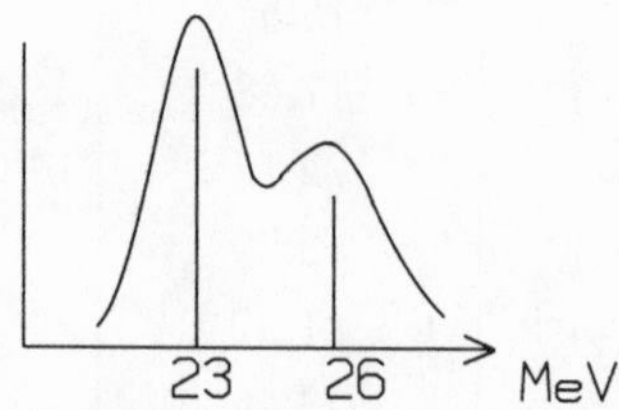

Figure 11.4: Schematic comparison of observed and calculated photoabsorption cross section in the giant resonance region for ^{16}O. Lines show location and relative strength of the calculated result; the integrated theoretical result is too high by about a factor of 2 (see text).

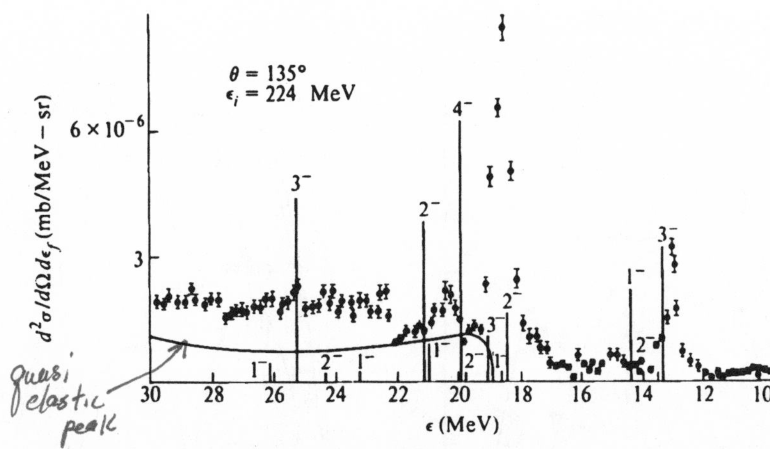

Figure 11.5: Experimentally observed spectrum of scattered electrons at $\theta = 135^o$ and $\epsilon_i = 224$ MeV compared with calculated spectrum for states in Fig. 11.3 (arbitrary overall normalization; the integrated areas for the various complexes are compared with theory in the next figure). From Refs. [N2, N71].

The theoretical results for the form factors are all too high, and they are reduced in amplitude by approximately the following numerical factors for each of the indicated complexes: 2/3 (13 MeV); 2/3 (17 MeV); 1 (19 MeV); 2/3 (20.4 MeV); $1/\sqrt{2}$ (Coulomb part of giant dipole resonance).

In *summary*, the shell model provides a basis for understanding the dominant features of the set of negative-parity $T = 1$ particle-hole excitations in this nucleus up to excitation energies of the order of 30 MeV. Linearization of the equations of motion for the collective particle-hole excitations provides a semiquantitative description of both the location of the levels and the spatial distribution of the transition current densities through which they are excited by the electromagnetic interaction.

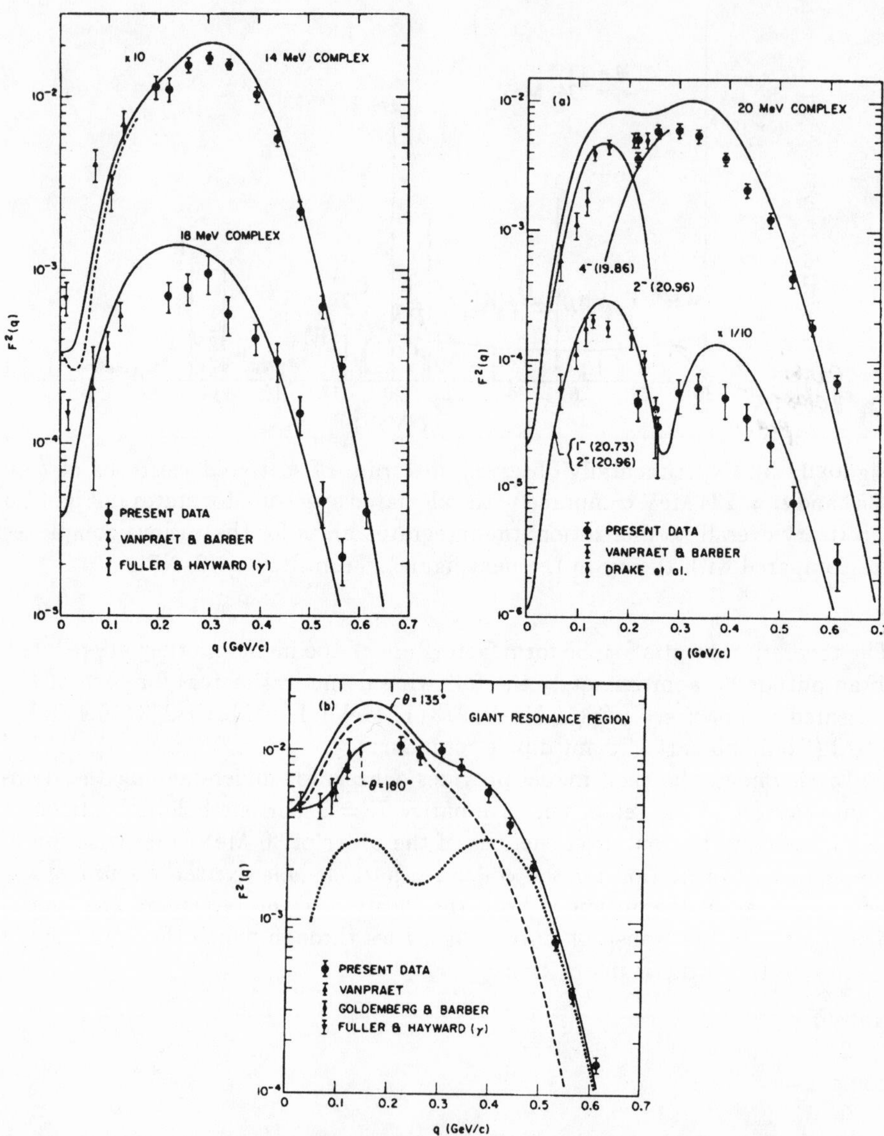

Figure 11.6: Calculated transverse inelastic form factor for (e, e') defined as $\mathcal{F}^2(q) \equiv (d\sigma/d\Omega)[4\pi\sigma_{\text{M}}(1/2 + \tan^2\theta/2)]^{-1}$ for negative-parity $T = 1$ states in ^{16}O and experimental values obtained from areas under the resonance curves for the following complexes of states (see Figs. 11.3 and 11.5): 13 MeV peak $(3^-2^-1^-)$; 17 MeV peak (2^-1^-); 19 MeV complex $(3^-2^-4^-1^-2^-)$; 20.4 MeV peak (2^-1^-); and 20.8-26.0 MeV giant resonance region $(3^-2^-1^-1^-)$. The latter includes the calculated quasielastic background. From Refs. [N62, N71]. Energies in figure are calculated values; see text for amplitude reduction of calculated curves.

12

CEBAF'S ROLE

In this section we examine the role that a high-energy, high-intensity, high-resolution, continuous electron beam with matched detectors can play in understanding basic nuclear structure.

Electron Scattering. The electron scattering process in one-photon exchange approximation with particle emission is illustrated in Fig. 12.1.[67] As discussed in Section 7, this process is governed by the Fourier transform of the transition matrix element of the local electromagnetic current density in the target

$$(\mathcal{J}_\mu)_{\mathrm{fi}} = \int d^3x \, \langle f|\hat{\mathcal{J}}_\mu(\mathbf{x})|i\rangle e^{i\mathbf{k}\cdot\mathbf{x}} \tag{12.1}$$

We start by considering some specific coincidence reactions and working to 0th order in the Coulomb interaction $(\mathcal{J}_0)_{\mathrm{fi}}$.

Single Proton Emission $(e, e'p)$. The process is illustrated schematically in Fig. 12.2. The basic Coulomb scattering process involves the interaction with a single nucleon as shown in Fig. 12.3. In this case the transition matrix element takes the form

$$(\mathcal{J}_0)_{\mathrm{fi}} \approx f(k^2) \int e^{i\mathbf{k}\cdot\mathbf{x}} \underbrace{[e^{-i\mathbf{q}\cdot\mathbf{x}} \psi_i(\mathbf{x})]}_{\rho_{\mathrm{fi}}(\mathbf{x})} d^3x = f(k^2)\, \tilde{\psi}_i(\mathbf{k}-\mathbf{q}) \tag{12.2}$$

Several comments are relevant:

1. Only the Coulomb interaction has been retained. Evidently the Fourier transform of the transition charge density $\rho_{\mathrm{fi}}(\mathbf{x})$ can be measured in this process.

2. Plane wave Born approximation (PWBA) has been assumed for the outgoing proton.

3. The single-particle shell model can be studied, for one can:

   a. Determine the energies of the hole states produced in the final nucleus from knowledge of the initial electron energy and measurement of the energies of all the final particles produced in the reaction;

   b. Measure the Fourier transform of the wave function of the hole states $\tilde{\psi}_i(\mathbf{k}-\mathbf{q})$; and

   c. Determine the widths of the hole states from the energy distribution of the final particles.

[67]Note the change in notation: for $(e, e'X)$ in this section k_μ is the momentum transfer and q_μ the momentum of the produced particle.

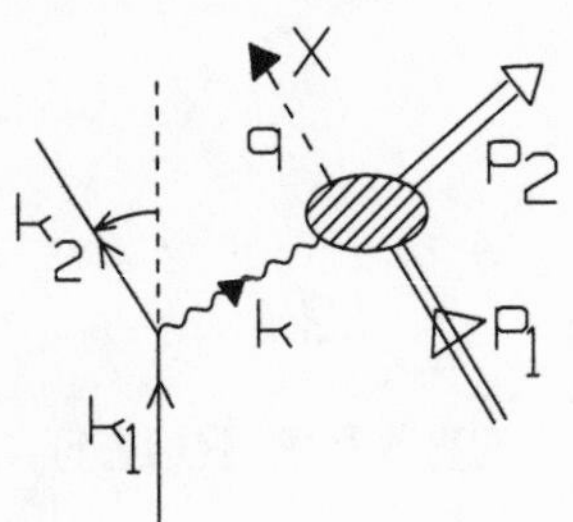

Figure 12.1: Basic electron scattering process in the one-photon-exchange approximation with particle emission.

4. The location of the initial participating nucleon in the nuclear interior can be varied by looking at different $\psi_i(\mathbf{x})$.

5. The reaction proceeds through the well-known electromagnetic interaction with the nucleon.

6. Once the nuclear structure is understood, modifications of the single-particle current in the nucleus can be examined.

Associated Production of Hypernuclei Through $(e, e'K^+)$. Strangeness is conserved in the strong interactions, and thus particles carrying strangeness are produced at least in pairs. A basic reaction on the nucleon is the electroproduction of K^+ mesons $e + p \to e' + K^+(S = +1) + \Lambda(S = -1)$ as illustrated in Fig. 12.3b. This leaves behind a Λ hypernucleus as illustrated schematically in Fig. 12.4. To lowest order the Coulomb matrix element governing this process is given by

$$\begin{aligned}
(\mathcal{J}_0)_{fi} &\approx A(k^2, \ldots) \int d^3x \, e^{i\mathbf{k}\cdot\mathbf{x}} \underbrace{[e^{-i\mathbf{q}\cdot\mathbf{x}} \psi^*_{f\Lambda}(\mathbf{x})\psi_{ip}(\mathbf{x})}_{\rho_{fi}(\mathbf{x})} \\
&= A(k^2, \ldots) \, \widetilde{\psi^*_{f\Lambda}\psi_{ip}}(\mathbf{k} - \mathbf{q}) \qquad (12.3)
\end{aligned}$$

Several comments are of interest:

1. PWBA for the outgoing K^+ has been assumed in writing this expression.

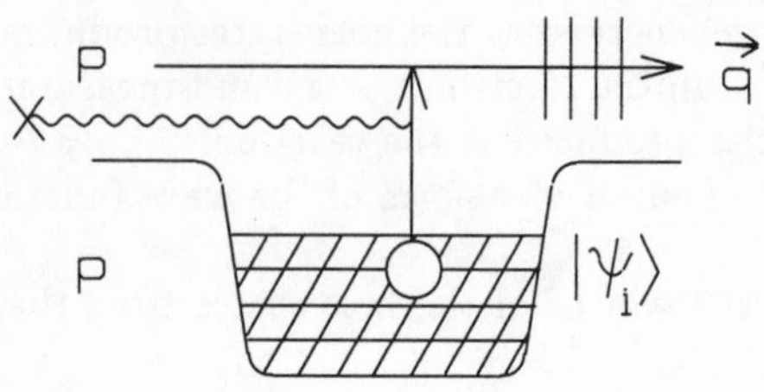

Figure 12.2: Some specific coincidence reactions: Single proton emission $(e, e'p)$.

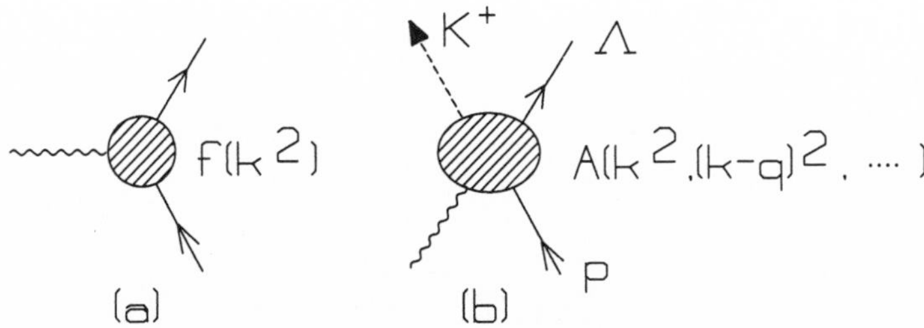

Figure 12.3: Basic elementary Coulomb interaction process with a nucleon in the nucleus in the coincidence reactions: (a) $(e, e'p)$; (b) $(e, e'K^+)$.

2. This reaction puts a distinguishable baryon, the Λ with $S = -1$, into any shell model state; there is no Pauli principle restriction on the states at the baryon level. Thus one is inserting a known "impurity," a hyperon, into a fundamental condensed matter system, the nucleus.

3. The wave function for the initial proton $\psi_{ip}(\mathbf{x})$ is in principle known from the previous discussion.

4. The single-particle shell model for the Λ can be studied; for example, one can study the

a. Location of all the states;

b. Spin structure of the states, in particular, the effects of the spin-orbit interaction;

c. Widths of the states.

5. The $\Lambda - N$ interaction can be examined through the depth of the potential V_Λ in which the Λ moves (see Probs. 5.5 and 5.6) and the spin-dependent splittings arising from $V_{\text{sp}}\boldsymbol{\sigma}_\Lambda \cdot \boldsymbol{\sigma}_N$.

6. The modification of the decay modes of the Λ produced by the nuclear environment can be studied. For example, the free decay mode $\Lambda \rightarrow N + \pi$ is inhibited by the effects of the Pauli principle on the final nucleon. Also, there is a new decay mode $\Lambda + N \rightarrow N + N$ present that depends, among other things, on the correlations between the Λ and nucleons in the nucleus.

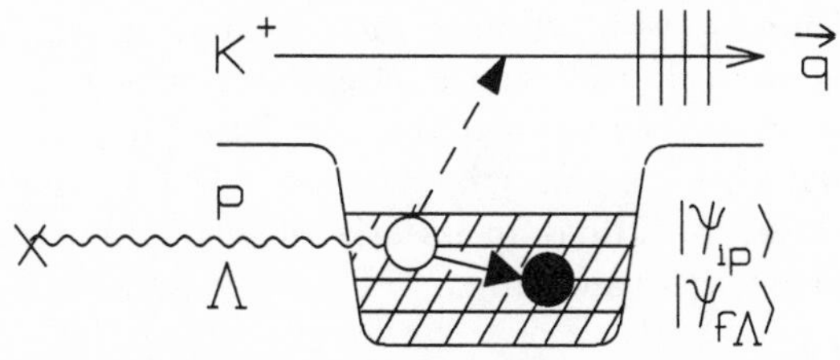

Figure 12.4: Some specific coincidence reactions: Associated production of hypernuclei $(e, e'K^+)$.

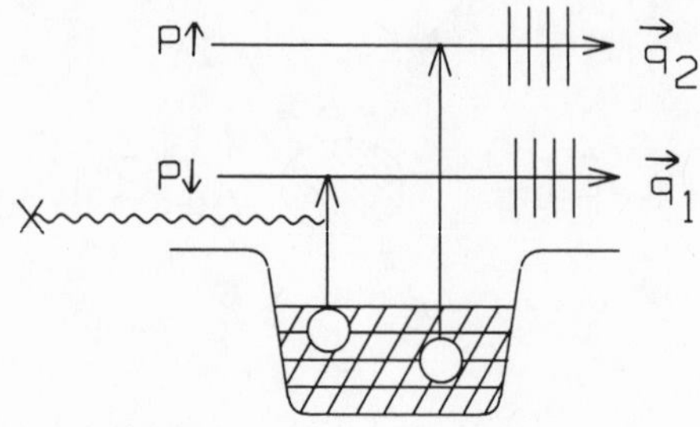

Figure 12.5: Some specific coincidence reactions: Two-proton emission $(e, e'pp)$.

7. Where the Λ goes in the nuclear interior can be controlled by choice of the final Λ state and wave function $\psi^*_{f\Lambda}(\mathbf{x})$.[68] [69]

Two Proton Emission $(e, e'pp)$. The basic Coulomb scattering process involves the nucleon as shown in Fig. 12.3a. The nuclear two-proton emission reaction is illustrated schematically in Fig. 12.5. The transition Coulomb matrix element for this process is given by

$$\begin{aligned}(\mathcal{J}_0)_{fi} \quad &\approx \quad f(k^2)\int(e^{i\mathbf{k}\cdot\mathbf{x}_1}+e^{i\mathbf{k}\cdot\mathbf{x}_2})\underbrace{[\Psi_f^*(\mathbf{x}_1,\mathbf{x}_2)\Psi_i(\mathbf{x}_1,\mathbf{x}_2)]}_{\approx\,[e^{-i\mathbf{q}_1\cdot\mathbf{x}_1}e^{-i\mathbf{q}_2\cdot\mathbf{x}_2}]\Psi_i(\mathbf{x}_1,\mathbf{x}_2)}d^3x_1\,d^3x_2 \\ &= \quad f(k^2)\,\{\tilde{\Psi}_i(\mathbf{k}-\mathbf{q}_1,-\mathbf{q}_2)+\tilde{\Psi}_i(-\mathbf{q}_1,\mathbf{k}-\mathbf{q}_2)\} \qquad (12.4)\end{aligned}$$

This result merits several comments:

1. The calculation here is for distinguishable nucleons, that is, for $\{p\uparrow, p\downarrow\}$.

2. This matrix element *vanishes* in the extreme independent-particle shell model where the two-nucleon wave functions in the first line of Eq. (12.4) are just the product of single-particle wave functions; either one or the other of the single-particle wave functions gives zero upon integration over all space by orthogonality. This result is evident — a one-body operator cannot simultaneously change the states of a pair of particles.

3. The existence of this process thus depends on some type of *correlation* in the two-nucleon wave function $\Psi(\mathbf{x}_1, \mathbf{x}_2)$.

4. In principle this process can be used to study the nature of the two body Bethe-Goldstone wave function in the nuclear medium. $\Psi_i(\mathbf{x}_1, \mathbf{x}_2)$ is the quantity discussed in Sections 4 and 5. Recall the sketch in Fig. 4.6; this is the expected structure of the two-nucleon correlation function. The predicted short-range correlation is of particular interest. The two-nucleon correlation function in the nuclear medium is an essential element of nuclear structure, and this quantity has never been measured.

[68] In this text N is used generically to indicate a nucleon, p or n.

[69] Reactions such as $e + n \to e' + K^+ + \Sigma^-$ are also possible; they occur at a significantly higher energy transfer.

Exact Expressions for These Processes. The previous discussion is just 0th order and based on only the Coulomb interaction $(\mathcal{J}_0)_{\mathrm{fi}}$. The *exact* expression governing the process in Fig. 12.1 is given by

$$(\mathcal{J}_\mu)_{\mathrm{fi}} = \int e^{i\mathbf{k}\cdot\mathbf{x}} \langle \Psi_f ; \mathbf{q}^{(-)} | \hat{\mathcal{J}}_\mu(\mathbf{x}) | \Psi_i \rangle \, d^3x \tag{12.5}$$

Here $|\Psi_i\rangle$ is the exact ground state of the target nucleus, and $|\Psi_f ; \mathbf{q}^{(-)}\rangle$ is the exact scattering state, with incoming wave boundary conditions, for the final nuclear target and the particle X of momentum $\mathbf{q}$. This state satisfies the *Lippman-Schwinger* equation

$$\begin{aligned} |\Psi_f ; \mathbf{q}^{(-)}\rangle &= [1 + \frac{1}{E_f - \hat{H}_0 - i\eta}\hat{H}_1 + \\ &+ \frac{1}{E_f - \hat{H}_0 - i\eta}\hat{H}_1 \frac{1}{E_f - \hat{H}_0 - i\eta}\hat{H}_1 + \cdots]|\Phi_f ; \mathbf{q}\rangle \end{aligned} \tag{12.6}$$

Here $\hat{H} = \hat{H}_0 + \hat{H}_1$ is the full nuclear hamiltonian and $|\Phi_f ; \mathbf{q}\rangle$ is the initial eigenstate of $\hat{H}_0$. The techniques for developing a graphic analysis of $(\hat{\mathcal{J}}_\mu)_{\mathrm{fi}}$ are developed in Ref. [N2]. Assume that in the absence of $\hat{H}_1$ the initial target ground state contains neither particles nor holes

$$a|\Phi_0\rangle = b|\Phi_0\rangle = 0 \tag{12.7}$$

Typical Feynman diagrams contributing to the three processes discussed above are then shown in Fig. 12.6. This graphic analysis provides a systematic procedure that allows one, in principle, to calculate the various processes to all orders in $\hat{H}_1$. We take this development no further at this point, referring the dedicated reader to Ref. [N2] for the technical details of such an analysis. Instead, we summarize a few interesting open questions concerning the use of the nuclear local electromagnetic current density, as probed by electron scattering, to study basic nuclear structure.

Some Open Questions. A few open questions are the following:

1. What is the role of nuclear final-state interactions in $(e, e'X)$ coincidence reactions? The first improvement one can think of over the 0th-order analysis presented above is to avoid the PWBA for an outgoing hadron and to use instead a solution in a nuclear optical potential[70]

$$e^{i\mathbf{q}\cdot\mathbf{x}} \longrightarrow \chi_{\mathbf{q}}^{(-)}(\mathbf{x}) \tag{12.8}$$

This procedure is commonly used, with great success, in the analysis of nuclear reactions (see Ref. [N3]). There are some open questions here: Is it justified to use an optical potential to summarize all the effects of the strong nuclear

[70] We discuss the nuclear optical potential in more detail in Part II.

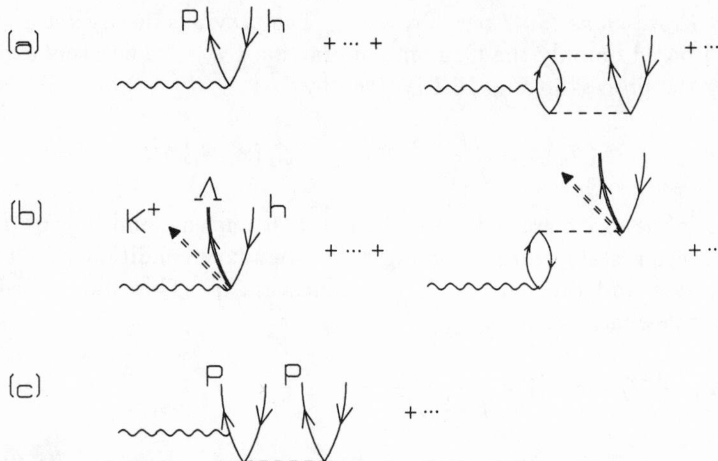

Figure 12.6: Typical Feynman diagrams contributing to the processes (a) $(e, e'p)$, (b) $(e, e'K^+)$, and (c) $(e, e'pp)$ in an expansion of the required current matrix elements $(\hat{\mathcal{J}}_\mu)_{\mathrm{f\,i}}$ in a series of Feynman diagrams. The meaning of the various elements is indicated schematically in the diagrams (see Ref. [N2]).

interactions on the electromagnetically ejected hadron X? Is a more detailed multiple scattering analysis required instead?

2. The discussion in Part I of this book is in terms of nonrelativistic quantum mechanics. What is the effect of a consistent relativistic treatment of the interacting nuclear many-body system? A relativistic treatment is essential when $k = O(m)$, that is, when the momentum transfer becomes comparable to the nucleon mass.

3. How good an approximation is the use of the free electromagnetic current interaction (Fig. 12.3) for nucleons in the nucleus? It has been assumed that the processes illustrated in Fig. 12.7a occurring in the nucleus behave just the way they do as in free space. This *cannot* be the case. Can the leading effects be described by a simple modification of the nucleon form factor $f(k^2)$, or the production amplitude $A(k^2, ...)$?

4. What is the role of two-body currents? Such currents must be present in the nuclear system. They arise, for example, from the exchange of charged mesons (Fig. 12.7b). This is an additional contribution to the electromagnetic current in the nuclear medium that is not present in a collection of isolated, free nucleons. What about three- (and more) body currents?

5. All these calculations should be carried out with the full nuclear wave functions $|\Psi_i\rangle$ and $|\Psi_f; \mathbf{q}\rangle$ [see Eqs. (12.5) and (12.6)]. These are solutions to the many-body Schrödinger equation. How does one construct and use these solutions as accurately as possible?

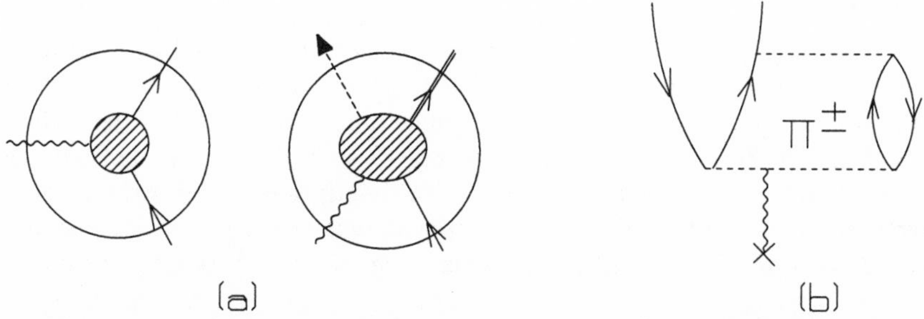

Figure 12.7: Some open questions concerning electromagnetic interactions with nuclei: (a) Modification of the interaction with free nucleons in the nuclear medium; (b) two-body electromagnetic currents.

6. What about neutron emission? The processes $(e, e'n)$, $(e, e'pn)$, $(e, e'nn)$ are as important as those involving only protons from the point of view of basic nuclear structure. How do these processes take place? (One possible mechanism is a charge-exchange reaction with an initially struck proton.)

7. A key open question is to understand the apparent suppression of the longitudinal (Coulomb) response in inelastic electron scattering (e, e') from medium and large size nuclei in the quasielastic region, as illustrated schematically in Fig. 12.8. Many possible explanations have been put forward. These include effects of

a. Correlations — both short and long range;

b. Modification of the nucleon form factor $f(k^2)$ in nuclei. For example, an increase in charge radius in the nuclear medium can explain the effect;

c. Strong-interaction vacuum polarization effects; and

d. Missing experimental cross section (perhaps under the radiative tail of the transverse peak?).

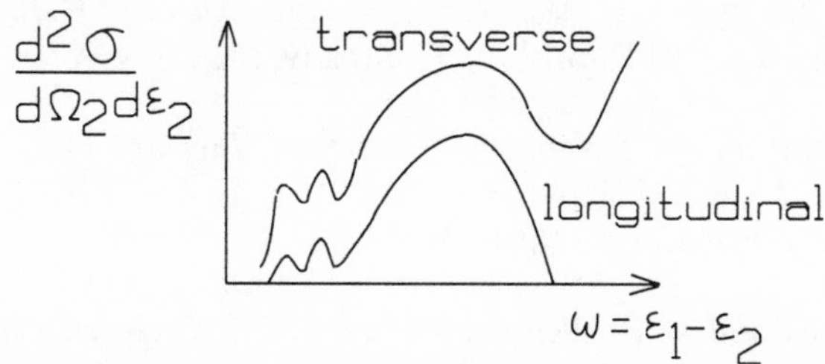

Figure 12.8: Sketch of transverse and longitudinal (Coulomb) response observed in inelastic electron scattering (e, e') from medium and large size nuclei in the quasielastic region. The measured response in the latter case appears systematically to lie below the calculated values.

The Coulomb interaction with the charges in the nucleus represents the most basic part of the electromagnetic interaction with nuclei, and the suppression of the Coulomb response is a significant mystery. There must be a definitive solution to this problem before we can make much progress in nuclear physics.

8. Finally, there is the deep underlying theoretical question. On what *level* should we do nuclear physics? With what description do we start to get the most economical explanation of nuclear phenomena and to provide a theoretical calculation of these observed phenomena to the desired degree of accuracy?

- With nucleons and static potentials, the basis for this discussion in Part I?
- With hadrons, mesons and baryons, as the underlying set of degrees of freedom with which we describe the nucleus?
- With quarks and gluons as the basic set of degrees of freedom, and nuclei as confined quark systems?

This last question is the most challenging. It is one we attempt to address, at least in part, in this text. In Part II we proceed to the second level, a relativistic description of the nucleus in terms of *hadrons*, baryons and mesons.

At whatever level the theoretical description of the nucleus is developed, it is essential to incorporate as constraints the known elements of basic nuclear structure presented in Part I.

References

[N1] M. A. Preston and R. K. Bhaduri, *Structure of the Nucleus*, Addison-Wesley Publishing Company, Reading, Massachusetts (1975); 2nd printing (1982)

[N2] A. L. Fetter and J. D. Walecka, *Quantum Theory of Many-Particle Systems*, McGraw-Hill Publishing Company, Inc., New York (1971)

[N3] A. de Shalit and H. Feshbach, *Theoretical Nuclear Physics Vol. I, Nuclear Structure*, John Wiley and Sons, Inc., New York (1974); H. Feshbach, *Vol. II, Nuclear Reactions* (1991)

[N4] J. M. Blatt and V. F. Weisskopf, *Theoretical Nuclear Physics*, John Wiley and Sons, Inc., New York (1952)

[N5] A. Bohr and B. R. Mottelson, *Nuclear Structure Vol. I, Single-Particle Motion* (1969) and *Vol. II, Nuclear Deformations* (1975), W. A. Benjamin, Inc., Reading, Massachusetts

[N6] R. Machleidt, *The Meson Theory of Nuclear Forces and Nuclear Structure*, in *Advances in Nuclear Physics* **19**, eds. J. W. Negele and E. Vogt, Plenum Press, New York (1989), Chap. 2

[N7] P. J. Siemens and A. S. Jensen, *Elements of Nuclei*, Addison-Wesley Publishing Co. Inc., Redwood City, California (1987)

[N8] G. Breit, I. E. Hoisington, S. S. Share, and H. M. Thaxon, *Phys. Rev.* **55**, 1103 (1939); R. E. Seamon, K. A. Friedman, G. Breit, R. D. Haracz, J. M. Holt, and A. Prakash, *Phys. Rev.* **165**, 1579 (1968)

[N9] R. Jastrow, *Phys. Rev.* **81**, 165 (1951)

[N10] M. Lacombe, B. Loiseau, J. M. Richard, R. Vinh Mau, J. Côté, P. Pirès, and R. de Tourreil, *Phys. Rev.* **C21**, 861 (1980)

[N11] R. V. Reid, Jr., *Ann. Phys.* **50**, 411 (1968)

[N12] H. Yukawa, *Proc. Phys.-Math. Soc. Jpn. Ser.3* **17**, 48 (1935)

[N13] W. Pauli, *Meson Theory of Nuclear Forces*, Interscience Publishers, Inc., New York (1946); 2nd Ed. (1948)

[N14] J. D. Walecka, *Lectures on Electron Scattering*, ANL-83-50, Argonne National Laboratory, Argonne, Illinois (1984); CEBAF, Newport News, Virginia (1987)

[N15] R. Hofstadter *Rev. Mod. Phys.* **28**, 214 (1956)

[N16] E. E. Chambers and R. Hofstadter, *Phys. Rev.* **103**, 1454 (1956)

[N17] C. F. Von Weizsäcker, *Z. Phys.* **96**, 431 (1935)

[N18] A. E. S. Green, *Phys. Rev.* **95**, 1006 (1954)

[N19] K. A. Brueckner, C. A. Levinson, and H. M. Mahmoud, *Phys. Rev.* **95**, 217 (1954)

[N20] H. A. Bethe, *Phys. Rev.* **103**, 1353 (1956)

[N21] K. A. Brueckner, "Theory of Nuclear Structure," in *The Many Body Problem*, ed. C. DeWitt, John Wiley and Sons, Inc., New York (1959), p. 47

[N22] B. Day, *Rev. Mod. Phys.* **39**, 719 (1967)

[N23] H. A. Bethe and J. Goldstone, *Proc. Roy. Soc. (London)* **A238**, 551 (1957)

[N24] V. F. Weisskopf, *Helv. Phys. Act.* **23**, 187 (1950); *Science* **113**, 101 (1951)

[N25] L. C. Gomes, J. D. Walecka, and V. F. Weisskopf, *Ann. Phys.* **3**, 241 (1958)

[N26] J. D. Walecka and L. C. Gomes, *Ann. Acad. Brasileira Ciêncas* **39**, 361 (1967)

[N27] H. A. Bethe, B. H. Brandow, and A. G. Petschek, *Phys. Rev.* **129**, 225 (1963)

[N28] R. Rajaraman and H. A. Bethe, *Rev. Mod. Phys.* **39**, 745 (1967)

[N29] B. D. Day, *Rev. Mod. Phys.* **50**, 495 (1978)

[N30] V. R. Pandharipande and R. B. Wiringa, *Rev. Mod. Phys.* **51**, 821 (1979)

[N31] J. G. Zabolitsky, *Advances in Nuclear Physics* **12**, eds. J. W. Negele and E. Vogt, Plenum Press, New York (1981), p. 1

[N32] R. Wiringa, in *Proc. CEBAF Summer Workshop, June (1992)*, eds. F. Gross and R. Holt, *A.I.P. Conf. Proc.* **269** A.I.P., New York (1993), p. 210

[N33] K. Huang and C. N. Yang, *Phys. Rev.* **105**, 767 (1957)

[N34] T. D. Lee and C. N. Yang, *Phys. Rev.* **105**, 1119 (1957)

[N35] C. DeDominicis and P. C. Martin, *Phys. Rev.* **105**, 1417 (1957); and private communication

[N36] V. M. Galitskii, *Sov. Phys.-JETP* **7**, 104 (1958)

[N37] M. G. Mayer and J. H. D. Jensen, *Elementary Theory of Nuclear Shell Structure*, John Wiley and Sons, Inc., New York (1955)

[N38] J. W. Negele, *Rev. Mod. Phys.* **54**, 913 (1982)

[N39] J. P. Svenne, *Adv. in Nucl. Phys.* **11**, eds. J. W. Negele and E. Vogt, Plenum Press, New York (1979), p. 179

[N40] T. H. R. Skyrme, *Phil. Mag.* **1**, 1043 (1956); *Nucl. Phys.* **9**, 615 (1959)

[N41] D. Gogny *Nuclear Physics with Electromagnetic Interactions*, eds. H. Arenhövel and D. Drechsel, Lecture Notes in Physics **108**, Springer, Berlin (1979), p. 88

[N42] L. I. Schiff, *Quantum Mechanics*, 3rd ed., McGraw-Hill Book Company, Inc., New York (1968)

[N43] P. M. Morse and H. Feshbach, *Methods of Theoretical Physics*, McGraw-Hill Book Company, Inc., New York (1953)

[N44] A. R. Edmonds, *Angular Momentum in Quantum Mechanics*, 3rd printing, Princeton University Press, Princeton, New Jersey (1974)

[N45] L. I. Schiff, *Phys. Rev.* **96**, 765 (1954)

[N46] K. Alder, A. Bohr, T. Huus, B. Mottelson, and A. Winther, *Rev. Mod. Phys.* **28**, 432 (1956)

[N47] T. deForest and J. D. Walecka, *Adv. Phys.* **15**, 1 (1966)

[N48] M. Jacob and G. C. Wick, *Ann. Phys.* **7**, 404 (1959)

[N49] T. W. Donnelly and W. C. Haxton, *Atomic Data and Nuclear Data Tables* **23**, 103 (1979)

[N50] T. W. Donnelly and W. C. Haxton, *Atomic Data and Nuclear Data Tables* **25**, 1 (1980)

[N51] M. Rotenberg, R. Bivens, N. Metropolis, and J. K. Wooten, Jr., *The 3-j and 6-j Symbols*, The Technology Press, M. I. T., Cambridge, Massachusetts (1959)

[N52] *Nuclear Data Sheets*

[N53] J. Rainwater, *Phys. Rev.* **79**, 432 (1950)

[N54] A. M. Lane, *Nuclear Theory*, W. A. Benjamin, Inc., New York (1964)

[N55] M. Baranger, *Phys. Rev.* **120**, 957 (1960)

[N56] G. E. Brown and M. Bolsterli, *Phys. Rev. Lett.* **3**, 472 (1959)

[N57] G. E. Brown, *Unified Theory of Nuclear Models*, North-Holland Publishing Company, Amsterdam (1964)

[N58] E. P. Wigner, *Phys. Rev.* **51**, 106 (1937)

[N59] H. Noya, A. Arima, and H. Horie, *Suppl. Prog. Theor. Phys. (Kyoto)* **8**, 33 (1959)

[N60] R. F. Casten, *Nuclear Structure from a Simple Perspective*, Oxford University Press, Oxford, England (1990)

[N61] M. Goldhaber and E. Teller, *Phys. Rev.* **74**, 1046 (1948)

[N62] T. W. Donnelly and J. D. Walecka, *Annu. Rev. Nucl. Sci.* **25**, 329 (1975)

[N63] V. Gillet and M. A. Melkanoff, *Phys. Rev.* **133**, B1190 (1964)

[N64] B. R. Barrett, ed., *Effective Interactions and Operators in Nuclei*, Springer-Verlag, Berlin (1975)

[N65] T.T.S. Kuo and E. Osnes, *Lecture Notes in Physics* **364**, Springer-Verlag, Berlin (1990)

[N66] B. A. Brown, *Proc. Int. Nucl. Phys. Conf., Harrogate, U. K.*, eds. J. L. Durell, *et al.*, Inst. of Phys. Conf. Series No. 86, Bristol (1987), p. 119

[N67] B. A. Brown and B. H. Wildenthal, *Annu. Rev. Nucl. Part. Sci.* **38**, 29 (1988)

[N68] J. Speth and T. Suzuki, *Nucl. Phys.* **A358**, 139c (1981)

[N69] K. Nakayama, S. Krewald, J. Speth, and W. G. Love, *Nucl. Phys.* **A431**, 419 (1984)

[N70] T. W. Donnelly and G. E. Walker, *Ann. Phys.* **60**, 209 (1970)

[N71] I. Sick, E. B. Hughes, T. W. Donnelly, J. D. Walecka, and G. E. Walker, *Phys. Rev. Lett.* **23**, 1117 (1969)

[N72] H. de Vries, C. W. de Jager, and C. de Vries, *Atomic Data and Nuclear Data Tables* **36**, 495 (1987)

[N73] P. Moller, J. R. Nix, W. D. Meyers, and W. J. Swiatecki, submitted to *Atomic Data and Nuclear Data Tables* (1993)

[N74] C. Mahaux, P. F. Bortignon, R. A. Broglia, and C. H. Dasso, *Phys. Rep.* **120**, 1 (1985); M. Jaminon and C. Mahaux, *Phys. Rev.* **C40**, 354 (1989)

[N75] C. Mahaux and R. Sartor, *Advances in Nuclear Physics* **20**, eds. J. W. Negele and E. Vogt, Plenum Press, New York (1991), p. 1

[N76] D. R. Tilley, H. R. Weller, and G. M. Hale, *Nucl. Phys.* **A541**, 1 (1992)

PROBLEMS: PART I

The first five problems review the analysis in nonrelativistic quantum mechanics of the scattering of a spinless particle by a spherically symmetric potential (see Refs. [N42], [N4], [R5]). In these problems the notation $x \equiv |\vec{x}| \equiv r$ is employed.

1.1. The Green's function for the scalar Helmholtz equation satisfies the differential equation $(\nabla^2 + k^2)G_k^{(+)}(\vec{x}-\vec{y}) = -\delta^{(3)}(\vec{x}-\vec{y})$ and is given by the limit $\eta \to 0$ of

$$G_k^{(+)}(\vec{x}-\vec{y}) = \int \frac{d^3t}{(2\pi)^3} \frac{e^{i\vec{t}\cdot(\vec{x}-\vec{y})}}{t^2 - k^2 - i\eta}$$

Show by contour integration that[71]

$$\begin{aligned} G_k^{(+)}(\vec{x}-\vec{y}) &= \frac{e^{ik|\vec{x}-\vec{y}|}}{4\pi|\vec{x}-\vec{y}|} \\ &= \frac{ik}{4\pi}\sum_{l=0}^{\infty}(2l+1)j_l(kx_<)h_l^{(1)}(kx_>)P_l(\cos\theta_{\vec{x},\vec{y}}) \end{aligned}$$

1.2. The Schrödinger equation for the scattering wave function with energy $E = \hbar^2k^2/2\mu_{\text{red}}$ and potential $V = \hbar^2 v(x)/2\mu_{\text{red}}$ can be rewritten as an inhomogeneous integral equation with outgoing wave boundary conditions

$$\psi_{\vec{k}}^{(+)}(\vec{x}) = e^{i\vec{k}\cdot\vec{x}} - \int G_k^{(+)}(\vec{x}-\vec{y})v(y)\psi_{\vec{k}}^{(+)}(\vec{y})d^3y$$

Verify this result. Show that as $x \to \infty$

$$\begin{aligned} \psi_{\vec{k}}^{(+)}(\vec{x}) &\overset{x\to\infty}{\longrightarrow} e^{i\vec{k}\cdot\vec{x}} + f(\vec{k}',\vec{k})\frac{e^{ikx}}{x} \\ f(\vec{k}',\vec{k}) &= -\frac{1}{4\pi}\int e^{-i\vec{k}'\cdot\vec{y}}v(y)\psi_{\vec{k}}^{(+)}(\vec{y})d^3y \end{aligned}$$

Here $\vec{k}' \equiv |\vec{k}|\,(\vec{x}/|\vec{x}|)$. Show by computing the ratio of scattered to incident flux that the cross section is given by $d\sigma/d\Omega = |f(\vec{k}',\vec{k})|^2$.

1.3. Substitute the following ansatz for the scattering wave function

$$\psi_{\vec{k}}^{(+)}(\vec{x}) = \sum_{l=0}^{\infty}(2l+1)i^l\psi_l^{(+)}(x;k)P_l(\cos\theta_{\vec{k},\vec{x}})$$

[71] Here $h_l^{(1)} = j_l + in_l$; and $x_>$ ($x_<$) is the greater (lesser) of $|\vec{x}|$ and $|\vec{y}|$.

Show that the integral equations decouple and that

$$\psi_l^{(+)}(x;k) \quad \stackrel{x\to\infty}{\longrightarrow} \quad \frac{1}{2kx}\{e^{-i[kx-(l+1)\pi/2]} + S_l(k)e^{i[kx-(l+1)\pi/2]}\}$$

$$S_l(k) \quad = \quad 1 - 2ik\int_0^\infty j_l(ky)v(y)\psi_l^{(+)}(y;k)y^2dy$$

$$f(\vec{k}',\vec{k}) \quad = \quad \frac{1}{2ik}\sum_{l=0}^{\infty}(2l+1)[S_l(k)-1]P_l(\cos\theta_{\vec{k}',\vec{k}})$$

Show that the radial wave functions everywhere satisfy

$$\left[\frac{1}{x}\frac{d^2}{dx^2}x - \frac{l(l+1)}{x^2} + k^2 - v(x)\right]\psi_l^{(+)}(x;k) \quad = \quad 0$$

1.4. Use $\psi_{\vec{k}}^{(+)}(\vec{x})$ to compute the net incoming flux through a large sphere of radius R. Prove $|S_l| = 1$ if there is only elastic scattering. Hence introduce the phase shift $S_l = \exp\{2i\delta_l\}$. [*Hint*: One must ultimately use superposition $\Psi \equiv \alpha\psi_{\vec{k}}^{(+)}(\vec{x}) + \beta\psi_{\vec{k}'}^{(+)}(\vec{x})$ with $|\vec{k}'| = |\vec{k}|$ in this proof.]

These last two problems reduce the analysis to a radial Schrödinger equation and scattering boundary condition

$$\psi_l^{(+)}(x;k) \quad \stackrel{x\to\infty}{\longrightarrow} \quad \frac{e^{i\delta_l}}{kx}\cos\left\{kx - \frac{(l+1)\pi}{2} + \delta_l\right\}$$

1.5. Suppose there is a nonvanishing incoming flux through the large sphere of radius R in Prob. 1.4. Show the reaction cross section is given by $\sigma_r = (\pi/k^2)\sum_l(2l+1)(1-|S_l|^2)$.

1.6. Derive the first Born aproximation for the scattering amplitude used in Eq. (1.12).

1.7. Show for a hard sphere potential that $\tan\delta_l = j_l(ka)/n_l(ka)$.

1.8. (a) Suppose an attractive square well potential of range d and depth V_0 has one bound state at zero energy. Show $V_0d^2 = \hbar^2\pi^2/8\mu_{\text{red}}$. Use the effective range expansion $k\cot\delta_0 = -1/a + r_0k^2/2$ to prove that $a = -\infty$, $r_0 = d$.
(b) Suppose the potential in (a) is an infinite barrier (hard core) to $|\vec{x}| = b$ and then an attractive square well to $|\vec{x}| = b + b_w$. Show $V_0b_w^2 = \hbar^2\pi^2/8\mu_{\text{red}}$, $a = -\infty$, and $r_0 = 2b + b_w$.

1.9. Derive the 1-π exchange potential in Eq. (A.9d).

2.1. Consider the scattering of a particle of charge ze from a charge distribution $\rho(|\vec{x}|)$ through the Coulomb interaction $H' = ze^2\int\rho(|\vec{y}|)d^3y/4\pi|\vec{x}-\vec{y}|$. Use the Born approximation. Show the cross section takes the form $d\sigma/d\Omega = \sigma_{\text{p}}|F(\vec{q}^2)|^2$ where σ_{p} is the cross section for scattering from a point charge and the form factor $F(\vec{q}^2)$ is the Fourier transform of the charge distribution with respect to the momentum transfer.

2.2. Calculate the form factor in Prob. 2.1 for the following charge distributions: (a) uniform to R_{C}; (b) gaussian; and (c) exponential.

2.3. Calculate the Coulomb interaction energy of Z charges uniformly distributed over a sphere of radius R_{C} and hence derive Eq. (2.12).

2.4. Consider the small oscillations of an incompressible liquid drop.[72] Write the general surface as $r = r_0[1+\sum_{l=1}^{\infty}\sum_m q_{lm}Y_{lm}(\theta,\phi)]$ with $q_{lm}^* = (-1)^m q_{l-m}$ and work to second order in q_{lm}.
(a) Show that the volume and surface area of the drop are given by $V = (4\pi r_0^3/3)[1 + (3/4\pi)\sum_{lm}|q_{lm}|^2] \equiv 4\pi a^3/3$ and $S - 4\pi a^2 = (a^2/2)\sum_{lm}(l-1)(l+2)|q_{lm}|^2$.
(b) Assume irrotational flow so that $\vec{v} = \vec{\nabla}\psi$. Hence show $\nabla^2\psi = 0$. Show that the kinetic energy is given by $T = (\rho a^5/2)\sum_{lm}|\dot{q}_{lm}|^2/l$ where $\rho = mn$ is the mass density and $\dot{q} \equiv dq/dt$. (*Hint*: What is the boundary condition at the surface of the drop?)
(c) Hence derive the lagrangian

$$L = \frac{1}{2}\rho a^5 \sum_{lm} \frac{|\dot{q}_{lm}|^2}{l} - \frac{\sigma a^2}{2}\sum_{lm}(l-1)(l+2)|q_{lm}|^2$$

Here σ is the surface tension. Derive the equations of motion for the normal modes and identify the normal mode frequencies. Plot the spectrum.

2.5 Consider the quantization of the system in Prob. 2.4. Introduce the canonical momenta, hamiltonian, and canonical commutation relations. Use $q_{lm} \equiv (\hbar/2\sqrt{B_lC_l})^{1/2}[a_{lm}+(-1)^m a_{l-m}^*]$ and $p_{lm} \equiv i(\hbar\sqrt{B_lC_l}/2)^{1/2}[a_{lm}^* - (-1)^m a_{l-m}]$ where the lagrangian is written $2L \equiv \sum_{lm} B_l|\dot{q}_{lm}|^2 - \sum_{lm} C_l|q_{lm}|^2$. Hence reduce the problem to the form

$$H = \frac{1}{2}\sum_{lm} \hbar\omega_l(a_{lm}^* a_{lm} + a_{lm}a_{lm}^*)$$

$$[a_{lm}, a_{l'm'}^*] = \delta_{ll'}\delta_{mm'}$$

Discuss this quantum system in detail.

2.6. Assume the drop in Probs. 2.4 and 2.5 is uniformly charged. Add the additional Coulomb interaction energy.
(a) Show the result is to replace $C_l \to C_l[1 - 10\gamma/(2l+1)(l+2)]$ where γ is the ratio of Coulomb to surface energy $\gamma \equiv [(3/5)Z^2e^2/4\pi a] \,/\, [4\pi\sigma a^2]$.
(b) Show that fission will occur (i.e., the restoring force will vanish) when $\gamma \geq 2$. What is the corresponding inequality for Z^2/A?

3.1. Prove that a two-body tensor force with Serber exchange $V = V_T S_{12}\frac{1}{2}(1+P_M)$ makes no contribution to the energy of a spin-$\frac{1}{2}$ isospin-$\frac{1}{2}$ Fermi gas (i.e., nuclear matter) in lowest order.

3.2. (a) Assume the nuclear interactions are equivalent to a slowly varying potential $-U(r)$. Within any small volume element, assume that the particles form a non-interacting Fermi gas with levels filled up to an energy $-B$. In equilibrium, B must be the same throughout the nucleus. From this description, derive the Thomas-Fermi expression for the nuclear density $n(r) = (2/3\pi^2)(2m/\hbar^2)^{3/2}[U(r) - B]^{3/2}$.
(b) Derive the results of part (a) by balancing the hydrostatic force $-\vec{\nabla}P$ and the force from the potential $n\vec{\nabla}U$ (Ref. [N2]).

4.1. The expectation value of a two-body operator $(1/2)\sum_{ij} O(\vec{x}_i, \vec{x}_j)$ for a system of identical particles involves knowledge of the two-body density $\rho^{(2)}(\vec{x}, \vec{y})$ computed

[72] Lord Rayleigh, *Proc. Roy. Soc. (London)* **A29**, 91 (1879); *Theory of Sound*, Dover, New York (1945).

from the full wave function. Use the analysis in Eqs. (3.9)-(3.15) to show that for a noninteracting Fermi gas with degeneracy g, one has $\rho^{(2)}_{\mathrm{FG}} = (1/2)A(A-1)[\rho^{(1)}]^2\{1 - (1/g)[3j_1(k_{\mathrm{F}}|\vec{x}-\vec{y}|)/k_{\mathrm{F}}|\vec{x}-\vec{y}|]^2\}$ where $\rho^{(1)} = 1/V$. Hence show that like fermions are anticorrelated in space by the Pauli principle. Compare this correlation length with the interparticle spacing.

4.2. Extend Prob. 4.1. How would one calculate $\rho^{(2)}$ for an interacting system in the independent-pair approximation? Sketch this quantity for a hard-core gas with parameters appropriate to nuclear matter; use the result in Fig. 4.6.

4.3.Attempt a partial wave decomposition of the B-G equation when $\vec{P} \neq 0$. Show the partial waves are coupled. Discuss.

4.4. To help understand Eq. (4.29), verify directly from Eq. (4.8) that $\psi_{\vec{P}\vec{k}}(\vec{x})$ has no Fourier components in $\overline{\Gamma}$ except for $\vec{k}$ itself (see Fig. 4.2).

4.5. Verify the claim made in writing Eq. (4.22).

5.1. The symmetry energy E_4/A in Eqs. (2.13) and (2.16) may be estimated as follows. Assume the nonsingular potential of Eq. (3.10) and compute the expectation value of $\hat{H}$ in the Fermi gas model for $A = Z + N$ nucleons with $\delta = (N-Z)/A \neq 0$.
(a) Use Eq. (3.9) to prove that

$$\frac{E_4}{A} = \frac{\delta^2}{3}\left\{\frac{\hbar^2 k_{\mathrm{F}}^2}{2m^*} - \frac{k_{\mathrm{F}}^3}{\pi}\int_0^\infty V(z)[a_M + a_W j_0^2(k_{\mathrm{F}}z)]z^2 dz\right\} \; ; \qquad \delta \to 0$$

where the effective mass at k_{F} is given by $\hbar^2 k_{\mathrm{F}}/m^* = \{d[\varepsilon_k^0 + U(k)]/dk\}_{k=k_{\mathrm{F}}}$ and $U(k)$ is defined by Eqs. (3.16)-(3.18).
(b) With $m^* \approx 0.65$ (Table 5.1) and the potential of Fig. 4.3 show $a_4 \approx 37$ MeV.

5.2. Prove that the leak contribution to the energy arising from $w(r)$ is of $O(c^4)$ for a hard-sphere gas.

5.3. Use the Bethe-Goldstone equation for a pure hard-core p-wave potential to prove $E^{(c)}_{l=1}/A = (\hbar^2 k_{\mathrm{F}}^2/2m)(c^3/\pi)$ for nuclear matter.

5.4. The compressibility of nuclear matter is defined by $K_V^{-1} \equiv k_{\mathrm{F}}^2[d^2(E/A)/dk_{\mathrm{F}}^2]_{\mathrm{equil}}$.
(a) Evaluate K_V approximately from Fig. 5.3 for the two cases shown. Relate K_V to the usual thermodynamic compressibility.
(b) Compare with the value for a noninteracting Fermi gas at the same density.

5.5. The Λ is a baryon of isospin 0 and strangeness -1; it is distinguishable from the nucleon. Show that the energy shift of a hard-sphere Λ in a nuclear matter gas of hard spheres is given by $E_c = (\hbar^2 k_{\mathrm{F}}^2/2m)\{16(k_{\mathrm{F}}a)/3\pi + 8(k_{\mathrm{F}}a)^2/\pi^2 + O(k_{\mathrm{F}}a)^3\}$. Here a is the range of the Λ-nucleon hard core and it is assumed for simplicity that $m_\Lambda = m$.

5.6. Consider the binding energy of a Λ in the nucleus. If the nucleus is a square well potential of depth U_0 and range $R = r_0 A^{1/3}$ show the binding energy is given for large s by $B_\Lambda = U_0 - (\hbar^2\pi^2/2\mu_\Lambda R^2)(1 - 2/s + \cdots)$ where $s = [(2\mu_\Lambda R^2/\hbar^2)U_0]^{1/2}$ and μ_Λ is the reduced mass of the Λ-nucleus system. Explain how to use these results to identify the binding energy of a Λ in nuclear matter.
(b) Suppose the Λ-nucleon potential is of the form in Fig. 5.1. Discuss the binding energy of the Λ in nuclear matter within the independent-pair approximation.

6.1. Consider the Hartree-Fock Eqs. (6.15).

(a) Use the hermiticity of T and V to prove the eigenvalues ε_δ are real.
(b) Prove that the solutions corresponding to different eigenvalues are orthogonal.

6.2. Consider a noninteracting Fermi gas in a big box with periodic boundary conditions. Now add a two-body interaction $V(|\vec{x}_i - \vec{x}_j|)$.
(a) Show the original single-particle wave functions solve the Hartree-Fock equations.
(b) Consider the new dispersion relation $\varepsilon(\vec{k}_\delta)$. Show that with short-range potentials the direct and exchange contributions are comparable, while with long-range potentials the direct term dominates. (The neglect of the exchange contribution results in the Hartree approximation.)

6.3. Consider the state of two indentical additional valence nucleons (n or p) in the same j shell $|j^2JM\rangle = (1/\sqrt{2})\sum_{m_1m_2}\langle jm_1jm_2|jjJM\rangle a^\dagger_{m_1}a^\dagger_{m_2}|0\rangle \equiv \xi^\dagger_{JM}|0\rangle$.
(a) Prove that J must be even.
(b) To investigate the level ordering, use a multipole expansion of the two-nucleon interaction $V(r_1, r_2, \cos\theta_{12}) = \sum_K f_k(r_1, r_2)P_K(\cos\theta_{12})$. Prove that (Ref. [N2, N44])

$$\langle j^2JM|V|j^2JM\rangle = \sum_{\text{even } K} F_K(-1)^{J+1}\begin{Bmatrix} j & j & J \\ j & j & K \end{Bmatrix}(2j+1)^2\begin{pmatrix} j & K & j \\ \frac{1}{2} & 0 & -\frac{1}{2} \end{pmatrix}^2$$

Here the Slater integrals are defined by $F_K \equiv \int\int R^2_{nl}(r_1)R^2_{nl}(r_2)f_K(r_1, r_2)r_1^2dr_1r_2^2dr_2$.
(c) For a potential $V = -g\delta^{(3)}(\vec{r}_1 - \vec{r}_2)$ show one has $f_K = (-g/4\pi)(2K+1)\delta(r_1 - r_2)/r_1r_2$ (Ref. [N2]). Plot the resulting spectrum for several values of j. Discuss.[73]

7.1. (a) Use the angular momentum commutation relations to prove that $\exp\{-i\beta\hat{J}_y\}\hat{J}_z\exp\{i\beta\hat{J}_y\} = \hat{J}_z\cos\beta + \hat{J}_x\sin\beta$.
(b) Prove $\vec{J}\cdot\vec{e}_{z'}[\exp\{-i\beta\hat{J}_y\}]|jm\rangle = m[\exp\{-i\beta\hat{J}_y\}]|jm\rangle$ where $\vec{e}_{z'} \equiv \vec{e}_z\cos\beta + \vec{e}_x\sin\beta$.
(c) Hence conclude that the operator $\hat{R}_{-\beta} = \exp\{-i\beta\hat{J}_y\}$ rotates the physical state vector by an angle $+\beta$ about the y-axis. Thus verify the result in Eq. (7.32).

7.2. Verify the integral vector identity in Eq. (7.51).

7.3. Equation (7.44) is the general expression to $O(\alpha)$ for photoemission.
(a) What is the analogous expression for the photoabsorption cross section integrated over the absorption line $\int_{\text{abs line}}\sigma_{fi}(\omega)\,d\omega$?
(b) The Wigner-Weisskopf theory of the line width in QED[74] results in the replacement $\delta(E_f - E_i - \hbar\omega) \to (\gamma/2\pi)[(E_f - E_i - \hbar\omega)^2 + \gamma^2/4]^{-1}$ in the transition rate. What is the effect on the answer in part (a)?

7.4. Show the effect of including nuclear recoil in the density of final states is to multiply the electron scattering cross section in Eq. (7.77) by a factor r where $r^{-1} = 1 + (2k_1/M_T)\sin^2\theta/2$ to $O(1/M_T)$. What is the analogous factor in Eq. (7.44)?

7.5. Consider photodisintegration of the deuteron ^{2}H. Work in the C-M system where $\vec{r}_p + \vec{r}_n = 0$. Neglect spin.
(a) Since ^{2}H is just bound, its wave function extends well outside the two-nucleon potential. Show that the wave function in this region is $\phi_{\text{out}} = N\exp\{-a\rho\}/\rho$ where

[73] An excellent discussion of the many-particle shell model is contained in A. de-Shalit and I. Talmi, *Nuclear Shell Theory*, Academic Press, New York (1963).
[74] E. Wigner and V. F. Weisskopf, *Z. Phys.* **63**, 54 (1930).

$\varepsilon_b \equiv \hbar^2 a^2/2\mu_{\text{red}}$. Here $\vec{\rho} = \vec{r}_p - \vec{r}_n$ is the relative coordinate. Sketch a comparison with the expected behavior of the actual wave function inside the potential. Show $N \approx (a/2\pi)^{1/2}$, and assume $\phi_i \approx \phi_{\text{out}}$.
(b) Make the Born approximation for the final state, assuming a plane wave in the relative coordinate $\phi_f \approx (1/\Omega)^{1/2} \exp\{i\vec{k}_f \cdot \vec{\rho}\}$.
(c) Start from Eq. (7.1), make the long wavelength dipole approximation $e^{i\vec{k}\cdot\vec{x}} \approx 1$ in the matrix element of the current, and derive the Bethe-Peierls' cross section for photodisintegration of the deuteron[75]

$$\frac{d\sigma}{d\Omega} = \frac{2\alpha}{a^2}\frac{y^{3/2}}{(1+y)^3}\cos^2\theta_{k_f}$$

Here $y \equiv k_f^2/a^2$ and polarized photons are assumed with $\cos\theta_{k_f} \equiv \vec{e}_\lambda \cdot \vec{k}_f/|\vec{k}_f|$. Plot your results.
(d) Sketch the final integrand of the required matrix element as a function of ρ, and use this as a basis for a discussion of the validity of the approximations made.

7.6. Derive the electron scattering cross section in Eq. (7.77). (see Refs. [N14, N47] and Part IV).

8.1. Rainwater (Ref. [N53]) pointed out that one can lower the energy of the system of a particle moving in a constant potential inside a liquid drop by allowing the drop to acquire a permanent deformation. The particle effectively exerts a pressure on the walls of the potential.
(a) Assume the deformation $\vec{r}' = \vec{r}(1+\sum_{\lambda\mu} q_{\lambda\mu}Y_{\lambda\mu})$ of Prob. 2.4. Assume the potential follows the drop so that $V'(\vec{r}') = V(\vec{r})$. Take $\Delta V \equiv V'(\vec{r}') - V(\vec{r}')$. Hence show the shift in the single-particle energy is $\Delta\varepsilon_{jm} = \langle ljm| -r(\partial V/\partial r)\sum_{\lambda\mu} q_{\lambda\mu}Y_{\lambda\mu}|ljm\rangle = -2T_{nl}\sum_{\lambda\mu} q_{\lambda\mu}\langle ljm|Y_{\lambda\mu}|ljm\rangle$.[76]
(b) Add the surface energy of Prob. 2.4 so that the total energy of the system is

$$E = \varepsilon_j - 2T_{nl}\sum_\lambda q_{\lambda 0}\langle jm|Y_{\lambda 0}|jm\rangle + \frac{1}{2}\sigma a^2 \sum_{\lambda\mu}(\lambda-1)(\lambda+2)|q_{\lambda\mu}|^2$$

Minimize this expression with respect to $q_{\lambda\mu}$. Show a permanent $\bar{q}_{\lambda 0}$ lowers the total energy of the system.
(c) Now add particles to fill the m states in the $|jm\rangle$ shell. Discuss what happens to $\bar{q}_{20}$ as the states are filled.

8.2. Calculate the quadrupole moment of the uniformly charged core as a function of the permanent deformation $\bar{q}_{20}$ in Prob. 8.1. Discuss how it varies with the filling of the $|jm\rangle$ shell.

8.3. Consider the single-particle shell model matrix elements of the multipole operators in Eqs. (7.25) and (7.79). Use the nonrelativistic quantum mechanical densities of Eq. (8.12) and also Eq. (7.16). Let $M_{JM} \equiv j_J(qx)Y_{JM}(\Omega_x)$ and $\vec{\mathbf{M}}_{JL1}^M \equiv j_L(qx)\vec{\mathcal{Y}}_{JL1}^M$. Prove the following relations (Refs. [N44, N47, N49]).

[75] *Hint*: If $H = p^2/2\mu_{\text{red}} + V(x)$, then $p/\mu_{\text{red}} = (i/\hbar)[H, x]$.
[76] *Hint*: Establish the virial relation $r(\partial V/\partial r) = [\vec{r}\cdot\vec{\nabla}, H_{\text{part}}] + 2T_{\text{part}}$ for a particle moving in the potential $V(r)$.

$$\langle n'l'\tfrac{1}{2}j'||M_J||nl\tfrac{1}{2}j\rangle = (-1)^{j+J+\frac{1}{2}}\frac{1}{\sqrt{4\pi}}[(2l'+1)(2l+1)(2j'+1)(2j+1)]^{1/2}$$
$$\times\sqrt{2J+1}\begin{Bmatrix} l' & j' & 1/2 \\ j & l & J \end{Bmatrix}\begin{pmatrix} l' & J & l \\ 0 & 0 & 0 \end{pmatrix}\langle n'l'|j_J(qr)|nl\rangle$$
$$\langle n'l'\tfrac{1}{2}j'||\vec{M}_{JL}\cdot\vec{\sigma}||nl\tfrac{1}{2}j\rangle = (-1)^{l'}\sqrt{\frac{6}{4\pi}}[(2l'+1)(2l+1)(2j'+1)(2j+1)]^{1/2}$$
$$\times\sqrt{(2L+1)(2J+1)}\begin{Bmatrix} l' & l & L \\ 1/2 & 1/2 & 1 \\ j' & j & J \end{Bmatrix}\begin{pmatrix} l' & L & l \\ 0 & 0 & 0 \end{pmatrix}\langle n'l'|j_L(qr)|nl\rangle$$

Here $\langle n'l'|j_L(qr)|nl\rangle = \int_0^\infty R^*_{n'l'}(r)j_L(qr)R_{nl}(r)r^2dr$.

8.4. Show

$$\langle n'l'\tfrac{1}{2}j'||\vec{M}_{JL}\cdot\vec{\nabla}||nl\tfrac{1}{2}j\rangle = (-1)^{l'+j-\frac{1}{2}}\frac{1}{\sqrt{4\pi}}[(2l'+1)(2l+1)(2j'+1)(2j+1)]^{1/2}$$
$$\times\sqrt{(2L+1)(2J+1)}\begin{Bmatrix} l' & j' & 1/2 \\ j & l & J \end{Bmatrix}$$
$$\times\left[\begin{Bmatrix} L & 1 & J \\ l & l' & l+1 \end{Bmatrix}\frac{\begin{pmatrix} l' & L & l+1 \\ 0 & 0 & 0 \end{pmatrix}}{\begin{pmatrix} l+1 & 1 & l \\ 0 & 0 & 0 \end{pmatrix}}\frac{l+1}{2l+1}\langle n'l'|j_L(qr)\left(\frac{d}{dr}-\frac{l}{r}\right)|nl\rangle\right.$$
$$\left.+\begin{Bmatrix} L & 1 & J \\ l & l' & l-1 \end{Bmatrix}\frac{\begin{pmatrix} l' & L & l-1 \\ 0 & 0 & 0 \end{pmatrix}}{\begin{pmatrix} l-1 & 1 & l \\ 0 & 0 & 0 \end{pmatrix}}\frac{l}{2l+1}\langle n'l'|j_L(qr)\left(\frac{d}{dr}+\frac{l+1}{r}\right)|nl\rangle\right]$$

8.5. Insert the explicit form of the Dirac spinors into the general form of the vertex, expand through $O(1/m^2)$, assume q_0 and F_2 are $O(1/m)$, and derive Eq. (8.21). Use the standard representation of the Dirac matrices where $\vec{\gamma} = i\vec{\alpha}\beta, \gamma_4 = \beta, \sigma_{\mu\nu} = [\gamma_\mu, \gamma_\nu]/2i$ and $\beta = \begin{pmatrix} 1 & 0 \\ 0 & -1 \end{pmatrix}$, $\vec{\alpha} = \begin{pmatrix} o & \vec{\sigma} \\ \vec{\sigma} & 0 \end{pmatrix}$ in 2×2 form.

9.1. Prove that a filled l-shell produces a spherically symmetric probability distribution. Repeat for a filled j-shell.

9.2. (a) Consider the TDA Eqs. (9.12). Use the assumed properties of the two-nucleon potential to prove that the eigenvalues are real and that the eigenfunctions satisfy the orthogonality relation of Eq. (9.15).
(b) Repeat for the RPA Eqs. (9.24) and the orthogonality relation of Eq. (9.26).

9.3. (a) Show that for a spin- and isospin-dependent two-nucleon potential the reduced TDA particle-hole interaction is (Ref. [N2])

$$v^{JT}_{ab;lm} = -\sum_{J'T'}(2J'+1)(2T'+1)\begin{Bmatrix} j_m & j_a & J' \\ j_b & j_l & J \end{Bmatrix}\begin{Bmatrix} 1/2 & 1/2 & T' \\ 1/2 & 1/2 & T \end{Bmatrix}$$

$$\times[\langle lbJ'T'|V|amJ'T'\rangle - (-1)^{\frac{1}{2}+\frac{1}{2}+T'}(-1)^{j_a+j_m+J'}\langle lbJ'T'|V|maJ'T'\rangle]$$

(b) Show the RPA adds the intraction $u^{JT}_{ab;lm} = (-1)^{\frac{1}{2}-\frac{1}{2}-T}(-1)^{j_m-j_l-J}\, v^{JT}_{ab;ml}$.

9.4. As a model for the large amplitude collective motion of deformed nuclei consider the quantum mechanics of the symmetric top.

(a) Introduce the Euler angles, construct the lagrangian, find the canonical momenta, and obtain the hamiltonian. Show $H = T = (\hbar^2/2I_1)\vec{J}^2 + \hbar^2(1/2I_3 - 1/2I_1)J_\gamma^2$ where $I_2 = I_1$, J_γ is the component of the angular momentum along the figure axis, and the square of the total angular momentum is $\vec{J}^2 = J_\beta^2 + (J_\alpha - J_\gamma \cos\beta)^2/\sin^2\beta + J_\gamma^2$.

(b) Introduce canonical quantization with the generalized coordinates $\{\alpha, \xi \equiv \cos\beta, \gamma\}$ and operators hermitian with respect to the volume element $d\tau = d\alpha d\gamma d\xi$. Hence show $J_\gamma = (1/i)\partial/\partial\gamma$ and

$$\vec{J}^2 = -\left[\frac{1}{\sin\beta}\frac{\partial}{\partial\beta}\sin\beta\frac{\partial}{\partial\beta} + \frac{\partial^2}{\partial\gamma^2} + \frac{1}{\sin^2\beta}\left(\frac{\partial}{\partial\alpha} - \cos\beta\frac{\partial}{\partial\gamma}\right)^2\right]$$

(c) Consider the Schrödinger equation $H\psi(\alpha\beta\gamma) = E\psi(\alpha\beta\gamma)$. Differentiate, use the commutation relations, and use the eigenstates of angular momentum to show that the rotation matrices $\mathcal{D}^j_{mk}(\alpha\beta\gamma) = \langle jm|\exp\{i\hat{J}_z\alpha\}\exp\{i\hat{J}_y\beta\}\exp\{i\hat{J}_z\gamma\}|jk\rangle$ are the eigenfunctions for the symmetric top. What are the eigenvalues? What is the normalization constant?

(d) Show that an object that has full azimuthal symmetry must have $k = 0$. (*Hint*: use superposition.) Hence derive the rotational spectrum observed for deformed even-even nuclei $E = (\hbar^2/2I)j(j+1)$.

10.1. Derive the reduced form of the additional RPA interaction $u^{[15]L}_{ab;lm}$ in Eq. (10.22).

10.2. Show $u^{[1]}_{ab;lm} = -3u^{[15]}_{ab;lm}$ for a delta-function potential [Eq. (10.33)].

10.3. It is a result from Prob. 6.3 that with a short-range attractive interaction $-g\delta^{(3)}(\vec{r})$ the paired state with $J = 0$ of two identical nucleons in the j-shell is by far the most tightly bound. The normal coupling scheme for the j^N configuration with N odd consists of a core with $(N-1)/2$ pairs coupled to zero, thus $|j^N jm\rangle = \mathcal{A}a^\dagger_{jm}(\xi^\dagger_0)^{(N-1)/2}|0\rangle$ where $\mathcal{A}$ is the appropriate normalization constant. If $\hat{T}_K$ is a one-body ITO, prove (Refs. [N37, N2])

$$\begin{aligned}\langle j^N j||\hat{T}_K||j^N j\rangle &= \langle j||T_K||j\rangle \quad ; \quad \text{K odd}\\ &= \frac{2j+1-2N}{2j-1}\langle j||T_K||j\rangle \quad ; \quad \text{K even and} \neq 0\end{aligned}$$

Hence conclude that in this normal coupling scheme odd moments are just the single-particle values and even moments change sign in going from particles to holes.

10.4. In the Goldhaber-Teller model of the GDR the protons are assumed to move as a unit against the neutrons. Consider the model hamiltonian $H = \vec{p}^2/2\mu_{\text{red}} + \mu_{\text{red}}\omega^2\vec{r}^2/2$ where $(\vec{r}, \vec{p})$ are the relative coordinate and momentum and the reduced mass is $1/\mu_{\text{red}} = 1/Zm + 1/Nm$.

(a) Quantize this system. Discuss the excitation spectrum and transition matrix elements of the operator $\vec{r}$.

(b) Show the charge density operator in the C-M system in this model is $\hat{\rho}_N(\vec{x}) = \rho_0(|\vec{x} - N\vec{r}/A|)$ where $\rho_0(x)$ is the ground-state proton density. Expand in $\vec{r}$ and show the Coulomb transition matrix element is related to the ground-state form factor by $|\langle 1^- || \hat{M}_1^{\text{coul}}(q) || 0 \rangle|^2 = (N/A)^2 (q^2/2\mu_{\text{red}})(\hbar/\omega) |\langle 0 || \hat{M}_0^{\text{coul}}(q) || 0 \rangle|^2$.
(c) Show the current density operator is $\vec{j}_N(\vec{x}) = \rho_0(x)(N/A) d\vec{r}/dct$ to first order in $\vec{r}$. Prove it is conserved to this order.
(d) Calculate the l.h.s. and show $|\langle 1^- || \hat{T}_1^{\text{el}}(q) || 0 \rangle|^2 = 2(\omega/qc)^2 |\langle 1^- || \hat{M}_1^{\text{coul}}(q) || 0 \rangle|^2$ for all q (Ref. [N47]).

10.5. Consider the quantized oscillating liquid drop model of Probs. 2.4-2.6.
(a) The charge density operator is given to order $\hat{q}_{lm}$ by $\hat{\rho}_N(\vec{x}) = (3Z/4\pi a^3)\theta(a[1 + \sum_{lm} \hat{q}_{lm} Y_{lm}(\theta, \phi)] - r)$. Show the Coulomb form factors for single surfon excitation are given by $|\langle L^\pi || \hat{M}_L^{\text{coul}}(q) || 0 \rangle|^2 = (9Z^2/16\pi^2)(2L+1)(\hbar/2\sqrt{B_L C_L}) |j_L(qa)|^2$. Plot as a function of qa and discuss.
(b) Show the conserved current density operator is given to order $\hat{q}_{lm}$ by $\vec{j}_N(\vec{x}) = (3Z/4\pi a) \sum_{lm} (1/l)(d\hat{q}_{lm}/dct)[\vec{\nabla}(r/a)^l Y_{lm}(\theta, \phi)]\theta(a - r)$.
(c) Show $|\langle L^\pi || \hat{T}_L^{\text{el}}(q) || 0 \rangle|^2 = [(L+1)/L](\omega/qc)^2 |\langle L^\pi || \hat{M}_L^{\text{coul}}(q) || 0 \rangle|^2$ for all q.

10.6. Let $\hat{D}$ be a sum of hermitian single-particle operators.
(a) Prove the identity

$$\langle \Psi_0 | [\hat{D}, [\hat{H}, \hat{D}]] | \Psi_0 \rangle = 2 \sum_n (E_n - E_0) |\langle \Psi_n | \hat{D} | \Psi_0 \rangle|^2$$

(b) If the r.h.s. is evaluated in the RPA, show that the result is the same as evaluating the l.h.s. in the Hartree-Fock shell model ground state. Whenever the double commutator is a c number, the RPA thus preserves this energy-weighted sum rule.[77]

11.1. The nucleus ${}_{49}\text{In}_{66}^{115}$ has a ground state with $j^\pi = (9/2)^+, \mu = 5.51$ n.m., and $Q = 83 \times 10^{-26}$ cm^2. The first excited state at 0.335 MeV has $j^\pi = (1/2)^-$ and a half-life of $\tau_{1/2} = 4.5$ hr. Can you quantitatively account for these properties using the shell model?[78]

11.2. The simplest application of the analysis in this section is to the nucleus ^{4}He. Assume s.h.o. wave functions with an oscillator parameter $b = 1.59$ fm from elastic electron scattering (Part IV). Retain the $(1s)^{-1}(1p)$ configurations and calculate the spectrum of the members of the [15] supermultiplet using the potential of Eq. (11.1). Take the spin-orbit splitting from experiment to be $\varepsilon_{1/2} - \varepsilon_{3/2} \approx 4$ MeV. Compare with the results in Section 10 and with experiment.

11.3. (a) The simplest model of ${}_8\text{O}_{10}^{18}$ is a $(1d_{5/2})_\nu^2$ configuration. Use the nonsingular potential of Eq. (11.1) and s.h.o. wave functions with $b = 1.77$ fm from elastic electron scattering to compute the spectrum. Compare with the experimental spectrum $0^+(-3.92 \text{ MeV}), 2^+(-1.94 \text{ MeV}), 4^+(-0.37 \text{ MeV})$.

[77] D. J. Thouless, *Nucl. Phys.* **22**, 78 (1961).
[78] Do not forget internal conversion. This is an additional contribution to the electromagnetic transition rate due to the direct ejection of atomic electrons; its importance increases with the multipolarity and Z, and decreases with energy. One writes $\omega_{\text{fi}}^{\text{tot}} = \omega_{\text{fi}}^\gamma + \omega_{\text{fi}}^{\text{int con}} \equiv (1+\xi)\omega_{\text{fi}}^\gamma$ and the internal conversion coefficients ξ are tabulated.

(b) Extend this calculation to include all states in the $(1p)^{-1}(2s, 1d)$ configurations. Compare with the result shown in Fig. 61.3 of Ref. [N2].

c) Discuss how one might carry out the calculation in the independent-pair approximation if V_{NN} is singular and contains a hard core (Ref. [N2]).

APPENDICES: PART I

A Meson Exchange Potentials

Consider the lowest-order scattering operator in nonrelativistic potential scattering

$$S^{(1)} = \frac{-i}{\hbar c} \int \mathcal{H}_I(x) d^4x \tag{A.1}$$

In second quantization the interaction hamiltonian with a potential V_{eff} and distinguishable fermions is given by

$$\int d^3x\, \mathcal{H}_I(x) = \int \int d^3x\, d^3y\, \psi_a^\dagger(x)\psi_b^\dagger(y) V_{\text{eff}}(\vec{y} - \vec{x})\psi_b(y)\psi_a(x) \tag{A.2}$$

The kinematic situation is illustrated in Fig. A.1. That part of the nonrelativistic field operator that contributes to the matrix element is the following

$$\psi \doteq \frac{1}{\Omega^{1/2}} \sum_{\vec{k}\lambda} a_{\vec{k}\lambda} \chi_\lambda e^{ik\cdot x} \tag{A.3}$$

Here Ω is the quantization volume. The matrix element of the scattering operator for the situation illustrated in Fig. A.1 thus takes the form

$$S_{fi}^{(1)} = \frac{-i}{\hbar c \Omega^2} (2\pi)^4 \delta^{(4)}(k_1 + k_2 - k_3 - k_4) \tilde{V}_{\text{eff}}(\vec{q}) \tag{A.4}$$

where $\vec{q} = \vec{k}_1 - \vec{k}_3 = \vec{k}_4 - \vec{k}_2$. Here spin indices have been suppressed.

The interaction lagrangian density for scalar meson exchange is given by

$$\mathcal{L}_I = g_s \bar{\psi}\psi\phi \tag{A.5}$$

The Feynman rules (Fig. A.1) then yield the following lowest-order S-matrix

$$\begin{aligned} S_{fi} &= \left(\frac{-i}{\hbar c}\right)^2 (-g_s)^2 (2\pi)^4 \delta^{(4)}(k_1 + k_2 - k_3 - k_4) \frac{1}{\Omega^2} \frac{\hbar}{ic} \frac{1}{q^2 + m_s^2} \\ &\times \bar{u}_a(k_3) u_a(k_1) \bar{u}_b(k_4) u_b(k_2) \end{aligned} \tag{A.6}$$

The limit $M \to \infty$ represents static sources; in this limit $q_0 = O(1/M)$ and $\bar{u}u \to \delta_{ss'}$. A comparison of Eqs. (A.4) and (A.6) then allows the identification

$$\tilde{V}_{\text{eff}}(\vec{q}) = -\frac{g_s^2}{c^2} \frac{1}{\vec{q}^2 + m_s^2} \tag{A.7}$$

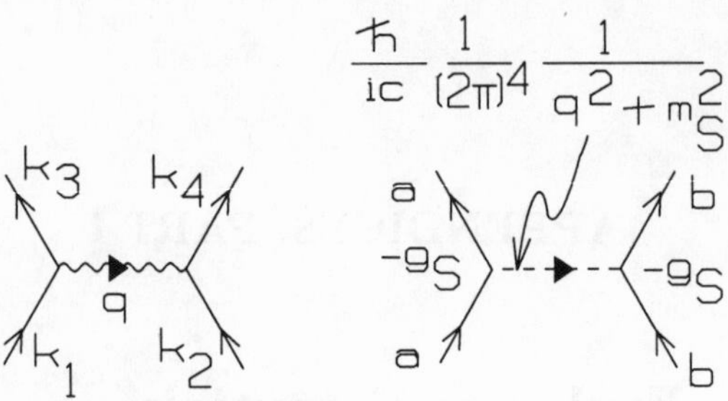

Figure A.1: Kinematics for scattering, and scalar meson exchange.

Note the sign. The Fourier transform of this relation then yields the celebrated Yukawa potential

$$
\begin{aligned}
V_{\text{eff}}(r) &= \int e^{i\vec{q}\cdot\vec{r}}\tilde{V}_{\text{eff}}(\vec{q})\frac{d^3q}{(2\pi)^3} \\
&= -\frac{g_s^2}{4\pi c^2}\frac{e^{-m_s r}}{r}
\end{aligned}
\tag{A.8}
$$

Here all masses are in units of inverse Compton wavelengths $mc/\hbar$.

We **summarize** the effective potentials obtained in this fashion from various meson exchanges and lagrangian densities. The respective vertices are indicated pictorially in Fig. A.2.

	$\mathcal{L}_I$	$V_{\text{eff}}(r)$	(A.9)
(a)	$g_s\bar{\psi}\psi\phi$	$-\frac{g_s^2}{4\pi c^2}\frac{e^{-m_s r}}{r}$	
(b)	$ig_v\bar{\psi}\gamma_\mu\psi\omega_\mu$	$\frac{g_v^2}{4\pi}\frac{e^{-m_v r}}{r}$	
(c)	$ig_\rho\bar{\psi}\gamma_\mu\frac{1}{2}\vec{\tau}\psi\cdot\vec{\rho}_\mu$	$\frac{g_\rho^2}{4\pi}\frac{1}{4}\vec{\tau}_1\cdot\vec{\tau}_2\frac{e^{-m_\rho r}}{r}$	

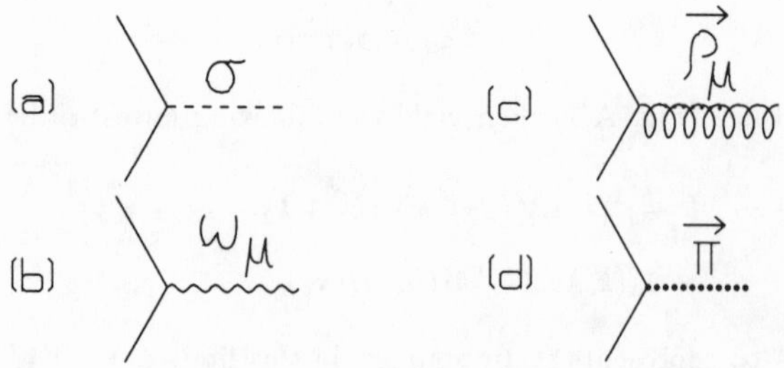

Figure A.2: Pictorial representation of vertices for various types of meson exchange: (a) neutral scalar σ; (b) neutral vector ω; (c) isovector vector $\vec{\rho}$; (d) isovector pseudoscalar $\vec{\pi}$.

(d) $\quad ig_\pi \bar{\psi}\gamma_5 \vec{\tau}\psi \cdot \vec{\pi} \quad$ $$\frac{g_\pi^2}{4\pi c^2}\left(\frac{m_\pi}{2M}\right)^2 \frac{1}{3}\vec{\tau}_1 \cdot \vec{\tau}_2 \times \left\{ \vec{\sigma}_1 \cdot \vec{\sigma}_2 + S_{12}\left[1 + \frac{3}{m_\pi r} + \frac{3}{(m_\pi r)^2}\right]\right\} \frac{e^{-m_\pi r}}{r}$$

Several comments are relevant here:

- Note the presence of the tensor interaction proportional to S_{12} in the one-pion exchange potential (d).
- The neutral vector meson result (b) assumes that the ω_μ is coupled to the conserved baryon current. Then the interaction is just like that in QED, only with a finite photon mass.
- The pion potential in (d) actually yields a divergent Fourier transform. The one pion exchange potential (OPEP) (d) is obtained in coordinate space if this Fourier transform is interpreted in the following fashion

$$\frac{(\vec{\sigma}^{(1)} \cdot \vec{q})(\vec{\sigma}^{(2)} \cdot \vec{q})}{q^2 + m_\pi^2} \equiv \int e^{-i\vec{q}\cdot\vec{r}}[-(\vec{\sigma}^{(1)} \cdot \vec{\nabla})(\vec{\sigma}^{(2)} \cdot \vec{\nabla})]\frac{e^{-m_\pi r}}{4\pi r} d^3r \tag{A.10}$$

- Note that the OPEP is still singular at the origin.

B $b_\alpha^\dagger$ Is a Rank-j Irreducible Tensor Operator

The canonical transformation to particles and holes for the situation illustrated in Fig. 6.1 is defined in Eq. (6.3)

$$\begin{aligned} a_\alpha^\dagger &\equiv c_\alpha^\dagger & \alpha > F \\ b_\alpha^\dagger &\equiv S_{-\alpha} c_{-\alpha} & \alpha < F \end{aligned} \tag{B.11}$$

Here $|-\alpha\rangle = |nlsj, -m_j\rangle \equiv |a, -m_\alpha\rangle$. The goal of this appendix is to show that the hole creation operator $b_\alpha^\dagger$ is an irreducible tensor operator (Ref. [N44]) and hence properly creates an eigenstate of angular momentum. The additional phase $S_\alpha \equiv (-1)^{j_\alpha - m_\alpha} = -S_{-\alpha}$ (recall j_α is half-integral) is essential to the argument. The first step in the proof is the construction of the angular momentum operator

$$\begin{aligned} \hat{\vec{J}} &= \sum_{\alpha\beta} c_\alpha^\dagger \langle \alpha|\vec{J}|\beta\rangle c_\beta = \sum_{nljmm'>F} a_{nljm}^\dagger \langle jm|\vec{J}|jm'\rangle a_{nljm'} \\ &+ \sum_{nljmm'<F} \langle jm|\vec{J}|jm'\rangle (-1)^{j-m} b_{nlj-m} (-1)^{j-m'} b_{nlj-m'}^\dagger \end{aligned} \tag{B.12}$$

Now use the anticommutation relations $(-1)^{j-m} b_{nlj-m}(-1)^{j-m'} b_{nlj-m'}^\dagger = (-1)^{m-m'} [\delta_{mm'} - b_{nlj-m'}^\dagger b_{nlj-m}]$ and the Wigner-Eckart Theorem (Ref. [N44]), which implies

$$\begin{aligned} \sum_m \langle jm|\vec{J}|jm\rangle &= 0 \\ \langle jm|J_{1q}|jm'\rangle &= (-1)^{m'-m+1} \langle j-m'|J_{1q}|j-m\rangle \end{aligned} \tag{B.13}$$

A change of dummy summation indices then gives

$$\hat{\vec{J}} = \sum_{nljmm'>F} a^\dagger_{nljm}\langle jm|\vec{J}|jm'\rangle a_{nljm'} + \sum_{nljmm'<F} b^\dagger_{nljm}\langle jm|\vec{J}|jm'\rangle b_{nljm'} \tag{B.14}$$

One now demonstrates that $b^\dagger_\alpha = b^\dagger_{am_\alpha}$ is an ITO by considering the commutator

$$\begin{aligned}[\hat{\vec{J}}, b^\dagger_{am_\alpha}] &= \sum_{\beta\gamma}\langle\beta|\vec{J}|\gamma\rangle[b^\dagger_\beta b_\gamma, b^\dagger_\alpha] = \sum_{\beta\gamma}\langle\beta|\vec{J}|\gamma\rangle\delta_{\alpha\gamma}b^\dagger_\beta \\ &= \sum_{m'_\alpha}\langle j_\alpha m'_\alpha|\vec{J}|j_\alpha m_\alpha\rangle b^\dagger_{am'_\alpha}\end{aligned} \tag{B.15}$$

This is the definition of an ITO (Ref. [N44]).

These arguments can readily be extended to include isospin by taking

$$\begin{aligned}|\alpha\rangle &= |nl\frac{1}{2}jm_j;\frac{1}{2}m_t\rangle \\ |-\alpha\rangle &= |nl\frac{1}{2}j-m_j;\frac{1}{2}-m_t\rangle \equiv |a,-m_{j\alpha}-m_{t\alpha}\rangle \\ S_\alpha &= (-1)^{j_\alpha-m_{j\alpha}}(-1)^{\frac{1}{2}-m_{t\alpha}}\end{aligned} \tag{B.16}$$

Then the same proof shows $b^\dagger_\alpha$ is an ITO with respect to both angular momentum and isospin.

Part II

THE RELATIVISTIC NUCLEAR MANY-BODY PROBLEM

13

INTRODUCTION

Part II of this book is based on Ref. [R1], which contains a detailed list of background references; only those individual references directly relevant to the discussion in the text will be cited as we proceed. Relativistic quantum mechanics and relativistic quantum field theory form the basis for the discussion in this part of the book. The best introduction to this subject is still to be found in the texts by Bjorken and Drell (Refs. [R2, R3]).[1] There will be one change from Ref. [R1]; here the following metric will be employed (Ref. [R4])

$$\begin{aligned} x_\mu &= (\mathbf{x}, ix_0) = (\mathbf{x}, it) \\ a \cdot b &= \mathbf{a} \cdot \mathbf{b} - a_0 b_0 \end{aligned} \tag{13.1}$$

In this metric, the gamma matrices are hermitian, and satisfy

$$\gamma_\mu \gamma_\nu + \gamma_\nu \gamma_\mu = 2\delta_{\mu\nu} \tag{13.2}$$

The conversion algorithm to go from the metric of Bjorken and Drell to this metric is given in Table XII of Ref. [R1] and Appendix N of this book. In the text we also employ units where $\hbar = c = 1$.

Motivation. The traditional many-body problem, as described in Part I and Ref. [N2], starts from static two-body potentials fit to two-body scattering and bound-state data. These potentials are inserted in the many-body Schrödinger equation and that equation is then solved under certain approximations; the few-body problem can now be solved exactly using modern computing techniques. Electroweak currents are then constructed from the properties of free nucleons, as in Section 8, and used to probe the properties of the system. Although this approach to nuclear physics has had a great many successes, as we have seen, it is clearly inadequate for obtaining a more detailed understanding of the nuclear system.

A more appropriate set of degrees of freedom for nuclear physics consists of the *hadrons*, the strongly interacting mesons and baryons. There are many arguments one can give for this: the long-range part of the most successful nucleon-nucleon potentials (see Section 1) consists of the exchange of mesons, including $\pi\,(0^-,1), \sigma\,(0^+,0), \rho\,(1^-,1), \omega\,(1^-,0)$; in addition, electron scattering from nuclei provides unambiguous evidence for the presence of exchange currents

[1] See also Ref. [R103].

in nuclei — these are additional electromagnetic currents arising from the flow of charged mesons back and forth between the nucleons; furthermore, mesons are produced from nuclei in copious amounts every day at meson factories such as LAMPF, SIN, and TRIUMF.

A current goal of nuclear physics is to describe the properties of nuclear matter under extreme conditions. The properties at high temperature and high density govern the behavior of astrophysical phenomena such as supernovas and neutron stars; these high-temperature, high-density properties will be probed through relativistic heavy-ion reactions at the Relativistic Heavy Ion Collider (RHIC). At CEBAF the response of the nucleus at high momentum transfers will be measured — momentum transfers much larger than the mass of the nucleon itself. In developing a theoretical framework that allows one to extrapolate from the known properties of terrestrial nuclei to these extreme conditions, it is essential to incorporate general principles of physics such as quantum mechanics, special relativity, and microscopic causality.

The only consistent theoretical framework we have for describing such a relativistic, interacting, many-body system is *relativistic quantum field theory based on a local lagrangian density.* In analogy with quantum electrodynamics (QED), it is convenient to refer to local relativistic quantum field theories based on hadronic degrees of freedom as quantum hadrodynamics (QHD).[2]

As with QED, we shall also concentrate on renormalizable field theories. Once the coupling constants and masses have been determined from experiment, such theories then provide a consistent framework in which one can in principle calculate unambiguous theoretical results to compare with experimental data. The theoretical results may, or may not, correctly describe the data; however, that is now a question to be answered by the comparison between calculation and experiment. It is hoped that the low-energy, large-distance properties of such a theory can be used to successfully describe much of nuclear physics. In the end, the requirement of renormalizability simply limits the class of local lagrangian densities that will be considered here.

Viewed as an effective low-energy limit of the underlying theory of quantum chromodynamics (QCD) describing hadrons as composites of interacting quarks and gluons, the above requirements may be too restrictive. We will return to such questions at the end of Part II; however, while the details of the hadronic lagrangian may change, the theoretical techniques employed, and much of the phenomenology, will not. In particular, as we shall see, there is compelling evidence that whatever the effective field theory for low-energy, large-distance QCD may be in the nucleus, it must be dominated by linear, isoscalar, scalar, and vector interactions. We proceed to a simple model problem in QHD.

[2]See Ref. [R98] for an alternative approach.

14

A SIMPLE MODEL WITH (σ, ω) AND MEAN FIELD THEORY

We start with the following fields

- A baryon field for the neutrons and protons

$$\psi = \begin{pmatrix} p \\ n \end{pmatrix} \tag{14.1}$$

- A neutral scalar field ϕ coupled to the scalar density $\bar{\psi}\psi$
- A neutral vector field V_λ coupled to the conserved baryon current $i\bar{\psi}\gamma_\lambda\psi$.

The choice is motivated by several considerations. First, we want to describe the bulk properties of nuclear matter; these fields and couplings provide the smoothest average nuclear interactions,[3] and as such, should describe the dominant features of the bulk properties. Second, large neutral scalar and vector contributions are observed empirically in a Lorentz-invariant analysis of the free nucleon-nucleon scattering amplitude — they dominate the amplitude. Third, as shown in Appendix A, in the static limit of infinitely heavy baryon sources (which will *not* be assumed in the subsequent discussion), these exchanges give rise to an effective nucleon-nucleon interaction of the form

$$V_{\text{static}} = \frac{g_v^2}{4\pi}\frac{e^{-m_v r}}{r} - \frac{g_s^2}{4\pi}\frac{e^{-m_s r}}{r} \tag{14.2}$$

With appropriate choices of coupling constants and masses this potential describes the main qualitative features of the nucleon-nucleon interaction: a short-range repulsion between baryons coming from ω exchange, and a long-range attraction coming from σ exchange.

Lagrangian. The lagrangian density for this system of fields and couplings is given by

$$\begin{aligned} \mathcal{L} &= -\frac{1}{4}F_{\mu\nu}F_{\mu\nu} - \frac{1}{2}m_v^2 V_\mu^2 - \frac{1}{2}\left[\left(\frac{\partial\phi}{\partial x_\mu}\right)^2 + m_s^2\phi^2\right] \\ &\quad -\bar{\psi}\left[\gamma_\mu\left(\frac{\partial}{\partial x_\mu} - ig_v V_\mu\right) + (M - g_s\phi)\right]\psi \end{aligned} \tag{14.3}$$

[3] Spin- and isospin-dependent interactions tend to average out in nuclear matter.

Here, as in QED, the field tensor is defined by

$$F_{\mu\nu} \equiv \frac{\partial V_\nu}{\partial x_\mu} - \frac{\partial V_\mu}{\partial x_\nu} \tag{14.4}$$

The field equations are derived from Hamilton's principle (Ref. [R5])

$$\delta \int \mathcal{L}\left(q, \frac{\partial q}{\partial x_\mu}\right) d^4x = 0 \tag{14.5}$$

Here q is any field variable. Lagrange's equations follow as

$$\frac{\partial}{\partial x_\mu} \frac{\partial \mathcal{L}}{\partial(\partial q/\partial x_\mu)} - \frac{\partial \mathcal{L}}{\partial q} = 0 \tag{14.6}$$

Taken in turn for the field variables $\{V_\mu, \phi, \bar{\psi}\}$ these give rise to the field equations

$$\begin{aligned} \frac{\partial}{\partial x_\nu} F_{\mu\nu} + m_v^2 V_\mu &= i g_v \bar{\psi} \gamma_\mu \psi \\ \left[\left(\frac{\partial}{\partial x_\mu}\right)^2 - m_s^2\right] \phi &= -g_s \bar{\psi} \psi \\ \left[\gamma_\mu \left(\frac{\partial}{\partial x_\mu} - i g_v V_\mu\right) + (M - g_s \phi)\right] \psi &= 0 \end{aligned} \tag{14.7}$$

The field equation for $\bar{\psi} \equiv \psi^\dagger \gamma_4$ follows from the adjoint of the last relation. The first of these field equations is just the relativistic form of Maxwell's equations with massive quanta and a *conserved baryon current* as source

$$B_\mu = i \bar{\psi} \gamma_\mu \psi \tag{14.8}$$

The second field equation is the Klein-Gordon equation for the scalar field with the baryon scalar density $\bar{\psi}\psi$ as source. The third field equation is the Dirac equation for the baryon field with the meson fields V_λ and ϕ included in a "minimal" fashion.

The stress tensor in continuum mechanics is given by[4]

$$T_{\mu\nu} \equiv \mathcal{L} \delta_{\mu\nu} - \frac{\partial \mathcal{L}}{\partial(\partial q/\partial x_\mu)} \frac{\partial q}{\partial x_\nu} \tag{14.9}$$

For a uniform system in equilibrium at rest the expectation value of the stress tensor must take the form (Ref. [R6])

$$\langle \hat{T}_{\mu\nu} \rangle = p\, \delta_{\mu\nu} + (p + \varepsilon) u_\mu u_\nu \tag{14.10}$$

Here p is the pressure, ε is the energy density, and $u_\mu = (\mathbf{0}, i)$ is the four-velocity of the fluid.

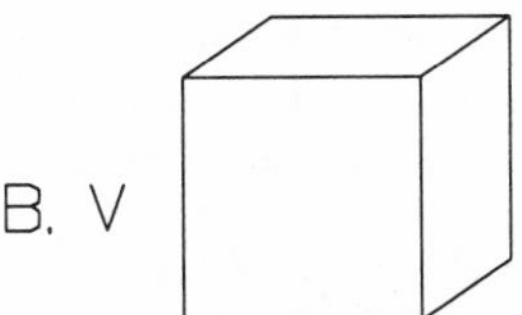

Figure 14.1: A large box of volume V filled uniformly with B baryons.

Mean Field Theory (MFT). Consider a large box of volume V filled uniformly with B baryons (Fig. 14.1). Since baryon number is conserved, so is the baryon density $\rho_{\rm B} \equiv B/V$. Now decrease the size of the box. As the baryon density gets large, so do the source terms on the right-hand side of the meson field equations in Eqs. (14.7). When the source gets strong, and there are many quanta present, one can attempt to replace the meson fields by classical fields and the sources by their expectation values — as in the theory of electromagnetism. In the limit of large $\rho_{\rm B}$ we replace

$$\begin{aligned} \hat{\phi} \rightarrow \langle \hat{\phi} \rangle &= \phi_0 \\ \hat{V}_\lambda \rightarrow \langle \hat{V}_\lambda \rangle &= i\delta_{\lambda 4} V_0 \end{aligned} \tag{14.11}$$

The vector field can develop only a fourth component since there is no spatial direction in the problem for a uniform system at rest. Furthermore, for a uniform system at rest the classical fields ϕ_0 and V_0 must be *constants* independent of space and time, which greatly simplifies the remaining problem. For example, the vector meson field equations in Eqs. (14.7) now reduce to the form

$$V_0 = \frac{g_v}{m_v^2} \rho_{\rm B} \tag{14.12}$$

Here the expectation value of the baryon current has been written as $B_\mu = (\mathbf{0}, i\rho_{\rm B})$. The classical vector meson field is thus given in terms of *conserved* quantities. We refer to this simplification of the description of the full interacting quantum system as mean field theory (MFT).

Lagrangian Density in MFT. The substitution of the constant fields ϕ_0, V_0 into the lagrangian density in Eq. (14.3) reduces it to the form

$$\mathcal{L}_{\rm MFT} = \frac{1}{2} m_v^2 V_0^2 - \frac{1}{2} m_s^2 \phi_0^2 - \bar{\psi}[\gamma_\mu \frac{\partial}{\partial x_\mu} + \gamma_4 g_v V_0 + M^*]\psi \tag{14.13}$$

Here the *effective mass* of the nucleon is defined by

$$M^* \equiv M - g_s \phi_0 \tag{14.14}$$

[4]Note the hamiltonian density is $\mathcal{H} = -T_{44} = \Pi_q(\partial q/\partial t) - \mathcal{L}$ with $\Pi_q = \partial \mathcal{L}/\partial(\partial q/\partial t)$.

Dirac Equation. If one looks for solutions of the form $\psi = \mathcal{U}(\mathbf{p})\exp(i\mathbf{p}\cdot\mathbf{x} - iEt)$, then the Dirac equation for the baryons in the constant fields ϕ_0, V_0 corresponding to the last of Eqs. (14.7) takes the form

$$(\boldsymbol{\alpha}\cdot\mathbf{p} + \beta M^*)\mathcal{U}(\mathbf{p}) = (E - g_v V_0)\mathcal{U}(\mathbf{p}) \tag{14.15}$$

Repeated application of this relation and use of the properties of the Dirac matrices $\boldsymbol{\alpha}, \beta$ yields the eigenvalue equation

$$E = g_v V_0 \pm (\mathbf{p}^2 + M^{*2})^{1/2} \tag{14.16}$$

We will refer to these eigenvalues as $E_\pm$. Note that familiar manipulations of the Dirac Eq. (14.15) lead to the relation

$$\bar{\mathcal{U}}(\mathbf{p})\mathcal{U}(\mathbf{p}) = \frac{M^*}{(\mathbf{p}^2 + M^{*2})^{1/2}}\mathcal{U}^\dagger(\mathbf{p})\mathcal{U}(\mathbf{p}) \tag{14.17}$$

We normalize our Dirac spinors to $\mathcal{U}^\dagger\mathcal{U} = 1$ (corresponding to unit baryon density in the laboratory frame).

Field Expansion. In the Schrödinger picture the baryon field operator can be expanded in terms of the complete set of solutions to the Dirac equation according to

$$\hat{\psi}(\mathbf{x}) = \frac{1}{\sqrt{V}}\sum_{\mathbf{k}\lambda}\left[\mathcal{U}(\mathbf{k}\lambda)A_{\mathbf{k}\lambda}e^{i\mathbf{k}\cdot\mathbf{x}} + \mathcal{V}(-\mathbf{k}\lambda)B^\dagger_{\mathbf{k}\lambda}e^{-i\mathbf{k}\cdot\mathbf{x}}\right] \tag{14.18}$$

The solution $\mathcal{V}$ corresponds to E_-. The theory is now quantized by imposing the equal time anticommutation relations

$$\begin{aligned}\left\{A_{\mathbf{k}\lambda}, A^\dagger_{\mathbf{k}'\lambda'}\right\} &= \delta_{\mathbf{k},\mathbf{k}'}\delta_{\lambda,\lambda'} \\ \left\{B_{\mathbf{k}\lambda}, B^\dagger_{\mathbf{k}'\lambda'}\right\} &= \delta_{\mathbf{k},\mathbf{k}'}\delta_{\lambda,\lambda'}\end{aligned} \tag{14.19}$$

Everything else anticommutes.

Hamiltonian Density. Insertion of this field expansion in the hamiltonian density derived from Eq. (14.13), use of the orthonormality of the wave functions, and use of the canonical anticommutation relations results in an expression of the form[5]

$$\hat{\mathcal{H}} = \hat{\mathcal{H}}_{\mathrm{MFT}} + \delta\mathcal{H} \tag{14.20}$$

Here the mean field theory hamiltonian is given by

$$\hat{\mathcal{H}}_{\mathrm{MFT}} = \frac{1}{2}m_s^2\phi_0^2 - \frac{1}{2}m_v^2V_0^2 + g_vV_0\hat{\rho}_{\mathrm{B}} +$$

[5]These manipulations are formally identical to those carried out in the free field case (Prob. 14.2).

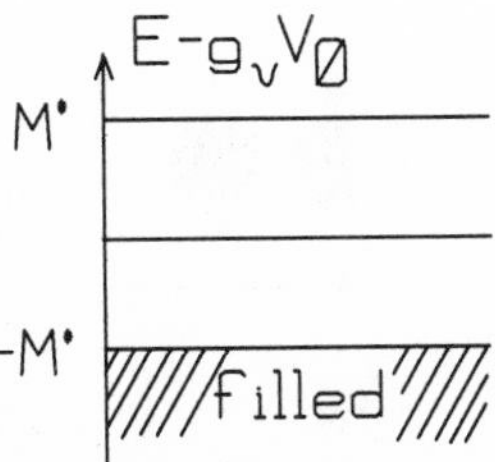

Figure 14.2: Filled negative energy Fermi sea of baryons in mean field theory.

$$
\begin{aligned}
& +\frac{1}{V}\sum_{\mathbf{k}\lambda}(\mathbf{k}^2 + M^{*2})^{1/2}(A^{\dagger}_{\mathbf{k}\lambda}A_{\mathbf{k}\lambda} + B^{\dagger}_{\mathbf{k}\lambda}B_{\mathbf{k}\lambda}) \\
\hat{\rho}_{\mathrm{B}} \quad = \quad & \frac{1}{V}\sum_{\mathbf{k}\lambda}(A^{\dagger}_{\mathbf{k}\lambda}A_{\mathbf{k}\lambda} - B^{\dagger}_{\mathbf{k}\lambda}B_{\mathbf{k}\lambda})
\end{aligned}
\tag{14.21}
$$

The additional term $\delta\mathcal{H}$ is defined by

$$
\delta\mathcal{H} \equiv -\frac{1}{V}\sum_{\mathbf{k}\lambda}[(\mathbf{k}^2 + M^{*2})^{1/2} - (\mathbf{k}^2 + M^2)^{1/2}] \tag{14.22}
$$

This is the zero-point energy; it represents the difference of energy of a filled negative energy Fermi sea of baryons with mass M^* and that of a filled negative energy Fermi sea of baryons of mass M (see Fig. 14.2). Its presence is familiar from Dirac hole theory. The baryon density appearing in Eq. (14.21) is the normal-ordered expression

$$
\begin{aligned}
\hat{\rho}_{\mathrm{B}} \quad & \equiv \quad \hat{\psi}^{\dagger}(\mathbf{x})\hat{\psi}(\mathbf{x}) - \langle 0|\hat{\psi}^{\dagger}(\mathbf{x})\hat{\psi}(\mathbf{x})|0\rangle \\
& = \quad \hat{\psi}^{\dagger}(\mathbf{x})\hat{\psi}(\mathbf{x}) - \frac{1}{V}\sum\nolimits_{\mathbf{k}\lambda} 1 \\
& \equiv \quad :\hat{\psi}^{\dagger}(\mathbf{x})\hat{\psi}(\mathbf{x}):
\end{aligned}
\tag{14.23}
$$

The vacuum expectation value subtracted off in the second line simply counts the number of filled states in the negative energy Dirac sea; it is independent of the interaction. The baryon density now counts the number of baryons minus the number of antibaryons relative to the vacuum.

Since both $\hat{\mathcal{H}}_{\mathrm{MFT}}$ and $\hat{\rho}_{\mathrm{B}}$ are now diagonal operators, the mean field theory has been *solved exactly*. The qualitative argument given at the beginning of this section indicates that mean field theory can be expected to be correct in the limit $\rho_{\mathrm{B}} \to \infty$.

Nuclear Matter. It is evident from Eq. (14.21) that the ground state of nuclear matter in the MFT is obtained by filling levels up to k_{F} with $\{p\uparrow, p\downarrow, n\uparrow, n\downarrow\}$; the degeneracy factor for nuclear matter is thus $\gamma = 4$ (Fig. 14.3). The energy density $\varepsilon \equiv E/V$ and baryon density obtained from the expectation

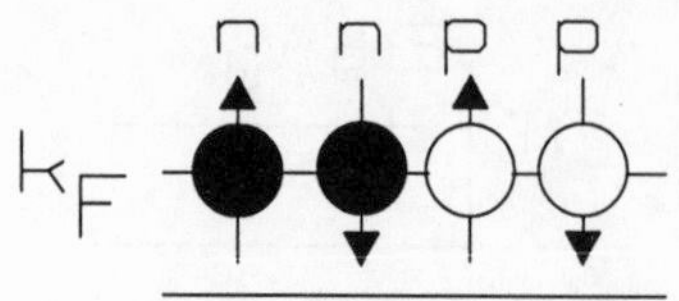

Figure 14.3: Ground state of nuclear matter in the MFT. The quantity γ is the spin-isospin degeneracy factor; here, for nuclear matter, $\gamma = 4$.

value of Eqs. (14.21) in this state, and the use of Eqs. (14.12) and (14.14), provide a parametric equation of state for nuclear matter in MFT[6]

$$\begin{aligned}
\varepsilon(\rho_{\rm B};\phi_0) &= \frac{g_v^2}{2m_v^2}\rho_{\rm B}^2 + \frac{m_s^2}{2g_s^2}(M-M^*)^2 + \frac{\gamma}{(2\pi)^3}\int_0^{k_{\rm F}} d^3k(\mathbf{k}^2+M^{*2})^{1/2} \\
p(\rho_{\rm B};\phi_0) &= \frac{g_v^2}{2m_v^2}\rho_{\rm B}^2 - \frac{m_s^2}{2g_s^2}(M-M^*)^2 + \frac{\gamma}{3(2\pi)^3}\int_0^{k_{\rm F}} d^3k\frac{\mathbf{k}^2}{(\mathbf{k}^2+M^{*2})^{1/2}} \\
\rho_{\rm B} &= \frac{\gamma}{(2\pi)^3}\int_0^{k_{\rm F}} d^3k = \frac{\gamma}{6\pi^2}k_{\rm F}^3
\end{aligned} \tag{14.24}$$

The value of the classical condensed scalar field $\phi_0 = (M - M^*)/g_s$ can be determined with the aid of thermodynamics. At a fixed volume and baryon number (V, B), the system will minimize its energy

$$\left(\frac{\partial E}{\partial \phi_0}\right)_{V,B} = 0 \tag{14.25}$$

Differentiation of the first of Eqs. (14.24) then gives

$$\begin{aligned}
\phi_0 &= \frac{g_s}{m_s^2}\rho_s \\
\rho_s &\equiv \frac{\gamma}{(2\pi)^3}\int_0^{k_{\rm F}} d^3k\frac{M^*}{(\mathbf{k}^2+M^{*2})^{1/2}}
\end{aligned} \tag{14.26}$$

Here the scalar density ρ_s is defined by the second equation. These relations provide a *self-consistency* equation for the scalar field, which appears under the integral in the form $M^* = M - g_s\phi_0$. They are recognized as the scalar meson field Eq. (14.7) in MFT where the source term is obtained by summing the scalar density in Eq. (14.17) over the occupied levels. Note that the scalar density ρ_s is damped at large $k_{\rm F}$ relative to the baryon density $\rho_{\rm B}$ due to Lorentz contraction [Eq. (14.17)].

[6] The expression for the pressure is derived in Appendix C.

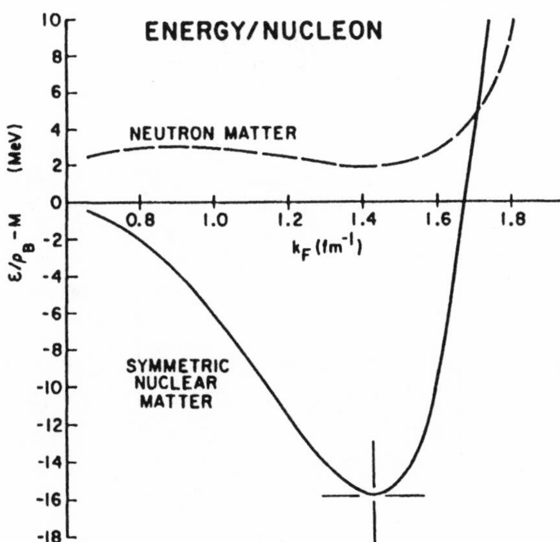

Figure 14.4: Saturation curve for nuclear matter. These results are calculated in the relativistic MFT with baryons and neutral scalar and vector mesons. The coupling constants are chosen to fit the value and position of the minimum. The prediction for neutron matter ($\gamma = 2$) is also shown. From Ref. [R1].

There are only two parameters in this MFT of nuclear matter. We choose to fit them to the two experimentally accessible properties of uniform nuclear matter, the binding energy and density (see Fig. 14.4); this yields

$$C_s^2 \equiv g_s^2 \left(\frac{M^2}{m_s^2}\right) = 267.1 \qquad\qquad C_v^2 \equiv g_v^2 \left(\frac{M^2}{m_v^2}\right) = 195.9 \qquad (14.27)$$

The mechanism for saturation in this relativistic MFT is the repulsion between like baryons and the damping of the scalar meson attraction with increasing baryon density. As shown in Section 3, a Hartree-Fock variational calculation with the static potential of Eq. (14.2) demonstrates that the corresponding non-relativistic many-body sytem is unstable against collapse. Thus, even though the binding energy is small compared to the nucleon mass, saturation here is entirely a relativistic phenomenon.

The solution to the self-consistency equation for M^* as a function of density is shown in Fig. 14.5. Note that at nuclear matter saturation density $M^*/M = 0.56$, and we clearly have a *new energy scale* in this problem as the scalar field energy is of the same order as the nucleon mass itself (see r.h.s. of Fig. 14.5).

Note that while the scalar meson density $\bar{\psi}\psi$ is the simplest thing one can write down relativistically, its nonrelativistic limit is complicated, and corre-

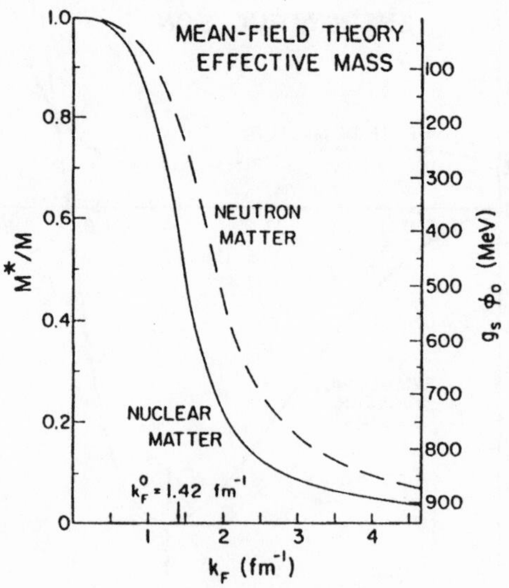

Figure 14.5: Effective mass as a function of density for nuclear ($\gamma = 4$) and neutron ($\gamma = 2$) matter based on Fig. 14.4. From Ref. [R1].

sponds to an infinite series of velocity-dependent terms, since [cf. Eq. (14.17)]

$$\frac{M^*}{(\mathbf{p}^2 + M^{*2})^{1/2}} = 1 - \frac{1}{2}\frac{\mathbf{p}^2}{M^{*2}} + \frac{3}{8}\frac{\mathbf{p}^4}{M^{*4}} + \cdots \tag{14.28}$$

Applications. All other properties of nuclear and neutron matter are now predicted, for example:

Neutron Matter Equation of State. Neutron matter in the MFT is obtained from the analysis of nuclear matter by simply replacing $\gamma = 4$ by $\gamma = 2$. Neutron matter is unbound (Fig. 14.4). The equation of state p vs. ε for neutron matter is shown in Fig. 14.6. Note the approach from below to the causal limit $\varepsilon = p$ (where $c_{\text{sound}} = c_{\text{light}}$) at high density. There is a phase separation in this model, similar to the gas-liquid transition in the van der Waal's equation of state.[7]

Neutron Star Mass vs. Central Density. Insertion of the neutron matter equation of state in the Tolman Oppenheimer Volkoff equations for a static spherically symmetric metric in general relativity (Ref. [R6]) allows one to compute the mass of a neutron star as a function of the central density (Fig. 14.7). One finds for the maximum mass of a neutron star $(M/M_{\odot})_{\text{max}} = 2.57$ with this equation of state; it is about as "stiff" as one can get and still be consistent with causality and the saturation properties of nuclear matter.

[7]The properties of the two phases of neutron matter are here determined by a Maxwell construction.

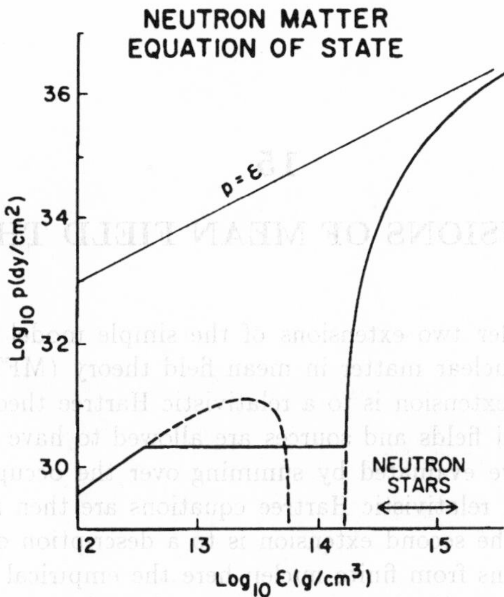

Figure 14.6: Predicted equation of state for neutron matter at all densities based on Fig. 14.4. A Maxwell construction is used to determine the equilibrium curve in the region of the phase transition. The density regime relevant to neutron stars is also indicated. From Ref. [R1].

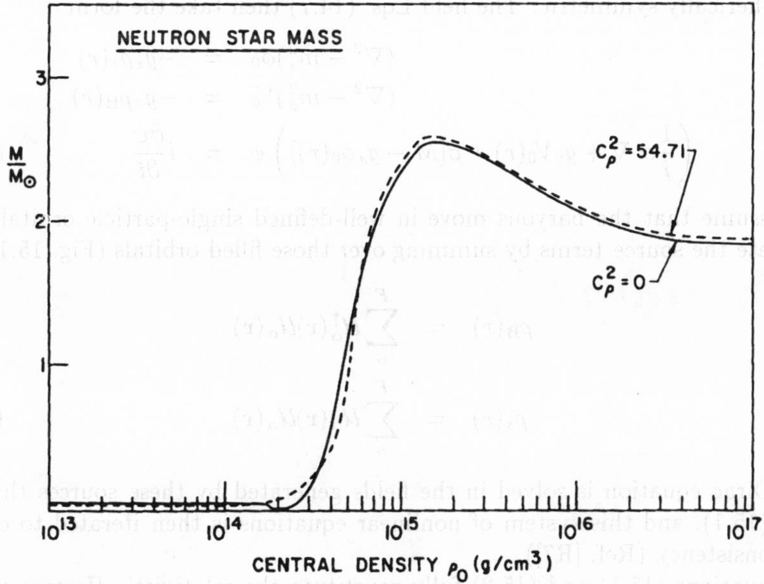

Figure 14.7: Calculated neutron star mass in units of the solar mass $M_\odot$ as a function of the central density based on Fig. 14.6 (solid curve). From Ref. [R1].

15

EXTENSIONS OF MEAN FIELD THEORY

We next consider two extensions of the simple model with (σ, ω) meson fields applied to nuclear matter in mean field theory (MFT) in the previous section. The first extension is to a relativistic Hartree theory of finite nuclei. Here the condensed fields and sources are allowed to have a spatial variation and the sources are evaluated by summing over the occupied orbitals. The coupled, nonlinear, relativistic Hartree equations are then solved by iteration (Ref. [R7, R1]). The second extension is to a description of the scattering of high energy nucleons from finite nuclei; here the empirical covariant nucleon-nucleon scattering amplitude and the Hartree densities are used to construct a Dirac optical potential from which the scattering amplitude is then generated (Refs. [R8, R9, R10, R11, R12, R1]).

Relativistic Hartree Theory of Finite Nuclei. In static finite systems, the condensed scalar and vector fields of the previous section and their source terms will acquire spatial variations. Closed shells are assumed so that these quantities are spherically symmetric. The field Eqs. (14.7) then take the form

$$\begin{aligned}
(\nabla^2 - m_s^2)\phi_0 &= -g_s\rho_s(r) \\
(\nabla^2 - m_v^2)V_0 &= -g_v\rho_B(r) \\
\left(\frac{1}{i}\boldsymbol{\alpha}\cdot\nabla + g_v V_0(r) + \beta[M - g_s\phi_0(r)]\right)\psi &= i\frac{\partial\psi}{\partial t}
\end{aligned} \qquad (15.1)$$

We assume that the baryons move in well-defined single-particle orbitals and evaluate the source terms by summing over those filled orbitals (Fig. 15.1).

$$\begin{aligned}
\rho_B(r) &= \sum_{\alpha}^{F} \mathcal{U}_\alpha^\dagger(\mathbf{r})\mathcal{U}_\alpha(\mathbf{r}) \\
\rho_s(r) &= \sum_{\alpha}^{F} \bar{\mathcal{U}}_\alpha(\mathbf{r})\mathcal{U}_\alpha(\mathbf{r})
\end{aligned} \qquad (15.2)$$

The Dirac equation is solved in the fields generated by these sources through Eqs. (15.1), and this system of nonlinear equations is then iterated to obtain self-consistency (Ref. [R7]).

Equations (15.1) and (15.2) fully constitute the relativistic Hartree theory in this spherically symmetric case. The reduction to nonlinear, coupled, one-dimensional radial differential equations is carried out in detail in Ref. [R1].

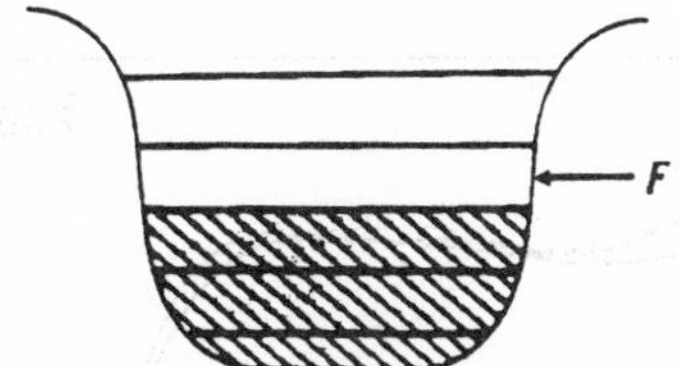

Figure 15.1: Filled orbitals in relativistic Hartree theory of finite nuclei.

While invaluable for numerical calculations, it serves no purpose to just reproduce that reduction here. Instead, we have included Probs. 15.1-15.5, which take the reader through that derivation and summarize the results. The generation of relativistic Hartree wave functions is an extremely useful tool for every nuclear physicist; fortunately, a computer program for solving the relativistic Hartree equations is now available for general distribution in Ref. [R13].

To make the model more realistic for application to nuclei with $N \neq Z$, and in anticipation of the broader formulation of QHD in Section 25, additional condensed fields are included here for the neutral $\rho_0^0(r)$ meson coupled to the isovector baryon density $g_\rho \psi^\dagger(\tau_3/2)\psi$, and for the electromagnetic field (Coulomb potential) $A_0(r)$. There are four parameters in this analysis $\{g_s, g_v, g_\rho, m_s\}$ that must be determined from experiment. The authors in Ref. [R7] choose to fit the following quantities $\{(E/B)_{\text{n.m.}}, (k_F)_{\text{n.m.}}, (a_4)_{\text{n.m.}}, (\sqrt{\langle r^2 \rangle})_{^{40}\text{Ca}}\}$— the binding energy, density, and symmetry energy of nuclear matter, and the root mean square radius of ^{40}Ca. The masses $\{m_v \equiv m_\omega, m_\rho\}$ are fixed at their observed values. The ground state properties of all nuclei are then determined.

Nuclear Charge Densities. Resulting nuclear charge densities are shown in Figs. 15.2,15.3, and 15.4 and compared with the results obtained from elastic electron scattering. The central density in ^{208}Pb determines $(k_F)_{\text{n.m.}}$; the mean square radius of ^{40}Ca is fit; the charge density of ^{16}O is then obtained for free. The quality of the description of the charge densities is comparable to the most sophisticated nonrelativistic density-dependent Hartree-Fock calculations that exist (also illustrated in these figures).

Single-Particle Spectrum. Figure 15.5 shows the calculated Hartree single-particle spectrum for ^{208}Pb. One obtains all the shell closures of the nuclear shell model (Section 6). Just as in atomic physics, a Dirac particle moving in spatially varying fields $\phi_0(r)$ and $V_0(r)$ will exhibit a spin-orbit splitting. Whereas the large effects of these condensed fields *cancel* in the total binding energy of the nucleus, they *add* with the correct sign in the spin-orbit interaction (see Ref. [R1]), bringing it to a level consistent with the empirical nuclear shell model. *One can thus, with this relativistic Hartree theory, derive the nuclear shell model by fitting only a few bulk properties of nuclear matter.*[8]

[8]References [R99, R100] summarize more extensive descriptions within this framework.

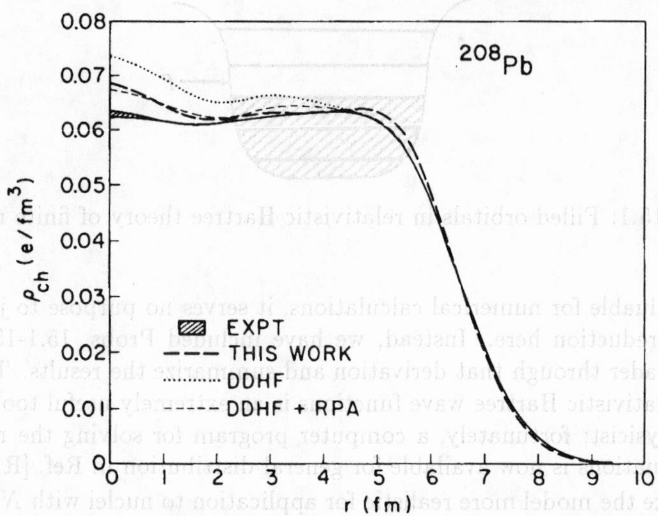

Figure 15.2: Charge density for ^{208}Pb. The solid curve and shaded area represent the fit to experimental data. Relativistic results are indicated by the long dashed lines (Refs. [R7, R1]). Taken from Ref. [R1].

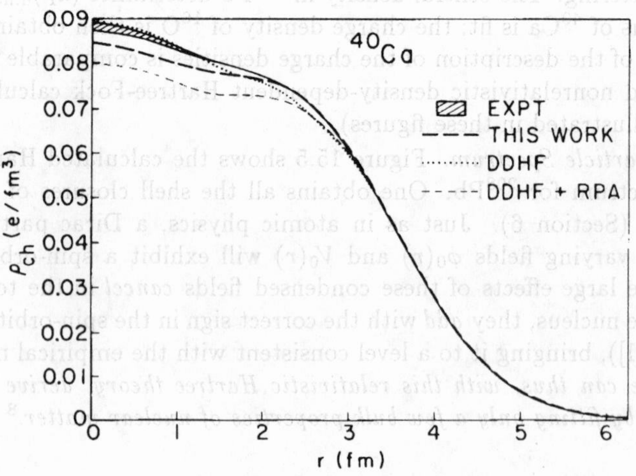

Figure 15.3: Same as Fig. 15.2 for ^{40}Ca.

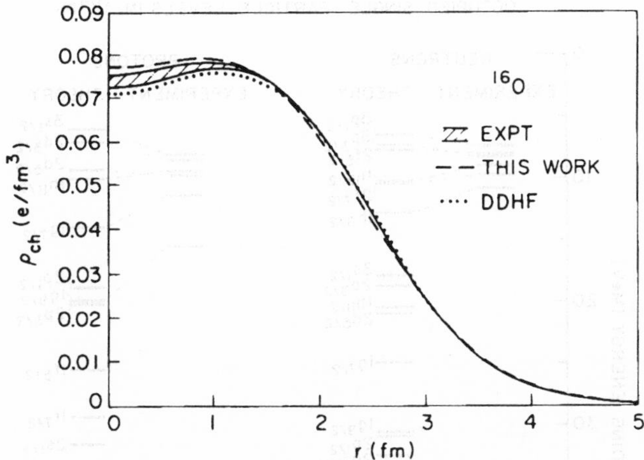

Figure 15.4: Same as Fig. 15.2 for ^{16}O.

Nucleon Scattering. This analysis can be extended to a description of those nucleons lying *above* the Fermi surface involved in scattering from the nucleus. The basic idea is to construct a Dirac optical potential from the relativistic Hartree densities and N-N scattering amplitude. The Dirac equation is then solved in this optical potential to generate the scattering amplitude.

The details of this construction are discussed in Ref. [R1] and in the original source Refs. [R8, R9, R10, R11, R12]. Simply reproducing those details here will take us too far afield from our principal purpose. The optical potential does, however, play a central role in nuclear physics, and it will play a crucial role in the interpretation of coincident $(e, e'X)$ experiments at CEBAF as seen in Section 12. Thus it is important for every nuclear physicist to have an understanding of it. The most thorough treatment of the subject of nuclear reactions is contained in Volume II of deShalit and Feshbach (Ref. [N3]), and the dedicated reader is referred to that reference. We shall be content here in Probs. 15.6-15.10 to guide the reader through a simple nonrelativistic model where the multiple scattering amplitude can be found exactly and the approximations in the optical potential clearly identified.

The (σ, ω) model is too simple to describe the detailed spin dependence of the nucleon-nucleon $(N-N)$ scattering. We therefore compromise and take the $N-N$ scattering amplitude

$$f_{NN} = f_s 1^{(1)} \cdot 1^{(2)} + f_v \gamma_\mu^{(1)} \cdot \gamma_\mu^{(2)} + \cdots \tag{15.3}$$

from experiment. Here the amplitude is written explicitly in terms of Lorentz invariant quantities. It is an empirical fact that when written in this fashion the scalar and vector terms are very large (several hundered MeV). The relativistic Hartree densities generated previously can be folded with the empirical

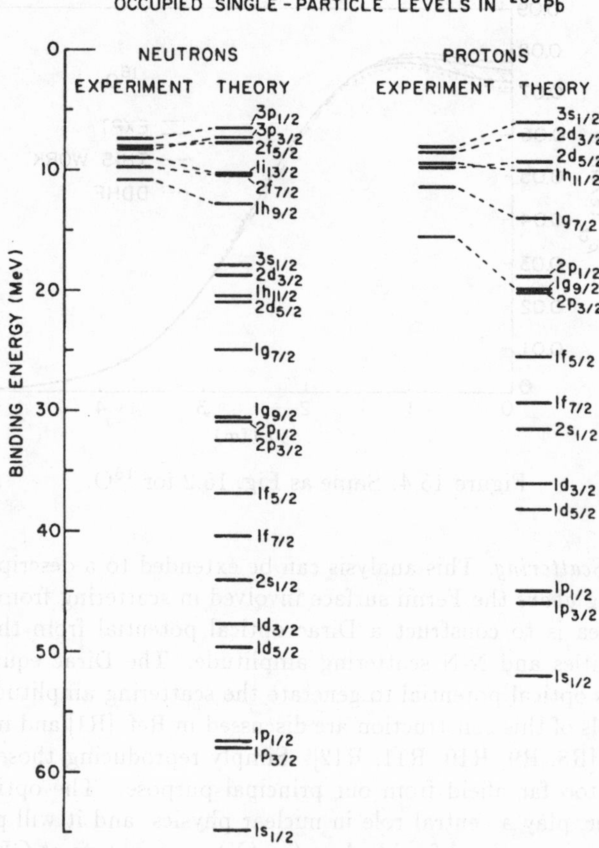

Figure 15.5: Predicted spectrum for occupied levels in ^{208}Pb. Experimental levels are from neighboring nuclei (Refs. [R7, R1]). Taken from Ref. [R1].

scattering amplitude to generate a Dirac optical potential (Prob. 15.10)

$$U_{\rm opt}(x) = -4\pi A \int \frac{d^3\Delta}{(2\pi)^3} e^{i\boldsymbol{\Delta}\cdot\mathbf{x}} \langle f_{NN}(\boldsymbol{\Delta})\tilde{\rho}^{(1)}(\Delta)\rangle \qquad (15.4)$$

This analysis is known as the "relativistic impulse approximation (RIA)" (Refs. [R8, R9, R10, R11, R12, R1]). One can then solve the Dirac equation for scattering in this potential. The results are illustrated in Fig. 15.6. *It is evident that the main features of nucleon-nucleus scattering, in particular the spin dependence, can be understood within this same relativistic framework.*

The source of the spin dependence responsible for these spin observables is again the interaction of a Dirac particle with spatially varying condensed scalar and vector fields; it is the same source of spin dependence as in the shell model derived above.

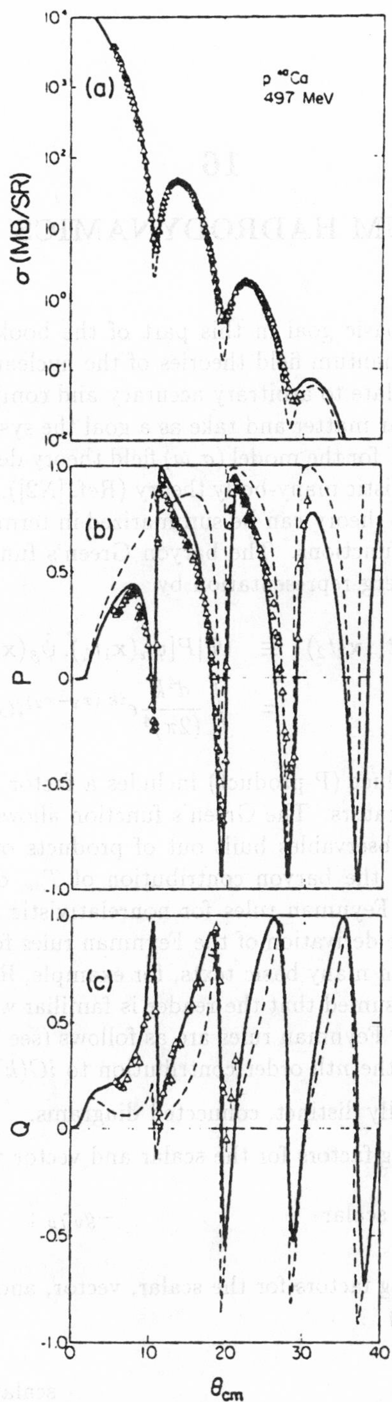

Figure 15.6: Calculated cross section, analyzing power, and spin rotation function for $\vec{p} + {}^{40}\mathrm{Ca}$ at $T_L = 497$ MeV using the Dirac (relativistic) impulse approximation (RIA, solid curve). (Figure prepared by Prof. B. C. Clark, from Ref. [R1].)

16

QUANTUM HADRODYNAMICS (QHD-I)

Motivation. The basic goal in this part of the book is to develop model hadronic relativistic quantum field theories of the nuclear system in which one can, in principle, calculate to arbitrary accuracy and compare with experiment. Let us return to nuclear matter and take as a goal the systematic calculation of the corrections to MFT for the model (σ, ω) field theory developed in Section 14. Just as with nonrelativistic many-body theory (Ref. [N2]), the content of the full relativistic many-body theory can be summarized in terms of a set of Feynman rules for the Green's functions. The baryon Green's function, for example, is defined in the Heisenberg representation by

$$\begin{aligned} iG_{\alpha\beta}(\mathbf{x}_1 t_1, \mathbf{x}_2 t_2) &\equiv \langle \Psi | P[\hat{\psi}_\alpha(\mathbf{x}_1 t_1), \hat{\bar{\psi}}_\beta(\mathbf{x}_2 t_2)] | \Psi \rangle \\ &= \int \frac{d^4k}{(2\pi)^4} e^{ik\cdot(x_1 - x_2)} iG_{\alpha\beta}(k) \end{aligned} \tag{16.1}$$

The time-ordered product (P-product) includes a factor of (-1) for the interchange of fermion operators. The Green's function allows one to calculate the expectation value of observables built out of products of field operators; this Green's function gives the baryon contribution of $\hat{T}_{\mu\nu}$ defined in Section 14. The derivation of the Feynman rules for nonrelativistic many body theory is given in Ref. [N2]. The derivation of the Feynman rules for relativistic theories is developed in detail in many basic texts, for example, Refs. [R1, R2, R3] (see also Ref. [R4]); it is assumed that the reader is familiar with this material. For the present theory, the Feynman rules are as follows (see Fig. 16.1):

Feynman rules for the nth order contribution to $iG(k)$:

1. Draw all topologically distinct, connected diagrams.
2. Include the following factors for the scalar and vector vertices respectively

$$ig_s 1 \; ; \qquad \text{scalar} \qquad\qquad -g_v\gamma_\mu \; ; \qquad \text{vector} \tag{16.2}$$

3. Include the following factors for the scalar, vector, and baryon propagators, respectively (Fig. 16.1)

$$\frac{1}{i}\frac{1}{k^2 + m_s^2} \; ; \qquad\qquad \text{scalar} \tag{16.3}$$

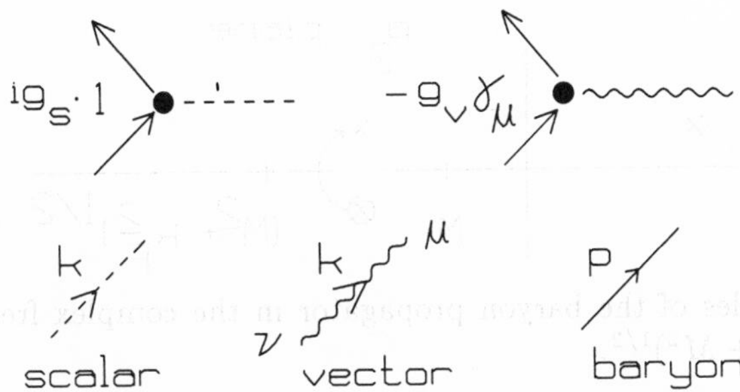

Figure 16.1: Elements of the Feynman rules for QHD-I (see text).

$$\frac{1}{i}\frac{1}{k^2+m_v^2}\left(\delta_{\mu\nu}+\frac{k_\mu k_\nu}{m_v^2}\right) \; ; \qquad \text{vector}$$

$$\frac{1}{i}\left[\frac{1}{i\gamma_\mu p_\mu + M} + 2\pi i(i\gamma_\mu p_\mu - M)\delta(p^2+M^2)\theta(p_0)\theta(k_F - |\mathbf{p}|)\right] \; ; \; \text{baryon}$$

4. Conserve four-momentum at each vertex.
5. Include a factor of $\int d^4q/(2\pi)^4$ for each independent internal line.
6. Take the Dirac matrix product along a fermion line.
7. Include a factor of $(-1)^F$ where F is the number of closed fermion loops.
8. Include a factor of δ_{ij} along a fermion line for isospin (here $i, j = p, n$).

There are several features of these rules that merit discussion:

1. The masses all carry a small negative imaginary part to give the proper Feynman singularities in the propagator.

2. The term proportional to $k_\mu k_\nu$ in the vector meson propagator goes out in any S-matrix element since the vector meson couples to the conserved baryon current.[9] The theory is analogous to massive QED with an additional scalar interaction; it is renormalizable.

3. It is the extra contribution to the baryon propagator, present at finite baryon density, which complicates finite-density, relativistic nuclear many-body theory. Its role is to move a finite number of poles from the 4th to the 1st quadrant so that when one evaluates expectation values by closing contours in the upper-1/2 p_0 plane, there will be a contribution from the occupied single-particle orbitals (see Fig. 16.2). Note that when contours are closed in the upper-1/2 p_0 plane, *one cannot avoid picking up the contribution of the negative frequency poles in the 2nd quadrant.* These contributions are an essential feature of the relativistic many-body problem and are completely absent in the nonrelativistic many-body problem where these antiparticle contributions are pushed off to infinity and ignored.

[9] The proof is similar to that for the analogous terms in the photon propagator in QED (see Refs. [R2, R3, R4]).

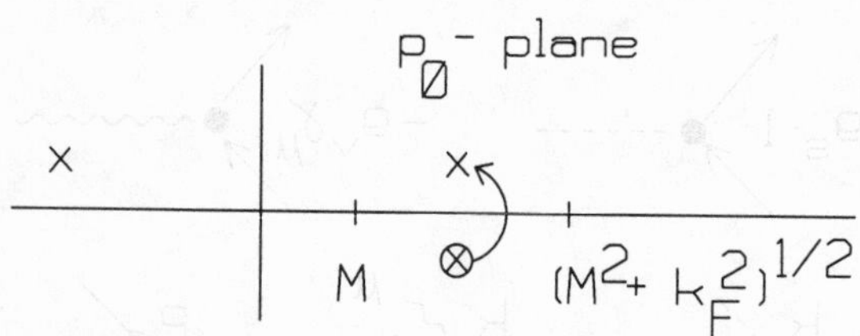

Figure 16.2: Poles of the baryon propagator in the complex frequency plane. Here $E_{\mathrm{F}} = (k_{\mathrm{F}}^2 + M^2)^{1/2}$.

4. The familiar expression $-1/(i\gamma_\mu p_\mu + M)$ can be used for the baryon propagator if one keeps track of the location of the singularities and the contour in Fig. 16.2.

An alternative way of writing the baryon propagator helps illustrate these points. From Fig. 16.2 one can write

$$
\begin{aligned}
iG_0(p) \quad = \quad & \frac{1}{i}\left[\frac{1}{i\gamma_\mu p_\mu + M - i\eta}\right. \qquad\qquad (16.4) \\
& \left. + \left(\frac{1}{i\gamma_\mu p_\mu + M + i\eta} - \frac{1}{i\gamma_\mu p_\mu + M - i\eta}\right)\theta(p_0)\theta(k_{\mathrm{F}} - |\mathbf{p}|)\right]
\end{aligned}
$$

The term in the second line can now be rewritten as

$$
\begin{aligned}
& \left(\frac{M - i\gamma_\mu p_\mu}{p^2 + (M + i\eta)^2} - \frac{M - i\gamma_\mu p_\mu}{p^2 + (M - i\eta)^2}\right)\theta(p_0)\theta(k_{\mathrm{F}} - |\mathbf{p}|) \\
& = 2\pi i\delta(P^2 + M^2)(i\gamma_\mu p_\mu - M)\theta(p_0)\theta(k_{\mathrm{F}} - |\mathbf{p}|) \quad (16.5)
\end{aligned}
$$

This is the result quoted in Eq. (16.3).

An Application — Relativistic Hartree Approximation (RHA). We present the results of a self-consistent one-baryon-loop calculation, done in detail in two different ways in Ref. [R1]. The self-consistent sum of "tadpole" diagrams for the baryon Green's function is illustrated diagrammatically in Fig. 16.3; it treats the forward scattering of a baryon in the medium from the other baryons in a self-consistent fashion.

The meson propagators are calculated by retaining just the disconnected contributions terminating in the baryon tadpoles, which are present in the nu-

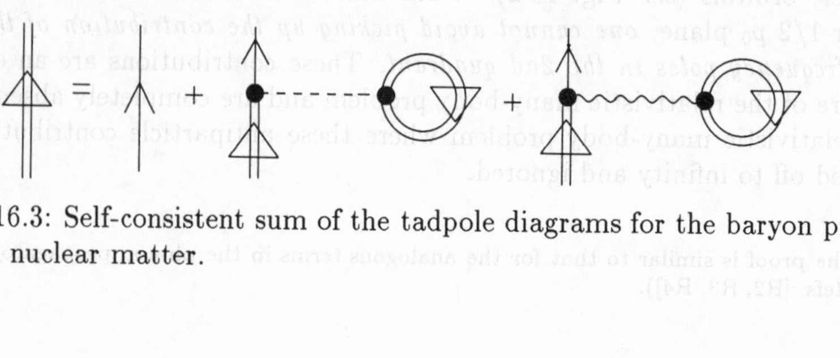

Figure 16.3: Self-consistent sum of the tadpole diagrams for the baryon propagator in nuclear matter.

clear medium at finite density.

In order to carry out the calculation of finite, physical quantities in this renormalizable field theory, *counter terms* involving the self-coupling of the scalar meson field up through powers of ϕ^4 must first be added to the lagrangian density

$$\delta\mathcal{L}_{\rm CTC} = \sum_{n=1}^{4} \frac{c_n}{n!}\phi^n \tag{16.6}$$

These counter terms are fixed in the *vacuum sector* by demanding that the appropriate calculated scalar meson amplitudes, including now the (divergent) loop contributions, take on specified physical values. To minimize the role played by many-body forces in nuclear matter, we assume here that the relevant amplitudes vanish when the four-momenta of the scalar mesons vanish ($q_i = 0$).[10]

With the familiar definition of the energy density $\varepsilon \equiv E/V$, we now find our previous MFT results plus a correction term

$$\varepsilon_{\rm RHA} = \varepsilon_{\rm MFT} + \Delta\varepsilon_{\rm 0-PT} \tag{16.7}$$

The "zero-point" correction term provides a proper evaluation of our previous result in Eqs. (14.20) and (14.22)

$$\Delta\varepsilon_{\rm 0-PT} = \delta\mathcal{H} - \delta\mathcal{L}_{\rm CTC} \tag{16.8}$$

It is given by (Ref. [R1])

$$\begin{aligned}\Delta\varepsilon_{\rm 0-PT} \quad &= \quad -\frac{\gamma}{16\pi^2}[M^{*4}\ln\left(\frac{M^*}{M}\right) + M^3(M-M^*) - \frac{7}{2}M^2(M-M^*)^2 \\ &\quad + \frac{13}{3}M(M-M^*)^3 - \frac{25}{12}(M-M^*)^4]\end{aligned} \tag{16.9}$$

The details of this calculation are given at the end of this section; we first present some numerical results.

Numerical Results. The modification of the MFT equation of state upon inclusion of the additional term in Eq. (16.9) is shown in Fig. 16.4. Note the MFT result remains correct at high density. This provides a partial justification of our initial derivation of MFT in Section 14. The modification of the MFT binding energy curve is shown in Fig. 16.5.

The additional term $\Delta\varepsilon_{\rm 0-PT}$ is a small shift on the new energy scale, but it is important for a quantitative description of the saturation properties of nuclear matter in this model. This additional contribution is *completely absent* in any nonrelativistic many-body problem where the negative frequency poles in Fig. 16.2 are pushed out to infinity and ignored; it is inherently a relativistic effect.

Relativistic Hartree Approximation (RHA). We proceed to rederive the MFT results for uniform nuclear matter by a self-consistent sum of an infinite class

[10] Other choices are possible and have been extensively investigated (Refs. [R1, R75]). The inclusion of $\Delta\varepsilon_{\rm 0-PT}(M^*)$ does not appear to improve the phenomenology.

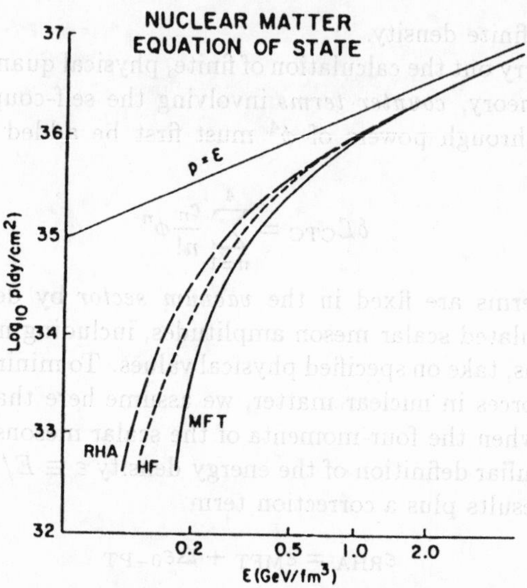

Figure 16.4: Nuclear matter equation of state. The mean-field theory (MFT) results are shown as the solid line. The relativistic Hartree approximation (RHA), which includes $\Delta\varepsilon_{0-\text{PT}}$, produces the long-dash line. (From Ref. [R1].)

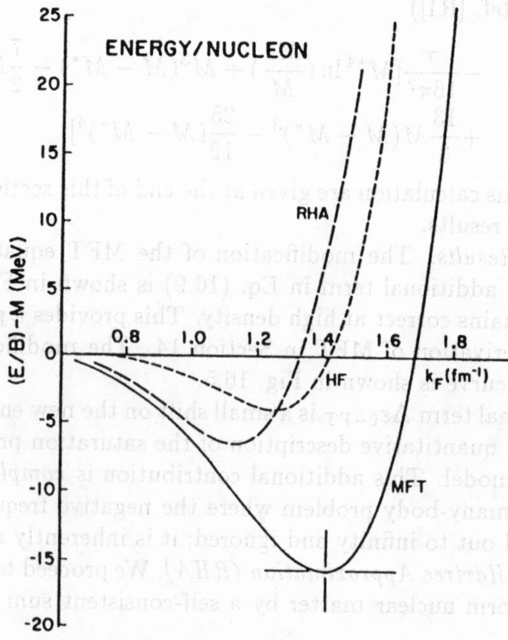

Figure 16.5: Energy/nucleon in nuclear matter. The curves are calculated and labeled as in Fig. 16.4. (From Ref. [R1].)

of Feynman diagrams. In so doing, we provide a framework for a proper treatment of the remaining term $\delta\mathcal{H}$ in Eqs. (14.20) and (14.22). Consider the self-consistent sum of tadpole diagrams for the baryon Green's function as illustrated in Fig. 16.3. Here the self-consistency enters in that the baryon loop, which represents the forward scattering ($q = 0$) interaction with the other baryons in the medium, is calculated from the full Green's function itself. The analytic expression of Dyson's equation for the baryon propagator takes the form (Refs. [N2], [R4])

$$G(p) = G_0(p) + G_0(p)\Sigma^*(p)G(p) \tag{16.10}$$

Here

$$G_0(p) \equiv -\frac{1}{i\gamma_\mu p_\mu + M} \tag{16.11}$$

The appropriate $\pm i\eta$ singularity structure here is given in Eqs. (16.4) and (16.5). The proper self-energy is calculated through the Feynman rules and can be written in the form

$$\Sigma^*(p) \quad = \quad -g_s\phi_0 - ig_v\gamma_\mu V_\mu^0 \tag{16.12}$$

Here $V_\lambda^0 = i\delta_{\lambda 4}V_0$. The meson fields ϕ_0, V_0 arising at finite baryon density are defined implicitly in terms of the appropriate baryon sources represented by closed loops of the baryon Green's function; they are constants for nuclear matter.[11] The solution to Dyson's equation can be written as

$$\begin{aligned} G(p)^{-1} &= G_0(p)^{-1} - \Sigma^*(p) \\ &= -(i\gamma_\mu p_\mu + M) + g_s\phi_0 + ig_v\gamma_\mu V_\mu^0 \end{aligned} \tag{16.13}$$

The baryon Green's function for nuclear matter in the RHA is thus given by

$$G(p) = -\frac{1}{i\gamma_\mu(p_\mu - g_v V_\mu^0) + M^*} \tag{16.14}$$

The change in singularity structure at finite baryon density from the usual Feynman Green's function is shown in Fig. 16.6.

In the renormalizable field theory QHD-I, the counter terms in Eq. (16.6) must be added to the lagrangian density. We choose to fix the counter terms in the vacuum sector to exactly *cancel* the one-baryon-loop contribution to the relevant scalar meson amplitudes at $q_i = 0$; the physical implications of this choice are described below. With $\varepsilon \equiv E/V$, we find our previous MFT results for nuclear matter plus the correction term in Eq. (16.7). The "zero point" correction term provides a proper evaluation of the additional terms in Eq. (16.8), and the result of the following calculation has already been given in Eq. (16.9). We proceed now to sketch the derivation.

[11] The signs and factors follow from the Feynman rules. The iteration of Dyson's equations in this case sums the tadpole diagrams for the baryon Green's function.

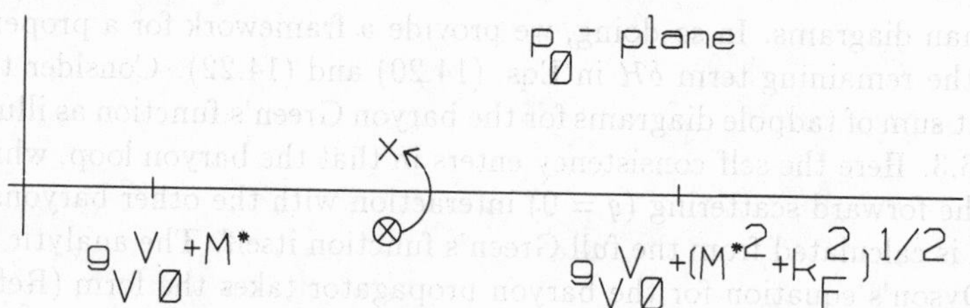

Figure 16.6: Shift in position of the pole in the Feynman propagator for the baryon at finite baryon density in RHA.

The stress tensor in quantum field theory is given by Eq. (14.9), where there is one contribution from each of the generalized coordinates. In QHD-I the stress tensor takes the form

$$T_{\mu\nu} \quad = \quad \mathcal{L}\delta_{\mu\nu} + \bar{\psi}\gamma_\mu \frac{\partial}{\partial x_\nu}\psi + \text{meson terms} \tag{16.15}$$

The baryon Green's function defined in Eq. (16.1) can be used to evaluate the baryon contribution to the expectation value of the stress tensor

$$\begin{aligned}\langle \bar{\psi}\gamma_\mu \frac{\partial}{\partial x_\nu}\psi\rangle &= -\text{Lim}_{t_2\to t_1^+}\,\text{Lim}_{\mathbf{x}_2\to\mathbf{x}_1}\,\text{tr}\,\gamma_\mu \frac{\partial}{\partial x_{1\nu}} iG(\mathbf{x}_1 t_1;\mathbf{x}_2 t_2) \\ &= \frac{\text{tr}}{(2\pi)^4}\int d^4k\;\gamma_\mu k_\nu G(k)\end{aligned} \tag{16.16}$$

Here tr indicates a trace over the matrix indices. Note that the baryon contribution to $\mathcal{L}$ in Eq. (14.3) *vanishes* when evaluated along the dynamic path in Eq. (14.7). Note also that for constant meson fields, there are no derivative terms involving the meson fields remaining in $T_{\mu\nu}$. The contribution $\mathcal{L}_{\text{CTC}}$ will be evaluated with the constant field ϕ_0.

It is always the difference of the expectation value of the stress tensor relative to vacuum that is of physical interest. This difference now takes the following form

$$\begin{aligned}&\langle \hat{T}_{\mu\nu}\rangle - \langle \hat{T}_{\mu\nu}\rangle_{\text{vac}} = \\ &\quad -\frac{\text{tr}}{(2\pi)^4}\int d^4k\;\gamma_\mu k_\nu \left[\frac{1}{i\gamma_\mu(k_\mu - g_v V_\mu^0) + M^*} - \frac{1}{i\gamma_\mu k_\mu + M}\right] \\ &\quad +\delta_{\mu\nu}\left(\frac{1}{2}m_v^2 V_0^2 - \frac{1}{2}m_s^2\phi_0^2 + \sum_{n=1}^{4}\frac{c_n}{n!}\phi_0^n\right)\end{aligned} \tag{16.17}$$

A shift of integration variables in the first integral leads to the result[12]

$$\langle\hat{T}_{\mu\nu}\rangle - \langle\hat{T}_{\mu\nu}\rangle_{\text{vac}} =$$
$$-\frac{\text{tr}}{(2\pi)^4} g_v V_\nu^0 \int d^4t \gamma_\mu \frac{1}{i\gamma_\mu t_\mu + M^*} + \delta_{\mu\nu}\left(\frac{1}{2}m_v^2 V_0^2 - \frac{1}{2}m_s^2\phi_0^2 + \sum_{n=1}^{4}\frac{c_n}{n!}\phi_0^n\right)$$
$$-\frac{\text{tr}}{(2\pi)^4}\int d^4t\, \gamma_\mu t_\nu \left[\frac{1}{i\gamma_\mu t_\mu + M^*} - \frac{1}{i\gamma_\mu t_\mu + M}\right] \tag{16.18}$$

The baryon Green's function $-(i\gamma_\mu t_\mu + M^*)^{-1}$ can be written with the aid of Eqs. (16.4) and (16.5) as

$$G(t) = -\left[\left(\frac{1}{i\gamma_\mu t_\mu + M^*}\right)_{\text{F}} + 2\pi i\delta(t^2 + M^{*2})(i\gamma_\mu t_\mu - M^*)\theta(t_0)\theta(k_{\text{F}} - |\mathbf{t}|)\right] \tag{16.19}$$

The first term is the familiar Feynman propagator, and the second term reproduces the MFT results, including the proper baryon density, as the reader can readily verify. Thus one can write[13]

$$\langle\hat{T}_{\mu\nu}\rangle - \langle\hat{T}_{\mu\nu}\rangle_{\text{vac}} = T_{\mu\nu}^{\text{MFT}} + \delta T_{\mu\nu}$$
$$\delta T_{\mu\nu} = -\frac{\text{tr}}{(2\pi)^4}\int d^4t\, \gamma_\mu t_\nu \left[\frac{1}{i\gamma_\mu t_\mu + M^*} - \frac{1}{i\gamma_\mu t_\mu + M}\right]_{\text{F}} + \delta_{\mu\nu}\sum_{n=1}^{4}\frac{c_n}{n!}\phi_0^n \tag{16.20}$$

Let us digress briefly to obtain some insight into this expression. As we shall demonstrate, a formal evaluation of the first term in $-\delta T_{44}$ shows that it is equal to $\delta\mathcal{H}$. Since the second term is just $-\delta\mathcal{L}_{\text{CTC}}$, one obtains Eq. (16.8). All the integrals make sense when defined through dimensional regularization (see Refs. [R14, R15, R16, R1]); a formal evaluation of the first term then gives

$$\begin{aligned}
-\delta T_{44}^{(1)} &= \frac{i\,\text{tr}}{(2\pi)^4}\int d^4t\, \gamma_4 t_0 \left[\frac{1}{i\gamma_\mu t_\mu + M^*} - \frac{1}{i\gamma_\mu t_\mu + M}\right]_{\text{F}} \\
&= \frac{8i}{(2\pi)^4}\int d^4t \left[\frac{t_0^2}{\mathbf{t}^2 + M^{*2} - t_0^2} - \frac{t_0^2}{\mathbf{t}^2 + M^2 - t_0^2}\right]_{\text{F}} \\
&= -\frac{8i}{(2\pi)^4}\int d^4t \left[\frac{\omega_t^{*2}}{t_0^2 - \omega_t^{*2}} - \frac{\omega_t^2}{t_0^2 - \omega_t^2}\right]_{\text{F}}
\end{aligned} \tag{16.21}$$

Here $\omega_t^2 \equiv \mathbf{t}^2 + M^2$ (similarly for ω_t^{*2}), and the second line follows after rationalizing the denominators and taking the trace [note that here $\text{tr}\,1 =$

[12] It is shown below that with the appropriate choice of counter terms this expression is finite. Dimensional regularization (see Ref. [R1]) can be used to consistently define all the integrals, and the limit $n \to 4$ is well defined. A shift of integration variables is thus justified.

[13] Note that $\text{tr}\int d^4t\gamma_\mu(i\gamma_\lambda t_\lambda + M^*)_{\text{F}}^{-1} = 0$ with dimensional regularization and symmetric integration.

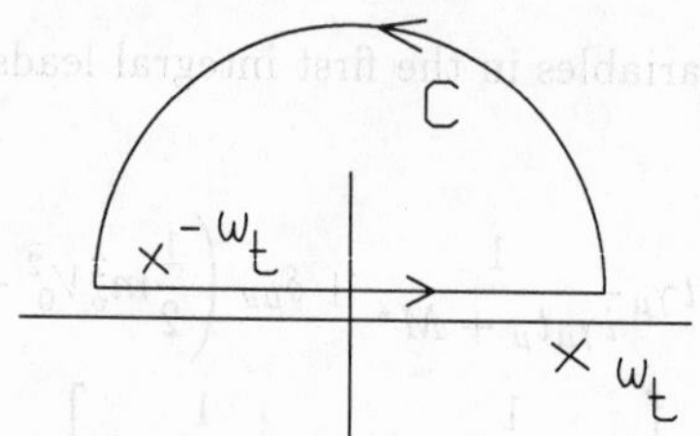

Figure 16.7: Location of Feynman singularities for evaluation of the first term in $-\delta T_{44}$.

4(Dirac) × 2(isospin)]. The integration over dt_0 can be done by noting the location of the Feynman singularities (Fig. 16.7) and using residues. This gives

$$\oint_C dt_0 \left[\frac{\omega_t^{*2}}{t_0^2-\omega_t^{*2}} - \frac{\omega_t^2}{t_0^2-\omega_t^2}\right]_{\mathrm{F}} = 2\pi i\left[\frac{\omega_t^{*2}}{-2\omega_t^*} - \frac{\omega_t^2}{-2\omega_t}\right] \tag{16.22}$$

Thus one finds

$$-\delta T_{44}^{(1)} = \frac{-4}{(2\pi)^3}\int d^3t\,\left[\sqrt{\mathbf{t}^2+M^{*2}} - \sqrt{\mathbf{t}^2+M^2}\right] = \delta\mathcal{H} \tag{16.23}$$

where $\delta\mathcal{H}$ was defined in Eqs. (14.20) and (14.22). After this brief digression, we return to an evaluation of the expression in Eq. (16.20), which is our goal.

Renormalization Prescription. The constants c_i with $i = 1, \ldots, 4$ in $\delta\mathcal{L}_{\mathrm{CTC}}$ are now determined with the renormalization prescription shown pictorially in Fig. 16.8. The interpretation of these requirements is as follows:

1. The constant c_1 is chosen to exactly cancel the vacuum baryon loop with scalar coupling. There is no linear term in ϕ_0 remaining in δT_{44} and the vacuum is therefore stable against the formation of a condensed scalar field (at least for small ϕ_0). There are now no scalar tadpoles remaining in the vacuum sector of the theory; the effect of this term is exactly the same as normal ordering the baryon scalar density $:\bar{\psi}\psi:$, the source of the scalar field, in the vacuum sector.

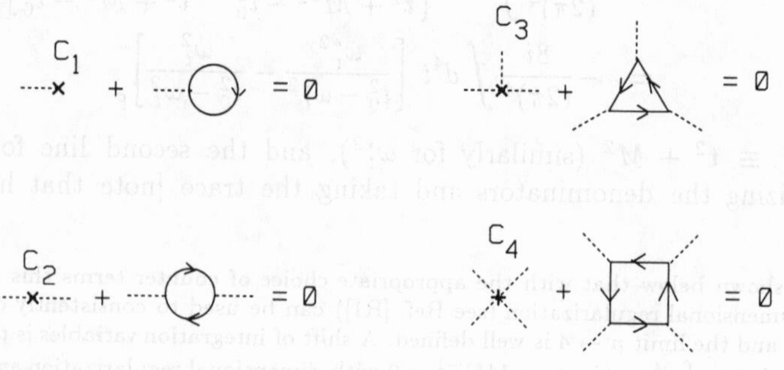

Figure 16.8: Renormalization prescription for the terms in $\delta\mathcal{L}_{\mathrm{CTC}}$ in RHA.

2. The constant c_2 is chosen to cancel the one-baryon-loop contribution to the scalar propagator. This implies that the mass associated with the scalar field is indeed m_s^2.

3. The constant c_3 is chosen to exactly cancel the one-baryon-loop contribution to the interaction of three scalar mesons. This eliminates any residual renormalized (scalar)^3 coupling.

4. The constant c_4 is chosen to cancel the one-baryon-loop contribution to the scalar-scalar scattering amplitude. This eliminates any residual renormalized (scalar)^4 coupling.

5. These conditions are imposed when all of the momenta of the scalar mesons in these processes vanish, or $q_i = 0$. This choice, which is arbitrary and here defines the renormalized model and the RHA, attempts to minimize (at least at low density) the role of many body forces in nuclear matter arising from multiple scalar meson exchange.

6. The first four terms in $\delta T_{\mu\nu} = \sum_{n=0}^{\infty} a_n \phi_0^n$ have now been specified by our renormalization presciption; all additional powers remain with specified, finite coefficients.

The correct numerical factors for the various contributions in Fig. 16.8 follow from the Feynman rules above. They are discussed in Ref. [R1], and are here left as a problem (Prob. 16.5). The result is

$$\delta T_{\mu\nu} = \frac{-\mathrm{tr}}{(2\pi)^4} \int d^4t \left\{ \gamma_\mu t_\nu \left[\frac{1}{i\gamma_\lambda t_\lambda + M^*} - \frac{1}{i\gamma_\lambda t_\lambda + M} \right]_{\mathrm{F}} + i\delta_{\mu\nu} \sum_{n=1}^{4} \frac{1}{(i\gamma_\lambda t_\lambda + M)^n} \frac{(g_s\phi_0)^n}{n} \right\} \tag{16.24}$$

Dimensional Regularization. As already stated, all the integrals are defined through dimensional regularization (Ref. [R14]). This permits one to carry out all the algebraic manipulations in the standard manner. The basic momentum integral in n dimensions in the Lorentz metric with $t^2 = \mathbf{t}^2 - t_0^2$ is given by (see Refs. [R14, R15, R16, R1] and Prob. 16.6).[14]

$$\int \frac{d^n t\,(t^2)^p}{(t^2 + a^2 - i\epsilon)^q} = \frac{i}{(a^2)^{q-p}} \frac{n(\pi a^2)^{n/2}}{\Gamma(n/2+1)} \frac{\Gamma(q - [p + n/2])\Gamma(p + n/2)}{2\Gamma(q)} \tag{16.25}$$

Go to dimensionless variables with

$$\frac{M^*}{M} = 1 - \frac{g_s\phi_0}{M} \equiv 1 - \eta \tag{16.26}$$

[14]Expressions in the Lorentz metric are first converted to the euclidian metric by rotating the contour in the complex t_0 plane. All algebraic manipulations with the momentum integrals are then in the euclidian metric.

Equation (16.24) then takes the form

$$\frac{\delta T_{\mu\nu}}{M^4} = \frac{-\text{tr}}{(2\pi)^4}\int d^n k \left\{\gamma_\mu k_\nu \left[\frac{1}{i\gamma_\lambda k_\lambda + 1 - \eta} - \frac{1}{i\gamma_\lambda k_\lambda + 1}\right]_{\text{F}} + i\delta_{\mu\nu}\sum_{p=1}^{4}\frac{\eta^p}{p}\frac{1}{(i\gamma_\lambda k_\lambda + 1)^p}\right\} \tag{16.27}$$

The expression for $\delta T_{\mu\nu}$ in Eq. (16.27) is now *explicitly finite*. To see this, first expand the term in square brackets in the first line in a power series

$$\left[\frac{1}{i\gamma_\lambda k_\lambda + 1 - \eta} - \frac{1}{i\gamma_\lambda k_\lambda + 1}\right] = \sum_{p=1}^{\infty}\frac{\eta^p}{(i\gamma_\lambda k_\lambda + 1)^{p+1}} \tag{16.28}$$

Now use the cyclic property of the tr to show that

$$\text{tr}\,\gamma_\mu k_\nu \left[\sum_{p=1}^{\infty}\frac{\eta^p}{(i\gamma_\lambda k_\lambda + 1)^{p+1}}\right] = -\text{tr}\left[k_\nu \frac{1}{i}\frac{\partial}{\partial k_\mu}\sum_{p=1}^{\infty}\frac{\eta^p}{p\,(i\gamma_\lambda k_\lambda + 1)^p}\right] \tag{16.29}$$

A partial integration on k_μ now reduces the integral in Eq. (16.27) to the form

$$\frac{\delta T_{\mu\nu}}{M^4} = \frac{-\text{tr}}{(2\pi)^4}\frac{\delta_{\mu\nu}}{i}\int d^n k \left[\sum_{p=1}^{\infty}\frac{\eta^p}{p(i\gamma_\lambda k_\lambda + 1)^p} - \sum_{p=1}^{4}\frac{\eta^p}{p(i\gamma_\lambda k_\lambda + 1)^p}\right] \tag{16.30}$$

The first four terms in the sum over p are now exactly cancelled by the second term coming from $\delta\mathcal{L}_{\text{CTC}}$. The terms with $p \geq 5$ have enough powers of k downstairs for *convergence* in the physical case of four dimensions ($n = 4$). Thus $\delta T_{\mu\nu}/M^4$ as a power series in η has finite coefficients and starts with η^5.

Evaluation of $\delta T_{\mu\nu}/M^4$. We proceed to analyze the expression in Eq. (16.27). A scaling out of the factor $(1-\eta)$ in the integration variable in the first term and a rationalization of the denominator in the last leads to

$$\frac{\delta T_{\mu\nu}}{M^4} = \frac{-\text{tr}}{(2\pi)^4}\int d^n k \left\{\gamma_\mu k_\nu[(1-\eta)^n - 1]\frac{1}{i\gamma_\lambda k_\lambda + 1} + i\delta_{\mu\nu}\sum_{p=1}^{4}\frac{\eta^p}{p}\frac{(1 - i\gamma_\lambda k_\lambda)^p}{1+k^2}\right\} \equiv I_1 + I_2 \tag{16.31}$$

With dimensional regularization, one can do the gamma matrix algebra as if in four dimensions. An evaluation of the first term in Eq. (16.31) then gives

$$\begin{aligned} I_1 &= \frac{-\text{tr}}{(2\pi)^4}\int d^n k\,\frac{\gamma_\mu k_\nu(1 - i\gamma_\lambda k_\lambda)}{1+k^2}[(1-\eta)^n - 1] \\ &= \frac{8i}{(2\pi)^4}\int d^n k\,\frac{k_\mu k_\nu}{1+k^2}[(1-\eta)^n - 1] \\ &= \frac{8i}{(2\pi)^4}\frac{\delta_{\mu\nu}}{n}\int d^n k\,\frac{k^2}{1+k^2}[(1-\eta)^n - 1] \end{aligned} \tag{16.32}$$

Now make use of Eq. (16.25). This gives

$$I_1 = \frac{8i}{(2\pi)^4}[(1-\eta)^n - 1]\frac{\delta_{\mu\nu}}{n}\left\{\frac{in\pi^{n/2}}{\Gamma(1+n/2)}\frac{\Gamma(-n/2)\Gamma(1+n/2)}{2\Gamma(1)}\right\} \tag{16.33}$$

Use $z\Gamma(z) = \Gamma(z+1)$ which holds for all z, and take out common overall factors

$$I_1 = \frac{-4}{(2\pi)^4}\delta_{\mu\nu}\frac{\pi^{n/2}\Gamma(-n/2+3)}{(-n/2)(-n/2+1)}\left\{\frac{[(1-\eta)^n - 1]}{(-n/2+2)}\right\} \tag{16.34}$$

Now take the limit $n \to 4$; to do this, write $n = 4+\epsilon$ and take the limit $\epsilon \to 0$. Consider the following contribution to this limit

$$I_1 \doteq \text{Lim}_{\epsilon\to 0}\frac{-2}{16\pi^2}\delta_{\mu\nu}\left\{\frac{(1-\eta)^{4+\epsilon} - 1}{-\epsilon/2}\right\} \tag{16.35}$$

Use

$$(1-\eta)^{4+\epsilon} = (1-\eta)^4 e^{\epsilon \ln(1-\eta)} \approx (1-\eta)^4[1+\epsilon \ln(1-\eta)] \tag{16.36}$$

This gives

$$\begin{aligned} I_1 &\doteq \text{Lim}_{\epsilon\to 0}\frac{-2}{16\pi^2}\delta_{\mu\nu}\left\{\frac{(1-\eta)^4 - 1}{-\epsilon/2} + \frac{\epsilon}{-\epsilon/2}(1-\eta)^4\ln(1-\eta)\right\} \\ &\equiv I_{1a} + I_{1b} \end{aligned} \tag{16.37}$$

Since the whole expression $I_1 + I_2$ in Eq. (16.31) must be finite as $\epsilon \to 0$ by Eq. (16.30), the term I_{1a} must be exactly cancelled by a corresponding term in the quartic polynomial in I_2. The term I_{1b} is finite as $\epsilon \to 0$ and gives for the remaining contribution

$$I_1 \doteq \frac{4}{16\pi^2}\delta_{\mu\nu}(1-\eta)^4\ln(1-\eta) \tag{16.38}$$

Now the term I_2 must yield additional finite terms in η, up through η^4, so that the entire result $I_1 + I_2$ starts off as η^5 in the power series expansion.[15] Thus one can write the final result as

$$\frac{\delta T_{\mu\nu}}{M^4} = \frac{\gamma}{16\pi^2}\delta_{\mu\nu}[(1-\eta)^4\ln(1-\eta) + \eta - \frac{7}{2}\eta^2 + \frac{13}{3}\eta^3 - \frac{25}{12}\eta^4] \tag{16.39}$$

where $\gamma = 4$ is the spin-isospin degeneracy in nuclear matter. This is the result quoted in Eq. (16.9) since

$$-\delta T_{44} \;=\; \delta\mathcal{H} - \delta\mathcal{L}_{\text{CTC}} \equiv \Delta\varepsilon_{0-\text{PT}} \tag{16.40}$$

[15] The explicit evaluation of I_2 to show that it indeed has all the required properties is left as an exercise for the reader (Prob. 16.7).

17

APPLICATIONS

In this section we present selected applications and extensions of QHD-I.[16]

Relativistic Random Phase Approximation (RRPA). The random phase approximation (RPA) for nuclear excitations was discussed in Sections 9 and 10. The relativistic generalization of the RPA for the excitation spectrum of nuclear matter in the (σ, ω) model (QHD-I) was first carried out by Chin (Ref. [R17]). The meson propagators are calculated by summing the ring diagrams as in the nonrelativistic RPA. These diagrams are indicated in Fig. 17.1.[17] We remind the reader that the excitations of the system can be obtained from the poles of the polarization propagator, which is just the self-energy insertion in Fig. 17.1.[18] The TDA is obtained by keeping only forward-going graphs, as indicated in Fig. 17.2a. Here there is only one particle-hole (p-h) pair present at any time. The RPA is obtained by treating these as Feynman diagrams, and hence keeping any possible time ordering as indicated in Fig. 17.2b. Here there may be any number of p-h pairs present at any time.

Several features of Chin's calculation of nuclear matter in QHD-I are worthy of comment:

1. Since the baryon propagator has both a Feynman and finite-density part,[19] these diagrams now include strong vacuum polarization.
2. The proper self-energies are renormalized in the vacuum sector with the appropriate counter terms.
3. At finite baryon density the vector and scalar mesons mix (see Fig. 17.1).
4. At high baryon density vector meson exchange dominates.
5. At high baryon density the excitation spectrum of nuclear matter in QHD-I in the RPA is that of *zero-sound*;[20] the velocity of zero-sound approaches the velocity of light from below as the baryon density increases.

RPA Calculation of Collective Excitations of Closed Shell Nuclei. An RPA calculation (Sections 9-11) of the excitation spectrum of ^{16}O using the particle-hole interaction of QHD-I fit to the properties of nuclear matter has been carried out by Furnstahl (Ref. [R18]). The interaction is illustrated in Fig. 17.3. Here

[16] Several other applications are discussed in Refs. [R1, R75]

[17] The sum of the ring diagrams gives the correct high density behavior of the electron gas where the interaction is $e^2/\vec{q}^{\,2}$ (Ref. [N2]).

[18] See Section 60 of Ref. [N2].

[19] We write $G = G_{\rm F} + G_{\rm D}$ in Eq. (16.19) and refer to $G_{\rm D}$ as the density-dependent part.

[20] The RPA here retains only those terms involving $G_{\rm D}$ in the baryon propagator.

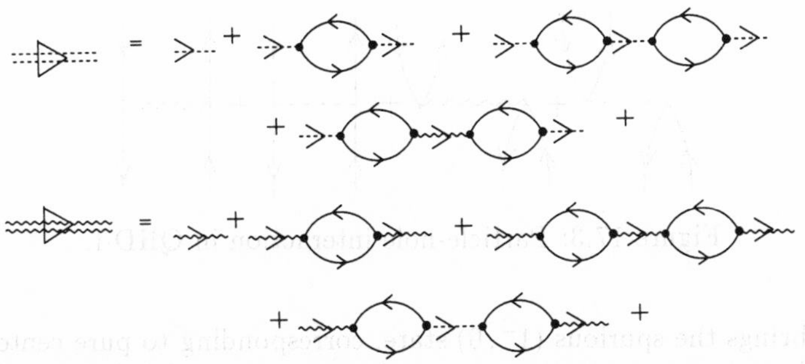

Figure 17.1: Sum of ring diagrams for the meson propagators in the relativistic random phase approximation (RRPA).

only those terms involving the density-dependent part of the baryon propagator are retained and retardation is neglected in the meson propagators; the calculation is otherwise relativistic. The p-h configuration energies and wave functions are taken from the relativistic Hartree calculations of Section 15. The resulting spectra are shown in Figs.17.4 and 17.5. Since the binding energy of nuclear matter results from a strong cancellation of two large contributions, one might wonder whether the nuclear excitation spectrum has any reality. This calculation demonstrates that the particle-hole interaction of QHD-I, with a minimal set of parameters fit to the bulk properties of nuclear matter, produces a realistic spectrum in ^{16}O.

Extension to RRPA. The inclusion of the negative energy states in the RPA calculation of nuclear excitations and of ground state properties of nuclei through the extension to the RRPA has three meritorious effects:

1. It cures the *magnetic moment problem* and the isoscalar convection current gets corrected back from $\mathbf{p}/M^*$ to $\mathbf{p}/M$. This contribution returns to that of the Schmidt lines of Section 8 (see Refs. [R19, R20, R21, R22, R23]).

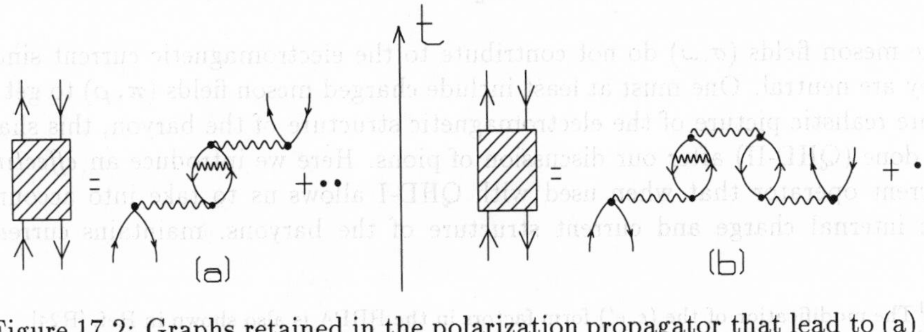

Figure 17.2: Graphs retained in the polarization propagator that lead to (a) the TDA, and (b) the RPA as discussed in Section 60 of Ref. [N2].

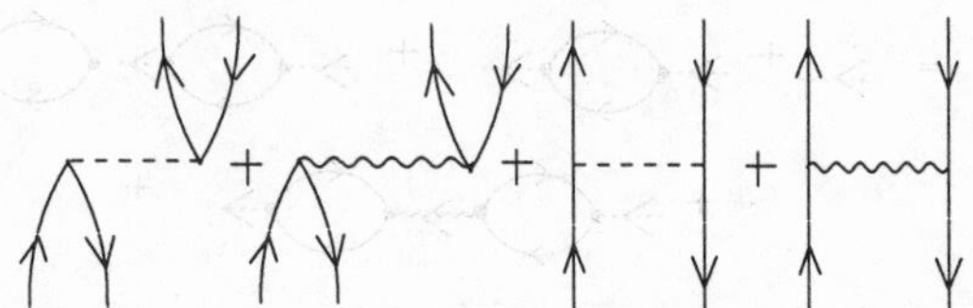

Figure 17.3: Particle-hole interaction in QHD-I.

2. It brings the spurious $(1^-, 0)$ state, corresponding to pure center-of-mass motion, down to zero excitation energy when enough particle-hole configurations are admixed (Ref. [R24]).

3. It preserves the conservation of the electromagnetic current (Ref. [R24]).[21]

On the other hand, both the scalar and vector meson propagators develop poles at $q_0 = 0$ and $|\mathbf{q}| \neq 0$ at finite baryon density in the RRPA (Refs. [R25, R26]); the values of $|\mathbf{q}|$ are typically a few M. Poles in the meson propagators at $q_0 = 0$ and $|\mathbf{q}| \neq 0$ indicate an instability of the MFT ground state against density oscillations of the corresponding wavelengths. There are several interesting possibilities here, for example:

1. The RRPA approximation to the propagators is inadequate.

2. Vertex modifications are important. (Phenomenological form factors signigicantly affect the numerical results.)[22]

3. The internal composite structure of the baryon must be taken into account well before these poles develop.

4. The instability is real.

Electromagnetic Interaction. The electromagnetic current operator in QHD-I is given by

$$\hat{J}_\mu(x) = i\bar{\psi}(x)\gamma_\mu \frac{1}{2}(1 + \tau_3)\psi(x) \tag{17.1}$$

The r.h.s. is expressed in terms of the field operators. The use of the equations of motion verifies that this current is conserved

$$\frac{\partial J_\mu}{\partial x_\mu} = 0 \tag{17.2}$$

The meson fields (σ, ω) do not contribute to the electromagnetic current since they are neutral. One must at least include charged meson fields $(\boldsymbol{\pi}, \boldsymbol{\rho})$ to get a more realistic picture of the electromagnetic structure of the baryon; this shall be done (QHD-II) after our discussion of pions. Here we introduce an *effective* current operator that when used with QHD-I allows us to take into account the internal charge and current structure of the baryons, maintains current

[21] The modification of the (e, e') form factors in the RRPA is also shown in Ref. [R24].

[22] The structure of the vertex in QHD-I is studied in Refs. [R27, R28] ; there is evidence for a damping at large spacelike q^2 with the vector interaction.

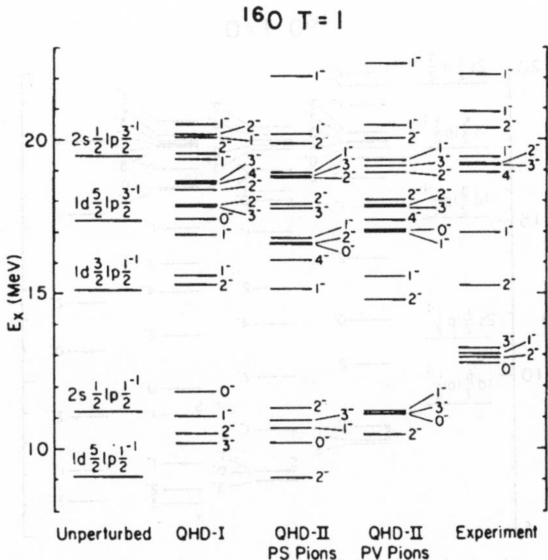

Figure 17.4: Negative parity $T = 1$ states in ^{16}O calculated in RPA using the particle-hole interaction of QHD-I with relativistic Hartree configuration energies and wave functions. Only $1d(1p)^{-1}$ and $2s(1p)^{-1}$ unperturbed levels are shown. The pion contribution is discussed in Sections 20-23. Taken from Ref. [R18].

conservation, and allows us to model the behavior of the interacting relativistic many-body system.

Effective Electromagnetic Current Operator in QHD-I. The general structure of the electromagnetic vertex of the nucleon was introduced in Section 8 (see Fig. 8.4)

$$\langle \mathbf{p}'\sigma'\rho'|J_\mu(0)|\mathbf{p}\sigma\rho\rangle = \frac{i}{\Omega}\bar{u}(\mathbf{p}'\sigma')\eta^\dagger_{\rho'}\left[F_1(q^2)\gamma_\mu + F_2(q^2)\sigma_{\mu\nu}q_\nu\right]\eta_\rho u(\mathbf{p}\sigma)$$

$$F_i \equiv \frac{1}{2}(F_i^S + \tau_3 F_i^V)$$

$$F_1^S(0) = 1 \qquad 2mF_2^S(0) = \lambda_p' + \lambda_n = -0.120$$

$$F_1^V(0) = 1 \qquad 2mF_2^V(0) = \lambda_p' - \lambda_n = +3.706 \qquad (17.3)$$

Here $p = p' + q$. The form factors possess a spectral representation. $F_2^V(q^2)$, for example, has the analytic properties in the complex q^2 plane indicated in Fig. 17.6 and can be expressed as (see Refs. [R29, R30, R31])

$$F_2^V(q^2) = \frac{1}{\pi}\int_{4m_\pi^2}^\infty \frac{\rho_2^V(\sigma^2)\,d\sigma^2}{\sigma^2 + q^2} \qquad (17.4)$$

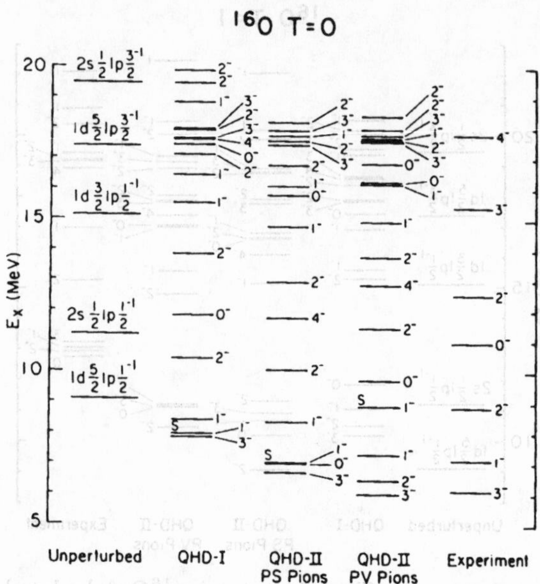

Figure 17.5: Same as Fig. 17.4 for $T = 0$ states. S denotes the spurious 1^- state.

The spectral weight function can be expressed as the absorptive part of the amplitude for a virtual time-like photon to produce an N–$\bar{N}$ pair as indicated schematically in Fig. 17.7. The physical region for electron scattering is space-like $q^2 > 0$. The singularities closest to the physical region come from the lowest-mass intermediate states; here it would be two pions. These are all exact statements and relations.

To evaluate the two-pion contribution to the spectral weight function for $F_2(q^2)$ in Born approximation (without pion rescattering) one can simply look at the Feynman diagram for the lowest order vertex correction illustrated in Fig. 17.8. The Feynman rules for pions will be developed later in this part of the book. Here we anticipate and calculate the contribution of this diagram to

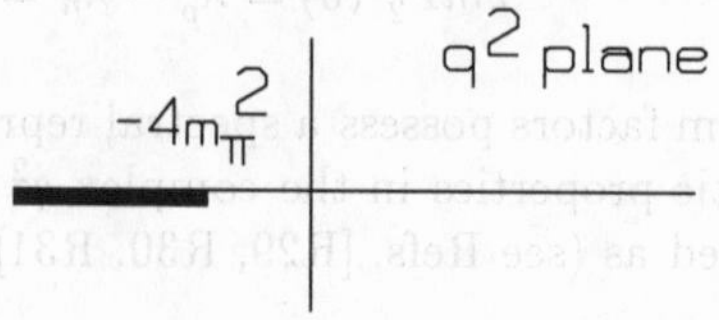

$-4m_\pi^2$ $\quad$ q^2 plane

Figure 17.6: Analytic properties of the nucleon form factor $F_2^V(q^2)$ in the complex q^2 plane.

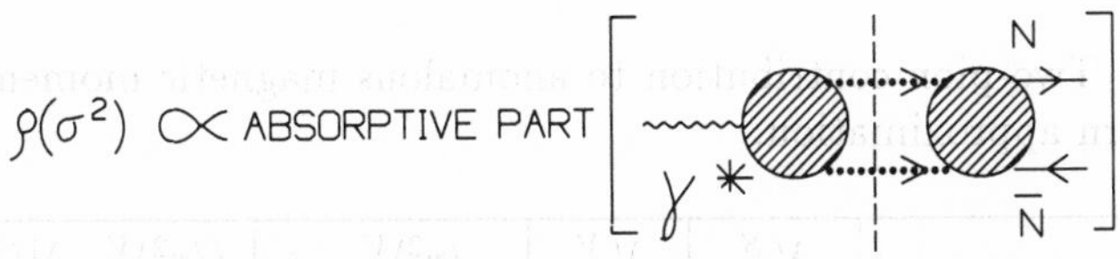

$\rho(\sigma^2) \propto$ ABSORPTIVE PART $\left[\gamma^* \; \text{N} \; \bar{\text{N}} \right]$

Figure 17.7: Schematic representation of spectral weight function as absorptive part.

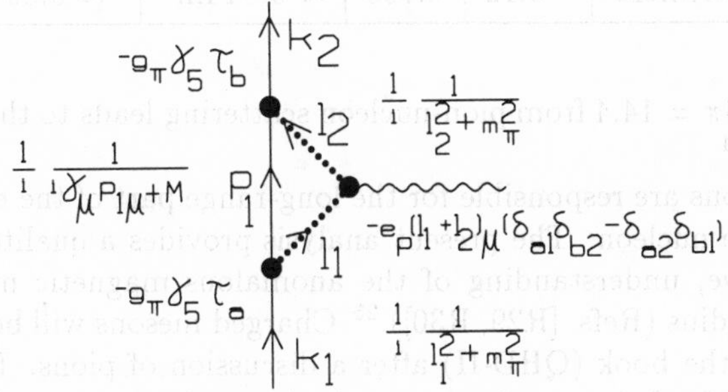

Figure 17.8: Two pion contribution to S_{fi} and $F_2(q^2)$ in Born approximation.

S_{fi} from the following pion-nucleon and pion-photon lagrangian densities[23]

$$
\begin{aligned}
\mathcal{L}_{\pi N} &= ig_\pi \bar{\psi}\gamma_5 \boldsymbol{\tau}\ \psi \cdot \boldsymbol{\pi} \\
\mathcal{L}_{\gamma\pi} &= -e_p \left[\boldsymbol{\pi} \times \frac{\partial \boldsymbol{\pi}}{\partial x_\mu} \right]_3 A_\mu
\end{aligned}
\tag{17.5}
$$

The component contributions to the diagram are then indicated in Fig. 17.8. The result from this diagram can be put into the following form (here $m \equiv M$)

$$
2MF_2(q^2) = \tau_3 \frac{g_\pi^2}{4\pi} \int_0^1 dx\,(1-x)^2 \int_0^x dy \frac{M^2}{M^2(1-x)^2 + m_\pi^2 x + q^2 y(x-y)}
\tag{17.6}
$$

The spectral representation and two-pion contribution to the spectral weight function follow directly. Note this contribution is entirely isovector.

The integral in Eq. (17.6) is well-defined, and one can use it to calculate the two-pion contribution to the anomalous magnetic moment of the nucleon by simply evaluating $2mF_2(0)$. One can get the two-pion contribution, that of longest range, to the mean-square radius of the isovector magnetic moment through

$$
\frac{F_2^V(q^2)}{F_2^V(0)} = 1 - \frac{q^2}{6}\langle r^2 \rangle_2^V + \cdots
\tag{17.7}
$$

[23]The absorptive part is independent of the particular form of the $\pi - N$ coupling used.

Table 17.1: Two-pion contribution to anomalous magnetic moment of the nucleon in Born approximation.

	λ'^{S}	λ'^{V}	$\langle r^2\rangle^V_{\text{mag}}$	$(\langle r^2\rangle^V_{\text{mag}})^{1/2}$
Theory	0	3.20	0.24 fm^2	0.49 fm
Experiment	−0.12	3.706	$\approx$ 0.64 fm^2	$\approx$ 0.80 fm

The use of $g_\pi^2/4\pi = 14.4$ from pion-nucleon scattering leads to the results shown in Table 17.1.[24]

Charged pions are responsible for the long-range part of the electromagnetic structure of the nucleon. The present analysis provides a qualitative, and even semiquantitative, understanding of the anomalous magnetic moment and its mean-square radius (Refs. [R29, R30]).[25] Charged mesons will be included later in this part of the book (QHD-II) after a discussion of pions. Here we will be content to include their contribution to the internal structure of the nucleon in QHD-I in a phenomenological fashion.

We assume[26]

$$\frac{F_1(q^2)}{F_1(0)} \approx f_{\text{SN}}(q^2) \approx \frac{F_2(q^2)}{F_2(0)} \tag{17.8}$$

Now take out the common overall form factor and define an *effective Møller potential* $f_{\text{SN}}(q^2)/q^2$ (Fig. 17.9). Since this is an overall factor in the scattering amplitude, the electron scattering cross sections will be simply expressed in terms of an effective Mott cross section $\sigma_{\text{M}}^{\text{eff}} = \sigma_{\text{M}}[f_{\text{SN}}(q^2)]^2$.

Now define an *effective current operator* in QHD-I

$$\begin{aligned} \hat{J}^{\gamma}_{\mu}(x) &= i\bar{\psi}\gamma_\mu \underset{\sim}{Q}\psi + \frac{1}{2m}\frac{\partial}{\partial x_\nu}(\bar{\psi}\sigma_{\mu\nu}\underset{\sim}{\lambda}'\psi) \\ \underset{\sim}{Q} &= \frac{1}{2}(1+\tau_3) \qquad\qquad \underset{\sim}{\lambda}' = \lambda'_p\frac{1}{2}(1+\tau_3) + \lambda_n\frac{1}{2}(1-\tau_3) \end{aligned} \tag{17.9}$$

Again the r.h.s. is expressed in terms of field operators. This effective current is to be used in lowest order. It takes into account the *internal* charge and magnetic structure of the baryons coming from the charged mesons in a phenomenological

[24] Note that the definition $F_i = (F_i^S + \tau_3 F_i^V)/2$ used here differs by a factor of 2 from that used in Eq. (10.10) of Ref. [R1] where $F_i = F_i^S + \tau_3 F_i^V$.

[25] It was argued before their discovery that vector mesons with $(1^-,1)$ and $(1^-,0)$, the (ρ,ω), must be present to make these results quantitative (Refs. [R32, R33]).

[26] Although this assumption serves our purpose over much of the domain of nuclear physics, one really must do better than this to have a quantitative calculation at high q^2. A more accurate representation of the single-nucleon form factors is given in Ref. [N14]; it is incorporated into the effective nuclear current in Part IV.

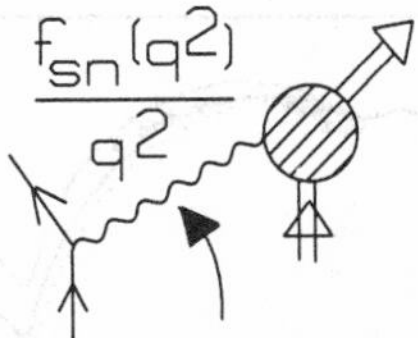

Figure 17.9: The effective Møller potential.

manner. Although very simple minded, this effective current has the following features to recommend its use in QHD-I:

- It is local.
- It is Lorentz covariant.
- It is conserved.
- It gives the correct (e, e) amplitude for a free, isolated nucleon.

High-q^2 Electron Scattering. Since we now have a completely relativistic framework, one can push the calculation of nuclear (e, e') processes to any q^2. There is nothing *inherent* in the calculations limiting them to low q^2. One goal is to push to very high q^2 to see where QHD breaks down – where one is forced to invoke another dynamic description of the internal structure of the hadrons.

The calculation of Kim (Ref. [R34]) for elastic magnetic scattering from ^{17}O is shown in Fig. 17.10. This calculation uses the solution to the Dirac equation for a $(1d_{5/2})_\nu$ in the relativistic Hartree potentials for ^{16}O (Section 15). It uses the current in Eq. (17.9). A center of mass (C-M) correction factor is also included.[27]

The wave functions, and other form factors, calculated by Kim in Ref. [R34] are shown in Figs. 17.11 and 17.12. The configuration assignments are ^{3}He $(1s_{1/2})_\nu^{-1}$, ^{17}O $(1d_{5/2})_\nu$, and ^{209}Bi $(1h_{9/2})_\pi$.

As a second application of this approach, consider quasielastic electron scattering (e, e'). Rosenfelder (Ref. [R35]) has studied this process for both ^{40}Ca and ^{208}Pb using a local Fermi gas with Dirac spinors and the quantities $M^*(r)$ and $\rho_{\rm B}(r)$ taken from a relativistic Thomas-Fermi calculation of the densities in QHD-I (Ref. [R1]). The solutions to the Dirac equation for both the initial bound nucleon and final continuum nucleon in the quaiselastic scattering process are thus generated consistently within this framework. Rosenfelder also uses the electromagnetic current operator of Eq. (17.9). The results are shown in Figs. 17.13 and 17.14. Particularly satisfying are the positions of the peaks and the shapes of the curves, which are obtained with no further parameters.

[27]The C-M correction factor here $f_{\rm CM} = \exp\{\vec{q}^{\,2}b^2/4A\}$ is taken from the simple harmonic oscillator (see Ref. [N14]) — it is the only nonrelativistic element in the calculation.

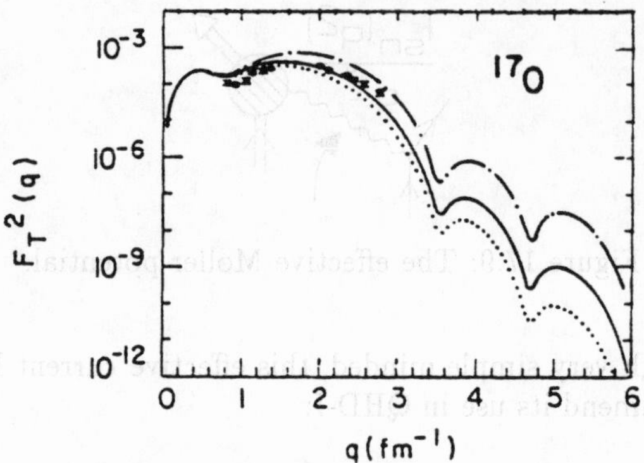

Figure 17.10: Magnetic form factor squared for ^{17}O. The dotted curve omits the C-M correction factor and the dashed-dot curve omits the single-nucleon form factor $f_{\rm SN}(q^2) = [1 + q^2/(855\,{\rm MeV})^2]^{-2}$. Calculated using relativistic Hartree wave functions and the current operator in Eq. (17.9). From Ref. [R34].

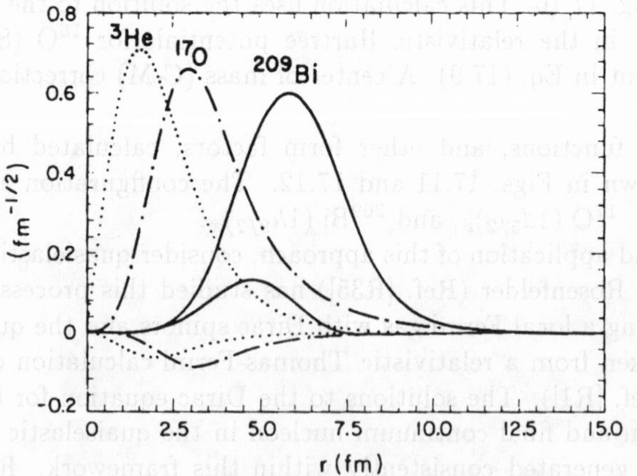

Figure 17.11: Upper and lower component Dirac radial wave functions $G(r)$ and $F(r)$ for the three cases discussed in the text. From Ref. [R34].

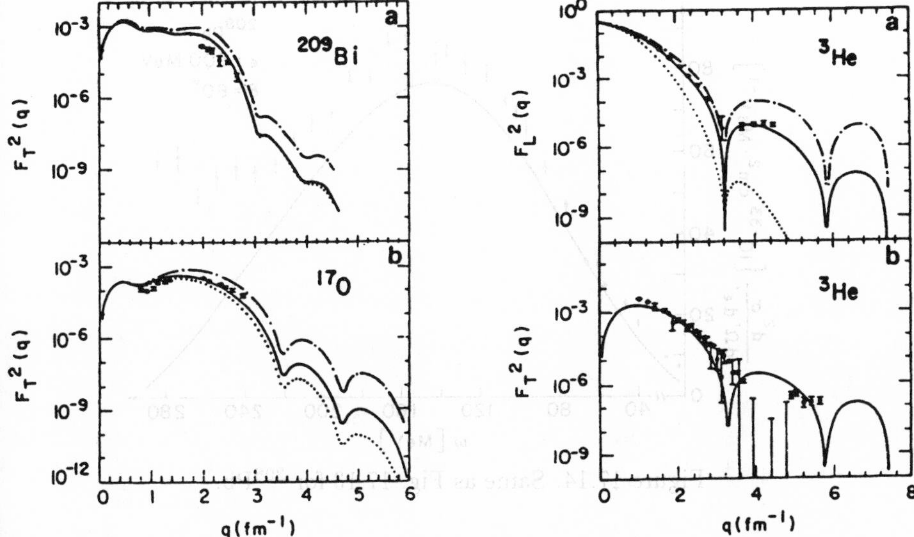

Figure 17.12: Magnetic and charge form factors squared for the other examples discussed in the text. The notation is the same as in Fig. 17.10. We use $d\sigma/d\Omega = 4\pi\sigma_M F^2 r$ where $F^2 = (q_\mu^2/\mathbf{q}^2)^2 F_L^2 + [(q_\mu^2/2\mathbf{q}^2) + \tan^2\theta/2]F_T^2$. Taken from Ref. [R34].

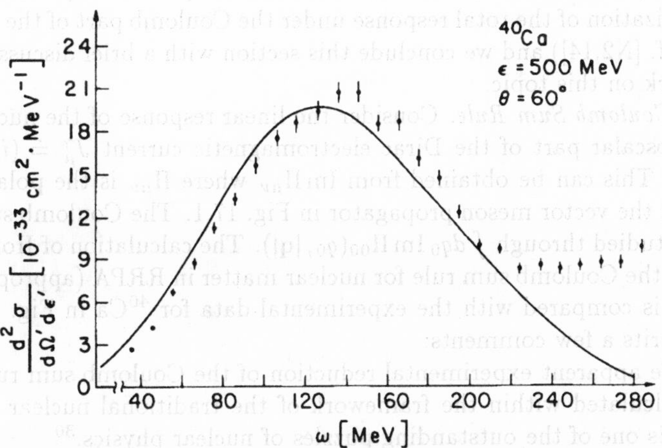

Figure 17.13: Quasielastic electron scattering from ^{40}Ca in the relativistic MFT compared with experimental values. The calculation assumes a local relativistic Fermi gas with the quantities $M^*(r)$ and $\rho_B(r)$ taken from a relativistic Thomas Fermi calculation of these quantities in QHD-I. From Refs. [R35, R1].

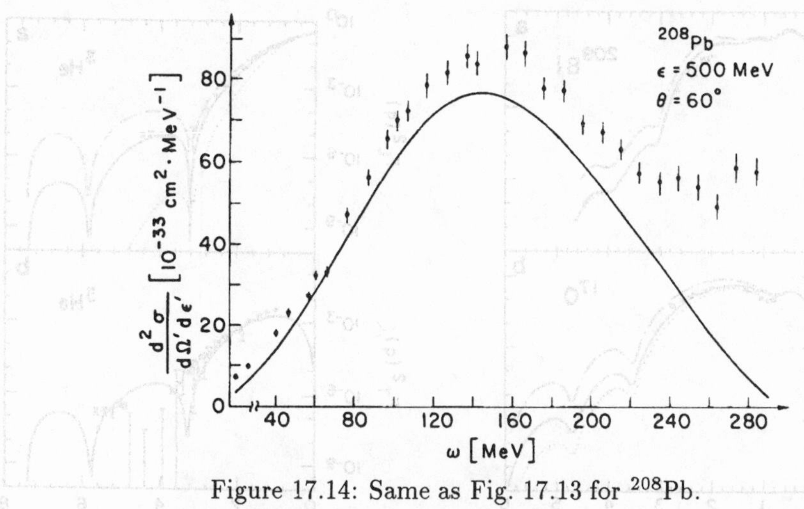

Figure 17.14: Same as Fig. 17.13 for ^{208}Pb.

This indicates that the MFT of Section 14 provides a consistent and reasonable description of the gross quasielastic nuclear response in this kinematic regime.

This nuclear response contains both longitudinal (Coulomb) and transverse contributions.[28] When, after a great deal of painstaking experimental work, the Coulomb response is separated, it is found that this calculation, as do almost all others, overestimates the Coulomb response in this kinematic regime by almost a factor of 2 (see Section 12). The Coulomb sum rule provides a particularly useful characterization of the total response under the Coulomb part of the quasielastic peak (Ref. [N2,14]) and we conclude this section with a brief discussion of some QHD work on this topic.

The Coulomb Sum Rule. Consider the linear response of the nuclear system to the isoscalar part of the Dirac electromagnetic current $J_\mu^\gamma \doteq (i/2)\bar{\psi}\gamma_\mu\psi$ in RRPA.[29] This can be obtained from $\text{Im}\,\Pi_{\mu\nu}$ where $\Pi_{\mu\nu}$ is the polarization insertion in the vector meson propagator in Fig. 17.1. The Coulomb sum rule can then be studied through $\int dq_0\, \text{Im}\,\Pi_{00}(q_0, |\mathbf{q}|)$. The calculation of Horowitz (Ref. [R36]) of the Coulomb sum rule for nuclear matter in RRPA (appropriately normalized) is compared with the experimental data for ^{40}Ca in Fig. 17.15. This result merits a few comments:

1. The apparent experimental reduction of the Coulomb sum rule from the values calculated within the framework of the traditional nuclear many-body problem is one of the outstanding puzzles of nuclear physics.[30]

[28] The quasielastic nuclear response is discussed in detail in Refs. [N2,N14] (see Probs.17.3-5); here we postpone further examination of this topic until Section 39.

[29] See Ref. [N2] for a discussion of linear response.

[30] Other suggestions for the resolution of the Coulomb sum rule puzzle include modification of the intrinsic properties of the nucleon, in particular, of $f_{SN}(q^2)$.

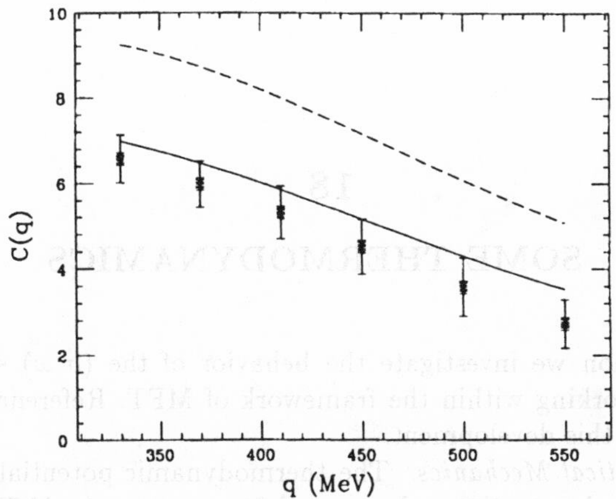

Figure 17.15: Coulomb sum rule C($|\mathbf{q}|$) in relativistic Hartree (dashed) or RRPA (solid) approximations versus momentum transfer (nucleon form factor left in). The Saclay ^{40}Ca electron scattering results are also shown. From Ref. [R36].

2. In the present calculation $\approx 15\%$ reduction comes from RPA and $\approx 15\%$ from RRPA. The first effect takes into account the actual excitation spectrum of the interacting system and comes from the density-dependent piece of the baryon propagator. The second arises from the modification of strong vacuum polarization in the nuclear medium due to the change in the baryon mass from M to M^*.

3. The rest of the current in Eq. 17.9 has been included in the results in Ref. [R36].

4. Other authors find results very similar to these (Refs. [R37, R38, R39]).

18

SOME THERMODYNAMICS

In this section we investigate the behavior of the (σ, ω) system at finite temperature, working within the framework of MFT. Reference [N2] provides background for this development.

Some Statistical Mechanics. The thermodynamic potential of a system at specified chemical potential, volume, and temperature (μ, V, T), as illustrated in Fig. 18.1, is given by

$$\Omega(\mu, V, T) = -\frac{1}{\beta} \ln Z_{\mathrm{G}} \tag{18.1}$$

The grand partition function appearing in this expression is defined by

$$Z_{\mathrm{G}} \equiv \mathrm{Tr}\left\{e^{-\beta(\hat{H}-\mu\hat{B})}\right\} \tag{18.2}$$

Here $\hat{B}$ is the baryon number operator and the Trace goes over a complete set of states in the many-baryon Hilbert space. As usual $\beta \equiv 1/k_{\mathrm{B}}T$. One has the thermodynamic relations

$$\begin{aligned} \Omega &= -pV \\ d\Omega &= -S\,dT - p\,dV - B\,d\mu \end{aligned} \tag{18.3}$$

where S is the entropy.

MFT. Consider nuclear matter in MFT in QHD-I. The hamiltonian and baryon number operators are given in Section 14

$$\begin{aligned} \hat{H}_{\mathrm{MFT}} &= V\left[\frac{1}{2}m_s^2\phi_0^2 - \frac{1}{2}m_v^2 V_0^2\right] + g_v V_0 \hat{B} \\ &\quad + \sum_{\mathbf{k}\lambda} \sqrt{\mathbf{k}^2 + M^{*2}}(A_{\mathbf{k}\lambda}^\dagger A_{\mathbf{k}\lambda} + B_{\mathbf{k}\lambda}^\dagger B_{\mathbf{k}\lambda}) \\ \hat{B} &= \sum_{\mathbf{k}\lambda}(A_{\mathbf{k}\lambda}^\dagger A_{\mathbf{k}\lambda} - B_{\mathbf{k}\lambda}^\dagger B_{\mathbf{k}\lambda}) \end{aligned} \tag{18.4}$$

These operators are diagonal in the basis of eigenstates of the baryon and anti-baryon number operators

$$\begin{aligned} A_{\mathbf{k}\lambda}^\dagger A_{\mathbf{k}\lambda}|n_{\mathbf{k}\lambda}\rangle &= n_{\mathbf{k}\lambda}|n_{\mathbf{k}\lambda}\rangle \\ B_{\mathbf{k}\lambda}^\dagger B_{\mathbf{k}\lambda}|\bar{n}_{\mathbf{k}\lambda}\rangle &= \bar{n}_{\mathbf{k}\lambda}|\bar{n}_{\mathbf{k}\lambda}\rangle \end{aligned} \tag{18.5}$$

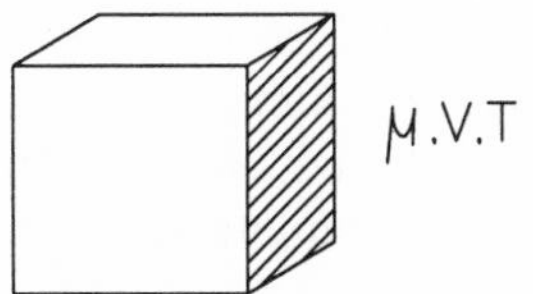

Figure 18.1: Many-body system at specified chemical potential, volume, and temperature (μ, V, T).

It is a straightforward matter to calculate the Trace in Eq. (18.2) in this basis; the calculation is very similar to that for a noninterating Fermi gas (Ref. [N2]) and is carried out in detail in Appendix D.[31]

At the end of the calculation, the vector field V_0 will be determined by the thermal average of the equation of motion. For a uniform system this gives

$$\begin{aligned} m_v^2 V_0 &= g_v \langle\langle \hat{\rho}_{\mathrm{B}} \rangle\rangle \\ &\equiv g_v \rho_{\mathrm{B}}(\mu, V, T; \phi_0, V_0) \end{aligned} \tag{18.6}$$

Here the additional explicit dependence on the condensed fields (ϕ_0, V_0) is also indicated (recall $M^* = M - g_s\phi_0$). The explicit dependence on V_0 in Eq. (18.4) allows one to immediately conclude from Eqs. (18.1) and (18.2) that

$$\left(\frac{\partial\Omega}{\partial V_0}\right)_{\mu,V,T;\phi_0} = 0 \tag{18.7}$$

The condensed scalar field is determined at the end of the calculation through the use of Gibbs' relation for thermodynamic equilibrium; a system in equilibrium at specified (μ, V, T) will minimize its thermodynamic potential

$$\left(\frac{\partial\Omega}{\partial \phi_0}\right)_{\mu,V,T} = 0 \tag{18.8}$$

Now the total differential of the thermodynamic potential is given by

$$\begin{aligned} d\Omega = & \left(\frac{\partial\Omega}{\partial \mu}\right)_{V,T;\phi_0,V_0} d\mu + \left(\frac{\partial\Omega}{\partial V}\right)_{\mu,T;\phi_0,V_0} dV + \left(\frac{\partial\Omega}{\partial T}\right)_{\mu,V;\phi_0,V_0} dT \\ & + \left(\frac{\partial\Omega}{\partial V_0}\right)_{\mu,V,T;\phi_0} dV_0 + \left(\frac{\partial\Omega}{\partial \phi_0}\right)_{\mu,V,T;V_0} d\phi_0 \end{aligned} \tag{18.9}$$

Keep (μ, V, T) constant and make use of Eq. (18.7). This allows one to write the equilibrium condition as

$$\left(\frac{\partial\Omega}{\partial \phi_0}\right)_{\mu,V,T} = \left(\frac{\partial\Omega}{\partial \phi_0}\right)_{\mu,V,T;V_0} = 0 \tag{18.10}$$

[31] We assume a single species of baryon in this simple model calculation. There will always be additional additive contributions to the thermodynamic potential arising from other species. For example, even in QHD-I in MFT, there will be contributions from the free (σ, ω) fields.

As a consequence, the condensed vector field can be held constant during the minimization procedure. The vanishing of the partial derivatives in Eqs. (18.7) and (18.10) is extremely useful, for it implies that the condensed meson fields (ϕ_0, V_0) can be held *constant* in computing thermodynamic variables. For example,

$$B = -\left(\frac{\partial\Omega}{\partial\mu}\right)_{V,T} = -\left(\frac{\partial\Omega}{\partial\mu}\right)_{V,T;\phi_0,V_0} \tag{18.11}$$

The equilibrium condition for ϕ_0 follows from Eq. (18.10) and Eqs. (18.1) and (18.2)

$$m_s^2\phi_0 = g_s\rho_s(\mu, V, T; \phi_0, V_0) \tag{18.12}$$

Here ρ_s is the thermodynamic average of the scalar density, written out in detail below.

At the end of the calculation it is useful to specify the baryon density $\rho_{\rm B}$, and adjust the chemical potential until one arrives at that specified value of $\rho_{\rm B}$ as the equilibrium value. The condensed vector field is then also specified by Eq. (18.6)

$$V_0 = \frac{g_v}{m_v^2}\rho_{\rm B} \tag{18.13}$$

Equation of State. The equation of state in parametric form follows directly from the thermodynamic potential; its derivation is given in Appendix D

$$\begin{aligned}
\varepsilon(\rho_{\rm B}, T) &\equiv \frac{1}{V}E = \frac{g_v^2}{2m_v^2}\rho_{\rm B}^2 + \frac{m_s^2}{2g_s^2}(M - M^*)^2 \\
&\quad + \frac{\gamma}{(2\pi)^3}\int d^3k\sqrt{\mathbf{k}^2 + M^{*2}}(n_k + \bar{n}_k) \\
p(\rho_{\rm B}, T) &= \frac{g_v^2}{2m_v^2}\rho_{\rm B}^2 - \frac{m_s^2}{2g_s^2}(M - M^*)^2 \\
&\quad + \frac{1}{3}\frac{\gamma}{(2\pi)^3}\int d^3k\frac{k^2}{(k^2 + M^{*2})^{1/2}}(n_k + \bar{n}_k) \\
\rho_{\rm B} &= \frac{\gamma}{(2\pi)^3}\int d^3k(n_k - \bar{n}_k)
\end{aligned} \tag{18.14}$$

At thermodynamic equilibrium, the self-consistency relation for $\phi_0 = (M - M^*)/g_s$ must be satisfied

$$\phi_0 = \frac{g_s}{m_s^2}\rho_s = \frac{g_s}{m_s^2}\frac{\gamma}{(2\pi)^3}\int d^3k\frac{M^*}{(k^2 + M^{*2})^{1/2}}(n_k + \bar{n}_k) \tag{18.15}$$

The thermal distribution functions in these expressions are defined by

$$n_k = \frac{1}{e^{\beta(E_k^* - \mu^*)} + 1} \qquad \bar{n}_k = \frac{1}{e^{\beta(E_k^* + \mu^*)} + 1} \tag{18.16}$$

Here

$$
\begin{aligned}
E_k^* &\equiv \sqrt{\mathbf{k}^2 + M^{*2}} \\
\mu^* &\equiv \mu - g_v V_0
\end{aligned}
\tag{18.17}
$$

In thermodynamic equilibrium, at the end of the calculation, one can replace

$$
\mu^* = \mu - \frac{g_v^2}{m_v^2}\rho_{\mathrm{B}} \tag{18.18}
$$

Note carefully the signs appearing in Eqs. (18.14) and (18.15); it is $(n_k + \bar{n}_k)$, which appears in (ε, p, ρ_s), and $(n_k - \bar{n}_k)$, which appears in ρ_{B}.

Some Limiting Results. Some limiting cases of this equation of state are of interest:

1. As $T \to 0$ at finite baryon density one has $n_k \to \theta(k_{\mathrm{F}} - |\mathbf{k}|)$ and $\bar{n}_k \to 0$. The system becomes a degenerate Fermi gas of baryons, and one recovers the MFT results of Section 14 at $T = 0$.

2. As $\rho_{\mathrm{B}} \to \infty$ for any finite T, the system again becomes degenerate.

3. As $T \to \infty$, baryon pairs are produced, and the self-consistent baryon mass $M^* \to 0$. The analytic relation at $\rho_{\mathrm{B}} = 0$ is

$$
\frac{M^*}{M} \longrightarrow \left[1 + \frac{g_s^2}{m_s^2}\frac{\gamma(k_{\mathrm{B}}T)^2}{12}\right]^{-1} \quad ; \qquad T \to \infty \tag{18.19}
$$

4. In the limit $T \to \infty$ the equation of state takes a form similar to that of a black body

$$
\varepsilon = \frac{7\pi^2\gamma}{120}(k_{\mathrm{B}}T)^4 \qquad\qquad p = \frac{1}{3}\varepsilon \tag{18.20}
$$

Numerical Results. A procedure for numerical analysis of the equation of state consists of the following:

1. Solve the self-consistency Eq. (18.15) for ϕ_0 (or equivalently $M^* = M - g_s\phi_0$) at fixed β and μ^*;
2. The distribution functions $n_k(\mu^*)$ and $\bar{n}_k(\mu^*)$ are now determined;
3. Compute the resultant ρ_{B} from the last of Eqs. (18.14);
4. Determine the corresponding chemical potential μ from Eq. (18.18);
5. Compute (ε, p) from Eqs. (18.14).

The resulting isotherms (constant temperature cuts of the equation of state) are shown for neutron matter with $\gamma = 2$ in Fig. 18.2. We use the coupling constants of Section 14. Several comments are of interest:

1. One sees a phase transition; it is similar here to the gas-liquid phase transition exhibited by the van der Waal's equation of state.

2. In the region of phase equilibrium, Gibbs' criteria for phase equilibrium are satisfied

$$
p_1 = p_2 \qquad\qquad \mu_1 = \mu_2 \qquad\qquad T = \text{constant} \tag{18.21}
$$

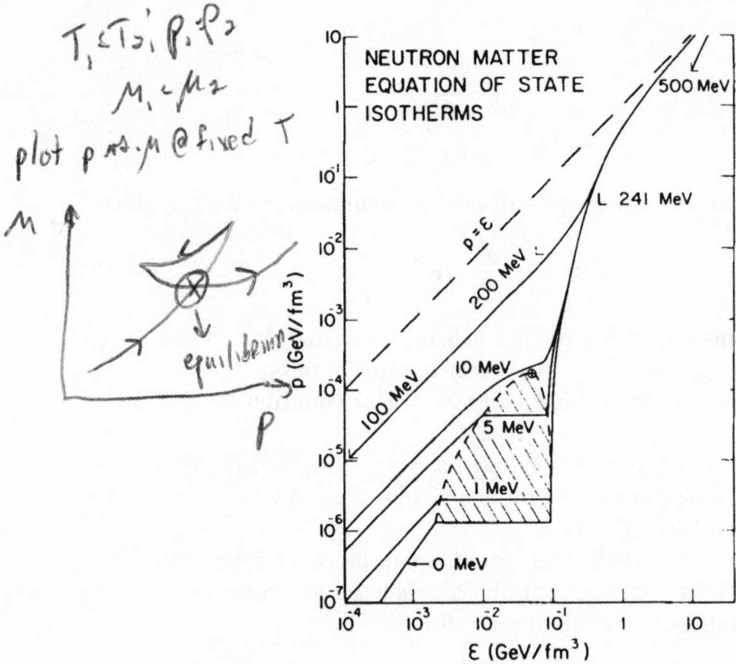

Figure 18.2: Isotherms of the neutron matter equation of state at finite temperature as calculated in the MFT of QHD-I. The curves are labeled by the value of $k_{\rm B}T$, and the left-hand end point of an isotherm corresponds to zero baryon density. The shaded area shows the region of phase separation, and the critical point at a temperature of $k_{\rm B}T_{\rm c} = 9.1 \pm 0.2$ MeV is indicated by $\odot$. From Ref. [R1].

3. The equilibrium conditions are determined numerically by plotting p vs. μ at fixed T and seeing where the curve crosses itself.

4. One has a critical region and a critical temperature, above which there is no phase transition.

5. The high temperature isotherms *terminate* as the energy density is decreased. There is a finite, limiting value of ε as $\rho_{\rm B} \to 0$. One simply has a vanishingly dilute solution of baryons in a sea of pairs.

Solution to the Self-Consistency Relation. The corresponding solution to the self-consistency equation for M^* as a function of temperature T is shown in Fig. 18.3 at $\rho_{\rm B} = \mu = 0$. The physics here is that pairs are produced as the temperature is increased. This does not change $\rho_{\rm B}$ but does change ρ_s; increasing ρ_s decreases M^*, which makes it easier to produce pairs. The equilibrium situation is obtained by solving the transcendental self-consistency relation at each temperature and density. *At high temperature the baryons are massless.* As one lowers the temperature, the baryons acquire a mass (at $k_{\rm B}T \ll$

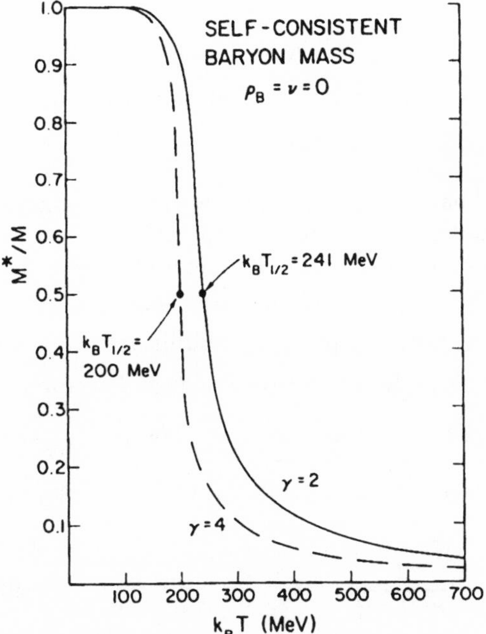

Figure 18.3: Self-consistent nucleon mass as a function of temperature at vanishing baryon density. Results are indicated both for neutron matter ($\gamma = 2$ — based on Fig. 18.2) and nuclear matter ($\gamma = 4$).

M) due to the self-consistent freezing out of the vacuum pairs; they then retain that mass down to $T = 0$.

Finite Temperature Field Theory in QHD-I. As at zero temperature, one can characterize the finite temperature field theory in terms of a set of Feynman rules. The thermal baryon Green's function, for example, is defined by (see Ref. [N2])

$$\mathcal{G}_{\alpha\delta}(\mathbf{x}_1\tau_1, \mathbf{x}_2\tau_2) \equiv -\text{Tr}\{\hat{\rho}_G P_\tau[\hat{\psi}_{\kappa\alpha}(\mathbf{x}_1\tau_1)\hat{\bar{\psi}}_{\kappa\delta}(\mathbf{x}_2\tau_2)]\}$$

$$\equiv \int \frac{d^3\kappa}{(2\pi)^3} e^{i\boldsymbol{\kappa}\cdot(\mathbf{x}_1-\mathbf{x}_2)} \frac{1}{\beta} \sum_{n\,(\text{odd})} e^{-i\omega_n(\tau_1-\tau_2)} \mathcal{G}_{\alpha\delta}(\boldsymbol{\kappa}, \omega_n) \quad (18.22)$$

The frequency is given by

$$\omega_n \equiv \frac{(2n+1)\pi}{\beta} \quad (18.23)$$

The statistical operator is defined by

$$\hat{\rho}_G \equiv \frac{e^{-\beta\hat{K}}}{\text{Tr}\{e^{-\beta\hat{K}}\}} \quad (18.24)$$

with

$$\hat{K} \equiv \hat{H} - \mu\hat{B} \tag{18.25}$$

The fields are in the Heisenberg representation with imaginary time

$$\hat{\psi}_K(\mathbf{x}\tau) \equiv e^{\hat{K}\tau}\hat{\psi}(\mathbf{x})e^{-\hat{K}\tau} \tag{18.26}$$

Given the thermal baryon Green's function, one can compute ensemble averages of bilinear combinations of baryon field operators.

Feynman Rules for $-\mathcal{G}(\boldsymbol{\kappa}\ ,\omega_n)$. We proceed to give the Feynman rules for the thermal baryon Green's function $-\mathcal{G}(\boldsymbol{\kappa}\ ,\omega_n)$.[32]

1. Draw all topologically distinct connected diagrams.
2. Include the following factors for the scalar and vector vertices, respectively

$$g_s 1\ ; \qquad \text{scalar} \qquad\qquad i g_v \gamma_\mu\ ; \qquad \text{vector} \tag{18.27}$$

3. Include the following factors for the scalar, vector, and baryon propagators, respectively

$$\begin{aligned} &\frac{1}{\kappa^2 + m_s^2} &; \qquad &\text{scalar} \\ &\frac{1}{\kappa^2 + m_v^2}\left(\delta_{\mu\nu} + \frac{\kappa_\mu \kappa_\nu}{m_v^2}\right) &; \qquad &\text{vector} \\ &\frac{1}{i\gamma_\mu p_\mu + M} &; \qquad &\text{baryon} \end{aligned} \tag{18.28}$$

Here, in the first two (boson) propagators

$$\begin{aligned} \kappa &= (\boldsymbol{\kappa}\ , i[i\omega_n]) \\ \omega_n &= \frac{2n\pi}{\beta} \end{aligned} \tag{18.29}$$

while in the last (fermion) propagator

$$\begin{aligned} p &= (\mathbf{p}, i[\mu + i\omega_n]) \\ \omega_n &= \frac{(2n+1)\pi}{\beta} \end{aligned} \tag{18.30}$$

4. Conserve frequency and wave number at each vertex.
5. For each internal line perform

$$\int \frac{d^3q}{(2\pi)^3}\frac{1}{\beta}\sum_n \tag{18.31}$$

[32] They are derived as in Ref. [N2]; see Ref. [R41]. Reference [R42] contains a much more extensive discussion of QHD at finite temperature.

6. Take the Dirac matrix product along fermion lines.[33]
7. Include a factor of $(-1)^F$ where F is the number of closed fermion loops.

Application. In Section 14 the MFT at zero temperature was extended to the RHA, which includes a proper treatment of the zero-point energy. For the extension to the RHA treatment of the finite-temperature equation of state, see Ref. [R43].

$$P = \frac{1}{2} m_v^2 V_0^2 - \frac{1}{2} m_s^2 \phi_0^2 + \frac{1}{\beta}\frac{\gamma}{(2\pi)^3}\int d^3k \left[\ln\left(1+e^{-\beta(E_k^*-\mu^*)}\right) + \ln\left(1+e^{-\beta(E_k^*+\mu^*)}\right)\right]$$

$$du = k^2 dk \qquad v = \ln\left(1+e^{-\beta(E_k^*-\mu^*)}\right)$$

$$u = \frac{1}{3}k^3 \qquad dv = \frac{e^{-\beta(E_k^*-\mu^*)}}{1+e^{-\beta(E_k^*-\mu^*)}}\left[-\beta\frac{k}{(k^2+m^{*2})^{1/2}}\right]$$

$$P = \frac{1}{2} m_v^2 V_0^2 - \frac{1}{2} m_s^2 \phi_0^2 + \frac{1}{3}\frac{\gamma}{(2\pi)^3}\int d^3k\,(n_k+\bar{n}_k)\frac{k^2}{(k^2+m^{*2})^{1/2}}$$

$$\varepsilon = \frac{1}{V}E = \frac{1}{V}\langle\langle \hat{H}_{MFT}\rangle\rangle = \frac{1}{2}m_s^2\phi_0^2 - \frac{1}{2}m_v^2 V_0^2 + g_v V_0 \rho_B + \frac{\gamma}{(2\pi)^3}\int d^3k\,(n_k+\bar{n}_k)$$

[33]We have again suppressed isospin; there is a factor δ_{ij} with $i, j = 1, 2$ in the baryon propagator.

19

QCD AND A PHASE TRANSITION

Quarks and Color.[34] There is now convincing evidence that hadrons are composed of a simpler substructure of *quarks*. The primary evidence for this is the following:

- If one assumes the baryons are composed of quark triplets (qqq) and the mesons are quark-antiquark pairs $(q\bar{q})$ then, with appropriate quantum numbers for the quarks (flavors), one can describe and predict the observed supermultiplets of hadrons.
- The assumption of interaction with point-like quarks provides a marvelously simple and accurate description of electroweak currents.
- Dynamic evidence for a point-like quark-parton substructure of hadrons is obtained from deep inelastic electron scattering (e, e') and neutrino reactions $(\nu_{\mathrm{l}}, \mathrm{l}^-)$.

Quarks come in many *flavors*; the quark field can be written as

$$\psi = \begin{pmatrix} u \\ d \\ s \\ c \\ \vdots \end{pmatrix} \tag{19.1}$$

One assigns quarks an additional intrinsic degree of freedom called *color*, which takes three values $i = R, G, B$. The quark field then becomes

$$\psi = \begin{pmatrix} u_R & u_{\mathrm{G}} & u_{\mathrm{B}} \\ d_R & d_{\mathrm{G}} & d_{\mathrm{B}} \\ s_R & s_{\mathrm{G}} & s_{\mathrm{B}} \\ c_R & c_{\mathrm{G}} & c_{\mathrm{B}} \end{pmatrix} = (\psi_R, \psi_{\mathrm{G}}, \psi_{\mathrm{B}}) \equiv \psi_i \; ; \qquad i = R, G, B \tag{19.2}$$

[34] The material on quarks, gluons, and quantum chromodynamics (QCD) will be developed in detail, with appropriate references, in Part III of this book. The present section is included as background for a simple model calculation of the phase diagram of nuclear matter.

It is convenient to construct a column vector from the color fields

$$\underline{\psi} \equiv \begin{pmatrix} \psi_R \\ \psi_G \\ \psi_B \end{pmatrix} \tag{19.3}$$

Matrices in this color space will be here denoted with a bar under a symbol. This is a very compact notation

- Each ψ_i has many flavors.
- Each flavor is a four-component Dirac field.

QCD. Quantum chromodynamics (QCD) is a theory of the strong interactions binding quarks into the observed hadrons. It is a Yang-Mills nonabelian gauge theory built on color and invariance under local $SU(3)_c$. We develop this theory in detail in Part III of this book. Here for the present purposes we anticipate that discussion and summarize some of the results:

- Introduce massless gauge boson fields, the gluons, $A_\mu^a(x)$ with $a = 1, \ldots, 8$; one for each generator.
- The lagrangian density then takes the form

$$\mathcal{L}_{\text{QCD}} = -\frac{1}{4}\mathcal{F}_{\mu\nu}^a \mathcal{F}_{\mu\nu}^a - \bar{\underline{\psi}}\gamma_\mu \left(\frac{\partial}{\partial x_\mu} - \frac{i}{2}g\underline{\lambda}^a A_\mu^a(x)\right)\underline{\psi} \tag{19.4}$$

  Repeated latin superscripts are summed $a = 1, \ldots, 8$.

- The $\underline{\lambda}^a$ are the $SU(3)$ matrices satisfying

$$[\frac{1}{2}\underline{\lambda}^a, \frac{1}{2}\underline{\lambda}^b] = if^{abc}\frac{1}{2}\underline{\lambda}^c \tag{19.5}$$

- The gluon field tensor is given by

$$\mathcal{F}_{\mu\nu}^a = \frac{\partial A_\nu^a}{\partial x_\mu} - \frac{\partial A_\mu^a}{\partial x_\nu} + gf^{abc}A_\mu^b A_\nu^c \tag{19.6}$$

- The lagrangian density is written for massless quarks; however, a mass term of the form

$$\delta\mathcal{L}_{\text{mass}} = -\bar{\underline{\psi}}\underline{M}\underline{\psi} \tag{19.7}$$

  where

$$\underline{M} = \begin{pmatrix} m & & \\ & m & \\ & & m \end{pmatrix} \tag{19.8}$$

  is the unit matrix with respect to color, leaves local $SU(3)_c$ invariance.

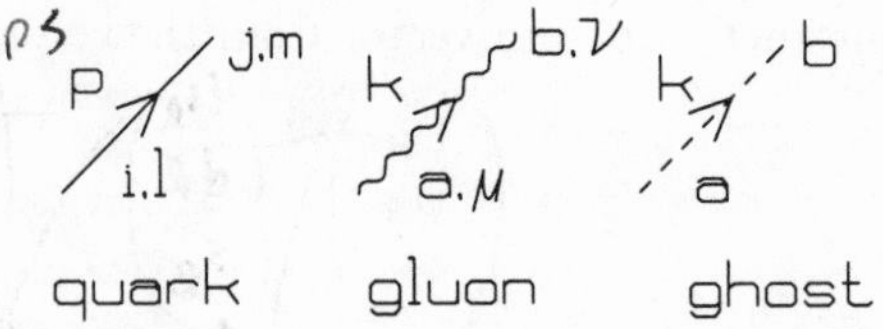

p, i.l, j.m — quark; k, a.μ, b.ν — gluon; k, a, b — ghost

Figure 19.1: Propagators in QCD.

Feynman Rules. The theory of QCD can again be characterized by a set of Feynman rules. The quark Green's function in the vacuum sector is defined by

$$
\begin{aligned}
iG_{\alpha\beta}(\mathbf{x}_1t_1, \mathbf{x}_2t_2) &\equiv \langle 0|P[\hat{\psi}_\alpha(\mathbf{x}_1t_1), \hat{\bar{\psi}}_\beta(\mathbf{x}_2t_2)]|0\rangle \\
&\equiv \int \frac{d^4k}{(2\pi)^4} e^{ik\cdot(x_1-x_2)} iG_{\alpha\beta}(k)
\end{aligned} \tag{19.9}
$$

The Feynman rules for $iG(k)$ are as follows (Refs. [R44, R45, R4]):[35]

1. Draw all topologically distinct, connected diagrams.

2. Include the following factors for the quark, gluon, and *ghost* lines, respectively (Fig. 19.1):[36]

$$
\begin{aligned}
&\frac{1}{i}\frac{1}{i\gamma_\mu p_\mu}\delta_{ij}\delta_{lm} \quad ; && \text{quark (massless)} \\
&\frac{1}{i}\delta^{ab}\frac{1}{k^2}\left(\delta_{\mu\nu} - \frac{k_\mu k_\nu}{k^2}\right) \quad ; && \text{gluon (Landau gauge)} \\
&\frac{1}{i}\delta^{ab}\frac{1}{k^2} \quad ; && \text{ghost}
\end{aligned} \tag{19.10}
$$

The ghost is an internal element, coupled to gluons, that is required to generate the correct S-matrix in a nonabelian gauge theory.

3. Include the following factors for the vertices indicated in Fig. 19.2:

$$
\begin{aligned}
&-g\frac{1}{2}\lambda^a_{ji}\delta_{lm}\gamma_\mu \quad ; && (\text{quark})^2\text{–gluon} \\
&gf^{abc}[(q-r)_\lambda\delta_{\mu\nu} + (p-q)_\nu\delta_{\lambda\mu} + (r-p)_\mu\delta_{\lambda\nu}] \quad ; && (\text{gluon})^3 \\
&-ig^2[f^{abe}f^{cde}(\delta_{\lambda\nu}\delta_{\sigma\mu} - \delta_{\lambda\sigma}\delta_{\mu\nu}) \\
&\quad + f^{ace}f^{bde}(\delta_{\lambda\mu}\delta_{\sigma\nu} - \delta_{\lambda\sigma}\delta_{\mu\nu}) \\
&\quad + f^{ade}f^{cbe}(\delta_{\lambda\nu}\delta_{\sigma\mu} - \delta_{\sigma\nu}\delta_{\lambda\mu})] \quad ; && (\text{gluon})^4
\end{aligned}
$$

[35] See Ref. [R45] for a much more extensive discussion, including Feynman rules with other choices of gauge.

[36] All quark indices are now explicit: $i, j = R, G, B$ for color; $l, m = u, d, s, c, \cdots$ for flavor.

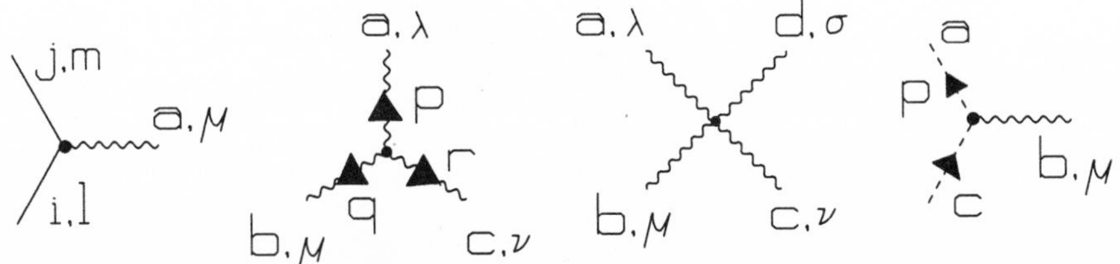

Figure 19.2: Vertices in QCD.

$$-gf^{abc}p_\mu \qquad ; \quad (\text{ghost})^2\text{–gluon} \tag{19.11}$$

4. Take the Dirac matrix product along fermion lines.
5. Conserve four-momentum at each vertex.
6. Include a factor $\int d^4q/(2\pi)^4$ for each independent internal line.
7. Include a factor of $(-1)^{F+G}$ where F is the number of closed fermion loops and G is the number of closed ghost loops.

Properties of QCD. QCD has two absolutely remarkable properties. The first is *confinement.* It is an empirical fact that quarks and color are confined to the interior of hadrons. There is evidence from lattice gauge theory calculations, discussed in detail in Part III of this book, that confinement is a dynamic property of QCD arising from the strong, nonlinear gluon couplings. The second property is *asymptotic freedom*, again arising from the nonlinear gluon couplings; this implies that at very large momenta, or equivalently at very short distances, the renormalized coupling constant gets very small and the theory is asymptotically free. The effect arises from the antishielding of the color charge (as opposed to the shielding one has in QED). When the effective coupling constant is small, one can do perturbation theory.

Relation Between QCD and QHD. What is the relationship of QHD to QCD? At this stage in the present development there are various possibilities:

- There is an approximate separation radius R in coordinate space for the hadron; and one can use QCD at short distances inside of R and QHD at large distance outside of R. This is the basis of bag models of the hadrons, also discussed in Part III.

- One can perform this separation in momentum space using spectral representations. The contributions from nearby singularities can be expressed in terms of observed hadron amplitudes; and the far-off, asymptotically free, contributions can be calculated in perturbation theory. These two contributions can then be joined in some manner. This is one of the basic concepts of QCD sum rules (Ref. [R46] and additional references in Part III).

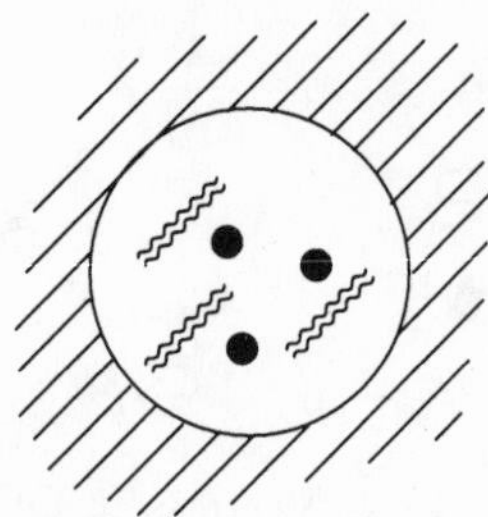

Figure 19.3: Vacuum bubble into which one inserts quarks and gluons as simple model of confinement property in QCD.

- A third possibility is that one has two models for two distinct phases of nuclear matter: QHD (treated in MFT) for a baryon-meson phase and QCD (treated as asymptotically free) for a quark-gluon phase. It is this third possibility that ties in directly with the development in this part of the book, and we proceed to carry out a simple model calculation of the phase diagram of nuclear matter.

Phase Diagram of Nuclear Matter. Nuclear matter will be modeled as consisting of two phases —a baryon-meson phase described using QHD-I in MFT (Section 18), and a quark-gluon phase described with asymptotically-free QCD. The discussion will be restricted to the nuclear domain consisting of u, d quarks (assumed massless) and their antiquarks. The quark field in the *nuclear domain* takes the form

$$\psi \doteq \begin{pmatrix} u \\ d \end{pmatrix} ; \qquad \text{nuclear domain} \tag{19.12}$$

The confinement property will be modeled by assuming that it takes a constant, finite energy per unit volume $+b$ to create a bubble into which the quarks and gluons can then be inserted (Fig. 19.3)[37]

$$\left(\frac{E}{V}\right)_{\rm vac} = +b \tag{19.13}$$

The following degeneracy factors will be used for the quark-gluon system

$$\begin{aligned} \gamma_{\rm Q} &= (3\,\text{colors}) \times (2\,\text{flavors}) \times (2\,\text{helicities}) = 12 \\ \gamma_{\rm G} &= (8\,\text{colors}) \times (2\,\text{helicities}) = 16 \end{aligned} \tag{19.14}$$

The equation of state for asymptotically free quarks and gluons (assumed massless) follows immediately (Ref. [N2]) [38]

$$\varepsilon = \frac{E}{V} = +b + \frac{\gamma_{\rm Q}}{(2\pi)^3}\int kd^3k(n_k + \bar{n}_k) + \frac{\gamma_{\rm G}}{(2\pi)^3}\int \frac{kd^3k}{e^{\beta k} - 1}$$

[37] This is the basic concept of the M.I.T. bag model (see Part III).

[38] Since gluons are not conserved, the gluon chemical potential vanishes (see Prob. 18.1).

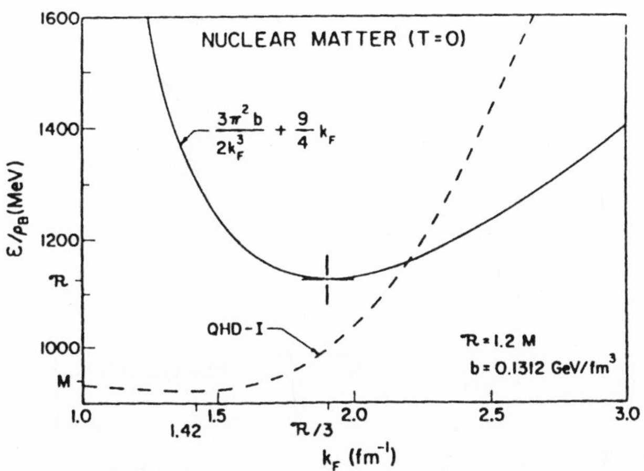

Figure 19.4: Saturation curve at $T = 0$ for nuclear matter. The solid curve denotes the quark-gluon result, with $\mathcal{R}$ from Eq. (19.22). The baryon density determines $k_{\rm F}$ through $\rho_{\rm B} = 2k_{\rm F}^3/3\pi^2$. The dashed curve is the result for nuclear matter from Section 14. From Ref. [R1].

$$
\begin{aligned}
p &= -b + \frac{1}{3}\left\{\frac{\gamma_{\rm Q}}{(2\pi)^3}\int kd^3k(n_k + \bar{n}_k) + \frac{\gamma_{\rm G}}{(2\pi)^3}\int \frac{kd^3k}{e^{\beta k}-1}\right\} \\
\rho_{\rm B} &= \frac{1}{3}\frac{\gamma_{\rm Q}}{(2\pi)^3}\int d^3k(n_k - \bar{n}_k)
\end{aligned} \tag{19.15}
$$

The fermion distribution functions here are given by (recall that quarks carry baryon number 1/3)

$$
n_k = \frac{1}{e^{\beta(k-\mu/3)}+1} \qquad \bar{n}_k = \frac{1}{e^{\beta(k+\mu/3)}+1} \tag{19.16}
$$

Some Analytical Results. Several analytical results follow from this simple parametric equation of state for the quark-gluon phase:

1. The equation of state at all T and $\rho_{\rm B}$ is given by

$$
3(p+b) = \varepsilon - b \tag{19.17}
$$

2. At finite baryon density $\rho_{\rm B} \equiv 2k_{\rm F}^3/3\pi^2$ (see Section 3) and zero temperature $T = 0$

$$
3(p+b) = \varepsilon - b = \frac{3}{2\pi^2}k_{\rm F}^4 \tag{19.18}
$$

Here the Fermi pressure of the quarks keeps the bubble from collapsing.

3. At finite temperature $T \neq 0$ and vanishing baryon density $\rho_{\rm B} = \mu = 0$ (vacuum)

$$
3(p+b) = \varepsilon - b = \frac{37}{30}\pi^2(k_{\rm B}T)^4 \tag{19.19}
$$

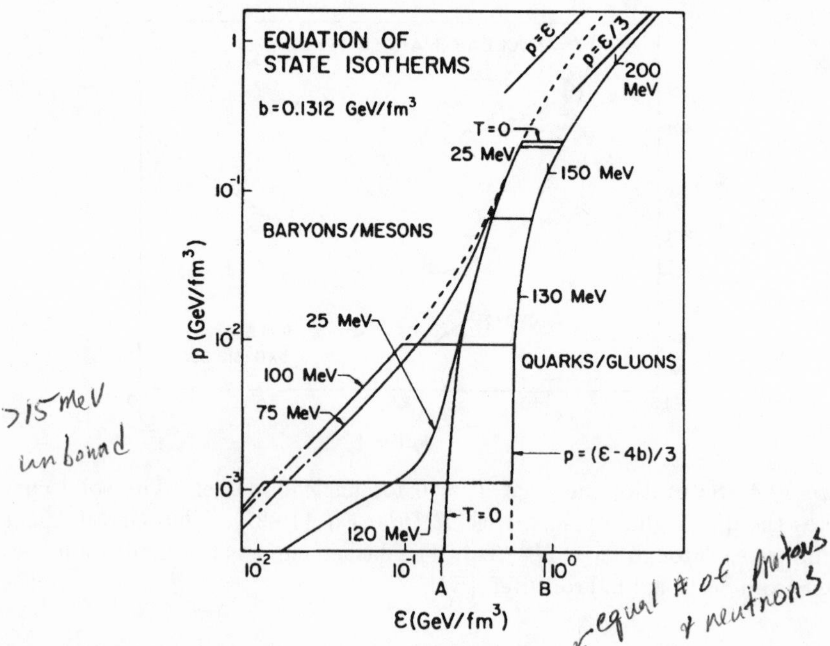

Figure 19.5: Equation of state isotherms for nuclear matter. Equilibrium between the baryon-meson and quark-gluon phases exists along the horizontal segments. The arrow 'B' indicates the energy density at the center of the most massive neutron star in Section 14. The endpoints of the highest T curves are $\rho_{\rm B} = 0$. From Ref. [R1].

Here the thermal pressure keeps the bubble from collapsing.

4. It follows from this last result that at zero baryon density, the pressure will *vanish* at a temperature T_0, which satisfies the condition

$$b = \frac{37}{90}\pi^2 (k_{\rm B}T_0)^4 \tag{19.20}$$

Above this temperature, the thermal pressure of the quarks and gluons causes the bubble to expand.

Phase Equilibrium. The equilibrium conditions for two phases described with the equations of state in Eqs. (18.14) and Eqs. (19.15) can be determined by demanding that Gibbs' criteria for phase equilibrium are again satisfied

$$p_1 = p_2 \qquad \mu_1 = \mu_2 \qquad T = \text{constant} \tag{19.21}$$

There is one free parameter b left in the calculation; it will be chosen *for purposes of illustration* so that quark-gluon matter at $T = 0$ saturates at baryon densities well above that of observed nuclear matter. The choice

$$\mathcal{R} \equiv 3(2\pi^2 b)^{1/4} \equiv 1.2\, M \; ; \qquad \text{arbitrary choice} \tag{19.22}$$

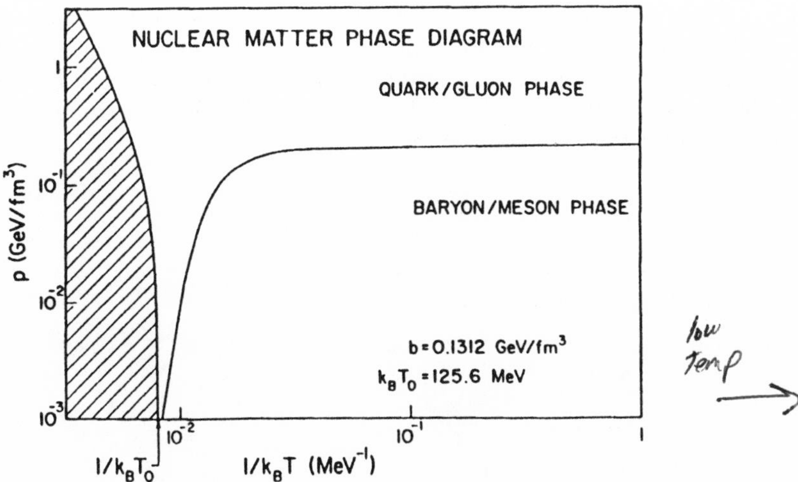

Figure 19.6: Phase diagram for nuclear matter based on Fig. 19.5. The equilibrium vapor pressure is ploted against $1/k_{\mathrm{B}}T$. The boundary of the shaded region is given by Eq. (19.19).

results in the situation illustrated in Fig. 19.4.

Numerical Results. The resulting isotherms for nuclear matter are shown in Fig. 19.5 for the indicated values of $k_{\mathrm{B}}T$. It is evident that at high enough baryon density ρ_{B}, or temperature T, the equilibrium phase is always quark-gluon.

The resulting phase diagram and vapor pressure curve for nuclear matter is shown in Fig. 19.6.

This is only a very simple model calculation, but it has the following features to recommend it:

- It is a completely relativistic model of the phase diagram.
- The QHD description of the baryon-meson phase is consistent with most observed properties of real nuclei.
- The QCD description of the quark-gluon phase is consistent with asymptotic freedom.
- The statistical mechanics has been done exactly.

Some important references for this phase transition are Refs. [R47, R48, R49, R50, R51]; others are given in Ref. [R1].

There are now regular international conferences on *quark matter*, and the reader is referred to the proceedings for the latest developments. A nice discussion of the detection of the quark-gluon plasma is contained in Ref. [R52]. The relativistic heavy-ion collider (RHIC), currently under construction, is designed to search for this new state of matter.

20

PIONS

The pion is the lightest mass quantum of the nuclear force (with mass here denoted by $m_\pi \equiv \mu$). It gives rise to the longest-range part of the two-nucleon interaction. It is a pseudoscalar meson with $J^\pi = 0^-$. It is also an isovector meson with three charge states (π^+, π^0, π^-) or in terms of hermitian components $\boldsymbol{\pi} = (\pi_1, \pi_2, \pi_3)$. The pion couples to the spin and isospin of the nucleon; it is the source of the long-range tensor force (see Section 1). The pion does not contribute to the binding energy of nuclear matter in the Hartree approximation since its contribution averages to zero in a spin, isospin saturated system.

Some General Considerations. Consider the general structure of the S-matrix for $\pi - N$ scattering (Fig. 20.1). Conservation of four-momentum implies

$$q_1 + p_1 = q_2 + p_2 \tag{20.1}$$

Here the symbols indicate four-vectors. Define the combinations

$$Q \equiv \frac{1}{2}(q_1 + q_2) \qquad\qquad P \equiv \frac{1}{2}(p_1 + p_2) \tag{20.2}$$

and the two independent Lorentz scalars

$$\nu = -\frac{P \cdot Q}{M} \qquad\qquad \kappa^2 = \frac{1}{4}(q_1 - q_2)^2 \tag{20.3}$$

Alternatively, one can work with two of the three scalars

$$s = -(p_1 + q_1)^2 \qquad u = -(p_2 - q_1)^2 \qquad t = -(q_1 - q_2)^2 \tag{20.4}$$

Note that these three quantities are linearly related[39]

$$s + t + u = 2M^2 + 2\mu^2 \tag{20.5}$$

In the center-of-momentum (C-M) system

$$\begin{aligned} s &= W^2 \\ t &= -2\mathbf{q}^2(1 - \cos\theta) \end{aligned} \tag{20.6}$$

[39] $s + t + u = -p_1^2 - p_2^2 - 3q_1^2 - q_2^2 - 2q_1 \cdot (p_1 - p_2 - q_2) = -p_1^2 - p_2^2 - q_1^2 - q_2^2 = 2M^2 + 2\mu^2$.

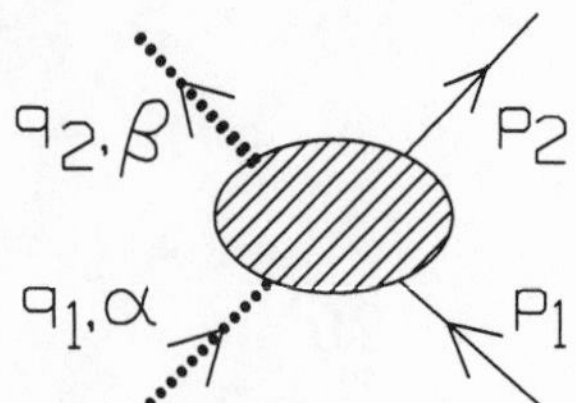

Figure 20.1: General structure of the S-matrix for $\pi - N$ scattering.

The general structure of the S-matrix is as follows:

$$
\begin{aligned}
S_{\mathrm{fi}} &= -\frac{(2\pi)^4}{\Omega^2} i\delta^{(4)}(p_1 + q_1 - p_2 - q_2)\left(\frac{M^2}{4E_1E_2\omega_1\omega_2}\right)^{1/2} T_{\mathrm{fi}} \\
T_{\mathrm{fi}} &= \bar{u}(p_2)\left[-A(s,t,u) + i\gamma_\mu Q_\mu B(s,t,u)\right] u(p_1) \qquad (20.7)
\end{aligned}
$$

Here periodic boundary conditions in a big box of volume Ω are assumed for the spatial wavefunctions (in the end $\Omega \to \infty$), the Dirac wavefunctions have invariant norm $\bar{u}u = 1$, and the scalar functions can alternatively be written in terms of the arguments $A(\nu, \kappa^2), B(\nu, \kappa^2)$. We suppress the nucleon isospinors. The general phenomenology of $\pi - N$ scattering is summarized in Appendix E.

Pseudoscalar Coupling and σ Exchange. Let us make a first attempt to include pions in our QHD lagrangian density in order to calculate some simple $\pi - N$ processes. We want covariance, parity invariance, and isospin invariance. We also ask for renormalizability. The only acceptable $\pi - N$ coupling is then $ig_\pi \bar{\psi}\gamma_5 \boldsymbol{\tau} \cdot \boldsymbol{\pi}\psi$. The simplest renormalizable $\pi - \phi$ coupling is $\frac{1}{2}g_\phi m_s \boldsymbol{\pi}^2\phi$. Here the coupling constants (g_π, g_ϕ) are dimensionless and real. The first attempt at a QHD-II lagrangian thus takes the form (Ref. [R1])

$$
\begin{aligned}
\mathcal{L} &= -\bar{\psi}\left[\gamma_\mu \frac{D}{Dx_\mu} + (M - g_s\phi) - ig_\pi\gamma_5\boldsymbol{\tau}\cdot\boldsymbol{\pi}\right]\psi \\
&\quad -\frac{1}{4}F_{\mu\nu}F_{\mu\nu} - \frac{1}{2}m_v^2 V_\mu^2 - \frac{1}{2}\left(\frac{\partial\phi}{\partial x_\lambda}\frac{\partial\phi}{\partial x_\lambda} + m_s^2\phi^2\right) \\
&\quad -\frac{1}{2}\left(\frac{\partial\boldsymbol{\pi}}{\partial x_\lambda}\cdot\frac{\partial\boldsymbol{\pi}}{\partial x_\lambda} + \mu^2\boldsymbol{\pi}^2\right) + \frac{1}{2}g_\phi m_s\boldsymbol{\pi}^2\phi \qquad (20.8)
\end{aligned}
$$

Here

$$
\begin{aligned}
\frac{D}{Dx_\mu} &\equiv \frac{\partial}{\partial x_\mu} - ig_v V_\mu(x) \\
F_{\mu\nu} &\equiv \frac{\partial V_\nu}{\partial x_\mu} - \frac{\partial V_\mu}{\partial x_\nu} \qquad (20.9)
\end{aligned}
$$

The lagrangian density thus takes the form

$$
\mathcal{L} = \mathcal{L}_{\mathrm{QHD}} + \mathcal{L}_\pi^0 + \mathcal{L}_\pi^{\mathrm{int}} \qquad (20.10)
$$

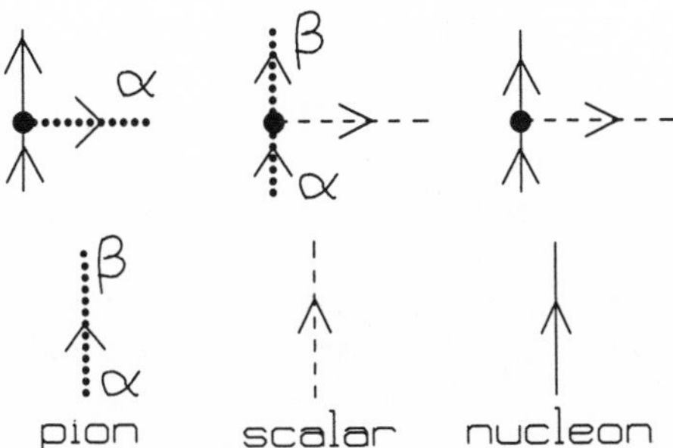

Figure 20.2: Some components of the Feynman rules for the S-matrix S_{fi} for the QHD-II lagrangian in Eq. (20.8).

The first term is the QHD-I lagrangian, the second is the free pion lagrangian, and the last is the interaction term

$$\mathcal{L}_{\pi}^{\text{int}} = ig_{\pi}\bar{\psi}\gamma_5\boldsymbol{\tau}\cdot\boldsymbol{\pi}\psi + \frac{1}{2}g_{\phi}m_s\boldsymbol{\pi}^2\phi \tag{20.11}$$

The Feynman rules for the S-matrix follow directly from this lagrangian density. We give a subset of the rules, useful for our immediate purposes.

Feynman Rules for Baryon, Scalar, and Pion Contributions to S_{fi}:[40]

1. Draw all topologically distinct connected diagrams.
2. Include the following vertex factors (Fig. 20.2).

$$\begin{aligned} -g_{\pi}\gamma_5\tau_{\alpha}\ ; &\qquad (\text{baryon})^2\text{–pion} \\ ig_{\phi}m_s\delta_{\alpha\beta}\ ; &\qquad (\text{pion})^2\text{–scalar} \\ ig_s\ ; &\qquad (\text{baryon})^2\text{–scalar} \end{aligned} \tag{20.12}$$

3. Include the following factors for the propagators (Fig. 20.2).

$$\begin{aligned} \frac{1}{i}\frac{1}{i\gamma_{\mu}p_{\mu}+M}\ ; &\qquad \text{baryon} \\ \frac{1}{i}\frac{1}{k^2+\mu^2}\delta_{\alpha\beta}\ ; &\qquad \text{pion} \\ \frac{1}{i}\frac{1}{k^2+m_s^2}\ ; &\qquad \text{scalar} \end{aligned} \tag{20.13}$$

4. Include the following factors for (incoming) external lines (Fig. 20.2).

$$\sqrt{\frac{M}{E}}\frac{1}{\sqrt{\Omega}}u\ ; \qquad \text{baryon}$$

[40] There is an additional (baryon)^2 – vector vertex, a vector propagator, and external vector lines (Prob. 20.6).

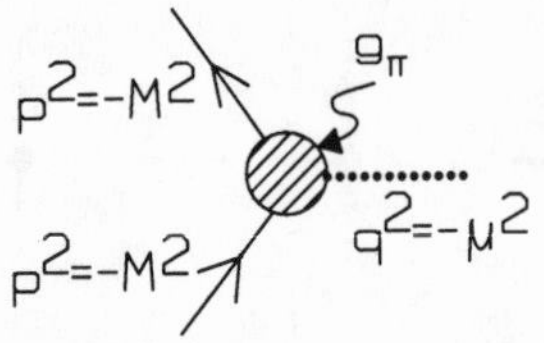

Figure 20.3: Renormalized coupling constant, with vertex corrections, appearing as the residue of the pole in particle-exchange amplitudes.

$$\frac{1}{\sqrt{2\omega\Omega}}\;; \qquad \text{pion} \qquad \frac{1}{\sqrt{2\omega\Omega}}\;; \qquad \text{scalar} \tag{20.14}$$

5. Include a factor $(2\pi)^4\delta^{(4)}(\Delta p)$ at each vertex.
6. Include a factor $\int d^4q/(2\pi)^4$ for each internal line.
7. Include a factor $(-1)^F$ where F is the number of closed baryon loops.

Particle-Exchange Poles. We now use these Feynman rules to calculate the contribution of the particle-exchange graphs to the π-N scattering amplitude. These exchanges give rise to *poles* in the amplitude. At the pole, the pole gives the entire contribution, and hence the exact answer. The pole will generally occur in an unphysical region of the kinematic variables. When close to the physical region, it will be a good approximation to keep just the pole contribution to the scattering amplitude. The coupling constants that appear are renormalized coupling constants. They are the *residues* at the poles. They include vertex corrections that remain when all of the particles at that vertex are on the mass shell (Fig. 20.3).

The particle-exchange graphs are shown in Fig. 20.4a,b,c. The s-channel

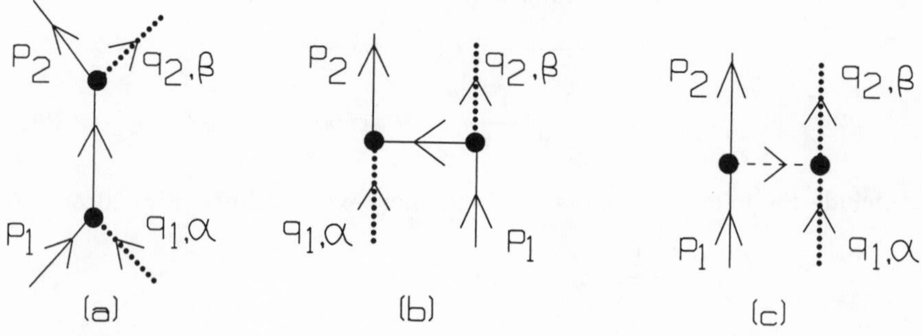

Figure 20.4: Particle-exchange graphs: (a) s-channel baryon; (b) u-channel baryon; (c) t-channel scalar.

baryon contribution to T_{fi} is

$$
\begin{aligned}
T_{(a)} &= g_\pi^2 \bar{u}(p_2)\left[\gamma_5\tau_\beta \frac{1}{i\gamma_\mu(p_1+q_1)_\mu + M}\gamma_5\tau_\alpha\right]u(p_1) \\
&= g_\pi^2 \bar{u}(p_2)\left[\tau_\beta \frac{M + i\gamma_\mu(p_1+q_1)_\mu}{M^2+(p_1+q_1)^2}\tau_\alpha\right]u(p_1) \qquad (20.15)
\end{aligned}
$$

The second line follows from the first by rationalizing and using the properties of γ_5.[41] Now use the Dirac equation[42]

$$
\begin{aligned}
(i\gamma_\mu p_{1\mu} + M)u(p_1) &= 0 \\
\frac{1}{2}\bar{u}(p_2)i\gamma_\mu(p_2+q_2-p_1)_\mu u(p_1) &= \frac{1}{2}\bar{u}(p_2)i\gamma_\mu q_{2\mu}u(p_1) \qquad (20.16)
\end{aligned}
$$

Also write $\tau_\beta\tau_\alpha \equiv \{\tau_\beta,\tau_\alpha\}/2+[\tau_\beta,\tau_\alpha]/2$. This allows one to identify [Eq. (E.3)]

$$
B^{\pm}_{(a)} = \frac{g_\pi^2}{M^2-s} \qquad\qquad A^{\pm}_{(a)} = 0 \qquad (20.17)
$$

The u-channel baryon pole comes from the graph in Fig. 20.4b. The contribution to the T-matrix is

$$
T_{(b)} = g_\pi^2\bar{u}(p_2)\left[\gamma_5\tau_\alpha\frac{1}{i\gamma_\mu(p_1-q_2)_\mu+M}\gamma_5\tau_\beta\right]u(p_1) \qquad (20.18)
$$

The analysis is the same as for the first term, only now $\tau_\alpha\tau_\beta = \delta_{\alpha\beta} - [\tau_\beta,\tau_\alpha]/2$. Thus one obtains

$$
B^{\pm}_{(b)} = \mp\frac{g_\pi^2}{M^2-u} \qquad\qquad A^{\pm}_{(b)} = 0 \qquad (20.19)
$$

The third term in Fig. 20.4c is a scalar exchange and contributes

$$
T_{(c)} = (-g_s g_\phi m_s)\bar{u}(p_2)\left[\frac{1}{(q_1-q_2)^2+m_s^2}\delta_{\alpha\beta}\right]u(p_1) \qquad (20.20)
$$

Thus

$$
A^{+}_{(c)} = \frac{g_s g_\phi m_s}{m_s^2 - t} \qquad\qquad B^{\pm}_{(c)} = A^{-}_{(c)} = 0 \qquad (20.21)
$$

In summary the pole terms are given by the sum of these contributions

$$
\begin{aligned}
B^{\pm} &= g_\pi^2\left[\frac{1}{M^2-s}\mp\frac{1}{M^2-u}\right] \\
A^{+} &= g_s g_\phi m_s\frac{1}{m_s^2-t}\,; \qquad\qquad A^{-} = 0 \qquad (20.22)
\end{aligned}
$$

[41]Note $\gamma_5 \equiv \gamma_1\gamma_2\gamma_3\gamma_4$ with $\{\gamma_5,\gamma_\mu\} = 0$ and $\gamma_5^2 = 1$.

[42]Thus between these Dirac spinors $q_1 \equiv (q_1+q_1)/2 = (q_1+p_2-p_1+q_2)/2 \doteq Q$.

Threshold Behavior. In the C-M system at threshold there is only s-wave scattering (Appendix E). The kinematic variables at threshold become

$$s = (M + \mu)^2 \qquad u = (M - \mu)^2 \qquad t = 0 \tag{20.23}$$

If one now takes the further limit of zero-mass pions, $\mu \to 0$, then the nucleon poles move to threshold. Thus at threshold, in the limit $\mu \to 0$, the nucleon poles give the *exact answer*!

The general expression for the s-wave scattering length is given by Eqs. (E.9) and (E.10)

$$a_{0+}^{\pm} = \frac{1}{4\pi}\frac{1}{(1+\mu/M)}(A^{\pm} + \mu B^{\pm}) \tag{20.24}$$

Substitution of the pole terms and the threshold values in Eqs. (20.22) and (20.23) then gives the following expressions for the contribution of the pole terms to the scattering lengths

$$\begin{aligned} a_{0+}^{-} &= 2\frac{g_\pi^2}{4\pi}\frac{\mu^2}{4M^2}\frac{1}{(1+\mu/M)}\frac{1}{(1-\mu^2/4M^2)}\frac{1}{\mu} \\ a_{0+}^{+} &= \left[\frac{g_s g_\phi}{4\pi}\frac{\mu}{m_s} - \frac{g_\pi^2}{4\pi}\frac{\mu}{M}\frac{1}{(1-\mu^2/4M^2)}\right]\frac{1}{(1+\mu/M)}\frac{1}{\mu} \end{aligned} \tag{20.25}$$

Now estimate the order of magnitude of these scattering lengths, using (Ref. [R1]) $\mu/M \approx 1/7$ and $g_\pi^2 = 14.4$. From the nucleon pole terms alone one has

$$\begin{aligned} a_{0+}^{-} &\approx \frac{g_\pi^2}{4\pi}2\frac{\mu^2}{4M^2}\frac{1}{\mu} \approx \frac{0.15}{\mu} \\ a_{0+}^{+} &\approx -\frac{g_\pi^2}{4\pi}\frac{\mu}{M}\frac{1}{\mu} \approx -\frac{2.1}{\mu} \end{aligned} \tag{20.26}$$

The experimental results are (Ref. [R1]; see also Ref. [R75])

$$\begin{aligned} a_{0+}^{-} &= \left(0.097 \begin{array}{c} +0.003 \\ -0.007 \end{array}\right)\frac{1}{\mu} \\ a_{0+}^{+} &= (-0.015 \pm 0.015)\frac{1}{\mu} \end{aligned} \tag{20.27}$$

The estimate for the isospin-odd scattering length a_{0+}^{-} is clearly in the right ballpark; the isospin-even scattering length a_{0+}^{+} is too large by two orders of magnitude!

The second estimate can be fixed by making use of the pole contribution from scalar exchange, which contributes only to a_{0+}^{+}. Arrange to have it *exactly cancel* the nucleon pole contribution. For reasons that will become clear as we proceed, this cancellation is arranged to take place at $\mu = 0$. From Eq. (20.25) this condition is

$$\frac{g_s g_\phi}{m_s} \equiv \frac{g_\pi^2}{M} \tag{20.28}$$

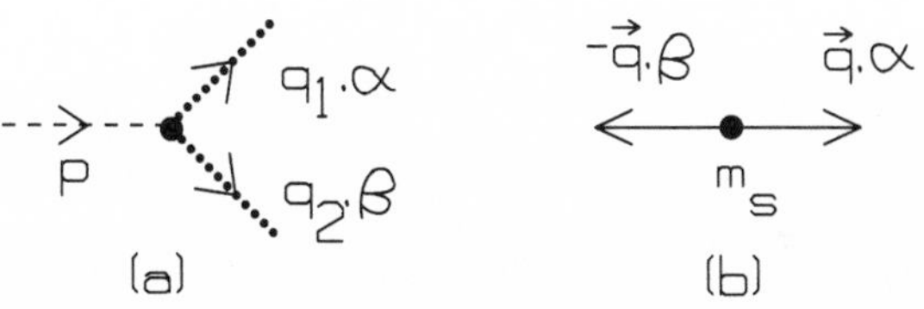

Figure 20.5: Decay amplitude for $\phi \to \pi + \pi$: (a) general structure; (b) in the C-M system.

It provides a constraint on coupling constants and masses. Substitution back into Eq. (20.25) then gives

$$a_{0+}^{+} \approx -\frac{g_\pi^2}{4\pi}\frac{\mu^2}{4M^2}\frac{\mu}{M}\frac{1}{(1+\mu/M)}\frac{1}{(1-\mu^2/4M^2)}\frac{1}{\mu} \tag{20.29}$$

An estimate similar to that above now gives

$$a_{0+}^{+} \quad \approx \quad -\frac{g_\pi^2}{4\pi}\frac{\mu^2}{4M^2}\frac{\mu}{M}\frac{1}{\mu} \approx -\frac{0.01}{\mu} \tag{20.30}$$

This scattering length is now *also* in the right ballpark; however, this arises by a cancellation of two orders of magnitude in the amplitude.[43]

These arguments are taken from Ref. [R1], which contains further references to original work. They were first developed within the framework of QHD in Ref. [R57].

Decay Rate for $\phi \to \pi + \pi$. Given g_π from π–N dispersion relations and (g_s, m_s) from MFT, Eq. (20.28) can be solved for g_ϕ. This in turn allows one to calculate the decay rate for the decay of the scalar meson (which we have referred to as the σ) into two pions $\phi \to \pi + \pi$ using the previously stated Feynman rules. The process is illustrated in Fig. 20.5. The S-matrix takes the form

$$\begin{aligned} S_{\rm fi} &= -\frac{(2\pi)^4}{\Omega^{3/2}} i\delta^{(4)}(p - q_1 - q_2)\tilde{T}_{\rm fi} \\ \tilde{T}_{\rm fi} &= -\frac{g_\phi m_s \delta_{\alpha\beta}}{\sqrt{8\omega_p \omega_{q_1}\omega_{q_2}}} \end{aligned} \tag{20.31}$$

The decay rate then follows from Fermi's Golden Rule

$$d\omega_{\rm fi} = \frac{2\pi}{\Omega}|\tilde{T}_{\rm fi}|^2 \delta(W_f - W_i)d\rho_f \tag{20.32}$$

Go to the C-M system (Fig. 20.5b). The rate is then calculated through the following series of steps:

[43]Note that the "natural" magnitude of the π–N scattering lengths is $a_0 \approx 1/\mu$.

1. In this frame

$$\begin{aligned} W_f &= 2\sqrt{\mathbf{q}^2+\mu^2}\,; \qquad\qquad W_i = m_s \\ \frac{\partial W_f}{\partial q} &= 2\frac{q}{\sqrt{\mathbf{q}^2+\mu^2}} = 2\frac{q}{\omega_q} \end{aligned} \tag{20.33}$$

2. The density of final states for a particle satisfying periodic boundary conditions in a box of volume Ω is

$$d\rho_f = \frac{\Omega d^3q}{(2\pi)^3} = \frac{\Omega q^2 d\Omega_q}{(2\pi)^3}\frac{\partial q}{\partial W_f}dW_f \tag{20.34}$$

3. The integral over final solid angles is $\int d\Omega_q = 2\pi$; since there are identical particles in the final state that come out back-to-back, the integration over 1/2 the total solid angle counts all processes.

4. The isopin sums are given by $\sum_{\alpha\beta}|\delta_{\alpha\beta}|^2 = 3$.

5. A combination of these factors yields $\omega_{\rm fi} = (3g_\phi^2/32\pi)(q/\omega_q)m_s$.

6. The kinematics give

$$\begin{aligned} m_s &= 2\omega_q = 2\sqrt{\mathbf{q}^2+\mu^2} \\ q &= \sqrt{\left(\frac{m_s}{2}\right)^2 - \mu^2} = \frac{m_s}{2}\sqrt{1-\frac{4\mu^2}{m_s^2}} \end{aligned} \tag{20.35}$$

7. The final result for the decay rate is

$$\omega_{\rm fi} \equiv \Gamma = \frac{3g_\phi^2}{32\pi}m_s\sqrt{1-\frac{4\mu^2}{m_s^2}} \tag{20.36}$$

Let us put some numbers in this expression. It is convenient to first summarize in one place the masses and coupling constants used so far (Section 14 and Ref. [R1]):

$$\begin{aligned} m_p &= 938.3\,\text{MeV} & \frac{g_\pi^2}{4\pi} &= 14.40 \\ m_{\pi^\pm} &= 139.6\,\text{MeV} & \frac{g_s^2}{4\pi} &= 7.303 \\ m_s &= 550\,\text{MeV} & \frac{g_v^2}{4\pi} &= 10.83 \\ C_s^2 \equiv g_s^2\left(\frac{M^2}{m_s^2}\right) &= 267.1 & \frac{g_\phi^2}{4\pi} &= 9.756 \\ C_v^2 \equiv g_v^2\left(\frac{M^2}{m_v^2}\right) &= 195.9 & \frac{g_s g_\phi}{m_s} &= \frac{g_\pi^2}{M} \\ m_v = m_\omega &= 782.0\,\text{MeV} \end{aligned} \tag{20.37}$$

In subsequent work we will make use of the combination

$$f_\pi^2 \equiv \frac{g_\pi^2}{4\pi}\left(\frac{\mu}{2M}\right)^2 = 0.0797 \tag{20.38}$$

These values give for the width of the scalar meson

$$\Gamma_{\phi\rightarrow\pi+\pi} = 1734\,\text{MeV} \tag{20.39}$$

There is no angular momentum barrier, and the scalar meson ϕ just falls apart into pions in free space. There is thus no simple interpretation of the degree of freedom of the scalar field as a particle (meson) since its width is so much larger than its mass in this model.

21

CHIRAL INVARIANCE

Introduction. We start from two observations. First, it was shown in the previous section that if the following condition is satisfied

$$\frac{g_s g_\phi}{m_s} = \frac{g_\pi^2}{M} \tag{21.1}$$

then the s-wave $\pi - N$ scattering lengths arising from the pole terms vanish in the limit of zero pion mass

$$a_0^{\pm} \stackrel{\mu \to 0}{\longrightarrow} 0 \tag{21.2}$$

Second, as we shall discuss in detail in Part IV, the charge-changing weak interactions couple to hadronic vector and axial vector currents

$$J_\mu^{\mathbf{V}}(x) \equiv \mathbf{V}_\mu(x) \qquad\qquad J_{\mu 5}^{\mathbf{V}}(x) \equiv \mathbf{A}_\mu(x) \tag{21.3}$$

The first current is a Lorentz four-vector and the second is an axial vector; both are isovectors. The vector current is conserved

$$\frac{\partial \mathbf{V}_\lambda}{\partial x_\lambda} = 0 \tag{21.4}$$

The axial vector current is *partially conserved*

$$\frac{\partial \mathbf{A}_\lambda}{\partial x_\lambda} \stackrel{\mu \to 0}{\longrightarrow} 0 \tag{21.5}$$

One knows from Noether's theorem that an invariance (symmetry) of the lagrangian leads directly to a conserved current

$$\delta \mathcal{L} \;=\; \frac{\partial}{\partial x_\lambda}\left[\frac{\partial \mathcal{L}}{\partial(\partial q_1/\partial x_\lambda)}\delta q_1 + \cdots\right] = 0 \tag{21.6}$$

Therefore, let us go back and examine the symmetry structure of the lagrangian.

Isospin Invariance — A Review. Recall the situation for isospin (Refs. [R2, R4]). Take the lagrangian density of Section 20 (with $g_\phi \equiv 0$ initially to simplify the illustration).

$$\begin{aligned}\mathcal{L} \;=\; & -\bar{\psi}\left[\gamma_\lambda \frac{D}{Dx_\lambda} + (M - g_s\phi) - ig_\pi\gamma_5\boldsymbol{\tau}\cdot\boldsymbol{\pi}\right]\psi \\ & -\frac{1}{2}\left[\frac{\partial \boldsymbol{\pi}}{\partial x_\lambda}\cdot\frac{\partial \boldsymbol{\pi}}{\partial x_\lambda} + \mu^2\boldsymbol{\pi}^2\right] + \mathcal{L}_\phi^0 + \mathcal{L}_v^0\end{aligned} \tag{21.7}$$

The last two terms are isoscalars and will not enter into the argument. In this expression the covariant derivative is defined by

$$\frac{D}{Dx_\lambda} \equiv \frac{\partial}{\partial x_\lambda} - ig_v V_\lambda(x) \tag{21.8}$$

The isospin transformation is defined by

$$e^{-i\boldsymbol{\omega}\cdot\hat{\mathbf{T}}}\underline{\psi}e^{+i\boldsymbol{\omega}\cdot\hat{\mathbf{T}}} = [\underline{e}^{\frac{1}{2}\boldsymbol{\tau}\cdot\boldsymbol{\omega}}]\underline{\psi}\;; \qquad \underline{\psi} \equiv \begin{pmatrix} \psi_p \\ \psi_n \end{pmatrix} \tag{21.9}$$

Here $\hat{\mathbf{T}}$ is an operator in the abstract occupation number Hilbert space and the $\boldsymbol{\tau}$ are 2×2 Pauli matrices. A bar under a symbol denotes a matrix. This relation can also be written as[44]

$$\hat{R}(\boldsymbol{\omega})\underline{\psi}\hat{R}(\boldsymbol{\omega})^{-1} \equiv \underline{\psi}' = \underline{\mathrm{r}}(\boldsymbol{\omega})\underline{\psi} \tag{21.10}$$

The corresponding transformation on the isovector pion field is just a rotation

$$\hat{R}(\boldsymbol{\omega})\pi_i\hat{R}(\boldsymbol{\omega})^{-1} \equiv \pi_i' = a_{ij}(\boldsymbol{\omega})\pi_j \tag{21.11}$$

These transformations leave the lagrangian in Eq. (21.7) invariant.

$$\mathcal{L} \rightarrow \mathcal{L}' = \mathcal{L} \tag{21.12}$$

It is often convenient to specify to the infinitesimal transformation $\boldsymbol{\omega} = \boldsymbol{\varepsilon} \rightarrow 0$ where the transformation becomes

$$\begin{aligned} \delta\underline{\psi} &= i\boldsymbol{\varepsilon}\cdot\frac{1}{2}\boldsymbol{\tau}\,\underline{\psi} \\ \delta\underline{\pi} &= i\boldsymbol{\varepsilon}\cdot\underline{\mathrm{t}}\,\underline{\pi} \end{aligned} \tag{21.13}$$

The quantities appearing in the second equation are defined by

$$\underline{\pi} = \begin{pmatrix} \pi_1 \\ \pi_2 \\ \pi_3 \end{pmatrix} \qquad (\underline{\mathrm{t}}_i)_{jk} \equiv -i\varepsilon_{ijk} \tag{21.14}$$

Equation (21.13) is a rewriting of the relation $\delta\vec{\pi} = -\vec{\varepsilon} \times \vec{\pi}$, which characterizes an infinitesimal rotation. Since they represent just the infinitesimal form of the transformation, Eqs. (21.13) still imply $\delta\mathcal{L} = 0$.[45]

[44] In subsequent developments in this book, carets will often be omitted from above operators in the abstract Hilbert space when the operator nature is evident from the context. The matrix notation of a bar under a symbol will similarly often be suppressed. For clarity, these indicators will generally be restored in the final result.

[45] Note the notation here, $\underline{\pi}$ is the pion field and *not* the canonical momentum, which will explicitly be denoted by $\dot{\underline{\pi}}, \dot{\phi}, \dot{\sigma}$, etc.

It now follows from Noether's theorem [Eq. (21.6)] that the following current is conserved (the constant arbitrary overall infinitesimal factor $-\varepsilon$ can be dropped)

$$\begin{aligned} \mathbf{V}_\mu &= i\bar{\underline{\psi}}\gamma_\mu \frac{1}{2}\boldsymbol{\tau}\,\underline{\psi} + i\frac{\partial \underline{\pi}^\dagger}{\partial x_\mu}\underline{\mathrm{t}}\,\underline{\pi} \\ &= i\bar{\underline{\psi}}\gamma_\mu \frac{1}{2}\boldsymbol{\tau}\,\underline{\psi} + \frac{\partial \boldsymbol{\pi}}{\partial x_\mu}\times\boldsymbol{\pi} \end{aligned} \qquad (21.15)$$

The integral over all space of the fourth component of the current is the total charge

$$\mathbf{T} \equiv \int d^3x\, \mathbf{V}_0 \qquad (21.16)$$

This has two properties:

1. If the current is conserved, then just as in the case of electrodynamics, the total charge is a *constant of the motion.*

2. The total charge is also the *generator of the transformation*; here the three charges form the isospin operators, which generate isospin transformations.[46]

The isospin operator is thus given by

$$\begin{aligned} \mathbf{T} &= \int d^3x \left[\underline{\psi}^\dagger \frac{1}{2}\boldsymbol{\tau}\,\underline{\psi} + \frac{1}{i}\dot{\underline{\pi}}^\dagger\underline{\mathrm{t}}\,\underline{\pi}\right] \\ &= \int d^3x \left[\underline{\psi}^\dagger \frac{1}{2}\boldsymbol{\tau}\,\underline{\psi} - \dot{\boldsymbol{\pi}}\times\boldsymbol{\pi}\right] \end{aligned} \qquad (21.17)$$

In quantum mechanics, the generators are characterized by their commutation relations. These can now be computed with the aid of the canonical (anti-) commutation rules

$$\begin{aligned} \left\{\psi_{\alpha\rho}(\mathbf{x}), \psi^\dagger_{\beta\sigma}(\mathbf{x}')\right\} &= \delta_{\alpha\beta}\delta_{\rho\sigma}\delta^{(3)}(\mathbf{x}-\mathbf{x}') \\ [\pi_i(\mathbf{x}), \dot{\pi}_j(\mathbf{x}')] &= i\delta_{ij}\delta^{(3)}(\mathbf{x}-\mathbf{x}') \end{aligned} \qquad (21.18)$$

It then follows from the defining Eqs. (21.17) that

$$[\hat{T}_i, \hat{T}_j] = i\varepsilon_{ijk}\hat{T}_k \qquad (21.19)$$

The generators of the isospin transformation form a *Lie algebra*, and just as with angular momentum, it is the algebra of $SU(2)$.

Chiral Transformation. What about the (partial) conservation of the axial vector current? To what symmetry principle does this correspond? We again define an operator transformation as in Eq. (21.10), only this time of the form

$$e^{-i\boldsymbol{\omega}\cdot\hat{\mathbf{T}}_5}\underline{\psi}\,e^{+i\boldsymbol{\omega}\cdot\hat{\mathbf{T}}_5} \equiv \underline{\psi}' = [\,e^{\frac{i}{2}\boldsymbol{\tau}\cdot\boldsymbol{\omega}\gamma_5}]\,\underline{\psi} \qquad (21.20)$$

[46] Just as the angular momentum operators $\mathbf{J}$ generate rotations.

This transformation mixes Dirac components as well as isospin. The matrices appearing in this expression are thus 8×8; they are unitary

$$\underline{r}^{\dagger} = [\underline{e}^{-\frac{i}{2}\boldsymbol{\tau}\cdot\boldsymbol{\omega}\gamma_5}] = \underline{r}^{-1} \tag{21.21}$$

Note also that

$$\begin{aligned} \underline{\psi}^{\dagger}\underline{r}^{\dagger}\gamma_4 &= \bar{\underline{\psi}}\,\underline{r} \\ \underline{r}\,\gamma_\mu &= \gamma_\mu\,\underline{r}^{-1} \end{aligned} \tag{21.22}$$

These follow from $\{\gamma_5, \gamma_\mu\} = 0$. The infinitesimal form of this chiral transformation is

$$\delta\underline{\psi} = [\, i\boldsymbol{\varepsilon}\cdot\frac{1}{2}\boldsymbol{\tau}\;\gamma_5\,]\,\underline{\psi} \tag{21.23}$$

Properties of the Transformation. It is proven in Prob. 21.2 that

$$\begin{aligned} e^{\frac{i}{2}\boldsymbol{\tau}\cdot\boldsymbol{\omega}} &= \cos\frac{\omega}{2} + i\mathbf{n}\cdot\boldsymbol{\tau}\,\sin\frac{\omega}{2} \\ e^{\frac{i}{2}\boldsymbol{\tau}\cdot\boldsymbol{\omega}\gamma_5} &= \cos\frac{\omega}{2} + i\mathbf{n}\cdot\boldsymbol{\tau}\gamma_5\,\sin\frac{\omega}{2} \end{aligned} \tag{21.24}$$

Here $\boldsymbol{\omega} \equiv \mathbf{n}\omega$ with $\mathbf{n}$ a unit vector. The first equation is a relation on the 2×2 $SU(2)$ matrices, and the second on the 8×8 chiral matrices.

Let us now look at the transformation properties of the various parts of the lagrangian in Eq. (21.7) under this chiral transformation:

1. The transformation of the baryon kinetic energy in the lagrangian goes as follows [note Eqs. (21.22)]

$$\bar{\underline{\psi}}'\,\gamma_\mu\frac{D}{Dx_\mu}\,\underline{\psi}' = \bar{\underline{\psi}}\,\gamma_\mu\frac{D}{Dx_\mu}\underline{r}^{-1}\underline{r}\,\underline{\psi} = \bar{\underline{\psi}}\,\gamma_\mu\frac{D}{Dx_\mu}\,\underline{\psi} \tag{21.25}$$

The kinetic energy is invariant under the chiral transformation because of the presence of the additional γ_μ. It is assumed in this analysis that $\boldsymbol{\omega}$ = constant independent of space-time so that it can be moved through the derivatives — it is a *global* (rather than a *local*) invariance.

2. Consider the transformation of the interaction term in Eq. (21.7) where for simplicity we initially assume $g_\pi = g_s$. The transformation of the baryon field alone leads to the following

$$\hat{R}\bar{\underline{\psi}}(\phi + i\gamma_5\boldsymbol{\tau}\cdot\boldsymbol{\pi})\underline{\psi}\hat{R}^{-1} = \bar{\underline{\psi}}\,\underline{r}(\phi + i\gamma_5\boldsymbol{\tau}\cdot\boldsymbol{\pi})\underline{r}\,\underline{\psi} \tag{21.26}$$

The finite transformation is analyzed in Prob. 21.3. Here we explicitly evaluate the infinitesimal form with $\boldsymbol{\omega} \equiv \boldsymbol{\varepsilon} \to 0$ [47]

$$\hat{R}\bar{\underline{\psi}}(\phi + i\gamma_5\boldsymbol{\tau}\cdot\boldsymbol{\pi})\underline{\psi}\hat{R}^{-1} \approx$$

[47] This is enough to identify the corresponding current through Noether's theorem.

$$
\begin{aligned}
&\approx \bar{\psi}(1 + \frac{i}{2}\boldsymbol{\varepsilon} \cdot \boldsymbol{\tau}\gamma_5)(\phi + i\gamma_5 \boldsymbol{\tau} \cdot \boldsymbol{\pi})(1 + \frac{i}{2}\boldsymbol{\varepsilon} \cdot \boldsymbol{\tau}\gamma_5)\psi \\
&\approx \bar{\psi}\{\phi + i\gamma_5 \boldsymbol{\tau} \cdot \boldsymbol{\pi} + i\boldsymbol{\varepsilon} \cdot \boldsymbol{\tau}\gamma_5 \phi \\
&\quad -\frac{1}{2}[(\boldsymbol{\varepsilon} \cdot \boldsymbol{\tau})(\boldsymbol{\tau} \cdot \boldsymbol{\pi}) + (\boldsymbol{\tau} \cdot \boldsymbol{\pi})(\boldsymbol{\varepsilon} \cdot \boldsymbol{\tau})]\}\psi \\
&= \bar{\psi}[(\phi - \boldsymbol{\varepsilon} \cdot \boldsymbol{\pi}) + i\gamma_5 \boldsymbol{\tau} \cdot (\boldsymbol{\pi} + \boldsymbol{\varepsilon}\phi)]\psi + O(\varepsilon^2)
\end{aligned}
\tag{21.27}
$$

This is not the full story, however, since one must also simultaneously transform the meson fields; for infinitesimal transformations these transformations can be written in the form

$$
\begin{aligned}
\hat{R}\phi\hat{R}^{-1} &= \phi' \equiv \phi + \delta\phi \\
\hat{R}\boldsymbol{\pi}\hat{R}^{-1} &= \boldsymbol{\pi}' \equiv \boldsymbol{\pi} + \delta\boldsymbol{\pi}
\end{aligned}
\tag{21.28}
$$

It is evident that the additional terms arising from the transformation of the baryon fields alone can be exactly cancelled by the transformation of the meson fields if that transformation is defined to satisfy

$$
\begin{aligned}
\delta\phi &= \boldsymbol{\varepsilon} \cdot \boldsymbol{\pi} \\
\delta\boldsymbol{\pi} &= -\boldsymbol{\varepsilon}\phi
\end{aligned}
\tag{21.29}
$$

The interaction lagrangian is then left invariant under this overall chiral transformation

$$
\delta\mathcal{L}_{\text{int}} = 0 \tag{21.30}
$$

If one starts from a more general interaction term that has different coupling constants for the scalar and pion fields [Eq. (21.7)]

$$
\mathcal{L}_{\text{int}} = \bar{\psi}(g_s\phi + ig_\pi\gamma_5\, \boldsymbol{\tau} \cdot \boldsymbol{\pi})\psi \tag{21.31}
$$

then the transformation of the meson fields that leaves this interaction term invariant depends on the strengths of the couplings

$$
\delta\phi = \left(\frac{g_\pi}{g_s}\right) \boldsymbol{\varepsilon} \cdot \boldsymbol{\pi} \qquad\qquad \delta\boldsymbol{\pi} = -\left(\frac{g_s}{g_\pi}\right) \boldsymbol{\varepsilon}\phi \tag{21.32}
$$

It is undesirable to have a symmetry property of the lagrangian depend on the strength of the coupling,[48] and therefore it shall henceforth be assumed that there is a single coupling constant

$$
g_\pi \;=\; g_s' \equiv g \tag{21.33}
$$

The possiblity $g_s \neq g_\pi$ is further ruled out by the symmetry properties of the meson kinetic energy, which we proceed to examine.

[48] Which is renormalized order by order.

3. The kinetic energy for the meson fields takes the form

$$\mathcal{L}_{\text{mes}}^{0} = -\frac{1}{2}\left[\left(\frac{\partial \phi}{\partial x_\lambda}\right)^2 + \left(\frac{\partial \boldsymbol{\pi}}{\partial x_\lambda}\right)^2\right] \tag{21.34}$$

Now with a transformation of the form in Eq. (21.29), the bilinear combination $\phi^2 + \boldsymbol{\pi}^2$ is invariant. This follows since

$$\begin{aligned} \delta\left(\phi^2 + \boldsymbol{\pi}^2\right) &= 2\left(\phi\,\delta\phi + \boldsymbol{\pi}\cdot\delta\boldsymbol{\pi}\right) \\ &= 2\left(\phi\boldsymbol{\varepsilon}\cdot\boldsymbol{\pi} - \boldsymbol{\pi}\cdot\boldsymbol{\varepsilon}\phi\right) = 0 \end{aligned} \tag{21.35}$$

It follows from this observation that the meson kinetic energy is invariant under the infinitesimal chiral transformation in Eq. (21.29)

$$\delta\mathcal{L}_{\text{mes}}^{0} = 0 \tag{21.36}$$

4. Consider next the meson mass and meson interaction terms. A meson potential of the form $-V(\phi^2 + \boldsymbol{\pi}^2)$ can now be added to the lagrangian. We choose to include the meson mass terms in this V. Note that the meson fields are assumed to enter in the bilinear combination $\phi^2 + \boldsymbol{\pi}^2$ to ensure invariance under the chiral transformation, as discussed above.

5. Finally, consider how the baryon mass term transforms

$$\hat{R}\,\bar{\psi}M\psi\,\hat{R}^{-1} \quad = \quad \bar{\psi}M\,\underline{\underline{\mathrm{rr}}}\,\psi \neq \bar{\psi}M\psi \tag{21.37}$$

The baryon mass term is *not* invariant under the chiral transformation. Since the baryon mass represents the largest energy in the problem, ones first reaction is simply to quit at this point and try another approach to pion physics. Alternatively, one can invoke the concept that the baryon mass term in the underlying lagrangian indeed vanishes, the lagrangian respects chiral symmetry, and the baryon mass is then generated through *spontaneous symmetry breaking*. This can arise through the nature of the meson mass and interaction term V, as we shall see. With an appropriate form of V, the scalar field will develop a *vacuum expectation value*

$$\langle\phi\rangle_{\text{vac}} \neq 0 \tag{21.38}$$

This term produces a baryon mass and spontaneously breaks the chiral symmetry. Conventionally, one denotes the scalar field appearing in the underlying chirally symmetric lagrangian, the field that develops the vacuum expectation value, as σ where

$$\langle\sigma\rangle_{\text{vac}} \equiv \sigma_0 \tag{21.39}$$

The field ϕ is then used to denote the fluctuations about this vacuum expectation value

$$\sigma \equiv \sigma_0 + \phi \tag{21.40}$$

6. In *summary*, the following is a lagrangian for baryons, isoscalar scalar and vector mesons, and pions that is invariant under the chiral transformation

$$\begin{aligned}\mathcal{L} &= -\bar{\psi}\left[\gamma_\lambda \frac{D}{Dx_\lambda} - g(\sigma + i\boldsymbol{\tau}\cdot\boldsymbol{\pi}\gamma_5)\right]\psi - \frac{1}{2}m_v^2 V_\mu^2 - \frac{1}{4}F_{\mu\nu}F_{\mu\nu} \\ &\quad -\frac{1}{2}\left[\left(\frac{\partial\sigma}{\partial x_\lambda}\right)^2 + \left(\frac{\partial\boldsymbol{\pi}}{\partial x_\lambda}\right)^2\right] - V(\sigma^2 + \boldsymbol{\pi}^2) \end{aligned} \tag{21.41}$$

The covariant derivative is defined in Eq. (21.8). This lagrangian is left invariant

$$\delta\mathcal{L} = 0 \tag{21.42}$$

under the chiral transformation

$$\begin{aligned}\delta\underline{\psi} &= [i\boldsymbol{\varepsilon}\cdot\frac{1}{2}\boldsymbol{\tau}\gamma_5\,]\,\underline{\psi} \\ \delta\sigma &= \boldsymbol{\varepsilon}\cdot\boldsymbol{\pi} \\ \delta\boldsymbol{\pi} &= -\boldsymbol{\varepsilon}\sigma\end{aligned} \tag{21.43}$$

Conserved Axial Current. Noether's theorem in Eq. (21.6) implies that this chiral invariance property of the lagrangian leads to a conserved current (the constant overall factor of $-\boldsymbol{\varepsilon}$ is again removed)

$$\mathbf{A}_\mu = i\bar{\psi}\gamma_\mu\gamma_5\frac{1}{2}\boldsymbol{\tau}\psi + \left(\frac{\partial\sigma}{\partial x_\mu}\boldsymbol{\pi} - \frac{\partial\boldsymbol{\pi}}{\partial x_\mu}\sigma\right) \tag{21.44}$$

The pion $\boldsymbol{\pi}$ is an isovector pseudoscalar field, and this is evidently an isovector, Lorentz axial-vector current.

The conservation of the axial vector current in Eq. (21.44) can also be proven explicitly from the equations of motion, and it will be of use in the subsequent developments to do so. The Euler-Lagrange equations derived from the lagrangian density in Eq. (21.41) for the $(\bar{\psi}, \psi, \sigma, \boldsymbol{\pi})$ fields, respectively, are as follows:

$$\begin{aligned}\left[\gamma_\lambda\frac{D}{Dx_\lambda} - g(\sigma + i\boldsymbol{\tau}\cdot\boldsymbol{\pi}\gamma_5)\right]\psi &= 0 \\ \bar{\psi}\left[\gamma_\lambda\frac{\overleftarrow{D}}{Dx_\lambda} + g(\sigma + i\boldsymbol{\tau}\cdot\boldsymbol{\pi}\gamma_5)\right] &= 0 \\ \left(\frac{\partial}{\partial x_\lambda}\right)^2\sigma = -g\bar{\psi}\psi + 2\sigma V'(\sigma^2 + \boldsymbol{\pi}^2) \\ \left(\frac{\partial}{\partial x_\lambda}\right)^2\boldsymbol{\pi} = -ig\bar{\psi}\gamma_5\boldsymbol{\tau}\psi + 2\boldsymbol{\pi}V'(\sigma^2 + \boldsymbol{\pi}^2)\end{aligned} \tag{21.45}$$

Here the derivative $\overleftarrow{D}/Dx_\mu$ acts to the left and is given by

$$\frac{\overleftarrow{D}}{Dx_\mu} \equiv \frac{\overleftarrow{\partial}}{\partial x_\mu} + ig_v V_\mu \; ; \qquad \text{acts to left} \tag{21.46}$$

Now use these relations to calculate the four-divergence of the axial vector current

$$\begin{aligned} \frac{\partial \mathbf{A}_\mu}{\partial x_\mu} &= i\left[\frac{\partial \bar{\psi}}{\partial x_\mu}\gamma_\mu\gamma_5 \frac{1}{2}\boldsymbol{\tau}\psi - \bar{\psi}\gamma_5 \frac{1}{2}\boldsymbol{\tau}\gamma_\mu \frac{\partial \psi}{\partial x_\mu}\right] + \boldsymbol{\pi}\left(\frac{\partial}{\partial x_\lambda}\right)^2 \sigma - \sigma\left(\frac{\partial}{\partial x_\lambda}\right)^2 \boldsymbol{\pi} \\ &= i\bar{\psi}\left[\gamma_\mu \frac{\overleftarrow{D}}{Dx_\mu}\gamma_5 \frac{1}{2}\boldsymbol{\tau} - \gamma_5 \frac{1}{2}\boldsymbol{\tau}\gamma_\mu \frac{D}{Dx_\mu}\right]\psi + \boldsymbol{\pi}\left(\frac{\partial}{\partial x_\lambda}\right)^2 \sigma - \sigma\left(\frac{\partial}{\partial x_\lambda}\right)^2 \boldsymbol{\pi} \end{aligned} \tag{21.47}$$

Substitution of Eqs. (21.45) now shows that this expression vanishes

$$\frac{\partial \mathbf{A}_\lambda}{\partial x_\lambda} = 0 \tag{21.48}$$

This relation cannot hold exactly if the pion mass is nonzero for then the pion would not decay (Prob. 21.5). Thus chiral symmetry must be an *approximate* one in nature where $\mu \neq 0$. We discuss a model for this breaking of chiral symmetry at the lagrangian level in the next section. For the present, we continue the discussion of the conserved current.

Generators of the Chiral Transformation. The integral over all space of the fourth component of the conserved axial vector current yields the following expression

$$\hat{\mathbf{T}}_5 = \int d^3x \left[\psi^\dagger \gamma_5 \frac{1}{2}\boldsymbol{\tau}\psi + \dot{\boldsymbol{\pi}}\sigma - \dot{\sigma}\boldsymbol{\pi}\right] \tag{21.49}$$

Just as in the discussion of the isospin transformation above, the three components of the vector operator $\hat{\mathbf{T}}_5$ have the following important properties:

- They are constants of the motion.
- They are the generators of the chiral transformation.[49]

Commutation Relations for the Generators. Just as with the generators of the isospin transformation in Eq. (21.17), the generators of the chiral transformation in Eq. (21.49) can again be characterized by their commutation relations. These are evaluated with the aid of the canonical (anti-)commutation relations

[49]The explicit proof, through use of the canonical (anti-)commutation relations, that the operators in Eq. (21.49) do indeed generate the transformation in Eq. (21.43) as the infinitesimal form of the defining Eqs. (21.20) and (21.28) is left as an exercise for the reader (Prob. 21.6).

satisfied by the $(\psi, \boldsymbol{\pi})$ fields in Eqs. (21.18) as well as that satisfied by the σ field

$$[\sigma(\mathbf{x}), \dot{\sigma}(\mathbf{x}')] = i\delta^{(3)}(\mathbf{x} - \mathbf{x}') \tag{21.50}$$

The result is (Appendix F)

$$\begin{aligned} [\hat{T}_i^5, \hat{T}_j^5] &= i\varepsilon_{ijk}\hat{T}_k \qquad\qquad [\hat{T}_i, \hat{T}_j] = i\varepsilon_{ijk}\hat{T}_k \\ [\hat{T}_i, \hat{T}_j^5] &= i\varepsilon_{ijk}\hat{T}_k^5 \end{aligned} \tag{21.51}$$

Note that it is only the entire set of isospin and chiral generators $(\hat{\mathbf{T}}, \hat{\mathbf{T}}^5)$ that is closed under commutation, and hence forms a Lie algebra. As shown in Appendix F, appropriate linear combinations of the generators reduce this algebra to that of two commuting $SU(2)$ subalgebras — $SU(2)_L \otimes SU(2)_R$.

Further progress depends on the specific form of the potential term $V(\sigma^2 + \boldsymbol{\pi}^2)$ and generation of the baryon mass by spontaneous symmetry breaking. We proceed to a discussion of the chiral σ-model of Schwinger and Gell-Mann and Lévy (Refs. [R58, R59]).

22

THE σ-MODEL

The development in this section is based on Refs. [R58, R59, R1]. We first summarize the argument up to this point. Consider the following lagrangian density built from isospinor baryon, isoscalar scalar and vector, and isovector pion fields $\{\psi, \sigma, V_\lambda, \boldsymbol{\pi}\}$

$$\begin{aligned} \mathcal{L} \;=\; & -\bar{\psi}\left[\gamma_\mu \frac{D}{Dx_\mu} - g(\sigma + i\boldsymbol{\tau}\cdot\boldsymbol{\pi}\gamma_5)\right]\psi - \frac{1}{2}m_v^2 V_\mu^2 - \frac{1}{4}F_{\mu\nu}F_{\mu\nu} \\ & -\frac{1}{2}\left[\left(\frac{\partial\sigma}{\partial x_\lambda}\right)^2 + \left(\frac{\partial\boldsymbol{\pi}}{\partial x_\lambda}\right)^2\right] - V(\boldsymbol{\pi}^2 + \sigma^2) \end{aligned} \tag{22.1}$$

Here

$$\frac{D}{Dx_\mu} \equiv \frac{\partial}{\partial x_\mu} - ig_v V_\mu(x) \tag{22.2}$$

The following statements then hold:

1. This lagrangian is invariant under isospin transformations. As a consequence, there is an isovector current $\mathbf{V}_\lambda(x)$ that is conserved $\partial\mathbf{V}_\lambda/\partial x_\lambda = 0$.

2. This lagrangian is invariant under chiral transformations. As a consequence, there is an isovector, axial-vector current $\mathbf{A}_\lambda$ that is conserved $\partial\mathbf{A}_\lambda/\partial x_\lambda = 0$.

3. This latter symmetry is observed to be only an approximate one in nature where the mass of the pion is nonzero $\mu \neq 0$ (PCAC); it can be expected to hold exactly only in the limit $\mu \to 0$.

4. This lagrangian is invariant under the symmetry group $SU(2)_L \bigotimes SU(2)_R$.

5. There is a conserved baryon current $B_\mu = i\bar{\psi}\gamma_\mu\psi$ to which the neutral vector meson V_μ couples.[50]

6. This lagrangian is invariant under parity, charge conjugation, and time reversal (P, C, T).

7. The quantum theory generated by this lagrangian is renormalizable if the potential V is at most quartic in the meson fields.

8. The baryon mass term is absent.

We must now discuss the form of V.

[50] This follows, as in the last section, from the invariance of the lagrangian under phase transformations of ψ and Noether's theorem (Prob. 22.3).

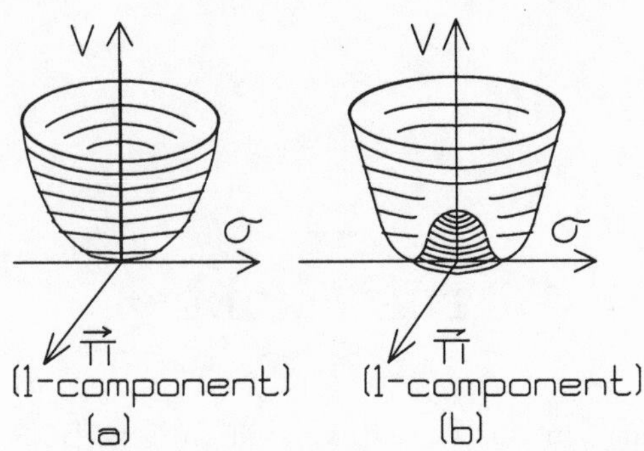

Figure 22.1: Meson potential surfaces V for (a) usual mass terms; (b) spontanteous symmetry breaking.

Spontaneous Symmetry Breaking. If the potential V represents just the usual meson mass term then

$$V^0_{\text{mass}}(\sigma^2 + \boldsymbol{\pi}^2) = \frac{1}{2}m^2(\sigma^2 + \boldsymbol{\pi}^2) \tag{22.3}$$

This potential surface is illustrated in Fig. 22.1a. Note that the meson masses must be equal $m_s^2 = m_\pi^2 \equiv m^2$ if chiral symmetry is to be preserved. This potential is clearly minimized if $\sigma_0 = \boldsymbol{\pi}_0 = 0$; hence there is no constant expectation value for the meson fields in the vacuum.

In contrast, assume the potential V has the following form, as illustrated in Fig. 22.1b

$$V^0 = \frac{\lambda}{4}\left[(\sigma^2 + \boldsymbol{\pi}^2) - v^2\right]^2 \tag{22.4}$$

Here $\lambda > 0$ and we will henceforth assume $g > 0$ and $v > 0$. The potential depends only on the combination $(\sigma^2 + \boldsymbol{\pi}^2)$ and respects chiral symmetry. Since it is at most quartic in the fields, it leads to a renormalizable theory. Now, however, it is clear that the origin is a local *maximum* rather than a true minimum, and the lowest energy state in V, the vacuum, will have a nonzero, constant expectation value for the meson fields. Thus, in this chirally symmetric theory, with this particular form for V, one will have

$$\langle\sigma^2 + \boldsymbol{\pi}^2\rangle_{\text{vac}} \neq 0 \tag{22.5}$$

If the vacuum is to have a definite parity, as appears to be the case in nature, then the vacuum expectation value of the pseudoscalar $\boldsymbol{\pi}$ field must vanish

$$\langle\boldsymbol{\pi}\rangle_{\text{vac}} \equiv \boldsymbol{\pi}_0 = 0 \tag{22.6}$$

How can these observations be reconciled? It has already been pointed out that chiral symmetry cannot be an exact symmetry of nature, but can be expected

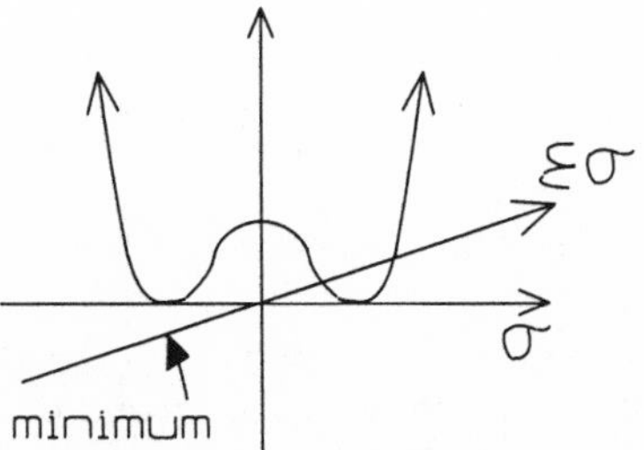

Figure 22.2: Contribution of additional chiral symmetry breaking term $\varepsilon\sigma$ to potential V as viewed along the σ axis.

to hold only in the limit that the pion mass $\mu \to 0$. Let us therefore augment the lagrangian in Eq. (22.1) with a chiral symmetry *breaking* term chosen so that the true minimum of the potential lies at a point satisfying Eq. (22.6). Evidently an additional term of the form

$$\delta V_{\text{csb}} \equiv \varepsilon\sigma \tag{22.7}$$

with $\varepsilon > 0$ will accomplish this objective. This additional term represents a plane in Fig. 22.1b whose addition tilts the potential so that a true minimum occurs along the negative σ axis (Fig. 22.2). Now in the true minimum of V, in the vacuum, there will be an additional constant field σ_0

$$\begin{aligned} \langle\sigma\rangle_{\text{vac}} &= \sigma_0 \\ \langle\boldsymbol{\pi}\rangle_{\text{vac}} &= 0 \end{aligned} \tag{22.8}$$

What does the additional term in Eq. (22.7) do to the equations of motion and the conservation of the axial vector current? It is clear from Eq. (21.45) that only the equation of motion for the σ field is modified; it now contains an additional term ε

$$\left(\frac{\partial}{\partial x_\lambda}\right)^2 \sigma = -g\bar{\psi}\psi + 2\sigma V'(\sigma^2 + \boldsymbol{\pi}^2) + \varepsilon \tag{22.9}$$

The modification of the four-divergence of the axial vector current follows immediately from Eq. (21.47)

$$\frac{\partial \mathbf{A}_\lambda}{\partial x_\lambda} = \varepsilon\boldsymbol{\pi} \tag{22.10}$$

It is proportional to the $\boldsymbol{\pi}$ field with constant of proportionality ε; evidently exact chiral symmetry is restored in the limit $\varepsilon \to 0$.

Now there are constant mean fields $(\sigma_0, \boldsymbol{\pi}_0)$ in the vacuum, and for the ground state, one just minimizes the total V

$$V = \frac{\lambda}{4}\left[(\sigma^2 + \boldsymbol{\pi}^2) - v^2\right]^2 + \varepsilon\sigma \tag{22.11}$$

The minimization conditions are

$$\begin{aligned}\frac{\partial V}{\partial \pi_0} &= \lambda\pi_0[(\sigma_0^2+\pi_0^2)-v^2]=0 \\ \frac{\partial V}{\partial \sigma_0} &= \lambda\sigma_0[(\sigma_0^2+\pi_0^2)-v^2]+\varepsilon=0 \end{aligned} \tag{22.12}$$

For positive ε, the second equation implies $[(\sigma_0^2+\pi_0^2)-v^2]\neq 0$; the first equation then implies that $\pi_0=0$, in accord with the second of Eqs. (22.8). The first of Eqs. (22.8) is now also satisfied with σ_0 given by the solution to

$$\lambda\sigma_0[\sigma_0^2-v^2]=-\varepsilon \tag{22.13}$$

It is evident from Fig. 22.2 that the absolute minimum of V occurs for negative σ_0. One can therefore introduce the following definitions

$$\sigma_0 \equiv -\frac{M}{g} \qquad \varepsilon\equiv\frac{M}{g}\mu^2 \qquad \lambda\equiv\frac{m_s^2-\mu^2}{2M^2}g^2 \tag{22.14}$$

The first relation defines a *baryon mass* M, for the presence of this term precisely serves the role of a mass in the Dirac lagrangian

$$-\bar{\psi}\left[\gamma_\mu\frac{D}{Dx_\mu}-g\sigma_0\right]\psi=-\bar{\psi}\left[\gamma_\mu\frac{D}{Dx_\mu}+M\right]\psi \tag{22.15}$$

The second of Eqs. (22.14) expresses the chiral symmetry breaking parameter ε in terms of what will turn out to be the pion mass $\mu^2\equiv m_\pi^2$. Note that now one has explicitly $\varepsilon\to 0$ as $\mu^2\to 0$, and chiral symmetry is restored in the limit that the pion mass goes to zero. The third of Eqs. (22.14) defines the scalar mass m_s^2; evidently $m_s^2>\mu^2$ here. These mass terms will be identified through the quadratic field terms in the lagrangian obtained in a new expansion about the true minimum of V.[51]

Now consider excitations built on this ground state. One looks for

$$\begin{aligned}\sigma &= \sigma_0+\phi \\ \boldsymbol{\pi} &= \boldsymbol{\pi}\end{aligned} \tag{22.16}$$

Here ϕ is the excitation of the scalar field and $\boldsymbol{\pi}$ that of the pion field; these take place about the minimum of the potential sketched in Fig. 22.3. Substitution of these relations into the lagrangian density in Eq. (22.1) leads to the following

$$\begin{aligned}\mathcal{L} &= -\bar{\psi}\left[\gamma_\mu\frac{D}{Dx_\mu}-g(-\frac{M}{g}+\phi+i\boldsymbol{\tau}\cdot\boldsymbol{\pi}\gamma_5)\right]\psi+\mathcal{L}_v^0 \\ &\quad -\frac{1}{2}\left[\left(\frac{\partial\phi}{\partial x_\lambda}\right)^2+\left(\frac{\partial\boldsymbol{\pi}}{\partial x_\lambda}\right)^2\right]-\frac{\lambda}{4}\left\{[(\sigma_0+\phi)^2+\boldsymbol{\pi}^2]-v^2\right\}^2 \\ &\quad -\varepsilon(\sigma_0+\phi)\end{aligned} \tag{22.17}$$

[51] Note that if one takes $\{g, M, \mu=m_\pi\}$ from experiment together with $\{g_v, m_v\}$, where $g=g_\pi$, then the only parameter left in the model is m_s, the mass of the scalar field.

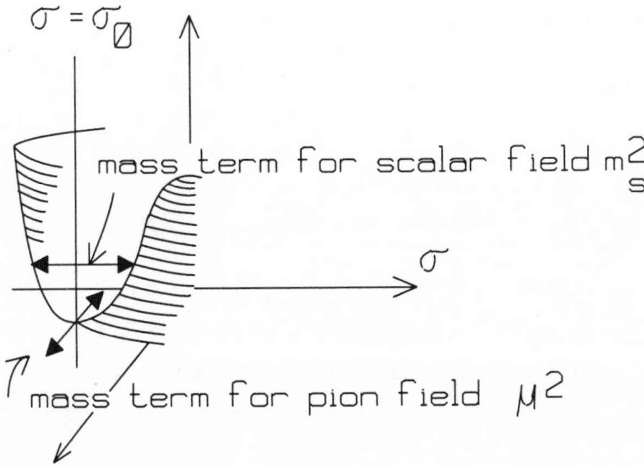

Figure 22.3: Sketch of the meson potential in which the excitations about the minimum take place.

Collect the terms in V (recall it is $-V$ that appears in $\mathcal{L}$):

1. The constant terms are

$$V_0 = \frac{\lambda}{4}(\sigma_0^2 - v^2)^2 + \varepsilon\sigma_0 \tag{22.18}$$

2. The terms linear in ϕ are

$$V_1 = \left[\frac{\lambda}{4}4\sigma_0(\sigma_0^2 - v^2) + \varepsilon\right]\phi = 0 \tag{22.19}$$

This expression vanishes at the minimum by Eq. (22.13).

3. The terms quadratic in $\boldsymbol{\pi}$ are

$$\begin{aligned} V_2 &= \frac{\lambda}{4}2(\sigma_0^2 - v^2)\boldsymbol{\pi}^2 = -\frac{\varepsilon}{2\sigma_0}\boldsymbol{\pi}^2 \\ V_2 &= \frac{1}{2}\mu^2\boldsymbol{\pi}^2 \end{aligned} \tag{22.20}$$

Here the minimum condition in Eq. (22.13) has again been used, as well as the definitions in Eq. (22.14). One can now identify μ^2 as the pion mass.

4. The terms quadratic in ϕ are analyzed in the same manner

$$\begin{aligned} V_3 &= \frac{\lambda}{4}\left[4\sigma_0^2 + 2(\sigma_0^2 - v^2)\right]\phi^2 = \frac{\lambda}{4}\left[4\frac{M^2}{g^2} + 2\left(\frac{-\varepsilon}{\lambda\sigma_0}\right)\right]\phi^2 \\ &= \left[\left(\frac{m_s^2 - \mu^2}{2M^2}\right)g^2\frac{M^2}{g^2} + \frac{1}{2}\mu^2\right]\phi^2 \\ V_3 &= \frac{1}{2}m_s^2\phi^2 \end{aligned} \tag{22.21}$$

This allows an identification of m_s^2 as the scalar mass.

5. The remaining cubic and quartic interactions of the meson fields $(\phi, \boldsymbol{\pi})$ are then given by

$$\begin{aligned} V_4 + V_5 &= \frac{\lambda}{4}\left\{4\sigma_0\phi(\phi^2+\boldsymbol{\pi}^2) + (\phi^2+\boldsymbol{\pi}^2)^2\right\} \\ &= \frac{g}{4M}(m_s^2-\mu^2)\left\{(\phi^2+\boldsymbol{\pi}^2)^2\frac{g}{2M} - 2\phi(\phi^2+\boldsymbol{\pi}^2)\right\} \\ V_4 + V_5 &= \frac{1}{2}(m_s^2-\mu^2)\left[\left(\frac{g}{2M}\right)^2(\phi^2+\boldsymbol{\pi}^2)^2 - 2\left(\frac{g}{2M}\right)\phi(\phi^2+\boldsymbol{\pi}^2)\right] \end{aligned} \tag{22.22}$$

A collection of these terms then yields the rewritten V

$$\begin{aligned} V &= \frac{1}{2}\mu^2\boldsymbol{\pi}^2 + \frac{1}{2}m_s^2\phi^2 \\ &+\frac{1}{2}(m_s^2-\mu^2)\left[\left(\frac{g}{2M}\right)^2(\phi^2+\boldsymbol{\pi}^2)^2 - 2\left(\frac{g}{2M}\right)\phi(\phi^2+\boldsymbol{\pi}^2)\right] + \text{constant} \end{aligned} \tag{22.23}$$

A constant term in V is irrelevant.

Summary. In summary, with the σ-model for the generation of the baryon mass by spontaneous symmetry breaking, the lagrangian density of Eq. (22.1) takes the form

$$\begin{aligned} \mathcal{L} &= -\bar{\psi}\left[\gamma_\mu\frac{D}{Dx_\mu} + M - g(\phi + i\boldsymbol{\tau}\cdot\boldsymbol{\pi}\gamma_5)\right]\psi - \frac{1}{2}m_v^2V_\mu^2 - \frac{1}{4}F_{\mu\nu}F_{\mu\nu} \\ &-\frac{1}{2}\left[\left(\frac{\partial\phi}{\partial x_\lambda}\right)^2 + m_s^2\phi^2\right] - \frac{1}{2}\left[\left(\frac{\partial\boldsymbol{\pi}}{\partial x_\lambda}\right)^2 + \mu^2\boldsymbol{\pi}^2\right] + \frac{g}{2M}(m_s^2-\mu^2)\phi(\phi^2+\boldsymbol{\pi}^2) \\ &-\frac{1}{2}\left(\frac{g}{2M}\right)^2(m_s^2-\mu^2)(\phi^2+\boldsymbol{\pi}^2)^2 + \text{constant} \end{aligned} \tag{22.24}$$

Several aspects of this result deserve comment:

1. This result, with the isoscalar vector field set equal to zero $V_\mu = 0$, is the σ-model of Schwinger (Ref. [R58]) and Gell-Mann and Lévy (Ref. [R59]). Reference [R60] is a very nice discussion of chiral dynamics, which includes many applications of the σ-model and further theoretical analysis of it.

2. The partially conserved axial vector current (PCAC) now follows as an operator relation

$$\frac{\partial\mathbf{A}_\lambda}{\partial x_\lambda} = \varepsilon\boldsymbol{\pi} = \left(\frac{M}{g}\right)\mu^2\boldsymbol{\pi} \tag{22.25}$$

Hence

$$\frac{\partial\mathbf{A}_\lambda}{\partial x_\lambda} \xrightarrow{\mu\to 0} 0 \tag{22.26}$$

Chiral symmetry is an exact symmetry of the lagrangian in Eq. (22.24) in the limit $\varepsilon = (M/g)\mu^2 \to 0$. Chiral symmetry is broken if $\mu^2 = (g/M)\varepsilon \neq 0$.

3. Equation (22.24) looks like the lagrangian we started with in Eq. (20.8), only now there is a single coupling constant g and an *additional prescribed set of nonlinear couplings.*[52] In particular, one can identify the previous coupling introduced on an *ad hoc* basis in Section 20

$$\mathcal{L}_\phi = \frac{1}{2}g_\phi m_s \phi \boldsymbol{\pi}^2 \tag{22.27}$$

From Eq. (22.24) one has

$$g_\phi m_s = g\frac{(m_s^2 - \mu^2)}{M} \tag{22.28}$$

In the chiral limit $\mu^2 = 0$ there is also a single coupling constant

$$g_s \;=\; g_\pi \;=\; g \tag{22.29}$$

In Section 20 the following relation was imposed

$$\frac{g_\phi g_s}{m_s} = \frac{g_\pi^2}{M} \tag{22.30}$$

to ensure a cancellation for the s-wave $\pi - N$ scattering lengths so that

$$a_0^\pm \xrightarrow{\mu \to 0} 0 \tag{22.31}$$

Now observe that in the chiral limit $\mu^2 = 0$ the relation in Eq. (22.30) follows identically from Eqs. (22.28) and (22.29)! The relation in Eq. (22.30) thus follows as a direct result of the underlying chiral symmetry of the lagrangian in Eq. (22.24).

4. Thus in the *soft pion limit* where $q_\lambda \to 0$, the pions *decouple* from the nucleons; this is a direct result of chiral symmetry.

5. Note that while $\mu = m_\pi = 139.6\,\text{MeV}$ may be small on the scale of particle physics, it is significant on the scale of nuclear physics, in which case going to the chiral limit $\mu \to 0$ is nontrivial.

6. Even when the relation in Eq. (22.30) is satisfied, one still has a cancellation of two orders of magnitude in the scattering amplitude required to obtain the small scattering lengths and Eq. (22.31). It would be much more satisfying to effect this cancellation *directly in the lagrangian*, and then have to deal only with scattering amplitudes that are already of the right order of magnitude. To accomplish this, we proceed to carry out in the next section a *chiral transformation* as a unitary transformation on the lagrangian. The argument in the next section is originally due to Weinberg (Refs. [R63, R64, R65]).

[52] What do these additional couplings do to the MFT of nuclear matter in Section 14? If one makes a preliminary identification of ϕ with the scalar field of QHD-I, then $\langle\phi\rangle_{\text{n.m.}} \equiv \phi_0$ and $M^* = M - g\phi_0$. The potential in QHD-I is just $V_0 = (1/2)m_s^2\phi_0^2 = (m_s^2/2g^2)(M-M^*)^2$. Now the potential, in the chiral limit $\mu^2 = 0$, is given by (Prob. 22.4) $V_{\text{chiral}} = V_0[(M+M^*)^2/4M^2]$. This changes the self-consistency relation for M^*; new *abnormal* solutions may be obtained (Refs. [R61, R62]). There is not much difference from QHD-I when $M^* \approx M$.

23

THE CHIRAL TRANSFORMATION

The goal of this section is to carry out a unitary transformation on Eq. (22.24) so that the cancellations that occur in the soft-pion limit in the chiral-invariant theory occur in the lagrangian itself. In order to do this, one uses a particular chiral transformation to generate the unitary transformation. The argument is due to Weinberg, and this section follows Refs. [R63, R64, R65, R66, R1].

The Chiral Transformation. The chiral transformation matrix, again denoted with a bar under the symbol, is discussed in Section 21 and Prob. 21.2. It is given by

$$\underline{e}^{\frac{i}{2}\boldsymbol{\omega}\cdot\boldsymbol{\tau}\gamma_5} = \cos\frac{\omega}{2} + i\mathbf{n}\cdot\boldsymbol{\tau}\gamma_5 \sin\frac{\omega}{2} \tag{23.1}$$

Here $\boldsymbol{\omega} = \omega\mathbf{n}$ with $\mathbf{n}^2 = 1$. This matrix is unitary since $(\boldsymbol{\tau}, \gamma_5)$ are hermitian.

The Meson-Nucleon Interaction. It is possible to write the interaction of the $(\boldsymbol{\pi}, \phi)$ in the σ-model as a chiral transformation; it can be expressed as

$$\begin{aligned} M - g\phi - ig\gamma_5\boldsymbol{\tau}\cdot\boldsymbol{\pi} &= \sqrt{(M-g\phi)^2 + g^2\boldsymbol{\pi}^2}\,\underline{e}^{-\frac{i}{2}\boldsymbol{\Omega}\cdot\boldsymbol{\tau}\gamma_5} \\ &= \sqrt{(M-g\phi)^2 + g^2\boldsymbol{\pi}^2}\left[\cos\frac{\Omega}{2} - i\mathbf{n}\cdot\boldsymbol{\tau}\gamma_5 \sin\frac{\Omega}{2}\right] \end{aligned} \tag{23.2}$$

Here $\boldsymbol{\pi} \equiv \mathbf{n}\pi$; note $\mathbf{n} \equiv \boldsymbol{\pi}/\pi$ with $\pi \equiv \sqrt{\boldsymbol{\pi}^2}$; also $\boldsymbol{\Omega} \equiv \mathbf{n}\Omega$, and the angle Ω is defined so as to make the second of Eqs. (23.2) an identity (see Fig. 23.1)

$$\begin{aligned} \cos\frac{\Omega}{2} &= \frac{M - g\phi}{\sqrt{(M-g\phi)^2 + g^2\boldsymbol{\pi}^2}} \\ \sin\frac{\Omega}{2} &= \frac{g\pi}{\sqrt{(M-g\phi)^2 + g^2\boldsymbol{\pi}^2}} \end{aligned} \tag{23.3}$$

The following unitary operator will now be used to transform the baryon field

$$\underline{e}^{-\frac{i}{4}\boldsymbol{\Omega}\cdot\boldsymbol{\tau}\gamma_5} = \cos\frac{\Omega}{4} - i\mathbf{n}\cdot\boldsymbol{\tau}\gamma_5 \sin\frac{\Omega}{4} \tag{23.4}$$

Note the presence of $\Omega/4$ in the argument. This is a local transformation since $\boldsymbol{\Omega}(x)$ depends on position. Define a new baryon field N through the relation

$$\underline{\mathrm{N}} \equiv \underline{e}^{-\frac{i}{4}\boldsymbol{\Omega}\cdot\boldsymbol{\tau}\gamma_5}\underline{\psi} \tag{23.5}$$

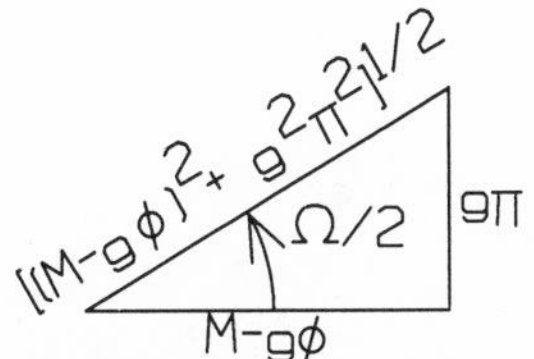

Figure 23.1: Definition of angle $\Omega/2$ in representation of the interaction with the baryon in the σ-model.

The interaction term then becomes

$$\begin{aligned}\bar{\psi}\left[M - g\phi - ig\gamma_5\boldsymbol{\tau}\cdot\boldsymbol{\pi}\right]\psi &= \bar{\psi}\sqrt{(M-g\phi)^2 + g^2\boldsymbol{\pi}^2}\, e^{-\frac{i}{2}\boldsymbol{\Omega}\cdot\boldsymbol{\tau}\gamma_5}\psi \\ &= \sqrt{(M-g\phi)^2 + g^2\boldsymbol{\pi}^2}\;\bar{N}N \qquad (23.6)\end{aligned}$$

The relation $\{\gamma_5, \gamma_4\} = 0$ has been used in obtaining this result [cf. Eq. (21.37)]. The pseudoscalar coupling of the $\boldsymbol{\pi}$ has now been *eliminated* in this new interaction term. This transformation can be analyzed in more detail starting from Eq. (23.4) using half-angle formulas and Fig. 23.1

$$\tan\frac{\Omega}{4} = \frac{\sin\Omega/2}{1+\cos\Omega/2} = \frac{g\pi}{\sqrt{(M-g\phi)^2 + g^2\boldsymbol{\pi}^2} + (M-g\phi)} \qquad (23.7)$$

Thus the angle $\Omega/4$ can be represented as in Fig. 23.2. Furthermore

$$\cos\frac{\Omega}{4} = \frac{1}{\sqrt{1+\tan^2\Omega/4}} \qquad (23.8)$$

Now define a new field

$$\boldsymbol{\xi} = \frac{1}{(M-g\phi) + \sqrt{(M-g\phi)^2 + g^2\boldsymbol{\pi}^2}}\, g\boldsymbol{\pi} \qquad (23.9)$$

A combination of Eqs. (23.4) and (23.7)-(23.9) allows one to express the new baryon field as

$$\begin{aligned}\underset{\sim}{\mathrm{N}} &= \underset{\sim}{\mathrm{e}}^{-\frac{i}{4}\boldsymbol{\Omega}\cdot\boldsymbol{\tau}\gamma_5}\underset{\sim}{\psi} \\ &= \frac{1}{\sqrt{1+\boldsymbol{\xi}^2}}\left[1 - i\boldsymbol{\tau}\cdot\boldsymbol{\xi}\gamma_5\right]\psi \qquad (23.10)\end{aligned}$$

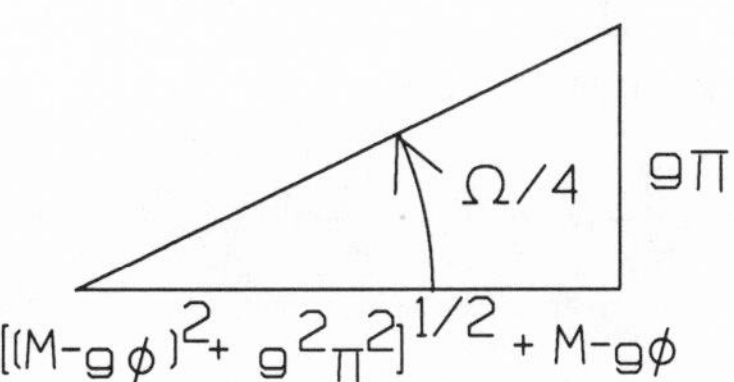

Figure 23.2: Represention of the angle $\Omega/4$.

Now introduce new pion and scalar fields as generalized coordinates through the relations

$$\begin{aligned}\sqrt{(M-g\phi)^2+g^2\boldsymbol{\pi}^2} &\equiv M-g\phi' \\ \boldsymbol{\xi} &\equiv \frac{g}{2M}\boldsymbol{\pi}' \equiv \frac{f}{\mu}\boldsymbol{\pi}'\end{aligned} \tag{23.11}$$

Here a new coupling constant has been introduced that satisfies

$$\frac{g}{2M} \equiv \frac{f}{\mu} \tag{23.12}$$

Its value is greatly reduced from that of the pseudoscalar $\pi - N$ coupling constant; from Eqs. (20.37)[53]

$$f^2 = \left(\frac{\mu}{2M}\right)^2 g_\pi^2 = 1.001 \tag{23.13}$$

The meson-nucleon interaction term in Eq. (23.6) then takes the form

$$\begin{aligned}\bar{\psi}\left[M-g\phi-ig\gamma_5\boldsymbol{\tau}\cdot\boldsymbol{\pi}\right]\psi &= \sqrt{(M-g\phi)^2+g^2\boldsymbol{\pi}^2}\,\bar{N}N \\ &= (M-g\phi')\,\bar{N}N\end{aligned} \tag{23.14}$$

It represent a new mass term and new scalar interaction for the baryons.

Baryon Kinetic Energy. Consider next the remaining baryon contributions to the lagrangian. First note that because of the presence of the extra γ_μ, the baryon current is invariant under the transformation in Eq. (23.5)

$$\bar{\psi}\gamma_\mu\psi = \bar{N}\gamma_\mu N \tag{23.15}$$

It is thus also true that the interaction with V_μ is unchanged

$$\bar{\psi}(ig_v\gamma_\mu V_\mu)\psi = \bar{N}(ig_v\gamma_\mu V_\mu)N \tag{23.16}$$

Now consider the transformation of the baryon kinetic energy

$$\begin{aligned}\bar{\psi}\gamma_\mu\frac{\partial}{\partial x_\mu}\psi &= \bar{N}\gamma_\mu e^{-\frac{i}{4}\boldsymbol{\Omega}\cdot\boldsymbol{\tau}\gamma_5}\frac{\partial}{\partial x_\mu}e^{\frac{i}{4}\boldsymbol{\Omega}\cdot\boldsymbol{\tau}\gamma_5}N \\ &= \bar{N}\gamma_\mu\frac{1}{\sqrt{1+\boldsymbol{\xi}^2}}(1-i\boldsymbol{\tau}\cdot\boldsymbol{\xi}\gamma_5)\frac{\partial}{\partial x_\mu}\frac{1}{\sqrt{1+\boldsymbol{\xi}^2}}(1+i\boldsymbol{\tau}\cdot\boldsymbol{\xi}\gamma_5)\,N\end{aligned} \tag{23.17}$$

The algebra can be performed to give

$$\bar{\psi}\gamma_\mu\frac{\partial}{\partial x_\mu}\psi = \bar{N}\gamma_\mu\frac{\partial}{\partial x_\mu}N + \bar{N}\gamma_\mu\frac{1}{\sqrt{1+\boldsymbol{\xi}^2}}(1-i\boldsymbol{\tau}\cdot\boldsymbol{\xi}\gamma_5)$$

[53]This coupling constant differs by $\sqrt{4\pi}$ from the f_π defined in Eq. (20.38). We will henceforth use the notation $\alpha_\pi \equiv f_\pi^2 = f^2/4\pi = (g_\pi^2/4\pi)(\mu/2M)^2 = 0.0797$.

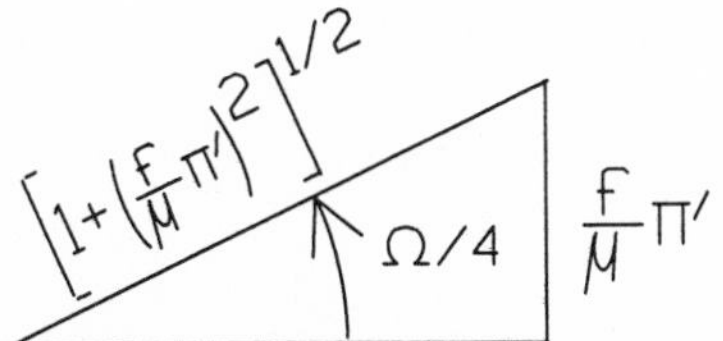

Figure 23.3: Redraw the angle $\Omega/4$.

$$
\begin{aligned}
&\times\left\{-\frac{1}{(1+\boldsymbol{\xi}^2)^{3/2}}\boldsymbol{\xi}\cdot\frac{\partial\boldsymbol{\xi}}{\partial x_\mu}(1+i\gamma_5\boldsymbol{\tau}\cdot\boldsymbol{\xi})+\frac{1}{(1+\boldsymbol{\xi}^2)^{1/2}}(i\gamma_5\boldsymbol{\tau}\cdot\frac{\partial\boldsymbol{\xi}}{\partial x_\mu})\right\}N \\
&= \bar{N}\gamma_\mu\left\{\frac{\partial}{\partial x_\mu}-\frac{1}{1+\boldsymbol{\xi}^2}\boldsymbol{\xi}\cdot\frac{\partial\boldsymbol{\xi}}{\partial x_\mu}\right. \\
&\left.+\frac{1}{1+\boldsymbol{\xi}^2}\left[i\gamma_5\boldsymbol{\tau}\cdot\frac{\partial\boldsymbol{\xi}}{\partial x_\mu}+\left(\boldsymbol{\xi}\cdot\frac{\partial\boldsymbol{\xi}}{\partial x_\mu}+i\boldsymbol{\tau}\cdot\boldsymbol{\xi}\times\frac{\partial\boldsymbol{\xi}}{\partial x_\mu}\right)\right]\right\}N \qquad (23.18)
\end{aligned}
$$

The second and fourth terms cancel, and one thus obtains the following result for the transformation

$$
\bar{\psi}\gamma_\mu\frac{D}{Dx_\mu}\psi = \bar{N}\left\{\gamma_\mu\frac{D}{Dx_\mu}+\frac{1}{1+\boldsymbol{\xi}^2}\left[i\gamma_\mu\gamma_5\boldsymbol{\tau}\cdot\frac{\partial\boldsymbol{\xi}}{\partial x_\mu}+i\gamma_\mu\boldsymbol{\tau}\cdot\boldsymbol{\xi}\times\frac{\partial\boldsymbol{\xi}}{\partial x_\mu}\right]\right\}N \qquad (23.19)
$$

It is understood from Eq. (23.11) that in this expression

$$
\boldsymbol{\xi} = \left(\frac{f}{\mu}\right)\boldsymbol{\pi}' \qquad (23.20)
$$

The baryon kinetic energy is now expressed in terms of the new generalized coordinates $(N, V_\mu, \boldsymbol{\pi}')$. Note that the original pseudoscalar Yukawa coupling to the pion with coupling constant g has been replaced by derivative couplings characterized by a coupling constant f/μ, multiplied by $[1+(f/\mu)^2\boldsymbol{\pi}'^2]^{-1}$, which generates an infinite series of nonlinear interactions.

Some Preliminaries. The next step is to transform the meson lagrangian to the new fields defined in Eqs. (23.11). It is first useful to perform some preliminary manipulations on these quantities. From Eqs. (23.7) and (23.9) one has

$$
\tan\frac{\Omega}{4} = \xi \qquad (23.21)
$$

This allows Fig. 23.2 to be redrawn as in Fig. 23.3. From Figs. 23.1 and 23.3 one finds

$$
\begin{aligned}
\cos\frac{\Omega}{2} &= \frac{M-g\phi}{\sqrt{(M-g\phi)^2+g^2\boldsymbol{\pi}^2}} \\
&\equiv 2\cos^2\frac{\Omega}{4}-1 = \frac{1-\boldsymbol{\xi}^2}{1+\xi^2} \qquad (23.22)
\end{aligned}
$$

The first line can be rewritten with the aid of Eq. (23.11) as

$$\cos\frac{\Omega}{2} = \frac{M - g\phi}{M - g\phi'} = \frac{1 - 2(f/\mu)\phi}{1 - 2(f/\mu)\phi'} \tag{23.23}$$

Equations (23.22) and (23.23) allow one to express the original scalar field in terms of the new generalized coordinates, that is to obtain $\phi(\phi', \boldsymbol{\pi}')$

$$\left(\frac{f}{\mu}\right)\phi = \frac{(f/\mu)\phi'(1 - \boldsymbol{\xi}^2) + \boldsymbol{\xi}^2}{1 + \xi^2} \tag{23.24}$$

In a similar fashion one can write

$$\begin{aligned} \sin\frac{\Omega}{2} &= \frac{g\pi}{\sqrt{(M - g\phi)^2 + g^2\boldsymbol{\pi}^2}} \\ &\equiv 2\sin\frac{\Omega}{4}\cos\frac{\Omega}{4} = \frac{2\xi}{1 + \xi^2} \end{aligned} \tag{23.25}$$

This is equivalent to [see Eqs. (23.9) and (23.11)]

$$\frac{1}{M - g\phi'}g\boldsymbol{\pi} = \frac{1}{1 + \xi^2}2\boldsymbol{\xi} \tag{23.26}$$

Hence one can also solve for $\boldsymbol{\pi}(\phi', \boldsymbol{\pi}')$

$$\boldsymbol{\pi} = \frac{1 - 2(f/\mu)\phi'}{1 + \xi^2}\boldsymbol{\pi}' \tag{23.27}$$

It is now necessary to calculate the meson contributions to the lagrangian using these nonlinear relations. Define

$$\mathcal{R} \equiv \frac{1 - 2(f/\mu)\phi'}{1 + \xi^2} \tag{23.28}$$

Then the above results can be rewritten as

$$\begin{aligned} \boldsymbol{\pi} &= \mathcal{R}\boldsymbol{\pi}' \\ \phi &= \phi' + \frac{f}{\mu}\mathcal{R}\boldsymbol{\pi}'^2 \end{aligned} \tag{23.29}$$

As the reader can easily demonstrate, it follows that

$$\phi^2 + \boldsymbol{\pi}^2 = \phi'^2 + \mathcal{R}\boldsymbol{\pi}'^2 \tag{23.30}$$

Meson Lagrangian. The algebra involved in transforming the meson lagrangian to the new field variables is nontrivial; it is developed in Appendix

G. Here we quote the result from that analysis. The meson kinetic energy is transformed to the following expression

$$-2T_{\text{mes}} = \left(\frac{\partial \phi}{\partial x_\lambda}\right)^2 + \left(\frac{\partial \boldsymbol{\pi}}{\partial x_\lambda}\right)^2 = \left(\frac{\partial \phi'}{\partial x_\lambda}\right)^2 + \mathcal{R}^2 \left(\frac{\partial \boldsymbol{\pi}'}{\partial x_\lambda}\right)^2 \tag{23.31}$$

The meson potential term takes the form

$$\begin{aligned} -V_{\text{mes}} &= -\frac{1}{2}(m_s^2\phi^2 + \mu^2\boldsymbol{\pi}^2) \\ &+ \frac{g}{2M}(m_s^2 - \mu^2)\phi(\boldsymbol{\pi}^2 + \phi^2) - \frac{1}{2}\left(\frac{g}{2M}\right)^2 (m_s^2 - \mu^2)(\pi^2 + \phi^2)^2 \\ &= -\frac{1}{2}(m_s^2\phi'^2 + \mathcal{R}\mu^2\boldsymbol{\pi}'^2) + (m_s^2 - \mu^2)\left[\frac{f}{\mu}\phi'^3 - \frac{1}{2}\left(\frac{f}{\mu}\right)^2 \phi'^4\right] \end{aligned} \tag{23.32}$$

Summary. In summary, after the chiral transformation to the new variables defined in Eqs. (23.10), (23.28), and (23.29), the lagrangian of Eq. (22.24) takes the form (see Refs. [R63, R64, R65, R1])

$$\begin{aligned} \mathcal{L} \;=\; & -\bar{N}\left\{\gamma_\mu \frac{D}{Dx_\mu} + (M - g\phi') + \right. \\ & \left. + \frac{1}{1+\xi^2}\left[i\gamma_\mu\gamma_5\boldsymbol{\tau}\cdot\frac{\partial \boldsymbol{\xi}}{\partial x_\mu} + i\gamma_\mu\boldsymbol{\tau}\cdot\boldsymbol{\xi}\times\frac{\partial \boldsymbol{\xi}}{\partial x_\mu}\right]\right\}N \\ & -\frac{1}{2}\left[\left(\frac{\partial \phi'}{\partial x_\lambda}\right)^2 + m_s^2\phi'^2\right] - \frac{1}{2}\mathcal{R}\left[\mathcal{R}\left(\frac{\partial \boldsymbol{\pi}'}{\partial x_\lambda}\right)^2 + \mu^2\boldsymbol{\pi}'^2\right] + \text{constant} \\ & +(m_s^2 - \mu^2)\left[\frac{f}{\mu}\phi'^3 - \frac{1}{2}\left(\frac{f}{\mu}\right)^2\phi'^4\right] - \frac{1}{2}m_v^2 V_\mu^2 - \frac{1}{4}F_{\mu\nu}F_{\mu\nu} \end{aligned} \tag{23.33}$$

Here

$$\boldsymbol{\xi} \equiv \left(\frac{f}{\mu}\right)\boldsymbol{\pi}' \tag{23.34}$$

The coupling constants are related by Eq. (23.13).

Discussion. There are several interesting aspects of this result:

1. Equation (23.33) gives the lagrangian density in terms of the new generalized coordinates $\mathcal{L}(N, V_\mu, \phi', \boldsymbol{\pi}')$.

2. Upon setting the coupling constants $f/\mu = 0$, $g_v = 0$ at fixed M one obtains the *free* lagrangian (to within a constant)

$$\begin{aligned} \mathcal{L}_0 \;=\; & -\bar{N}\left(\gamma_\lambda\frac{\partial}{\partial x_\lambda} + M\right)N - \frac{1}{2}m_v^2 V_\lambda^2 - \frac{1}{4}F_{\mu\nu}F_{\mu\nu} \\ & -\frac{1}{2}\left[\left(\frac{\partial \phi'}{\partial x_\lambda}\right)^2 + m_s^2\phi'^2\right] - \frac{1}{2}\left[\left(\frac{\partial \boldsymbol{\pi}'}{\partial x_\lambda}\right)^2 + \mu^2\boldsymbol{\pi}'^2\right] \end{aligned} \tag{23.35}$$

This enables us to identify the *particle content* of the theory: there is a baryon field $\underset{\sim}{N} = \begin{pmatrix} N_p \\ N_n \end{pmatrix}$ with mass M, a vector meson field V_μ with mass m_v, a pion field $\boldsymbol{\pi}'$ with mass $\mu \equiv m_\pi$, and a scalar field ϕ' with mass m_s.

3. The result obtained in Eq. (23.33) is fully equivalent to the original theory in Eq. (22.24). Since the original theory is renormalizable, so is the transformed theory (see Ref. [R66]).

4. The underlying lagrangian possesses exact chiral symmetry when $\mu^2 = 0$.

5. One has gone from a pseudoscalar pion coupling to the baryon of the form $ig\bar{\psi}\gamma_5\boldsymbol{\tau}\cdot\boldsymbol{\pi}\psi$ with a coupling constant $g^2 = g_\pi^2 = (4\pi)\,14.4$ or $g = 13.45$, to pseudovector coupling $-i(f/\mu)\bar{N}\gamma_\lambda\gamma_5\boldsymbol{\tau}\cdot(\partial\boldsymbol{\pi}'/\partial x_\lambda)N$ with coupling constant $f = 1.000$. This was the original goal — to effect the cancellation in the lagrangian itself.

6. The lagrangian now has an infinite series of nonlinear couplings — it contains all powers of f/μ.[54]

7. It is possible to go one step further and define still *another* new scalar field

$$m_s\phi' \equiv \chi \tag{23.36}$$

Rewrite the lagrangian in terms of χ

$$\begin{aligned}
\mathcal{L} \;=\; & -\bar{N}\left\{\gamma_\lambda\frac{D}{Dx_\lambda} + (M - \frac{g}{m_s}\chi) + \right. \\
& \left. + \frac{1}{1+\boldsymbol{\xi}^2}\left[i\gamma_\lambda\gamma_5\boldsymbol{\tau}\cdot\frac{\partial\boldsymbol{\xi}}{\partial x_\lambda} + i\gamma_\lambda\boldsymbol{\tau}\cdot\boldsymbol{\xi}\times\frac{\partial\boldsymbol{\xi}}{\partial x_\lambda}\right]\right\}N \\
& -\frac{1}{2}\left[\frac{1}{m_s^2}\left(\frac{\partial\chi}{\partial x_\lambda}\right)^2 + \chi^2\right] - \frac{1}{2}\mathcal{R}\left[\mathcal{R}\left(\frac{\partial\boldsymbol{\pi}'}{\partial x_\lambda}\right)^2 + \mu^2\boldsymbol{\pi}'^2\right] + \text{constant} \\
& + \left(1 - \frac{\mu^2}{m_s^2}\right)\left[\frac{f}{\mu}\frac{1}{m_s}\chi^3 - \frac{1}{2}\left(\frac{f}{\mu}\right)^2\frac{1}{m_s^2}\chi^4\right] - \frac{1}{2}m_v^2V_\mu^2 - \frac{1}{4}F_{\mu\nu}F_{\mu\nu}
\end{aligned} \tag{23.37}$$

Here

$$\begin{aligned}
\boldsymbol{\xi} &\equiv \left(\frac{f}{\mu}\right)\boldsymbol{\pi}' \\
\mathcal{R} &= \frac{1 - 2(f/\mu)(1/m_s)\chi}{1+\boldsymbol{\xi}^2}
\end{aligned} \tag{23.38}$$

Weinberg (Refs. [R63, R64, R65]) was interested in obtaining an *effective* pion lagrangian that would reproduce the results of current algebra based on chiral symmetry (Ref. [R67]). Since the above is chirally symmetric (when $\mu^2 = 0$ at

[54]Note that canonical quantization starting directly from this form of $\mathcal{L}$ is complicated by the derivative couplings and nonlinearities.

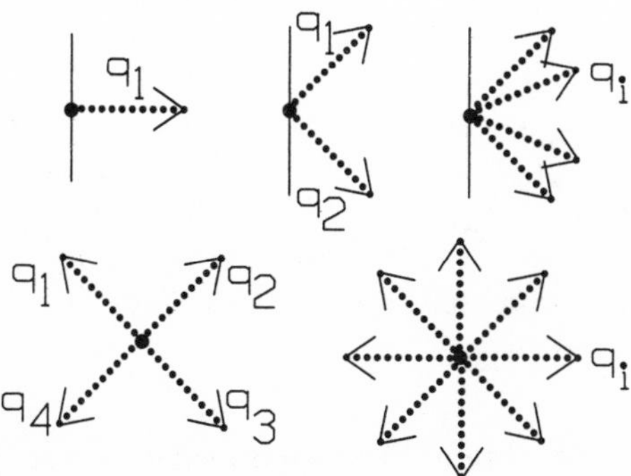

Figure 23.4: Some pion vertices contained in the transformed lagrangian.

fixed $f/\mu = g/2M$) for *any* value of m_s, he took the formal limit $m_s \to \infty$. In this case

$$\mu^2 = 0 \qquad\qquad m_s^2 \to \infty \qquad\qquad \mathcal{R} = \frac{1}{1+\boldsymbol{\xi}^2}$$

$$\begin{aligned}\mathcal{L}(N, V_\mu, \boldsymbol{\pi}', \chi) &= -\bar{N}\left\{\gamma_\lambda \frac{D}{Dx_\lambda} + M + \right. \\ &\left. + \frac{1}{1+\xi^2}\left[i\gamma_\lambda\gamma_5\boldsymbol{\tau}\cdot\frac{\partial\boldsymbol{\xi}}{\partial x_\lambda} + i\gamma_\lambda\boldsymbol{\tau}\cdot\boldsymbol{\xi}\times\frac{\partial\boldsymbol{\xi}}{\partial x_\lambda}\right]\right\}N \\ &-\frac{1}{2}\mathcal{R}\left[\mathcal{R}\left(\frac{\partial\boldsymbol{\pi}'}{\partial x_\lambda}\right)^2\right] - \frac{1}{2}m_v^2 V_\mu^2 - \frac{1}{4}F_{\mu\nu}F_{\mu\nu} - \frac{1}{2}\chi^2 + \text{const.}\end{aligned} \tag{23.39}$$

The scalar field χ now appears only in the last line and evidently *decouples* from the problem! With $V_\mu = 0$, this is the nonlinear σ model of Weinberg.[55]

8. All the pion couplings in Eqs. (23.37) and (23.39) now involve at least one derivative of the pion field. [The denominators can be expanded in a power series in $(f/\mu)^2$.] The effective pion vertices contained in these lagrangians are illustrated in Fig. 23.4. All these vertices now vanish in the soft-pion limit where all $q_i \to 0$! This condition is now built into the lagrangian — this was our goal.

The Equivalence Theorem. The equivalence theorem states that for fermions on the mass shell, the pseudoscalar and pseudovector couplings are *identical.* The situation is illustrated in Fig. 23.5. To prove this result, use the Dirac equation $(i\gamma_\mu p_\mu + M)u(p) = 0$ on the vertex in Fig. 23.5b

$$\begin{aligned}\bar{u}(p')\gamma_5(-i\gamma_\mu q_\mu)u(p) &= \bar{u}(p')\gamma_5(-i\gamma_\mu p_\mu + i\gamma_\mu p'_\mu)u(p) \\ &= \bar{u}(p')[(-i\gamma_\mu p'_\mu)\gamma_5 + \gamma_5(-i\gamma_\mu p_\mu)]u(p) \\ &= 2M\bar{u}(p')\gamma_5 u(p)\end{aligned} \tag{23.40}$$

[55]The theory is renormalizable for any finite value of m_s.

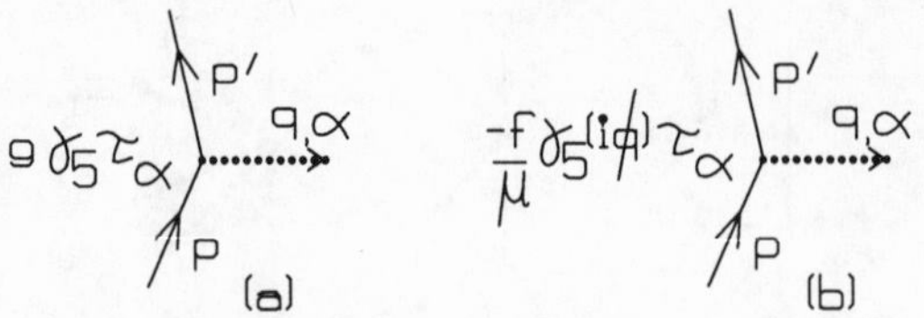

Figure 23.5: Equivalence of pseudoscalar (a) and pseudovector (b) couplings for fermions on their mass shell.

Since $(2M/\mu)f = g$, one obtains the pseudoscalar vertex in Fig. 23.5a. This proves the theorem.

In the pseudoscalar case, the vertex is $g\gamma_5$ where g is large. There is also no explicit dependence on q_λ; hence one can build the soft-pion limit into the amplitude only by a cancellation of large terms. In the pseudovector case, the vertex is proportional to $(f/\mu)q_\lambda$ where f is much smaller. In addition, the vertex now vanishes *explicitly* in the soft-pion limit where $q_\lambda \to 0$ at fixed $f/\mu \equiv g/2M$.

24

DYNAMIC RESONANCES

In Section 22 the fields $\{\psi, V_\mu, \sigma, \boldsymbol{\pi}\}$ were used to construct a chiral-invariant lagrangian, and the mass of the baryon was then generated by choosing a meson interaction potential V that leads to spontanteously broken chiral symmetry. The outcome of that discussion is the following lagrangian density

$$
\begin{aligned}
\mathcal{L} = & -\bar{\psi}\left[\gamma_\mu \frac{D}{Dx_\mu} + M - g(\varphi + i\gamma_5 \boldsymbol{\tau}\cdot\boldsymbol{\pi})\right]\psi - \frac{1}{2}m_v^2 V_\mu^2 - \frac{1}{4}F_{\mu\nu}F_{\mu\nu} \\
& -\frac{1}{2}\left[\left(\frac{\partial\varphi}{\partial x_\lambda}\right)^2 + m_\sigma^2\varphi^2\right] - \frac{1}{2}\left[\left(\frac{\partial\boldsymbol{\pi}}{\partial x_\lambda}\right)^2 + \mu^2\boldsymbol{\pi}^2\right] + \text{constant} \\
& +\frac{g}{2M}(m_\sigma^2 - \mu^2)\varphi(\varphi^2 + \boldsymbol{\pi}^2) - \frac{1}{2}\left(\frac{g}{2M}\right)^2(m_\sigma^2 - \mu^2)(\varphi^2 + \boldsymbol{\pi}^2)^2 \qquad (24.1)
\end{aligned}
$$

Here

$$
\frac{D}{Dx_\mu} = \frac{\partial}{\partial x_\mu} - ig_v V_\mu \qquad (24.2)
$$

We change notation slightly for clarity in the following discussion; the scalar field is here denoted φ, and its mass by m_σ. Several features of the above result are of interest:

1. The underlying lagrangian is chiral invariant when $\varepsilon \equiv (M/g)\mu^2 = 0$. PCAC is thus satisfied as are the soft-pion theorems (Section 23). If one identifies $\{M;\ g = g_\pi;\ \mu = m_\pi;\ m_v = m_\omega\}$ with experimental values, and takes g_v from Section 14, then there is only one parameter left — the mass m_σ^2 of the scalar field.

2. The last line in this chiral-invariant lagrangian represents strong, nonlinear couplings in the fields $\{\varphi, \boldsymbol{\pi}\}$. We have seen that this scalar field *decouples* in the limit that its mass becomes very large $m_\sigma^2 \to \infty$.

3. If this chiral scalar field is identified with the low mass scalar field of Sections 14 and 15 with $m_\sigma = m_s = 550\,\text{MeV}$, then the nonlinear terms in the last line are such as to destroy the successful phenomenological MFT description of nuclear matter and finite nuclei discussed there (Refs. [R1, R75]). We have also noted that the empirical $N - N$ scattering amplitude exhibits evidence of strong, low-mass isoscalar scalar (and vector) exchange.

4. We are thus faced with the following *problem*: Can one take the meson-baryon dynamics to be described by this underlying chiral-invariant lagrangian

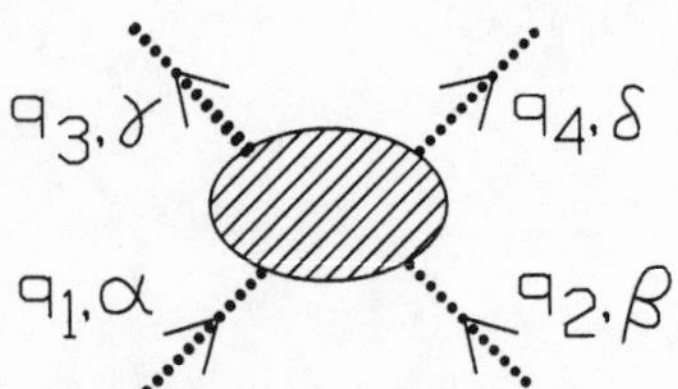

Figure 24.1: $\pi - \pi$ scattering.

with a very large (but finite) m_σ, and then generate a resonant low-mass scalar *dynamically* through the strong nonlinear couplings, which would then play the required nuclear role? Can we understand a *light* σ *as of dynamic origin*? We proceed to discuss the calculation of Ref. [R68] (see other references cited there).

A Low-Mass σ. In this section we demonstrate that the strong, nonlinear, chiral-invariant couplings involving the $\{\varphi, \boldsymbol{\pi}\}$ present in the lagrangian in Eq. (24.1) will give rise to a low-mass, broad, (near-)resonant amplitude in the $(0^+, 0)$ $\pi - \pi$ scattering channel. The argument follows Ref. [R68].

The process of $\pi - \pi$ scattering is illustrated in Fig. 24.1. The analysis of this process is discussed in Appendix H, where it is shown that the general form of the S-matrix is given by

$$S_{\rm fi} = -\frac{(2\pi)^4 i}{\Omega^2}\delta^{(4)}(q_1 + q_2 - q_3 - q_4)\frac{1}{\sqrt{2^4\omega_1\omega_2\omega_3\omega_4}}T_{\rm fi} \tag{24.3}$$

One introduces the same kinematic variables $\{s, t, u\}$ as in $\pi - N$ scattering (Appendix H). The differential cross section in the C-M system is given by

$$\begin{aligned} \left(\frac{d\sigma}{d\Omega}\right)_{\rm CM} &= |f_{\rm CM}|^2 \\ f_{\rm CM} &= -\frac{T_{\rm fi}}{8\pi W} \end{aligned} \tag{24.4}$$

The Feynman rules are generated directly from Eq. (24.1) (see Section 20); they can be used to compute the pole contributions to the scattering amplitude arising from the meson interaction processes illustrated in Fig. 24.2. The $\boldsymbol{\pi}^4$ contact term in Eq. (24.1) contributes to $\pi - \pi$ scattering and its contribution must also be included; thus we here work at "tree level" (no loops). The T-matrix arising from this sum of graphs follows immediately

$$\begin{aligned} T_{\rm fi}^{\rm tree} &= \left[\frac{g}{2M}(m_\sigma^2 - \mu^2)(2i)\right]^2 \\ &\times \left[\delta_{\alpha\beta}\delta_{\gamma\delta}\frac{1}{m_\sigma^2 - s} + \delta_{\alpha\gamma}\delta_{\beta\delta}\frac{1}{m_\sigma^2 - t} + \delta_{\alpha\delta}\delta_{\beta\gamma}\frac{1}{m_\sigma^2 - u}\right] \\ &+ \left[\frac{1}{2}\left(\frac{g}{2M}\right)^2(m_\sigma^2 - \mu^2)\right][2\times 2\times 2][\delta_{\alpha\beta}\delta_{\gamma\delta} + \delta_{\alpha\gamma}\delta_{\beta\delta} + \delta_{\alpha\delta}\delta_{\beta\gamma}] \end{aligned} \tag{24.5}$$

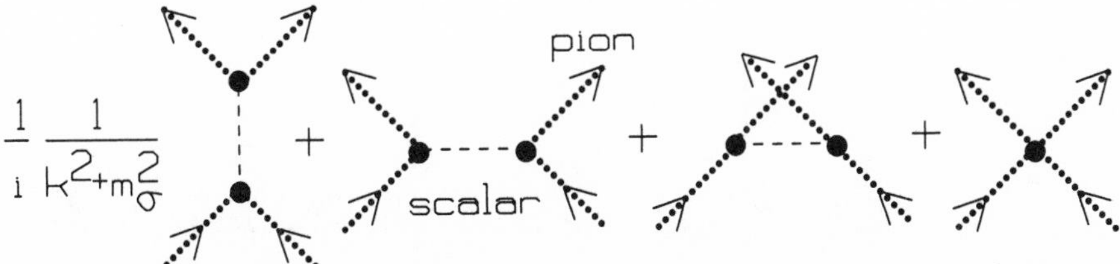

Figure 24.2: Tree-level contributions to $\pi-\pi$ scattering amplitude in this theory.

A combination of these results gives

$$T_{\text{fi}}^{\text{tree}} = \left(\frac{g}{2M}\right)^2 (m_\sigma^2 - \mu^2)4\left\{(\delta_{\alpha\beta}\delta_{\gamma\delta} + \delta_{\alpha\gamma}\delta_{\beta\delta} + \delta_{\alpha\delta}\delta_{\beta\gamma}) \right.$$
$$\left. -(m_\sigma^2 - \mu^2)\left(\delta_{\alpha\beta}\delta_{\gamma\delta}\frac{1}{m_\sigma^2 - s} + \delta_{\alpha\gamma}\delta_{\beta\delta}\frac{1}{m_\sigma^2 - t} + \delta_{\alpha\delta}\delta_{\beta\gamma}\frac{1}{m_\sigma^2 - u}\right)\right\} \tag{24.6}$$

Now consider the $(J^\pi, T) = (0^+, 0)$ channel (the channel of the low-mass sigma). First note that since the quantity $\boldsymbol{\pi}^2 = \pi_\alpha\pi_\alpha$ is an obvious isoscalar, we can identify the state $\frac{1}{\sqrt{3}}|\alpha\alpha>$ as the isoscalar channel.[56] Hence we examine

$$\frac{1}{3}T_{\gamma\gamma;\alpha\alpha}^{\text{tree}} = \frac{1}{3}\frac{g^2}{M^2}(m_\sigma^2 - \mu^2)\left\{(9+3+3) \right.$$
$$\left. -(m_\sigma^2 - \mu^2)\left(\frac{9}{m_\sigma^2 - s} + \frac{3}{m_\sigma^2 - t} + \frac{3}{m_\sigma^2 - u}\right)\right\}$$
$$= \frac{g^2}{M^2}(m_\sigma^2 - \mu^2)\left\{3\left(\frac{s-\mu^2}{s-m_\sigma^2}\right) + \left(\frac{t-\mu^2}{t-m_\sigma^2} + \frac{u-\mu^2}{u-m_\sigma^2}\right)\right\} \tag{24.7}$$

Note that this amplitude is symmetric under the interchange $t \rightleftharpoons u$, as it should be for bosons.

Consider next the s-wave amplitude. First note that in the C-M system one has (Appendix H)

$$t = -2\mathbf{q}^2(1 - \cos\theta) \qquad s = W^2 = 4(\mathbf{q}^2 + \mu^2)$$
$$u = -2\mathbf{q}^2(1 + \cos\theta) \qquad s + t + u = 4\mu^2 \tag{24.8}$$

The s-wave scattering amplitude can now be projected by simply taking the angular average

$$t_0 \equiv \frac{1}{2}\int_{-1}^{+1} d\cos\theta \frac{1}{3}T_{\gamma\gamma;\alpha\alpha} \tag{24.9}$$

[56] Repeated indices are summed.

One thus needs to evaluate

$$\begin{aligned} I &= \frac{1}{2}\int_{-1}^{+1} dx \left\{ 3\left(\frac{s-\mu^2}{s-m_\sigma^2}\right) + 2 \right. \\ & \left. -(m_\sigma^2 - \mu^2)\left[\frac{1}{m_\sigma^2 + 2\mathbf{q}^2(1-x)} + \frac{1}{m_\sigma^2 + 2\mathbf{q}^2(1+x)}\right]\right\} \end{aligned} \tag{24.10}$$

The result is

$$t_0^{\text{tree}} = \frac{g^2}{M^2}(m_\sigma^2 - \mu^2)\left\{3\left(\frac{s-\mu^2}{s-m_\sigma^2}\right) + 2 - \frac{(m_\sigma^2-\mu^2)}{2\mathbf{q}^2}\ln\left(1 + \frac{4\mathbf{q}^2}{m_\sigma^2}\right)\right\} \tag{24.11}$$

Consider the *soft-pion limit* of this result. First go to thereshold by setting $s = 4\mu^2$ and $\mathbf{q}^2 = 0$

$$t_0^{\text{tree}} \to \frac{g^2}{M^2}(m_\sigma^2 - \mu^2)\left\{\frac{9\mu^2}{4\mu^2 - m_\sigma^2} + 2 - \frac{2(m_\sigma^2 - \mu^2)}{m_\sigma^2}\right\} \tag{24.12}$$

Note the cancelation of the terms of $O(1)$ inside the brackets; the result is

$$\left(t_0^{\text{tree}}\right)_{\text{th}} = -\frac{7g^2\mu^2}{M^2} + O(\mu^4) \tag{24.13}$$

The soft-pion limit is again built into this amplitude, which was calculated from the chiral-invariant lagrangian in Eq. (24.1).

$$\left(t_0^{\text{tree}}\right)_{\text{th}} \overset{\mu\to 0}{\longrightarrow} 0 \tag{24.14}$$

Note that the inclusion of the $\boldsymbol{\pi}^4$ contact term was essential in arriving at this result. Thus the tree amplitude has the correct pole structure and satisfies the soft-pion constraint imposed by chiral symmetry.

It is now necessary to develop a framework in which one can describe *resonance dynamics.* The method that shall be used is to take the Born, or tree amplitude, as the driving term and then introduce a procedure to *unitarize* that amplitude. Forcing it to satisfy unitarity is equivalent to summing a selected part of an infinite series of Feynman diagrams. This infinite summation of graphs allows for repeated interaction, and with an appropriate Born term, one can build up a dynamic resonance.[57]

We initially assume two equal mass distinguishable spinless particles and elastic scattering in this discussion for clarity and to illustrate the method. One knows from quantum mechanics that

$$\begin{aligned} \left(\frac{d\sigma}{d\Omega}\right)_{\text{CM}} &= |f_{\text{CM}}|^2 \\ f_{\text{CM}} &= \sum_{l=0}^{\infty}(2l+1)P_l(\cos\theta)\, f_l \end{aligned} \tag{24.15}$$

[57] In the present case, the Born amplitude is given by the set of graphs in Fig. 24.2.

It follows from unitarity that the partial wave amplitudes f_l must have the form

$$f_l = \frac{e^{i\delta_l}\sin\delta_l}{q} \tag{24.16}$$

For elastic scattering, the phase shifts δ_l are real. Equation (24.16) can be rewritten as

$$f_l = \frac{1}{q}\frac{\sin\delta_l}{e^{-i\delta_l}} = \frac{1}{q}\frac{\sin\delta_l}{\cos\delta_l - i\sin\delta_l} = \frac{1}{q}\frac{1}{\cot\delta_l - i} \tag{24.17}$$

The imaginary part of $1/f_l$ is thus a *known* quantity

$$\text{Im}\frac{1}{f_l} = -q \tag{24.18}$$

It follows from Eqs. (24.4) and (24.9) that for s-waves

$$\begin{aligned} f_{\text{CM}} &= -\frac{T_{\text{fi}}}{8\pi W} \\ f_0 &= \frac{1}{2}\int_{-1}^{1} dx\, f_{\text{CM}} \equiv -\frac{t_0}{8\pi W} \end{aligned} \tag{24.19}$$

Hence

$$e^{i\delta_0}\sin\delta_0 = \frac{1}{\cot\delta_0 - i} = -\frac{qt_0}{8\pi W} \equiv K(s)t_0(s) \tag{24.20}$$

Here

$$K(s) = -\frac{q}{8\pi W} = -\frac{1}{16\pi}\left(1 - \frac{4\mu^2}{s}\right)^{1/2} \tag{24.21}$$

Suppose one has some real first approximation to Eq. (24.20), such as a set of pole contributions (Born or tree amplitude) $K(s)t_0^{\text{tree}}(s)$. (The situation is illustrated in Fig. 24.3.) This amplitude can be forced to satisfy Eq. (24.18), and hence *unitarized* by making the replacement

$$K(s)t_0^{\text{tree}} \longrightarrow \frac{K(s)t_0^{\text{tree}}(s)}{1 - iK(s)t_0^{\text{tree}}(s)} \tag{24.22}$$

This clearly gives $\text{Im}(1/e^{i\delta_0}\sin\delta_0) = -1$, as should be. In *summary*, we shall take as the unitarized amplitude

$$e^{i\delta_0}\sin\delta_0 = \frac{K(s)t_0^{\text{tree}}(s)}{1 - iK(s)t_0^{\text{tree}}(s)} \tag{24.23}$$

To interpret this result, simply expand the denominator; one finds a repeated application of the driving term (Fig. 24.3a), with just enough of the sum over

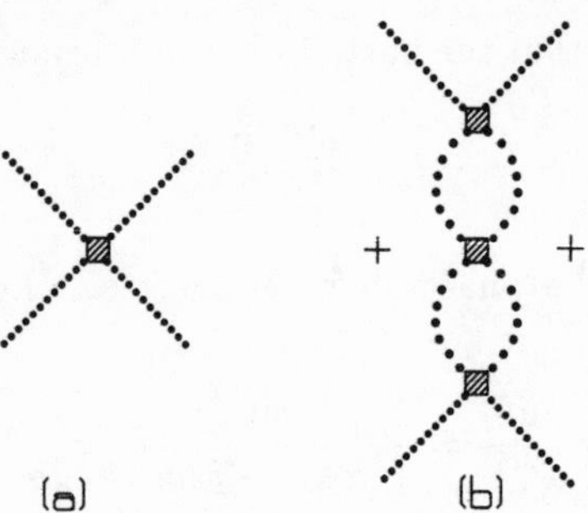

Figure 24.3: (a) Driving term; (b) unitarized (partial) summation of graphs in description of resonance dynamics.

intermediate states to ensure unitarity (Fig. 24.4b). Of course this is an approximation scheme, and one expects these results to be quantitatively modified in any strong coupling theory; however, the qualitative features of repeated application of a driving term leading to resonance behavior might be expected to remain.[58]

It remains to take into account the *identity* of the particles — here bosons. The required modification in the $T = 0$ (symmetric) channel in Eq. (24.15) is (Probs. 24.1-3)

$$\begin{aligned} f_{\text{CM}} &\rightarrow f(\theta) + f(\pi - \theta) \\ f_{\text{CM}} &= \sum\nolimits_{l \text{ even}} (2l+1) P_l(\cos\theta)\, 2f_l \end{aligned} \tag{24.24}$$

Hence in Eqs. (24.19) and (24.21)

$$\begin{aligned} 2f_0 &= -\frac{t_0}{8\pi W} \\ K(s) &= -\frac{q}{16\pi W} = -\frac{1}{32\pi}\left(1 - \frac{4\mu^2}{s}\right)^{1/2} \end{aligned} \tag{24.25}$$

The results of Serot and Lin in Ref. [R68] for $\pi - \pi$ scattering in the $(0^+, 0)$ channel starting from the driving term in Eq. (24.11) are shown in Fig. 24.4. The scalar strength required in QHD-I is reproduced at the appropriate energy for arbitrarily large values of the chiral σ mass.[59]

The $\Delta(1232)$. The $\pi - N$ resonance in the $(\frac{3}{2}^+, \frac{3}{2})$ channel is the first excited state of the nucleon (Fig. 24.5) and this $\Delta(1232)$ state, which lies in the $\pi - N$

[58] One can give a more rigorous basis for this discussion by invoking the N/D solution to dispersion relations where only a certain set of nearby singularities is retained. See Refs. [R70, R71].

[59] See also the figure in that reference where there is an actual peaked contribution to the driving term in the N–N interaction.

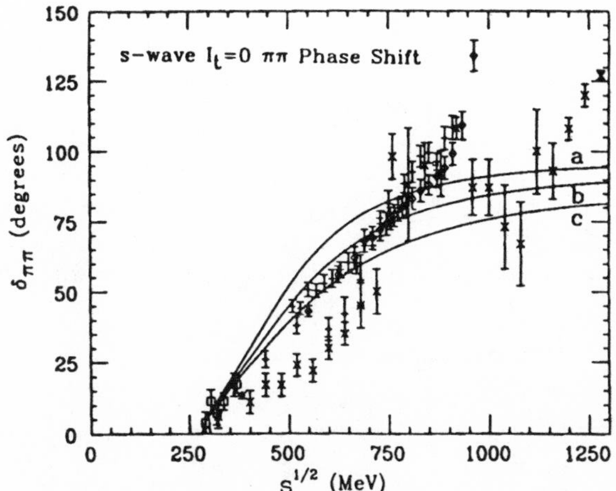

Figure 24.4: The calculated unitarized s-wave $\pi - \pi$ phase shift as a function of (twice) the pion C-M energy compared with the experimental data. The chiral σ masses used here are $m_\sigma = 950\,\text{MeV}$ (a), 1400 MeV (b), and 14 GeV (c). From Ref. [R68].

continuum, dominates the interaction of intermediate energy pions with nuclei. Its effects are all the more important because of the suppression of the $\pi - N$ s-wave interaction, which we have been discussing. Any model that claims a connection with nuclear physics must contain the $\Delta(1232)$. It is not possible to include a field for this particle in a simple renormalizable theory, and so the only possibility, if one starts from the lagrangian of Eq. (24.1), is that this state also occurs as a *dynamic resonance.* Fortunately, one knows from the work of Chew and Low in the static model (Ref. [R69]) and on the relativistic extensions of this work (Refs. [R53, R55, R72]) that this will indeed be the case (see also Ref. [R70]). The driving mechanism is nucleon exchange.

We retain the nucleon pole terms illustrated in Fig. 24.6. The pole contri-

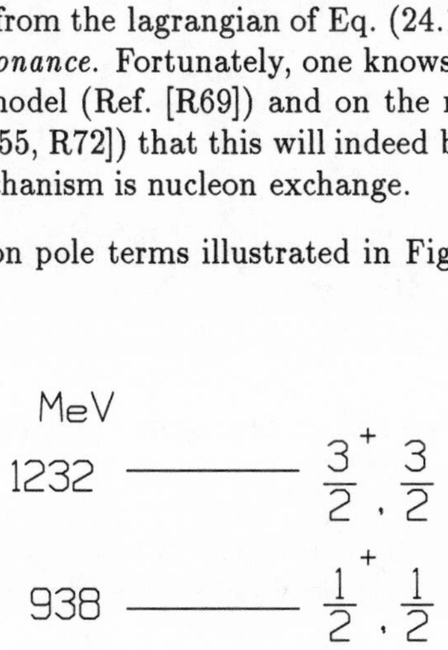

MeV

1232 ———— $\frac{3}{2}^+, \frac{3}{2}$

938 ———— $\frac{1}{2}^+, \frac{1}{2}$

Figure 24.5: First excited state of the nucleon.

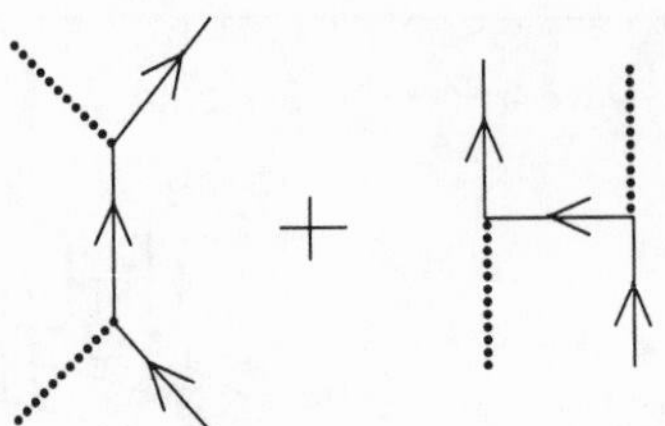

Figure 24.6: Nucleon pole terms as driving mechanism for the $\Delta(1232)$.

bution was evaluated in Section 20.

$$B^{\pm} = g^2 \left[\frac{1}{M^2 - s} \mp \frac{1}{M^2 - u} \right] \qquad A^{\pm} = 0 \tag{24.26}$$

The isospin 3/2 amplitude is given by (Appendix E) $B^{3/2} = B^+ - B^-$, and hence

$$B^{3/2} = -\frac{2g^2}{M^2 - u} \tag{24.27}$$

Now project out the $J^\pi = \frac{3}{2}^+$ amplitude; the procedure is given in Appendix E. To simplify the discussion, consider first just the static limit (as did Chew and Low).[60] Write

$$W \equiv M + \omega \qquad \omega^2 \equiv \mathbf{q}^2 + \mu^2 \tag{24.28}$$

Now let $M \to \infty$. In this case one has

$$\begin{aligned} u &= 2M^2 + 2\mu^2 - W^2 + 2\mathbf{q}^2(1 - x) \\ M^2 - u &= W^2 - M^2 - 2\omega^2 + 2\mathbf{q}^2 x \\ &\overset{M\to\infty}{\longrightarrow} 2M\omega + 2\mathbf{q}^2 x + \cdots \end{aligned} \tag{24.29}$$

Note the linear dependence on $x = \cos\theta$ in the last expression. An expansion in inverse powers of M gives[61]

$$B^{3/2} = -\frac{2g^2}{M^2 - u} \overset{M\to\infty}{\longrightarrow} -\frac{2g^2}{2M\omega}\left(1 - \frac{2\mathbf{q}^2 x}{2M\omega} + \cdots\right) \tag{24.30}$$

From Appendix E one has

$$f_{1^+} = \frac{1}{32\pi M^2}\left\{4M^2\omega B_1 + 2M\mathbf{q}^2 B_2\right\} \tag{24.31}$$

[60] The scalar exchange graph is omitted here for simplicity; it is included in the calculations in Refs. [R55, R72]. As $m_\sigma \to \infty$, scalar meson exchange does not contribute in the $J^\pi = \frac{3}{2}^+$ channel.

[61] One must keep the $x = \cos\theta$ dependence in these expressions to get a nonzero result when projecting the amplitude in the $J^\pi = \frac{3}{2}^+$ channel.

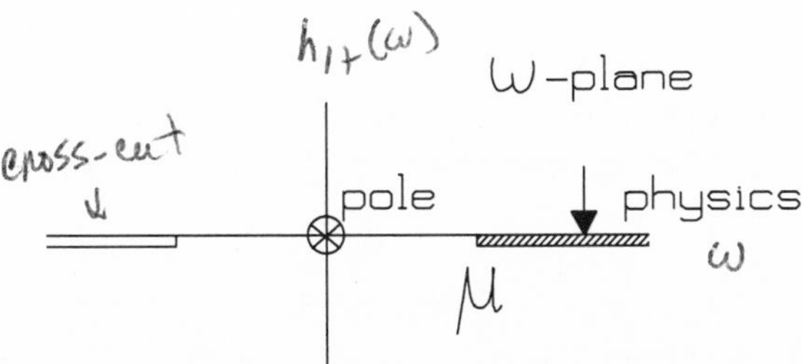

Figure 24.7: Analytic properties of the amplitude $f_{1^+}/\mathbf{q}^2$ as a function of ω. Here the left-hand crossed cut is neglected for simplicity; its effects can easily be included (Refs. [R69, R53, R55, R71, R72]).

To leading order in $1/M$ only the first term contributes and

$$B_1^{3/2} \equiv \int_{-1}^{1} x\,dx\,B^{3/2} = \frac{2g^2}{3}\frac{\mathbf{q}^2}{(M\omega)^2} \tag{24.32}$$

Hence the nucleon pole contribution to the scattering amplitude in the $(\frac{3}{2}^+, \frac{3}{2})$ channel is

$$\begin{aligned} f_{1^+}^{\text{pole}} &= \left(\frac{g^2}{4\pi}\frac{\mu^2}{4M^2}\right)\frac{4}{3}\frac{\mathbf{q}^2}{\omega}\frac{1}{\mu^2} \\ &= \frac{4f_\pi^2}{3}\frac{\mathbf{q}^2}{\mu^2}\frac{1}{\omega} \end{aligned} \tag{24.33}$$

Note that this amplitude $f_{1^+}/\mathbf{q}^2$ has a *simple pole in the variable* ω at $\omega = 0$.

Consider now the *dynamics* in this channel. The full amplitude $f_{1^+}(\omega)/\mathbf{q}^2$ must have the following properties:

1. It must have the *analytic properties* in the variable ω indicated in Fig. 24.7. In particular, it must have the pole in Eq. (24.33).
2. The scattering amplitude $f_{1^+}/\mathbf{q}^2 = e^{i\delta_{1^+}}\sin\delta_{1^+}/q^3$ must satisfy *unitarity* on the right-hand physical cut

$$\text{Im}\left(\frac{f_{1^+}}{\mathbf{q}^2}\right)^{-1} = -q^3 \tag{24.34}$$

The *solution* to this problem can be written in the form (Ref. [R71])

$$\frac{f_{1^+}}{\mathbf{q}^2} = \frac{N(\omega)}{D(\omega)} \tag{24.35}$$

Here $N(\omega)$ has all the *left-hand singularities* (in this case just the pole at the origin) and $N(\omega)$ is real on the right-hand unitarity cut. We will take

$$N(\omega) = \frac{f_{1^+}^{\text{pole}}}{\mathbf{q}^2} = \frac{4f_\pi^2}{3\mu^2}\frac{1}{\omega} \tag{24.36}$$

The denominator $D(\omega)$ has the right-hand unitarity cut. Since the exact value of the amplitude is *known at the pole*, which is contained in $N(\omega)$, one can subtract D there and impose the condition $D(0) \equiv 1$. Unitarity then dictates the form

$$D(\omega) = 1 - \frac{\omega}{\pi}\int_{\mu}^{\infty} \frac{d\omega'}{\omega'} \frac{q'^3 N(\omega')}{\omega' - \omega - i\eta} \tag{24.37}$$

Evidently, on the right-hand cut

$$\mathrm{Im}\frac{\mathbf{q}^2}{f_{1+}} = \frac{1}{N(\omega)}\mathrm{Im}D(\omega) = \frac{1}{N(\omega)}[-q^3 N(\omega)] = -q^3 \tag{24.38}$$

Thus we have a solution to the problem posed.

A combination of these results gives the explicit expression

$$D(\omega) = 1 - \frac{\omega}{\pi}\int_{\mu}^{\infty} \frac{d\omega'}{\omega'} \frac{q'^3}{\mu^2} \frac{4f_\pi^2}{3\omega'} \frac{1}{\omega' - \omega - i\eta} \tag{24.39}$$

The integral does not converge in this static limit. Chew and Low put in a cut-off (Ref. [R69]); the relativistic version, retaining all the correct kinematic factors converges (Ref. [R55, R72]). Since the integral gets most of its contribution from high ω', one can neglect ω in the denominator in the region of interest. In this case $D(\omega)$ takes the approximate form[62]

$$D(\omega) \approx 1 - \frac{\omega}{\omega_{\mathrm{R}}} - iq^3 N(\omega) \tag{24.40}$$

Now

$$\frac{\mathbf{q}^2}{f_{1+}} = q^3(\cot\delta_{1+} - i) \approx \frac{1}{N(\omega)}\left[\left(1 - \frac{\omega}{\omega_{\mathrm{R}}}\right) - iq^3 N(\omega)\right] \tag{24.41}$$

The term $-iq^3$ cancels, and the result is

$$\frac{4q^3}{3\omega\mu^2}\cot\delta_{1+} = \frac{1}{f_\pi^2}\left(1 - \frac{\omega}{\omega_{\mathrm{R}}}\right) \tag{24.42}$$

This is the Chew-Low effective range formula (Ref. [R69, R53]).

At $\omega = \omega_{\mathrm{R}}$, the phase shift passes through $\pi/2$, and one has a resonance.[63] In quantum mechanics a resonance can be represented as

$$e^{i\delta_{1+}}\sin\delta_{1+} = \frac{-\Gamma/2}{\omega - \omega_{\mathrm{R}} + i\Gamma/2} \tag{24.43}$$

[62] Here

$$\frac{4}{3\pi}\left(\frac{f_\pi}{\mu}\right)^2 \int_{\mu}^{\omega_c} \frac{q'^3}{\omega'^3} d\omega' \equiv \frac{1}{\omega_{\mathrm{R}}}$$

Note that $N(\omega)$ must have the correct sign (positive) to get a resonance with $\omega_{\mathrm{R}} > 0$.

[63] It was shown by Chew and Low in Ref. [R69] (also in Ref. [R55]) that the pole term $N(\omega)$ is of the wrong sign, or too small, to lead to low-energy resonances in the other p-waves $(\frac{3}{2}^+, \frac{1}{2}), (\frac{1}{2}^+, \frac{3}{2}), (\frac{1}{2}^+, \frac{1}{2})$ in $\pi - N$ scattering.

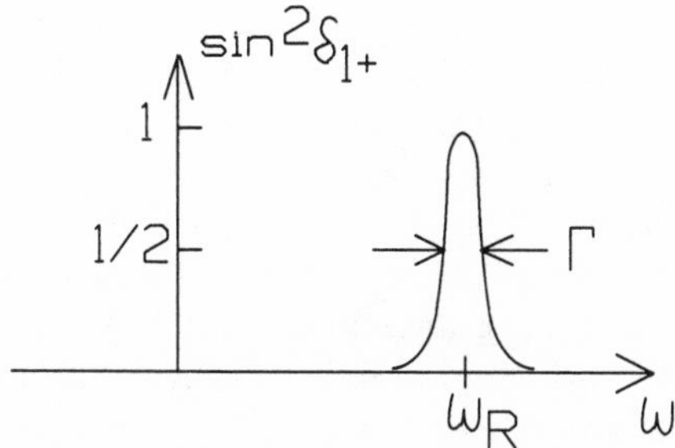

Figure 24.8: Resonant phase shift.

The situation is illustrated in Fig. 24.8. It follows from the above that

$$e^{i\delta_{1+}} \sin \delta_{1+} \approx \frac{q^3 N(\omega)}{1 - \omega/\omega_{\rm R} - iq^3 N(\omega)} \tag{24.44}$$

One can thus identify $\Gamma/2 = q^3 \omega_{\rm R} N(\omega)$, and evaluation at resonance gives

$$\left(\frac{\Gamma}{2}\right)_{\rm res} = \frac{4f_\pi^2}{3} \frac{q_{\rm R}^3}{\mu^2} \tag{24.45}$$

The following numbers are relevant:

$$\begin{array}{lll} f_\pi^2 = 0.0797 & M = 938.3\,{\rm MeV} & \mu = 139.6\,{\rm MeV} \\ W = 1232\,{\rm MeV} & \omega_{\rm R} = 2.104\,\mu & q_{\rm R} = 1.851\,\mu \end{array} \tag{24.46}$$

Insertion into Eq. (24.45) gives

$$\Gamma_{\rm res} = 188.2\,{\rm MeV} \qquad \Gamma_{\rm expt} = 120\,{\rm MeV} \tag{24.47}$$

The second number is the experimental value.

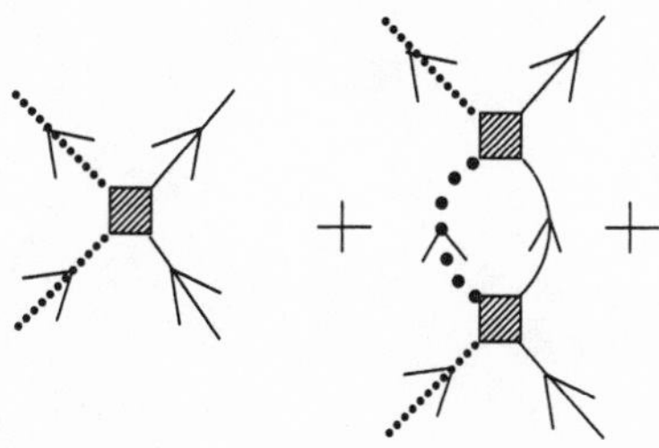

Figure 24.9: Driving term and its iteration in dynamical calculation of the $\Delta(1232)$.

This calculation can again be interpreted as a partial summation of the infinite set of diagrams in Fig. 24.9; just enough of the infinite summation is kept to unitarize the driving term, here generated from the nucleon pole.

The result of a relativistic dispersion calculation keeping all the close lying left-hand singularities in Fig. 24.7 is given in Ref. [R55]; this calculation is convergent and requires no cutoff. A modern more-detailed version of this calculation, starting from the lagrangian in Eq. (24.1) is contained in Ref. [R72]. The calculation is finite in this renormalizable theory, and the $\Delta(1232)$ is again found as a dynamic resonance.

25

A MODEL WITH $(\sigma, \omega, \pi, \rho)$ — QHD-II

In this section the discussion is extended to include the charged $\boldsymbol{\rho}$ vector meson with (ρ^+, ρ^0, ρ^-) in a renormalizable theory (QHD-II) containing the baryon field ψ and $(\sigma, \omega, \boldsymbol{\pi}, \boldsymbol{\rho})$ meson fields. The inclusion of the $\boldsymbol{\rho}$ is through a Yang-Mills nonabelian gauge theory based on local isospin symmetry; this is, in fact, the original application considered by Yang and Mills in Ref. [R73]. The mass of the ρ is then generated through a coupling to a new scalar field (a Higgs field) and spontaneous symmetry breaking; the procedure is the same as in the standard model of the electroweak interactions. Yang-Mills theories are discussed in detail in Part III of this book, and the standard model is discussed in Part IV. We here simply get ahead of ourselves and summarize the construction of QHD-II. The discussion is based on Refs. [R57, R1].

For illustration, start from the lagrangian density of Section 20 that describes QHD-I plus a pion field

$$\begin{aligned}\mathcal{L} &= -\bar{\psi}\left[\gamma_\lambda\left(\frac{\partial}{\partial x_\lambda} - ig_v V_\lambda\right) + (M - g_s\phi) - ig_\pi\gamma_5\boldsymbol{\tau}\cdot\boldsymbol{\pi}\right]\psi \\ &\quad -\frac{1}{2}m_v^2 V_\mu^2 - \frac{1}{4}F_{\mu\nu}F_{\mu\nu} - \frac{1}{2}\left[\left(\frac{\partial\phi}{\partial x_\lambda}\right)^2 + m_s^2\phi^2\right] \\ &\quad -\frac{1}{2}\left[\left(\frac{\partial\boldsymbol{\pi}}{\partial x_\lambda}\right)^2 + \mu^2\boldsymbol{\pi}^2\right] + \frac{1}{2}g_\phi m_s\boldsymbol{\pi}^2\phi \end{aligned} \tag{25.1}$$

With $g_s g_\phi/m_s = g_\pi^2/M$ one reproduces the soft-pion limit for $\pi - N$ scattering; however, the pion has not yet been included in a chiral-invariant fashion; for that one needs the lagrangian in Section 22 (and the discussion in Sections 23 and 24).

Global Isospin Invariance. It was shown in Section 21 that this lagrangian is invariant under global isospin transformations. A global invariance involves a transformation of the fields that is the same at all points in space-time. From Noether's theorem, this invariance implies that the isovector current

$$\begin{aligned}\mathbf{V}_\mu(x) &= i\bar{\underline{\psi}}\gamma_\mu\frac{1}{2}\underline{\tau}\,\underline{\psi} + i\frac{\partial\underline{\pi}^\dagger}{\partial x_\mu}\mathbf{t}\,\underline{\pi} \\ &= i\bar{\underline{\psi}}\gamma_\mu\frac{1}{2}\underline{\tau}\,\underline{\psi} + \frac{\partial\boldsymbol{\pi}}{\partial x_\mu}\times\boldsymbol{\pi}\end{aligned} \tag{25.2}$$

is conserved

$$\frac{\partial \mathbf{V}_\lambda}{\partial x_\lambda} = 0 \tag{25.3}$$

Here $(t_i)_{jk} \equiv -i\varepsilon_{ijk}$.

Local Isospin Invariance. The Yang-Mills theory of the $\boldsymbol{\rho}$ converts this symmetry into *local* isospin invariance, where the isospin transformation can vary from point to point in space-time. This is accomplished through the following series of steps:[64]

1. Introduce *gauge boson* fields, one for each generator

$$\boldsymbol{\rho}_\mu : (\rho^1_\mu, \rho^2_\mu, \rho^3_\mu) \tag{25.4}$$

2. Use the *covariant derivative*

$$\begin{aligned} \frac{D}{Dx_\mu}\underline{\psi} &\rightarrow \left[\frac{D}{Dx_\mu} - ig_\rho \frac{1}{2}\boldsymbol{\tau} \cdot \boldsymbol{\rho}_\mu(x)\right]\underline{\psi} \equiv \frac{\mathcal{D}}{\mathcal{D}x_\mu}\underline{\psi}\,; && \text{on isospinors} \\ \frac{\partial \underline{\pi}}{\partial x_\mu} &\rightarrow \left[\frac{\partial \underline{\pi}}{\partial x_\mu} - ig_\rho \underline{\mathbf{t}} \cdot \boldsymbol{\rho}_\mu(x)\right]\underline{\pi} \,; && \text{on isovectors} \end{aligned} \tag{25.5}$$

The second equation can be rewritten in vector form as

$$\frac{\partial \boldsymbol{\pi}}{\partial x_\mu} \rightarrow \left(\frac{\partial}{\partial x_\mu} + g_\rho \boldsymbol{\rho}_\mu \times\right)\boldsymbol{\pi} \equiv \frac{\mathcal{D}\boldsymbol{\pi}}{\mathcal{D}x_\mu}\,; \qquad \text{on isovectors} \tag{25.6}$$

3. Introduce the *field tensor* for the kinetic energy[65]

$$\begin{aligned} \rho^i_{\mu\nu} &= \frac{\partial \rho^i_\nu}{\partial x_\mu} - \frac{\partial \rho^i_\mu}{\partial x_\nu} + g_\rho \varepsilon_{ijk} \rho^j_\mu \rho^k_\nu \\ \boldsymbol{\rho}_{\mu\nu} &= \frac{\partial \boldsymbol{\rho}_\nu}{\partial x_\mu} - \frac{\partial \boldsymbol{\rho}_\mu}{\partial x_\nu} + g_\rho \boldsymbol{\rho}_\mu \times \boldsymbol{\rho}_\nu \end{aligned} \tag{25.7}$$

In analogy to QED the kinetic energy term for the gauge fields is taken as

$$\mathcal{L}_{\text{KE}} = -\frac{1}{4}\boldsymbol{\rho}_{\mu\nu} \cdot \boldsymbol{\rho}_{\mu\nu} \tag{25.8}$$

In contrast to QED, this lagrangian now contains specific cubic and quartic self-couplings of the $\boldsymbol{\rho}$.

4. One can add a mass term

$$\mathcal{L}_{\text{mass}} = -\frac{1}{2}m^2_\rho \boldsymbol{\rho}^2_\mu \tag{25.9}$$

This term, however, breaks local gauge invariance. The mass has to be generated by another mechanism. The procedure used is to add another scalar field with a

[64] For a discussion of Yang-Mills theories see also Ref. [R4].

[65] The structure constants are defined through $\left[\frac{1}{2}\tau_i, \frac{1}{2}\tau_j\right] = i\varepsilon_{ijk}\frac{1}{2}\tau_k$.

locally gauge-invariant coupling and to then generate the mass by spontaneous symmetry breaking as in Section 22. The procedure is exactly the same as that used in the standard model of the electroweak interactions, and we postpone a discussion of that until Part IV. Here we anticipate the result, which is to add a term to the lagrangian of the form

$$\mathcal{L}_{\text{Higgs}} = -\frac{1}{2}m_\rho^2 \boldsymbol{\rho}_\mu^2 + \delta\mathcal{L}_{\text{Higgs}} \tag{25.10}$$

The last term contains cubic and quartic couplings involving the $\boldsymbol{\rho}$ and a real scalar Higgs field η (whose mass may be taken very large); it is given in Refs. [R57, R1]). The inclusion of the $\boldsymbol{\rho}$ in the lagrangian in Eq. (25.1) in this renormalizable theory thus takes the form

$$\begin{aligned}
\mathcal{L}_{\text{QHD-II}} &= -\bar{\psi}\left[\gamma_\lambda \frac{\mathcal{D}}{\mathcal{D}x_\lambda} + (M - g_s\phi) - ig_\pi\gamma_5 \boldsymbol{\tau}\cdot\boldsymbol{\pi}\right]\psi \\
&\quad -\frac{1}{2}m_v^2 V_\mu^2 - \frac{1}{4}F_{\mu\nu}F_{\mu\nu} - \frac{1}{2}m_\rho^2\boldsymbol{\rho}_\mu^2 - \frac{1}{4}\boldsymbol{\rho}_{\mu\nu}\cdot\boldsymbol{\rho}_{\mu\nu} \\
&\quad -\frac{1}{2}\left[\left(\frac{\partial\phi}{\partial x_\lambda}\right)^2 + m_s^2\phi^2\right] + \frac{1}{2}g_\phi m_s \phi \boldsymbol{\pi}^2 \\
&\quad -\frac{1}{2}\left[\frac{\mathcal{D}\boldsymbol{\pi}}{\mathcal{D}x_\lambda}\cdot\frac{\mathcal{D}\boldsymbol{\pi}}{\mathcal{D}x_\lambda} + \mu^2\boldsymbol{\pi}^2\right] + \delta\mathcal{L}_{\text{Higgs}}
\end{aligned} \tag{25.11}$$

This lagrangian contains cubic and quartic meson couplings involving the $\boldsymbol{\rho}_\mu$ in the terms $-\boldsymbol{\rho}_{\mu\nu}\cdot\boldsymbol{\rho}_{\mu\nu}/4$ and $(\mathcal{D}\boldsymbol{\pi}/\mathcal{D}x_\lambda)^2$; they are specified in terms of a single coupling constant g_ρ.

Let us identify the couplings linear in $\boldsymbol{\rho}_\mu$ in Eq. (25.11). These are

$$\begin{aligned}
\mathcal{L}^{\text{int}} &= ig_\rho\bar{\psi}\gamma_\mu\frac{1}{2}\boldsymbol{\tau}\cdot\boldsymbol{\rho}_\mu\psi - \frac{g_\rho}{2}\left[\frac{\partial\boldsymbol{\pi}}{\partial x_\mu}\cdot(\boldsymbol{\rho}_\mu\times\boldsymbol{\pi}) + (\boldsymbol{\rho}_\mu\times\boldsymbol{\pi})\cdot\frac{\partial\boldsymbol{\pi}}{\partial x_\mu}\right] \\
&= g_\rho\boldsymbol{\rho}_\mu\cdot\mathbf{V}_\mu
\end{aligned} \tag{25.12}$$

This piece of the lagrangian gives the linear coupling of the $\boldsymbol{\rho}_\mu$ to the $(N, \boldsymbol{\pi})$ system. The last equality arises from the identification in Eq. (25.2); note that this is *not* the full isovector current in this locally isospin invariant theory, since the $\boldsymbol{\rho}_\mu$ now has self couplings that contribute to it.

Some Ingredients in Feynman Diagrams. Some ingredients in Feynman diagrams for the S-matrix (as in Section 20) arising from this lagrangian are shown in Fig. 25.1; the analytical expressions are

$$\begin{aligned}
&\text{NN}\rho - \text{vertex} && -g_\rho\gamma_\mu\frac{1}{2}\tau_a \\
&\pi\pi\rho - \text{vertex} && -g_\rho(q_1+q_2)_\mu\varepsilon_{abc} \\
&\rho - \text{propagator} && \frac{1}{i}\frac{1}{q^2+m_\rho^2}\delta_{ab}(\delta_{\mu\nu} + \frac{q_\mu q_\nu}{m_\rho^2})
\end{aligned} \tag{25.13}$$

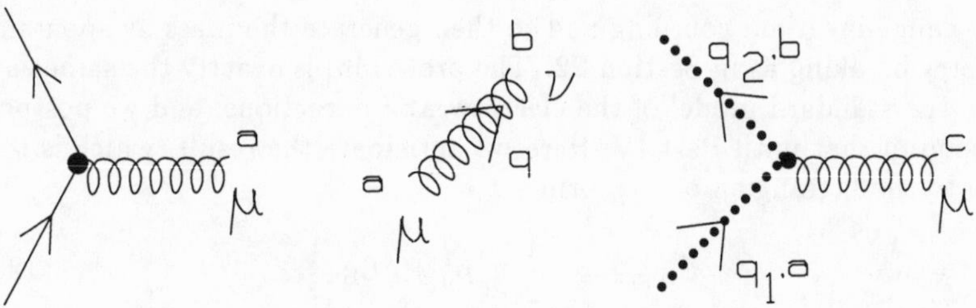

Figure 25.1: Some ingredients in Feynman diagrams for the S-matrix (as in Section 20) involving the ρ.

This last expression is written in the unitary gauge where the particle content of the theory is manifest. Despite the appearance of the last term in the propagator, this theory is renormalizable, as in the standard model of the electroweak interactions (Part IV).

Contribution of $\boldsymbol{\rho_\mu}$ *Exchange to* π*–N Scattering.* Let us use the above results to calculate the contribution of the ρ pole to π–N scattering; the process is illustrated in Fig. 25.2. The Feynman rules readily permit an identification of the contribution to the T-matrix (Section 20).

$$T_{\mathrm{fi}} = g_\rho^2 \bar{u}(p_2)\left[\gamma_\mu \frac{1}{2}\tau_c \frac{1}{m_\rho^2 - t}\left(\delta_{\mu\nu} + \frac{q_\mu q_\nu}{m_\rho^2}\right)\varepsilon_{abc}(q_1+q_2)_\nu\right]u(p_1) \tag{25.14}$$

Note that the term $q_\mu q_\nu/m_\rho^2$ in the ρ propagator drops out at this level from current conservation on either the $\boldsymbol{\pi}$ vertex where $q\cdot(q_1+q_2) = (q_1-q_2)\cdot(q_1+q_2) = 0$ or on the nucleon vertex where $\bar{u}(p_2)\gamma_\mu(p_2-p_1)_\mu u(p_1) = 0$. The use of $\varepsilon_{abc}\tau_c = (i/2)[\tau_b, \tau_a]$ permits the above result to be written as

$$T_{\mathrm{fi}} = \bar{u}(p_2)\left[i\gamma_\mu Q_\mu \frac{1}{2}[\tau_b, \tau_a]\right]u(p_1)\frac{g_\rho^2}{m_\rho^2 - t} \tag{25.15}$$

Hence the ρ pole contribution to π–N scattering can be identified as

$$B_\rho^{(-)} = \frac{g_\rho^2}{m_\rho^2 - t} \tag{25.16}$$

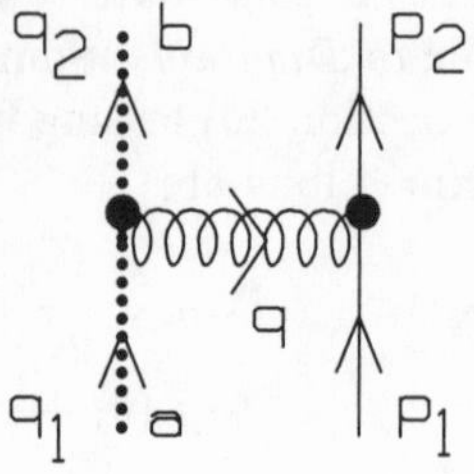

Figure 25.2: Contribution of the ρ pole to π–N scattering.

This exchange does not contribute to the isospin even (+) amplitude.

Further discussion of QHD-II is contained in Refs. [R57, R1].

Chiral Symmetry. The question arises as to whether chiral symmetry can be imposed on QHD with vector mesons. In Ref. [R74] a chiral-invariant Yang-Mills theory based on the local gauge symmetry $\mathrm{SU}(2)_R \bigotimes \mathrm{SU}(2)_L$ (Appendix F) is constructed. The baryon mass is again generated through the spontaneous symmetry breaking of the linear σ-model and the vector meson masses are produced as above through the Higgs mechanism. The theory is parity conserving. Two baryon isodoublets with opposite hypercharge y are necessary to eliminate chiral anomalies and leave one with a meaningful renormalizable quantum field theory. The minimal set of hadrons required thus consists of $\{N, \Xi; \sigma, \omega, \boldsymbol{\pi}, \boldsymbol{\rho}, \mathbf{a}; \eta, \xi\}$ where $\mathbf{a}$ is the chiral partner of the $\boldsymbol{\rho}$ (the $\mathbf{a}$ naturally occurs at a higher mass in the model), and the η and ξ represent scalar and pseudoscalar Higgs particles. The parameters in this minimal theory consist of eight coupling constants and and one mass $(g_\omega, g_{0\pi} + y g_{1\pi}, g_\rho, \mu_{\mathrm{M}}^2, \lambda_{\mathrm{M}}, \mu_{\mathrm{H}}^2, \lambda_{\mathrm{H}}; m_\omega)$, where μ^2 and λ define the meson interaction potentials that lead to spontaneous symmetry breaking.

The lagrangian for this model $\mathcal{L}_{\mathrm{QHD-III}}$ is obtained by using covariant derivatives containing right- and left-handed gauge bosons treated on an equal footing in the lagrangian in Eq. (F.11)[66]

$$\frac{D}{Dx_\lambda}\psi_R = \left(\frac{\partial}{\partial x_\lambda} - i g_v V_\lambda - \frac{i}{2} G \boldsymbol{\tau} \cdot \mathbf{r}_\lambda\right) \psi_R$$
$$\frac{D}{Dx_\lambda}\psi_L = \left(\frac{\partial}{\partial x_\lambda} - i g_v V_\lambda - \frac{i}{2} G \boldsymbol{\tau} \cdot \mathbf{l}_\lambda\right) \psi_L$$
$$\frac{D}{Dx_\lambda}\chi = \frac{\partial}{\partial x_\lambda}\chi - \frac{i}{2} G(\boldsymbol{\tau} \cdot \mathbf{r}_\lambda)\chi + \frac{i}{2} G\chi(\boldsymbol{\tau} \cdot \mathbf{l}_\lambda) \tag{25.17}$$

The meson potential is taken to have the form of the σ-model in Section 22

$$V(\mathrm{tr}\,\chi^\dagger\chi) = -\frac{\mu^2}{2}(\mathrm{tr}\,\chi^\dagger\chi) + \frac{\lambda}{4}(\mathrm{tr}\,\chi^\dagger\chi)^2 \tag{25.18}$$

The kinetic energy from the field tensors for the gauge fields $(\mathbf{r}_\lambda, \mathbf{l}_\lambda)$ is also included as in Eq. (25.8).

Two complex doublets of Higgs fields (ϕ_R, ϕ_L), treated on an equal footing, are added; a Higgs potential of the form of Eq. (25.18) then gives rise to mass terms for the gauge bosons by spontaneous symmetry breaking exactly as in the standard model of electroweak interactions in Part IV.

The physical vector meson fields are obtained as linear combinations of the gauge fields[67]

$$\boldsymbol{\rho}_\lambda = \frac{1}{\sqrt{2}}(\mathbf{r}_\lambda + \mathbf{l}_\lambda) \qquad\qquad \mathbf{a}_\lambda = \frac{1}{\sqrt{2}}(\mathbf{r}_\lambda - \mathbf{l}_\lambda) \tag{25.19}$$

[66] Note the order in the last expression.

[67] One more diagonalization of $\vec{a}_\lambda$ and $\partial\vec{\pi}/\partial x_\lambda$ is still required.

The reader will thus be able to reproduce the lagrangian of Ref. [R74]. Applications do not exist at this time, and we are content to leave the inclusion of vector mesons in a chiral symmetric renormalizable QHD with the statement that it can be accomplished.

26

CEBAF'S ROLE

In this section we consider the relation of the relativistic nuclear many-body problem to CEBAF. First the discussion in Part II of this book is briefly summarized.

The Relativistic Nuclear Many-Body Problem —Summary/Review. One desires a theoretical description of nuclei under *extreme conditions* of density, temperature, and momentum transfer $k^2 \gg M^2$. In developing this description it is essential to incorporate general principles of physics: quantum mechanics, special relativity, and microscopic causality. We work here with hadronic degrees of freedom since they appear to provide the most economical description and extrapolation of observed nuclear properties.[68] The only consistent theoretical framework one has for describing such a relativistic interacting many-body system is lagrangian field theory based on a local lagrangian density. The condition of renormalizability is imposed so that one has a *theory* and can consistently investigate the consequences. This requirement can be taken to define a *purely hadronic theory* (QHD). Such theories provide a guide into unexplored regions of nuclear structure, and as such provide a theoretical framework for predicting and interpreting CEBAF experimental results, as well as those from the Relativistic Heavy Ion Collider (RHIC), and the intense, high-energy meson factory (KAON). A recent review of QHD is given in Ref. [R75].

Mean Field Theory of Nuclear Matter. A basic coincidence reaction to be studied at CEBAF is $(e, e'p)$ (Section 12). The relativistic MFT of nuclear matter developed in Section 14 and the effective electromagnetic current of Section 17 provide a basis for making a relativistic calculation of this process. The kinematic situation is illustrated in Fig. 26.1.[69]

The S-matrix for the $(e, e'X)$ coincidence process has the general form

$$S_{fi} = -(2\pi)^4\delta^{(4)}(p_1 + k - p_2 - q)\frac{ee_p}{\sqrt{\Omega^2}}\bar{u}(k_2)\gamma_\mu u(k_1)\frac{1}{k_\nu^2}\langle p_2\, q^{(-)}|J_\mu(0)|p_1\rangle \quad (26.1)$$

[68]QCD based on quark and gluon degrees of freedom appears to be the underlying theory of the strong interactions. Hence the present analysis must be viewed as an economical description of the consequences of QCD in the strong-coupling regime — a subject to which we turn our attention in Part III of this book.

[69]Note the change in notation: for $(e, e'X)$ in this section k_μ is the momentum transfer and q_μ the momentum of the produced particle.

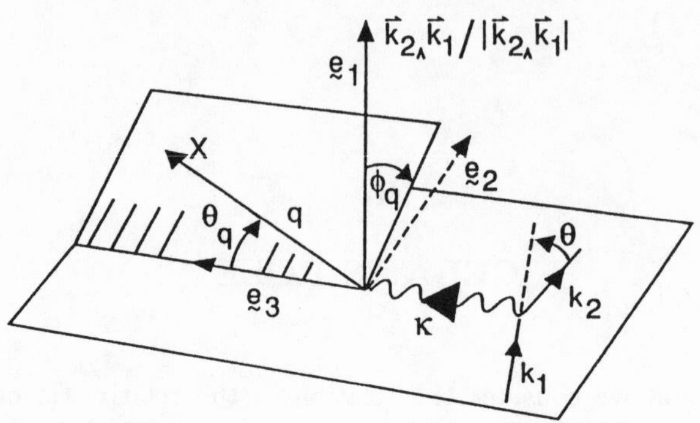

Figure 26.1: Kinematic situation for a (e, e′X) coincidence experiment.

Here $k \equiv k_1 - k_2$, and (p_1, p_2) are the initial and final four-momenta of the target with $(p_1^2 = -M_1^2, p_2^2 = -M_2^2)$. The cross section follows as (Ref. [N14])[70]

$$\frac{d^5\sigma}{d\varepsilon_2 d\Omega_2 d\Omega_q} = \sigma_{\rm M}\left(\frac{M_1|\mathbf{q}|}{\pi W}\right)\left\{\frac{k_\mu^4}{k^{*4}}|\mathcal{J}_C|^2\right.$$

$$+\left(\frac{k_\mu^2}{2k^{*2}}+\frac{W^2}{M_1^2}\tan^2\frac{\theta}{2}\right)(|\mathcal{J}^{+1}|^2+|\mathcal{J}^{-1}|^2)+\frac{k_\mu^2}{2k^{*2}}2{\rm Re}\,(\mathcal{J}^{+1})^*(\mathcal{J}^{-1})\cos 2\phi_q$$

$$\left.+\frac{k_\mu^2}{k^{*2}}\left(\frac{k_\mu^2}{k^{*2}}+\frac{W^2}{M_1^2}\tan^2\frac{\theta}{2}\right)^{1/2}\sqrt{2}\,{\rm Im}\,\mathcal{J}_C^*(\mathcal{J}^{+1}+\mathcal{J}^{-1})\sin\phi_q\right\} \tag{26.2}$$

Here the ϕ_q dependence has been made explicit. The transition matrix element of the current appearing in this expression is defined by

$$\mathcal{J}_\mu \equiv \frac{\sqrt{M_1 M_2}}{4\pi W}\left(\frac{2\omega_q E_1 E_2 \Omega^3}{M_1 M_2}\right)^{1/2}\langle p_2\, q^{(-)}|J_\mu(0)|p_1\rangle \tag{26.3}$$

The quantities $(W, \mathbf{k}^*)$ are the total energy and photon momentum in the C-M system, $\mathbf{q}$ is in the C-M system, $k_\mu^2 = \mathbf{k}^{*2} - k_0^{*2}$, and $\mathcal{J}^\lambda = \mathbf{e}_\lambda \cdot \mathcal{J}$ are the spherical components of the current $\mathcal{J}_\mu = (\mathcal{J}, \mathcal{J}_0)$; the unit vectors are defined so that $\mathbf{e}_1$ lies along $\mathbf{k}_2 \times \mathbf{k}_1$ and $\mathbf{e}_3$ along $\mathbf{k}$.

The relativistic MFT calculation is carried out in Ref. [R40]. The four response functions in Eq. (26.2) obtained from the transition matrix elements of the current $\mathcal{J}_\mu$ are shown in Fig. 26.2. This calculation has the following features:

1. The relativistic MFT provides a realistic model of nuclear matter (Section 14).

[70] See Prob. 26.1.

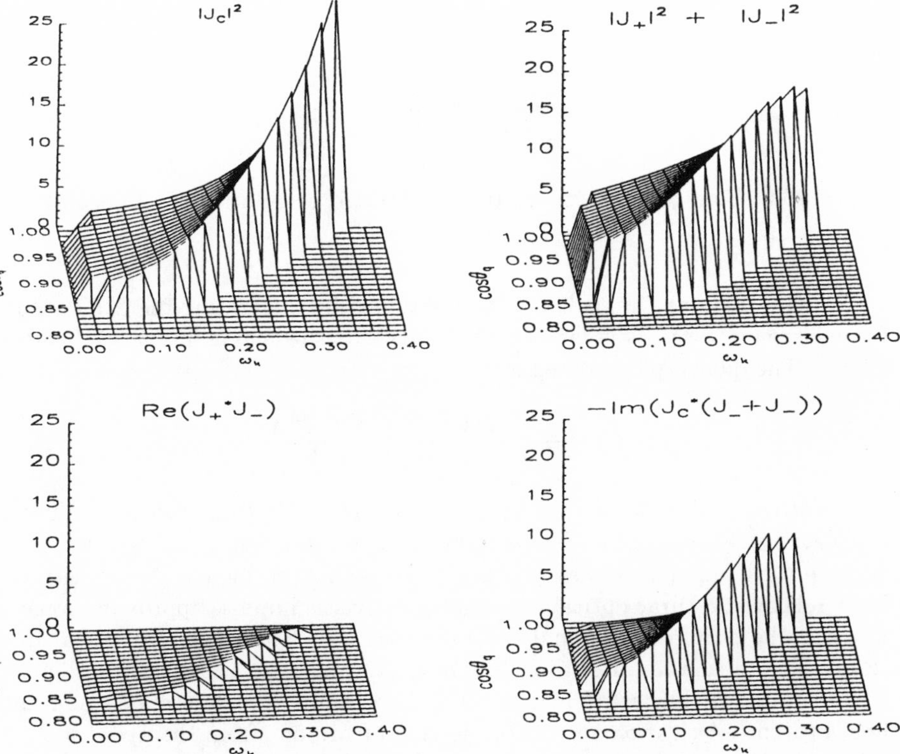

Figure 26.2: The four proton response functions in MFT obtained from the transition matrix elements of the current $\mathcal{J}_\mu$ and evaluated per proton are plotted as functions of energy loss and $\cos\theta_q$. Here $|\mathbf{k}| = 0.5\,\text{GeV}$ and $\phi_q = \pi/2$; the ϕ_q dependence is now in the response. Also $k_\text{F} = 0.28\,\text{GeV}$ and $M^*/M = 0.56$ (appropriate for nuclear matter). The vertical scale is $25.0\,\text{GeV}^{-2}$ for all four response functions, and ω_k is in GeV. See Ref. [R40]. The author would like to thank S. Pollock for preparing this figure.

2. The full nucleon vertex $\Gamma_\mu = F_1\gamma_\mu - F_2\sigma_{\mu\nu}k_\nu$ has been used (see Prob. 17.5); the current is conserved and gives the correct result for a free nucleon.

3. The calculation is completely relativistic.

4. The resulting response surfaces in Fig. 26.2 map out the complete Fermi sphere, weighted by the appropriate electromagnetic interaction; one can examine any part of the Fermi sphere, including the deeply bound states, by looking at the appropriate region of the response surface. Correlations (Sections 4 and 5) will modify the Fermi sphere and add a tail to the momentum distribution.

5. The $(e, e'n)$ response surfaces are also worth looking at (Ref. [R40]).

6. The inelastic $^{40}\text{Ca}(e, e')$ and $^{208}\text{Pb}(e, e')$ cross sections for the finite system obtained from relativistic MFT have already been shown in Figs. 17.13 and 17.14 (Refs. [R35, R1]).

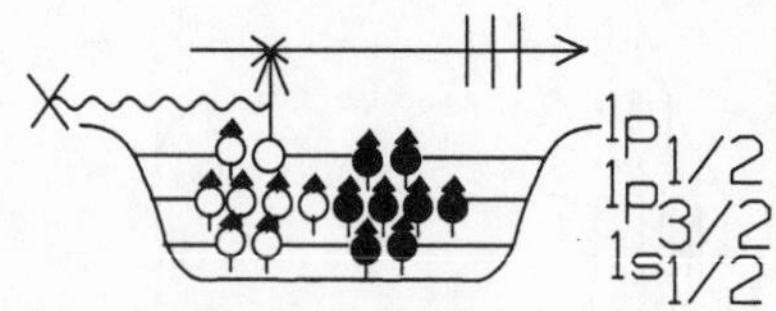

Figure 26.3: Configuration for relativistic (DWIA) calculation of $^{16}O(e,e'\vec{p})^{15}N$.

7. The Coulomb sum rule for ^{40}Ca obtained by integrating the MFT result for nuclear matter is indistinguishable from the dashed curve in Fig. 17.15 (Ref. [R40]). The quantity calculated is

$$C(k) \equiv \int d\omega \frac{\mathbf{k}^4}{k_\mu^4} \frac{1}{\sigma_{\mathrm{M}}} \left(\frac{d^3\sigma^{\mathrm{Coul}}}{d\Omega_2 d\varepsilon_2} \right) \tag{26.4}$$

Relativistic Hartree Wave Functions and Dirac Optical Potential. These calculations can be extended to a more realistic situation by using relativistic Hartree wave functions to describe the initial bound proton and then constructing a relativistic Dirac optical potential (relativistic impulse approximation) to describe the proton in the final state (Section 15). This is carried out in Ref. [R76]. The calculation uses the effective electromagnetic current of Section 17. The situation is illustrated in Fig. 26.3. The results for the cross section and final proton polarization vector for ejection of protons from the $1p_{1/2}$ shell in $^{16}O(e,e'\vec{p})^{15}N$ are shown in Fig. 26.4. Some features of these results are of interest:

1. This is a realistic model for ^{16}O (Section 15).

2. The calculation is fully relativistic.

3. The results are shown as the solid line (DWIA) in Fig. 26.4; they are compared with relativistic plane wave (PWIA) and nonrelativistic distored wave and plane wave calculations.

4. Interesting spin effects show up for the final proton in the relativistic (DWIA) calculation; this might be anticipated from the discussion in Section 15.

Scattering from Discrete Nuclear Levels at High Momentum Transfer. One can use the relativistic Hartree wave functions of Section 15 and the effective current of Section 17 to extrapolate electron scattering cross sections from discrete nuclear levels out to high momentum transfer. In Ref. [R34] elastic magnetic scattering is calculated for three nuclei. The assumed configurations are shown in Fig. 26.5. The relativistic Hartree wave functions in each case are shown in Fig. 17.11. The squares of the magnetic form factors are shown in Figs. 17.10 and 17.12. Several features of these results are of interest:

1. This is a completely relativistic calculation of the electron scattering response (except for the center-of-mass correction).

2. The calculations are roughly consistent with all existing data.

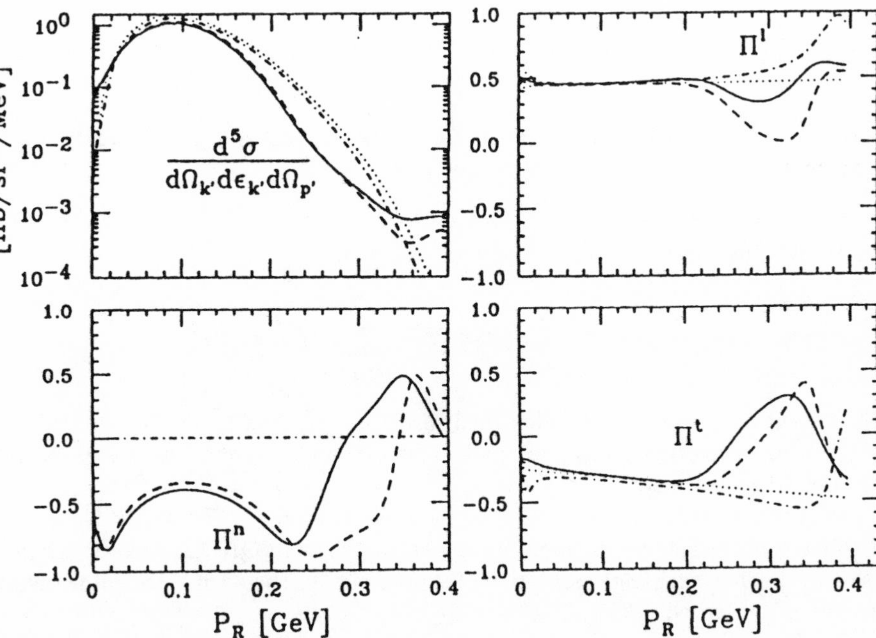

Figure 26.4: Cross section and polarization vector for the ejection of a 500 – MeV proton from the $1p_{1/2}$ shell of ^{16}O in constant-k in-plane kinematics. The incident electron energy is 2 GeV. Calculations are shown for Dirac DWIA (solid line), nonrelativistic DWIA (dashed line), Dirac PWIA (dot-dashed line), and nonrelativistic PWIA (dotted line). Here $\mathbf{p}_R = \mathbf{k} - \mathbf{q}$. From Ref. [R76].

3. These calculations provide a guide, within the framework of QHD, for an extrapolation of the observed nuclear response to high momentum transfers.

4. It is at very large momentum transfer that one might expect to see an explicit transition from meson-baryon to quark-gluon degrees of freedom.

5. One should include a relativistic exchange-current contribution here — and we proceed to discuss this topic.

Exchange Currents. A prime example of the need for explicit hadronic descriptions of nuclei is provided by the additional two-body currents arising from the exchange of charged mesons between nucleons. Although many exchange current calculations exist, we briefly describe those of Dubach, Koch, and Donnelly in Ref. [R77] for concreteness. These authors keep the static limit [leading $O(1/M)$] of the time-ordered Feynman diagrams shown in Fig. 26.6. Each of these processes clearly represents an additional contribution to the current in

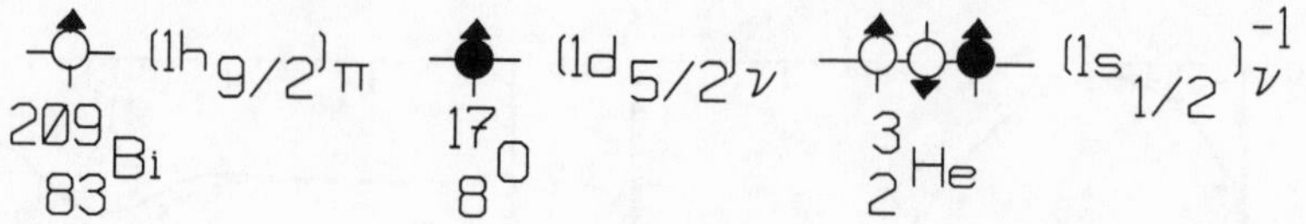

$(1h_{9/2})_\pi$ $^{209}_{83}$Bi $(1d_{5/2})_\nu$ $^{17}_{8}$O $(1s_{1/2})^{-1}_\nu$ $^{3}_{2}$He

Figure 26.5: Assumed configurations for relativistic calculation of elastic magnetic scattering.

the traditional picture, which is now extended to

$$\hat{J}_\mu(\mathbf{x}) = \sum_{i=1}^{A} J_\mu^{(1)}(\mathbf{x}_i;\mathbf{x}) + \sum_{i<j=1}^{A} J_\mu^{(2)}(\mathbf{x}_i,\mathbf{x}_j;\mathbf{x}) \tag{26.5}$$

The two-body current can be identified through reproduction of the S-matrix; it is constructed explicitly in Prob. 26.4. This exchange current has the following features to recommend it:

1. If the two-nucleon potential is of the form $V = V^{\text{neutral}} + V^{\text{OPEP}}$ where the first term arises from neutral meson exchange as in QHD-I, and the second term is the 1-π exchange potential of Appendix A, then this electromagnetic current is conserved (Prob. 26.5).

2. The threshold π electroproduction graphs satisfy the Kroll-Ruderman (soft-pion) theorem.

3. This 1-π contribution represents the longest-range part of the two-body exchange current; it is exact as $|\mathbf{x}_1 - \mathbf{x}_2| \to \infty$.

4. The charge density operator is unmodified to leading $O(1/M)$; hence transition matrix elements of the charge density can be used to calibrate the nuclear structure in exchange-current calculations.

5. Assume that ^{3}He can be described by a $(1s_{1/2})_\nu^{-1}$ harmonic oscillator shell model configuration. The magnetic moment calculated with the inclusion of these exchange currents is $\mu = -2.078$ n.m., now closer to the experimental

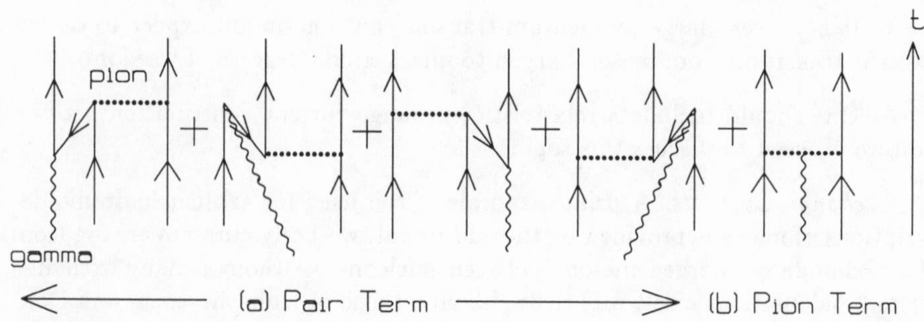

Figure 26.6: Time-ordered Feynman diagrams retained in the 1$-\pi$ exchange current calculation in Ref. [R77].

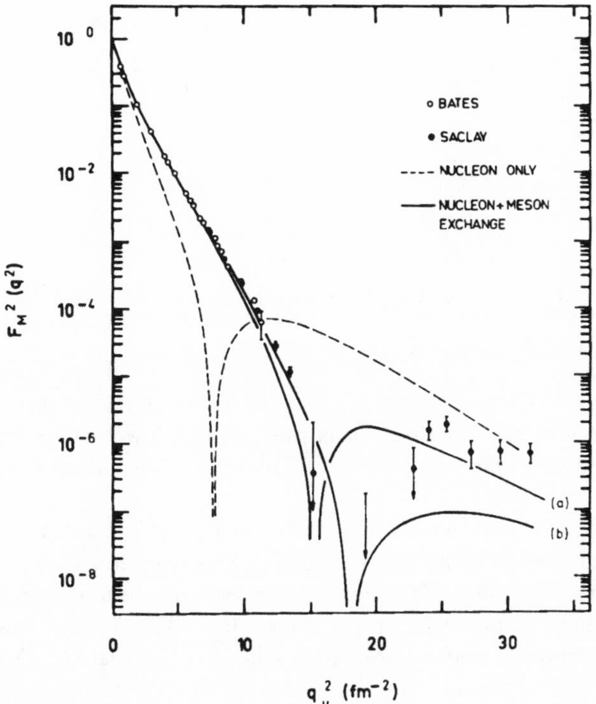

Figure 26.7: Elastic magnetic form factor for ^{3}He(e, e') out to high k^2 from Ref. [R79] (Here $q_\mu^2 \equiv k_\mu^2$). Two exchange-current theories are shown: (a) from Ref. [R80]; (b) from Ref. [R81].

value $\mu = -2.127$ n.m. than is the Schmidt value $\mu = -1.913$ n.m., indicating that one is in the right ballpark.

6. The effect on elastic magnetic electron scattering at modest momentum transfers, say $k^2 \leq 6\ \text{fm}^{-2}$, is not large (Fig. 26.7), indicating the *marginal* role of exchange currents in the traditional nuclear physics domain.

7. A relativistic QHD calculation of this exchange current, without the $1/M$ expansion, is contained in Ref. [R78].

Figure 26.7 taken from Ref. [R79] illustrates the state of the art with elastic magnetic scattering from ^{3}He. The dashed line shows the result obtained from the best three-body calculation done in the traditional picture; the three-body wave function is obtained by solving the Faddeev equations with potentials fit to two-body data, and the current is obtained from the properties of free nucleons. The best three-body calulation in the traditional picture clearly fails at high k^2. Also shown in Fig. 26.7 are two meson exchange current calculations (Refs. [R80, R81]) that include the pion exchange current discussed above, as well as other hadronic contributions (see also Ref. [R101]). The difference between

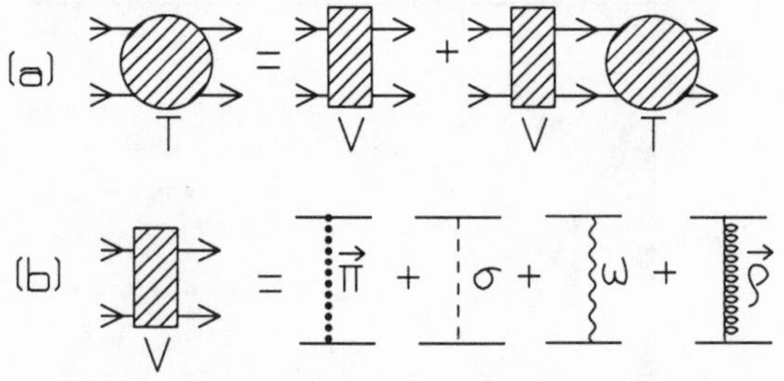

Figure 26.8: (a) T-matrix, and (b) potential for relativistic calculation of N–N scattering in Ref. [R82].

these two curves is a good measure of the present theoretical uncertainty. While the exchange current contribution is marginal at low momentum transfers, it is now a *dominant* effect at large k^2. This application illustrates the need for QHD.

Relativistic Description of Two-Body System. It is essential to have a fully relativistic calculation of the properties of the two-nucleon system, formulated in terms of hadronic degrees of freedom, to project the observed electron scattering response to high momentum transfer (see Ref. [R97]). One such calculation is currently being carried out by the authors in Ref. [R82]. A diagrammatic representation of the relativistic T-matrix for N–N scattering is shown in Fig. 26.8a. The relativistic potential appearing in this expression is calculated in terms of meson exchanges as illustrated in Fig. 26.8b. Some interesting features of this calculation are the following:

1. The calculation uses the Gross reduction of the Bethe-Salpeter equation.
2. The calculation is fully relativistic.
3. The calculation goes beyond QHD in that *empirical* form factors are used at the meson-nucleon vertices to model the short-distance behavior.
4. There are 12 parameters in Model I in this calculation: coupling constants, form factor parameters, and the ps - pv ratio for the pion-nucleon coupling.[71]
5. A good fit is obtained for the N–N phase shifts (Fig. 26.9).
6. To compute electron scattering, the electromagnetic interaction must now be included in a gauge-invariant fashion. That work is in progress.

Relativistic two-body calculations of the properties of nuclear matter within the framework of QHD exist and extend the discussion of Section 4 (Refs. [R83, R84, R1]), as do other such calculations (Ref. [R85, R86, R102]). A discussion of relativistic transport theory (Ref. [R87]), while extremely interesting and crucial for developments such as RHIC, would take us too far afield.

Excitation of the Nucleon. Electron excitation of the nucleon will form a central focus at CEBAF. Coincidence studies $(e, e'X)$ will explore the structure

[71] The best fit in Model I gives $\{m_s = 516.0\,\text{MeV}, g_s^2/4\pi = 5.513, g_v^2/4\pi = 9.851\}$ with $m_v \equiv m_\omega$ [compare Eqs. (20.37)].

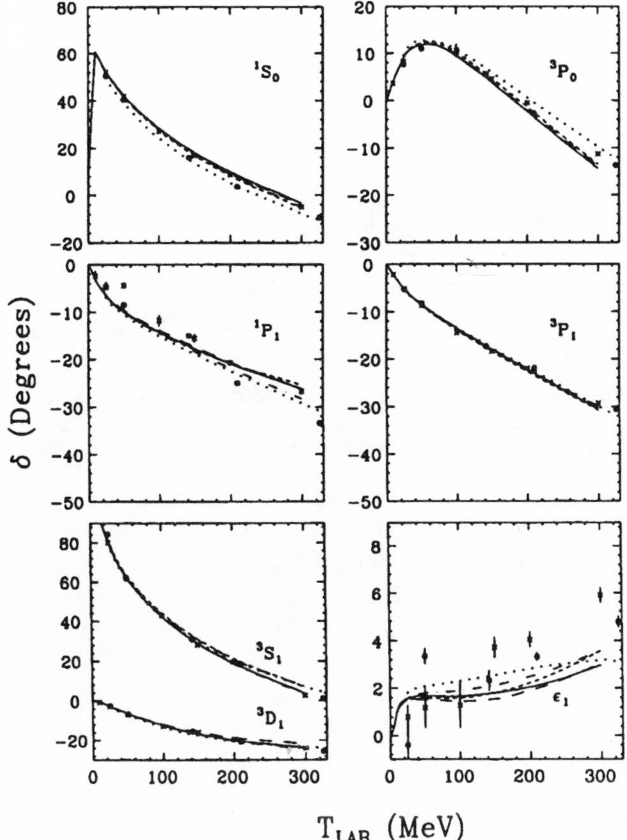

Figure 26.9: Fits to the $J = 0$ and $J = 1$ N–N phase parameters. Solid line is Model I, long dashed line is Model II, and short dashed line is the full Bonn result. From Ref. [R82].

of the nucleon in unprecedented detail (Ref. [N14]). Figure 26.10 shows existing data from SLAC on $^1\mathrm{H}(e, e')$ taken in the resonance region. The assumption of Breit-Wigner resonances and a smooth polynomial background allows one to extract the inelastic (e, e') cross section as the area under the peak (Ref. [R89]); the ratio $(d\sigma_{\mathrm{in}}/d\sigma_{\mathrm{el}})_{\theta=6^\circ}$ is plotted as a function of k^2 for the first nucleon resonance in Fig. 26.11.

We have seen in Section 25 that the $\Delta(1232)$ is obtained as a dynamic resonance in QHD. Electron excitation of this resonance (Fig. 26.12) can be viewed as an excitation process into the proper π–N channel followed by a dynamic final-state enhancement that builds up the resonance (Figs. 24.6 and 24.9). As a model for $\mathrm{N}(e, e')\Delta$ consider the following

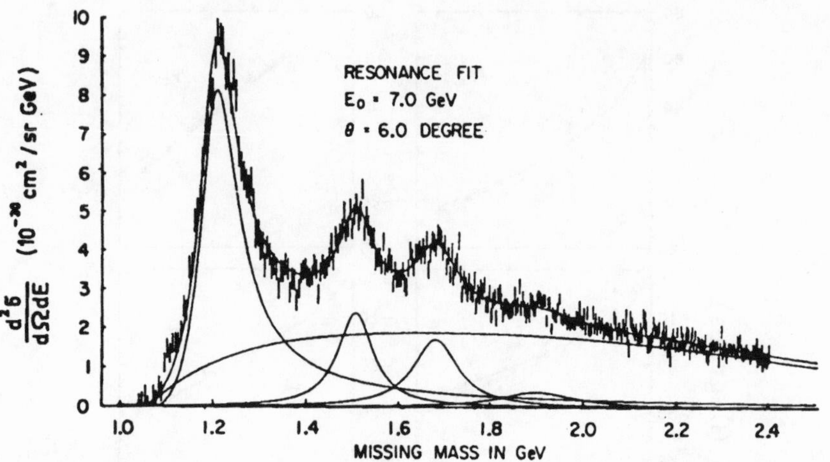

Figure 26.10: SLAC experimental inelastic spectrum for ${}^1\mathrm{H}(e,e')$ at $\varepsilon_1 = 7\ \mathrm{GeV}$, $\theta = 6^0$ resolved into Breit-Wigner resonances (see text). The elastic peak has been suppressed. From Refs. [R88, R90].

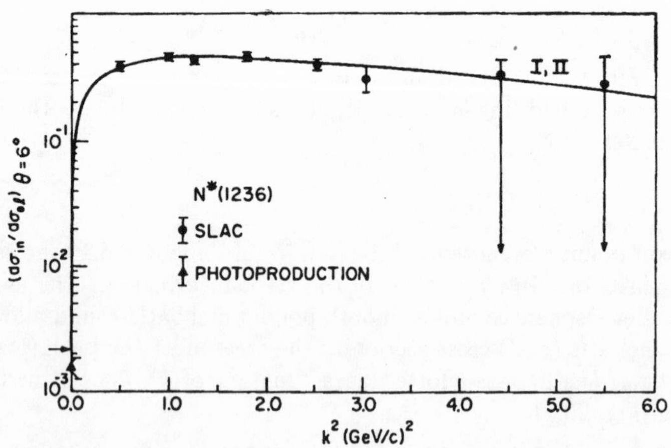

Figure 26.11: $d\sigma_{\text{in}}/d\sigma_{\text{el}}$ at $\theta = 6^o$ for the $3/2^+, 3/2(1232)$ resonance. Experimental points are from SLAC Group A (Ref. [R88]) and the resonance analysis of Breidenbach (Ref. [R89]). The point at $k^2 = 0$ is from photoproduction. The predictions of the model described in the text are also shown. From Ref. [R90].

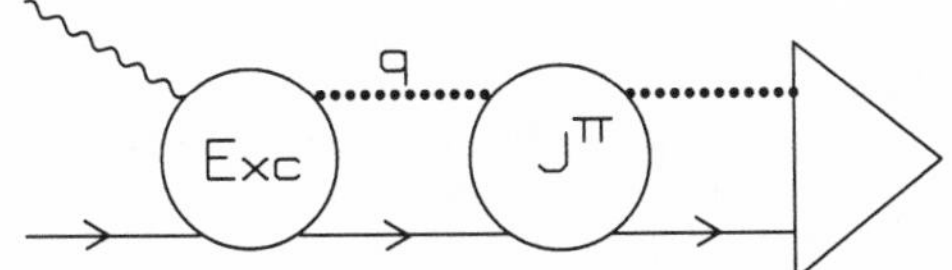

Figure 26.12: Electroexcitation of the first nucleon resonance.

$$\begin{aligned} a(W,k^2) &= \frac{a^{\text{lhs}}(W,k^2)}{D(W)} \\ D(W) &= \exp\left\{-\frac{1}{\pi}\int_{W_0}^{\infty}\frac{\delta(W')dW'}{W'-W-i\varepsilon}\right\} \end{aligned} \tag{26.6}$$

Here $a^{\text{lhs}}(W,k^2)$ is the appropriate multipole projection (Prob. 26.2) of a set of Feynman graphs thought to play an important role in the excitation of the resonance and $D(W)$ is a final-state enhancement factor. The excitation graphs are treated as a generalized Feynman amplitude in that renormalized coupling constants and electromagnetic form factors $F(k^2)$ are used at the vertices; the justification for this procedure is that this amplitude has the correct left-hand singularity structure arising from the pole terms in a dispersion treatment of this process (Ref. [R91]). The graphs used in the present calculation are shown in Fig. 26.13. The excitation amplitude arising from the first three graphs is constructed explicitly in Prob. 26.9.

Division of numerator and denominator in Eq. (26.6) by $D(M_s)$ changes nothing and

$$\frac{D(W)}{D(M_s)} = \exp\left\{-\frac{(W-M_s)}{\pi}\int_{W_0}^{\infty}\frac{\delta(W')dW'}{(W'-M_s)(W'-W-i\varepsilon)}\right\} \tag{26.7}$$

This relation provides additional convergence. If it is assumed that $a(W,k^2) \equiv a^{\text{lhs}}(W,k^2)$ at a given point, as in Section 24 where in the static limit one has a simple pole at $W = M$, then the quantity M_s is determined.[72]

This model has several features to recommend it (Refs. [R90], [N14]):

1. It has the correct analytic properties since $a^{\text{lhs}}(W,k^2)$ has the correct left hand singularities and $D(W)$ has the right-hand unitarity cut.
2. It has the correct threshold behavior in both $|\mathbf{k}^*|$ and $|\mathbf{q}|$.
3. The current is conserved (see Prob. 26.9).
4. It satisfies Watson's theorem $a = |a|e^{i\delta}$ on the physical cut (Prob. 26.6); here δ is the strong interaction π–N phase shift.
5. It is an approximate solution to Omnès integral equation (Prob. 26.7).
6. The electroproduction amplitude resonates at the same W_{R} as elastic scattering (Prob. 26.8).
7. The calculation is completely relativistic.

[72]The calculation shown uses $M_s = 0.95M$ and $\text{Re}\,\delta(W)$ everywhere in the integral.

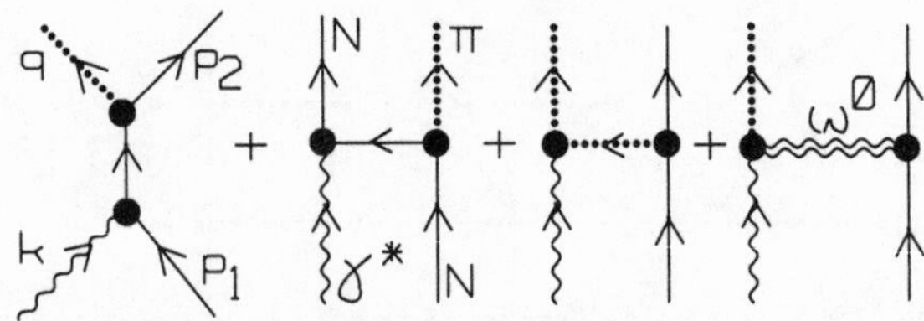

Figure 26.13: Generalized Feynman amplitude used as the excitation mechanism in the $|\pi N\rangle$ channel for the $\Delta(1232)$. It is assumed that $F_\pi \approx F_1^V$. [For the last graph, which plays a minor role for the $\Delta(1232)$, $g_{\omega\pi\gamma}$ is obtained from $\omega \to \pi + \gamma$, and $g_{\omega\pi\gamma}g_{\omega NN}$ from an overall fit to the inelastic resonance spectra; it is assumed that $F_{\omega\pi\gamma} \approx F_2^V/F_2^V(0)$.] From Refs. [R95, R90].

8. The k^2 dependence is explicit.

This model is a simple synthesis and summary of a great deal of theoretical work on $\mathrm{N}(e,e')\Delta(1232)$ within an hadronic framework (Refs. [R91, R92, R93, R94, R95]). The result is shown as the theoretical curve in Fig. 26.11. Note that this QHD picture of the first excited state of the nucleon holds out to $k^2 \approx 4\,\mathrm{GeV}^2 = 100\,\mathrm{fm}^{-2}$. A coupled channel extension of this model exists that describes the inelastic form factors in the higher resonance regions in Fig. 26.10 (Refs. [R95, R90]). Precise coincidence studies and measurement of all the amplitudes for all the excited states out to high k^2 will challenge our understanding of the internal dynamics of the nucleon.

Open Questions. There are many questions one can ask about the structure of the nucleus within the framework of QHD:

- Can one find *dynamic* evidence for the condensed scalar field ϕ_0 (or equivalently the effective mass M^*) and condensed scalar field V_0?

- What are the modifications of the properties of the nucleon in the nuclear medium? As one simple example, there is a Pauli principle restriction on the allowed intermediate nucleon states in the process illustrated in Fig. 17.8 that contributes to the anomalous magnetic moment of the nucleon. A calculation of this effect indicates a few percent quenching of the isovector moment in nuclei (Ref. [R96]). What is the nuclear effect on the nucleon form factors?

- Are there strong, nonlinear meson couplings present in nuclei? Do these give rise to effective many-body forces? Can one probe various aspects of the chiral-invariant lagrangian of Section 24?

- In Section 24, a relativistic description of the first excited state of the nucleon, the $\Delta(1232)$, was developed within the framework of QHD. What are the modifications of this excited state in the nuclear medium? How is the electromagnetic transition matrix element modified in the nucleus?

- Can one calculate the vertex corrections and form factors in QHD? The nucleon is a complicated object in QHD, with a meson and N–$\bar{N}$ cloud. One has to have a reliable estimate of this internal structure before it is possible to ascertain where this purely hadronic picture breaks down.
- Can one find procedures for solving QHD-I (or QHD-II) in the strong coupling limit? Perhaps a lattice calculation?
- Is there a derivation of QHD-I as an effective theory starting from the chiral-invariant lagrangian of Section 24?
- Is there a derivation of QHD-I as an effective theory starting from QCD?[73]

It is to QCD, in particular the nuclear domain of strong-coupling QCD, that we now turn our attention in the next part of this book.

References

[R1] B. D. Serot and J. D. Walecka, *Advances in Nuclear Physics* **16**, eds. J. W. Negele and E. Vogt, Plenum Press, New York (1986)

[R2] J. D. Bjorken and S. D. Drell, *Relativistic Quantum Mechanics*, McGraw-Hill Book Company, Inc., New York (1964)

[R3] J. D. Bjorken and S. D. Drell, *Relativistic Quantum Fields*, McGraw-Hill Book Company, Inc., New York (1965)

[R4] J. D. Walecka, *Lectures on Advanced Quantum Mechanics and Field Theory*, CEBAF, Newport News, VA (1991-92) unpublished

[R5] A. L. Fetter and J. D. Walecka, *Theoretical Mechanics of Particles and Continua*, McGraw-Hill Book Company, Inc., New York (1980)

[R6] S. Weinberg, *Gravitation and Cosmology*, John Wiley and Sons, Inc., New York (1972)

[R7] C. J. Horowitz and B. D. Serot, *Nucl. Phys.* **A368**, 503 (1981)

[R8] B. C. Clark, S. Hama, R. L. Mercer, L. Ray, and B. D. Serot, *Phys. Rev. Lett.* **50**, 1644 (1983)

[R9] B. C. Clark, S. Hama, R. L. Mercer, L. Ray, G. W. Hoffmann, and B. D. Serot, *Phys. Rev.* **C28**, 1421 (1983)

[R10] J. A. McNeil, J. R. Shepard, and S. J. Wallace, *Phys. Rev. Lett.* **50**, 1439 (1983)

[73]See, in this regard, Ref. [Q100].

[R11] J. R. Shepard, J. A. McNeil, and S. J. Wallace, *Phys. Rev. Lett.* **50**, 1443 (1983)

[R12] S. J. Wallace, *Annu. Rev. Nucl. Part. Sci.* **37**, 267 (1987)

[R13] C. J. Horowitz, D. P. Murdock, and B. D. Serot, in *Computational Nuclear Physics I: Nuclear Structure*, eds. K. Langanke, J. A. Maruhn, and S. E. Koonin, Springer-Verlag, Berlin (1991), p. 128

[R14] G. 't Hooft and M. Veltman *Nucl. Phys.* **B44**, 189 (1972)

[R15] G. 't Hooft, *Nucl. Phys.* **B62**, 444 (1973)

[R16] G. Leibbrandt, *Rev. Mod. Phys.* **47**, 849 (1975)

[R17] S. A. Chin, *Ann. Phys.* **108**, 301 (1977)

[R18] R. J. Furnstahl, *Phys. Lett.* **152B**, 313 (1985)

[R19] T. Matsui, *Nucl. Phys* **A370**, 365 (1981)

[R20] W. Bentz, A. Arima, H. Hyuga, K. Shimizu, and K. Yazaki, *Nucl. Phys.* **A436**, 593 (1985)

[R21] H. Kurasawa and T. Suzuki, *Phys. Lett.* **165B**, 234 (1985)

[R22] J. A. McNeil, R. D. Amado, C. J. Horowitz, M. Oka, J. R. Shepard, and D. A. Sparrow, *Phys, Rev.* **C34**, 746 (1986)

[R23] R. J. Furnstahl and B. D. Serot, *Nucl. Phys.* **A468**, 539 (1987)

[R24] J. F. Dawson and R. J. Furnstahl, *Phys. Rev.* **C42**, 2009 (1990)

[R25] R. J. Furnstahl and C. J. Horowitz, *Nucl. Phys.* **A485**, 632 (1988)

[R26] T. D. Cohen, M. K. Banerjee, and C.-Y. Ren, *Phys. Rev.* **C36**, 1653 (1987)

[R27] J. Milana, *Phys. Rev.* **C44**, 527 (1991)

[R28] M. P. Allendes and B. D. Serot, *Phys. Rev.* **C45**, 2975 (1992)

[R29] G. F. Chew, R. Karplus, S. Gasiorowicz, and F. Zachariasen, *Phys. Rev.* **110**, 265 (1958)

[R30] P. Federbush, M. L. Goldberger, and S. B. Treiman, *Phys. Rev.* **112**, 642 (1958)

[R31] S. D. Drell and F. Zachariasen, *Electromagnetic Structure of Nucleons*, Oxford University Press, Oxford, England (1961)

[R32] W. R. Frazer and J. R. Fulco, *Phys. Rev.* **117**, 1609 (1960)

[R33] Y. Nambu, *Phys. Rev.* **106**, 1366 (1957)

[R34] E.-J. Kim, *Phys. Lett.* **B174**, 233 (1986)

[R35] R. Rosenfelder, *Ann. Phys.* **128**, 188 (1980)

[R36] C. J. Horowitz, *Phys. Lett.* **B208**, 8 (1988)

[R37] X. Ji, *Phys. Lett.* **B219**, 143 (1989)

[R38] H. Kurasawa and T. Suzuki, *Nucl. Phys.* **A490**, 571 (1988)

[R39] K. Wehrberger and F. Beck, *Nucl. Phys.* **A491**, 587 (1989)

[R40] S. J. Pollock, *Acta Phys. Pol.* **B19**, 419 (1988); and private communication

[R41] J. D. Walecka, *New Vistas in Nuclear Dynamics*, eds. P. J. Brussard and J. H. Koch, Plenum Press, New York (1986), p. 229

[R42] R. J. Furnstahl and B. D. Serot, *Phys. Rev.* **C44**, 2141 (1991)

[R43] R. A. Freedman, *Phys. Lett.* **71B**, 369 (1977)

[R44] C. Quigg, *Gauge Theories of the Strong, Weak, and Electromagnetic Interactions*, Benjamin Cummings Publ. Co., Inc., Reading, Massachusetts (1983)

[R45] T.-P. Cheng and L.-F. Li, *Gauge Theory of Elementary Particle Physics*, Clarendon Press, Oxford, England (1984)

[R46] A. Radyushkin, *Lectures on QCD Sum Rules*, CEBAF, Newport News, VA (1991), to be published

[R47] J. C. Collins and M. J. Perry, *Phys. Rev. Lett.* **34**, 1353 (1975)

[R48] G. Baym and S. A. Chin, *Phys. Lett.* **62B**, 241 (1976)

[R49] S. A. Chin, *Phys. Lett.* **78B**, 552 (1978)

[R50] J. Kuti, B. Lukács, J. Polónyi, and K. Szlachányi, *Phys. Lett.* **95B**, 75 (1980)

[R51] J. Kuti, J. Polónyi, and K. Szlachányi, *Phys. Lett.* **98B**, 199 (1981)

[R52] G. Bertsch, in *Trends in Theoretical Physics Vol. I*, eds. P. J. Ellis and Y. C. Tang, Addison-Wesley Publ. Co., Reading, Massachusetts (1990), p. 79

[R53] G. F. Chew, M. L. Goldberger, F. E. Low, and Y. Nambu, *Phys. Rev.* **106**, 1337 (1957)

[R54] S. Fubini and J. D. Walecka, *Phys. Rev.* **116**, 194 (1959)

[R55] S. C. Frautschi and J. D. Walecka, *Phys. Rev.* **120**, 1486 (1960)

[R56] M. Jacob and G. C. Wick, *Ann. Phys.* **7**, 404 (1959)

[R57] B. D. Serot, Ph.D. Thesis, Stanford University, Stanford, California (1979) unpublished

[R58] J. Schwinger, *Ann. Phys.* **2**, 407 (1957)

[R59] M. Gell-Mann and M. Lévy, *Nuovo Cimento* **16**, 705 (1960)

[R60] B. W. Lee, *Chiral Dynamics*, Gordon and Breach, New York (1972)

[R61] T. D. Lee and G. C. Wick, *Phys. Rev.* **D9**, 2291 (1974)

[R62] T. D. Lee, *Rev. Mod. Phys.* **47**, 267 (1975)

[R63] S. Weinberg, *Phys. Rev. Lett.* **18**, 188 (1967)

[R64] S. Weinberg, *Phys. Rev.* **166**, 1568 (1968)

[R65] S. Weinberg, *Physica* **A96**, 327 (1979)

[R66] T. Matsui and B. D. Serot, *Ann. Phys.* **144**, 107 (1982)

[R67] V. de Alfaro, S. Fubini, G. Furlan, C. Rossetti, *Currents in Hadron Physics*, North-Holland Publ. Co., Amsterdam (1973)

[R68] W. Lin and B. D. Serot, *Phys. Lett.* **B233**, 23 (1989); *Nucl. Phys.* **A512**, 637 (1990)

[R69] G. F. Chew and F. E. Low, *Phys. Rev.* **101**, 1570 (1956)

[R70] S. Gasiorowicz, *Elementary Particle Physics*, John Wiley and Sons, Inc., New York (1966)

[R71] G. F. Chew, *S-Matrix Theory of Strong Interactions*, 2nd printing, W. A. Benjamin, Inc., New York (1962)

[R72] W. Lin and B. D. Serot, *Nucl. Phys.* **A524**, 601 (1991)

[R73] C. N. Yang and R. L. Mills, *Phys. Rev.* **96**, 191 (1954)

[R74] B. D. Serot and J. D. Walecka, *Acta Phys. Pol.* **B23**, 655 (1992)

[R75] B. D. Serot, *Rep. Prog. Phys.* **55**, 1855 (1992)

[R76] J. W. Van Orden, *Proc. CEBAF 1988 Summer Workshop*, CEBAF, Newport News, Virginia (1988)

[R77] J. Dubach, J. H. Koch, and T. W. Donnelly, *Nucl. Phys.* **A271**, 279 (1976)

[R78] P. G. Blunden and E. J. Kim, *Nucl. Phys.* **A531**, 461 (1991)

[R79] J. M. Cavedon, B. Frois, D. Goutte, M. Huet, Ph. Leconte, J. Martino, X-H. Phan, S. K. Platchkov, S. E. Williamson, W. Boeglin, I. Sick, P. De Witt-Huberts, L. S. Cardman, and C. N. Papanicolas, *Phys. Rev. Lett.* **49**, 986 (1982)

[R80] E. Hadjimichael, B. Goulard, and R. Bornais, *Phys. Rev.* **C27**, 831 (1983)

[R81] D. O. Riska, *Nucl. Phys.* **A350**, 227 (1980)

[R82] F. Gross, J. W. Van Orden, and K. Holinde, *Phys. Rev.* **C45**, 2094 (1992)

[R83] C. J. Horowitz and B. D. Serot, *Phys.Lett.* **B137**, 287 (1984)

[R84] C. J. Horowitz and B. D. Serot, *Nucl. Phys.* **A464**, 613 (1987); **A473**, 760(E) (1987)

[R85] M. R. Anastasio, L. S. Celenza, W. S. Pong, and C. M. Shakin, *Phys. Rep.* **100**, 327 (1983)

[R86] R. Brockmann and R. Machleidt, *Phys. Lett.* **B149**, 283 (1984); *Phys. Rev.* **C42**, 1965 (1990)

[R87] R. Malfliet, *Relativistic Transport Theory*, preprint KVI-952, Groningen, The Netherlands (1992)

[R88] E. Bloom *et al.*, SLAC Group A, reported by W. K. Panofsky in *Int. Conf. on High Energy Phys., Vienna, 1968*, CERN, Geneva (1968), p. 23

[R89] M. Breidenbach, Ph. D. Thesis, M.I.T. (1971) unpublished

[R90] J. D. Walecka, *Acta Phys. Pol.* **B3**, 117 (1972)

[R91] S. Fubini, Y. Nambu, and V. Wataghin, *Phys. Rev.* **111**, 329 (1958)

[R92] P. Dennery, *Phys. Rev.* **124**, 2000 (1961)

[R93] N. Zagury, *Phys. Rev.* **145**, 1112 (1966)

[R94] S. L. Adler, *Ann. Phys.* **50**, 189 (1968)

[R95] P. L. Pritchett, J. D. Walecka, and P. A. Zucker, *Phys. Rev.* **184**, 1825 (1969)

[R96] S. D. Drell and J. D. Walecka, *Phys. Rev.* **120**, 1069 (1960)

[R97] J. Tjon, in *Hadronic Physics with Multi-GeV Electrons*, Les Houches Series, Nova Science Publ., New York (1991), p. 89

[R98] B. D. Keister and W. N. Polyzou, *Adv. Nucl. Phys.* **20**, 225 (1991)

[R99] P. G. Reinhard, *Rep. Prog. Phys.* **52**, 439 (1989)

[R100] Y. K. Gambhir, P. Ring, and A. Thinet, *Ann. Phys.* **511**, 129 (1990)

[R101] J. Carlson, V. R. Pandharipande, and R. Schiavilla, in *Modern Topics in Electron Scattering*, eds. B. Frois and I. Sick, World Scientific, Singapore (1991), p. 177

[R102] B. ter Haar and R. Malfliet, *Phys. Rep.* **149**, 207 (1987)

[R103] F. Gross, *Relativistic Quantum Mechanics and Field Theory*, John Wiley and Sons, Inc., New York (1993)

PROBLEMS: PART II

The first seven problems review some basic results from advanced quantum mechanics: properties of the Dirac equation and introduction to the relativistic quantum theory of fields. We use $\vec{\gamma} = i\vec{\alpha}\beta, \gamma_4 = \beta$ with $\vec{\alpha} = \begin{pmatrix} 0 & \vec{\sigma} \\ \vec{\sigma} & 0 \end{pmatrix}, \beta = \begin{pmatrix} 1 & 0 \\ 0 & -1 \end{pmatrix}$. Also $\sigma_{\mu\nu} = [\gamma_\mu, \gamma_\nu]/2i$ and the Maxwell field tensor is $F_{\mu\nu} = \partial A_\nu/\partial x_\mu - \partial A_\mu/\partial x_\nu$. The relation to the conventions of Bjorken and Drell is discussed in detail in Appendix N.

13.1. The Dirac equation for a free particle is $i\hbar\partial\Psi/\partial t = [(\hbar c/i)\vec{\alpha}\cdot\vec{\nabla} + \beta m_0 c^2]\Psi$.
(a) Find stationary-state plane-wave solutions of the form $\Psi = \exp\{\frac{i}{\hbar}(\vec{p}\cdot\vec{x} - Et)\}\psi$. Show the spinors ψ form the columns of the modal matrix

$$\begin{aligned} \mathcal{M} &= \left(\frac{E_p + m_0c^2}{2E_p}\right)^{1/2} \begin{bmatrix} 1 & -c\vec{\sigma}\cdot\vec{p}/(E_p + m_0c^2) \\ c\vec{\sigma}\cdot\vec{p}/(E_p + m_0c^2) & 1 \end{bmatrix} \\ &= \left(\frac{E_p + m_0c^2}{2E_p}\right)^{1/2} \left[1 - \frac{c\beta\vec{\alpha}\cdot\vec{p}}{E_p + m_0c^2}\right] \end{aligned}$$

(b) Show this modal matrix is unitary with $\mathcal{M}^\dagger = \mathcal{M}^{-1}$. Here $E_p = c(\vec{p}^{\,2} + m_0^2c^2)^{1/2}$.
(c) The ith column is obtained from $\mathcal{M}\eta_i$ where η_i is a unit spinor with 1 in the ith row and zeros elsewhere. Hence relate the modal matrix to the Lorentz transformation from the particle's rest frame to one where it moves with momentum $\vec{p}$.

13.2. The Dirac equation $[\gamma_\mu(\partial/\partial x_\mu - ieA_\mu/\hbar c) + m_0c/\hbar]\Psi = 0$ describes a particle of spin 1/2 and magnetic moment $e\hbar/2m_0c$. Pauli observed that the equation remains gauge invariant and covariant if a term $(-e\delta/4m_0c^2)\sigma_{\mu\nu}F_{\mu\nu}\Psi$ is added to the above.
(a) Show that the Dirac hamiltonian is then $H = c\vec{\alpha}\cdot(\vec{p} - e\vec{A}/c) + \beta m_0c^2 + e\Phi - (e\hbar\delta/2m_0c)\beta(\vec{\sigma}\cdot\vec{\mathcal{H}} - i\vec{\alpha}\cdot\vec{\mathcal{E}})$. Identify $\vec{\sigma}$.
(b) Show that in the non-relativistic limit and with vanishingly small fields, this describes a particle of spin 1/2 and magnetic moment $e\hbar(1+\delta)/2m_0c$.
(c) Recall that $\vec{s} = \vec{\sigma}/2$ and $\vec{\pi} = \vec{p} - e\vec{A}/c$ are the Dirac spin and kinetic momentum. Show the equation of motion of a particle with anomalous magnetic moment δ in a uniform magnetic field $\vec{\mathcal{H}} = \mathcal{H}\vec{z}$ is $d(\vec{s}\cdot\vec{\pi})/dt = -\omega_l\delta\beta(\vec{s}\times\vec{\pi})\cdot\vec{z}$ where $\omega_l = e\mathcal{H}/m_oc$.

13.3. Show the energy levels of a Dirac particle in a point Coulomb potential are given by (Ref. [N42]) $E = m_0c^2[1 + \alpha^2Z^2/(n' + \sqrt{(j+1/2)^2 - \alpha^2Z^2})^2]^{-1/2}$ with $n' = 0, 1, \ldots$.

14.1.[74] Consider a real scalar field with equations of motion $[(\partial/\partial x_\mu)^2 - m_s^2]\phi = 0$ and lagrangian density $\mathcal{L} = (-c^2/2)[(\partial\phi/\partial x_\mu)^2 + m_s^2\phi^2]$. Here $m_s = m_{s0}c/\hbar$ is the inverse Compton wavelength.

[74] An excellent initial introduction to relativistic quantum field theory is contained in G. Wentzel, *Quantum Theory of Fields*, Interscience, New York (1949).

(a) Show the canonical momentum density is $\pi = \partial\phi/\partial t$ and hamiltonian density is $\mathcal{H} = (1/2)[\pi^2 + c^2(\vec{\nabla}\phi)^2 + m_s^2c^2\phi^2]$.
(b) Construct the stress tensor and the four-momentum $P_\mu = (1/ic)\int d^3x T_{4\mu} = (\vec{P}, iH/c)$. Prove that P_μ is both a four-vector and a constant of the motion. Show $T_{44} = -\mathcal{H}$ and $\vec{P} = -\int d^3x\, \pi\vec{\nabla}\phi$.
(c) Work in a big box of volume Ω and use p.b.c. First, treat the problem as one in classical continuum mechanics and introduce the general field expansions

$$\phi(\vec{x},t) = \frac{1}{\sqrt{\Omega}}\sum_{\vec{k}}(\frac{\hbar}{2\omega_k})^{1/2}(c_{\vec{k}}e^{i\vec{k}\cdot\vec{x}-i\omega_k t} + c_{\vec{k}}^\dagger e^{-i\vec{k}\cdot\vec{x}+i\omega_k t})$$

$$\pi(\vec{x},t) = \frac{1}{i\sqrt{\Omega}}\sum_{\vec{k}}(\frac{\hbar\omega_k}{2})^{1/2}(c_{\vec{k}}e^{i\vec{k}\cdot\vec{x}-i\omega_k t} - c_{\vec{k}}^\dagger e^{-i\vec{k}\cdot\vec{x}+i\omega_k t})$$

Show this reduces both H and $\vec{P}$ to normal modes with $H = \sum_{\vec{k}}\hbar\omega_k(c_{\vec{k}}^\dagger c_{\vec{k}} + c_{\vec{k}}c_{\vec{k}}^\dagger)/2$ and $\vec{P} = \sum_{\vec{k}}\hbar\vec{k}(c_{\vec{k}}^\dagger c_{\vec{k}} + c_{\vec{k}}c_{\vec{k}}^\dagger)/2$. Here $\omega_k = c(\vec{k}^2 + m_s^2)^{1/2}$. The problem is now reduced to a system of uncoupled harmonic oscillators.
(d) Now quantize the system. Work in the Schrödinger picture. Show Ehrenfest's theorem and the canonical commutation relations $[\pi(\vec{x}), \phi(\vec{x}')] = (\hbar/i)\delta^{(3)}(\vec{x}-\vec{x}')$ generate the correct equations of motion. Hence show one recovers the correct classical limit of the theory.
(e) The fields are now operators. Show the canonical commutation relations of (d) imply $[c_{\vec{k}}, c_{\vec{k}'}^\dagger] = \delta_{\vec{k}\vec{k}'}$ with all other commutators vanishing. Interpret H and $\vec{P}$ in terms of the particle content of the theory.
(f) Explain how one recovers the explicit time dependence of the operators in (c) in quantum mechanics.

14.2. Consider a complex Dirac field with equations of motion $(\gamma_\mu\partial/\partial x_\mu + M)\psi = 0$ and lagrangian density $\mathcal{L} = -\hbar c\bar{\psi}(\gamma_\mu\partial/\partial x_\mu + M)\psi$. Here $\bar{\psi} \equiv \psi^\dagger\gamma_4$ and $M = m_0c/\hbar$.
(a) Show the canonical momentum density is $\pi_\psi = i\hbar\psi^\dagger$ and hamiltonian is $H = \int \psi^\dagger(x)(c\vec{\alpha}\cdot\vec{p} + \beta m_0c^2)\psi(x)d^3x$. Here $\vec{p} = (\hbar/i)\vec{\nabla}$.
(b) Construct the stress tensor and the four-momentum $P_\mu = (1/ic)\int d^3x T_{4\mu} = (\vec{P}, iH/c)$. Show $T_{44} = -\mathcal{H}$ and $\vec{P} = \int d^3x\psi^\dagger(x)\vec{p}\psi(x)$.
(c) Show that the current $j_\mu = ie\bar{\psi}(x)\gamma_\mu\psi(x)$ is conserved and hence conclude the total charge $Q = e\int d^3x\psi^\dagger(x)\psi(x)$ is a constant of the motion.
(d) Work in a big box of volume Ω and use p.b.c. First, treat the problem as one in classical continuum mechanics[75] and introduce the general field expansions

$$\psi(\vec{x},t) = \frac{1}{\sqrt{\Omega}}\sum_{\vec{k}\lambda}(a_{\vec{k}\lambda}u(\vec{k}\lambda)e^{i\vec{k}\cdot\vec{x}-i\omega_k t} + B_{\vec{k}\lambda}v(\vec{k}\lambda)e^{i\vec{k}\cdot\vec{x}+i\omega_k t})$$

Show this reduces both H and $\vec{P}$ to normal modes with $H = \sum_{\vec{k}\lambda}\hbar\omega_k(a_{\vec{k}\lambda}^\dagger a_{\vec{k}\lambda} - B_{\vec{k}\lambda}^\dagger B_{\vec{k}\lambda})$, $\vec{P} = \sum_{\vec{k}\lambda}\hbar\vec{k}(a_{\vec{k}\lambda}^\dagger a_{\vec{k}\lambda} + B_{\vec{k}\lambda}^\dagger B_{\vec{k}\lambda})$, and $Q = e\sum_{\vec{k}\lambda}(a_{\vec{k}\lambda}^\dagger a_{\vec{k}\lambda} + B_{\vec{k}\lambda}^\dagger B_{\vec{k}\lambda})$. Here $\omega_k = c(\vec{k}^2 + M^2)^{1/2}$.

[75] Keep track of the proper order of the factors, however, so that the calculation is still correct after one introduces anticommutation rules.

(e) Now quantize the system. Show that the usual commutation relations for the creation and destruction operators lead to a system with no lowest energy ground state.[76] Hence introduce anticommutation relations for the fermion creation and destruction operators. Work in the Schrödinger picture. Show Ehrenfest's theorem and the canonical anticommutation relations $\{\psi(\vec{x}), \pi_\psi(\vec{x}')\} = i\hbar\delta^{(3)}(\vec{x} - \vec{x}')$ generate the correct equations of motion.
(f) Now make a canonical transformation to $B_{\vec{k}\lambda} \equiv b^\dagger_{-\vec{k}\lambda}$ so that the field becomes

$$\psi(\vec{x}, t) = \frac{1}{\sqrt{\Omega}}\sum_{\vec{k}\lambda}(a_{\vec{k}\lambda}u(\vec{k}\lambda)e^{i\vec{k}\cdot\vec{x}-i\omega_k t} + b^\dagger_{\vec{k}\lambda}v(-\vec{k}\lambda)e^{-i\vec{k}\cdot\vec{x}+i\omega_k t})$$

The quantities in (d) then take the form

$$H = \sum_{\vec{k}\lambda}\hbar\omega_k(a^\dagger_{\vec{k}\lambda}a_{\vec{k}\lambda} + b^\dagger_{\vec{k}\lambda}b_{\vec{k}\lambda} - 1)$$

$$\vec{P} = \sum_{\vec{k}\lambda}\hbar\vec{k}(a^\dagger_{\vec{k}\lambda}a_{\vec{k}\lambda} + b^\dagger_{\vec{k}\lambda}b_{\vec{k}\lambda} - 1)$$

$$Q = e\sum_{\vec{k}\lambda}(a^\dagger_{\vec{k}\lambda}a_{\vec{k}\lambda} - b^\dagger_{\vec{k}\lambda}b_{\vec{k}\lambda} + 1)$$

Interpret H, $\vec{P}$, and Q in terms of the particle content of the theory. Interpret and discuss the additive constants in these expressions.
(g) Show the canonical anticommutation relations for the creation and destruction operators are equivalent to the canonical anticommutation relations on the fields.
(h) Explain how one recovers the explicit time dependence of the field operator in (f) in quantum mechanics.

14.3. Carry out the canonical formalism for a real spin 1 field with mass. The field equations are $[(\partial/\partial x_\mu)^2 - m_v^2]\psi_\nu = 0$ with $\nu = 1, \ldots, 4$ and $\partial\psi_\nu/\partial x_\nu = 0$. With the introduction of the field tensor $\psi_{\mu\nu} = \partial\psi_\nu/\partial x_\mu - \partial\psi_\mu/\partial x_\nu$ these become $\partial\psi_{\mu\nu}/\partial x_\nu = -m_v^2\psi_\mu$.
(a) Show the lagrangian density $\mathcal{L} = -(1/4)\psi_{\mu\nu}\psi_{\mu\nu} - (m_v^2/2)\psi_\mu\psi_\mu$ gives the correct equations of motion.
(b) Use the canonical procedure to find the hamiltonian.
(c) What is the stress tensor? The momentum?
(d) Show the following expansions reduce the problem to normal modes

$$\vec{\psi}_T = \frac{1}{\sqrt{\Omega}}\sum_{\vec{k}}\sum_{\lambda=1,2}\left(\frac{\hbar c^2}{2\omega_k}\right)^{1/2}(a_{\vec{k}\lambda}\vec{e}_{\vec{k}\lambda}e^{ik\cdot x} + a^\dagger_{\vec{k}\lambda}\vec{e}_{\vec{k}\lambda}e^{-ik\cdot x})$$

$$\vec{\psi}_L = \frac{1}{\sqrt{\Omega}}\sum_{\vec{k}}\left(\frac{\hbar\omega_k}{2m_v^2}\right)^{1/2}(a_{\vec{k}0}\vec{e}_{\vec{k}0}e^{ik\cdot x} + a^\dagger_{\vec{k}0}\vec{e}_{\vec{k}0}e^{-ik\cdot x})$$

Here $\vec{e}_{\vec{k}0} \equiv \vec{k}/|\vec{k}|$.
(e) Quantize the system. What are the commutation rules?

[76] This is the first half of Pauli's theorem on the connection between spin and statistics.

(f) Compare with the free electromagnetic field where $m_v^2 = 0$.

14.4. (a) Include an interaction of the Dirac field in Prob. 14.2 with an electromagnetic field by making the minimal gauge invariant substitution $\partial/\partial x_\mu \rightarrow \partial/\partial x_\mu - ieA_\mu/\hbar c$. What is the new H?
(b) Generalize Probs. 14.1 and 14.3 to complex (charged) fields. How would you then include with minimal coupling an interaction with the electromagnetic field?

14.5. Derive the expression for the pressure in Eq. (14.24) directly from the MFT stress tensor obtained from Eq. (14.13).

15.1.[77] Consider the hamiltonian $h = -i\vec{\alpha}\cdot\vec{\nabla} + g_v V_0(r) + \beta[M - g_s\phi_0(r)]$ for a Dirac particle moving in spherically symmetric vector and scalar fields. Define the angular momentum by $\vec{J} = \vec{L} + \vec{S} = -i\vec{r}\times\vec{\nabla} + \vec{\Sigma}/2$. Here $\vec{\Sigma} = \begin{pmatrix} \vec{\sigma} & 0 \\ 0 & \vec{\sigma} \end{pmatrix}$ and $\psi = \begin{pmatrix} \psi_A \\ \psi_B \end{pmatrix}$.
(a) Prove $[h, J_i] = [h, \vec{J}^2] = [h, \vec{S}^2] = 0$. Note $[h, \vec{L}^2] \neq 0$. Introduce $K = \beta(\vec{\Sigma}\cdot\vec{L} + 1) = \beta(\vec{\Sigma}\cdot\vec{J} - 1/2)$. Show $[h, K] = 0$.
(b) Label the eigenvalues of K by $K\psi = -\kappa\psi$. Show the states can be characterized by the eigenvalues $\{j, s = 1/2, -\kappa, m\}$. Show $K^2 = \vec{L}^2 + \vec{\Sigma}\cdot\vec{L} + 1 = \vec{J}^2 + 1/4$. Hence conclude that $\kappa = \pm(j + 1/2)$.
(c) Show $-\kappa\psi_A = (\vec{\sigma}\cdot\vec{L} + 1)\psi_A$ and $-\kappa\psi_B = -(\vec{\sigma}\cdot\vec{L} + 1)\psi_B$. Use (b) to show that

$$\begin{aligned} \vec{L}^2\psi_A &= \left[\left(j + \frac{1}{2}\right)^2 + \kappa\right]\psi_A = l_A(l_A + 1)\psi_A \\ \vec{L}^2\psi_B &= \left[\left(j + \frac{1}{2}\right)^2 - \kappa\right]\psi_B = l_B(l_B + 1)\psi_B \end{aligned}$$

Thus, although ψ is not an eigenstate of $\vec{L}^2$, the upper and lower components are separately eigenstates with eigenvalues determined from these relations. They also have fixed j and $s = 1/2$.
(d) Introduce spin spherical harmonics $\Phi_{\kappa m} = \sum_{m_l m_s} \langle l m_l \frac{1}{2} m_s | l \frac{1}{2} j m\rangle Y_{l m_l}(\theta, \phi)\chi_{m_s}$. Here $j = |\kappa| - 1/2$. Hence show the solutions to this Dirac equation take the form

$$\psi_{n\kappa m} = \frac{1}{r}\begin{pmatrix} iG(r)_{n\kappa}\Phi_{\kappa m} \\ -F(r)_{n\kappa}\Phi_{-\kappa m} \end{pmatrix}$$

Here $l = \kappa$ if $\kappa > 0$ and $l = -(\kappa + 1)$ if $\kappa < 0$. Write out the first few wave functions.

15.2. Consider the relativistic Hartree Eqs. (15.1). Label the baryon states by $\{\alpha\} = \{n\kappa t, m_\alpha\} \equiv \{a, m_\alpha\}$. Here $t = 1/2\,(-1/2)$ for protons (neutrons). Look for stationary state solutions, and insert the form of Dirac wave functions in Prob. 15.1.
(a) Show $\vec{\sigma}\cdot\vec{\nabla}(G/r)\Phi_{\kappa m} = -(1/r)(d/dr + \kappa/r)G\Phi_{-\kappa m}$. What is the relation for F?
(b) Show the coupled radial Dirac equations reduce to

$$\begin{aligned} &\frac{d}{dr}G_a(r) + \frac{\kappa}{r}G_a(r) - [E_a - g_v V_0(r) + M - g_s\phi_0(r)]F_a(r) = 0 \\ &\frac{d}{dr}F_a(r) - \frac{\kappa}{r}F_a(r) + [E_a - g_v V_0(r) - M + g_s\phi_0(r)]G_a(r) = 0 \end{aligned}$$

[77] Since it is now clear where all the factors go, we shall here and henceforth also set $\hbar = c = 1$ in the problems.

(c) Show the normalization condition is $\int_0^\infty dr(|G_a(r)|^2 + |F_a(r)|^2) = 1$.

15.3. Consider the relativistic Hartree Eqs. (15.1) for the meson fields

(a) Show $\sum_{m=-j}^{m=j} \Phi^\dagger_{\kappa m}\Phi_{\kappa' m} = \delta_{\kappa\kappa'}(2j+1)/4\pi$ for $\kappa = \pm\kappa'$.

(b) Hence show the meson field equations become

$$\frac{d^2}{dr^2}\phi_0(r) + \frac{2}{r}\frac{d}{dr}\phi_0(r) - m_s^2\phi_0(r) = -g_s \sum_a^{\text{occ}} \left(\frac{2j_a+1}{4\pi r^2}\right)[|G_a(r)|^2 - |F_a(r)|^2]$$

$$\frac{d^2}{dr^2}V_0(r) + \frac{2}{r}\frac{d}{dr}V_0(r) - m_v^2 V_0(r) = -g_v \sum_a^{\text{occ}} \left(\frac{2j_a+1}{4\pi r^2}\right)[|G_a(r)|^2 + |F_a(r)|^2]$$

15.4. How would you solve the relativistic Hartree equations in Probs. 15.2-3?

15.5. Enlarge the set of equations in Probs. 15.2-3 to include a condensed neutral ρ field $b_0(r)$ coupled to the third component of the isovector baryon density $g_\rho \psi^\dagger \frac{1}{2}\tau_3\psi$ and Coulomb field $A_0(r)$ coupled to the charge density $e_p\psi^\dagger\frac{1}{2}(1+\tau_3)\psi$ (Ref. [R1]).

15.6.[78] Consider nonrelativistic potential scattering. Enlarge the concept of a potential to include nonlocal interactions $v\psi \to \int v(\vec{x},\vec{y})\psi(\vec{y})d^3y$, and then separable potentials $v(\vec{x},\vec{y}) = \sum_{lm} 4\pi\lambda_l v_l(x)v_l(y)Y_{lm}(\Omega_x)Y^*_{lm}(\Omega_y)$.

(a) Show the scattering amplitude for a single fixed scatterer at the origin is

$$f_l(k) = \frac{e^{i\delta_l}\sin\delta_l}{k} = -\frac{\lambda_l}{4\pi}|v_l(k)|^2\left[1 + \lambda_l\int\frac{d^3t}{(2\pi)^3}|v_l(t)|^2\frac{1}{t^2-k^2-i\eta}\right]^{-1}$$

Here $v_l(k) = 4\pi\int v_l(x)j_l(kx)x^2dx$.

(b) Consider A fixed scatterers at positions $\{\vec{x}_1,\ldots,\vec{x}_A\}$ and an interaction $v(\vec{x},\vec{y}) = \sum_{i=1}^A v(\vec{x}-\vec{x}_i,\vec{y}-\vec{x}_i)$. Show the multiple scattering problem reduces to a set of matrix equations (recall that a bar under a symbol denotes a matrix)

$$\begin{aligned}[\underline{1}+\underline{\mathcal{G}}](\underline{\lambda\psi}) &= (\underline{Ve}) \\ -4\pi f &= (\underline{e'v'})^\dagger[\underline{1}+\underline{\mathcal{G}}]^{-1}(\underline{Ve})\end{aligned}$$

The matrix indices are $\{ilm\}$. The Green's function is

$$\mathcal{G}^{ij}_{lm;l'm'} = -\frac{4\pi f_l(k)}{|v_l(k)|^2}\int\frac{d^3t}{(2\pi)^3}\frac{v_{lm}(\vec{t})v^*_{l'm'}(\vec{t})}{t^2-k^2-i\eta}e^{i\vec{t}\cdot(\vec{x}_i-\vec{x}_j)} \tag{26.1}$$

Here $v_{lm}(\vec{k}) = (4\pi)^{1/2}i^lY^*_{lm}(\Omega_k)v_l(k)$ and $\mathcal{G}^{(i=j)} \equiv 0$. The other quantities appearing in these equations are defined by

$$\begin{aligned}(\underline{\lambda\psi})^i_{lm} &= \lambda_l\psi^i_{lm}(\vec{k}) \\ (\underline{Ve})^i_{lm} &= V_{lm}e^i = -\left(\frac{4\pi f_l(k)}{|v_l(k)|^2}\right)v_{lm}(\vec{k})e^{i\vec{k}\cdot\vec{x}_i} \\ (\underline{e'v'})^i_{lm} &= e'_iv'_{lm} = e^{i\vec{k}'\cdot\vec{x}_i}v_{lm}(\vec{k}')\end{aligned}$$

[78] Probs. 15.6-15.10 are from L. L. Foldy and J. D. Walecka, *Ann. Phys.* **54**, 447 (1969).

These relations provide an exact solution for the multiple scattering amplitude $f \equiv f_{\vec{k}'\vec{k}}(\vec{x}_1, \ldots, \vec{x}_A)$.

15.7. Make the multiple scattering expansion $[\underline{1} + \underline{\mathcal{G}}]^{-1} = \underline{1} - \underline{\mathcal{G}} + \underline{\mathcal{G}}^2 + \cdots$ in Prob. 15.6. Derive the following rules for the nth order contribution to $-4\pi f$:
(a) Set down A points $\{\vec{x}_1, \ldots, \vec{x}_A\}$;
(b) Draw $n+1$ connected directed line segments connecting the points with one incoming and one outgoing line. Include all possibilities;
(c) Assign factors $e^{i\vec{k}\cdot\vec{x}_i}$ and $e^{-i\vec{k}'\cdot\vec{x}_j}$ for the incoming and outgoing lines;
(d) Include a factor $-e^{i\vec{t}\cdot(\vec{x}_j - \vec{x}_i)}/(t^2 - k^2 - i\eta)$ for each propagator between points;
(e) Include the following factor for each vertex

$$-4\pi f(\vec{p}, \vec{q}) = -4\pi \sum_l f_l(k)(2l+1)P_l(\cos\theta_{\vec{p}\vec{q}})\frac{v_l(p)v_l(q)}{|v_l(k)|^2}$$

(f) Integrate $\int d^3t/(2\pi)^3$ over each internal line.

15.8. For a quantum mechanical target one must integrate over the probability of finding the target particles in any particular configuration

$$f(\vec{k}'\vec{k}) = \int \rho^A(\vec{x}_1, \cdots, \vec{x}_A) f_{\vec{k}'\vec{k}}(\vec{x}_1, \cdots, \vec{x}_A) d^3x_1 \cdots d^3x_A$$

This probability is given by the square of the target ground-state wave function $\rho^A = |\Psi_0|^2$. Introduce the the following expansion for this ground-state density

$$\rho^A(\vec{x}_1, \cdots, \vec{x}_A) = \rho^{(1)}(\vec{x}_1)\cdots\rho^{(1)}(\vec{x}_A) + \sum_{\text{contractions}} [\rho^{(1)}(\vec{x}_1)\cdots\rho^{(1)}(\vec{x}_A)] + \cdots$$

$$\rho^{(1)}(\vec{x})^\bullet \rho^{(1)}(\vec{y})^\bullet \equiv \Delta(\vec{x}, \vec{y}) \equiv \rho^{(2)}(\vec{x}, \vec{y}) - \rho^{(1)}(\vec{x})\rho^{(1)}(\vec{y})$$

The sum goes over all possible pairs of contractions. Here $\rho^{(1)}$ and $\rho^{(2)}$ are the one and two particle densities (compare Probs. 4.1 and 4.2). Note $\int d^3y \Delta(\vec{x}, \vec{y}) = 0$. Demonstrate the validity of this expansion by showing:
(a) The density ρ^A is symmetric under particle interchange;
(b) It satisifies the consistency relation

$$\int d^3x_A \, \rho^{(A)}(\vec{x}_1, \cdots, \vec{x}_A) = \rho^{(A-1)}(\vec{x}_1, \cdots, \vec{x}_{A-1})$$

(c) It gives the correct expectation value for any one-body operator $\langle \sum_i O(i) \rangle$;
(d) It gives the correct expectation value for any two-body operator $\langle \sum_{i<j} O(ij) \rangle$.

15.9. A single-particle optical potential allows one to simply solve the appropriate wave equation to generate the scattering amplitude. With the results in Probs. 15.7-8, one can state precisely when the equivalent single-particle potential will reproduce the full scattering amplitude. In general this one-body potential will be nonlocal $U(\vec{x}, \vec{y})$ and has a double Fourier transform $\tilde{U}(\vec{p}, \vec{q}) = \int\int e^{-i\vec{p}\cdot\vec{x}} U(\vec{x}, \vec{y}) e^{i\vec{q}\cdot\vec{y}} d^3x d^3y$.

Make the following assumptions: (1) The effective number of scatterings n_{eff} satisfies $n_{\text{eff}} \ll A$; (2) no target particle is multiply struck; (3) retain just the first term in the expansion in Prob. 15.8 [with $\rho^{(1)}(|\vec{x}|)$]. Demonstrate the following:

(a) The lowest order optical potential is given by

$$\tilde{U}_0(\vec{p},\vec{q}) = -4\pi A\tilde{\rho}^{(1)}(\vec{p}-\vec{q})\sum_l f_l(k)\frac{v_l(p)v_l(q)}{|v_l(k)|^2}(2l+1)P_l(\cos\theta_{\vec{p}\vec{q}})$$

Here $\tilde{\rho}^{(1)}(\vec{p}-\vec{q}) = \int e^{-i(\vec{p}-\vec{q})\cdot\vec{x}}\rho^{(1)}(x)d^3x$.
(b) For an extended system, the Fourier transform of the one-body density implies $\vec{p} \approx \vec{q} \approx \vec{k}$. show that U_0 then has the approximate limiting form

$$\begin{aligned} U_0(\vec{p},\vec{q}) &\approx -4\pi A f(0)\tilde{\rho}^{(1)}(\Delta)\,; \qquad \vec{\Delta} \equiv \vec{p}-\vec{q} \\ U_0(\vec{x},\vec{y}) &\approx -4\pi A f(0)\rho^{(1)}(x)\,\delta^{(3)}(\vec{x}-\vec{y}) \end{aligned}$$

Here $f(0)$ is the forward scattering amplitude from a single target particle. In this limit the optical potential becomes local with $U_0(x) = -4\pi A f(0)\rho^{(1)}(x)$.

15.10. Assume the interaction $v_l(x)$ in Probs. 15.6-9 vanishes outside some radius a.
(a) Show more precisely in Prob. 15.9 that if the target particles do not overlap so that $2a < |\vec{x}_i - \vec{x}_j|$, and if the energy is high enough $k|\vec{x}_i - \vec{x}_j| \to \infty$, then

$$\begin{aligned} \tilde{U}_0(\vec{p},\vec{q}) &\to -4\pi A\tilde{\rho}^{(1)}(\Delta)\sum_l f_l(k)(2l+1)P_l\left(1-\frac{\Delta^2}{2k^2}\right) \\ U_0(x) &\to -4\pi A\int e^{i\vec{\Delta}\cdot\vec{x}}f(\vec{\Delta})\tilde{\rho}^{(1)}(\Delta)\frac{d^3\Delta}{(2\pi)^3} \end{aligned}$$

Evidently the limit of Prob. 15.9 $U_0(x) \approx -4\pi A f(0)\rho^{(1)}(x)$ is recovered here.
(b) Discuss some classes of corrections to this lowest-order optical potential.

16.1. The Feynman propagator for the free scalar meson is defined by

$$\frac{1}{i}\Delta_F(\vec{x}_1t_1,\vec{x}_2t_2) \equiv \langle 0|T[\hat{\phi}(\vec{x}_1t_1),\hat{\phi}(\vec{x}_2t_2)]|0\rangle$$

Use the field expansion in Prob. 14.1 to derive the result in Eq. (16.3).

16.2. (a) Use the field expansion in Prob. 14.2 to calculate the Feynman propagator for the free baryon in Eq. (16.1); hence establish the first part of the result in Eq. (16.3).
(b) Repeat for the noninteracting system at finite baryon density to derive the full result in Eq. (16.3).
(c) Compare with the nonrelativistic propagator in Ref. [N2].

16.3. Use the field expansion in Prob. 14.3 to derive the Feynman propagator for the free massive vector meson in Eq. (16.3).

16.4. Show the retention of the finite density part of the baryon propagator [second term in Eq. (16.19)] and the meson mass terms in the stress tensor in Eq. (16.18) give rise to the MFT of Section 14.

16.5. Use the Feynman rules and renormalization conditions discussed in the text to compute the numerical factors in the counter terms in Eq. (16.24).

16.6. (a) The Dirichlet integral $I_n(R) = \int\cdots\int dx_1\cdots dx_n$ with $x_1^2+\cdots+x_n^2 \leq R^2$ is the volume element of a sphere in n-dimensional euclidian space. Show $I_n(R) = nC_n\int_0^R r^{n-1}dr = (\sqrt{\pi})^n R^n\,/\,\Gamma(1+n/2)$. This makes the n-dependence explicit.

(b) Euler's integral is $\int_0^1 t^{x-1}(1-t)^{y-1}dt = \Gamma(x)\Gamma(y)/\Gamma(x+y)$ for $\text{Re}\, x > 0$ and $\text{Re}\, y > 0$ (Ref. [R5]). Use this to establish Eq. (16.25). (*Hint*: first rotate the t_0 contour.)

16.7. Evaluate the second integral in Eq. (16.31) to validate the claims made in arriving at $\delta T_{\mu\nu}/M^4$ in Eq. (16.39).

16.8. Compute the lowest-order baryon self-energy in QHD-I retaining just the density-dependent part of the baryon propagator. Make a nonrelativistic reduction and reproduce the usual exchange contribution to the energy (Refs. [N2], [R1]).

16.9. Compute the lowest-order meson self-energies in QHD-I retaining only the terms involving the density-dependent part of the baryon propagator. Compare with the nonrelativistic result for the polarization propagator in Ref. [N2].

17.1. Define an effective N–N potential that in lowest order gives the same S-matrix as QHD (Appendix A). Neglect retardation in the meson propagators. Show the effective potential to be used with relativistic Hartree wave functions to compute nuclear spectra is $\nu(1,2) = (-g_s^2/4\pi r_{12})e^{-m_s r_{12}} + \gamma_\mu^{(1)}\gamma_\mu^{(2)}(g_v^2/4\pi r_{12})e^{-m_v r_{12}}$ (Ref. [R18]).

17.2. (a) Use the effective current in Eq. (17.9). Show the multipole operators to be used with the relativistic Hartree wave functions for elastic magnetic scattering take the form

$$T_{JM}^{\text{mag}}(q) = \begin{pmatrix} (iq\lambda'/2m)\Sigma'_{JM} & Q\Sigma_{JM} \\ Q\Sigma_{JM} & (-iq\lambda'/2m)\Sigma'_{JM} \end{pmatrix}$$

Here $\Sigma_{JM} = j_J(qx)\vec{\mathcal{Y}}_{JJ1}^M \cdot \vec{\sigma}$ and $\Sigma'_{JM} = (-i/q)[\vec{\nabla} \times j_J(qx)\vec{\mathcal{Y}}_{JJ1}^M] \cdot \vec{\sigma}$.
(b) Generalize to the other multipoles and inelastic transitions (Ref. [R34]).

17.3. (a) Retain just the Coulomb interaction. Show the electron scattering cross section can be written as $d^2\sigma/d\Omega_2 d\varepsilon_2 = \sigma_{\text{Mott}}^{\text{eff}}(q_\mu^2/q^2)^2 R(q,\omega)$ where

$$R(q,\omega) = \overline{\sum_i}\sum_f |\langle f| \int e^{-i\vec{q}\cdot\vec{x}}\hat{\rho}(\vec{x})d^3x|i\rangle|^2\, \delta(E_f - E_i - \omega)$$

(b) Show that for a uniform system of nonrelativistic charged point nucleons

$$\int e^{-i\vec{q}\cdot\vec{x}}\hat{\rho}(\vec{x})d^3x = \sum_{\vec{k}\lambda} a^\dagger_{\vec{k}-\vec{q}\lambda} a_{\vec{k}\lambda}$$

(c) Compute the quasielastic response $R(q,\omega)$ for a nonrelativistic noninteracting Fermi gas (Section 3). Show

$$R(q,\omega) = \frac{3Z}{4\pi}\frac{m}{k_{\text{F}}^2}\int_0^1 d^3x\, \theta(|\vec{x}-\vec{\Delta}|-1)\delta\left(\xi + \vec{\Delta}\cdot\vec{x} - \frac{\Delta^2}{2}\right)$$

Here the dimensionless variables are defined by $\vec{\Delta} = \vec{q}/k_{\text{F}}$, $\xi = m\omega/k_{\text{F}}^2$, $\vec{x} = \vec{k}/k_{\text{F}}$.
(d) Evaluate the integral in part (c). Show (Refs. [N2,N14])

$$\begin{aligned}
\left(\frac{3Z}{4\pi}\frac{m}{k_{\text{F}}^2}\right)^{-1} R(q,\omega) &= \frac{\pi}{\Delta}[1-(\frac{\xi}{\Delta}-\frac{\Delta}{2})^2]\,; && \Delta > 2 \;\; ; \frac{\Delta}{2}+1 \geq \frac{\xi}{\Delta} \geq \frac{\Delta}{2}-1 \\
&= \frac{\pi}{\Delta}[1-(\frac{\xi}{\Delta}-\frac{\Delta}{2})^2]\,; && \Delta < 2 \;\; ; \frac{\Delta}{2}+1 \geq \frac{\xi}{\Delta} \geq 1-\frac{\Delta}{2} \\
&= 2\pi\frac{\xi}{\Delta} \,; && \Delta < 2 \;\; ; 1-\frac{\Delta}{2} \geq \frac{\xi}{\Delta} \geq 0
\end{aligned}$$

(e) Plot these results as a function of ξ for fixed Δ. Discuss.

17.4. Use the results of Prob. 17.3 to derive the Coulomb sum rule for the noninteracting Fermi gas $C(q) \equiv (1/Z) \int_0^\infty d\omega\, R^{\text{in}}(q,\omega)$

$$\begin{aligned} C(q) &= 1 & ; \qquad & q \geq 2k_{\text{F}} \\ &= \frac{3}{2}\left(\frac{q}{2k_{\text{F}}}\right) - \frac{1}{2}\left(\frac{q}{2k_{\text{F}}}\right)^3 & ; \qquad & 2k_{\text{F}} \geq q \end{aligned}$$

17.5. Calculate the cross section for inelastic scattering from nuclear matter in the relativistic MFT of Section 14 retaining the full relativistic single-nucleon electromagnetic vertex $\Gamma_\mu = F_1\gamma_\mu + F_2\sigma_{\mu\nu}q_\nu$. Discuss (Ref. [R40]).

17.6. Derive Eq. (17.6), and verify the first line of Table 17.1.

18.1. (a) Use the condition of thermodynamic equilibrium to prove that conservation of baryon number implies the chemical potential of an antibaryon is the negative of the chemical potential of a baryon.
(b) Show that if there is no such conservation law, the chemical potential of an additional species must vanish.

18.2. Consider the noninteracting, relativistic system of fermions in Prob. 14.2; let $F \equiv Q/e$ be the fermion number.
(a) Compute the thermodynamic potential and parametric equation of state $\varepsilon(\rho_{\text{F}}, T)$ and $p(\rho_{\text{F}}, T)$.
(b) Give analytic expressions for the limiting cases $\rho_{\text{F}} \to \infty$ and $T \to \infty$.
(c) Formulate a numerical procedure for arbitrary (ρ_{F}, T).

18.3. Verify the high temperature limits in Eqs. (18.19) and (18.20).

18.4. Include the contribution of the noninteracting, noncondensed (σ, ω) fields in $\hat{H}_{\text{MFT}}$ (Probs. 14.1 and 14.3).
(a) Compute the additional contribution to Ω (see Prob. 18.1).
(b) Compute the new very high T energy density and equation of state.

18.5. Use the Feynman rules for the temperature Green's function and the self-consistent Hartree approximation of Section 16 to rederive the finite temperature MFT results of this section.

19.1. Verify the limiting results in Eqs. (19.18)-(19.20) for the model quark-gluon equation of state.

19.2. (a) Include a contribution for noninteracting pions in the baryon-meson thermodynamic potential and equation of state. What is the pion chemical potential?
(b) Recompute a few of the isotherms in Fig. 19.5.
(c) Discuss the impact of this, and higher mass hadron contributions.

19.3. Draw the Feynman diagrams and use the Feynman rules for QCD to obtain an expression for the second-order contribution to the quark self energy.[79] You need not yet evaluate the integrals.

19.4. Repeat Prob. 19.3 for the gluon self-energy. Remember the ghost loop and all the gluon loops.

[79] Second order in g.

19.5. Repeat Prob. 19.3 for the quark-gluon vertex.

The next five problems review the relativistic analysis of an arbitrary two-body scattering or reaction process $a+b \to c+d$; this can include massless participants as well as $(e, e'X)$ through one-photon exchange (Ref. [N14]). The analysis is from the classic paper of Jacob and Wick (Ref. [R56]), which uses helicity states for the particles with $\vec{J} \cdot (\vec{p}/p)\,|\vec{p}\lambda\rangle = \lambda|\vec{p}\lambda\rangle$. The helicity is unchanged under rotation or Lorentz transformations along $\vec{p}$ (as long as it is not reversed). Through the use of general properties of the scattering operator $\hat{S}$, the angular distribution can be exhibited, and unitarity, as well as symmetry properties of the S-matrix, readily imposed in each subspace of total J.

20.1. The direct product state $|\vec{p}_a\vec{p}_b\lambda_a\lambda_b\rangle$ can also be denoted by $|P_\mu\theta_p\phi_p\lambda_a\lambda_b\rangle$ where $P_\mu = (\vec{P}, iE)$ is the total four-momentum and (θ_p, ϕ_p) are the direction of the relative momentum $\vec{p} = (m_b\vec{p}_a - m_a\vec{p}_b)/(m_a + m_b)$. Since $[\hat{S}, \hat{P}_\mu] = 0$ one can define

$$\langle\vec{p}_c\vec{p}_d\lambda_c\lambda_d|\hat{S}|\vec{p}_a\vec{p}_b\lambda_a\lambda_b\rangle \equiv \frac{(2\pi)^6\sqrt{vv'}}{\Omega^2 pp'}\delta^{(4)}(P_\mu - P'_\mu)\langle\theta'\phi'\lambda_c\lambda_d|\hat{S}(P_\mu)|\theta\phi\lambda_a\lambda_b\rangle$$

Here p and the relative velocity $v = p/\sqrt{p^2 + m_a^2} + p/\sqrt{p^2 + m_b^2}$ are C-M values.
(a) Set $\hat{S} = 1$ and use the normalization of the single-particle states to show that in the C-M system $\langle\theta'\phi'\lambda_c\lambda_d|\theta\phi\lambda_a\lambda_b\rangle = \delta(\cos\theta - \cos\theta')\delta(\phi - \phi')\delta_{\lambda_c\lambda_a}\delta_{\lambda_d\lambda_b}\delta_{ca}\delta_{db}$.
(b) Write $\hat{S} = 1 + i\hat{T}$ and show the cross section in the C-M frame is

$$\frac{d\sigma}{d\Omega_{p'}} = \frac{(2\pi)^2}{p^2}\,|\langle\theta\phi\lambda_c\lambda_d|\hat{T}(E)|00\lambda_a\lambda_b|^2$$

(c) Unitarity states $\hat{S}^\dagger\hat{S} = 1$. Assume only two-body states are accessible. Show

$$\sum_{\lambda_1\lambda_2 \text{ particles}} \int d\Omega\langle\theta\phi\lambda_1\lambda_2|\hat{S}(E)|\theta_f\phi_f\lambda_c\lambda_d\rangle^*\langle\theta\phi\lambda_1\lambda_2|\hat{S}(E)|\theta_i\phi_i\lambda_a\lambda_b\rangle$$
$$= \delta_{ca}\delta_{db}\delta_{\lambda_c\lambda_a}\delta_{\lambda_d\lambda_b}\delta^{(2)}(\theta_f\phi_f|\theta_i\phi_i)$$

20.2. The single-particle state can be constructed by rotation (see Section 7) $|\vec{p}\lambda\rangle = \hat{R}_{-\phi-\theta\phi}|p_+\lambda\rangle$ where p_+ lies along the positive z axis; the last angle is a phase convention. Define $|p_-\lambda\rangle \equiv (-1)^{s-\lambda}\hat{R}_{0-\pi 0}|p_+\lambda\rangle$ and the two-particle state by $|p_+\lambda_1\lambda_2\rangle \equiv |p_+\lambda_1\rangle|p_-\lambda_2\rangle$. Let $(\alpha\beta\gamma)$ be Euler angles and $\lambda \equiv \lambda_1 - \lambda_2$. A basic theorem then provides the eigenstates of angular momentum

$$|pJM\lambda_1\lambda_2\rangle = \frac{\mathcal{N}}{2\pi}\int_0^{2\pi} d\alpha \int_0^{\pi} \sin\beta d\beta \int_0^{2\pi} d\gamma\, \mathcal{D}^J_{M\lambda}(-\alpha-\beta-\gamma)^*\hat{R}_{-\alpha-\beta-\gamma}|p_+\lambda_1\lambda_2\rangle$$

(a) Prove this is an eigenstate of $\hat{J}_z$ with eigenvalue M. (*Hint*: Try $e^{i\omega\hat{J}_z}$.)

(b) Prove this is an eigenstate of $\hat{\vec{J}}^2$ with eigenvalue $J(J+1)$. (*Hint*: Insert a complete set of eigenstates.)
(c) Identify the angles $(\alpha, \beta) = (\phi, \theta)$ and $|p\theta\phi\lambda_1\lambda_2\rangle = \hat{R}_{-\phi-\theta\phi}|p_+\lambda_1\lambda_2\rangle$. Use the normalization condition in Prob. 20.1 to show $\mathcal{N} = [(2J+1)/4\pi]^{1/2}$. Hence conclude

that the transformation coefficients to eigenstates of angular momentum are just the rotation matrices

$$\langle\theta\phi\lambda_1\lambda_2|JM\lambda_1'\lambda_2'\rangle = \left(\frac{2J+1}{4\pi}\right)^{1/2} \mathcal{D}^J_{M\lambda}(-\phi-\theta\phi)^*\delta_{\lambda_1'\lambda_1}\delta_{\lambda_2'\lambda_2}$$

20.3. Use the results of Probs. 20.1-2 and the rotational invariance of the scattering operator $[\hat{\vec{J}},\hat{S}]=0$ to exhibit the general angular dependence of the cross section (here $\lambda\equiv\lambda_a-\lambda_b$ and $\mu\equiv\lambda_c-\lambda_d$, and we use p and p_0)

$$\begin{aligned}\frac{d\sigma}{d\Omega_p} &= |f_{\lambda_c\lambda_d|\lambda_a\lambda_b}(\theta,\phi)|^2 \\ f_{\lambda_c\lambda_d|\lambda_a\lambda_b}(\theta,\phi) &= \frac{1}{2p_0}\sum_J(2J+1)\langle\lambda_c\lambda_d|\hat{T}^J(E)|\lambda_a\lambda_b\rangle\mathcal{D}^J_{\lambda\mu}(-\phi-\theta\phi)^*\end{aligned}$$

20.4. (a) Use the unitarity of the scattering operator in Prob. 20.1 and the transformation in Prob. 20.2 to show the finite submatrices $\underline{S}^J(E)$ satisfy

$$\sum_{\lambda_1\lambda_2\ \text{particles}}\langle\lambda_1\lambda_2|\hat{S}^J(E)|\lambda_c\lambda_d\rangle^*\langle\lambda_1\lambda_2|\hat{S}^J(E)|\lambda_a\lambda_b\rangle = \delta_{ac}\delta_{bd}\delta_{\lambda_a\lambda_c}\delta_{\lambda_d\lambda_b}$$

(b) It is shown in Ref.[R56] that the parity operator can be defined so that $\hat{P}|JM\lambda_1\lambda_2\rangle = (-1)^{J-s_1-s_2}\eta_1\eta_2|JM-\lambda_1-\lambda_2\rangle$ where η is the intrinsic parity. Show that if parity is a good symmetry with $[\hat{P},\hat{S}]=0$ then the number of independent helicity amplitudes is reduced

$$\langle-\lambda_c-\lambda_d|\hat{S}^J(E)|-\lambda_a-\lambda_b\rangle = \eta_a\eta_b\eta_c^*\eta_d^*(-1)^{s_c+s_d-s_a-s_b}\langle\lambda_c\lambda_d|\hat{S}^J(E)|\lambda_a\lambda_b\rangle$$

(c) It is also shown in Ref. [R56] that the antiunitary time reversal operator and phase conventions for the states (note!) can be defined so that $\hat{T}|JM\lambda_1\lambda_2\rangle = (-1)^{J-M}|J-M\lambda_1\lambda_2\rangle$. Demonstrate that if time reversal is a good symmetry with $\hat{T}\hat{S}=\hat{S}^\dagger\hat{T}$ then the submatrices are symmetric

$$\langle\lambda_c\lambda_d|\hat{S}^J(E)|\lambda_a\lambda_b\rangle = \langle\lambda_a\lambda_b|\hat{S}^J(E)|\lambda_c\lambda_d\rangle$$

20.5. As an application of the results in Probs. 20.1-4 consider relativistic elastic scattering of strongly interacting spin-1/2 and spin-0 particles (e.g., $\pi+N$ or $\alpha+N$).
(a) What are the conditions imposed on the matrix $\underline{S}^J(E)$ by unitarity and the symmetry conditions in Prob. 20.4?
(b) Demonstrate that eigenstates of parity will diagonalize these matrices. Express the diagonal elements in terms of phase shifts. The standard notation is $\delta_{l\pm}$ where $J=l\pm1/2$ and the parity is $\eta_1\eta_2(-1)^l$.
(c) Show that the scattering amplitude between helicity states $\lambda=\pm1/2$ can be expressed in the form

$$\underline{f} = \begin{pmatrix} f_{++} & f_{+-} \\ f_{-+} & f_{--}\end{pmatrix} = \begin{bmatrix}(f_1+f_2)\cos\theta/2 & (f_1-f_2)e^{-i\phi}\sin\theta/2 \\ -(f_1-f_2)e^{i\phi}\sin\theta/2 & (f_1+f_2)\cos\theta/2\end{bmatrix}$$

Here $f_1=\sum_l(f_{l+}P'_{l+1}-f_{l-}P'_{l-1})$ and $f_2=\sum_l(f_{l-}-f_{l+})P'_l$ with $f_l\equiv e^{i\delta_l}\sin\delta_l/p_0$.

(d) One can always introduce another basis by making a unitary transformation on the helicity states $|\beta\rangle = \sum_{\alpha'} U_{\alpha'\beta}|\alpha'\rangle$. Chose the following transformation $U_{\alpha\beta} = \mathcal{D}^{1/2}_{\lambda'\lambda}(-\phi\theta\phi)$ on the final state. Show one then reproduces the form of the scattering amplitude given in Eq. (E.1).
(e) Interpret the transformation in (c) in terms of a rotation of the spin of the particle at rest from the z-axis to the direction of $\vec{p}$. Use the Lorentz transformation properties of the helicity states to relate the C-M scattering amplitude in (c) to the spin of the final particle in its rest frame.

20.6. Deduce from $\mathcal{L}$ in Eq. (20.8) the Feynman rules for the additional contributions to S_{fi} from the neutral vector meson V_μ.

20.7. Substitute the explicit representation of the Dirac spinors in Prob. 13.1 into Eq. (20.7) to derive Eq. (E.2). (Remember to renormalize to $\bar{u}u = 1$).

20.8. Take appropriate matrix elements and use the theory of angular momenta to derive the isospin relations in Eq. (E.4).

20.9. Derive Eq. (E.11).

21.1. One of the most useful operator identities in quantum mechanics is

$$e^{i\hat{S}}\hat{O}e^{-i\hat{S}} = \hat{O} + i[\hat{S},\hat{O}] + \frac{i^2}{2!}[\hat{S},[\hat{S},\hat{O}]] + \frac{i^3}{3!}[\hat{S},[\hat{S},[\hat{S},\hat{O}]]] + \cdots$$

This algebraic identity also holds for matrices.
(a) Verify the first few terms.
(b) Prove to all orders by making a Taylor series expansion of $\hat{F}(\lambda) = e^{i\lambda\hat{S}}\hat{O}e^{-i\lambda\hat{S}} = \sum_n (\lambda^n/n!)(\partial^n \hat{F}/\partial\lambda^n)_{\lambda=0}$ and then setting $\lambda = 1$ (Ref. [R2]).

21.2. The τ matrices satisfy $[\frac{1}{2}\tau_i, \frac{1}{2}\tau_j] = i\epsilon_{ijk}\frac{1}{2}\tau_k$ and $\{\tau_i, \tau_j\} = 2\delta_{ij}$.
(a) Show $\exp\{\frac{i}{2}\vec{\omega}\cdot\vec{\tau}\} = \cos\omega/2 + i\vec{n}\cdot\vec{\tau}\sin\omega/2$ where $\vec{\omega} = (\omega_1, \omega_2, \omega_3) = \vec{n}\omega$.
(b) Extend the proof to show $\exp\{\frac{i}{2}\vec{\omega}\cdot\vec{\tau}\gamma_5\} = \cos\omega/2 + i\vec{n}\cdot\vec{\tau}\gamma_5\sin\omega/2$.

21.3. (a) Let $\exp\{\frac{i}{2}\vec{\omega}\cdot\vec{\tau}\gamma_5\} \equiv \underline{\mathrm{r}}(\vec{\omega})$. Show the finite chiral transformation is $\underline{\mathrm{r}}(\phi + i\vec{\tau}\cdot\vec{\pi}\gamma_5)\underline{\mathrm{r}} = \phi(\cos\omega + i\vec{n}\cdot\vec{\tau}\gamma_5\sin\omega) + \vec{\pi}\cdot[i\vec{\tau}\gamma_5 - \vec{n}\sin\omega - 2\vec{n}(i\vec{n}\cdot\vec{\tau}\gamma_5)\sin^2\omega/2]$.
(b) Let $\vec{\omega} \equiv \vec{\varepsilon} \to 0$ and verify the infinitesimal chiral transformation in Eq. (21.27).

21.4. Nambu and Jona-Lasinio have proposed a dynamic model for spontanteous breaking of chiral symmetry.[80] Their model involves a four-fermion coupling with lagrangian density

$$\mathcal{L}_{\mathrm{NJL}} = -\bar{\psi}\left(\gamma_\mu\frac{\partial}{\partial x_\mu}\right)\psi + G[(\bar{\psi}\psi)^2 + (\bar{\psi}i\gamma_5\psi)^2]$$

Prove this lagrangian is invariant under the global chiral transformation $\psi \to e^{i\gamma_5\theta/2}\psi$.

21.5. The matrix element needed for pion decay is (Section 42) $\sqrt{2\omega_p\Omega}\langle 0|J^{(+)}_{\lambda 5}(0)|\pi^-, p\rangle = iF_\pi(p^2)p_\lambda$. Prove that if the axial vector current is conserved and $p^2 = -\mu^2 \neq 0$, then $F_\pi = 0$ and the pion cannot decay.

21.6. Use the canonical (anti-)commutation relations to show that the operator $\hat{\vec{T}}_5$ in Eq. (21.49) is indeed the generator of the chiral transformation in Eqs. (21.43).

[80] Y. Nambu and G. Jona-Lasinio, *Phys. Rev.* **122**, 345 (1961); **124**, 246 (1961).

21.7. (a) Verify Eq. (F.11); (b) Verify Eqs. (F.15) and (F.16)

22.1. Consider QCD in the nuclear domain with massless quarks $m_u = m_d = 0$ [Eqs. (19.1)-(19.4) and 19.12)]. Show $\mathcal{L}_{\text{QCD}}$ is invariant under the chiral transformation.

22.2. Prove Noether's theorem in Eq. (21.6).

22.3. Use the invariance of the lagrangian in Eq. (22.1) under global phase transformations of the baryon field ψ to deduce the conserved baryon current.

22.4. Consider the chiral symmetric σ-model of nuclear matter in MFT.
(a) Show $V_{\text{chiral}} = V_0[(M+M^*)^2/4M^2]$ where $V_0 = (1/2)m_s^2\phi_0^2 = (m_s^2/2g^2)(M-M^*)^2$ is the scalar meson potential of QHD-I.
(b) How is the self-consistency relation for M^* modified?
(c) Discuss the solution to this new self-consistency relation as a function of density (see Refs. [R61, R62, R1]).

22.5. (a) Derive the lowest order three-body nucleon force in the chiral symmetric σ-model.
(b) Repeat for the four-body force.
(c) Estimate the contribution of these two interactions to the binding energy of ^{4}He. Use s.h.o. wave functions and assume a $(1s_{1/2})^4$ configuration.

23.1. (a) Start from the lagrangian in Eq. (22.24) and construct the S-matrix for $N+N \to N+N+\pi$ at tree level (no internal loops). Verify that the production amplitude vanishes in the chiral soft-pion limit $q_\lambda \to 0$.
(b) Repeat starting from the lagrangian in Eq. (23.33). Compare.

23.2. Repeat Prob. 23.1 for the decay of the scalar meson with mass m_s into 4 pions.

24.1. Consider scattering in nonrelativistic quantum mechanics (Probs. 1.1-6). Let $f(\theta,\phi) = \sum_l (2l+1) f_l\, P_l(\cos\theta)$ with $f_l = e^{i\delta_l}\sin\delta_l/k$ be the scattering amplitude calculated for two distinguishable particles. Now assume the two particles are identical, implying either a symmetric or antisymmetric spatial wave function. Show the effect on the scattering amplitude is to replace $f \to f(\theta) \pm f(\pi-\theta)$ (see Ref. [N42]).

24.2. Consider the scattering of two particles with isotopic spin $|t_1 m_1 t_2 m_2\rangle$. Assume the scattering operator is invariant under isospin rotations so that $[\hat{\vec{T}}, \hat{S}] = 0$.
(a) Show the scattering amplitude can be written

$$f_{m_1' m_2';\, m_1 m_2} = \sum_{TM_T} \langle t_1 m_1' t_2 m_2' | t_1 t_2 T M_T\rangle \langle t_1 m_1 t_2 m_2 | t_1 t_2 T M_T\rangle f^T(\theta,\phi)$$

(b) Show the scattering amplitude of Prob. 20.3 is calculated from $\langle \lambda_c \lambda_d | T^{JT}(E) | \lambda_a \lambda_b\rangle$ which satisfies the unitarity condition in Prob. 20.4.

24.3. Consider the relativistic analysis of the scattering of two identical 0^+ bosons. Here there is only one two-particle state $|\vec{k}_1 \vec{k}_2\rangle = c^\dagger_{\vec{k}_1} c^\dagger_{\vec{k}_2}|0\rangle$; it is automatically symmetric under particle interchange.
(a) Show the transformation coefficients in Prob. 20.2 with the correct symmetry are now given by $\langle\theta,\phi|JM\rangle = Y_{JM}(\theta,\phi)\,[1+(-1)^J]\,/\,\sqrt{2}$.
(b) Show these coefficients are properly normalized with respect to the volume element $\int d\Omega\,/\,2\,!$, which counts all the independent states.
(c) Show the unitarity relation, taking into account the volume element in (b), again has the form in Prob. 20.4.

(d) What is the corresponding form for the scattering amplitude in Prob. 20.3? Compare with the result in Prob. 24.1.
(e) Extend the analysis to include integer isospin. Show the transformation coefficients are now $\langle\theta,\phi|JM\rangle^T = Y_{JM}(\theta,\phi)\,[1+(-1)^{J+T}]/\sqrt{2}$.

24.4. (a) The substitution rule allows one to turn around an external leg on a Feynman diagram by reversing the sign of the four-momentum and making an appropriate wave function replacement. Show that a pion leg can be turned around with the replacement $q_i, \alpha \rightarrow -q_i, \alpha$. Hence show that the π–N scattering amplitude is invariant under the substitution $q_1, \alpha \rightleftharpoons -q_2, \beta$.
(b) If the scattering amplitude is an analytic function, the substitution rule implies crossing relations that relate the function in different regions of the variable(s). Prove the following crossing relations for $\pi - N$ scattering

$$A^{\pm}(s,t,u) = \pm A^{\pm}(u,t,s) \qquad\qquad B^{\pm}(s,t,u) = \mp B^{\pm}(u,t,s)$$

What are the crossing relations in terms of (ν, κ^2)?

24.5. The analyticity properties for π–N scattering shown in Fig. E.1 can be established from Feynman diagrams or axiomatic field theory (Ref. [R3]). Write a Cauchy integral, expand the contour, use the crossing relations of Prob. 24.4, and use the nucleon pole contributions of Eqs. (20.22) to derive the fixed κ^2 dispersion relations in Eqs. (E.7) and (E.8).

24.6. Derive the results in Eqs. (H.12).

24.7. What are the crossing relations for $\pi - \pi$ scattering?

24.8. Consider π–π scattering in the $J^\pi, T = 1^-, 1$ channel.
(a) Calculate the T-matrix in tree approximation starting from $\mathcal{L}$ in Eq. (24.1).
(b) Unitarize the amplitude as discussed in the text for the $0^+, 0$ channel.
(c) Calculate and plot the phase shift for the values of m_σ^2 in Fig. 24.4. Do you get anything that looks like the ρ-meson?
(d) The g_ϕ and m_s^2 from Eqs. (20.37) represent a phenomenological determination of the effects of the exchange of a correlated $0^+, 0$ pair of pions as discussed in this section. Repeat part (c) with these parameters. Discuss.

24.9. (a) Include the electromagnetic interaction through the minimal gauge invariant substitution in $\mathcal{L}$ in Eq. (24.1) and derive the conserved electromagnetic current.
(b) Show the electric charge is given by $Q = T_3 + B/2$.

25.1. Use Eq. (25.11) to derive the contribution of a condensed neutral ρ_μ^0 field in the relativistic Hartree theory of finite nuclei in Prob. 15.5.

25.2. (a) Use the Feynman rules in Eq. (25.13) to calculate the decay rate $\Gamma_{\rho\rightarrow\pi+\pi} = (g_\rho^2/4\pi)(m_\rho/12)[1-(2m_\pi/m_\rho)^2]^{3/2}$.
(b) Use $m_\rho = 768.1\,\text{MeV}$ and $\Gamma_{\rho\rightarrow\pi+\pi} = 151.5\,\text{MeV}$ to find $g_\rho^2/4\pi$ and compare with the value $g_\rho^2/4\pi = 5.19$ used in the relativistic Hartree calculations to get the correct symmetry energy (Refs. [R7, R1]).

25.3. Use Eqs. (25.17) and (25.19) and the analysis of Appendix A to derive the static N–N potential arising from the exchange of a single axial vector meson with mass m_a^2. Relate the coupling to that in Eq. (25.16).

26.1. Start from the S-matrix in Eq. (26.1) and current matrix element in Eq. (26.3) and derive the $(e, e'X)$ coincidence cross section in Eq. (26.2) (see Ref. [N14]).

26.2. Consider the $(e, e'X)$ coincidence cross section in Eq. (26.2) and Fig. 26.1.
(a) Introduce C-M unit vectors $(\vec{e}_x, \vec{e}_y, \vec{e}_z) = (-\vec{k}_2 \times \vec{k}_1/|\vec{k}_2 \times \vec{k}_1|,\ \vec{e}_{\vec{k}2},\ -\vec{k}/|\vec{k}|)$; note the transverse unit vectors are identical to those in the lab since they are unchanged under a Lorentz transformation along $\vec{k}$. Use the Jacob and Wick analysis in Probs. 20.1-3 to show that the transition matrix element can be parameterized, and the general angular dependence of particle 2 exhibited, through the following relations (Ref. [N14])

$$
\begin{aligned}
(\mathcal{J}_C)_{\lambda_f\lambda_i} &= \frac{k^*}{k_0^*}\frac{1}{\sqrt{4k^*q}}\sum\nolimits_J (2J+1)\langle\lambda_2\lambda_x|T^J(W,k^2)|\lambda_1\lambda_k\rangle \mathcal{D}^J_{\lambda_i\lambda_f}(-\phi_p, -\theta_p, \phi_p)^* \\
(\mathcal{J}^{\lambda_k})_{\lambda_f\lambda_i} &= \frac{1}{\sqrt{4k^*q}}\sum\nolimits_J (2J+1)\langle\lambda_2\lambda_x|T^J(W,k^2)|\lambda_1\lambda_k\rangle \mathcal{D}^J_{\lambda_i\lambda_f}(-\phi_p, -\theta_p, \phi_p)^*
\end{aligned}
$$

Here the first relation holds for $\lambda_k = 0$ and the second for $\lambda_k = \pm 1$; also $\lambda_i \equiv \lambda_1 - \lambda_k$ and $\lambda_f \equiv \lambda_2 - \lambda_x$; in addition, $\mathcal{J}^\lambda = \vec{e}_\lambda \cdot \vec{\mathcal{J}}$ with $\vec{e}_\lambda = \mp(\vec{e}_1 \pm i\vec{e}_2)/\sqrt{2}$ (Fig. 26.1).
(b) It is often convenient to measure the distribution of the particle X in the coordinate system with $(\vec{e}_1, \vec{e}_2, \vec{e}_3) = (-\vec{e}_x, \vec{e}_y, -\vec{e}_z) = (\vec{k}_2 \times \vec{k}_1/|\vec{k}_2 \times \vec{k}_1|,\ \vec{e}_{\vec{k}2},\ \vec{k}/|\vec{k}|)$ (Fig.26.1). Here $\theta_q = \theta_p$ and $\phi_q + \phi_p = 2\pi$. Show $\mathcal{D}^J_{\lambda_i\lambda_f}(-\phi_p, -\theta_p, \phi_p)^* = \mathcal{D}^J_{\lambda_f\lambda_i}(\phi_q, \theta_q, -\phi_q)$.

26.3. (a) Show the free baryon propagator can be decomposed as

$$
\frac{1}{i\gamma_\mu p_\mu + M} \equiv \left[\frac{1}{2E_p}\frac{\vec{\alpha}\cdot\vec{p} + \beta M + E_p}{E_p - p_0 - i\eta} + \frac{1}{2E_p}\frac{\vec{\alpha}\cdot\vec{p} + \beta M - E_p}{E_p + p_0 - i\eta}\right]\beta
$$

(b) Show the first term yields the usual non-relativistic result in Ref. [N2].
(c) Show the second term gives rise to backward propagation in time. Interpret.

26.4. Use the Feynman rules from the lagrangian in Prob. 24.9 to evaluate the contribution to the S-matrix from the graphs in Fig. 26.6. Retain just the second piece of the baryon propagator in Prob. 26.3. Construct the equivalent S-matrix from the current in Eq. (26.5); hence identify the additional two-body current. Write $J_\mu(\vec{x}_1, \vec{x}_2; \vec{x}) = \int e^{i\vec{k}\cdot\vec{x}} J_\mu(\vec{x}_1, \vec{x}_2; \vec{k}) d^3k/(2\pi)^3$. Work to leading order in $1/M$ and assume $k_0 = 0$.
(a) Show the pair contribution to the pion exchange current in Fig. 26.6a is given by

$$
\begin{aligned}
\vec{J}^{\text{pair}}(\vec{x}_1, \vec{x}_2; \vec{k}) &= -e_p f_\pi^2 [\vec{\tau}^{(1)} \times \vec{\tau}^{(2)}]_3 \left\{\left(\frac{\vec{\sigma}_1\cdot\vec{r}}{r}\right)\vec{\sigma}_2 e^{-i\vec{k}\cdot\vec{x}_2} + \vec{\sigma}_1\left(\frac{\vec{\sigma}_2\cdot\vec{r}}{r}\right)e^{-i\vec{k}\cdot\vec{x}_1}\right\} \\
&\quad\times\left(\frac{1+x_\pi}{x_\pi^2}\right)e^{-x_\pi}
\end{aligned}
$$

Here $x_\pi = \mu\vec{r}$ with $\vec{r} = \vec{x}_1 - \vec{x}_2$ and $\vec{R} = (\vec{x}_1 + \vec{x}_2)/2$. Also $\mu \equiv m_\pi$.
(b) Show the pion contribution in Fig. 26.6b is

$$
\begin{aligned}
\vec{J}^{\text{pion}}(\vec{x}_1, \vec{x}_2; \vec{k}) &= e_p\left(\frac{f_\pi}{\mu}\right)^2 [\vec{\tau}^{(1)} \times \vec{\tau}^{(2)}]_3 (\vec{\sigma}_1\cdot\vec{\nabla}_1)(\vec{\sigma}_2\cdot\vec{\nabla}_2)\int_{-1/2}^{1/2} dv \\
&\quad\times(-irv\vec{k} + \vec{y})\left(\frac{e^{-y}}{y}\right)\exp\{-i\vec{k}\cdot(\vec{R} - v\vec{r})\}
\end{aligned}
$$

Here $\vec{y} = [\mu^2 + (\vec{k}^2/4)(1-4v^2)]^{1/2}\vec{r}$. (*Hint*: Use $(ab)^{-1} = \int_0^1 dz[az + b(1-z)]^{-2}$.) (c) Show there is no exchange contribution to the charge density to this order in $1/M$. These results are from Ref. [R77].

26.5. Take the current to be the sum of the usual one-body current in Section 8 plus the exchange current in Prob. 26.4. Assume a two-nucleon potential $V = V^{\text{neutral}} + V^{\text{OPEP}}$ where the last term is the 1–π exchange potential of Appendix A. Prove the current is conserved $\partial \hat{J}_\mu/\partial x_\mu = \vec{\nabla}\cdot\hat{\vec{J}} + i[\hat{H},\hat{\rho}] = 0$ (Ref. [R77]).

26.6. Consider a 2-channel process where the first channel $a + b \rightleftharpoons a + b$ is elastic scattering through the strong interaction in a given partial wave, the transition amplitude is weak, say of $O(e)$ $\gamma + a \rightleftharpoons a + b$, and the scattering in the second channel $\gamma + a \rightleftharpoons \gamma + a$ is of $O(e^2)$. With time reversal invariance (Prob. 20.4c), the S-matrix for this process then takes the following form $\mathcal{S} = \begin{pmatrix} e^{2i\delta} & 2it \\ 2it & 1 \end{pmatrix}$ to $O(e)$. Use unitarity to prove Watson's theorem $t = |t|e^{i\delta}$; that is, the phase of the weak transition amplitude is that of the strong-interaction phase shift.[81]

26.7. The problem of constructing an analytic function $a(W,k^2)$ with a specified set of left-hand singularities in W given by $a^{\text{lhs}}(W,k^2)$ (real on the physical real axis) and obeying Watson's theorem along the right-hand physical cut (Prob. 26.6) was formulated by Omnès as an integral equation[82]

$$a(W,k^2) = a^{\text{lhs}}(W,k^2) + \frac{1}{\pi}\int_{W_0}^{\infty} \frac{e^{-i\delta(W')}\sin\delta(W')a(W',k^2)}{W'-W-i\varepsilon}dW'$$

The solution to this integral equation for $W_0 \le W \le \infty$ was also given by Omnès

$$a(W,k^2) = e^{i\delta(W)}\left[a^{\text{lhs}}(W,k^2)\cos\delta(W) + e^{\rho(W)}\frac{\mathcal{P}}{\pi}\int_{W_0}^{\infty}\frac{a^{\text{lhs}}(\xi,k^2)\sin\delta(\xi)e^{-\rho(\xi)}}{\xi - W}d\xi\right]$$

Here $\mathcal{P}$ is the Cauchy principal value and $\rho(W) = (\mathcal{P}/\pi)\int_{W_0}^{\infty}\delta(\zeta)d\zeta\,/\,(\zeta - W)$.
(a) Assume that $a^{\text{lhs}}(W,k^2)$ varies only slowly over the region where $\sin\delta(W) \ne 0$ on the physical cut, and factor it out of the integral. Show $a(W,k^2) \approx a^{\text{lhs}}(W,k^2)\chi(W)$ where $\chi(W) = \exp\{(1/\pi)\int_{W_0}^{\infty}\delta(W')dW'/(W'-W-i\varepsilon)\}\psi(W)$ and

$$\psi(W) = \exp\left[-\frac{1}{\pi}\int_{W_0}^{\infty}\frac{\delta(W')dW'}{W'-W-i\varepsilon}\right] + \frac{1}{\pi}\int_{W_0}^{\infty}\frac{\sin\delta(\xi)d\xi}{\xi - W - i\varepsilon}\exp\left[-\frac{\mathcal{P}}{\pi}\int_{W_0}^{\infty}\frac{\delta(\zeta)d\zeta}{\zeta-\xi}\right]$$

(b) Use the analytic properties of $\psi(W)$ and the observation that $\psi \to 1$ as $|W| \to \infty$ to write an unsubtracted dispersion relation for $\psi(W) - 1$. Show that on the right-hand physical cut the discontinuity of this function vanishes; hence conclude that $\psi(W) \equiv 1$!
(c). Thus derive the relation

$$a(W,k^2) = \frac{a^{\text{lhs}}(W,k^2)}{D(W)}$$

$$D(W) = \exp\left[-\frac{1}{\pi}\int_{W_0}^{\infty}\frac{\delta(W')dW'}{W'-W-i\varepsilon}\right]$$

[81] K. M. Watson, *Phys. Rev.* **88**, 1163 (1952).
[82] R. Omnès, *Nuovo Cimento* **8**, 316 (1958).

Here $D(W)$ serves as a final-state enhancement factor.

26.8. Evidently in Prob. 26.7(c) the final-state enhancement factor satisfies $D(W) = |D(W)|e^{-i\delta(W)}$ for $W \geq W_0$ and is purely imaginary at a resonance $\delta(W_R) = \pi/2$ in the elastic scattering channel. Hence a Taylor series gives $D(W) \approx (W - W_R)\text{Re}'D(W_R) + i\text{ImD}(W_R)$. Show the electroproduction amplitude in this channel then resonates at the same W_R and has a Breit-Wigner form.

26.9. Consider the Feynman diagrams in Fig. 26.13 as the excitation mechanism for the production of the low-lying nucleon resonances in Fig. 26.12. Treat these as generalized Feynman amplitudes using renormalized coupling constants and physical electromagnetic form factors $F(k^2)$ at the vertices; the justification for this procedure is that these terms give the correct pole contributions in the dispersion relations for these processes (Ref. [R91]).

(a) Show the contribution of the nucleon and pion pole terms takes the form (here $\bar{u}u = 1$)

$$\left(\frac{4\pi W}{M}\right) \mathcal{J}_\lambda^{\text{pole}} \varepsilon_\lambda = -g_\pi \bar{u}(p_2)\{\tau_\alpha M_\lambda^{(0)} + \delta_{\alpha 3} M_\lambda^{(+)} + \frac{1}{2}[\tau_\alpha, \tau_3] M_\lambda^{(-)}\} u(p_1)\varepsilon_\lambda$$

$$M_\lambda^{(i)} = \gamma_5 \frac{1}{i(p_1 + k)_\sigma \gamma_\sigma + M}[F_1^{(i)}\gamma_\lambda - F_2^{(i)}\sigma_{\lambda\rho}k_\rho]$$

$$+ s_i[F_1^{(i)}\gamma_\lambda - F_2^{(i)}\sigma_{\lambda\rho}k_\rho]\frac{1}{i(p_2 - k)_\sigma\gamma_\sigma + M}\gamma_5 - if_i\gamma_5\frac{(2q-k)_\lambda}{(q-k)^2 + \mu^2}F_\pi$$

Here $\{F^{(0)} = F^S/2, s_0 = +1, f_0 = 0\}$, $\{F^{(+)} = F^V/2, s_+ = +1, f_+ = 0\}$, and finally $\{F^{(-)} = F^V/2, s_- = -1, f_- = 1\}$.

(b) Assume that $F_\pi(k^2) \approx F_1^V(k^2)$ in the region of interest. Show that the replacement $\varepsilon_\lambda \to k_\lambda$ gives zero; hence conclude that this current is explicitly conserved.

Talk

APPENDICES: PART II

C Pressure in MFT

The pressure p can be determined from Eq. (14.10) and the field expansions, or one can use thermodynamics; these two approaches give identical results (Ref. [R1]). In this appendix the thermodynamic argument is summarized.

Start from the first law of thermodynamics

$$dE = -pdV\,; \qquad B\text{ fixed} \tag{C.1}$$

Consider the expression

$$\frac{\partial\varepsilon}{\partial\rho_{\rm B}} = \frac{\partial(E/V)}{\partial V}\frac{\partial V}{\partial\rho_{\rm B}} = \left(-\frac{E}{V^2}+\frac{1}{V}\frac{\partial E}{\partial V}\right)\left(-\frac{V^2}{B}\right) = \frac{\varepsilon}{\rho_{\rm B}}+\frac{p}{\rho_{\rm B}} \tag{C.2}$$

Thus

$$p = \rho_{\rm B}\frac{\partial\varepsilon}{\partial\rho_{\rm B}} - \varepsilon = \rho_{\rm B}^2\frac{\partial}{\partial\rho_{\rm B}}\left(\frac{\varepsilon}{\rho_{\rm B}}\right) \tag{C.3}$$

One can keep ϕ_0 fixed here since from Eq. (14.25)

$$\left(\frac{\partial\varepsilon}{\partial\phi_0}\right)_{V,B} = 0 \tag{C.4}$$

One then has from Eq. (C.3) and the first of Eqs. (14.24)

$$\begin{aligned} p &= \frac{g_v^2\rho_{\rm B}^2}{2m_v^2} - \left\{\frac{m_s^2}{2g_s^2}(M-M^*)^2 + \frac{\gamma}{(2\pi)^3}\int_0^{k_{\rm F}} d^3k(\vec{k}^{\,2}+M^{*2})^{1/2}\right\} \\ &\quad + \rho_{\rm B}\left\{\frac{\gamma}{(2\pi)^3}4\pi k_{\rm F}^2(k_{\rm F}^2+M^{*2})^{1/2}\frac{\partial k_{\rm F}}{\partial\rho_{\rm B}}\right\} \end{aligned} \tag{C.5}$$

Note $\rho_{\rm B} = \gamma k_{\rm F}^3/6\pi^2$; thus the third and fourth terms can be written in the form

$$\begin{aligned} &\frac{\gamma}{(2\pi)^3}4\pi\left[\frac{1}{3}k_{\rm F}^3(k_{\rm F}^2+M^{*2})^{1/2} - \int_0^{k_{\rm F}} k^2dk(k^2+M^{*2})^{1/2}\right] \\ &= \frac{\gamma}{(2\pi)^3}4\pi\int_0^{k_{\rm F}}\frac{1}{3}\frac{k^4}{(k^2+M^{*2})^{1/2}}dk \end{aligned} \tag{C.6}$$

This last equality follows upon a partial integration with $u = k^3/3$, $du = k^2dk\,; dv = kdk/(k^2+M^{*2})^{1/2}$, $v = (k^2+M^{*2})^{1/2}$. Thus we arrive at the following expression for the pressure

$$p = \frac{g_v^2\rho_{\rm B}^2}{2m_v^2} - \frac{m_s^2}{2g_s^2}(M-M^*)^2 + \frac{\gamma}{(2\pi)^3}\frac{1}{3}\int_0^{k_{\rm F}} d^3k\frac{k^2}{(k^2+M^{*2})^{1/2}} \tag{C.7}$$

This is the result quoted in Eq. (14.24).

D Thermodynamic Potential and Equation of State

First introduce some notation. Order all the single-particle modes and write

$$\begin{aligned}\{n_{\vec{k}\lambda}\} &\equiv \{n_1, n_2, n_3, \ldots, n_\infty\} \\ &\equiv \{n_i\}\,; \qquad\qquad i = 1, 2, \ldots, \infty\end{aligned} \tag{D.1}$$

The grand partition function is then given by

$$\begin{aligned}Z_G &= \sum_{n_1}\cdots\sum_{n_\infty}\sum_{\bar{n}_1}\cdots\sum_{\bar{n}_\infty} \\ &\times\langle n_1, \ldots, n_\infty; \bar{n}_1, \ldots, \bar{n}_\infty|e^{-\beta(\hat{H}-\mu\hat{B})}|n_1, \ldots, n_\infty; \bar{n}_1, \ldots, \bar{n}_\infty\rangle\end{aligned} \tag{D.2}$$

Use $\hat{H}_{\text{MFT}}$, $\hat{B}$, and the factorization of exponentials

$$\begin{aligned}Z_G &= \exp\{-\beta V(\frac{1}{2}m_s^2\phi_0^2 - \frac{1}{2}m_v^2V_0^2)\} \\ &\times\prod_i\sum_{n_i}\langle n_i|e^{-\beta(E_i^*-\mu^*)n_i}|n_i\rangle \times \prod_j\sum_{\bar{n}_j}\langle\bar{n}_j|e^{-\beta(E_j^*+\mu^*)\bar{n}_j}|\bar{n}_j\rangle\end{aligned} \tag{D.3}$$

Here

$$\begin{aligned}E_k^* &\equiv \sqrt{\vec{k}^{\,2} + M^{*2}} \\ \mu^* &\equiv \mu - g_vV_0\end{aligned} \tag{D.4}$$

There are just two values of the occupation numbers for fermions $n_i, \bar{n}_i = 0, 1$. Thus

$$\begin{aligned}Z_G &= \exp\{-\beta V(\frac{1}{2}m_s^2\phi_0^2 - \frac{1}{2}m_v^2V_0^2)\} \\ &\times\prod_i\{1 + e^{-\beta(E_i^*-\mu^*)}\}\prod_j\{1 + e^{-\beta(E_j^*+\mu^*)}\}\end{aligned} \tag{D.5}$$

Thus the thermodynamic potential in Eq. (18.1) is given by

$$\begin{aligned}\Omega &= V(\frac{1}{2}m_s^2\phi_0^2 - \frac{1}{2}m_v^2V_0^2) - \frac{1}{\beta}\sum\ln\{1 + e^{-\beta(E_i^*-\mu^*)}\} \\ &\quad -\frac{1}{\beta}\sum_j\ln\{1 + e^{-\beta(E_j^*+\mu^*)}\}\end{aligned} \tag{D.6}$$

Now compute the thermodynamic variables, for example

$$B = -\left(\frac{\partial\Omega}{\partial\mu}\right)_{T,V} = -\left(\frac{\partial\Omega}{\partial\mu}\right)_{T,V;\phi_0,V_0} \tag{D.7}$$

This gives, with $\sum_i \to [\gamma V/(2\pi)^3] \int d^3k$

$$\rho_{\rm B} = \frac{\gamma}{(2\pi)^3} \int d^3k (n_k - \bar{n}_k) \tag{D.8}$$

Here the thermal occupation numbers are defined by

$$n_k = \left[e^{\beta(E_k^* - \mu^*)} + 1 \right]^{-1} \qquad \bar{n}_k = \left[e^{\beta(E_k^* + \mu^*)} + 1 \right]^{-1} \tag{D.9}$$

The pressure can be obtained from $p = -\Omega/V$ which gives

$$\begin{aligned} p &= \frac{1}{2} m_v^2 V_0^2 - \frac{1}{2} m_s^2 \phi_0^2 \\ &+ \frac{1}{\beta} \frac{\gamma}{(2\pi)^3} \int d^3k \left[\ln\{1 + e^{-\beta(E^* - \mu^*)}\} + \ln\{1 + e^{-\beta(E^* + \mu^*)}\} \right] \end{aligned} \tag{D.10}$$

Now integrate by parts. In the first term, for example, define

$$\begin{aligned} dv &= k^2\,dk \; ; & u &= \ln\{1 + e^{-\beta(E^* - \mu^*)}\} \\ v &= \frac{1}{3} k^3 \; ; & du &= -\frac{e^{-\beta(E^* - \mu^*)}}{1 + e^{-\beta(E^* - \mu^*)}} \beta \frac{k\,dk}{E^*} \end{aligned} \tag{D.11}$$

Then

$$p = \frac{1}{2} m_v^2 V_0^2 - \frac{1}{2} m_s^2 \phi_0^2 + \frac{\gamma}{(2\pi)^3} \int d^3k \frac{1}{3} (n_k + \bar{n}_k) \frac{k^2}{(k^2 + M^{*2})^{1/2}} \tag{D.12}$$

The energy is obtained from

$$\frac{1}{V} E = \frac{1}{V} \langle\langle \hat{H} \rangle\rangle = \frac{1}{V} \frac{\partial(\beta\Omega)}{\partial\beta} + \mu\,\rho_{\rm B} \tag{D.13}$$

This gives

$$\frac{1}{V} E = \frac{1}{2} m_s^2 \phi_0^2 - \frac{1}{2} m_v^2 V_0^2 + g_v V_0 \rho_{\rm B} + \frac{\gamma}{(2\pi)^3} \int d^3k \sqrt{\vec{k}^2 + M^{*2}} (n_k + \bar{n}_k) \tag{D.14}$$

The value of the scalar field ϕ_0 is obtained from the minimization of the thermodynamic potential at fixed (μ, T, V)

$$\left(\frac{\partial\Omega}{\partial\phi_0} \right)_{\mu,V,T} = \left(\frac{\partial\Omega}{\partial\phi_0} \right)_{\mu,V,T;V_0} = 0 \tag{D.15}$$

This leads to the self-consistency condition (recall $M^* = M - g_s\phi_0$)

$$\phi_0 = \frac{g_s}{m_s^2} \frac{\gamma}{(2\pi)^3} \int d^3k \frac{M^*}{(\vec{k}^2 + M^{*2})^{1/2}} (n_k + \bar{n}_k) \equiv \frac{g_s}{m_s^2} \rho_s \tag{D.16}$$

This is identical to the thermal average of the scalar meson field equation for a uniform system. At the end of the calculation, one can replace the vector field by the baryon density

$$V_0 = \frac{g_v}{m_v^2} \rho_{\rm B} \tag{D.17}$$

This is the thermal average of the vector meson field equation for a uniform medium.

E $\pi - N$ Scattering

This material is from Refs. [R53, R54, R55].

In the C-M system one can write (in this appendix $q \equiv |\vec{q}|$)

$$\frac{d\sigma}{d\Omega} = \sum_f \overline{\sum_i} |\langle f| f_1 + f_2 \frac{(\vec{\sigma}\cdot\vec{q}_2)(\vec{\sigma}\cdot\vec{q}_1)}{\vec{q}^{\,2}} |i\rangle|^2 \tag{E.1}$$

Here the matrix element is taken between two-component Pauli spinors. The quantities f_1, f_2 can be related to the amplitudes defined in the text through substitution of the explicit representation for the four-component Dirac spinors in Prob. 13.1 (multiplied by $(E/M)^{1/2}$ to get $\bar{u}u = 1$).

$$\begin{aligned} f_1 &= \frac{[(W+M)^2 - \mu^2]}{16\pi W^2}[A + (W-M)B] \\ f_2 &= \frac{[(W-M)^2 - \mu^2]}{16\pi W^2}[-A + (W+M)B] \end{aligned} \tag{E.2}$$

The isospin structure of the amplitudes is given by

$$A_{\beta\alpha} = A^+ \delta_{\beta\alpha} + A^- \frac{1}{2}[\tau_\beta, \tau_\alpha] \tag{E.3}$$

with a similar relation for B. The indices refer to the hermitian components. The amplitude must be a second rank tensor in isospin space, and these are the only two tensors available. These amplitudes are related to those in the channels of total isospin by

$$A^+ = \frac{1}{3}(2A_{3/2} + A_{1/2}) \qquad\qquad A^- = \frac{1}{2}(A_{1/2} - A_{3/2}) \tag{E.4}$$

Similar relations hold for B.[83]

One can carry out a partial wave analysis in the C-M system. The conventional notation denotes the parity of the channel by $(-1)^{l+1}$ and the angular momentum in the channel by $j = l \pm 1/2$. Introduce

$$f_{l\pm} \equiv \frac{e^{2i\delta_{l\pm}} - 1}{2iq} = \frac{e^{i\delta_{l\pm}}\sin\delta_{l\pm}}{q} \tag{E.5}$$

Then[84]

$$\begin{aligned} f_1 &= \sum_{l=0}^{\infty} f_{l+} P'_{l+1}(x) - \sum_{l=2}^{\infty} f_{l-} P'_{l-1}(x) \\ f_2 &= \sum_{l=1}^{\infty} (f_{l-} - f_{l+}) P'_l(x) \end{aligned} \tag{E.6}$$

Here $P'_l(x) \equiv dP_l/dx$ with $x \equiv \cos\theta$.

[83] These relations are proved by taking appropriate matrix elements and using the theory of angular momenta (Prob. 20.8).

[84] These relations are derived using the analysis of Jacob and Wick (Ref. [R56]) in Probs. 20.1-5.

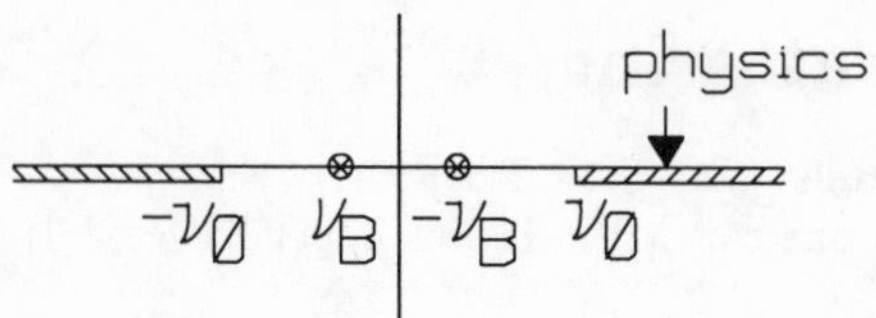

Figure E.1: Singularity structure of $\pi - N$ scattering amplitudes in the complex ν - plane at fixed κ^2.

The scalar functions A, B satisfy fixed momentum transfer dispersion relations[85]

$$
\begin{aligned}
A^{\pm}(\nu,\kappa^2) &= \frac{1}{\pi}\int_{\nu_0}^{\infty} d\nu' \mathrm{Im}\, A^{\pm}(\nu',\kappa^2)\left[\frac{1}{\nu'-\nu} \pm \frac{1}{\nu'+\nu}\right] \\
B^{\pm}(\nu,\kappa^2) &= \frac{g^2}{2M}\left[\frac{1}{\nu_B-\nu} \mp \frac{1}{\nu_B+\nu}\right] \\
&\quad +\frac{1}{\pi}\int_{\nu_0}^{\infty} d\nu' \mathrm{Im}\, B^{\pm}(\nu',\kappa^2)\left[\frac{1}{\nu'-\nu} \mp \frac{1}{\nu'+\nu}\right]
\end{aligned}
\tag{E.7}
$$

Here

$$
\nu_0 = \mu - \frac{\kappa^2}{M} \qquad\qquad \nu_B = -\left(\frac{\mu^2}{2M} + \frac{\kappa^2}{M}\right) \tag{E.8}
$$

The singularity structure of the amplitudes in the complex ν - plane at fixed κ^2 is shown in Fig. E.1.

At threshold in the C-M system $q = 0$ and $W = M + \mu$, and there is only s-wave scattering so $f_1^{\mathrm{th}} = f_{0+}$ and $f_2^{\mathrm{th}} = 0$. Thus from Eq. (E.2)

$$
f_{0+}^{\mathrm{th}} = f_1^{\mathrm{th}} = \frac{1}{4\pi(1+\mu/M)}(A + \mu B) \tag{E.9}
$$

The s-wave scattering length in $\pi - N$ scattering is defined by (note the sign)

$$
f_{0+}^{\mathrm{th}} \equiv a_{0+} \tag{E.10}
$$

Inversion of the defining relations for the partial wave amplitudes in the C-M system [Eqs. (E.2)-(E.6)] gives

$$
\begin{aligned}
f_{l\pm} &= \frac{1}{32\pi W^2}\left\{[(W+M)^2 - \mu^2][A_l + (W-M)B_l]\right. \\
&\quad \left. +[(W-M)^2 - \mu^2][-A_{l\pm1} + (W+M)B_{l\pm1}]\right\}
\end{aligned}
\tag{E.11}
$$

Here

$$
A_l(s) \equiv \int_{-1}^{1} P_l(x)A(s,t,u)dx\ ; \qquad\qquad \text{etc.} \tag{E.12}
$$

Also in the C-M system

$$
\vec{q}^{\,2} = \frac{1}{4W^2}[(W+M)^2 - \mu^2][(W-M)^2 - \mu^2] \tag{E.13}
$$

[85] See Prob. 24.5.

F The Symmetry $SU(2)_L \otimes SU(2)_R$

The generators for the isospin and chiral transformations in Section 21 are

$$\begin{aligned}\hat{T}_i &= \int d^3x \left[\psi^\dagger \frac{1}{2}\tau_i\psi - \varepsilon_{ijk}\dot{\pi}_j\pi_k\right] \\ \hat{T}_i^5 &= \int d^3x \left[\psi^\dagger \frac{1}{2}\tau_i\gamma_5\psi + \dot{\pi}_i\sigma - \dot{\sigma}\pi_i\right]\end{aligned} \tag{F.1}$$

Equations (21.18) and (21.50) give the canonical (anti-)commutation relations for the fields. It is then a basic exercise in quantum mechanics to show that the isospin generators form an $SU(2)$ algebra (Ref. [R4])

$$[\hat{T}_i, \hat{T}_j] = i\varepsilon_{ijk}\hat{T}_k \tag{F.2}$$

Now evaluate

$$[\hat{T}_i, \hat{T}_j^5] = \int d^3x \left\{\psi^\dagger[\frac{1}{2}\tau_i, \frac{1}{2}\tau_j]\gamma_5\psi - i\varepsilon_{ilj}\dot{\pi}_l\sigma - i\varepsilon_{ijl}\pi_l\dot{\sigma}\right\} = i\varepsilon_{ijk}\hat{T}_k^5 \tag{F.3}$$

In a similar fashion, for $i \neq j$, one has

$$[\hat{T}_i^5, \hat{T}_j^5] = \int d^3x \left\{\psi^\dagger[\frac{1}{2}\tau_i, \frac{1}{2}\tau_j]\gamma_5^2\psi - i\dot{\pi}_i\pi_j + i\pi_i\dot{\pi}_j\right\} = i\varepsilon_{ijk}\hat{T}_k \tag{F.4}$$

Define the linear combinations

$$\vec{T}_L \equiv \frac{1}{2}(\vec{T} + \vec{T}_5) \qquad \vec{T}_R \equiv \frac{1}{2}(\vec{T} - \vec{T}_5) \tag{F.5}$$

These may be said to correspond to left- and right-handed isospin, respectively.

It follows immediately from Eqs. (F.2)-(F.4) that

$$\begin{aligned}[\hat{T}_i^L, \hat{T}_j^L] &= i\varepsilon_{ijk}\hat{T}_k^L \\ [\hat{T}_i^R, \hat{T}_j^R] &= i\varepsilon_{ijk}\hat{T}_k^R \\ [\hat{T}_i^R, \hat{T}_j^L] &= 0\end{aligned} \tag{F.6}$$

The operators $\hat{\vec{T}}_L$ form an $SU(2)$ algebra $SU(2)_L$; the operators $\hat{\vec{T}}_R$ form an $SU(2)$ algebra $SU(2)_R$; and these generators commute with each other. The entire new set of generators thus forms the Lie algebra $SU(2)_L \bigotimes SU(2)_R$.

The following quantities are the projection operators for left- and right-handed massless Dirac particles, respectively (note Appendix N)

$$P_\downarrow = \frac{1}{2}(1 + \gamma_5) \qquad P_\uparrow = \frac{1}{2}(1 - \gamma_5) \tag{F.7}$$

They satisfy

$$\begin{aligned}P_\downarrow^2 &= P_\downarrow \qquad\qquad P_\uparrow^2 = P_\uparrow \\ P_\downarrow P_\uparrow &= P_\uparrow P_\downarrow = 0\end{aligned} \tag{F.8}$$

Define left- and right-handed Dirac fields by

$$\psi_L \equiv \frac{1}{2}(1+\gamma_5)\psi \qquad \psi_R \equiv \frac{1}{2}(1-\gamma_5)\psi \tag{F.9}$$

In addition, define the following combination of meson fields

$$\underset{\sim}{\chi} = \frac{1}{\sqrt{2}}(\sigma + i\vec{\tau}\cdot\vec{\pi}) \tag{F.10}$$

It is then a matter of straightforward algebra to verify that the lagrangian in Eq. (21.41) can be rewritten as (Prob. 21.7)

$$\begin{aligned}\mathcal{L} \;=\; & -\left[\bar{\psi}_L\gamma_\lambda\frac{D}{Dx_\lambda}\psi_L + \bar{\psi}_R\gamma_\lambda\frac{D}{Dx_\lambda}\psi_R\right] + \sqrt{2}g\left(\bar{\psi}_R\underset{\sim}{\chi}\psi_L + \bar{\psi}_L\underset{\sim}{\chi}^\dagger\psi_R\right) \\ & -V(\mathrm{tr}[\underset{\sim}{\chi}^\dagger\underset{\sim}{\chi}]) - \frac{1}{2}\mathrm{tr}\left[\left(\frac{\partial\underset{\sim}{\chi}}{\partial x_\lambda}\right)^\star\left(\frac{\partial\underset{\sim}{\chi}}{\partial x_\lambda}\right)\right] - \frac{1}{2}m_v^2V_\mu^2 - \frac{1}{4}F_{\mu\nu}F_{\mu\nu}\end{aligned} \tag{F.11}$$

Here $v_\mu^\star \equiv (\vec{v}^\dagger, +iv_0^\dagger)$.

Define the $SU(2)$ matrix $\underset{\sim}{r}$ and its infinitesimal form with $\vec{\omega} = \vec{\varepsilon} \to 0$ as

$$\underset{\sim}{r} \equiv \exp\{\frac{i}{2}\vec{\omega}\cdot\vec{\tau}\} \;\to\; 1 + \frac{i}{2}\vec{\varepsilon}\cdot\vec{\tau} \tag{F.12}$$

Since $\underset{\sim}{r}^\dagger\underset{\sim}{r} = 1$ is unitary and the trace is invariant under cyclic permutations, it follows by inspection that the above lagrangian is invariant under the global $SU(2)_R$ transformation defined by

$$\begin{aligned}\psi_R &\to \underset{\sim}{r}\psi_R \qquad\qquad \psi_L \to \psi_L \\ \underset{\sim}{\chi} &\to \underset{\sim}{r}\underset{\sim}{\chi}\end{aligned} \tag{F.13}$$

Similarly, it is invariant under the independent $SU(2)_L$ transformation defined by[86]

$$\begin{aligned}\psi_L &\to \underset{\sim}{r}\psi_L \qquad\qquad \psi_R \to \psi_R \\ \underset{\sim}{\chi} &\to \underset{\sim}{\chi}\underset{\sim}{r}^\dagger\end{aligned} \tag{F.14}$$

The infinitesimal forms of these transformations are easily seen to be (Prob. 21.7)

$$\begin{aligned}&\psi_R \to \underset{\sim}{r}\psi_R \to \left(1 + \frac{i}{2}\vec{\varepsilon}\cdot\vec{\tau}\right)\psi_R \\ &\underset{\sim}{\chi} \to \underset{\sim}{r}\underset{\sim}{\chi}(\sigma,\vec{\pi}) \to \underset{\sim}{\chi}(\sigma - \frac{1}{2}\vec{\varepsilon}\cdot\vec{\pi},\; \vec{\pi} + \frac{1}{2}\vec{\varepsilon}\sigma - \frac{1}{2}\vec{\varepsilon}\times\vec{\pi})\end{aligned} \tag{F.15}$$

This is the infinitesimal form of a particular combination of chiral and isospin transformations, both of which leave the lagrangian invariant. In the second case

$$\begin{aligned}&\psi_L \to \underset{\sim}{r}\psi_L \to \left(1 + \frac{i}{2}\vec{\varepsilon}\cdot\vec{\tau}\right)\psi_L \\ &\underset{\sim}{\chi} \to \underset{\sim}{\chi}(\sigma,\vec{\pi})\underset{\sim}{r}^\dagger \to \underset{\sim}{\chi}(\sigma + \frac{1}{2}\vec{\varepsilon}\cdot\vec{\pi},\; \vec{\pi} - \frac{1}{2}\vec{\varepsilon}\sigma - \frac{1}{2}\vec{\varepsilon}\times\vec{\pi})\end{aligned} \tag{F.16}$$

[86]Note the matrix multiplication on the right by the adjoint in the second expression.

G Chiral Transformation of Meson Lagrangian

In this appendix the algebra is developed that expresses the meson lagangian of the σ-model in Eq. (22.24) in terms of the chirally transformed meson fields. The chiral transformation for the meson fields is defined by

$$\begin{aligned}\vec{\pi} &= \mathcal{R}\vec{\pi}' \\ \phi &= \phi' + \left(\frac{f}{\mu}\right)\mathcal{R}\vec{\pi}'^2 \\ \mathcal{R} &\equiv \frac{1-2(f/\mu)\phi'}{1+(f/\mu)^2\vec{\pi}'^2}\end{aligned} \tag{G.1}$$

Here $f/\mu \equiv g/2M$. It has already been established in the text that

$$\phi^2 + \vec{\pi}^2 = \phi'^2 + \mathcal{R}\vec{\pi}'^2 \tag{G.2}$$

The result to be proven for the meson potential terms is

$$\begin{aligned}V &\equiv \frac{1}{2}(m_s^2\phi^2 + \mu^2\vec{\pi}^{\,2}) - \frac{g}{2M}(m_s^2-\mu^2)\phi(\vec{\pi}^{\,2}+\phi^2) \\ &\quad + \frac{1}{2}\left(\frac{g}{2M}\right)^2(m_s^2-\mu^2)(\vec{\pi}^{\,2}+\phi^2)^2 \\ &= \frac{1}{2}(m_s^2\phi'^2 + \mathcal{R}\mu^2\vec{\pi}'^2) + (m_s^2-\mu^2)\left[-\frac{f}{\mu}\phi'^3 + \frac{1}{2}\left(\frac{f}{\mu}\right)^2\phi'^4\right]\end{aligned} \tag{G.3}$$

In this appendix, the shorthand $f/\mu \equiv f$ will be employed in the algebra.
i. Consider first the mass terms

$$\begin{aligned}\frac{1}{2}m_s^2\phi^2 + \frac{1}{2}\mu^2\vec{\pi}^{\,2} &\equiv \frac{1}{2}m_s^2\phi'^2 + \frac{1}{2}\mathcal{R}\mu^2\vec{\pi}'^2 \\ &\quad + \underbrace{\frac{1}{2}m_s^2(\phi^2-\phi'^2) + \frac{1}{2}\mu^2(\vec{\pi}^{\,2}-\mathcal{R}\vec{\pi}'^2)} \\ &= \frac{1}{2}\mu^2\mathcal{R}(\mathcal{R}-1)\vec{\pi}'^2 + \underbrace{\frac{1}{2}m_s^2[2f\phi'\mathcal{R}\vec{\pi}'^2 + f^2\mathcal{R}^2(\vec{\pi}'^2)^2]} \\ &= \frac{1}{2}m_s^2\mathcal{R}\vec{\pi}'^2(1-\mathcal{R})\end{aligned} \tag{G.4}$$

Thus

$$\frac{1}{2}m_s^2\phi^2 + \frac{1}{2}\mu^2\vec{\pi}^{\,2} = \frac{1}{2}m_s^2\phi'^2 + \frac{1}{2}\mathcal{R}\mu^2\vec{\pi}'^2 + \frac{1}{2}(m_s^2-\mu^2)\vec{\pi}'^2\mathcal{R}(1-\mathcal{R}) \tag{G.5}$$

ii. Now combine all terms in $(m_s^2-\mu^2)$; their coefficient c_1 is

$$\begin{aligned}c_1 &= \frac{1}{2}\mathcal{R}(1-\mathcal{R})\vec{\pi}'^2 \underbrace{-f(\phi'+f\mathcal{R}\vec{\pi}'^2)(\phi'^2+\mathcal{R}\vec{\pi}'^2) + \frac{1}{2}f^2(\phi'^2+\mathcal{R}\vec{\pi}'^2)^2} \\ &= -f\phi'^3 + \frac{1}{2}f^2\phi'^4 \underbrace{-f\mathcal{R}\vec{\pi}'^2\phi' - \frac{1}{2}f^2\mathcal{R}^2(\pi'^2)^2} \\ &= -\frac{1}{2}\mathcal{R}(1-\mathcal{R})\vec{\pi}'^2\end{aligned} \tag{G.6}$$

This proves Eq. (G.3).

The result to be proven for the meson kinetic energy term is

$$-2T = \left(\frac{\partial\phi}{\partial x_\lambda}\right)^2 + \left(\frac{\partial\vec{\pi}}{\partial x_\lambda}\right)^2 = \left(\frac{\partial\phi'}{\partial x_\lambda}\right)^2 + \mathcal{R}^2\left(\frac{\partial\vec{\pi}'}{\partial x_\lambda}\right)^2 \tag{G.7}$$

i. First differentiate the defining relations in Eq. (G.1)

$$\begin{aligned}\frac{\partial\phi}{\partial x_\lambda} &= \frac{\partial\phi'}{\partial x_\lambda} + 2f\mathcal{R}\vec{\pi}'\cdot\frac{\partial\vec{\pi}'}{\partial x_\lambda} \\ &\quad + f\vec{\pi}'^2\left[\frac{-2f}{1+f^2\vec{\pi}'^2}\frac{\partial\phi'}{\partial x_\lambda} - \frac{(1-2f\phi')}{(1+f^2\vec{\pi}'^2)^2}2f^2\vec{\pi}'\cdot\frac{\partial\vec{\pi}'}{\partial x_\lambda}\right] \\ &= \left[\frac{1-f^2\vec{\pi}'^2}{1+f^2\vec{\pi}'^2}\right]\frac{\partial\phi'}{\partial x_\lambda} + \frac{2f\mathcal{R}}{1+f^2\vec{\pi}'^2}\vec{\pi}'\cdot\frac{\partial\vec{\pi}'}{\partial x_\lambda}\end{aligned} \tag{G.8}$$

Also

$$\begin{aligned}\frac{\partial\vec{\pi}}{\partial x_\lambda} &= \mathcal{R}\frac{\partial\vec{\pi}'}{\partial x_\lambda} + \vec{\pi}'\left[-\frac{2f}{1+f^2\vec{\pi}'^2}\frac{\partial\phi'}{\partial x_\lambda} - \frac{(1-2f\phi')}{(1+f^2\vec{\pi}'^2)^2}2f^2\vec{\pi}'\cdot\frac{\partial\vec{\pi}'}{\partial x_\lambda}\right] \\ &= \mathcal{R}\frac{\partial\vec{\pi}'}{\partial x_\lambda} - \frac{2f}{1+f^2\vec{\pi}'^2}\left(\frac{\partial\phi'}{\partial x_\lambda} + f\mathcal{R}\vec{\pi}'\cdot\frac{\partial\vec{\pi}'}{\partial x_\lambda}\right)\vec{\pi}'\end{aligned} \tag{G.9}$$

ii. Now compute the kinetic energy by collecting the various terms in the square of Eqs. (G.8) and (G.9). In addition to the terms appearing in the answer on the right-hand side of Eq. (G.7), there are two cross terms whose coefficients vanish. The coefficients (c_2, c_3) of the terms in $\frac{\partial\phi'}{\partial x_\lambda}\vec{\pi}'\cdot\frac{\partial\vec{\pi}'}{\partial x_\lambda}$ and $(\vec{\pi}'\cdot\frac{\partial\vec{\pi}'}{\partial x_\lambda})(\vec{\pi}'\cdot\frac{\partial\vec{\pi}'}{\partial x_\lambda})$ respectively are

$$\begin{aligned}c_2 &= \frac{1}{(1+f^2\vec{\pi}'^2)^2}\left[4f\mathcal{R}(1-f^2\vec{\pi}'^2) - 4f\mathcal{R}(1+f^2\vec{\pi}'^2) + 8f^3\mathcal{R}\vec{\pi}'^2\right] \equiv 0 \\ c_3 &= \frac{1}{(1+f^2\vec{\pi}'^2)^2}\left[4f^2\mathcal{R}^2 - 4f^2\mathcal{R}^2(1+f^2\vec{\pi}'^2) + 4f^2\vec{\pi}'^2(f^2\mathcal{R}^2)\right] \equiv 0\end{aligned} \tag{G.10}$$

Since both of these coefficients vanish identically, the result in Eq. (G.7) holds.

H $\pi - \pi$ Scattering

In this appendix we discuss some general phenomenology of $\pi - \pi$ scattering. The process is illustrated in Fig. H.1 The general form of the S-matrix is given by

$$S_{fi} = -\frac{(2\pi)^4 i}{\Omega^2}\delta^{(4)}(q_1 + q_2 - q_3 - q_4)\frac{1}{\sqrt{2^4\omega_1\omega_2\omega_3\omega_4}}T_{fi} \tag{H.1}$$

As before, the Lorentz invariant kinematic variables are defined by

$$s = -(q_1+q_2)^2 \qquad t = -(q_1-q_3)^2 \qquad u = -(q_1-q_4)^2 \tag{H.2}$$

They satisfy

$$s + t + u = 4\mu^2 \tag{H.3}$$

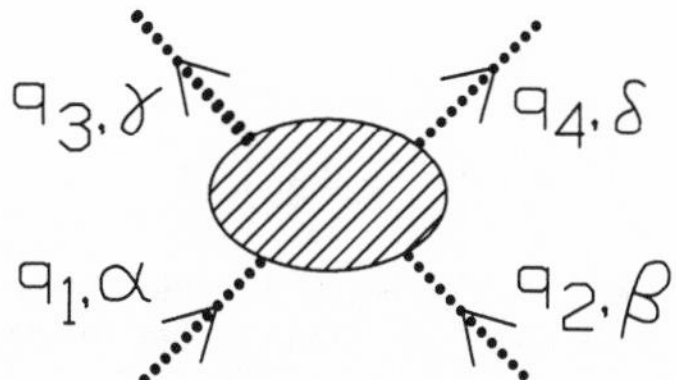

Figure H.1: $\pi - \pi$ scattering.

In the C-M system (Fig. H.2)

$$\begin{aligned} s &\equiv W^2 = 4(\vec{q}^{\,2} + \mu^2) = 4\omega^2 \\ t &= -2\vec{q}^{\,2}(1 - \cos\theta) \end{aligned} \tag{H.4}$$

The cross section is given by

$$\begin{aligned} d\sigma_{\rm fi} &= \frac{d\omega_{\rm fi}}{\rm Flux} \\ &= \frac{2\pi}{\Omega^2}\delta(W_f - W_i)\frac{|T_{\rm fi}|^2}{(2\omega)^4}\frac{d\rho_f}{\rm Flux} \end{aligned} \tag{H.5}$$

The incident flux is given by

$$\text{Flux} = \frac{1}{\Omega}2\frac{q}{\omega} \tag{H.6}$$

The density of states is

$$d\rho_f = \frac{\Omega}{(2\pi)^3}d^3q \tag{H.7}$$

Since the total energies are $W_f = 2(\vec{q}^{\,2} + \mu^2)^{1/2}$ and $W_i = 2\omega$, one has

$$\begin{aligned} \frac{dW_f}{dq} &= 2\frac{q}{\omega} \\ \frac{d\rho_f}{dW_f} &= \frac{\Omega q^2 d\Omega_q}{(2\pi)^3}\frac{1}{2q/\omega} = \frac{\Omega}{2}\frac{q\omega}{(2\pi)^3}d\Omega_q \end{aligned} \tag{H.8}$$

Thus the differential cross section in the C-M system is given by

$$\left(\frac{d\sigma}{d\Omega}\right)_{\rm CM} = \frac{1}{(2\pi)^2}\frac{1}{2^6\omega^2}|T_{\rm fi}|^2 \equiv |f_{\rm CM}|^2 \tag{H.9}$$

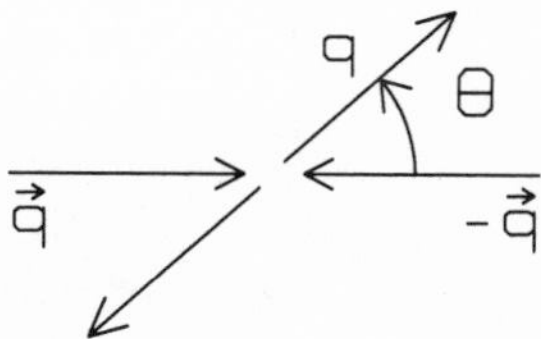

Figure H.2: $\pi - \pi$ scattering in the center-of-momentum (C-M) system.

The scattering amplitude is thus[87]

$$f_{\rm CM} = -\frac{1}{8\pi W} T_{\rm fi} \tag{H.10}$$

One still has a second rank tensor in isospin space $T_{\gamma\delta;\alpha\beta}$ from which appropriate matrix elements must be evaluated. In terms of hermitian components the scattering amplitude in Fig. H.1 must have the form

$$T_{\gamma\delta;\,\alpha\beta} = T^{(s)}\delta_{\gamma\delta}\delta_{\alpha\beta} + T^{(u)}\delta_{\gamma\beta}\delta_{\delta\alpha} + T^{(t)}\delta_{\gamma\alpha}\delta_{\delta\beta} \tag{H.11}$$

The amplitudes in states of given total isospin are then

$$\begin{aligned} T^0 &= 3T^{(s)} + T^{(u)} + T^{(t)} \\ T^1 &= T^{(u)} - T^{(t)} \\ T^2 &= T^{(u)} + T^{(t)} \end{aligned} \tag{H.12}$$

[87]The sign follows from unitarity (Probs.20.1-4).

Part III
STRONG COUPLING QCD

27

QCD — A REVIEW

Quantum chromodynamics (QCD) is the theory of the strong interactions binding quarks and gluons into observed hadrons (baryons and mesons), and, in turn, into observed nuclei. As such, it is truly the theory of the structure of matter. A brief introduction to QCD was given in Section 19. In this part of the book we turn our attention to developing the implications of the theory of QCD. Of particular interest is the strong-coupling regime appropriate to nuclear physics. First, however, it is necessary to further develop a basic understanding of the theory. We do that in this section with a review and summary, starting with the classic work of Yang and Mills (Ref. [Q1]) on nonabelian local gauge theories.

Yang-Mills Theory — A Review. Start with isospin invariance [$SU(2)$], which is the case originally studied by Yang and Mills in Ref. [Q1]. A discussion of isospin invariance is contained in Section 21. In direct analogy with angular momentum, the isospin operator $\hat{\mathbf{T}}$ is the generator of isospin transformations, and the operator $\hat{R}$ producing the finite, global isospin transformation through the angle $\boldsymbol{\theta} = \mathbf{n}\theta$ is obtained through exponentiation. Its effect on the field $\underline{\phi}$ depends on the particular isospin representation to which that field belongs.

$$
\begin{aligned}
\hat{R} &= e^{i\hat{\mathbf{T}}\cdot\boldsymbol{\theta}} \\
\hat{R}\underline{\phi}\hat{R}^{-1} \equiv \underline{\phi}' &= [e^{-i\underline{\mathbf{T}}\cdot\boldsymbol{\theta}}]\underline{\phi}
\end{aligned}
\tag{27.1}
$$

Here $\underline{\mathbf{T}}$ is a hermitian matrix representation of the generators $\hat{\mathbf{T}}$.[1] $\underline{\phi}$ is a column vector composed of a set of fields that mix among themselves under isospin transformations. Two examples consist of the previously studied nucleon and pion fields

$$
\underline{\psi} = \begin{pmatrix} \psi_p \\ \psi_n \end{pmatrix} = \begin{pmatrix} p \\ n \end{pmatrix}; \qquad \underline{\mathbf{T}} = \frac{1}{2}\boldsymbol{\tau}
$$

$$
\underline{\pi} = \begin{pmatrix} \pi_1 \\ \pi_2 \\ \pi_3 \end{pmatrix}; \qquad \underline{\mathbf{T}} = \underline{\mathbf{t}} \tag{27.2}
$$

[1]The generators form a Lie algebra

$$
[\hat{T}_i, \hat{T}_j] = i\varepsilon_{ijk}\hat{T}_k
$$

The quantities ε_{ijk} are the *structure constants* of the Lie algebra [here $SU(2)$].

Here the matrices $\underline{\mathrm{t}}$ are defined by $(t_i)_{jk} \equiv -i\varepsilon_{ijk}$. Global isospin invariance implies that the lagrangian is left invariant under this transformation.

$$\mathcal{L} \longrightarrow \mathcal{L}' = \mathcal{L} \tag{27.3}$$

Specification to infinitesimal transformations $\boldsymbol{\theta} \to 0$ reduces Eqs. (27.1) and (27.3) to

$$\begin{aligned} \delta\underline{\phi} &= -i\boldsymbol{\theta}\cdot\underline{\mathbf{T}}\,\underline{\phi} \\ \delta\mathcal{L} &= 0 \end{aligned} \tag{27.4}$$

The goal is to now convert this into a *local* gauge invariance where the isospin transformation angle $\boldsymbol{\theta}$ instead of being an overall constant (global invariance) may be a *function of the position in space-time* $\boldsymbol{\theta}(x)$. The transformation of the fields now takes the form

$$\begin{aligned} \underline{\phi}(x) \to \underline{\phi}'(x) &= [e^{-i\underline{\mathbf{T}}\cdot\boldsymbol{\theta}(x)}]\,\underline{\phi}(x) \\ &\equiv \underline{\mathrm{U}}[\boldsymbol{\theta}(x)]\,\underline{\phi}(x) \end{aligned} \tag{27.5}$$

The matrix $\underline{\mathrm{U}}[\boldsymbol{\theta}(x)]$ defined in this expression is unitary. The nucleon with isospin 1/2 plays a special role here since it forms a basis for the *fundamental* representation of the group $SU(2)$

$$\hat{R}\underline{\psi}\hat{R}^{-1} = [e^{-\frac{i}{2}\underline{\boldsymbol{\tau}}\cdot\boldsymbol{\theta}}]\,\underline{\psi} \tag{27.6}$$

In this case $\underline{\mathrm{U}}[\boldsymbol{\theta}(x)]$ is a three parameter, unitary, unimodular, 2×2 matrix.

Now if the lagrangian is built out of bilinear forms $\underline{\phi}^\dagger\underline{\phi}$, it will be invariant under the local transformation in Eq. (27.5)

$$\underline{\phi}'^\dagger\underline{\phi}' = \underline{\phi}^\dagger\underline{\mathrm{U}}^\dagger\underline{\mathrm{U}}\underline{\phi} = \underline{\phi}^\dagger\underline{\phi} \tag{27.7}$$

What about the gradient terms? To make them invariant under this local gauge transformation one introduces a *covariant derivative* such that

$$\frac{D'}{Dx_\mu}\underline{\phi}' = \underline{\mathrm{U}}(\boldsymbol{\theta})\frac{D}{Dx_\mu}\underline{\phi} \tag{27.8}$$

The same argument for invariance as in Eq. (27.7) can then be employed; if the kinetic energy term in $\partial\underline{\phi}/\partial x_\mu$ appears in the lagrangian $\mathcal{L}$ only in a bilinear combination of $D\underline{\phi}/Dx_\mu$, then $\mathcal{L}$ will be *invariant* under local gauge transformations.

To ensure Eq. (27.8), introduce *vector fields*, one for each generator

$$A_\mu^i(x)\;; \qquad\qquad i = 1, 2, 3 \tag{27.9}$$

Define the covariant derivative as

$$\frac{D}{Dx_\mu}\underline{\phi} \equiv \left[\frac{\partial}{\partial x_\mu} - ig\underline{\mathbf{T}}\cdot\mathbf{A}_\mu(x)\right]\underline{\phi} \tag{27.10}$$

Now let these new vector fields also transform under the local gauge transformation. How must these fields transform to ensure Eq. (27.8)? Use

$$\frac{\partial}{\partial x_\mu}\underline{\phi}'(x) = \underline{U}(\boldsymbol{\theta})\frac{\partial \underline{\phi}}{\partial x_\mu} + \left[\frac{\partial}{\partial x_\mu}\underline{U}(\boldsymbol{\theta})\right]\underline{\phi} \tag{27.11}$$

one wants

$$\left[\frac{\partial}{\partial x_\mu} - ig\underline{\mathbf{T}}\cdot\mathbf{A}'_\mu(x)\right]\underline{\phi}'(x) = \underline{U}\left[\frac{\partial}{\partial x_\mu} - ig\underline{\mathbf{T}}\cdot\mathbf{A}_\mu(x)\right]\underline{\phi}(x) \tag{27.12}$$

Substitution of Eq. (27.11) into Eq. (27.12) yields

$$\left[\frac{\partial}{\partial x_\mu}\underline{U}(\boldsymbol{\theta}) - ig\underline{\mathbf{T}}\cdot\mathbf{A}'_\mu(x)\underline{U}\right]\underline{\phi} = \underline{U}\left[-ig\underline{\mathbf{T}}\cdot\mathbf{A}_\mu(x)\right]\underline{\phi} \tag{27.13}$$

or

$$\frac{\partial}{\partial x_\mu}\underline{U}(\boldsymbol{\theta}) - ig\underline{\mathbf{T}}\cdot\mathbf{A}'_\mu\underline{U} = \underline{U}(-ig\underline{\mathbf{T}}\cdot\mathbf{A}_\mu) \tag{27.14}$$

or, finally

$$\underline{\mathbf{T}}\cdot\mathbf{A}'_\mu = \underline{U}(\underline{\mathbf{T}}\cdot\mathbf{A}_\mu)\underline{U}^{-1} - \frac{i}{g}\left[\frac{\partial}{\partial x_\mu}\underline{U}(\boldsymbol{\theta})\right]\underline{U}^{-1} \tag{27.15}$$

This result appears to depend on the representation $\underline{\mathbf{T}}$; in fact, it depends only on the commutation rules, that is, on the particular Lie algebra under consideration. To understand this, go to infinitesimals $\boldsymbol{\theta} \to 0$

$$\begin{aligned}\underline{\mathbf{T}}\cdot\mathbf{A}'_\mu &= (1 - i\underline{\mathbf{T}}\cdot\boldsymbol{\theta} + \cdots)(\underline{\mathbf{T}}\cdot\mathbf{A}_\mu)(1 + i\underline{\mathbf{T}}\cdot\boldsymbol{\theta} + \cdots) - \frac{i}{g}\left[-i\underline{\mathbf{T}}\cdot\frac{\partial\boldsymbol{\theta}}{\partial x_\mu}\right] \\ &= \underline{\mathbf{T}}\cdot\mathbf{A}_\mu - i[\underline{T}^i, \underline{T}^j]\theta^i A^j_\mu - \frac{1}{g}\underline{\mathbf{T}}\cdot\frac{\partial\boldsymbol{\theta}}{\partial x_\mu}\end{aligned} \tag{27.16}$$

The matrices $\underline{\mathbf{T}}$ provide a representation of the commutation rules, which in the case of $SU(2)$ gives $[\underline{T}^i, \underline{T}^j] = i\varepsilon_{ijk}\underline{T}^k$. Since the matrices $\underline{T}^i$ are linearly independent, one can equate coefficients and hence the infinitesimal transformation law for the vector fields becomes

$$A'^i_\mu = A^i_\mu - \frac{1}{g}\frac{\partial\theta^i}{\partial x_\mu} + \varepsilon_{ijk}\theta^j A^k_\mu \tag{27.17}$$

More generally, the ε_{ijk} are replaced by the structure constants of the particular group under discussion. For $SU(2)$ with nucleon and pion fields, the covariant derivatives and transformation law for the vector fields can be conveniently expressed in a vector notation

$$\begin{aligned}\frac{D}{Dx_\mu}\underline{\psi} &= \left[\frac{\partial}{\partial x_\mu} - \frac{i}{2}g\underline{\tau}\cdot\mathbf{A}_\mu(x)\right]\underline{\psi} \\ \frac{D}{Dx_\mu}\boldsymbol{\phi} &= \left(\frac{\partial}{\partial x_\mu} + g\mathbf{A}_\mu\times\right)\boldsymbol{\phi} \\ \delta\mathbf{A}_\mu &= -\frac{1}{g}\frac{\partial\boldsymbol{\theta}}{\partial x_\mu} + \boldsymbol{\theta}\times\mathbf{A}_\mu\,; \qquad \boldsymbol{\theta}\to 0\end{aligned} \tag{27.18}$$

What about the kinetic energy term for the vector mesons? Can we construct a term bilinear in derivatives of the vector meson field that is also locally gauge invariant? The answer to this question was given by Yang and Mills. Define a vector meson field tensor by the following relation

$$\mathcal{F}^i_{\mu\nu} = \frac{\partial A^i_\nu}{\partial x_\mu} - \frac{\partial A^i_\mu}{\partial x_\nu} + g\varepsilon_{ijk}A^j_\mu A^k_\nu \tag{27.19}$$

Again, more generally, the ε_{ijk} of $SU(2)$ are replaced by the structure constants of the group. Take, in analogy to QED, a kinetic energy term of the form

$$\mathcal{L}_{\text{KE}} = -\frac{1}{4}\mathcal{F}^i_{\mu\nu}\mathcal{F}^i_{\mu\nu} \tag{27.20}$$

In vector notation these relations can be written

$$\begin{aligned}\mathbf{F}_{\mu\nu} &= \frac{\partial \mathbf{A}_\nu}{\partial x_\mu} - \frac{\partial \mathbf{A}_\mu}{\partial x_\nu} + g\mathbf{A}_\mu \times \mathbf{A}_\nu \\ \mathcal{L}_{\text{KE}} &= -\frac{1}{4}\mathbf{F}_{\mu\nu}\cdot\mathbf{F}_{\mu\nu}\end{aligned} \tag{27.21}$$

With these definitions, it is then true that for infinitesimal transformations the change in the vector meson field tensor is perpendicular to the field tensor itself[2]

$$\delta\mathbf{F}_{\mu\nu} = \boldsymbol{\theta}\times\mathbf{F}_{\mu\nu}\ ; \qquad \boldsymbol{\theta}\to 0 \tag{27.22}$$

This implies that the kinetic energy term is unchanged under this infinitesimal transformation

$$\delta\mathcal{L}_{\text{KE}} = 0 \tag{27.23}$$

One would normally proceed to add a mass term for the vector mesons to the lagrangian

$$\delta\mathcal{L}_{\text{mass}} = -\frac{1}{2}m_A^2\mathbf{A}_\mu\cdot\mathbf{A}_\mu \tag{27.24}$$

Such a term clearly changes under the local gauge transformation in Eq. (27.18), and hence the only way to preserve local gauge invariance is to demand that the additional vector meson fields be *massless*

$$m_A^2 = 0 \tag{27.25}$$

These results can be combined to construct, for example, a model lagrangian for pions and nucleons that is locally gauge invariant under isospin transformations [a nonabelian gauge theory built on the group $SU(2)$]

$$\begin{aligned}\mathcal{L} &= -\bar{\underline{\psi}}\left(\gamma_\mu\frac{D}{Dx_\mu} + M - ig_\pi\gamma_5\boldsymbol{\tau}\cdot\boldsymbol{\phi}\right)\underline{\psi} - \frac{1}{4}\mathbf{F}_{\mu\nu}\cdot\mathbf{F}_{\mu\nu} \\ &\quad -\frac{1}{2}\left[\left(\frac{D\boldsymbol{\phi}}{Dx_\mu}\right)^\star\cdot\left(\frac{D\boldsymbol{\phi}}{Dx_\mu}\right) + m_\pi^2\boldsymbol{\phi}\cdot\boldsymbol{\phi}\right]\end{aligned} \tag{27.26}$$

[2]We leave the proof of this relation in $SU(2)$ as an exercise for the reader in Prob. 27.1 (see Ref. [R4]).

Here $v_\mu^\star \equiv (\mathbf{v}^\dagger, iv_0^\dagger)$; the metric is not complex conjugated under this * operation.

The discussion of nonabelian Yang-Mills theories can be specialized to the case of the abelian theory of QED where the invariance is to local phase transformations of the fields and the invariance group is simply $U(1)$. In this case one has the correspondence:

1. The finite transformation operator and unitary representation become

$$\hat{R} = e^{i\hat{Q}\theta} \; ; \qquad\qquad U = e^{-iq\theta} \tag{27.27}$$

Here $\hat{Q}$ is the electromagnetic charge operator and q is the charge carried by the field. Since there is only one generator $\hat{Q}$, the structure constants vanish.

2. A single vector field A_μ is introduced, and the covariant derivative is defined by

$$\frac{D\psi}{Dx_\mu} = \left[\frac{\partial}{\partial x_\mu} - ieqA_\mu(x)\right]\psi \tag{27.28}$$

3. Under local gauge transformations the vector field transforms as

$$A'_\mu = A_\mu - \frac{1}{e}\frac{\partial\theta}{\partial x_\mu} \tag{27.29}$$

4. The field tensor for this vector field is defined by

$$F_{\mu\nu} = \frac{\partial A_\nu}{\partial x_\mu} - \frac{\partial A_\mu}{\partial x_\nu} \tag{27.30}$$

5. The vector field must be massless $m_A^2 = 0$ to maintain local gauge invariance.

Remarkably enough, these arguments, used to construct a lagrangian that is invariant under local phase transformations, lead to quantum electrodynamics, the most accurate physical theory we have.

Quarks and Color. As discussed in Section 19, quarks come in several *flavors* $[u, d, s, c, b, (t), \ldots]$, each of which has a set of quantum numbers out of which the quantum numbers of the observed baryons are deduced from (qqq) triplets and the observed mesons from $(q\bar{q})$ pairs. In addition, the quarks are assigned an additional internal quantum number called *color*, which can take three values $(red, green, blue)$. The lightest mass quark fields will be represented as follows:[3]

$$\begin{aligned} \psi \; &= \; \begin{pmatrix} u \\ d \\ s \\ c \end{pmatrix} \longrightarrow \begin{pmatrix} u_R & u_G & u_B \\ d_R & d_G & d_B \\ s_R & s_G & s_B \\ c_R & c_G & c_B \end{pmatrix} \\ &\equiv \; (\psi_R, \psi_G, \psi_B) \equiv \psi_i \; ; \qquad i = R, G, B \end{aligned} \tag{27.31}$$

[3]The extension to any number of flavors is evident.

Define the column vector $\underline{\psi}$ by

$$\underline{\psi} \equiv \begin{pmatrix} \psi_R \\ \psi_G \\ \psi_B \end{pmatrix} \tag{27.32}$$

This is a very compact notation; each ψ_i contains many flavors as indicated in Eq. (27.31), and each quark flavor in turn represents a four-component Dirac field, for example

$$\psi_R = \begin{pmatrix} u_R \\ d_R \\ s_R \\ c_R \end{pmatrix} \quad ; \quad u_R = \begin{pmatrix} u_1 \\ u_2 \\ u_3 \\ u_4 \end{pmatrix}_R \quad ; \quad \text{etc.} \tag{27.33}$$

The lagrangian for the free quark fields can now be written compactly as

$$\mathcal{L} = -\bar{\underline{\psi}} \left(\gamma_\mu \frac{\partial}{\partial x_\mu} + \underline{\mathrm{M}} \right) \underline{\psi} \tag{27.34}$$

Here the mass term is the unit matrix with respect to color

$$\underline{\mathrm{M}} = \begin{pmatrix} \underline{\mathrm{m}} & & \\ & \underline{\mathrm{m}} & \\ & & \underline{\mathrm{m}} \end{pmatrix} \tag{27.35}$$

It may be *anything* with respect to flavor, for example,

$$\underline{\mathrm{m}} = \begin{pmatrix} m_u & & & \\ & m_d & & \\ & & m_s & \\ & & & m_c \end{pmatrix} \tag{27.36}$$

The lagrangian in Eq. (27.34) has a *global* invariance with respect to unitary transformations mixing the three internal color variables $[SU(3)]$. We denote the generators of this transformation by $\hat{G}^a$ with $a = 1, \ldots, 8$ and the eight parameters characterizing a three-by-three unitary, unimodular matrix by θ^a with $a = 1, \ldots, 8$. There are eight three-by-three, traceless, hermitian, Gell-Mann matrices $\underline{\lambda}_a$ — the analogues of the Pauli matrices.[4] The operator producing the finite color transformation is then given by

$$\hat{R} = e^{i\theta^a \hat{G}^a} \tag{27.37}$$

[4]These matrices satisfy the Lie algebra of $SU(3)$, the same algebra as satisfied by the generators

$$[\frac{1}{2}\underline{\lambda}^a, \frac{1}{2}\underline{\lambda}^b] = if^{abc}\frac{1}{2}\underline{\lambda}^c$$

It has the following effect on the quark field

$$\hat{R}\underline{\psi}\hat{R}^{-1} \;=\; \underline{U}(\theta)\underline{\psi} \;=\; \left[e^{-\frac{i}{2}\underline{\lambda}^a\theta^a}\right]\underline{\psi} \qquad (27.38)$$

Latin indices will now run from $1, \ldots, 8$, and repeated Latin indices are summed. The transformation in Eq. (27.38) with constant, finite θ^a leaves the lagrangian in Eq. (27.34) unchanged. Here $\underline{U}(\theta)$ is a unitary, unimodular three-by-three matrix, and the quark field in Eq. (27.38) forms a basis for the fundamental representation of $SU(3)$. The symmetry is with respect to color.

One can now make this global color invariance a *local* invariance where the transformation $\theta^a(x)$ can vary from point to point in space-time by using the theory developed by Yang and Mills:

1. Introduce massless vector meson fields, one for each generator

$$A_\mu^a(x)\;; \qquad\qquad a = 1, \ldots, 8 \qquad (27.39)$$

These vector mesons are known as *gluons*.

2. Define the covariant derivative by

$$\frac{D}{Dx_\mu}\underline{\psi} = \left[\frac{\partial}{\partial x_\mu} - \frac{i}{2}g\underline{\lambda}^a A_\mu^a(x)\right]\underline{\psi} \qquad (27.40)$$

3. Define the field tensor for the vector meson fields as

$$\mathcal{F}_{\mu\nu}^a = \frac{\partial A_\nu^a}{\partial x_\mu} - \frac{\partial A_\nu^a}{\partial x_\mu} + g f^{abc} A_\mu^b A_\nu^c \qquad (27.41)$$

Here f^{abc} are the structure constants of $SU(3)$.

4. Under infinitesimal local gauge transformations $\theta^a \to 0$ the vector meson fields and the field tensor transform according to

$$\begin{aligned} \delta A_\mu^a &= -\frac{1}{g}\frac{\partial\theta^a}{\partial x_\mu} + f^{abc}\theta^b A_\mu^c \\ \delta\mathcal{F}_{\mu\nu}^a &= f^{abc}\theta^b\mathcal{F}_{\mu\nu}^c\;; \qquad\qquad \theta^a \to 0 \end{aligned} \qquad (27.42)$$

Here the f^{abc} are the structure constants of the group; they are antisymmetric in the indices (abc). The matrices $(\lambda^a)_{ij}$ for $a = 1, \ldots, 8$ are given in order by

$$\begin{pmatrix} & 1 & \\ 1 & & \\ & & \end{pmatrix}\begin{pmatrix} & -i & \\ i & & \\ & & \end{pmatrix}\begin{pmatrix} 1 & & \\ & -1 & \\ & & \end{pmatrix}\begin{pmatrix} & & 1 \\ & & \\ 1 & & \end{pmatrix}\begin{pmatrix} & & -i \\ & & \\ i & & \end{pmatrix}$$

$$\begin{pmatrix} & & \\ & & 1 \\ & 1 & \end{pmatrix}\begin{pmatrix} & & \\ & & -i \\ & i & \end{pmatrix}\begin{pmatrix} 1/\sqrt{3} & & \\ & 1/\sqrt{3} & \\ & & -2/\sqrt{3} \end{pmatrix}$$

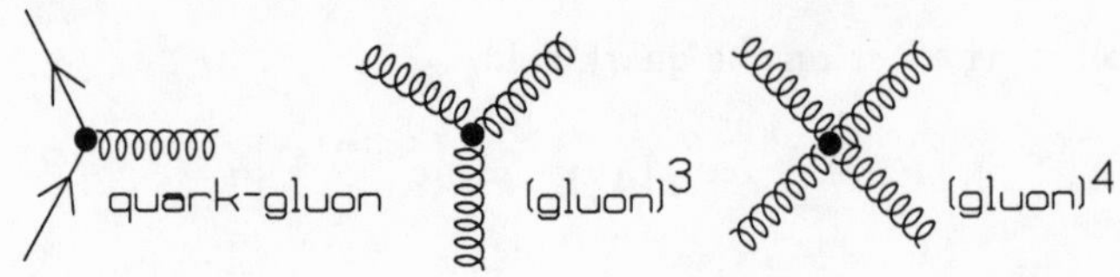

Figure 27.1: Processes described by the interaction term in the QCD lagrangian.

5. A combination of these results leads to the *lagrangian of QCD*

$$\mathcal{L}_{\rm QCD} = -\bar{\underset{\sim}{\psi}}\left\{\gamma_\mu\left[\frac{\partial}{\partial x_\mu} - \frac{i}{2}g\underset{\sim}{\lambda}^a A^a_\mu(x)\right] + \underset{\sim}{\rm M}\right\}\underset{\sim}{\psi} - \frac{1}{4}\mathcal{F}^a_{\mu\nu}\mathcal{F}^a_{\mu\nu} \quad (27.43)$$

The result that this lagrangian leads to asymptotic freedom is due to Gross and Wilczek (Refs. [Q2, Q3]) and Politzer (Refs. [Q4, Q5]); see also Gell-Mann and Fritzsch (Refs. [Q6, Q7]). References [Q8, Q9, Q10] contain good background material on QCD.

The lagrangian in Eq. (27.43) can be written out explicitly in powers of the coupling constant g

$$\begin{aligned}
\mathcal{L}_{\rm QCD} &= \mathcal{L}_0 + \mathcal{L}_1 + \mathcal{L}_2 \\
\mathcal{L}_0 &= -\bar{\underset{\sim}{\psi}}\left(\gamma_\mu\frac{\partial}{\partial x_\mu} + \underset{\sim}{\rm M}\right)\underset{\sim}{\psi} - \frac{1}{4}F^a_{\mu\nu}F^a_{\mu\nu} \\
\mathcal{L}_1 &= \frac{i}{2}g\bar{\underset{\sim}{\psi}}\gamma_\mu\underset{\sim}{\lambda}^a\underset{\sim}{\psi}A^a_\mu(x) - \frac{g}{2}f^{abc}F^a_{\mu\nu}A^b_\mu A^c_\nu \\
\mathcal{L}_2 &= -\frac{g^2}{4}f^{abc}f^{ade}A^b_\mu A^c_\nu A^d_\mu A^e_\nu
\end{aligned} \quad (27.44)$$

Here

$$F^a_{\mu\nu} \equiv \frac{\partial A^a_\nu}{\partial x_\mu} - \frac{\partial A^a_\mu}{\partial x_\nu} \quad (27.45)$$

The various processes described by the interaction terms in this lagrangian are illustrated in Fig. 27.1.

To obtain further insight into these results, it is useful to write the Yukawa interaction between the quarks and gluons in more detail. Recall, for example, the structure of the first two $\underset{\sim}{\lambda}^a$ matrices

$$\underset{\sim}{\lambda}^1 = \begin{pmatrix} & 1 & \\ 1 & & \\ & & \end{pmatrix} \qquad \underset{\sim}{\lambda}^2 = \begin{pmatrix} & -i & \\ i & & \\ & & \end{pmatrix} \quad (27.46)$$

These matrices connect the (R, G) quarks, and with explicit identification of the flavor components of the color fields, it is evident that this interaction contains the individual processes illustrated in Fig. 27.2. The quarks interact here by

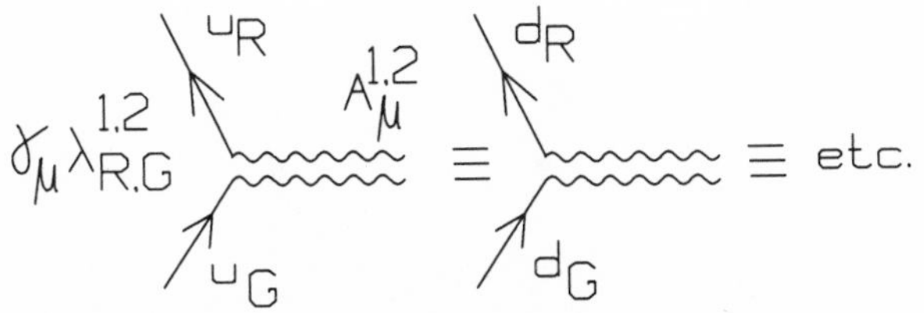

Figure 27.2: Individual processes described by the quark-gluon Yukawa coupling in QCD.

changing their color, which in turn is carried off by the gluons; the flavor of the quarks is unchanged and all flavors of quarks have an identical color coupling. If the gluons are represented with double lines connected to the incoming and outgoing quark lines, respectively, and a color assigned to each line as indicated in this figure, then color can be viewed as running continuously through a Feynman diagram built from these components. The Euler-Lagrange equations following from the QCD lagrangian provide further insight. They are readily derived to be the following

$$\left\{\gamma_\mu \left[\frac{\partial}{\partial x_\mu} - \frac{i}{2} g \underset{\sim}{\lambda}^a A_\mu^a(x)\right] + \underset{\sim}{\mathrm{M}}\right\} \underset{\sim}{\psi} = 0$$

$$\bar{\underset{\sim}{\psi}} \left\{\gamma_\mu \left[\frac{\overleftarrow{\partial}}{\partial x_\mu} + \frac{i}{2} g \underset{\sim}{\lambda}^a A_\mu^a(x)\right] - \underset{\sim}{\mathrm{M}}\right\} = 0$$

$$\frac{\partial \mathcal{F}_{\mu\nu}^a}{\partial x_\nu} = \frac{i}{2} g \bar{\underset{\sim}{\psi}} \gamma_\mu \underset{\sim}{\lambda}^a \underset{\sim}{\psi} + g f^{abc} \mathcal{F}_{\mu\nu}^b A_\nu^c \tag{27.47}$$

It follows immediately from these equations of motion that any current built out of quark fields and a unit matrix with respect to color is *conserved.*

$$\frac{\partial}{\partial x_\mu}\left(\frac{i}{3} \bar{\underset{\sim}{\psi}} \gamma_\mu \underset{\sim}{\psi}\right) = 0 \; ; \qquad \text{baryon current}$$

$$\frac{\partial}{\partial x_\mu}\left(i \bar{\underset{\sim}{\psi}} \gamma_\mu \underset{\sim}{\Sigma} \underset{\sim}{\psi}\right) = 0 \; ; \qquad \text{flavor current} \tag{27.48}$$

Here $\underset{\sim}{\Sigma}$ is a unit matrix with respect to color, but anything with respect to flavor, satisfying $[\underset{\sim}{\Sigma}, \underset{\sim}{\lambda}^a] = 0$

$$\underset{\sim}{\Sigma} = \begin{pmatrix} \underset{\sim}{\sigma} & & \\ & \underset{\sim}{\sigma} & \\ & & \underset{\sim}{\sigma} \end{pmatrix} \qquad \underset{\sim}{\sigma} = \begin{pmatrix} \sigma_u & & & X \\ & \sigma_d & & \\ & & \sigma_s & \\ X & & & \sigma_c \end{pmatrix} \tag{27.49}$$

It follows from the four-divergence of the third of Eqs. (27.47) and the antisymmetry of $\mathcal{F}_{\mu\nu}^a = -\mathcal{F}_{\nu\mu}^a$ that the color current, the source of the color field, is also

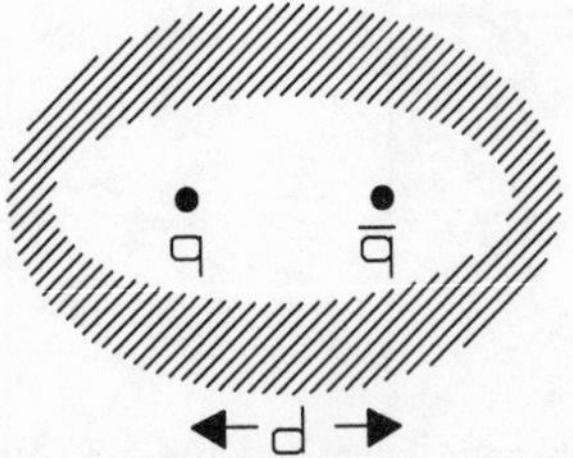

Figure 27.3: Confinement in QCD. Lattice gauge theory calculations indicate that the separation energy grows linearly with d.

conserved.

$$\frac{\partial}{\partial x_\mu}\left(\frac{i}{2}g\bar{\underline{\psi}}\gamma_\mu\underline{\lambda}^a\underline{\psi} + gf^{abc}\mathcal{F}^b_{\mu\nu}A^c_\nu\right) = 0 \tag{27.50}$$

The Feynman rules for the S-matrix following from the lagrangian density of QCD are derived, for example, in Refs. [Q11, Q12, Q13], [R4]; they lead to those for the Green's functions stated in Section 19.

QCD has two absolutely remarkable properties, confinement and asymptotic freedom.

Confinement. Colored quarks and gluons, the basic underlying degrees of freedom in the strong interactions, are evidently never observed as free asymptotic scattering states in the laboratory; you cannot hold an isolated quark or gluon in your hand. Quarks and gluons are confined to the interior of hadrons. There are strong indications from lattice gauge theory calculations, which we discuss in some detail in the next several sections, that confinement is indeed a dynamic property of QCD arising from the strong, nonlinear gluon couplings in the lagrangian. One can show in these calculations, for example, that the energy of a static $(q\bar{q})$ pair grows linearly with the distance d separating the pair (see Fig. 27.3). What actually happens then as the $(q\bar{q})$ pair is separated is that another $(q\bar{q})$ pair is formed, completely shielding the individual color charges of the first pair, and producing two mesons from one.

Asymptotic Freedom. The second remarkable property is asymptotic freedom. Recall from QED that vacuum polarization shields a point electric charge e_0 as indicated in Fig. 27.4a. The renormalized charge e_2^2 changes with the

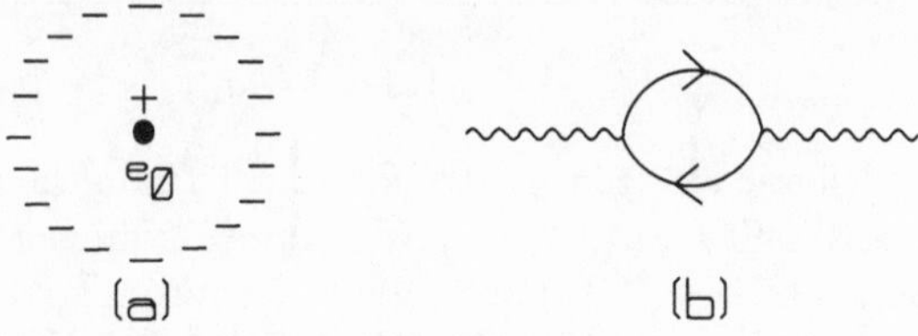

Figure 27.4: (a) Shielding of point charge by (b) vacuum polarization in QED.

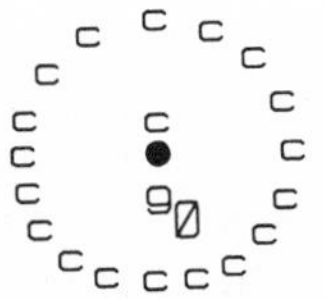

Figure 27.5: Antishielding of color charge in QCD by strong vacuum polarization.

distance scale, or momentum transfer λ^2, at which one measures the interior charge. The mathematical statement of this fact is the renormalization group equation of Gell-Mann and Low (Ref. [Q14])

$$\frac{de_2^2}{d\ln(\lambda^2/M^2)} = \psi(e_2^2) \tag{27.51}$$

The lowest order modification of the charge in QED arises from the vacuum polarization graph indicated in Fig. 27.4b. The renormalization group equations can be used to sum the leading logarithmic corrections to the renormalized charge to all orders. The result is that the renormalized charge measured at large $\lambda^2 \gg M^2$ is related to the usual value of the total charge e_1^2 by

$$e_2^2 \approx \frac{e_1^2}{1-(e_1^2/12\pi^2)\ln(\lambda^2/M^2)} \tag{27.52}$$

The first term in the expansion of the denominator arises from the graph in Fig. 27.4b.[5] The renormalized electric charge in QED is evidently *shielded* by vacuum polarization; the measured charge *increases* as one goes to shorter and shorter distances, or higher and higher λ^2.

Similar, although somewhat more complicated, arguments can be made in QCD. An isolated color charge g_0 is modified by strong vacuum polarization and surrounded with a corresponding cloud of color charge as indicated schematically in Fig. 27.5. In this case, the renormalization group equations lead to a sum of the leading ln corrections for $\lambda^2 \gg \lambda_1^2$ of the form (Refs. [Q2, Q3, Q4, Q5])

$$g_2^2 \approx \frac{g_1^2}{1+(g_1^2/16\pi^2)(33/3-2N_f/3)\ln(\lambda^2/\lambda_1^2)} \tag{27.53}$$

Here N_f is the number of quark flavors.[6] An expansion of the denominator again gives the result obtained by combining the lowest-order perturbation theory corrections to the quark and gluon propagators and quark vertex (Prob. 27.7).

[5]Prove this statement (Prob. 27.6).

[6]With $N_f = 1$, no gluon contribution of 33/3, and the observation $\text{tr}(\frac{1}{2}\lambda^a\frac{1}{2}\lambda^b) = \frac{N_f}{2}\delta^{ab}$, one recovers the result in Eq. (27.52). It is the gluon contribution that changes the sign in the denominator.

The plus sign in the denominator in this expression is crucial. One now draws the conclusion that there is *antishielding*; the charge *decreases* at shorter distances, or with larger λ^2.[7] The implications are enormous, for one now concludes that it is consistent to do *perturbation theory* at very short distances, or high momentum transfer. The renormalization group equations then provide a tool for summing the leading ln's of perturbation theory. This powerful result of asymptotic freedom in QCD is due to Politzer, Gross, and Wilczek in Refs. [Q2, Q3, Q4, Q5].

[7] The vacuum in QCD thus acts like a *paramagnetic* medium, where a moment surrounds itself with like moments, rather than the *dielectric* medium of QED where a charge surrounds itself with opposite charges.

28

PATH INTEGRALS

Nuclear physics is the study of the structure and dynamics of hadronic systems with baryon number $B \geq 1$. Such systems are composed of a confined quark/gluon substructure. The large-distance confinement of color, and the evolution into the large-distance hadronic structure, is governed by a regime where the coupling constant g is large and the nonlinear interactions of QCD are crucial. A central goal of nuclear physics is to deduce the consequences of QCD in this strong-coupling regime.[8] The subsequent developments are most conveniently presented in terms of a path integral formulation of quantum mechanics and field theory (Ref. [Q15]). This approach permits one to readily incorporate explicit local gauge invariance, and as a formulation of field theory in terms of multiple integrals over paths, provides a basis for carrying out large-scale numerical Monte Carlo evaluations of physical quantities. We start the discussion with a review of the basic concepts of path integrals. The material in this section is taken from Refs. [Q15, Q16] and [R1, R4]; it is meant as a *review*. We start with the problem of a single nonrelativistic particle in a potential.

Propagator and Path Integral. The quantum mechanical amplitude for finding a particle at position q_f at time t_f if it started at q_i at time t_i is given by[9]

$$\langle q_f t_f | q_i t_i \rangle = \int \mathcal{D}(q) \exp\left\{\frac{i}{\hbar} S(f,i)\right\} \tag{28.1}$$

The action appearing in this expression is defined by

$$S(f,i) = \int_{t_i}^{t_f} L(q,\dot{q})dt \tag{28.2}$$

Here $L = T - V$ is the lagrangian. The path integral appearing in Eq. (28.1) is illustrated in Fig. 28.1. It is defined in the following manner:[10]

1. Split the time interval $t_f - t_i$ into n subintervals Δt such that $t_f - t_i = n\Delta t$. Label the coordinate at time t_p by q_p. Fix $q_i \equiv q_0$ and $q_f \equiv q_n$.

2. Write the action in Eq. (28.2) as a finite sum over these intervals.

[8] As opposed to the very short-distance, high-momentum regime where one can do perturbative QCD.

[9] We restore $\hbar$ until the end of this section for reasons that will become evident.

[10] The derivation of these results for a T and V of the form in Eq. (28.15) proceeds in a manner similar to that given in the text for the partition function (Prob. 28.1).

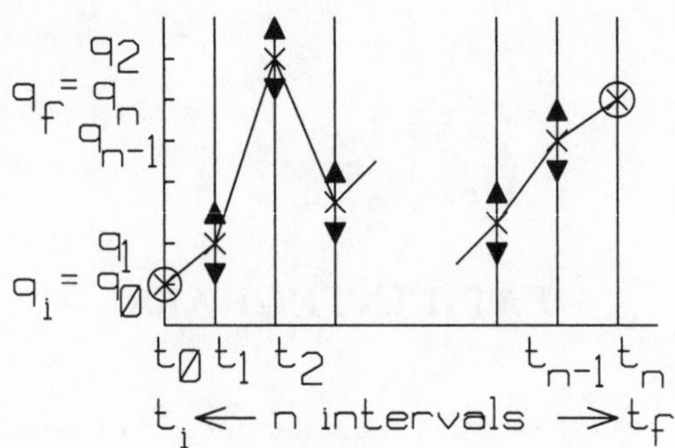

Figure 28.1: Definition of the path integral for a single nonrelativistic particle.

3. Define the time derivative of the coordinate appearing in that sum according to

$$\dot{q}_p = \frac{q_{p+1} - q_p}{\Delta t} \tag{28.3}$$

4. To evaluate the expression in Eq. (28.1), integrate over each coordinate at each intermediate time (see Fig. 28.1)

$$\int dq_1 \cdots \int dq_{n-1} \tag{28.4}$$

5. For the measure of integration, assign the following factor for each interval

$$\left(\frac{me^{-i\pi/2}}{2\pi\hbar\Delta t}\right)^{1/2} \quad ; \qquad \text{one factor for each interval} \tag{28.5}$$

6. Finally, take the limit $n \to \infty$ [which implies $\Delta t = (t_f - t_i)/n \to 0$].

This procedure generates the exact quantum mechanical transition amplitude (Ref. [Q15]). It involves only the *classical lagrangian*; however, one has to integrate over *all possible paths* connecting the initial and final points as illustrated in Fig. 28.1. The classical limit of this expression can be obtained by taking $\hbar \to 0$. In this limit, the method of stationary phase (Ref. [R5]) implies that the integral in Eqs. (28.1) and (28.2) will be determined by that path (or paths) where

$$\delta \int_{t_i}^{t_f} L(q, \dot{q})dt = 0 \tag{28.6}$$

Here the variation about the actual path is precisely that defined in classical mechanics, and the endpoints are held fixed. The reader will recognize Eq. (28.6) as Hamilton's principle. One immediately obtains classical mechanics as the classical limit of this path integral formulation of quantum mechanics.

Partition Function and the Path Integral. The partition function in the canonical ensemble, and hence statistical mechanics, bears an intimate relation

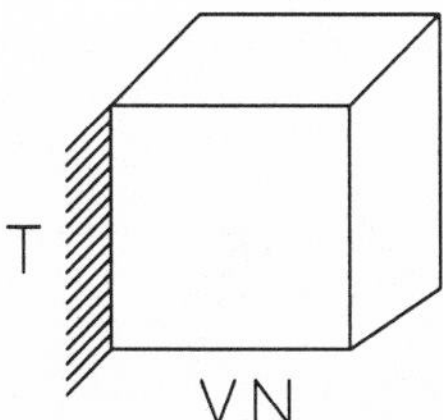

Figure 28.2: System at given (N, V, T) in the canonical ensemble.

to the propagator and path integral described above. Given a system in thermodynamic equilibrium at given (N, V, T) as illustrated in Fig. 28.2, the canonical partition function is defined by

$$Z(N,V,T) \;=\; \mathrm{Tr}\left[e^{-\beta\hat{H}}\right] = \sum_n e^{-\beta E_n} \tag{28.7}$$

Here $\beta \equiv 1/k_{\mathrm{B}}T$ where k_{B} is Boltzmann's constant and the Trace (Tr) indicates a sum over the diagonal elements evaluated for a complete set of states in the Hilbert space for given N; the second expression is obtained by evaluating the trace in the basis of eigenstates of the hamiltonian $\hat{H}$ satisfying $\hat{H}|E_n\rangle = E_n|E_n\rangle$. The Helmholtz free energy is then given in terms of the canonical partition function by

$$F(N,V,T) = -\frac{1}{\beta}\ln Z \tag{28.8}$$

Recall that the Helmholtz free energy is related to the energy and entropy by

$$\begin{aligned} F &= E - TS \\ dF &= dE - TdS - SdT \end{aligned} \tag{28.9}$$

The first and second laws of thermodynamics state

$$dE = TdS - PdV + \mu dN \tag{28.10}$$

Here μ is the chemical potential, and the last term contributes only for an open system. A combination of these equations yields

$$dF = -SdT - PdV + \mu dN \tag{28.11}$$

Hence if one knows the function $F(T, V, N)$, the entropy, pressure, and chemical potential can be obtained by differentiation

$$S = -\left(\frac{\partial F}{\partial T}\right)_{V,N} \qquad P = -\left(\frac{\partial F}{\partial V}\right)_{T,N} \qquad \mu = \left(\frac{\partial F}{\partial N}\right)_{V,T} \tag{28.12}$$

With N *uncoupled subsystems*, the hamiltonian is simply additive

$$\hat{H} = \sum_{\sigma=1}^{N} \hat{h}_{\sigma} \tag{28.13}$$

If the N subsystems are all *identical and localized*, the canonical partition function simplifies to

$$\begin{aligned} Z &= z^N \\ z &= \text{Tr}\left[e^{-\beta\hat{h}}\right] \end{aligned} \tag{28.14}$$

In the last expression the trace is now over a complete set of single-particle states and one works in the *microcanonical ensemble.* As a concrete example one can work with a one-dimensional problem of a particle in a potential where

$$\hat{h} = \frac{1}{2m}\hat{p}^2 + V(\hat{q}) \tag{28.15}$$

To convert the partition function to a *path integral* first introduce a basis of eigenstates of position where

$$\begin{aligned} \hat{q}|q\rangle &= q|q\rangle \\ \langle q'|q\rangle &= \delta(q'-q) \end{aligned} \tag{28.16}$$

Eigenstates of momentum will also be employed

$$\begin{aligned} \hat{p}|p\rangle &= p|p\rangle \\ \langle p'|p\rangle &= \delta(p'-p) \end{aligned} \tag{28.17}$$

The inner product between the coordinate and momentum eigenstates follows from general principles

$$\langle q|p\rangle = \frac{1}{\sqrt{2\pi\hbar}} \exp\left\{\frac{i}{\hbar} p\, q\right\} \tag{28.18}$$

The completeness relation states[11]

$$\int dp|p\rangle\langle p| = \int dq|q\rangle\langle q| = 1 \tag{28.19}$$

Now evaluate the trace in Eq. (28.14) using these *eigenstates of position*

$$z = \text{Tr}[e^{-\beta\hat{h}}] = \int dq\langle q|e^{-\beta\hat{h}}|q\rangle \tag{28.20}$$

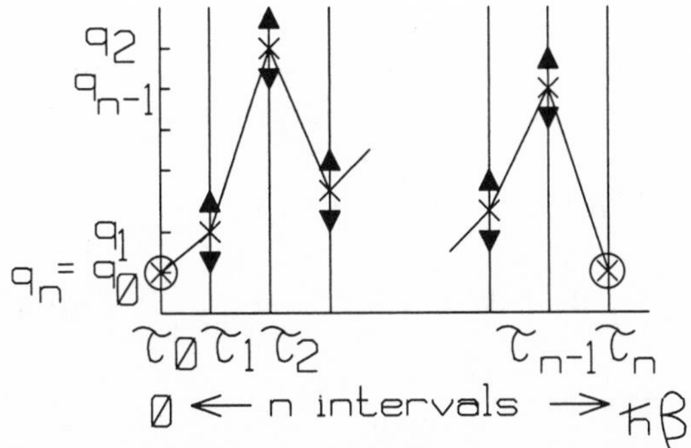

Figure 28.3: Evaluation of partition function as a path integral.

To proceed, divide $\hbar\beta$ into n intervals of width $\Delta\tau = \varepsilon$ and write $\hbar\beta = n\varepsilon$ as indicated in Fig. 28.3. Introduce the coordinate q_p at each intermediate τ_p as indicated in Fig. 28.3. Define the initial and final coordinates to have the same value $q_0 \equiv q_n = q$. This corresponds to a *cyclic boundary condition* on the coordinate and is needed for the evaluation of the trace. The exponential in the partition function can be factored into n terms

$$e^{-\beta\hat{h}} = e^{-(\varepsilon/\hbar)\hat{h}}e^{-(\varepsilon/\hbar)\hat{h}}\cdots e^{-(\varepsilon/\hbar)\hat{h}}\ ; \qquad \text{n terms} \tag{28.21}$$

The completeness relation $\int dq_p|q_p\rangle\langle q_p| = 1$ can be inserted between each of these factors to give

$$\begin{aligned} z &= \int dq_n \prod_{p=1}^{n-1} dq_p \langle q_0|e^{-(\varepsilon/\hbar)\hat{h}}|q_1\rangle\langle q_1|e^{-(\varepsilon/\hbar)\hat{h}}|q_2\rangle\cdots\langle q_{n-1}|e^{-(\varepsilon/\hbar)\hat{h}}|q_n\rangle \\ z &= \int \prod_{p=1}^{n} dq_p \prod_{l=0}^{n-1}\langle q_l|e^{-(\varepsilon/\hbar)\hat{h}}|q_{l+1}\rangle \end{aligned} \tag{28.22}$$

Now expand the matrix elements in this expression and work to *first order in* ε.

$$\begin{aligned} \langle q_l|e^{-(\varepsilon/\hbar)\hat{h}}|q_{l+1}\rangle &\approx \langle q_l|\left(1-\frac{\varepsilon}{\hbar}\hat{h}\right)|q_{l+1}\rangle \\ &\approx \left(1-\frac{\varepsilon}{\hbar}V(q_l)\right)\langle q_l|\left(1-\frac{\varepsilon}{\hbar}\frac{\hat{p}^2}{2m}\right)|q_{l+1}\rangle \\ &\approx \exp\left\{-\frac{\varepsilon}{\hbar}V(q_l)\right\}\langle q_l|\exp\left\{-\frac{\varepsilon}{\hbar}\frac{\hat{p}^2}{2m}\right\}|q_{l+1}\rangle \end{aligned} \tag{28.23}$$

[11] With a big box of length L and periodic boundary conditions one should really use

$$\langle p'|p\rangle = \delta_{p',p} \qquad \langle q|p\rangle = \frac{1}{\sqrt{L}}e^{\frac{i}{\hbar}pq}$$

Then at the end of the calculation the limit $\sum_p \to (L/2\pi\hbar)\int dp$ is taken.

This result is exact as $\varepsilon \to 0$.

Next insert eigenstates of momentum so that the $\hat{p}$ in the exponential can be replaced by its eigenvalue in the last factor (which we label as m.e.)

$$
\begin{aligned}
\text{m.e.} &\equiv \langle q_l | \exp\left\{-\frac{\varepsilon}{\hbar}\frac{\hat{p}^2}{2m}\right\} | q_{l+1}\rangle \\
&= \int dp \int dp' \langle q_l | p\rangle \exp\left\{-\frac{\varepsilon p^2}{2m\hbar}\right\} \langle p | p'\rangle\langle p' | q_{l+1}\rangle \\
&= \int dp \exp\left\{-\frac{\varepsilon p^2}{2m\hbar}\right\} \langle q_l | p\rangle\langle p | q_{l+1}\rangle \\
&= \int \frac{dp}{2\pi\hbar} \exp\left\{\frac{i}{\hbar}p(q_l - q_{l+1})\right\} \exp\left\{-\frac{\varepsilon p^2}{2m\hbar}\right\}
\end{aligned}
\tag{28.24}
$$

Complete the square in the exponent

$$
\begin{aligned}
\text{m.e.} &= \frac{1}{2\pi\hbar} \exp\left\{-\frac{m}{2\varepsilon\hbar}(q_l - q_{l+1})^2\right\} I \\
I &\equiv \int_{-\infty}^{\infty} dp \exp\left\{-\frac{\varepsilon}{2m\hbar}\left[p - \frac{im}{\varepsilon}(q_l - q_{l+1})\right]^2\right\}
\end{aligned}
\tag{28.25}
$$

A change of integration variables, and the use of Cauchy's theorem to shift the integral $\int_C e^{-z^2}$ up to the real axis yields

$$
I = \left(\frac{2m\hbar}{\varepsilon}\right)^{1/2} \int_{-\infty}^{\infty} dx\, e^{-x^2} = \left(\frac{2m\pi\hbar}{\varepsilon}\right)^{1/2}
\tag{28.26}
$$

A combination of these results yields

$$
\langle q_l | e^{-(\varepsilon/\hbar)\hat{h}} | q_{l+1}\rangle \approx \left(\frac{m}{2\pi\varepsilon\hbar}\right)^{1/2} \exp\left\{-\frac{m}{2\varepsilon\hbar}(q_l - q_{l+1})^2\right\} \exp\left\{-\frac{\varepsilon}{\hbar}V(q_l)\right\}
\tag{28.27}
$$

This expression is exact as $\varepsilon \to 0$; it allows us to express the partition function in this same limit as

$$
z = \text{Lim}_{\varepsilon \to 0} \left(\frac{m}{2\pi\varepsilon\hbar}\right)^{n/2} \int \prod_{l=0}^{n-1} dq_l \exp\left\{-\frac{\varepsilon}{\hbar}\sum_{p=0}^{n-1}\left[\frac{m}{2\varepsilon^2}(q_p - q_{p+1})^2 + V(q_p)\right]\right\}
\tag{28.28}
$$

The exponent appearing in this expression can be related to the *action* in the following manner. Define $\varepsilon \equiv d\tau$. Then as $\varepsilon \to 0$, one has

$$
\begin{aligned}
\frac{q_{p+1} - q_p}{\varepsilon} &= \frac{dq}{d\tau} \\
\varepsilon \sum_{p=0}^{n-1}\left[\frac{m}{2\varepsilon^2}(q_{p+1} - q_p)^2 + V(q_p)\right] &= \int_0^{\hbar\beta} d\tau \left[\frac{m}{2}\left(\frac{dq}{d\tau}\right)^2 + V(q)\right]
\end{aligned}
\tag{28.29}
$$

Recall the action is defined by

$$S(f,i) \equiv \int_{t_i}^{t_f} L(q, \frac{dq}{dt})dt \tag{28.30}$$

Now make the following substitution

$$t \equiv -i\tau \qquad\qquad dt = -id\tau \tag{28.31}$$

The action then becomes

$$S(f,i) \;=\; -i\int_{\tau_1}^{\tau_2} d\tau L(q, i\frac{dq}{d\tau}) \;\equiv\; i\bar{S}(\tau_2, \tau_1) \tag{28.32}$$

Hence

$$\bar{S}(\tau_2, \tau_1) = -\int_{\tau_1}^{\tau_2} L(q, i\frac{dq}{d\tau})d\tau \tag{28.33}$$

In *summary* the partition function in the microcanonical ensemble can be written as a path integral according to

$$z = \int \bar{\mathcal{D}}(q) \exp\left\{-\frac{1}{\hbar}\bar{S}(\hbar\beta, 0)\right\} \tag{28.34}$$

Here the action is evaluated for imaginary time, and from Eq. (28.33)

$$\bar{S}(\hbar\beta, 0) = \int_0^{\hbar\beta} \left[\frac{1}{2}m\left(\frac{dq}{d\tau}\right)^2 + V(q)\right] d\tau \tag{28.35}$$

This expression for the partition function as a path integral has the following interpretation:

1. Divide the τ integration for the action at imaginary times in Eq. (28.35) into n intervals of size $\Delta\tau$ with $\hbar\beta = n\Delta\tau$. Label the coordinate at the intermediate time τ_p by q_p as indicated in Fig. 28.3.
2. Write the action in Eq. (28.35) as the corresponding finite sum.
3. Define the τ derivative appearing in that expression by

$$\frac{dq_p}{d\tau} \equiv \frac{q_{p+1} - q_p}{\Delta\tau} \tag{28.36}$$

4. Impose cyclic boundary conditions to recover the trace

$$q_0 \equiv q_n \tag{28.37}$$

5. Carry out the multiple integral over the coordinates at all intermediate taus (Fig. 28.3); include $\int dq_0$ to recover the trace

$$\int dq_0 \int dq_1 \cdots \int dq_{n-1} \tag{28.38}$$

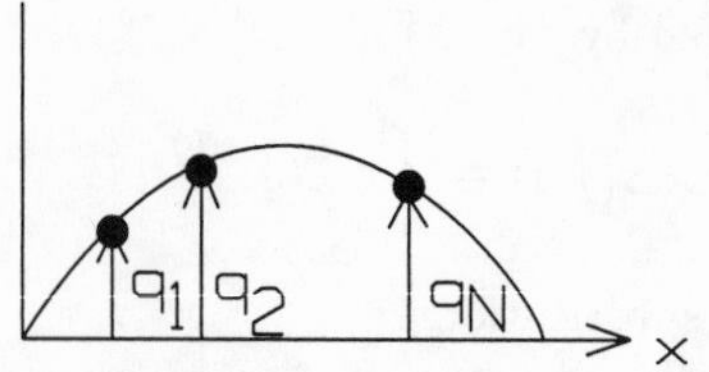

Figure 28.4: System with many degrees of freedom: mass points on a massless string with tension T moving in the transverse direction.

6. As a measure for the integration, include the following factor for each interval

$$\left(\frac{m}{2\pi\hbar\Delta\tau}\right)^{1/2} \quad ; \qquad \text{one for each interval} \tag{28.39}$$

7. At the end of the calculation, take the limit $n \rightarrow \infty$ (which implies $\Delta\tau = \hbar\beta/n \rightarrow 0$).

The resulting expression gives an exact representation of the partition function in the microcanonical ensemble.

Many Degrees of Freedom and Continuum Mechanics. Let us extend this discussion to systems with many degrees of freedom. Consider, for example, many mass points on a massless string with tension T moving in the transverse direction as illustrated in Fig. 28.4. Denote the transverse displacements by q_α with $\alpha = 1, \ldots, N$. Here the index α merely labels the coordinates. Then extend the notion of the path integral to include integrations over all possible configurations of *each* of the coordinates

$$\mathcal{D}(q) \rightarrow \prod_{\alpha=1}^{N} \mathcal{D}(q_\alpha) \tag{28.40}$$

The quantum mechanical amplitude for finding the coordinates in a configuration $\{q'_\alpha\}$ at time t_f if they started at $\{q_\alpha\}$ at time t_i is

$$\langle q'_1 q'_2 \cdots q'_N, t_f | q_1 q_2 \cdots q_N, t_i \rangle = \int \mathcal{D}(q) \exp\left\{\frac{i}{\hbar} S(f,i)\right\} \tag{28.41}$$

where $S(f,i)$ is the many-body action. The canonical partition function is evidently

$$Z = \int \bar{\mathcal{D}}(q) \exp\left\{-\frac{1}{\hbar}\bar{S}(\hbar\beta, 0)\right\} \tag{28.42}$$

It is now straightforward to proceed to the continuum limit; in the previous example it would be that of a continuous string.

$$q_\alpha \rightarrow q(\mathbf{x}, t) \qquad\qquad L \rightarrow \int d\mathbf{x} \mathcal{L}(q, \frac{\partial q}{\partial x_\mu}) \tag{28.43}$$

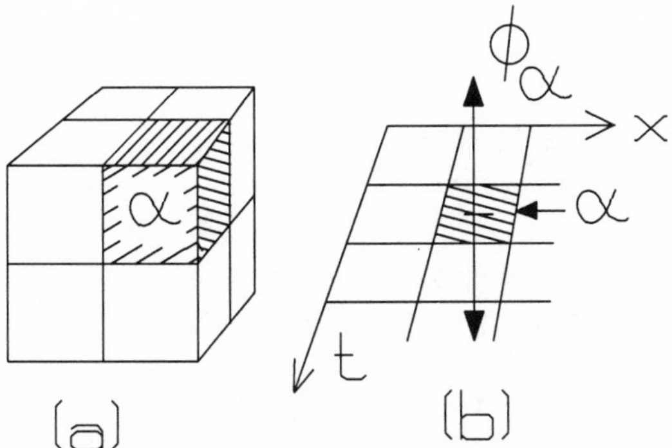

Figure 28.5: Path integral in field theory: (a) Subdivision of space-time into cells of volume ε^4; each cell is labeled by α. (b) Representation of the field in each cell.

The coordinate $\mathbf{x}$ now merely serves as a *label* to indicate the position of the dynamic variable q. The quantity $\mathcal{L}$ is the lagrangian density.[12]

Field Theory. We have now arrived at field theory.[13] Consider the path integral associated with a field variable ϕ in three-dimensional space.

$$\int \mathcal{D}(\phi) \exp\left\{\frac{i}{\hbar}S\right\} \tag{28.44}$$

This expression means the following:

1. Subdivide space-time into volumes of size ε^4. Label each cell with the index α as indicated in Fig. 28.5a.
2. Write the action S as a finite sum over these cells.
3. Include the measure for each interval

$$\prod_\alpha m_\alpha \tag{28.45}$$

Since all the subsequent developments will depend only on *ratios* of path integrals over the fields, the precise value of m_α is immaterial here.

4. Perform a multiple integration over the value of the field variable in each cell as illustrated in Fig. 28.5b.

$$\prod_\alpha \int d\phi_\alpha \tag{28.46}$$

[12]For generality, these expressions have been written for any number of spatial dimensions; they would apply, for example, to the dynamics of a one-dimensional string, a two-dimensional membrane, or a three-dimensional solid.

[13]We could be discussing, for example, sound waves in a gas (cf. Ref. [R5]).

5. Take the limit $\varepsilon \to 0$ where $\phi_\alpha \to \phi(\mathbf{x}, t)$. The path integral has now been exactly evaluated.[14]

Relativistic Quantum Field Theory. On the basis of the preceding discussion, it follows that the *partition function* in relativistic quantum field theory is given by[15]

$$Z = \int \bar{\mathcal{D}}(\phi) \exp\left\{\frac{1}{\hbar}\int_0^{\hbar\beta} d\tau \int d^3x\, \mathcal{L}(\phi, i\partial\phi/\partial\tau, \nabla\phi)\right\} \tag{28.47}$$

Cyclic boundary conditions required for the trace are imposed by demanding

$$\phi(\mathbf{x}, \hbar\beta) = \phi(\mathbf{x}, 0) \tag{28.48}$$

To conclude this section, we briefly *review* some additional results of relativistic quantum field theory, derived for example in Ref. [Q16] (see also Ref. [R1]). The *generating functional* for a scalar field theory is defined by

$$\tilde{W}(J) \equiv \frac{\int \mathcal{D}(\phi) \exp\{(i/\hbar)\int d^4x[\mathcal{L} + J\phi]\}}{\int \mathcal{D}(\phi) \exp\{(i/\hbar)\int d^4x\, \mathcal{L}\}} \tag{28.49}$$

It is a ratio of path integrals where the denominator is calculated from the numerator by simply setting the source current $J = 0$. The generating functional yields the Green's functions, or propagators, of the theory by variational differentiation (see Prob. 28.3) with respect to the source current $J(x)$. Here x denotes a four-vector.

$$\left(\frac{\hbar}{i}\right)^n \frac{\delta^n \tilde{W}(0)}{\delta J(x_1)\delta J(x_2)\ldots\delta J(x_n)} = \frac{\langle\Psi_0|T[\hat{\phi}(x_1)\cdots\hat{\phi}(x_n)]|\Psi_0\rangle}{\langle\Psi_0|\Psi_0\rangle} \tag{28.50}$$

The *crucial theorem* of Abers and Lee (Ref. [Q16]) now states the following: the correct limiting conditions on the times for extracting the ground-state expectation values from the generating functional are

$$\tilde{W}(J) = \mathrm{Lim}_{T' \to -i\infty}\mathrm{Lim}_{T \to +i\infty} \frac{\int \mathcal{D}(\phi) \exp\left\{(i/\hbar)\int_T^{T'} d^4x[\mathcal{L} + J\phi]\right\}}{[\cdots]_{J=0}} \tag{28.51}$$

This result can be combined with the *euclidicity postulate* that the Green's functions for certain problems can be analytically continued in the time.[16] Let

$$t \equiv -i\tau \qquad\qquad i\int_{i\infty}^{-i\infty} dt \to \int_{-\infty}^{\infty} d\tau \tag{28.52}$$

[14] This limit should be taken only at the very end of the calculation of the quantity of interest.

[15] This is not a ratio, but we will show in the next section how thermodynamic averages will involve the partition function in the form of a ratio.

[16] This can be validated in perturbation theory for the vacuum Green's functions.

The Green's functions are then generated in the euclidian metric where

$$\begin{aligned} x_\mu &= (\mathbf{x}, it) = (\mathbf{x}, \tau) \\ x_\mu^2 &= \mathbf{x}^2 + \tau^2 \end{aligned} \tag{28.53}$$

The generating functional that provides these euclidian Green's functions is evidently

$$\tilde{W}_E(J) \equiv \frac{\int \mathcal{D}(\phi) \exp\left\{(1/\hbar)\int d^4x[\mathcal{L}(\phi, i\partial\phi/\partial\tau, \nabla\phi) + J\phi]\right\}}{[\cdots]_{J=0}} \tag{28.54}$$

Here $d^4x \equiv d^3x d\tau$. The evaluation of this expression gives the Green's functions in the euclidian metric, which may then be analytically continued back to real time.[17]

This generating functional has an interpretation as the amplitude to go from the ground state to the ground state in the presence of external sources (Ref. [Q16]).

Now note the great similarity of the partition function in Eq. (28.47) and the generating functional for the Green's functions in the euclidian metric in Eq. (28.54).

We proceed to use these results as a basis for analyzing strong-coupling QCD.

[17]Alternatively, one can evaluate Eq. (28.49) in Minkowski space and build in the correct Feynman boundary conditions for the propagators by introducing an adiabatic damping factor in the action. Consider the mass term $\mathcal{L}_{\text{mass}} = -(1/2)m^2\phi^2$ in the lagrangian density, and take $m^2 \to m^2 - i\eta$ where η is a positive infinitesimal. See Prob. 28.7.

29

LATTICE GAUGE THEORY

The material in this section is based on Refs. [Q17, Q18, Q19, Q20]. We start by reviewing the motivation.

Motivation. The goal is to solve a locally gauge-invariant, nonabelian, strong-coupling field theory. We seek to understand confinement in QCD and the structure of hadrons and nuclei. Since it is the local gauge invariance that dictates the nature of the nonlinear couplings in the lagrangian, and since it is these nonlinear couplings that are presumably responsible for confinement, it is important that the approach incorporate local gauge invariance.

The method of solution, due to Wilson (Ref. [Q17]), puts the theory on a finite lattice of space-time points with separation a. This reduces the problem to a large, but finite, set of degrees of freedom. A natural momentum cut-off of $\Lambda \approx 1/a$ now appears in the theory. Various expectation values, which allow one to probe the consequences of QCD, can be related to the partition function. The resulting path integral ratios can be evaluated numerically with Monte Carlo techniques — in some cases, such as mean-field theory and strong-coupling theory, analytic results can be obtained. At the end, the continuum limit must be taken (or at least discussed). Asymptotic freedom, whereby the renormalized coupling constant becomes vanishingly small at short distances [Eq. (27.53)], facilitates the continuum limit and permits one to tie on to perturbative QCD.

Some Preliminaries. Recall from Section 28 that the canonical partition function is defined by

$$Z = \text{Tr}\left(e^{-\beta\hat{H}}\right) \tag{29.1}$$

The energy of the system is then given by[18]

$$E = \frac{\text{Tr}\left(\hat{H}e^{-\beta\hat{H}}\right)}{\text{Tr}\left(e^{-\beta\hat{H}}\right)} = \frac{\text{Tr}\left(\hat{H}e^{-\beta\hat{H}}\right)}{Z} \tag{29.2}$$

[18] From Eqs. (28.7)-(28.12)

$$E = F + \beta\left(\frac{\partial F}{\partial\beta}\right)_{V,N} = \left[\frac{\partial(\beta F)}{\partial\beta}\right]_{V,N}$$

$$E = \left[\frac{\partial}{\partial\beta}(-\ln Z)\right]_{V,N} = \frac{1}{Z}\text{Tr}\left(\hat{H}e^{-\beta\hat{H}}\right)$$

These observations allow one to introduce the *statistical operator* (see Ref. [N2])

$$\hat{\rho}_{\text{th}} = \frac{e^{-\beta \hat{H}}}{\text{Tr}\left(e^{-\beta \hat{H}}\right)} = \frac{e^{-\beta \hat{H}}}{Z} \tag{29.3}$$

The thermal average of a quantity is then given in terms of the statistical operator by

$$\langle\langle \hat{O} \rangle\rangle = \frac{\text{Tr}\left(\hat{O} e^{-\beta \hat{H}}\right)}{Z} = \text{Tr}\left(\hat{O} \hat{\rho}_{\text{th}}\right) \tag{29.4}$$

From the form of the partition function in Eq. (28.47), the statistical operator in relativistic quantum field theory can be identified as

$$\begin{aligned} \rho_{\text{th}} &= \frac{\exp\{-(1/\hbar)\bar{S}(\hbar\beta, 0)\}}{\int \bar{\mathcal{D}}(\phi) \exp\{-(1/\hbar)\bar{S}(\hbar\beta, 0)\}} = \frac{\exp\{-(1/\hbar)\bar{S}(\hbar\beta, 0)\}}{Z} \\ \langle\langle \hat{O}(\phi) \rangle\rangle &= \frac{\int \bar{\mathcal{D}}(\phi) O(\phi) \exp\{-(1/\hbar)\bar{S}(\hbar\beta, 0)\}}{\int \bar{\mathcal{D}}(\phi) \exp\{-(1/\hbar)\bar{S}(\hbar\beta, 0)\}} \end{aligned} \tag{29.5}$$

When the numerator and denominator are evaluated through path integrals in these thermal averages, the measure drops out in the *ratio* as advertised.

Now that the $\hbar$ has served its purpose, we shall henceforth return to the system of units where $\hbar = c = 1$.

As an introduction to lattice gauge theory, we start with the example of QED in one space and one time dimension. This theory is locally gauge invariant; however, the group here is abelian — it is just that of local phase transformations. This simple example is worth studying, though, since it provides insight into the approach of putting the theory on a lattice while maintaining exact gauge invariance. We will then discuss the extension to the more complicated nonabelian case. It is worthwhile to separate these concepts.

QED in One Space and One Time Dimension. Start with a system with *no fermions*—just the electromagnetic field

$$\begin{aligned} \mathcal{L} &= -\frac{1}{4} F_{\mu\nu} F_{\mu\nu} \\ F_{\mu\nu} &= \frac{\partial A_\nu}{\partial x_\mu} - \frac{\partial A_\mu}{\partial x_\nu} \end{aligned} \tag{29.6}$$

The lagrangian is gauge invariant under the transformation

$$A_\mu \rightarrow A_\mu + \frac{\partial \Lambda}{\partial x_\mu} \tag{29.7}$$

Now construct the appropriate action and partition function

$$\begin{aligned} \bar{S}(\beta,0) &= -\int_0^\beta d\tau \int dx \mathcal{L}(A_\mu, \frac{\partial A_\mu}{\partial x_\nu}) \\ &= -\int d^2x\, \mathcal{L}(A_\mu, \frac{\partial A_\mu}{\partial x_\nu}) \end{aligned} \tag{29.8}$$

Here $x_\mu = (x, \tau)$, and the cyclic boundary condition implies $A_\mu(x,\beta) = A_\mu(x,0)$. The partition function is then given by

$$Z = \int \bar{\mathcal{D}}(A_\mu) \exp\{-\bar{S}(\beta,0)\} \tag{29.9}$$

The volume element $\int \bar{\mathcal{D}}(A_\mu)$ in the path integral must be defined so that it is also gauge invariant.

In the path integral we are required to integrate over all field configurations at a given point. After the analytic continuation to imaginary time (temperature) it is convenient to define $A_\mu = (A_1, A_2)$ with real components and integrate over these. Now *everything* is euclidian. This change of variables amounts to a rotation of the contour of integration on A_2 from the imaginary to the real axis. This contour rotation is evidently justified, and defines a unique partition function $Z(\beta)$, if the resulting integrals are convergent. In the present case, after the rotation the exponential in the partition function takes the form $\exp\{-\frac{1}{2}\int_0^\beta d\tau \int dx(\partial A_1/\partial\tau - \partial A_2/\partial x)^2\}$ and the exponentially decreasing weight function will indeed give convergent integrals. Although we do not prove this contour rotation in detail here, a simple example of the mathematical manipulations involved in the analytic continuation and contour rotation is given in Prob. 29.1.[19]

We proceed to discuss the evaluation of the partition function for QED in one space and one time dimension, in the euclidian metric, using the technique of lattice gauge theory.

Lattice Gauge Theory. Divide space-time into a set of discrete points, or lattice sites as indicated in Fig. 29.1. The elementary square with neighboring points at the corners is called a *plaquette*; the length of its side is the point separation a. Each site may be associated with a plaquette in a one-to-one manner in this two-dimensional problem.[20] Each side of the square is of length a.

[19] In QED in three dimensions one can guarantee physical configurations in the path integral by carrying out the integrations in a particular gauge. For example, one can enforce the Coulomb gauge by adding delta functions in the integrand $\bar{\mathcal{D}}(A_\mu)\delta_P(\nabla\cdot\mathbf{A})\delta_P(\phi)$. Here δ_P represents a product over the volume elements α of size ε^4 (see Ref. [Q16]). The weighting function now takes the form $\exp\{-\frac{1}{2}\int_0^\beta d\tau \int d^3x[(\nabla\times\mathbf{A})^2 + (\partial\mathbf{A}/\partial\tau)^2]\}$. The partition function is then the same result one gets by working in the euclidian metric with $x_\mu = (\mathbf{x},\tau)$; $A_\mu = (\mathbf{A},\phi)$, and the decreasing exponentials again lead to convergent path integrals.

[20] For example, choose the point in the lower left-hand corner of each plaquette.

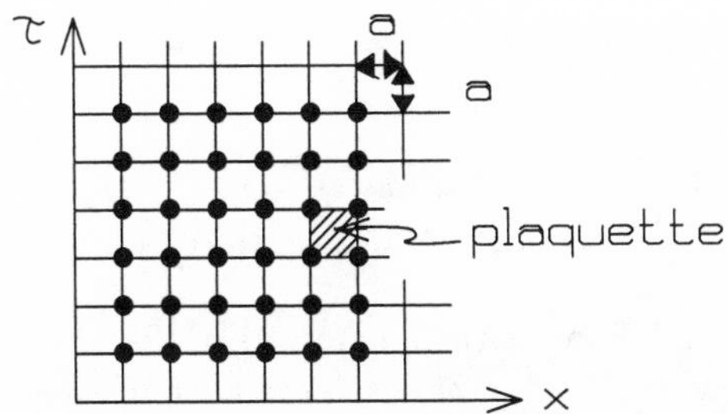

Figure 29.1: Division of space-time into a set of discrete points, or lattice sites, in one space and one time dimension. The elementary square with neighboring points at the corners is called a *plaquette*; the length of its side is the point separation a.

The sites around a given plaquette will be labeled by (i, j, k, l) as indicated in Fig. 29.2. The two-vector $(x_i)_\mu = (x_{i1}, x_{i2})$ will now denote the location of the ith site.

The sites are said to be connected by *links*, and the links are vectors having two directions. An electromagnetic field variable is now *assigned to each link* according to the following prescription

$$U_{ji}(A_\mu) \equiv \exp\left\{ie_0(x_j - x_i)_\mu A_\mu\left(\frac{1}{2}(x_j + x_i)\right)\right\} \tag{29.10}$$

The vector $(x_j - x_i)_\mu$ points from the point i to the point j; it is dotted into the field A_μ, which is evaluated at the center of the link $\frac{1}{2}(x_j + x_i)$. Note that

$$U_{ji}(A_\mu)^* = U_{ij}(A_\mu) \tag{29.11}$$

Thus the U_{ij} form an hermitian matrix. Each plaquette is assigned a direction running around it; there are evidently two possibilites, counterclockwise ($\hookrightarrow$) and clockwise ($\hookleftarrow$).

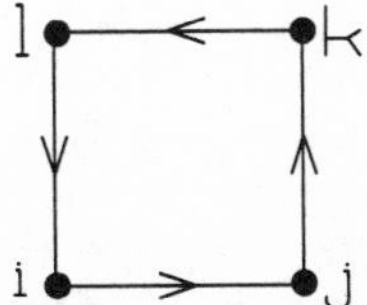

Figure 29.2: Label the sites around a given plaquette by (i, j, k, l). The sites are said to be connected by *links*.

The action is now assumed to receive the following contribution *from each plaquette*[21]

$$
\begin{aligned}
S_{\Box} &\equiv \sigma\left\{\left[1-(U_{il}U_{lk}U_{kj}U_{ji})_{\hookrightarrow}\right]+\left[1-(U_{jk}U_{kl}U_{li}U_{ij})_{\hookleftarrow}\right]\right\} \\
S_{\Box} &= 2\sigma\left[1-\mathrm{Re}\,(U_{il}U_{lk}U_{kj}U_{ji})_{\hookrightarrow}\right]
\end{aligned} \tag{29.12}
$$

The second line follows with the aid of Eq. (29.11).

Let us now discuss the *gauge invariance* of this result. Under a change in gauge the field transforms according to

$$
A_\mu \quad \rightarrow \quad A_\mu + \partial\Lambda/\partial x_\mu \equiv A'_\mu \tag{29.13}
$$

The exponent in Eq. (29.10) then changes to

$$
\begin{aligned}
ie_0(x_j-x_i)_\mu A'_\mu &= ie_0(x_j-x_i)_\mu A_\mu + ie_0(x_j-x_i)_\mu\,\partial\Lambda/\partial x_\mu \\
&\approx ie_0(x_j-x_i)_\mu A_\mu + ie_0\left[\Lambda(x_j)-\Lambda(x_i)\right]
\end{aligned} \tag{29.14}
$$

The second line follows from the definition of the gradient for small separation a; this is the whole trick, for now the quantity U_{ji} in Eq. (29.10) transforms under a gauge transformation according to

$$
U_{ji}(A'_\mu) = e^{ie_0\Lambda(x_j)}U_{ji}(A_\mu)e^{-ie_0\Lambda(x_i)} \tag{29.15}
$$

The change is simply an initial phase factor depending only on the initial point x_i and a final phase factor depending on the final point x_j. The contribution to the action from a plaquette in Eq. (29.12), since it depends on the product of the U's around the plaquette, is unchanged under this transformation and hence gauge invariant

$$
(U_{il}U_{lk}U_{kj}U_{ji})'_{\hookrightarrow} \quad = \quad (U_{il}U_{lk}U_{kj}U_{ji})_{\hookrightarrow} \ ; \qquad \text{invariant} \tag{29.16}
$$

The additional phase factors cancel in pairs in this expression.[22]

Consider next the *continuum limit* of these expressions. One wants to show that in the limit $a \rightarrow 0$ the correct continuum results are recovered. The phases appearing in the contribution of a single plaquette to the action for finite a in Eq. (29.12) are additive; define their sum as follows

$$
(x_j-x_i)_\mu A_\mu\left[\frac{1}{2}(x_j+x_i)\right]+\ldots+(x_i-x_l)_\mu A_\mu\left[\frac{1}{2}(x_i+x_l)\right] \equiv \oint_{\hookrightarrow} A_\mu dx_\mu \tag{29.17}
$$

[21] This form of the action ensures both gauge invariance and the correct continuum limit — see the following discussion.

[22] The introduction of the notion of point splitting to generate gauge-invariant currents is due to Schwinger (Ref. [Q21]).

This expression has the important property that it reverses sign with a change in direction around the plaquette

$$\oint_{\hookrightarrow} A_\mu dx_\mu = -\oint_{\hookleftarrow} A_\mu dx_\mu \tag{29.18}$$

With the definition in Eq. (29.17), the contribution of a given plaquette to the action in Eq. (29.12) can be written

$$S_\Box = \sigma\left\{\left[1 - \exp\left(ie_0 \oint_{\hookrightarrow} A_\mu dx_\mu\right)\right] + \left[1 - \exp\left(ie_0 \oint_{\hookleftarrow} A_\mu dx_\mu\right)\right]\right\} \tag{29.19}$$

The exponent is of order a^2, and for small a it can now be expanded to give

$$\begin{aligned} S_\Box &= \sigma\left\{1 - \left[1 + ie_0 \oint_{\hookrightarrow} A_\mu dx_\mu + \frac{1}{2!}\left(ie_0 \oint_{\hookrightarrow} A_\mu dx_\mu\right)^2 + \cdots\right]\right. \\ &\quad \left. +1 - \left[1 + ie_0 \oint_{\hookleftarrow} A_\mu dx_\mu + \frac{1}{2!}\left(ie_0 \oint_{\hookleftarrow} A_\mu dx_\mu\right)^2 + \cdots\right]\right\} \end{aligned} \tag{29.20}$$

Use of Eq. (29.18) leads to

$$S_\Box = \sigma e_0^2 \left(\oint_{\hookrightarrow} A_\mu dx_\mu\right)^2 + \cdots \tag{29.21}$$

In the continuum limit $a \to 0$, the expression in Eq. (29.17) is the usual *line integral*, and this limiting result for the contribution to the action of an individual plaquette in Eq. (29.21) is exact.

Now use Stokes' theorem on the line integral

$$\oint_{\hookrightarrow} dx_\mu A_\mu = \int_{\text{enclosed surface}} dS_\mu (\nabla \times \mathbf{A})_\mu \tag{29.22}$$

This relation is also exact as $a \to 0$. Evaluation in the present two-dimensional case gives

$$\oint_{\hookrightarrow} dx_\mu A_\mu \approx a^2 (\nabla \times \mathbf{A})_3 = a^2 \left(\frac{\partial A_2}{\partial x_1} - \frac{\partial A_1}{\partial x_2}\right) + O(a^3) \tag{29.23}$$

Thus from Eq. (29.21) one finds

$$S_\Box = \sigma e_0^2 a^4 \left(\frac{\partial A_2}{\partial x_1} - \frac{\partial A_1}{\partial x_2}\right)^2 + O(a^5) \tag{29.24}$$

Now define the overall coefficient to be

$$\sigma \equiv \frac{1}{2e_0^2 a^2} \tag{29.25}$$

Then

$$\text{Lim}_{a\to 0}\, S_{\Box} = a^2\frac{1}{4}F_{\mu\nu}F_{\mu\nu} = -a^2\mathcal{L} \tag{29.26}$$

Here $\mathcal{L}$ is the lagrangian density.

The total action is defined to be the *sum over all the plaquettes*

$$\bar{S} \equiv \sum_{\Box} S_{\Box} \tag{29.27}$$

Equation (29.26) then yields

$$\text{Lim}_{a\to 0}\, \bar{S} = -\int d^2x\mathcal{L}(A_\mu, \frac{\partial A_\mu}{\partial x_\nu}) \tag{29.28}$$

The right-hand side is just the classical action evaluated in the euclidian metric. Thus one has achieved the correct continuum limit of the theory.

We next discuss the *boundary conditions* to be employed in the evaluation of the partition function. For the partition function one must have periodic boundary conditions on fields in the direction of temperature (to produce the Trace)

$$A_\mu(x,\beta) = A_\mu(x,0) \tag{29.29}$$

This implies that the fields must have period $\beta = 1/k_BT$ in the (imaginary) time direction. For a finite system confined to a box of dimension L (Fig. 29.3a), we take periodic boundary conditions with period L in the space direction

$$A_\mu(x+L,\tau) = A_\mu(x,\tau) \tag{29.30}$$

The physical realization of periodic boundary conditions in space is achieved by joining the two ends of the line. The construction of the square lattice of side a in (x,τ) space then corresponds to a uniform division of the surface of a cylinder as illustrated in Fig. 29.3a. The size of the lattice is (n,n) in units of a. Evidently[23]

$$\beta = na \qquad\qquad L = na \tag{29.31}$$

The physical realization of periodic boundary conditions in the τ direction in (x,τ) space is achieved by joining the two ends of the cylinder as in Fig. 29.3b to form a *torus*. The lattice then divides the surface of the torus into small squares of side a; integrals $\int\int dx\, d\tau$ are obtained by summing over the squares.[24]

It is now nessesary to discuss the *measure* for the path integral, that is, the volume element $\bar{\mathcal{D}}(A_\mu)$ for the integration over all field configurations. The action $\bar{S}$ has been constructed to be gauge invariant. We want the measure to

[23] Recall $\hbar = c = 1$. Clearly one can generalize to different lattice sizes (m,n).

[24] For the generating functional $\tilde{W}_{\mathrm{E}}(J)$ one must integrate $\int_{-\infty}^{+\infty}\int_{-\infty}^{+\infty} d^2x$.

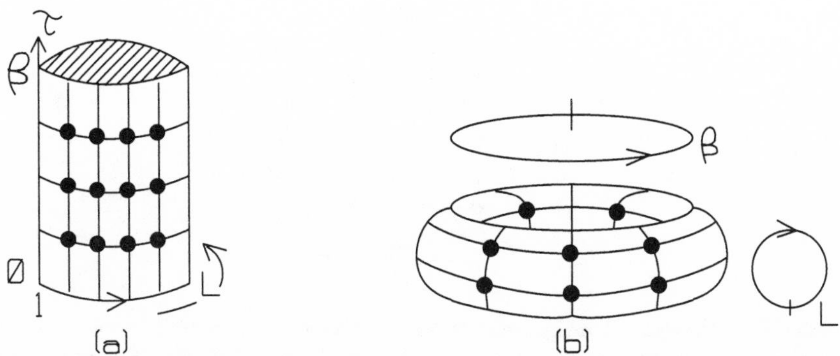

Figure 29.3: (a) One-dimensional spatial configuration, and (b) physical realization of periodic boundary conditions in one spatial dimension.

be gauge invariant so that the partition function Z will also have this property. The partition function should arise from summing only over physical field configurations. Recall that any *overall* factors in the measure are immaterial since they cancel in the ratio required in the thermal averages $\langle\langle O\rangle\rangle$ in Eq. (29.5).[25]

To construct a gauge-invariant measure, first examine the exponent of the matrices U_{ji}

$$\phi_{ji} \equiv e_0(x_j - x_i)_\mu A_\mu \left[\frac{1}{2}(x_j + x_i)\right] \tag{29.32}$$

This quantity is now a pure phase. The field variables in this lattice gauge theory enter only through the phase in

$$U_{ji} = \exp\{i\phi_{ji}\} \qquad\qquad \phi_{ij} = -\phi_{ji} \tag{29.33}$$

As before, ϕ depends on the direction of the link, reversing sign when the direction is reversed. Let us simplify the notation for the purposes of the present discussion and let $\{\phi_1, \phi_2, \phi_3, \phi_4\}$ stand for the phases of the links surrounding a given plaquette. Since the phase depends on the direction of the link, we will adopt the convention that the phase in the positive τ or positive x direction will be denoted by ϕ_i. This is illustrated in Fig. 29.4. The phase in the opposite direction is then $-\phi_i$. The contribution to the action from this plaquette is

$$\begin{aligned} S_\Box &= \sigma\{[1 - e^{i(\phi_1+\phi_2-\phi_3-\phi_4)}]_{\hookrightarrow} + [1 - e^{-i(\phi_1+\phi_2-\phi_3-\phi_4)}]_{\hookleftarrow}\} \\ &= 2\sigma[1 - \cos(\phi_1 + \phi_2 - \phi_3 - \phi_4)] \end{aligned} \tag{29.34}$$

What values can the phases ϕ_i take? Each component of A_μ along the link is an independent variable and ϕ can in principle take all values. Note, however, that

[25] Overall factors in the measure add a constant to $\ln Z$ and since $\beta F = \beta(E - TS) = -\ln Z$, they add a constant to S_{vac}, the entropy of the vacuum, which is unobservable.

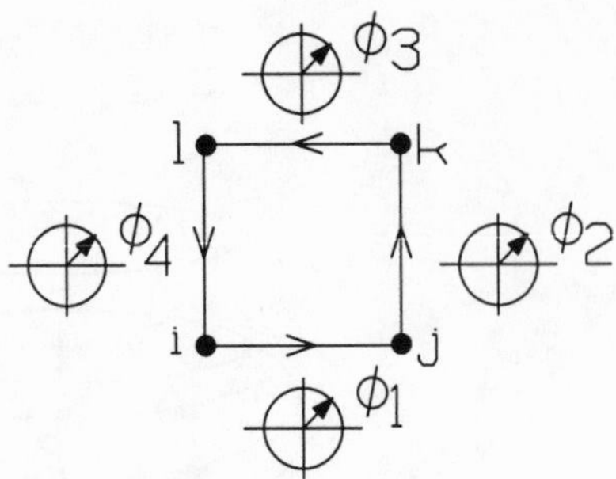

Figure 29.4: Phases on links surrounding a given plaquette. The convention is that the phase in the positive τ or positive x direction is denoted by ϕ_i. The phase in the opposite direction is then $-\phi_i$.

the function $e^{i\phi}$ is *periodic* with period 2π and thus any field configuration that corresponds to ϕ translated by 2π in any direction does not make an independent contribution to the action. *We will thus simply retain one complete period of ϕ_i — all other values of the field give rise to a value of $U = e^{i\phi}$ already counted.* Thus the measure will be taken as an integration over phase ϕ such that $0 \leq \phi \leq 2\pi$

$$I = \frac{1}{2\pi}\int_0^{2\pi} d\phi f(e^{i\phi}) \tag{29.35}$$

This measure is now *locally gauge invariant*! To see this, note that a local gauge transformation changes the phase along the link by some constant λ, thus $\phi \to \phi + \lambda$. The contribution of this link to the partition function is modified to

$$I_\lambda = \frac{1}{2\pi}\int_0^{2\pi} d\phi f(e^{i[\phi+\lambda]}) \tag{29.36}$$

Now change the dummy integration variable $\phi \to \phi + \lambda \equiv \phi'$ with $d\phi = d\phi'$

$$\begin{aligned} I_\lambda &= \frac{1}{2\pi}\int_\lambda^{2\pi+\lambda} d\phi' f(e^{i\phi'}) \\ &= \frac{1}{2\pi}\int_0^{2\pi} d\phi f(e^{i\phi}) = I \end{aligned} \tag{29.37}$$

The last line follows since $f(e^{i\phi})$ is *periodic in ϕ*.

Thus the gauge-invariant measure in this lattice gauge theory will be taken to be

$$\int \bar{\mathcal{D}}(A_\mu) \equiv \prod_{\text{links}} \int_0^{2\pi} \frac{d\phi_l}{2\pi} \tag{29.38}$$

Here ϕ_l is the phase associated with a given link in the positive τ or positive x direction.

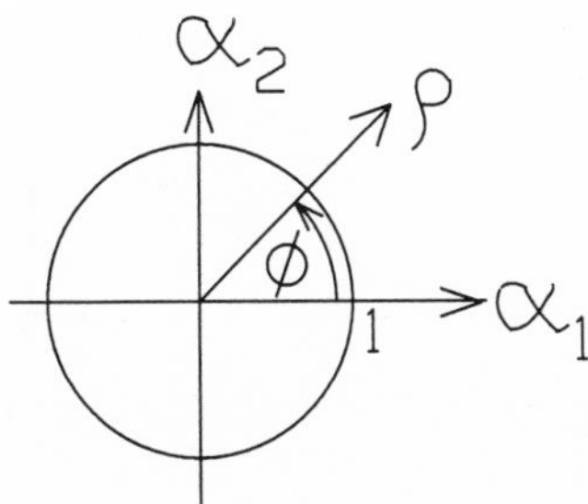

Figure 29.5: Cartesian coordinates in the field parameter space for a given link.

Note that in the continuum limit as the lattice spacing $a \to 0$, the field A_μ must cover an infinite range of values so that the phase ϕ defined in Eq. (29.32) covers the range $0 \to 2\pi$.

It is convenient for this and future discussions to write the measure in *another way*. Instead of integrating over the phase ϕ_{ji}, one can integrate over the matrix $U_{ji} = e^{i\phi_{ji}}$ itself. To see this, go to cartesian coordinates in this field parameter space as illustrated in Fig. 29.5. The integral in Eq. (29.35) can be written in the following fashion

$$I = \frac{1}{\pi}\int\int d\alpha_1 d\alpha_2 \delta(\alpha_1^2 + \alpha_2^2 - 1) f(\alpha_1 + i\alpha_2) \tag{29.39}$$

Evaluation in polar coordinates establishes the equality

$$\begin{aligned} I &= \frac{1}{\pi}\int\int d^2\alpha\,\delta(\boldsymbol{\alpha}^{\,2} - 1) f(\alpha_1 + i\alpha_2) \\ &= \frac{1}{\pi}\int\int \rho d\rho d\phi\,\delta(\rho^2 - 1) f(\rho e^{i\phi}) \\ &= \frac{1}{2\pi}\int\int d\rho^2\delta(\rho^2 - 1) d\phi f(\rho e^{i\phi}) = \frac{1}{2\pi}\int_0^{2\pi} d\phi f(e^{i\phi}) \end{aligned} \tag{29.40}$$

Hence one can take either of the following expressions as the measure for the path integral over a given link

$$\frac{1}{2\pi}\int_0^{2\pi} d\phi f(e^{i\phi}) = \frac{1}{\pi}\int\int d^2\alpha\,\delta(\boldsymbol{\alpha}^{\,2} - 1) f(\alpha_1 + i\alpha_2) \tag{29.41}$$

Summary. In summary, the partition function in lattice gauge theory for QED in $1+1$ dimensions is calculated as follows:

1. Temperature corresponds to imaginary time ($t = -i\tau$) and rotation of a contour in the path integrals over the fields converts the entire problem to the euclidian metric.

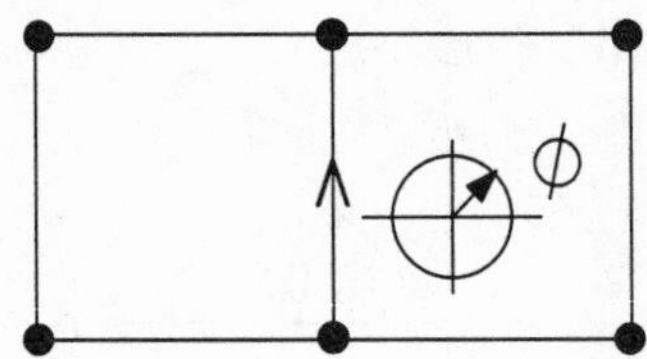

Figure 29.6: Each link appears in the contribution to the action of two neighboring plaquettes.

2. Space-tau is divided into an (n, n) lattice of dimension a. Squares with neighboring lattice sites at the four corners are *plaquettes*; connections between nearest neighbors (the edges of the plaquettes) are *links* (Figs. 29.1 and 29.2). Periodic boundary conditions are imposed in both the space and tau directions [recall Eqs. (29.31) and Fig. 29.3].

3. The field variables are associated with the links and enter through a phase $U_{ji} = e^{i\phi_{ji}}$ [Eqs. (29.10) and (29.32)]. The phase changes sign with direction (Fig. 29.4).

4. The action receives a contribution from each plaquette; it is obtained from the product of the field contributions around the plaquette [Eq. (29.12)] $S_\Box = \sigma\{[1 - U_{\circlearrowleft}] + [1 - U_{\circlearrowright}]\} \equiv \sigma\{[1 - U_\Box] + [1 - U_\Box^*]\}$.

5. The partition function is obtained by integrating the exponential of the action with the measure of Eq. (29.38). Thus

$$\begin{aligned} S_\Box &= 2\sigma(1 - \mathrm{Re}\, U_\Box) \\ U_\Box &= \exp\{i(\phi_1 + \phi_2 - \phi_3 - \phi_4)\} \\ Z &= \prod_{\text{links}} \int_0^{2\pi} \frac{d\phi_l}{2\pi} \exp\{-\sum_\Box S_\Box\} \end{aligned} \tag{29.42}$$

The situation is illustrated in Fig. 29.4. The convention here is that the phase ϕ_i is associated with the positive x or τ direction along the link.

6. Equation (29.41) presents an alternate form of the measure.

7. The problem is coupled since each link occurs in the contribution to the action of two neighboring plaquettes (see Fig. 29.6).

8. The action is gauge invariant [Eqs. (29.15) and (29.16)].

9. The measure is also gauge invariant [Eqs. (29.36) and (29.37)].

10. The theory has the correct continuum limit [Eq. (29.28)]; σ is defined in Eq. (29.25).

This model is now completely defined. One can at this stage, for example, put the calculation of the partition function on a computer. Before launching on large-scale numerical calculations, however, it is always useful to first obtain some physical insight into the theory. We shall solve this model, as well as its subsequent generalizations, analytically in some simple limiting cases. We proceed to discuss mean field theory.

30

MEAN FIELD THEORY

Equations (29.42) summarize the content of lattice gauge theory for QED in $1+1$ dimensions; they are a self-contained set of expressions. In this section we solve those equations analytically under certain simplifying assumptions to get some physical insight into their implications. We work in mean field theory (MFT) where the basic idea is to reduce the coupled many-body problem to a one-body problem in a mean field coming from the average interaction with all the other degrees of freedom.

Although the exact problem clearly becomes more complicated, it is possible to carry out the MFT in any number of *dimensions*. In fact, one expects MFT to become more valid as the number of nearest neighbors increases. We first need some elementary considerations in different numbers of dimensions d.

As an aid in the counting, divide the lattice into basic building blocks from which the entire lattice can be constructed by simple repetition. In two dimensions ($d = 2$), the elementary building block is the square, and in three dimensions ($d = 3$) it is the cube. This is easily seen (Fig. 30.1) and extended to d dimensions. At each site draw the positive orthogonal coordinate axes and place the building block between the positive axes. The site can then be associated in a one-to-one fashion with the origin of this coordinate system, and the lattice constructed by repetition of the basic building block at each site. Some elementary results then follow immediately.

The *volume* per site is a^d. The volume of the lattice is then given by summing that volume over all the sites

$$\begin{aligned} \text{volume/site} &= a^d \\ \text{volume} &= \sum_{\text{sites}} a^d \rightarrow \int d^d x \end{aligned} \tag{30.1}$$

The number of *links* per site is given by counting the number of positive coordinate axes, which is just the dimension of the problem

$$\text{number of links/site} = d \tag{30.2}$$

The number of *plaquettes* per site is just the number of pairs of positive coordinate axes, for each pair determines an independent plane; the number of these pairs is $d(d-1)/2$

$$\text{number of plaquettes/site} = \frac{1}{2}d(d-1) \tag{30.3}$$

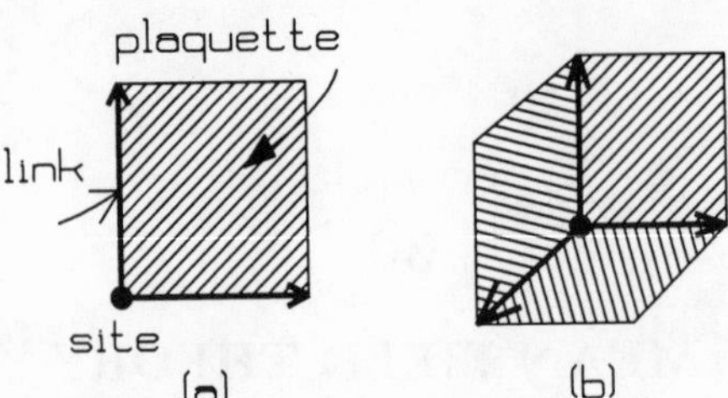

Figure 30.1: Elementary building blocks of the lattice in different number of dimensions: (a) $d = 2$; (b) $d = 3$.

A complete nonoverlapping enumeration of the terms in various sums is thus as follows:

$$\sum_{\text{sites}}$$
$$\sum_{\text{links}} = \sum_{\text{sites}} \sum_{\text{links/site}} \quad ; \qquad \text{e.g. } \sum_{\text{links}} 1 = d \sum_{\text{sites}} 1$$
$$\sum_{\square} = \sum_{\text{sites}} \sum_{\text{plaquettes/site}} \quad ; \qquad \text{e.g. } \sum_{\square} 1 = \frac{1}{2} d(d-1) \sum_{\text{sites}} 1 \qquad (30.4)$$

In the MFT the problem is reduced to a one-body problem in a mean field. This shall be done by concentrating on each of the above independent individual units, the basic building blocks, at a given site. We start the discussion of MFT by recalling a similar analysis of a more familiar physical system; the Ising model (see Refs. [Q22, Q23]).

Ising Model — Review. The Ising model consists of a set of spins on a lattice where the spins can take two values, up and down, and there is a constant interaction, either attractive or repulsive, between nearest neighbors. The physical situation is indicated in Fig. 30.2. The hamiltonian is

$$H = -J \sum_{\langle ij \rangle} s_i s_j \qquad s_i, s_j = \pm 1 \qquad (30.5)$$

Here $\langle ij \rangle$ indicates nearest neighbors on the lattice. Evidently a sum over nearest neighbors is identical to a sum over links

$$\sum_{\langle ij \rangle} = \sum_{\text{links}} \qquad (30.6)$$

The constant J can have either sign. If $J > 0$ the interaction between nearest neighbors is attractive and the spins tend to align.

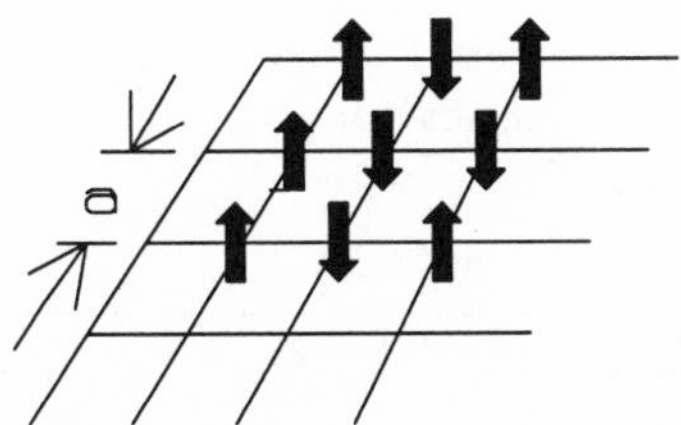

Figure 30.2: Two-dimensional Ising model.

The partition function for this system is given by

$$\begin{aligned} Z &= \sum_{s_1=\pm1}\sum_{s_2=\pm1}\cdots \exp\{-\beta H\} \equiv \sum_{\{s\}} \exp\{-\beta H\} \\ &= \sum_{\{s\}} \exp\{\beta J \sum_{\text{links}} s_i s_j\} \end{aligned} \tag{30.7}$$

With an external field H_{ext} with which the spins interact, the partition function is modified to

$$Z = \sum_{\{s\}} \exp\left\{\beta\left[J\sum_{\text{links}} s_i s_j + H_{\text{ext}}\sum_{\text{sites}} s_i\right]\right\} \tag{30.8}$$

The spins in this problem are clearly coupled; each spin enters in the exponent through an additive term coupling it to its nearest neighbors.

MFT.[26] The object is to replace the coupled problem by an effective one-body problem where a given spin can move dynamically in a mean field created by the average interaction with its neighbors. With periodic boundary conditions all sites are equivalent. Consider then the ith site and denote the expectation value of the spin at that site by $\langle s\rangle \equiv m$. Through the use of the partition function and the definition of thermal averages discussed in the previous section one has

$$\langle s\rangle \;\equiv\; m \;=\; \frac{\sum_{\{s\}} s_i \exp\{\beta J \sum_{\text{links}} s_i s_j\}}{\sum_{\{s\}} \exp\{\beta J \sum_{\text{links}} s_i s_j\}} \tag{30.9}$$

The problem will now be *decoupled* by replacing the two-body interaction terms by the following average value for the interaction with the spin at the ith site

$$\langle s_i s_j\rangle_i \;=\; s_i\langle s_j\rangle_i \;\equiv\; s_i\, m \tag{30.10}$$

The spin s_i is still dynamic, but it sees only the average value of the spin at the neighboring site. Since s_i is decoupled from all the other spins, the dependence

[26] This is the author's own version of MFT; it is not Bragg-Williams.

on s_i now *factors* in the exponential. The remaining sums over all the other spins in the lattice are now identical in the numerator and denominator and *cancel in the ratio.* Hence one is left with

$$m = \frac{\sum_{\{s_i\}} s_i \exp\{\beta J \sum_{\text{links at } i\text{th site}} s_i m\}}{\sum_{\{s_i\}} \exp\{\beta J \sum_{\text{links at } i\text{th site}} s_i m\}} \tag{30.11}$$

The goal of reduction to an effective one-body problem has been accomplished. It is still a nontrivial problem since the expectation value of the spin m itself depends on the average field, which appears in the exponential.

It remains to evaluate

$$\begin{aligned} \sum_{\text{links at } i\text{th site}} \langle s_i s_j \rangle_i &= (s_i\, m) \times (\text{number links/site}) \\ &\equiv s_i m \gamma_l \end{aligned} \tag{30.12}$$

Here γ_l is the *effective coordination number*, the average number of other spins with which the ith spin interacts; this quantity is discussed in more detail below. With this definition the problem is reduced to

$$\begin{aligned} m &= \frac{\sum_{s_i=\pm 1} s_i \exp\{(\beta J \gamma_l m) s_i\}}{\sum_{s_i=\pm 1} \exp\{(\beta J \gamma_l m) s_i\}} \\ &= \frac{e^x - e^{-x}}{e^x + e^{-x}}\,; \qquad\qquad x \equiv \beta J \gamma_l m \\ m &= \tanh\{(\beta J \gamma_l) m\} \end{aligned} \tag{30.13}$$

This is a self-consistency equation for the magnetization m in the Ising model in MFT. With an external field, the same calculation yields

$$m = \tanh\{\beta(J\gamma_l m + H_{\text{ext}})\} \tag{30.14}$$

Let us now concentrate on γ_l, which measures the average number of nearest neighbor spins with which the ith spin interacts. One cannot determine this quantity unambiguously at this stage of the argument without a more powerful principle such as the minimization of the total free energy.[27] To see this, note that the term $s_i s_j$ in the hamiltonian in the exponential could be replaced by ms_i where it serves as a mean field for the ith spin; or it could be replaced by ms_j where it serves as a mean field for the jth spin and is no longer seen by the ith spin; or it could be replaced by any combination of these expressions. In a later section the MFT equations will be derived in a global and more rigorous manner. Here we simply use intuition to gain physical insight. It was shown in Eqs. (30.4) that the sum over links can be decomposed into a complete and

[27] Recall in quantum mechanics the Hartree (or Hartree-Fock) equations are determined by minimizing the expectation value of the total hamiltonian with a product (or determinant) of single-particle wave functions.

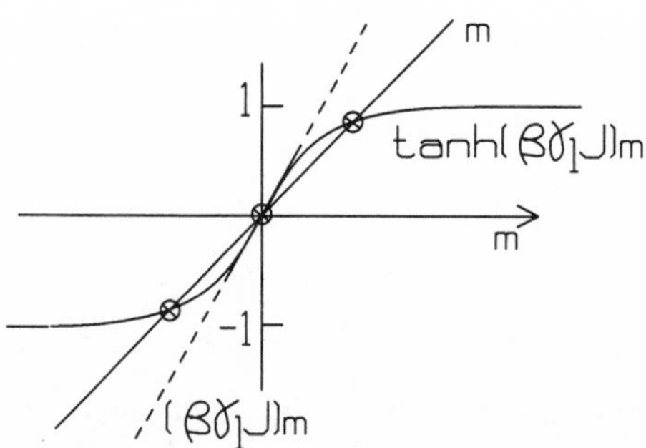

Figure 30.3: Graphic solution to self-consistency equation for magnetization in Ising model in MFT.

nonoverlapping sum over sites if one first sums over the independent links at that site. It will be assumed that it is the independent links at the ith site that are coupled in the mean field approximation to the spin at the ith site. Equations (30.4) then provide a simple expression for the effective coordination number

$$\gamma_l \equiv \text{number links/site} = d \tag{30.15}$$

Equation (30.13) can now be solved graphically as indicated in Fig. 30.3. It is evident from the figure that if the slope $(\beta J\gamma_l) < 1$ the only solution is $m = 0$. On the other hand, if $(\beta J\gamma_l) > 1$, there is an additional intersection point with $m > 0$.[28] There is thus a critical value of the slope at the origin above which one begins to obtain a macroscopic value of $m \neq 0$.

$$\beta_C J\gamma_l = 1 \qquad\qquad T_C = \frac{J\gamma_l}{k_B} \tag{30.16}$$

Insertion of Eq. (30.15) determines the *critical temperature* to be

$$T_C = \frac{Jd}{k_B} \tag{30.17}$$

This is the critical temperature below which one finds a macroscopic magnetization $m \neq 0$ in the Ising model in d dimensions in MFT.

The two-dimensional Ising model was solved analytically by Onsager in a tour de force calculation, and indeed for positive J there is a phase transition to a ferromagnetic phase at a transition temperature very close to that given

[28] There are actually two solutions with $m \neq 0$ corresponding to the two possible directions for the total spin of the lattice. One solution may be selected by starting with finite H_{ext} and then letting $H_{\text{ext}} \to 0$. To show it is the stable solution, it is also necessary to show that the solution with $m \neq 0$ has lower free energy that the one with $m = 0$ (see Ref. [Q23]).

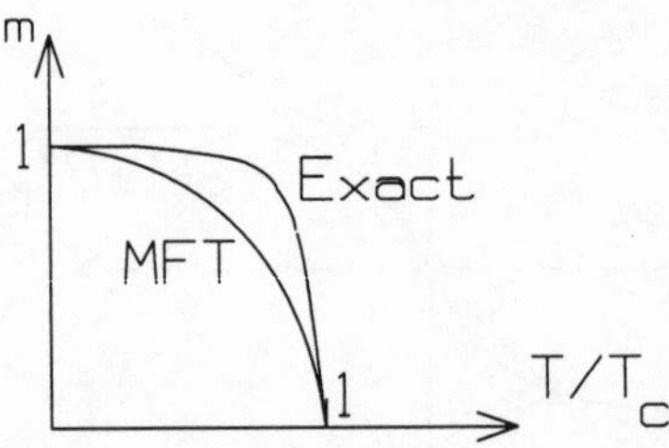

Figure 30.4: Sketch of behavior of magnetization in two-dimensional Ising model as a function of temperature in MFT compared with exact result (see Refs. [Q22, Q23]).

by this MFT (see Refs. [Q22, Q23])

$$T_{\mathrm{C}} = \frac{2J}{k_{\mathrm{B}}} \; ; \qquad \text{MFT}$$
$$T_{\mathrm{C}} = \frac{2.269J}{k_{\mathrm{B}}} \; ; \qquad \text{Exact} \tag{30.18}$$

The behavior of the magnetization as a function of temperature is qualitatively correct when compared with the exact answer in this case as sketched in Fig. 30.4.

Lattice Gauge Theory for QED in MFT. Lattice gauge theory for QED in d dimensions will be treated in MFT using analogous arguments (Ref. [Q17]).

> *We again focus the discussion on the basic building block, or cell, from which the entire system is constructed by repetition.*

The cell contains one site, d links in the direction of the positive coordinate axes, and $d(d-1)/2$ plaquettes — planes formed by the positive coordinate axes (see Fig. 30.1).

Concentrate first on the dynamics of a single link variable connected to the ith site. Pick a gauge. Now suppose all other link variables coupled to this one through plaquettes appearing in the sum over plaquettes in the action were to be replaced by a mean value

$$\langle e^{i\phi}\rangle \equiv me^{i\chi} \qquad\qquad \langle e^{-i\phi}\rangle = me^{-i\chi} \tag{30.19}$$

Here a simple parameterization has been introduced for this complex vector; m is its modulus and χ is its phase. In order to have physics, one must deal with a *gauge-invariant* quantity; the contribution to the action from a given plaquette is a suitable candidate.[29] Substitution of the MFT result for each link into $S_{\Box}$

[29] Since we worked hard to make that gauge invariant.

leads to

$$\begin{aligned}\langle S_{\square}\rangle &= 2\sigma\{1-\mathrm{Re}[(me^{i\chi})^2(me^{-i\chi})^2]\} \\ &= 2\sigma(1-m^4)\end{aligned} \tag{30.20}$$

Thus the "magnetization" m represents a gauge-invariant quantity.

With periodic boundary conditions all sites, links, and plaquettes are equivalent. Again work within a given gauge. Consider a link variable connected to the ith site. Denote this generically as the ith link with $U_i = e^{i\phi_i}$. The MFT value of the contribution to the action in the exponential from each plaquette containing this link to the ith site leaves this link as a dynamic variable and replaces all the other links in that plaquette by their mean value

$$\begin{aligned}\langle S_{\square}\rangle_i &= 2\sigma\{1-\mathrm{Re}[m^3e^{i(\phi_i-\chi)}]\} \\ &= 2\sigma[1-m^3\cos(\phi_i-\chi)]\end{aligned} \tag{30.21}$$

As before, introduce $\gamma_{\square}$ as the effective coordination number. Here

> $\gamma_{\square}$ *is the average number of plaquettes in the ith cell that contain a given link to the ith site.*

One then has

$$\overline{\langle S_{\square}\rangle_i} = \frac{\int_0^{2\pi} d\phi_i \langle S_{\square}\rangle_i \exp\{-\langle S_{\square}\rangle_i\gamma_{\square}\}}{\int_0^{2\pi} d\phi_i \exp\{-\langle S_{\square}\rangle_i\gamma_{\square}\}} \tag{30.22}$$

Substitution of Eqs. (30.21) and (30.20) then yields

$$2\sigma(1-m^4) = \frac{\int_0^{2\pi} d\phi_i 2\sigma[1-m^3\cos(\phi_i-\chi)]\exp\{-2\sigma[1-m^3\cos(\phi_i-\chi)]\gamma_{\square}\}}{\int_0^{2\pi} d\phi_i \exp\{-2\sigma[1-m^3\cos(\phi_i-\chi)]\gamma_{\square}\}} \tag{30.23}$$

Now change the variable of integration in the integrals. Let $\phi \equiv \phi_i - \chi$ with $d\phi = d\phi_i$. The limits of integration can be restored to be $\int_0^{2\pi}$ since the integrand is periodic in ϕ. (Recall this is the argument used originally to justify the choice of gauge-invariant measure.) The result is (cf. Fig. 30.5)

$$2\sigma(1-m^4) = \frac{\int_0^{2\pi} d\phi 2\sigma(1-m^3\cos\phi)\exp\{-2\sigma(1-m^3\cos\phi)\gamma_{\square}\}}{\int_0^{2\pi} d\phi \exp\{-2\sigma(1-m^3\cos\phi)\gamma_{\square}\}} \tag{30.24}$$

A conceptual difficulty with this discussion is that although in the end it produces a relation between gauge-invariant quantities, it proceeds through the link variables, which are themselves gauge dependent.[30] A more satisfying approach is to use this discussion as *motivation* and simply make the *MFT ansatz*

[30] See, e.g., Ref. [Q94].

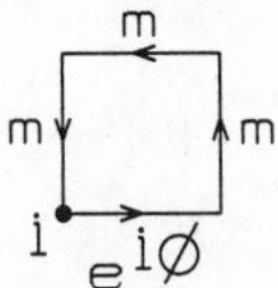

Figure 30.5: MFT ansatz for the contribution to the action from a plaquette in the ith cell attached to the ith link, a gauge-invariant quantity; it is parameterized in terms of a "magnetization" m and a single overall phase as $S_\Box = 2\sigma(1 - m^3 \cos\phi)$.

for the dynamic form of the contribution to the action from a plaquette in the ith cell attached to the ith link; this is $\langle S_\Box \rangle_i$ given in Eq. (30.21). Thus one assigns one common phase $(\phi_i - \chi)$ to each of these plaquettes and then evaluates the mean value of $\langle S_\Box \rangle_i$. The development of a magnetization m in this mean value $2\sigma(1 - m^4)$ can then be used to signal a phase transition in this MFT.[31]

Let us proceed then with the analysis of the MFT self-consistency relation in Eq. (30.24). Additive constants in the exponentials in the integrands lead to constant factors that can be cancelled in the *ratio* of integrals in this expression. In analogy with the Ising model, the final result is a transcendental equation for the magnetization m

$$2\sigma(1 - m^4) = \frac{\int_0^{2\pi} d\phi 2\sigma(1 - m^3 \cos\phi) \exp\{2\sigma\gamma_\Box m^3 \cos\phi\}}{\int_0^{2\pi} d\phi \exp\{2\sigma\gamma_\Box m^3 \cos\phi\}} \tag{30.25}$$

Cancellation of common terms on both sides leads to

$$m = \frac{\int_0^{2\pi} d\phi \cos\phi \exp\{2\sigma\gamma_\Box m^3 \cos\phi\}}{\int_0^{2\pi} d\phi \exp\{2\sigma\gamma_\Box m^3 \cos\phi\}} = \frac{I_1(2\sigma\gamma_\Box m^3)}{I_0(2\sigma\gamma_\Box m^3)} \tag{30.26}$$

The last equality identifies an integral representation of the modified Bessel function of imaginary argument (Ref. [R5]).

The effective coordination number $\gamma_\Box$ remains to be discussed. We proceed exactly as in the Ising model. Equations (30.4) express the sum over links and plaquettes as a complete set of nonoverlapping contributions obtained from a sum over sites where one first sums over the independent links and plaquettes at that site.

> *It will be assumed here that all of the plaquettes in the ith cell connected to a given link in the ith cell are coupled through a common term in the action of the form in Eq. (30.21).*

[31] The exact expression for the contribution to the action from the ith plaquette is $\langle S_\Box \rangle_i = 2\sigma[1 - \cos(\phi_1 + \phi_2 - \phi_3 - \phi_4)]$. The MFT ansatz discussed here assigns a single angle $(\phi_i - \chi)$ and multiplicative constant m to the angular dependence; the phase transition then signals an alignment of the contributions from the plaquettes.

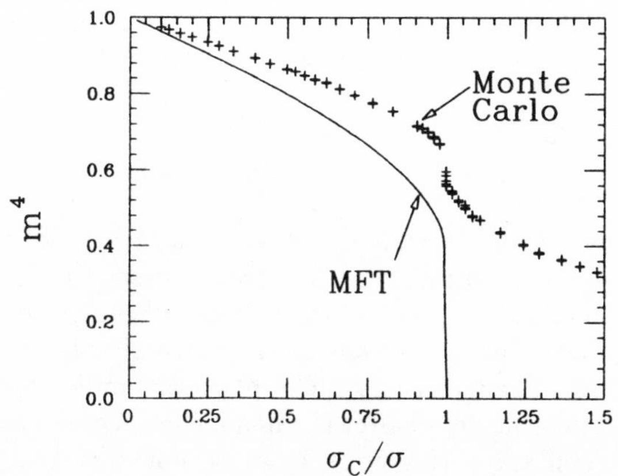

Figure 30.6: Sketch of the "magnetization" m^4 for QED in lattice gauge theory in 3+1 dimensions ($d = 4$) in MFT compared with the essentially exact Monte Carlo calculation on a 5^4 lattice. Here $S_\square \equiv 2\sigma(1 - m^4)$. (From Refs. [Q26, Q20]). The author is grateful to J. Dubach for preparing this figure.

The number of plaquettes in the ith cell passing through a link in the ith cell is equal to the number of positive coordinate axes orthogonal to a given positive axis; in d dimensions this number is just $(d - 1)$. Thus we choose

$$\gamma_\square = \text{number plaquettes /link} \equiv d - 1 \tag{30.27}$$

Numerical solution of Eq. (30.26) yields a universal result for the critical value σ_C above which there is a solution $m \neq 0$ (Ref. [Q20])

$$2\sigma_C\gamma_\square = 2\,(2.7878) \tag{30.28}$$

Combination with Eq. (30.27) yields

$$\sigma_C(d - 1) = 2.7878 \tag{30.29}$$

In 1+1 dimensions ($d = 2$) this MFT predicts a phase transition at $\sigma_C = 2.7878$. The problem with $d = 2$ can be solved exactly and contains no phase transition. Thus

$$\begin{aligned} \sigma_C = 2.7878\,; &\qquad \text{MFT} \\ \text{no phase transition}\,; &\qquad \text{Exact} \end{aligned} \tag{30.30}$$

This should not be so surprising since the same thing happens with the one-dimensional Ising model. Equation (30.17) predicts $T_C = J/k_B$ for $d = 1$ while the exact theory has no phase transition in one dimension.

In 3+1 dimensions ($d = 4$) the results of this MFT are more realistic. Comparison with the Monte Carlo results of Refs. [Q24, Q25, Q26, Q20] gives

$$\begin{aligned} \sigma_C &= 0.9293\,; && \text{MFT} \\ &= 0.4975\,; && \text{Monte Carlo on } 5^4 \text{ lattice} \end{aligned} \tag{30.31}$$

The resulting magnetization m is sketched in the two cases in Fig. 30.6. Although quantitatively incorrect, the MFT result is striking in its simple qualitative description of the observed exact numerical behavior of $S_\Box$.

The developments in this section are intended simply to provide some familiarity with lattice gauge theory. The physical implication of these results will be discussed in some detail after the nonabelian theory has been developed. The reader may well ask at this stage, however, how this theory of QED, which has as its continuum limit *the free field theory where the coupling constant e_0 never appears*, can ever give rise to a phase transition. The answer is that this lattice gauge theory is really a model field theory describing an entirely different physical situation. In this model, the coupling constant enters in the phase of the field variables U_{ji} and, thus, for finite a, the action contains an infinite series in e_0! For finite a it is a fully interacting field theory and may have many rich and interesting properties. In the continuum limit it is, indeed, constructed to reduce to the free field theory. It is thus essential to understand how one goes to the continuum limit in lattice gauge theory. With this in mind, we turn to a discussion of the nonabelian case.

31

NONABELIAN THEORY — SU(2)

We turn next to a development of lattice gauge theory for Yang-Mills nonabelian gauge groups. $SU(2)$ is considered as a specific example since in this case it is relatively simple to deal explicitly with all of the required matrices. The discussion is based on Refs. [Q17, Q18, Q19, Q20].

One now has an *additional internal space* for each field variable. The quantity $\mathbf{A}_\mu$ is a vector in this internal isospin space with three components $\mathbf{A}_\mu$: $(A_\mu^1, A_\mu^2, A_\mu^3)$. The situation for the basic plaquette is illustrated in Fig. 31.1. The phase angle $\boldsymbol{\theta}_{ji}$ is also defined to be a vector in this internal space

$$\boldsymbol{\theta}_{ji} \equiv g_0(x_j - x_i)_\mu \mathbf{A}_\mu \left[\frac{1}{2}(x_i + x_j)\right] \tag{31.1}$$

As before the subscript (ji) indicates the connected sites.

Recall that the modification of the covariant derivative required to go from the abelian QED theory to this Yang-Mills theory is

$$\frac{\partial}{\partial x_\mu} - ie_0 A_\mu(x) \rightarrow \frac{\partial}{\partial x_\mu} - \frac{i}{2} g_0 \boldsymbol{\tau} \cdot \mathbf{A}_\mu(x) \tag{31.2}$$

In contrast to the simple phases of the abelian theory of QED, we are thus motivated to introduce the link variables as 2×2 $SU(2)$ matrices

$$\underline{U}_{ji}(\mathbf{A}_\mu) \equiv \exp\{\frac{i}{2}\boldsymbol{\tau} \cdot \boldsymbol{\theta}_{ji}\} \tag{31.3}$$

Here the internal matrix structure is again denoted by a bar under the symbol. Substitution of Eq. (31.1) leads to

$$\underline{U}_{ji}(\mathbf{A}_\mu) = \exp\left\{\frac{i}{2} g_0 \boldsymbol{\tau} \cdot \left[(x_j - x_i)_\mu \mathbf{A}_\mu \left(\frac{1}{2}(x_i + x_j)\right)\right]\right\} \tag{31.4}$$

The contribution to the action is defined to be the trace (tr) of the matrix product in this internal space taken around a plaquette (Fig. 31.1)

$$(U_\Box)_{\hookrightarrow} \equiv \frac{1}{2}\text{tr}\left(\underline{U}_{ji}\underline{U}_{il}\underline{U}_{lk}\underline{U}_{kj}\right) \equiv \frac{1}{2}\text{tr}\underline{U}_\Box \tag{31.5}$$

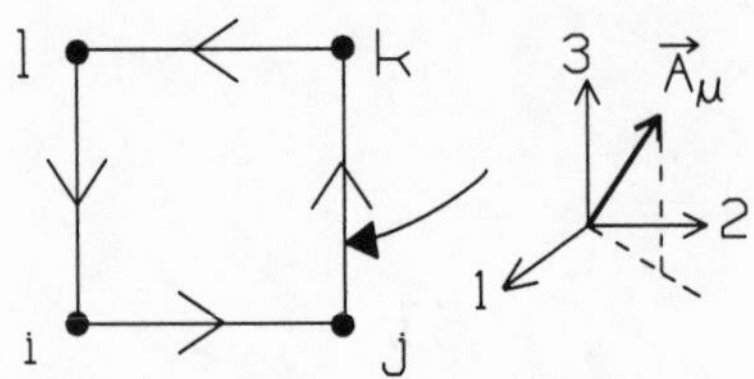

Figure 31.1: Basic plaquette and illustration of field $\mathbf{A}_\mu$ as vector in the internal isospin space in $SU(2)$ nonabelian lattice gauge theory.

The ordering of the matrices is now important; the final result after the trace is taken is just a c-number. If one goes around the plaquette in the opposite direction, the expression becomes

$$(U_\square)_{\hookleftarrow} \equiv \frac{1}{2}\mathrm{tr}\left(\mathsf{U}_{jk}\mathsf{U}_{kl}\mathsf{U}_{li}\mathsf{U}_{ij}\right) \tag{31.6}$$

Since the $\boldsymbol{\tau}$ matrices are hermitian, one evidently has[32]

$$\mathsf{U}_{ji}(\mathbf{A}_\mu)^\dagger = \mathsf{U}_{ij}(\mathbf{A}_\mu) \tag{31.7}$$

Substitution of this relation into Eq. (31.6) yields

$$\begin{aligned} (U_\square)_{\hookleftarrow} &= \frac{1}{2}\mathrm{tr}\left(\mathsf{U}_{kj}^\dagger\mathsf{U}_{lk}^\dagger\mathsf{U}_{il}^\dagger\mathsf{U}_{ji}^\dagger\right) = \frac{1}{2}\mathrm{tr}\left(\mathsf{U}_{ji}\mathsf{U}_{il}\mathsf{U}_{lk}\mathsf{U}_{kj}\right)^\dagger \\ &= \frac{1}{2}\mathrm{tr}\mathsf{U}_\square^\dagger = (U_\square)^*_{\hookrightarrow} \end{aligned} \tag{31.8}$$

The contribution of a plaquette to the action will be taken to be

$$\begin{aligned} S_\square &\equiv \sigma\{[1-(U_\square)_{\hookrightarrow}] + [1-(U_\square)_{\hookleftarrow}]\} \\ &= 2\sigma[1-\mathrm{Re}(U_\square)_{\hookrightarrow}] = 2\sigma\left(1-\frac{1}{2}\mathrm{Re}\left[\mathrm{tr}\mathsf{U}_\square\right]\right) \end{aligned} \tag{31.9}$$

The total action is obtained from a sum over plaquettes

$$\bar{S} \equiv \sum_\square S_\square \tag{31.10}$$

So far these are just ad hoc definitions. They assume importance if one can show, as in the abelian case, that

- The action is locally gauge invariant;
- The action has the correct continuum limit.

[32] Thus considered as matrices with respect to the site indices (j,i), they are again hermitian.

This we shall proceed to do.

Gauge Invariance. The goal is to show that the action as defined above is gauge invariant. We shall be content here to show invariance under an *infinitesimal* gauge transformation with $\boldsymbol{\theta} \to 0$ [Eq. (27.18)]; we work to first order in $\boldsymbol{\theta}$.

$$\mathbf{A}_\mu \to \mathbf{A}'_\mu = \mathbf{A}_\mu - \frac{1}{g_0}\frac{\partial\boldsymbol{\theta}}{\partial x_\mu} + \boldsymbol{\theta} \times \mathbf{A}_\mu \tag{31.11}$$

The link variables in Eq. (31.4) are then transformed into

$$\mathrm{U}_{ji}(\mathbf{A}'_\mu) = \tag{31.12}$$
$$\exp\left\{\frac{i}{2}g_0\boldsymbol{\tau}\cdot\left[(x_j - x_i)_\mu\mathbf{A}_\mu - \frac{1}{g_0}(x_j - x_i)_\mu\frac{\partial\boldsymbol{\theta}}{\partial x_\mu} + (x_j - x_i)_\mu\boldsymbol{\theta}\times\mathbf{A}_\mu\right]\right\}$$

The first term gives the original result. The definition of the gradient allows one to rewrite the second as

$$(x_j - x_i)_\mu\frac{\partial\boldsymbol{\theta}}{\partial x_\mu} = \boldsymbol{\theta}(x_j) - \boldsymbol{\theta}(x_i) \tag{31.13}$$

This expression is exact as $a \to 0$; we also work to first order in a. For the third term in Eq. (31.12) write

$$\begin{aligned} \frac{i}{2}\boldsymbol{\tau}\cdot[\boldsymbol{\theta}\times\mathbf{A}_\mu] &= \frac{i}{2}\epsilon_{ijk}\tau_k\theta^i A^j_\mu = \frac{1}{4}[\tau_i,\tau_j]\theta^i A^j_\mu \\ &= [\frac{1}{2}\boldsymbol{\tau}\cdot\boldsymbol{\theta}, \frac{1}{2}\boldsymbol{\tau}\cdot\mathbf{A}_\mu] \end{aligned} \tag{31.14}$$

Thus

$$\begin{aligned} \mathrm{U}_{ji}(\mathbf{A}'_\mu) &= \exp\left\{\frac{i}{2}g_0\boldsymbol{\tau}\cdot[(x_j - x_i)_\mu\mathbf{A}_\mu]\right. \\ &\left. -\frac{i}{2}\boldsymbol{\tau}\cdot[\boldsymbol{\theta}(x_j) - \boldsymbol{\theta}(x_i)] + g_0(x_j - x_i)_\mu[\frac{1}{2}\boldsymbol{\tau}\cdot\boldsymbol{\theta}, \frac{1}{2}\boldsymbol{\tau}\cdot\mathbf{A}_\mu]\right\} \end{aligned} \tag{31.15}$$

We now claim that with the neglect of terms of $O(a^2\boldsymbol{\theta})$ and $O(a\boldsymbol{\theta}^2)$ in the exponent, the relation in Eq. (31.15) can be rewritten as

$$\begin{aligned} \mathrm{U}_{ji}(\mathbf{A}'_\mu) &= [\exp\{-\frac{i}{2}\boldsymbol{\tau}\cdot\boldsymbol{\theta}(x_j)\}][\mathrm{U}_{ji}(\mathbf{A}_\mu)][\exp\{\frac{i}{2}\boldsymbol{\tau}\cdot\boldsymbol{\theta}(x_i)\}] \\ &\equiv \mathrm{g}^{-1}(x_j)\mathrm{U}_{ji}(\mathbf{A}_\mu)\mathrm{g}(x_i) \end{aligned} \tag{31.16}$$

Here $\mathrm{g}(x) = \exp\{\frac{i}{2}\boldsymbol{\tau}\cdot\boldsymbol{\theta}(x)\}$ is a local $SU(2)$ transformation. This result states that a gauge transformation in the nonabelian theory multiplies the link variables by a local gauge transformation, and its inverse, at the *sites at the ends of the links.*[33] The action is again gauge invariant if this result holds, for the matrices g cancel when the trace is taken around a plaquette[34]

$$\begin{aligned}\frac{1}{2}\mathrm{tr}\underset{\sim}{\mathrm{U}}_{\Box}(\mathbf{A}'_\mu) &= \frac{1}{2}\mathrm{tr}[\underset{\sim}{\mathrm{g}}^{-1}(x_j)\underset{\sim}{\mathrm{U}}_{ji}\underset{\sim}{\mathrm{g}}(x_i)\underset{\sim}{\mathrm{g}}^{-1}(x_i)\underset{\sim}{\mathrm{U}}_{il}\underset{\sim}{\mathrm{g}}(x_l) \\ &\quad \times\underset{\sim}{\mathrm{g}}^{-1}(x_l)\underset{\sim}{\mathrm{U}}_{lk}\underset{\sim}{\mathrm{g}}(x_k)\underset{\sim}{\mathrm{g}}^{-1}(x_k)\underset{\sim}{\mathrm{U}}_{kj}\underset{\sim}{\mathrm{g}}(x_j)] \\ &= \frac{1}{2}\mathrm{tr}\underset{\sim}{\mathrm{U}}_{\Box}(\mathbf{A}_\mu) \end{aligned} \tag{31.17}$$

It remains to demonstrate Eq. (31.16). To do this one can invoke the Baker-Haussdorf formula

$$\begin{aligned} e^A e^B &= e^{A+B+\frac{1}{2}[A,B]} \\ \text{if} \quad [A,[A,B]] &= [B,[A,B]] = 0 \end{aligned} \tag{31.18}$$

This is an *algebraic identity* holding for both operators and matrices. The derivation of this relation is discussed in Probs. 31.1-2. A combination of the first two exponentials in Eq. (31.16) then gives

$$\begin{aligned}\underset{\sim}{\mathrm{g}}^{-1}\underset{\sim}{\mathrm{U}} &= \exp\left\{\frac{i}{2}g_0\boldsymbol{\tau}\cdot(x_j-x_i)_\mu\mathbf{A}_\mu - \frac{i}{2}\boldsymbol{\tau}\cdot\boldsymbol{\theta}(x_j)\right. \\ &\quad \left.+\frac{1}{2}g_0(x_j-x_i)_\mu[\frac{1}{2}\boldsymbol{\tau}\cdot\boldsymbol{\theta}(x_j),\frac{1}{2}\boldsymbol{\tau}\cdot\mathbf{A}_\mu\left(\frac{1}{2}(x_i+x_j)\right)]\right\}\end{aligned} \tag{31.19}$$

The last term in the exponential commutes with the other two through the order to which we are working. Also, $\boldsymbol{\theta}$ can be evaluated at the midpoint of the link in the last term since it is already of $O(a\theta)$. Now combine with the last exponential in Eq. (31.16)

$$\begin{aligned}\underset{\sim}{\mathrm{g}}^{-1}\underset{\sim}{\mathrm{U}}\underset{\sim}{\mathrm{g}} &= \exp\left\{\frac{i}{2}g_0\boldsymbol{\tau}\cdot(x_j-x_i)_\mu\mathbf{A}_\mu - \frac{i}{2}\boldsymbol{\tau}\cdot\boldsymbol{\theta}(x_j) + \frac{i}{2}\boldsymbol{\tau}\cdot\boldsymbol{\theta}(x_i)\right. \\ &\quad +\frac{1}{2}g_0(x_j-x_i)_\mu[\frac{1}{2}\boldsymbol{\tau}\cdot\boldsymbol{\theta},\frac{1}{2}\boldsymbol{\tau}\cdot\mathbf{A}_\mu] \\ &\quad \left.-\frac{1}{2}g_0(x_j-x_i)_\mu[\frac{1}{2}\boldsymbol{\tau}\cdot\mathbf{A}_\mu\left(\frac{1}{2}(x_i+x_j)\right),\frac{1}{2}\boldsymbol{\tau}\cdot\boldsymbol{\theta}(x_i)]\right\}\end{aligned} \tag{31.20}$$

One can again verify that the conditions of Eq. (31.18) are satisfied through the order to which we are working, and, again, $\boldsymbol{\theta}$ can be evaluated at the midpoint of the link in the last term. This expression in Eq. (31.20) is immediately rewritten

[33] The result can actually be established for finite gauge transformations through more general considerations; to illustrate the concepts we are content to prove it here explicitly for infinitesimals.

[34] Compare Eqs. (29.15) and (29.16).

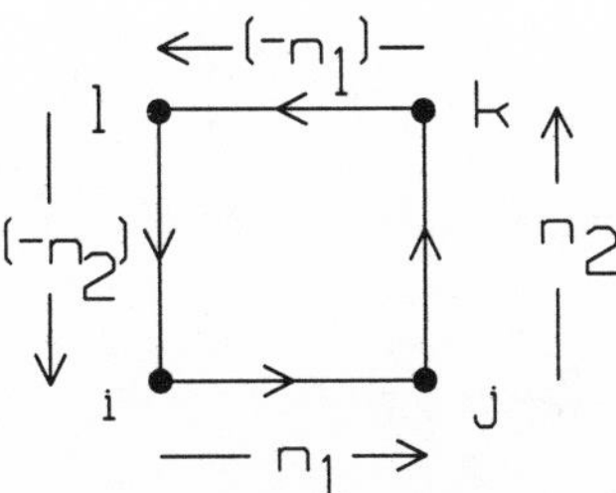

Figure 31.2: Quantities in basic plaquette used in evaluation of continuum limit $a \to 0$.

as

$$\exp\left\{\frac{i}{2}g_0\boldsymbol{\tau}\cdot[(x_j-x_i)_\mu\mathbf{A}_\mu]-\frac{i}{2}\boldsymbol{\tau}\cdot[\boldsymbol{\theta}(x_j)-\boldsymbol{\theta}(x_i)]\right.$$
$$\left.+g_0(x_j-x_i)_\mu[\frac{1}{2}\boldsymbol{\tau}\cdot\boldsymbol{\theta},\frac{1}{2}\boldsymbol{\tau}\cdot\mathbf{A}_\mu]\right\}=\mathfrak{U}_{ji}(\mathbf{A}'_\mu) \tag{31.21}$$

The last equality is just Eq. (31.15), the result we set out to prove. Thus the gauge invariance of the action has been established.[35]

Continuum Limit. Consider next the continuum limit of the model as $a \to 0$. The goal is to show that the action takes the correct form in this limit. The link variables are defined in Eq. (31.4). The action is obtained from a product of the link variables around a plaquette, as illustrated in Fig. 31.2. The relation in Eq. (31.18) can first be used to evaluate the product of two link variables

$$\mathfrak{U}_{kj}\mathfrak{U}_{ji}= \tag{31.22}$$
$$\exp\left\{\frac{i}{2}g_0\boldsymbol{\tau}\cdot\left[(x_k-x_j)_\mu\mathbf{A}_\mu\left(\frac{1}{2}(x_k+x_j)\right)+(x_j-x_i)_\mu\mathbf{A}_\mu\left(\frac{1}{2}(x_i+x_j)\right)\right]\right.$$
$$\left.-\frac{g_0^2}{2}(x_k-x_j)_\nu(x_j-x_i)_\mu\left[\frac{\boldsymbol{\tau}}{2}\cdot\mathbf{A}_\nu\left(\frac{1}{2}(x_k+x_j)\right),\frac{\boldsymbol{\tau}}{2}\cdot\mathbf{A}_\mu\left(\frac{1}{2}(x_i+x_j)\right)\right]\right\}$$

We neglect terms of $O(a^3)$ in the exponent; to this order the conditions of Eq. (31.18) are satisfied. In addition, to this order, the $\mathbf{A}_\mu$ in the last term can be evaluated at the center of the plaquette $x \equiv (x_i+x_j+x_k+x_l)/4$. Now write the small displacement vectors along the links as (Fig. 31.2)

$$(x_k-x_j)_\nu=a(n_2)_\nu \qquad\qquad (x_j-x_i)_\mu=a(n_1)_\mu \tag{31.23}$$

The result in Eq. (31.22) then takes the form

$$\mathfrak{U}_{kj}\mathfrak{U}_{ji}= \tag{31.24}$$

[35] At least to this order.

$$\exp\left\{\frac{i}{2}g_0\boldsymbol{\tau}\cdot\left(\int_i^k\right)_{\hookrightarrow}dx_\mu\mathbf{A}_\mu-\frac{g_0^2a^2}{2}(n_1)_\mu(n_2)_\nu\left[\frac{\boldsymbol{\tau}}{2}\cdot\mathbf{A}_\nu,\frac{\boldsymbol{\tau}}{2}\cdot\mathbf{A}_\mu\right]\right\}$$

As in Eq. (29.17), the integral is defined by the first two terms in Eq. (31.22). Now repeat, noting that

$$\begin{aligned}(n_1)_\mu(n_1)_\nu\left[\frac{\boldsymbol{\tau}}{2}\cdot\mathbf{A}_\nu,\frac{\boldsymbol{\tau}}{2}\cdot\mathbf{A}_\mu\right] &= 0\\ (n_2)_\mu(n_2)_\nu\left[\frac{\boldsymbol{\tau}}{2}\cdot\mathbf{A}_\nu,\frac{\boldsymbol{\tau}}{2}\cdot\mathbf{A}_\mu\right] &= 0\end{aligned}\tag{31.25}$$

This gives

$$\begin{aligned}\mathrm{U}_{lk}\mathrm{U}_{kj}\mathrm{U}_{ji} = \exp&\left\{\frac{i}{2}g_0\boldsymbol{\tau}\cdot\left(\int_i^l\right)_{\hookrightarrow}dx_\mu\mathbf{A}_\mu\right.\\ &\left.-\frac{g_0^2a^2}{2}[(n_1)_\mu(n_2)_\nu+(-n_1)_\nu(n_2)_\mu]\left[\frac{\boldsymbol{\tau}}{2}\cdot\mathbf{A}_\nu,\frac{\boldsymbol{\tau}}{2}\cdot\mathbf{A}_\mu\right]\right\}\end{aligned}\tag{31.26}$$

Once more yields[36]

$$\begin{aligned}&\mathrm{U}_{il}\mathrm{U}_{lk}\mathrm{U}_{kj}\mathrm{U}_{ji} =\\ &\exp\left\{\frac{i}{2}g_0\boldsymbol{\tau}\cdot\oint_{\hookrightarrow}dx_\mu\mathbf{A}_\mu+g_0^2a^2\left[\frac{\boldsymbol{\tau}}{2}\cdot\mathbf{A}_\mu,\frac{\boldsymbol{\tau}}{2}\cdot\mathbf{A}_\nu\right]\frac{1}{2}\alpha_{\mu\nu}\right\}\end{aligned}\tag{31.27}$$

Here $\alpha_{\mu\nu}$ is the antisymmetric tensor characterizing the *plane* of the plaquette (see Ref. [R4])

$$\alpha_{\mu\nu}\equiv(n_1)_\mu(n_2)_\nu-(n_1)_\nu(n_2)_\mu\tag{31.28}$$

Define the surface area associated with the plaquette to be

$$dS_{\mu\nu}\equiv a^2\alpha_{\mu\nu}\tag{31.29}$$

The result in Eq. (31.27) can be rewritten as

$$\mathrm{U}_{il}\mathrm{U}_{lk}\mathrm{U}_{kj}\mathrm{U}_{ji}=\exp\left\{\frac{i}{2}g_0\boldsymbol{\tau}\cdot\left[\oint_{\hookrightarrow}dx_\mu\mathbf{A}_\mu+g_0\mathbf{A}_\mu\times\mathbf{A}_\nu\frac{1}{2}dS_{\mu\nu}\right]\right\}\tag{31.30}$$

Stokes' theorem can now be used on the first term[37]

$$\oint_{\hookrightarrow}dx_\mu\mathbf{A}_\mu = \int_{\text{enclosed surface}}\frac{1}{2}dS_{\mu\nu}\left(\frac{\partial\mathbf{A}_\nu}{\partial x_\mu}-\frac{\partial\mathbf{A}_\mu}{\partial x_\nu}\right)\tag{31.31}$$

[36] Note $(-n_2)_\nu[(n_1)_\mu+(-n_1)_\mu]\equiv 0$.

[37] This relation is readily verified in three dimensions. With reference to Fig. 31.2 one has

$$\begin{aligned}\oint_{\hookrightarrow}\mathbf{A}\cdot d\mathbf{x} &= \int_{\text{enclosed surface}}(\nabla\times\mathbf{A})\cdot d\mathbf{S}\approx a^2(\nabla\times\mathbf{A})_3\\ &= a^2\left(\frac{\partial A_2}{\partial x_1}-\frac{\partial A_1}{\partial x_2}\right)=\frac{1}{2}dS_{\mu\nu}\left(\frac{\partial\mathbf{A}_\nu}{\partial x_\mu}-\frac{\partial\mathbf{A}_\mu}{\partial x_\nu}\right)\end{aligned}$$

Here Eq. (31.28) has been used. This is the stated result.

Hence Eq. (31.30) becomes

$$\begin{aligned} \mathrm{U}_{il}\mathrm{U}_{lk}\mathrm{U}_{kj}\mathrm{U}_{ji} &= \exp\left\{\frac{i}{2}g_0\boldsymbol{\tau}\cdot\left(\frac{\partial \mathbf{A}_\nu}{\partial x_\mu}-\frac{\partial \mathbf{A}_\mu}{\partial x_\nu}+g_0\mathbf{A}_\mu\times\mathbf{A}_\nu\right)\frac{1}{2}dS_{\mu\nu}\right\} \\ &= (\mathrm{U}_\square)_{\hookrightarrow} \end{aligned} \tag{31.32}$$

If one goes around the plaquette in the opposite direction the result is

$$(\mathrm{U}_\square)_{\hookleftarrow} = \exp\left\{-\frac{i}{2}g_0\boldsymbol{\tau}\cdot\left(\frac{\partial \mathbf{A}_\nu}{\partial x_\mu}-\frac{\partial \mathbf{A}_\mu}{\partial x_\nu}+g_0\mathbf{A}_\mu\times\mathbf{A}_\nu\right)\frac{1}{2}dS_{\mu\nu}\right\} \tag{31.33}$$

The action is obtained from the sum of these two contributions

$$S_\square = \sigma\left\{\left[1-\frac{1}{2}\mathrm{tr}\,(\mathrm{U}_\square)_{\hookrightarrow}\right]+\left[1-\frac{1}{2}\mathrm{tr}\,(\mathrm{U}_\square)_{\hookleftarrow}\right]\right\} \tag{31.34}$$

Now let $a \to 0$. As before, the odd terms in the exponentials *cancel*, and in this limit the action becomes[38]

$$S_\square = \frac{\sigma g_0^2}{8}\mathrm{tr}\left[\boldsymbol{\tau}\cdot\left(\frac{\partial \mathbf{A}_\nu}{\partial x_\mu}-\frac{\partial \mathbf{A}_\mu}{\partial x_\nu}+g_0\mathbf{A}_\mu\times\mathbf{A}_\nu\right)\frac{1}{2}dS_{\mu\nu}\right]^2 \tag{31.35}$$

Use $[\boldsymbol{\tau}\cdot\mathbf{v}]^2 = \mathbf{v}^2$ and $\mathrm{tr}\,1 = 2$.

For illustration, we now specialize to the case of 1+1 dimensions. Define[39]

$$\sigma \equiv \frac{2}{g_0^2 a^2} \tag{31.36}$$

The result is

$$S_\square = \frac{1}{2a^2}\left[\frac{1}{2}dS_{\mu\nu}\left(\frac{\partial \mathbf{A}_\nu}{\partial x_\mu}-\frac{\partial \mathbf{A}_\mu}{\partial x_\nu}+g_0\mathbf{A}_\mu\times\mathbf{A}_\nu\right)\right]^2 \tag{31.37}$$

For the case of 1+1 dimensions one can simply refer to Fig. 31.2. The contribution of a plaquette to the action becomes

$$\begin{aligned} S_\square &= \frac{1}{2a^2}a^4\left(\frac{\partial \mathbf{A}_2}{\partial x_1}-\frac{\partial \mathbf{A}_1}{\partial x_2}+g_0\mathbf{A}_1\times\mathbf{A}_2\right)^2 \\ &= \frac{1}{4}a^2\,\mathbf{F}_{\mu\nu}\cdot\mathbf{F}_{\mu\nu} \end{aligned} \tag{31.38}$$

This is the continuum action with the *full field tensor* $\mathbf{F}_{\mu\nu}$ of the nonabelian theory [see Eq. (27.21)]. Thus, in the continuum limit

$$\begin{aligned} S_\square &\overset{a\to 0}{\longrightarrow} \frac{1}{4}\mathbf{F}_{\mu\nu}\cdot\mathbf{F}_{\mu\nu}\,a^2 = -\mathcal{L}\left(\mathbf{A}_\mu,\frac{\partial \mathbf{A}_\mu}{\partial x_\nu}\right)a^2 \\ \sum_\square S_\square &\overset{a\to 0}{\longrightarrow} -\int d^2x\,\mathcal{L}\left(\mathbf{A}_\mu,\frac{\partial \mathbf{A}_\mu}{\partial x_\nu}\right) \end{aligned} \tag{31.39}$$

[38] See also Prob. 31.3.

[39] The definition in d dimensions is $\sigma \equiv 2/g_0^2 a^{4-d}$ (see Prob. 31.4).

This is the correct continuum limit for the gauge theory $SU(2)$.

Gauge-Invariant Measure. The remaining issue is to develop a gauge invariant measure for the path integrals in the partition function in the nonabelian theory. For each link $U = e^{i\phi}$ in the abelian $U(1)$ theory we took [Eq. (29.41)]

$$\frac{1}{2\pi}\int_0^{2\pi} d\phi f(e^{i\phi}) = \frac{1}{\pi}\int d^2\alpha\, \delta(\boldsymbol{\alpha}^2 - 1) f(\alpha_1 + i\alpha_2) \tag{31.40}$$

This is illustrated in Fig. 29.5. There is one such term for each *link* [Eq. (29.42)]. Now note that in this $U(1)$ case one can write

$$\begin{aligned} e^{i\phi} &= \cos\phi + i\sin\phi &&= \alpha_1 + i\alpha_2 \\ & \qquad\quad \alpha_1^2 + \alpha_2^2 &&= 1 \end{aligned} \tag{31.41}$$

A gauge transformation $A_\mu \to A_\mu - (1/e_0)\partial\Lambda/\partial x_\mu$ changes the link variable to

$$U_{ji} \to e^{-i\Lambda(x_j)} e^{i\phi_{ji}} e^{+i\Lambda(x_i)} \tag{31.42}$$

Or, equivalently

$$\begin{aligned} e^{i\phi} \to e^{i\phi'} &= e^{i(\phi+\lambda)} = \alpha_1' + i\alpha_2' \\ (\alpha_1')^2 + (\alpha_2')^2 &= 1 \end{aligned} \tag{31.43}$$

This is just a *rotation* on the unit circle in the two-dimensional internal space illustrated in Fig. 29.5. Under a rotation $d^2\alpha = d^2\alpha'$. With periodic boundary conditions, the integral around the unit circle is unchanged

$$\frac{1}{\pi}\int d^2\alpha\, \delta(\boldsymbol{\alpha}^2 - 1) f(\alpha_1' + i\alpha_2') = \frac{1}{\pi}\int d^2\alpha\, \delta(\boldsymbol{\alpha}^2 - 1) f(\alpha_1 + i\alpha_2) \tag{31.44}$$

Or, equivalently

$$\frac{1}{2\pi}\int_0^{2\pi} d\phi f(e^{i\phi'}) = \frac{1}{2\pi}\int_0^{2\pi} d\phi f(e^{i\phi}) \tag{31.45}$$

We argue by analogy in the nonabelian case of $SU(2)$. Recall that the 2×2 $SU(2)$ matrices can always be represented as

$$\exp\left\{\frac{i}{2}\boldsymbol{\tau}\cdot\boldsymbol{\theta}\right\} = \cos\frac{\theta}{2} + i\mathbf{n}\cdot\boldsymbol{\tau}\sin\frac{\theta}{2} \equiv \alpha_0 + i\boldsymbol{\tau}\cdot\boldsymbol{\alpha} \tag{31.46}$$

Here

$$\begin{aligned} \alpha_0 &= \cos\frac{\theta}{2} \qquad\qquad \boldsymbol{\alpha} = \mathbf{n}\sin\frac{\theta}{\mathbf{2}} \\ \alpha_0^2 + \boldsymbol{\alpha}^2 &= 1 \end{aligned} \tag{31.47}$$

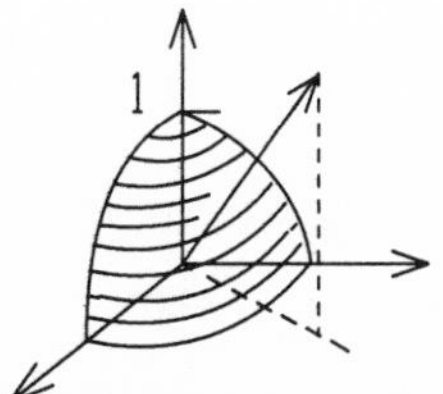

Figure 31.3: Schematic illustration of unit sphere in four-dimensional internal parameter space for each link in $SU(2)$.

This suggests that we now

1. Work in the four-dimensional internal parameter space $\alpha_\mu \equiv (\alpha_0, \boldsymbol{\alpha})$ for each link. The condition $\alpha_\mu^2 = \alpha_0^2 + \boldsymbol{\alpha}^2 = 1$ defines the *unit sphere* in this four-dimensional internal space. This is illustrated schematically in Fig. 31.3.
2. Impose periodic boundary conditions on this unit sphere.
3. Take the *measure* for each link to be

$$\begin{aligned} I &= \frac{1}{\pi^2}\int d^4\alpha\,\delta(\alpha_\mu^2 - 1) f(\alpha_0 + i\boldsymbol{\alpha}\cdot\boldsymbol{\tau}) \\ &= \frac{1}{\pi^2}\int d^4\alpha\,\delta(\alpha_0^2 + \boldsymbol{\alpha}^2 - 1) f(\alpha_0 + i\boldsymbol{\alpha}\cdot\boldsymbol{\tau}) \end{aligned} \tag{31.48}$$

The normalization constant can be verified by performing the integration. In the four-dimensional euclidian internal space the volume element can be written in spherical coordinates as (Ref. [R4])

$$\begin{aligned} \alpha_0 &= \rho\cos\psi \\ \alpha_1 &= \rho\sin\psi\sin\theta\cos\phi \\ \alpha_2 &= \rho\sin\psi\sin\theta\sin\phi \\ \alpha_3 &= \rho\sin\psi\cos\theta \\ d^4\alpha &= \rho^3\sin^2\psi\sin\theta\,d\psi\,d\theta\,d\phi\,d\rho \end{aligned} \tag{31.49}$$

Here (θ, ϕ) are the usual three-dimensional polar and azimuthal angles and the additional polar angle satisfies $0 \leq \psi \leq \pi$. The required normalization integral is then

$$\begin{aligned} &\frac{1}{\pi^2}\int d^4\alpha\,\delta(\alpha_0^2 + \boldsymbol{\alpha}^2 - 1) = \frac{1}{2\pi^2}\int (2\rho d\rho)\rho^2\delta(\rho^2 - 1)\sin^2\psi\sin\theta\,d\psi\,d\theta\,d\phi \\ &= \frac{1}{2\pi^2}\int_0^\pi \sin^2\psi\,d\psi\int_0^\pi \sin\theta\,d\theta\int_0^{2\pi} d\phi = 1 \end{aligned} \tag{31.50}$$

Consider now the *gauge invariance* of the measure. We have shown that under a (infinitesimal) gauge transformation the link variables transform according

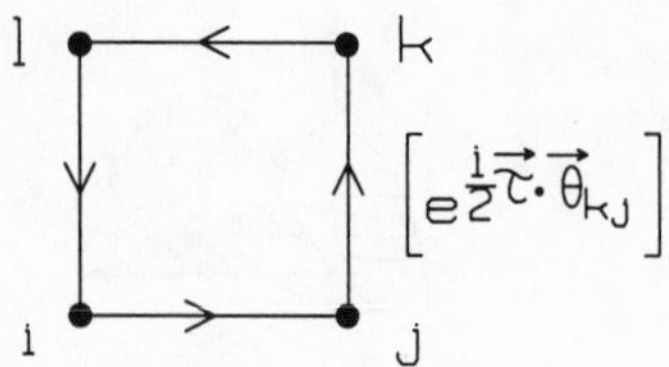

Figure 31.4: Basic link, link variable, and plaquette for the lattice gauge model of the nonabelian Yang-Mills theory with internal $SU(2)$ symmetry.

to Eq. (31.16)

$$
\begin{aligned}
&\mathrm{U}_{ji} \to \mathrm{U}'_{ji} \\
&\mathrm{U}'_{ji} = e^{-\frac{i}{2}\boldsymbol{\tau}\cdot\boldsymbol{\theta}(x_j)}\mathrm{U}_{ji}e^{\frac{i}{2}\boldsymbol{\tau}\cdot\boldsymbol{\theta}(x_i)}
\end{aligned}
\tag{31.51}
$$

Since this is simply another 2×2 $SU(2)$ matrix, it can again be expressed in the form

$$
\begin{aligned}
\mathrm{U}'_{ji} &= \alpha'_0 + i\boldsymbol{\alpha}'\cdot\boldsymbol{\tau} \\
(\alpha'_0)^2 + (\boldsymbol{\alpha}')^2 &= 1
\end{aligned}
\tag{31.52}
$$

This transformation is just a *rotation* on the surface of the unit sphere in the four-dimensional internal space $(\alpha_0, \boldsymbol{\alpha})$ (Fig. 31.3). Under a rotation $d^4\alpha = d^4\alpha'$. With periodic boundary conditions [Eqs. (31.49)] the integral on the surface of the unit sphere in the four-dimensional internal space is unchanged under this rotation. Thus

$$
\frac{1}{\pi^2}\int d^4\alpha\,\delta(\alpha_\mu^2 - 1)f(\alpha'_0 + i\boldsymbol{\alpha}'\cdot\boldsymbol{\tau}) = \frac{1}{\pi^2}\int d^4\alpha\,\delta(\alpha_\mu^2 - 1)f(\alpha_0 + i\boldsymbol{\alpha}\cdot\boldsymbol{\tau})
\tag{31.53}
$$

Hence this integral over the parameter set $(\alpha_0, \boldsymbol{\alpha})$ is unchanged under a change in gauge and the measure is gauge invariant.

Summary. In summary, the lattice gauge model for the Yang-Mills nonabelian gauge theory based on an internal $SU(2)$ symmetry is constructed as follows:

1. Assign to each link a phase angle and link variable (see Fig. 31.4)

$$
\begin{aligned}
\boldsymbol{\theta}_{ji} &= g_0(x_j - x_i)_\mu \mathbf{A}_\mu\left(\frac{1}{2}(x_i + x_j)\right) \\
\mathrm{U}_{ji}(\mathbf{A}_\mu) &= \exp\left\{\frac{i}{2}\boldsymbol{\tau}\cdot\boldsymbol{\theta}_{ji}\right\} \\
\mathrm{U}_{ji}(\mathbf{A}_\mu) &= (\alpha_0 + i\boldsymbol{\alpha}\cdot\boldsymbol{\tau})_{ji}
\end{aligned}
\tag{31.54}
$$

2. Impose periodic boundary conditions on $(\alpha_0, \boldsymbol{\alpha})$ that lie on the unit sphere in the four-dimensional internal parameter space for each link.

3. Take the contribution to the action from each plaquette to be (Fig. 31.4)[40]

$$\begin{aligned} \mathrm{U}_{\Box} &= \mathrm{U}_{ji}\mathrm{U}_{il}\mathrm{U}_{lk}\mathrm{U}_{kj} \\ S_{\Box} &= 2\sigma\left(1 - \mathrm{Re}\frac{1}{2}\mathrm{tr}\ \mathrm{U}_{\Box}\right) \end{aligned} \tag{31.55}$$

The total action is the sum over all plaquettes $\bar{S} = \sum_{\Box} S_{\Box}$.

4. The partition function is obtained from an integration over all links with the measure of Eq. (31.53)

$$Z = \prod_{\text{links}} \int \frac{d^4\alpha_l}{\pi^2}\,\delta(\alpha_{0l}^2 + \boldsymbol{\alpha}_l^2 - 1)\exp\{-\sum_{\Box} S_{\Box}\} \tag{31.56}$$

5. It has been demonstrated that the action is invariant under (infinitesimal) local gauge transformations, that the measure is also gauge invariant, and that this model has the proper continuum limit.

6. Equations (31.55),(31.56), and the last of Eqs. (31.54) are self-contained; they constitute $SU(2)$ lattice gauge theory.

[40] σ is given in 1+1 dimensions by Eq. (31.36) and in d dimensions by Prob. 31.4.

32

MEAN FIELD THEORY — $SU(n)$

To obtain some insight into nonabelian lattice gauge theory, and to gain some familiarity with it, consider again mean field theory (MFT). This time we take a more systematic approach than in Section 30 and start from the partition function for the entire many-body system, from which an analysis of its free energy will follow. The discussion is based on Refs. [Q27, Q28, Q29, Q30]. This section is taken from Ref. [Q30].

We first briefly repeat a summary of the results of the previous section. The basic plaquette is illustrated in Fig. 32.1. The partition function is obtained from a path integral over all link variables of the action, which is obtained from the sum over all plaquettes of a term formed from the product of the link variables around the plaquette[41]

$$\begin{aligned} Z &= \int (dU) e^{-S(U)} \\ S(U) &= \sum_{\Box} 2\sigma \left(1 - \frac{1}{n} \mathrm{Re}\,[\mathrm{tr} \underset{\sim}{U}_{\Box}]\right) \\ \underset{\sim}{U}_{\Box} &= \underset{\sim}{U}_{il} \underset{\sim}{U}_{lk} \underset{\sim}{U}_{kj} \underset{\sim}{U}_{ji} \end{aligned} \tag{32.1}$$

Stated in this form, the equations constitute lattice gauge theory for *any* internal symmetry group $SU(n)$.

For $SU(2)$, as we have seen, the link variables are expressed as

$$\begin{aligned} \underset{\sim}{U}_{ji} &= \exp\left\{ i g_0 (x_j - x_i)_\mu \frac{1}{2} \underset{\sim}{\tau} \cdot \mathbf{A}_\mu \left[\frac{1}{2}(x_i + x_j)\right]\right\} \\ &= (\alpha_0 + i \boldsymbol{\alpha} \cdot \boldsymbol{\tau})_{ji} \end{aligned} \tag{32.2}$$

For $SU(3)$ the link variables are expressed as

$$\underset{\sim}{U}_{ji} = \exp\left\{ i g_0 (x_j - x_i)_\mu \frac{1}{2} \lambda^a A_\mu^a \left[\frac{1}{2}(x_i + x_j)\right]\right\} \tag{32.3}$$

An explicit representation for the $SU(3)$ (and higher n) matrices will not be needed for the developments in this section.

[41] We here and henceforth simply use $\bar{S} \equiv S$ for the required action.

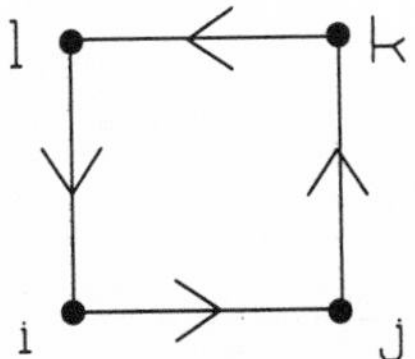

Figure 32.1: Basic plaquette in lattice gauge theory for $SU(n)$.

The measure for $SU(2)$ is

$$\begin{aligned}\int(dU) &= \prod_{\text{links}} \frac{1}{\pi^2}\int d^4\alpha_l\,\delta(\alpha_{0l}^2+\boldsymbol{\alpha}_l^2-1) \\ f(U) &= f(\alpha_0 + i\boldsymbol{\alpha}\cdot\boldsymbol{\tau}) \end{aligned} \tag{32.4}$$

This can also be generalized to $SU(n)$; the specific form will again not be required for the present developments.

Mean-Field Approach. Concentrate again on the link variables. Pick a gauge. Now *add and subtract a mean field contribution to the action*

$$\sum_{<ij>} \frac{H}{n}\text{Re}\,[\text{tr}\underset{\sim}{U}_{ji}] = \sum_{\text{links}} \frac{H}{n}\text{Re}\,[\text{tr}\underset{\sim}{U}_l] \tag{32.5}$$

Here H is a constant to be determined.[42] Since this expression is a *sum* over link variables, the exponential of this MFT contribution to the action *factors*. The product of the integrations of this contribution over the link variables then also factors. The path integrals over the link variables of this MFT contribution to the action are thus *decoupled*

$$\prod_{\text{links}}\int(dU_l)\exp\{\sum_{\text{links}} U_l\} = \prod_{\text{links}}\int(dU_l)e^{U_l} \tag{32.6}$$

After the addition and subtraction of this MFT term in the action, the partition function can be written identically as

$$\begin{aligned} Z &= \exp\{-2\sigma\sum_{\Box}1\}\int(dU)\exp\left\{\sum_{\text{links}}\frac{H}{n}\text{Re}[\text{tr}\underset{\sim}{U}_l]\right\} \\ &\quad\times\exp\left\{\sum_{\Box}\frac{2\sigma}{n}\text{Re}[\text{tr}\underset{\sim}{U}_{\Box}] - \sum_{\text{links}}\frac{H}{n}\text{Re}[\text{tr}\underset{\sim}{U}_l]\right\} \\ &\equiv \exp\{-2\sigma\sum_{\Box}1\}\,\tilde{Z} \end{aligned} \tag{32.7}$$

[42]The constant H will be treated as a variational parameter determined by a variational principle.

The first factor in the last expression contributes only a constant to the entropy of the vacuum and drops out of thermal averages.

Now define the MFT statistical operator

$$\rho_{\text{MFT}} \equiv \frac{\exp\left\{\sum_{\text{links}}(H/n)\,\text{Re}[\text{tr}\underset{\sim}{U}_l]\right\}}{\int(dU)\exp\left\{\sum_{\text{links}}(H/n)\,\text{Re}[\text{tr}\underset{\sim}{U}_l]\right\}} \tag{32.8}$$

With this one can calculate MFT thermal averages

$$\langle O\rangle_{\text{MFT}} \equiv \int(dU)O\,\rho_{\text{MFT}} \tag{32.9}$$

With this definition, Eq. (32.7) becomes

$$\frac{\tilde{Z}}{\int(dU)\exp\left\{\sum_{\text{links}}(H/n)\,\text{Re}[\text{tr}\underset{\sim}{U}_l]\right\}} = \left\langle \exp\left\{\sum_{\square}\frac{2\sigma}{n}\text{Re}[\text{tr}\underset{\sim}{U}_{\square}] - \sum_{\text{links}}\frac{H}{n}\text{Re}[\text{tr}\underset{\sim}{U}_l]\right\}\right\rangle_{\text{MFT}} \tag{32.10}$$

To proceed with the analysis one invokes the very useful and powerful *Peierls' inequality*

$$\langle \exp f(x)\rangle \;\geq\; \exp\langle f(x)\rangle \tag{32.11}$$

This inequality holds whenever the average value is computed with any positive measure (or weighting function);[43] it is proven in Appendix I. The great utility of this inequality is that it gets the mean value up into the exponent, and the MFT value of the action itself is readily computed. The inequality will be used to establish a variational principle for the free energy.

The use of Eq. (32.11) on Eq. (32.10) leads to

$$\begin{aligned}
&\frac{\tilde{Z}}{\int(dU)\exp\left\{\sum_{\text{links}}(H/n)\,\text{Re}[\text{tr}\underset{\sim}{U}_l]\right\}} \\
&\quad\geq \exp\left\{\left\langle\left[\sum_{\square}\frac{2\sigma}{n}\,\text{Re}[\text{tr}\underset{\sim}{U}_{\square}] - \sum_{\text{links}}\frac{H}{n}\text{Re}[\text{tr}\underset{\sim}{U}_l]\right]\right\rangle_{\text{MFT}}\right\} \\
&\quad\equiv \exp\left\{\int(dU)\rho_{\text{MFT}}\left[\sum_{\square}\frac{2\sigma}{n}\text{Re}[\text{tr}\underset{\sim}{U}_{\square}] - \sum_{\text{links}}\frac{H}{n}\text{Re}[\text{tr}\underset{\sim}{U}_l]\right]\right\}
\end{aligned} \tag{32.12}$$

The integrations on the right can now be performed because they *factor*. The integral in the second term is

$$I_2 = \frac{\int(dU)\left\{\sum_{\text{links}}\frac{H}{n}\text{Re}[\text{tr}\underset{\sim}{U}_l]\right\}\exp\left\{\sum_{\text{links}}\frac{H}{n}\text{Re}[\text{tr}\underset{\sim}{U}_l]\right\}}{\int(dU)\exp\left\{\sum_{\text{links}}\frac{H}{n}\text{Re}[\text{tr}\underset{\sim}{U}_l]\right\}} \tag{32.13}$$

[43] As is the case when one uses Eqs. (32.8) and (32.9).

The volume element factors $\int(dU) = \prod_{\text{links}} \int(dU_l)$ and each term being averaged in $\sum_{\text{links}}$ gives an identical result. We again choose to work in d dimensions and use the counting procedure of Section 30.[44] Thus

$$\begin{aligned} I_2 &= N_{\text{sites}} d_{\text{links/site}} H \frac{\int(dU_l)\frac{1}{n}\text{Re}[\text{tr}\underset{\sim}{U}_l] \exp\{\frac{H}{n}\text{Re}[\text{tr}\underset{\sim}{U}_l]\}}{\int(dU_l)\exp\{\frac{H}{n}\text{Re}[\text{tr}\underset{\sim}{U}_l]\}} \\ &= N_{\text{sites}} d\, H \left\langle \frac{1}{n}\text{Re}[\text{tr}\underset{\sim}{U}_l]\right\rangle_{\text{MFT}} \end{aligned} \qquad (32.14)$$

The expression being averaged in the final term is a one-body operator.

Consider next the first integral on the right-hand-side of Eq. (32.12). This is the heart of the matter. The problem at this stage has been reduced to calculating the mean value of $\text{Re}[\text{tr}\underset{\sim}{U}_\Box]$; now one is again dealing with a physical, gauge-invariant quantity (see the discussion in Section 30). But this integral can now be carried out since $\underset{\sim}{U}_\Box$ also factors! Thus

$$\begin{aligned} I_1 &= \frac{\int(dU)\{\sum_\Box \frac{2\sigma}{n}\text{Re}[\text{tr}\underset{\sim}{U}_\Box]\}\exp\{\sum_{\text{links}}\frac{H}{n}\text{Re}[\text{tr}\underset{\sim}{U}_l]\}}{\int(dU)\exp\{\sum_{\text{links}}\frac{H}{n}\text{Re}[\text{tr}\underset{\sim}{U}_l]\}} \\ \underset{\sim}{U}_\Box &= \underset{\sim}{U}_{il}\underset{\sim}{U}_{lk}\underset{\sim}{U}_{kj}\underset{\sim}{U}_{ji} \end{aligned} \qquad (32.15)$$

Now assume [as will be verified explicitly for $SU(2)$ below] that upon doing the path integral of a link variable over a link $\langle\underset{\sim}{U}_l\rangle_{\text{MFT}}$, only $(1/n)\,\text{Re}[\text{tr}\underset{\sim}{U}_l]$ remains; thus

$$\frac{\int(dU_l)\underset{\sim}{U}_l\exp\{\frac{H}{n}\text{Re}[\text{tr}\underset{\sim}{U}_l]\}}{\int(dU_l)\exp\{\frac{H}{n}\text{Re}[\text{tr}\underset{\sim}{U}_l]\}} = \frac{\int(dU_l)\frac{1}{n}\text{Re}[\text{tr}\underset{\sim}{U}_l]\exp\{\frac{H}{n}\text{Re}[\text{tr}\underset{\sim}{U}_l]\}}{\int(dU_l)\exp\{\frac{H}{n}\text{Re}[\text{tr}\underset{\sim}{U}_l]\}} \qquad (32.16)$$

Then all the contributions in Eqs. (32.15) are identical and[45]

$$I_1 = N_{\text{sites}}\frac{1}{2}d(d-1)_{\text{plaquettes/site}}2\sigma\left\langle\frac{1}{n}\text{Re}[\text{tr}\underset{\sim}{U}_l]\right\rangle^4_{\text{MFT}} \qquad (32.17)$$

The remaining mean value is identical to that appearing in Eq. (32.14).

Define

$$\begin{aligned} c(H) &\equiv \int(dU)\exp\left\{\frac{H}{n}\text{Re}[\text{tr}\underset{\sim}{U}]\right\} \\ c(H)t(H) &\equiv \int(dU)\frac{1}{n}\text{Re}[\text{tr}\underset{\sim}{U}]\exp\left\{\frac{H}{n}\text{Re}[\text{tr}\underset{\sim}{U}]\right\} \end{aligned} \qquad (32.18)$$

The previous results can then be summarized simply as

$$\begin{aligned} I_2 &= N_{\text{sites}}\, d\, Ht(H) \\ I_1 &= N_{\text{sites}}\frac{1}{2}d(d-1)2\,\sigma\, t^4(H) \end{aligned} \qquad (32.19)$$

[44]Note that with a lattice of N intervals in each of d dimensions and with periodic boundary conditions the number of sites is $N_{\text{sites}} = N^d$.

[45]Note $(1/n)\,\text{Re}[\text{tr}1] = 1$.

For the denominator of the l.h.s of Eq. (32.12) one also needs the relation

$$\begin{aligned}\int (dU) \exp\left\{\sum_{\text{links}} \frac{H}{n}\text{Re}[\text{tr}\underset{\sim}{U}_l]\right\} &= \left[\int (dU_l) \exp\{\frac{H}{n}\text{Re}[\text{tr}\underset{\sim}{U}_l]\}\right]^{N_{\text{sites}}\,d} \\ &= [c(H)]^{N_{\text{sites}}\,d} \end{aligned} \tag{32.20}$$

A combination of these relations allows us to rewrite the basic inequality for the partition function in Eq. (32.12) as

$$\frac{\tilde{Z}}{[c(H)]^{N_{\text{sites}}\,d}} \geq \exp\{N_{\text{sites}}[\sigma t^4(H)d(d-1) - Ht(H)d]\} \tag{32.21}$$

Now take the logarithm of both sides

$$\ln \tilde{Z} \geq N_{\text{sites}}\{\sigma t^4(H)d(d-1) - Ht(H)d + [\ln c(H)]d\} \tag{32.22}$$

Recall the relation between the partition function and the free energy (to within an additive constant $\langle S\rangle_{\text{vac}}$)

$$\tilde{Z} = e^{-\beta F} \tag{32.23}$$

Both are gauge-invariant quantities. Upon taking logarithms, and reversing the sense of the inequality in Eq. (32.22), one finds a *variational principle for the free energy*

$$\begin{aligned} F &\leq F_{\text{MFT}} \\ \frac{\beta F_{\text{MFT}}}{N_{\text{sites}}} &= -\sigma t^4(H)d(d-1) + Ht(H)d - [\ln c(H)]d \end{aligned} \tag{32.24}$$

The last expression is just the free-energy/site. The MFT expression for the free energy can now be minimized with respect to the parameter H to obtain the best bound on the actual free energy F. One has

$$\begin{aligned} &\left(\frac{dF}{dH}\right)_{\text{MFT}} = 0 \\ &-4\sigma t^3(H)t'(H)d(d-1) + t(H)d + Ht'(H)d - [c'(H)/c(H)]d = 0 \end{aligned} \tag{32.25}$$

Note from Eqs. (32.18) that

$$\frac{c'(H)}{c(H)} \equiv t(H) \tag{32.26}$$

Hence Eq. (32.25) for the optimal choice of H becomes

$$H = 4\sigma t^3(H)(d-1) \tag{32.27}$$

Define a "magnetization" M by

$$M^3 \equiv \frac{H}{4\sigma(d-1)} \tag{32.28}$$

Then the self-consistency equation for the magnetization is given by

$$M = t[4\sigma M^3(d-1)] \tag{32.29}$$

Here $t(H) = \langle \frac{1}{n}\text{Re}[\text{tr}\underset{\sim}{U}]\rangle_{\text{MFT}}$ is the one-body MFT average in Eq. (32.18). This is a very powerful result; it describes the nonabelian theory $SU(n)$ in d dimensions for arbitrary n and d!

Evaluation of Required Integrals for $SU(2)$. Let us specialize to the case of $SU(2)$. In this case the link variable takes the form

$$\underset{\sim}{U} = \alpha_0 + i\boldsymbol{\alpha}\cdot\boldsymbol{\tau} \tag{32.30}$$

The integral over link variables is performed according to

$$\int d\underset{\sim}{U} f(\underset{\sim}{U}) = \frac{1}{\pi^2}\int d^4\alpha\,\delta(\alpha_0^2+\boldsymbol{\alpha}^2-1)f(\alpha_0+i\boldsymbol{\alpha}\cdot\boldsymbol{\tau}) \tag{32.31}$$

First note that

$$\frac{1}{2}\text{Re}[\text{tr}\underset{\sim}{U}] = \alpha_0 \tag{32.32}$$

Thus

$$\begin{aligned}\int (dU)\underset{\sim}{U}e^{(H/2)\,\text{Re}[\text{tr}\underset{\sim}{U}]} &= \frac{1}{\pi^2}\int d^4\alpha\,\delta(\alpha_0^2+\boldsymbol{\alpha}^2-1)e^{H\alpha_0}(\alpha_0+i\boldsymbol{\alpha}\cdot\boldsymbol{\tau}) \\ &= \int (dU)\frac{1}{2}\text{Re}[\text{tr}\underset{\sim}{U}]e^{(H/2)\,\text{Re}[\text{tr}\underset{\sim}{U}]}\end{aligned} \tag{32.33}$$

The second equality follows since the angular average of the final term in the first line vanishes by symmetry. Equation (32.33) is the result that was to be verified in the discussion of Eq. (32.16).

The quantity $c(H)$ can be evaluated as follows for $SU(2)$

$$\begin{aligned}c(H) &= \frac{1}{\pi^2}\int d^4\alpha\,\delta(\alpha_0^2+\boldsymbol{\alpha}^2-1)e^{H\alpha_0} \\ &= \frac{1}{\pi^2}\int_{-1}^{1}d\alpha_0\int_0^\infty \alpha^2 d\alpha\,\delta[\alpha^2-(1-\alpha_0^2)]e^{H\alpha_0}d\Omega_\alpha \\ &= \frac{4\pi}{\pi^2}\int_{-1}^{1}d\alpha_0\int_0^\infty \alpha^2 d\alpha\frac{1}{2\alpha}[\delta(\alpha-\sqrt{1-\alpha_0^2})+\delta(\alpha+\sqrt{1-\alpha_0^2})]e^{H\alpha_0} \\ c(H) &= \frac{2}{\pi}\int_{-1}^{1}d\alpha_0\sqrt{1-\alpha_0^2}e^{H\alpha_0}\end{aligned} \tag{32.34}$$

It follows from Eq. (32.18) that

$$t(H) = \frac{c'(H)}{c(H)} = \frac{(2/\pi)\int_{-1}^{1}\alpha_0 d\alpha_0\sqrt{1-\alpha_0^2}e^{H\alpha_0}}{(2/\pi)\int_{-1}^{1}d\alpha_0\sqrt{1-\alpha_0^2}e^{H\alpha_0}} \tag{32.35}$$

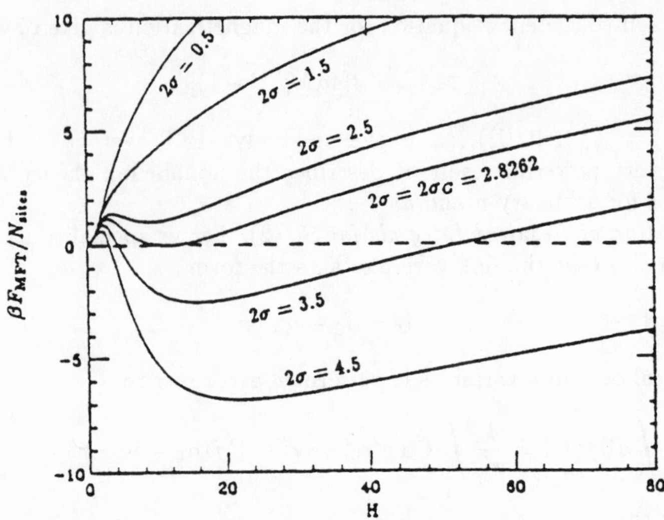

Figure 32.2: Values of $\beta F_{\mathrm{MFT}}(H)/N_{\mathrm{sites}}$ versus H for various values of 2σ for $SU(2)$ and $d = 4$. From Ref. [Q30].

This is the basic relation needed for the self-consistency Eq. (32.29).

Numerical results for $\beta F_{\mathrm{MFT}}(H)/N_{\mathrm{sites}}$ versus H for $d = 4$ are shown in Fig. 32.2 (from Ref. [Q30]).[46] It is evident from this figure that there is a critical value of σ

$$\sigma_{\mathrm{C}} = 1.4131 \tag{32.36}$$

For values of $\sigma < \sigma_{\mathrm{C}}$ the phase with $H = 0$ has the lowest free energy; however, for values of $\sigma > \sigma_{\mathrm{C}}$ a second phase exists with lower free energy and finite H [and hence finite M by Eq. (32.28)]. The calculation of M^4 as a function

[46]Some integrals are useful in this analysis (Ref. [Q30]). A change of variable $\alpha_0 \equiv \cos\psi$ and a subsequent integration by parts yields

$$\begin{aligned} c(H) &= \frac{2}{\pi}\int_0^{\pi} \sin^2\psi e^{H\cos\psi}\,d\psi \\ Hc(H) &= \frac{2}{\pi}\int_0^{\pi} \cos\psi e^{H\cos\psi}\,d\psi = 2I_1(H) \\ \sqrt{\pi}\Gamma(\nu+\frac{1}{2})I_\nu(z) &\equiv \left(\frac{z}{2}\right)^{\nu}\int_{-1}^{1}(1-t^2)^{\nu-\frac{1}{2}}e^{\pm zt}\,dt \end{aligned}$$

The last equation identifies a modified Bessel function. It follows that $I_1'(H) = I_0(H) - I_1(H)/H$; this is of use in Eq. (32.26). The first of these equations is the result one obtains starting from the four-dimensional euclidian volume element in spherical coordinates in Section 31.

of σ_C/σ for $SU(2)$ lattice gauge theory in $d = 4$ dimensions in MFT, obtained from the solution to Eq. (32.29), is assigned as Prob. 32.1.

Results in other dimensions, and comparison with some Monte Carlo calculations, are discussed in Probs. 32.2-3.

33

OBSERVABLES IN LGT

In this section we consider the calculation of observables in lattice gauge theory (LGT). These will include such quantities as the interaction potential between a static quark-antiquark pair, where a linear rise in the potential with separation distance will provide strong evidence for the complete screening of the strong color charge and the *confinement of color*; the corresponding "string tension," or force required to separate the two static color charges; and the mass of a "glueball," a particle without quarks arising entirely from the nonlinear gluon interactions. References [Q17, Q18, Q19, Q31, Q32, Q33, Q34, Q35, Q36, Q37] provide the relevant background; this section is based on Ref. [Q37].

The $(l\bar{l})$ *Interaction in QED.* For clarity we start the discussion with the abelian $U(1)$ theory and then generalize to the nonabelian case. Put a (heavy) charged lepton pair with charges $\pm e_0$ at two fixed points in space with world lines as shown in Fig. 33.1. The charges interact through the electromagnetic field. The interaction with the background electromagnetic field can be included exactly in LGT. Consider the partition function and the generating functional for the gauge fields (Sections 28 and 29)

$$
\begin{aligned}
Z &= \int (dU) \exp\{-S(U)\} \\
\tilde{W}_E(J) &= \frac{\int (dU) \exp\{-S(U) - S_J\}}{\int (dU) \exp\{-S(U)\}} \equiv \langle e^{-S_J} \rangle
\end{aligned} \tag{33.1}
$$

Everything is now in the euclidian metric, and the correct boundary conditions are implied: for Z one takes $\int_0^\beta d\tau$ with periodic boundary conditions in τ; and for $\tilde{W}_E(J)$ one computes $\int_{-\infty}^{\infty} d\tau$ — the Green's functions derived from $\tilde{W}_E(J)$ can be analytically continued back to Minkowski space at the end of the calculation.

Now compute the additional source term in the action S_J

$$
-S_J = \left[i \int d^4x \, \mathcal{L}_J \right]_E \tag{33.2}
$$

Start in Minkowski space

$$
\mathcal{L} = J_\mu A_\mu \tag{33.3}
$$

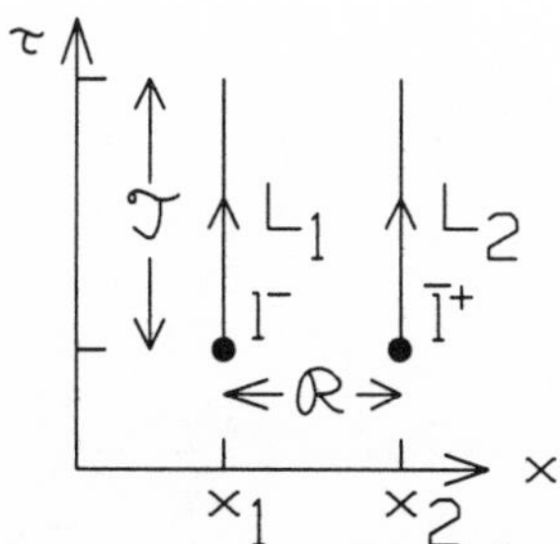

Figure 33.1: World lines for a (heavy) charged lepton pair with charges $\pm e_0$ at two fixed points in space; basis for calculating their interaction potential.

Here, for a pair of static charges, the source of the electromagnetic field takes the form $J_\mu = (\mathbf{0}, i\rho)$ and $J_\mu A_\mu = -\rho\,\Phi$ with

$$\rho = e_0\left[\delta^{(3)}(\mathbf{x}-\mathbf{x}_2) - \delta^{(3)}(\mathbf{x}-\mathbf{x}_1)\right] \tag{33.4}$$

Hence

$$\int d^4x\,\mathcal{L}_J = -e_0\int dt\,[\Phi(\mathbf{x}_2) - \Phi(\mathbf{x}_1)] \tag{33.5}$$

Now go to the euclidian metric (Section 29)

$$\begin{array}{lll} t \to -i\tau & x_\mu \to (\mathbf{x}, \tau) & x_\mu^2 = \mathbf{x}^2 + \tau^2 \\ \Phi \to -iA_0 & A_\mu \to (\mathbf{A}, A_0) & A_\mu^2 = \mathbf{A}^2 + A_0^2 \end{array} \tag{33.6}$$

Then

$$\begin{aligned} -S_J \equiv \left[i\int d^4x\,\mathcal{L}_J\right]_E &= ie_0\int d\tau\,[A_0(\mathbf{x}_2) - A_0(\mathbf{x}_1)] \\ &= ie_0\left[\int_{L_2} dx_\mu A_\mu - \int_{L_1} dx_\mu A_\mu\right] \end{aligned} \tag{33.7}$$

Here the integrals go along the world lines in Fig. 33.1.

Now for a static problem, the contribution to the line integral from a segment in the $\mathbf{x}$ direction at fixed t must be independent of t. Thus after the substitution in Eq. (33.6) the contribution to the line integral from the two segments shown in Fig. 33.2a must be equal.[47]

$$\int_{L_1'} dx_\mu A_\mu = \int_{L_2'} dx_\mu A_\mu \tag{33.8}$$

[47]In fact, here, both contributions vanish since $A_\mu = (\mathbf{0}, A_0)$.

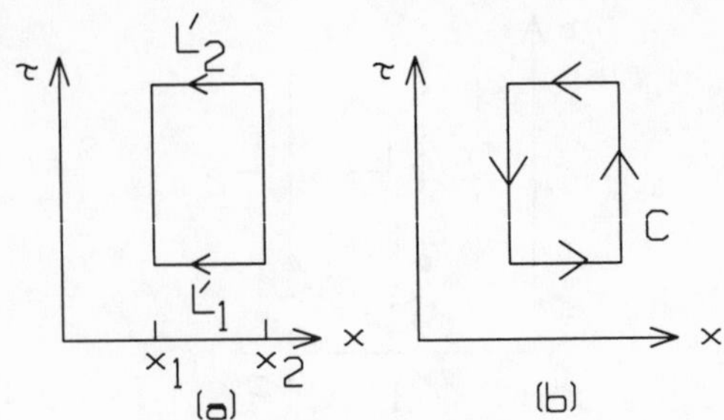

Figure 33.2: (a) Equal contributing end segments in line integral of vector potential in static case. (b) Resulting completion of line integral of vector potential around the closed curve C in space-time.

These segments can be combined with the expression in Eq. (33.7) to give a line integral around a closed curve C in space-time as shown in Fig. 33.2b.

$$S_J = -ie_0\left[\left(\int_{L_2} + \int_{L_2'} - \int_{L_1} - \int_{L_1'}\right) dx_\mu A_\mu\right] = -ie_0 \oint_C dx_\mu A_\mu \quad (33.9)$$

The resulting integral is *gauge invariant*. To see this, make a gauge transformation $A_\mu \to A_\mu + \partial\Lambda/\partial x_\mu$. Then, by the definition of the gradient, and since Λ is single-valued

$$\oint_C \left(\frac{\partial \Lambda}{\partial x_\mu}\right) dx_\mu = 0 \quad (33.10)$$

Hence the line integral of the vector potential around the closed curve in space-time is unchanged by the gauge transformation.

The euclidian action in Eq. (33.1) can thus be written

$$\tilde{W}_E(J) = \left\langle \exp\left\{ie_0 \oint_C A_\mu dx_\mu\right\}\right\rangle \quad (33.11)$$

Here the expectation value is evaluated with the statistical operator in Eq. (33.1)

$$\rho = \frac{e^{-S(U)}}{\int (dU) e^{-S(U)}} \quad (33.12)$$

Interpretation as a $V_{l\bar{l}}(R)$ Potential. The goal is to now make a connection with physics through the concept of an interaction potential between the two fixed charges. In order to make this connection introduce the concept of *effective degrees of freedom* for this system, and concentrate entirely on the space-time coordinates of the charges themselves. Return to the original discussion of the path integral in Section 28 in terms of particle coordinates. For a single particle the probability amplitude for finding the particle at (q_2, t_2) if it started at (q_1, t_1) is

$$\langle q_2 t_2 | q_1 t_1 \rangle = \int \mathcal{D}(q) \exp\left\{ i \int_{t_1}^{t_2} L(q, \dot{q}) dt \right\} \tag{33.13}$$

This probabilty amplitude, according to the general principles of quantum mechanics, can be written in the Schrödinger and Heisenberg pictures as

$$\begin{aligned} \langle q_2 t_2 | q_1 t_1 \rangle &= \langle q_2 | \Psi_{q_1}(t_2) \rangle \ ; & \text{S} - \text{Rep} \\ &= \langle q_2 | \exp\{-iH(t_2 - t_1)\} | q_1 \rangle \ ; & \text{H} - \text{Rep} \end{aligned} \tag{33.14}$$

This analysis is readily extended to two (or more) particles. Consider two charged particles, and take as the initial and final states the above charges at their fixed positions in space

$$|q_1\rangle \to |l\bar{l}\rangle \qquad\qquad |q_2\rangle \to |l\bar{l}\rangle \tag{33.15}$$

Now go to the euclidian metric and define the tau-interval

$$(t_2 - t_1) \equiv -i\mathcal{T} \tag{33.16}$$

Equation (33.14) then implies that the l.h.s. of Eq. (33.13) can be written as

$$\text{l.h.s.} = \langle l\bar{l} | e^{-H\mathcal{T}} | l\bar{l} \rangle \tag{33.17}$$

Define an effective hamiltonian for the two charged particles by $H = T + V$; for heavy (static) charges, one can neglect the kinetic energy T. Since the particles are heavy and fixed there is no dynamics and the states do not change; Eq. (33.17) then reduces to

$$\text{l.h.s.} = e^{-V(\mathcal{R})\mathcal{T}} \tag{33.18}$$

Here $V(\mathcal{R})$ is just the interaction potential of the two heavy (fixed) charges separated in coordinate space by a distance $\mathcal{R}$. We have now calculated the same amplitude in two different ways. A combination of Eqs. (33.18) and (33.13) and (33.11) then provides a definition of the *interaction potential of the two charges entirely in terms of the field variables.*[48]

$$\exp\{-V(\mathcal{R})\mathcal{T}\} = \left\langle \exp\left\{ ie_0 \oint_C A_\mu dx_\mu \right\} \right\rangle = \tilde{W}_E(J) \tag{33.19}$$

The logarithm of this relation provides an explicit expression for the potential

$$\begin{aligned} V(\mathcal{R}) &= -\frac{1}{\mathcal{T}} \ln \tilde{W}_E(J) \\ \tilde{W}_E(J) &= \frac{\int (dU) e^{-S(U)} \exp\{ie_0 \oint_C dx_\mu A_\mu\}}{\int (dU) e^{-S(U)}} \end{aligned} \tag{33.20}$$

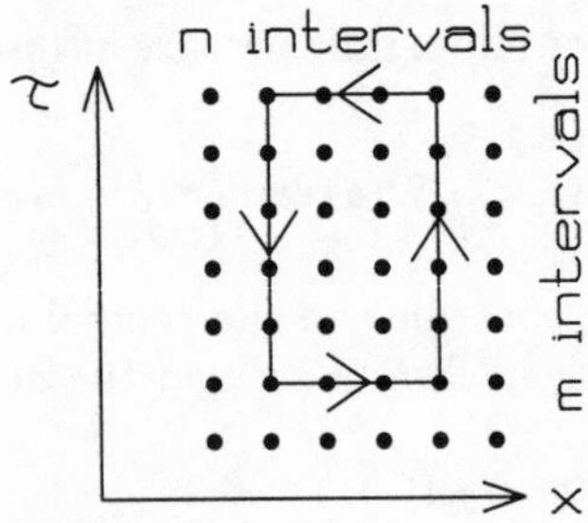

Figure 33.3: Contour C for the evaluation of Wilson loops in LGT.

The required integrals in Eq. (33.20) can be immediately evaluated in lattice gauge theory in terms of *Wilson loops.* These are defined by utilizing the specific contour C indicated in Fig. 33.3; the quantity $\tilde{W}_E(J)$ evaluated for this contour C will be denoted by $W_J(m,n)$. Here

$$\mathcal{T} = ma \qquad\qquad \mathcal{R} = na \tag{33.21}$$

Hence

$$V(\mathcal{R}) \quad = \quad -\frac{1}{ma}\ln W_J(m,n) \tag{33.22}$$

This expression is readily evaluated using the LGT techniques that have been developed, since the paths in the Wilson loops now just involve the link variables along the path — we shall do so analytically in the next section.

Nonabelian Theory. Analogous arguments indicate that these results may be taken over to the nonabelian LGT if one makes the replacements[49]

- For $SU(2)$

$$\oint_C dx_\mu A_\mu \rightarrow \oint_C dx_\mu \frac{1}{2}\underline{\tau}\cdot\mathbf{A}_\mu \quad \equiv \quad \oint_C dx_\mu \tilde{A}_\mu \tag{33.23}$$

- For $SU(3)$

$$\oint_C dx_\mu A_\mu \rightarrow \oint_C dx_\mu \frac{1}{2}\lambda^a A_\mu^a \quad \equiv \quad \oint_C dx_\mu \tilde{A}_\mu \tag{33.24}$$

The generalization of Eq. (33.20) for $SU(\mathcal{N})$ is

$$\tilde{W}_E(J) \quad = \quad \frac{\int(dU)e^{-S(U)}\,\mathrm{Re}(1/\mathcal{N})\mathrm{tr}\left[\exp\{ig_0\oint_C dx_\mu\tilde{A}_\mu\}\right]}{\int(dU)e^{-S(U)}}$$

$$\exp\{ig_0\oint_C dx_\mu\tilde{A}_\mu\} \quad \equiv \quad \prod\nolimits_{\text{links around C}} \mathrm{U}_l \tag{33.25}$$

[48]Recall that even in QED, with finite lattice spacing a, these are self-interacting fields.

[49]Note $[S(U),\oint_C dx_\mu\tilde{A}_\mu] = 0$ since the action $S(U) = 2\sigma\sum_\square\left(1-\frac{1}{\mathcal{N}}\mathrm{Re}[\mathrm{tr}\underline{\mathrm{U}}_\square]\right)$ is just a c-number. Thus the exponentials still factor $e^{-S(U)}e^{-S_J} = e^{-S(U)-S_J}$.

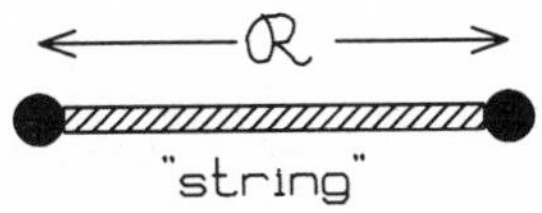

Figure 33.4: Confinement modeled by two heavy charges connected with a string.

One simply includes the appropriate link variables around the Wilson loop. The arguments in Section 31 indicate that this functional is again gauge invariant.

Confinement. It is of great interest to see if the nonabelian gauge theories based on an internal color symmetry can provide forces sufficiently strong so that an isolated color charge is completely shielded by strong vacuum polarization (a possibility first envisioned by Schwinger in another context (Ref. [Q38])). We seek a situation where one cannot pull the static charges apart in the absence of additional pairs, where they behave as if they were tied together with a "string" (Fig. 33.4). We look for

$$V(\mathcal{R}) = \bar{\sigma}\mathcal{R} \qquad\qquad \bar{\sigma} \equiv \text{string tension} \tag{33.26}$$

The goal is to see if this indeed happens in the above calculation, and, if it does, to calculate $\bar{\sigma}$.[50]

The anticipated form of the confining potential is sketched in Fig. 33.5. If Eq. (33.26) holds then $e^{-V(\mathcal{R})\mathcal{T}} = e^{-\bar{\sigma}\mathcal{R}\mathcal{T}}$ and one can rewrite Eq. (33.22) as

$$\bar{\sigma} \;=\; -\frac{1}{\mathcal{R}\mathcal{T}}\ln \tilde{W}_E(J) \;=\; -\frac{1}{a^2}\left[\frac{1}{mn}\ln W_J(m,n)\right] \tag{33.27}$$

Here Eq. (33.21) has been used and the result written in terms of the Wilson loop. The first of these equations states that the $\ln \tilde{W}_E(J)$ must be proportional to the *area* $\mathcal{R}\mathcal{T}$ of the Wilson loop in order to find a constant $\bar{\sigma}$; this proportionality to the area can be taken as a signal for confinement.[51]

Continuum Limit. Consider the continuum limit of this calculation where $a \to 0$ for a fixed lattice size N^d.[52] The situation is illustrated in Fig. 33.6. For a given lattice size, one needs to keep a *large* enough so the fixed static charges are far enough apart in space that the asymptotic expression (large $\mathcal{R}$) in Eq. (33.26) holds. Conversely, a must be *small* enough so that the result approximates the true continuum limit.

[50] In the physical world what presumably happens is that the string breaks and a pair of quarks is created from the vacuum to take the places at the ends of the string fragments — a meson turns into two mesons.

[51] Suppose $W_J(m,n) = \exp\{-\bar{\sigma}\mathcal{R}\mathcal{T} - \rho_1\mathcal{T} - \rho_2\mathcal{R} - b\}$ where the last three terms represent possible "transients" in the LGT calculation. Recall $\mathcal{R} = na, \mathcal{T} = ma$. Then

$$R(m,n) \;\equiv\; \frac{W_J(m,n)W_J(m-1,n-1)}{W_J(m,n-1)W_J(m-1,n)} \;=\; e^{-\bar{\sigma}a^2}$$

In this case the transients can be eliminated by taking $\bar{\sigma} = -(1/a^2)\ln R(m,n)$.

[52] Limited, for example, by computing power.

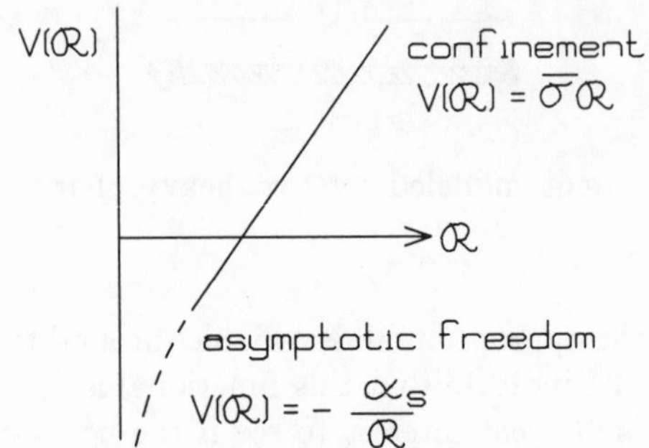

Figure 33.5: Anticipated form of confining static $(q\bar{q})$ potential. At short distances one has the asymptotic freedom result $V(\mathcal{R}) = -\alpha_s/\mathcal{R}$; at large distances is the anticipated string value $V(\mathcal{R}) = \bar{\sigma}\mathcal{R}$.

One test that can be used on the lattice dimension a is to see if one *reproduces the scaling of the coupling constant predicted by perturbation theory.* How does this work? *Asymptotic freedom* is the key ingredient here. It says that the size of the renormalized coupling constant decreases as one goes to smaller and smaller distance scales. Let g be the (renormalized) coupling constant one uses to describe physics on the distance scale of a as illustrated in Fig. 33.7a. For illustration, imagine the coupling constant to be associated with each site and let the size of the dot illustrate the strength of the coupling. Now suppose one halves the physical dimension of the lattice by letting $a \to a/2$ as illustrated in Fig. 33.7b. To maintain the same total strength of the coupling in a given physical region in space, the coupling on each site must be decreased in strength as illustrated. In the asymptotic domain the analytical relation between these scaled renormalized coupling constants is given in $SU(3)$ (that is, QCD) by the relation (Refs. [Q4, Q2], see Section 27)

$$g_2^2(\lambda^2) \quad \approx \quad \frac{g_1^2}{1 + (g_1^2/16\pi^2)\beta_0 \ln\{\lambda^2/\lambda_1^2\}}$$

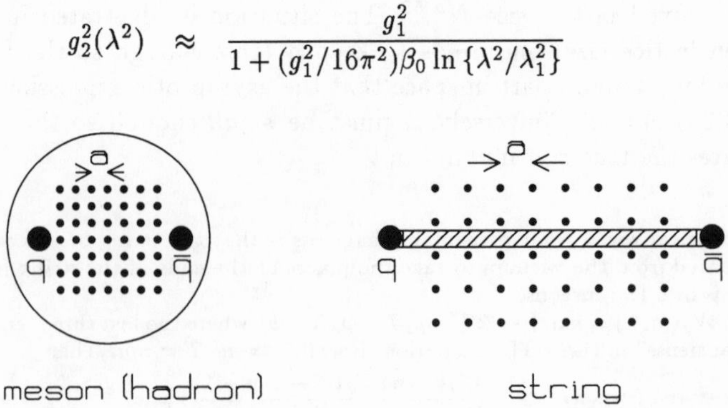

Figure 33.6: Continuum limit $a \to 0$ of calculation of the $(q\bar{q})$ potential for fixed lattice size N^d.

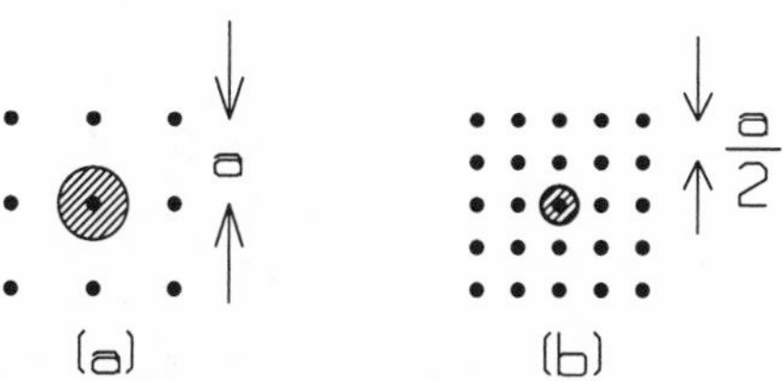

Figure 33.7: Schematic illustration of variation of strength of renormalized coupling constant g on the lattice with the distance scale of the lattice a. (a) For spacing a; (b) for spacing $a/2$.

$$\beta_0 = \frac{33}{3} - \frac{2N_f}{3} \qquad (= \frac{33}{3} \text{ here}) \tag{33.28}$$

In this expression $g_2^2(\lambda^2)$ is the renormalized coupling constant as measured at a momentum transfer λ^2 (Fig. 33.8a).[53] The initial coupling constant g_1^2 is fixed by experiment; it defines the starting point for the analysis. To visualize the situation, imagine this coupling constant g_1^2 has a certain physical distribution in space. The lattice then looks at this coupling on a certain physical distance scale characterized by the lattice spacing a as illustrated in Fig. 33.8b.

Equation (33.28) is based on the renormalization group; it sums the leading terms in $\ln q^2$. In LGT there is a natural distance scale a at which one wants to determine the new renormalized coupling constant, or, correspondingly, a natural momentum scale $q^2 \equiv \lambda^2 = 1/a^2$. Asymptotically in this quantity one can write Eq. (33.28) as

$$g_2^2(\frac{1}{a^2}) \approx \frac{16\pi^2}{\beta_0 \ln\{1/a^2\lambda_1^2\}} \equiv g^2 \tag{33.29}$$

Here it has been assumed that $(g_1^2\beta_0/16\pi^2)\ln\{1/a^2\lambda_1^2\} \gg 1$; we are interested in the limit $a \to 0$. The quantity in Eq. (33.29) will be defined as g^2, the renormalized coupling constant to be used at the distance scale a^2.[54] Equation (33.29) can be inverted to solve for a^2

$$a^2 = \frac{1}{\lambda_1^2}\exp\{-\frac{16\pi^2}{\beta_0 g^2}\} \equiv \frac{1}{\lambda_1^2} f_0^2(g^2) \tag{33.30}$$

Now substitute this relation in Eq. (33.27)

$$\bar{\sigma} = -\lambda_1^2\left[\frac{1}{f_0^2(g^2)}\frac{1}{mn}\ln W_J(m,n)\right] \tag{33.31}$$

[53]The corresponding result for $SU(\mathcal{N})$ from Refs. [Q4, Q2] is $\beta_0 = 11\mathcal{N}/3 - 2N_f/3$; here N_f is the number of "flavors" of fermions, that is, the number of different types of fermions belonging to the fundamental representation of $SU(\mathcal{N})$.

[54]Note that one *cannot* take the same limit in a nonasymptotically free theory, such as QED where (Section 27) $e_2^2 \approx e_1^2/[1-(e_1^2/12\pi^2)\ln(\lambda^2/M^2)]$ because here e_2^2 would change sign!

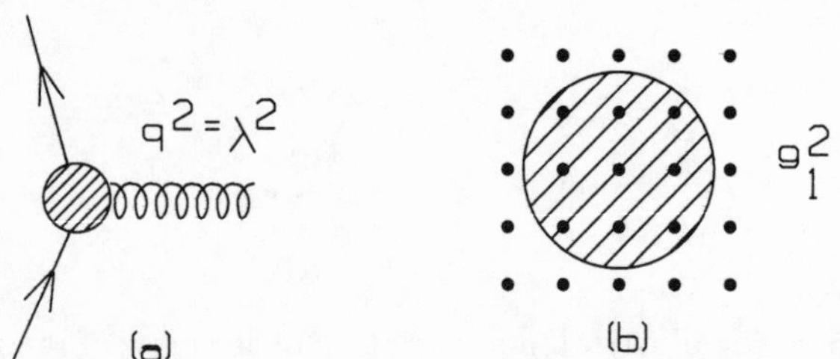

Figure 33.8: (a) Physical determination of the renormalized coupling constant $g_2^2(\lambda^2)$; (b) lattice view of the starting renormalized coupling constant g_1^2 with lattice spacing a; it has some overall strength and spatial extent.

The *scaling test* implies that the physical quantity expressed by the term in brackets on the r.h.s. of this equation should be independent of a [and hence of g^2 by Eq. (33.29)] for small enough a. If one has a proper description, then physical quantities such as that illustrated schematically in Fig. 33.8b should be unchanged if the underlying lattice with which one chooses to describe them decreases in size and the coupling constant used on that lattice is scaled appropriately. If the physical distance scale of the underlying lattice is small enough, one can use the asymptotic relation for the dependence of coupling constant on distance scale — and the coupling constant is becoming vanishingly small in this asymptotically free theory.

For SU($\mathcal{N}$) and dimension $d = 4$ define[55]

$$2\sigma \equiv \bar{\beta} \equiv \frac{2\mathcal{N}}{g^2} \tag{33.32}$$

This is now the only parameter left in the calculation [except for $1/\lambda_1$, which simply defines the unit of length through Eq. (33.30)]. Then from Eq. (33.30)

$$f_0^2(g^2) = \exp\left\{-\left(\frac{16\pi^2}{2\mathcal{N}}\right)\left(\frac{\bar{\beta}}{\beta_0}\right)\right\} \tag{33.33}$$

Thus the result in brackets in Eq. (33.31) should be *independent of $\bar{\beta}$ for large enough $\bar{\beta}$* — this is the scaling test.[56]

Results for $V_{\bar{q}q}$. We show some results of numerical calculations carried out by Otto and Stack for QCD in Fig. 33.9 (Ref. [Q32]).[57] These authors:

1. Work on a lattice of size 16^4;

[55] Recall from Section 31 for $SU(2)$ and $d = 2$ we had $2\sigma = 4/g^2a^2$ and with $d = 4$ the a^2 dependence disappears from this expression (see Prob. 31.4).

[56] According to Ref. [Q37] the authors in Ref. [Q32] include the next order correction to Eq. (33.33), which results in $f_0(g) \rightarrow f_1(g) = (8\pi^2\bar{\beta}/3\beta_0)^{51/121} f_0(g)$, but the factor in front of $f_0(g)$ is "purely decorative."

[57] The basis for the large-scale Monte Carlo numerical calculations is developed in Section 35.

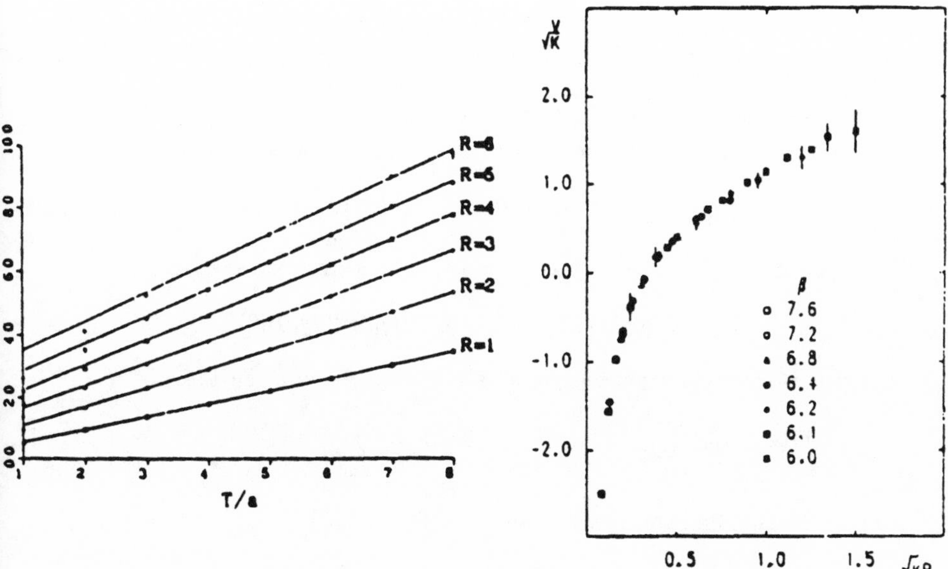

Figure 33.9: (a) $-\ln W(\mathcal{R}, \mathcal{T})$ vs $\mathcal{T}/a$ for $\bar{\beta} = 6$ and various R $= \mathcal{R}/a$; (b) the quark potential for all the data with $\bar{\beta} \geq 6$. Here $\kappa \equiv \bar{\sigma}$. From Refs. [Q32, Q37].

2. Assume

$$-\ln W(\mathcal{R}, \mathcal{T}) = \bar{\sigma}\mathcal{R}\mathcal{T} + \text{corrections} \tag{33.34}$$

3. Check the linearity in $\mathcal{T}/a$ for fixed $\mathcal{R}/a$ (Fig. 33.9a);
4. Check the scaling with $\bar{\beta}$ and find it works for $\bar{\beta} \geq 6$;
5. Fit their results for $V(\mathcal{R})$ in Fig. 33.9b to

$$V(\mathcal{R}) = \bar{\sigma}\mathcal{R} + B - \frac{\alpha}{\mathcal{R}} \tag{33.35}$$

They find

$$\sqrt{\bar{\sigma}} \approx (106 \pm 3)\lambda_1 \tag{33.36}$$

From fits to the spectra of heavy quarkonium [$(\bar{c}c)$, $(\bar{b}b)$, etc.] the string tension appearing in the potential $V(\mathcal{R})$ is determined to be

$$\bar{\sigma} \approx 1\,\text{GeV/fm} \tag{33.37}$$

Thus these authors determine

$$\lambda_1 \approx 4\text{ MeV} \qquad\qquad \sqrt{\bar{\sigma}} \approx 400\text{ MeV} \tag{33.38}$$

The key result from this work is that *the quark potential $V_{\bar{q}q}(\mathcal{R})$ does indeed appear to rise linearly with $\mathcal{R}$ for large $\mathcal{R}$ as seen in Fig. 33.9b — QCD does appear to lead to confinement.*

Determination of the Glueball Mass. Start back in Minkowski space. Let $\hat{Q}(t)$ be any hermitian operator in the Heisenberg representation

$$\hat{Q}(t) = e^{i\hat{H}t}\hat{Q}(0)e^{-i\hat{H}t} \; ; \qquad \text{H} - \text{Rep} \tag{33.39}$$

Consider the vacuum matrix element of this operator expressed as

$$G(t) \equiv \langle 0|\hat{Q}(t)\hat{Q}(0)|0\rangle - \langle 0|\hat{Q}(t)|0\rangle\langle 0|\hat{Q}(0)|0\rangle \tag{33.40}$$

This is a *correlation function.* Insert a complete set of eigenstates of $\hat{H}$

$$G(t) \;=\; \sum_{n\neq 0}\langle 0|\hat{Q}(t)|n\rangle\langle n|\hat{Q}(0)|0\rangle \;=\; \sum_{n\neq 0}|\langle n|\hat{Q}(0)|0\rangle|^2 e^{-iE_n t} \tag{33.41}$$

Now go to euclidian space with $t \to -i\mathcal{T}$ [Eq. (33.6)]

$$G(\mathcal{T}) = \sum_{n\neq 0}|\langle n|\hat{Q}(0)|0\rangle|^2 e^{-E_n\mathcal{T}} \tag{33.42}$$

Take the limit $\mathcal{T} \to \infty$. Let m_0 be the mass of the lightest "glueball."[58] Suppose the operator $\hat{Q}$ has *any* overlap with this state. Then

$$G(\mathcal{T}) \overset{\mathcal{T}\to\infty}{\longrightarrow} |\langle g_0|\hat{Q}(0)|0\rangle|^2 e^{-m_0\mathcal{T}} \tag{33.43}$$

Hence

$$m_0 = -\frac{1}{\mathcal{T}}\ln G(\mathcal{T}) \tag{33.44}$$

Here the limit $\mathcal{T} \to \infty$ is implied. With $\mathcal{T} = ma$ one can write instead

$$m_0 = -\frac{1}{a}\ln\frac{G(m)}{G(m-1)} \tag{33.45}$$

The use of Eq. (33.30) allows one to again check scaling to see if the lattice result is believable

$$m_0 = -\lambda_1\left[\frac{1}{f_0(g)}\ln\frac{G(m)}{G(m-1)}\right] \tag{33.46}$$

The scaling test implies that the physical quantity in brackets should be independent of $\bar{\beta}$ [Eqs. (33.32) and (33.33)] as $\bar{\beta} \to \infty$.

[58] An eigenstate of $\hat{H}$ formed from the nonlinear gluon couplings in the absence of valence quarks.

Clearly the operator $\hat{Q}$ must have the same quantum numbers as the state $|g_0\rangle$ to produce some overlap in Eq. (33.43). One picks $\hat{Q}$ to have as definite rotation properties as possible on the lattice in order to select states of definite angular momentum (see Refs. [Q35, Q36]). The latest result quoted in Ref. [Q37] is

$$m_{g_0}(0^{++}) \approx 375\lambda_1 \approx 1.5\,\mathrm{GeV} \tag{33.47}$$

Here the value of λ_1 from Eq. (33.38) has been used. Thus the mass of the lightest glueball appears to be about 1.5 GeV.

The calculation of a *correlation function* on the lattice amounts to an evaluation of the following expression (see Prob. 33.1).

$$\langle Q(\mathcal{T})Q(0)\rangle = \frac{\int(dU)e^{-S(U)}Q(\mathcal{T})Q(0)}{\int(dU)e^{-S(U)}} \tag{33.48}$$

34

STRONG-COUPLING LIMIT

It is always useful to have limiting analytical solutions to any theory. In this section LGT is solved analytically in the strong-coupling limit where the constant σ in the action becomes very *small.* Thus we are here interested in the following limit of the theory

$$\text{Fix a}^2\ ; \qquad \sigma \to 0 \quad \text{or} \quad \frac{1}{\sigma} \to \infty \tag{34.1}$$

From Eq. (33.32), for example, for SU($\mathcal{N}$) in $d = 4$ dimensions, $2\sigma = 2\mathcal{N}/g^2$. Small σ corresponds to large g^2; hence the name strong-coupling limit. References [Q17, Q18, Q34] provide basic background here; this section is based on Ref. [Q39].

Since the action appears in the exponent of the statistical operator, small σ is equivalent to *high temperature* in the usual partition function in statistical mechanics. One can think of $1/\sigma \equiv T_{\text{eff}}$ as an effective temperature here. If we recall the plot of the magnetization m^4 vs σ_C/σ in Fig. 30.6 and Prob. 32.2, then the strong-coupling limit corresponds to the far right hand side of the figure.

> *Basic Observation: In the limit $\sigma \to 0$, one can expand the exponential of the action in the statistical operator and keep the first nonvanishing term in a power series in σ.*

Let us start with $U(1)$ and then generalize. This calculation is readily carried out because the path integral over the individual link variables takes a very simple form

$$\begin{aligned}
\int (dU_l)1 &= \int_0^{2\pi} \frac{d\phi_l}{2\pi} = 1 \\
\int (dU_l)U_l &= \int_0^{2\pi} \frac{d\phi_l}{2\pi} e^{i\phi_l} = 0 \\
\int (dU_l)U_l^2 &= \int_0^{2\pi} \frac{d\phi_l}{2\pi} e^{2i\phi_l} = 0\ ; \qquad \text{etc.} \\
\int (dU_l)U_l U_l^* &= \int_0^{2\pi} \frac{d\phi_l}{2\pi} = 1
\end{aligned} \tag{34.2}$$

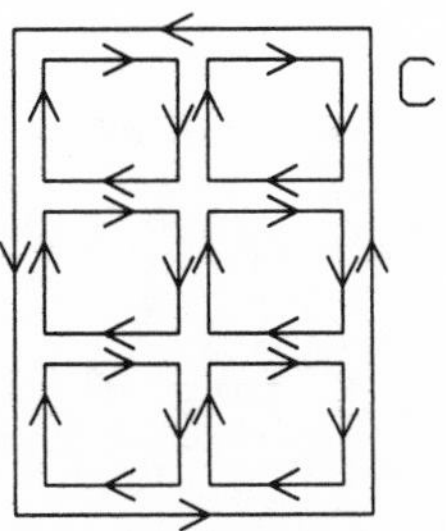

Figure 34.1: Contour C of Wilson loop (see Fig. 33.3) and tiling plaquettes.

As $a \to 0$ one can write

$$\exp\left\{ie_0 \oint_C dx_\mu A_\mu\right\} = \prod_{\text{links on } C} U_l \tag{34.3}$$

Here C indicates the contour of the Wilson loop (Fig. 34.1). This follows from the definition of the link variables

$$U_{ji} = \exp\left\{ie_0(x_j - x_i)_\mu A_\mu\left[\frac{1}{2}(x_j + x_i)\right]\right\} \tag{34.4}$$

The statistical average of Eq. (34.3) then forms the Wilson loop (Section 33)

$$W_J(m,n) = \frac{\int(dU)e^{-S(U)}\prod_{\text{links on C}} U_l}{\int(dU)e^{-S(U)}} \tag{34.5}$$

In this expression (Section 29)

$$\begin{aligned} S &= \sum\nolimits_{\Box} S_{\Box} \\ S_{\Box} &= \sigma\{[1-(U_{\Box})_{\hookrightarrow}] + [1-(U_{\Box})_{\hookleftarrow}]\} \\ (U_{\Box})_{\hookrightarrow} &= U_{il}U_{lk}U_{kj}U_{ji} = e^{-i\phi_4}e^{-i\phi_3}e^{i\phi_2}e^{i\phi_1} \end{aligned} \tag{34.6}$$

The situation for the basic plaquette is illustrated in Fig. 34.2.

Now expand $e^{-S(U)}$, and employing Eqs. (34.2), keep the minimum number of terms necessary to get a nonzero result. One must *pair all links in opposite directions* as illustrated in Fig. 34.3 (producing $U_l U_l^*$); the path integral over unpaired links *vanishes*. The minimum number of plaquettes necessary to pair all links starting from a given contour C in the Wilson loop must just *tile the area* as illustrated in Fig. 34.1. The number of such plaquettes p is clearly

$$p = mn \qquad \text{minimum number of plaquettes to tile Wilson loop} \tag{34.7}$$

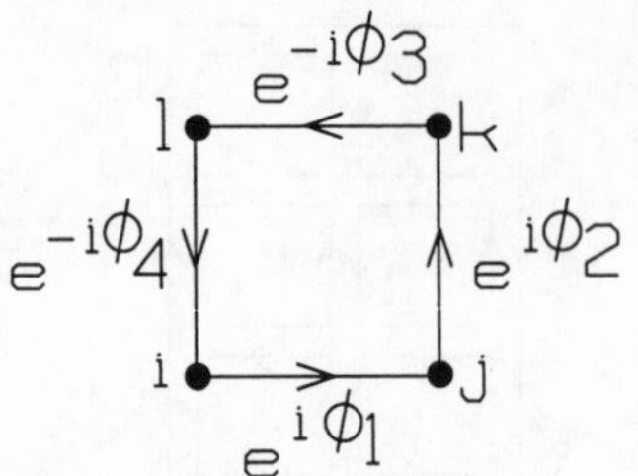

Figure 34.2: Link variables around the basic plaquette in $U(1)$. Recall the convention that the phase in the direction of the positive coordinate axes is taken as $+\phi$ (the phase in the opposite direction is $-\phi$).

Expand

$$e^{-S} = 1 - S + \frac{1}{2!}(-S)^2 + \frac{1}{3!}(-S)^3 + \cdots \tag{34.8}$$

We must go to pth order

$$\frac{(-1)^p}{p!}\left(\sum_{\Box} S_{\Box}\right)^p = \frac{(-1)^p}{p!}(\ldots + S_{\Box 1} + S_{\Box 2} + \ldots + S_{\Box p} + \ldots)^p \tag{34.9}$$

Here the tiling plaquettes have been explicitly indicated. How many ways can the required product of the p tiling plaquettes be obtained from this expression? The first term can be chosen from p factors; the second from $p-1$ factors, etc. — the answer is $p!$ ways. Hence the first contributing term in the statistical average arises from

$$e^{-S} \doteq \frac{(-1)^p}{p!} p!\, S_{\Box 1} S_{\Box 2} \ldots S_{\Box p} \tag{34.10}$$

Evidently $e^{-S} \doteq 1$ in the denominator of Eq. (34.5) for the same reasons. Hence this expression becomes

$$\begin{aligned} W_J(m,n) &= \sigma^p(-1)^p \int (dU)\,\{[1-(U_{\Box})_{\hookrightarrow}] + [1-(U_{\Box})_{\hookleftarrow}]\}_1 \times \cdots \\ &\quad \times \{[1-(U_{\Box})_{\hookrightarrow}] + [1-(U_{\Box})_{\hookleftarrow}]\}_p \prod_{\text{links on } C} U_l \end{aligned} \tag{34.11}$$

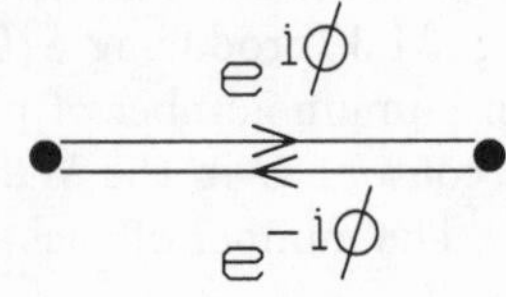

Figure 34.3: Pairing of link variables required to get first nonvanishing result in strong-coupling theory.

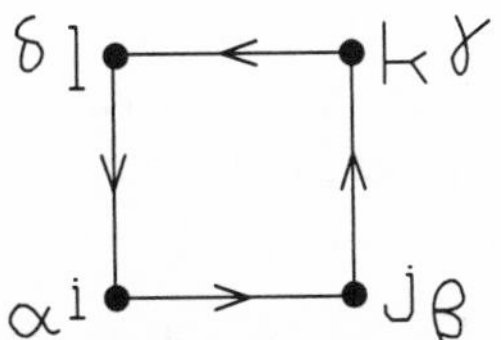

Figure 34.4: Sites and matrix indices for the basic plaquette in $SU(2)$.

Now note that the only nonvanishing contribution from this product of $S_\Box$ arises from the product of factors $-(U_\Box)_{\hookleftarrow}$, which produces the pairing illustrated in Fig. 34.1.

$$W_J(m,n) = \sigma^p \int (dU)(U_{\Box 1})_{\hookleftarrow}(U_{\Box 2})_{\hookleftarrow}\cdots(U_{\Box p})_{\hookleftarrow} \prod_{\text{links on } C} U_l \tag{34.12}$$

Equations (34.2) state that all the remaining integrals are unity. Hence one derives the lovely, simple, analytical result that in the strong-coupling limit $\sigma \to 0$ the Wilson loop is given by

$$W_J(m,n) = \sigma^p = \sigma^{mn} \tag{34.13}$$

We return later to a discussion of this result.

Nonabelian Theory: $SU(2)$. Let us now extend these arguments to the nonabelian case and take $SU(2)$ as a specific example. The same basic arguments will be employed, only now the link variables $\underset{\sim}{U}_{ji}$ are 2×2 matrices, and the contribution of a plaquette to the action involves the matrix product around the plaquette

$$(\underset{\sim}{U}_\Box)_{\hookrightarrow} = \underset{\sim}{U}_{il}\underset{\sim}{U}_{lk}\underset{\sim}{U}_{kj}\underset{\sim}{U}_{ji} \tag{34.14}$$

For the present discussion, matrix indices will be denoted with Greek superscripts; repeated indices are summed from 1 to 2. One needs

$$\frac{1}{2}\text{tr}\ (\underset{\sim}{U}_\Box)_{\hookrightarrow} = \frac{1}{2}\,(U_{il})^{\alpha\delta}\,(U_{lk})^{\delta\gamma}\,(U_{kj})^{\gamma\beta}\,(U_{ji})^{\beta\alpha} \tag{34.15}$$

Note that the initial and final matrix indices are tied together by the tr. The sites and matrix indices for the basic plaquette are illustrated in Fig. 34.4. In the opposite direction one has

$$(\underset{\sim}{U}_\Box)_{\hookleftarrow} = \underset{\sim}{U}_{kl}\underset{\sim}{U}_{li}\underset{\sim}{U}_{ij}\underset{\sim}{U}_{jk} \tag{34.16}$$

Since $\underset{\sim}{U}_{ij} = \underset{\sim}{U}_{ji}^\dagger$ Eqs. (34.14) and (34.16) can be rewritten

$$\begin{aligned}(\underset{\sim}{U}_\Box)_{\hookrightarrow} &= \underset{\sim}{U}_{li}^\dagger\underset{\sim}{U}_{kl}^\dagger\underset{\sim}{U}_{kj}\underset{\sim}{U}_{ji} \\ (\underset{\sim}{U}_\Box)_{\hookleftarrow} &= \underset{\sim}{U}_{kl}\underset{\sim}{U}_{li}\underset{\sim}{U}_{ji}^\dagger\underset{\sim}{U}_{kj}^\dagger\end{aligned} \tag{34.17}$$

Now all the link indices refer to the direction of the positive coordinate axes.

We claim that $\mathrm{tr}\,(\mathrm{U}_\Box)_{\hookleftarrow}$, by itself, is real. The same result holds for the other direction. The proof follows from the general representation of the $SU(2)$ link variables[59]

$$\mathrm{U}_l \;=\; (\alpha_0 + i\boldsymbol{\alpha}\cdot\boldsymbol{\tau})_l \tag{34.18}$$

Here the parameters $(\boldsymbol{\alpha}, \alpha_0)_l$ with $\boldsymbol{\alpha}_l^2 + \alpha_{0l}^2 = 1$ depend on the link and are real. Then

$$\begin{aligned} &\mathrm{tr}\;(\mathrm{U}_\Box)_{\hookleftarrow} = \\ &\quad \mathrm{tr}\,(\alpha_0 + i\boldsymbol{\alpha}\cdot\boldsymbol{\tau})_1\,(\alpha_0 + i\boldsymbol{\alpha}\cdot\boldsymbol{\tau})_2\,(\alpha_0 - i\boldsymbol{\alpha}\cdot\boldsymbol{\tau})_3\,(\alpha_0 - i\boldsymbol{\alpha}\cdot\boldsymbol{\tau})_4 \end{aligned} \tag{34.19}$$

The result follows since all the required traces $\mathrm{tr}\;(i\tau_k), \mathrm{tr}\;(i^2\tau_k\tau_l)$, etc. are real.

Thus for $SU(2)$

$$\begin{aligned} S_\Box &= 2\sigma\left[1 - \frac{1}{2}\mathrm{Re}\;\mathrm{tr}\,(\mathrm{U}_\Box)_{\hookrightarrow}\right] = 2\sigma\left[1 - \frac{1}{2}\mathrm{tr}\,(\mathrm{U}_\Box)_{\hookrightarrow}\right] \\ &= 2\sigma\left[1 - \frac{1}{2}\mathrm{tr}\,(\mathrm{U}_\Box)_{\hookleftarrow}\right] \end{aligned} \tag{34.20}$$

Basic Observation. The analysis again depends on the path integrals over the individual link variables

$$\begin{aligned} \int(d\mathrm{U}_l) &= 1 \qquad\qquad \int(d\mathrm{U}_l)\mathrm{U}_l = 0 \\ \int(d\mathrm{U}_l)(U_l)^{\beta\alpha}(U_l^\dagger)^{\gamma\delta} &= \frac{1}{2}\delta_{\alpha\gamma}\delta_{\beta\delta} \end{aligned} \tag{34.21}$$

The proof of these relations starts from the measure previously introduced for $SU(2)$

$$\int(d\mathrm{U}_l) \;\equiv\; \frac{1}{\pi^2}\int d^4\alpha\,\delta(\alpha_0^2 + \boldsymbol{\alpha}^2 - 1) = 1 \tag{34.22}$$

The terms linear in α give zero by symmetry.

$$\int(d\mathrm{U}_l)(U_l)^{\beta\alpha} \;=\; \frac{1}{\pi^2}\int d^4\alpha\,\delta(\alpha_0^2 + \boldsymbol{\alpha}^2 - 1)(\alpha_0\delta^{\beta\alpha} + i\boldsymbol{\alpha}\cdot\boldsymbol{\tau}^{\beta\alpha}) = 0 \tag{34.23}$$

Define

$$\frac{1}{\pi^2}\int d^4\alpha\,\delta(\alpha_0^2 + \boldsymbol{\alpha}^2 - 1)f(\alpha) \;\equiv\; \langle f(\alpha)\rangle \tag{34.24}$$

[59]Note α_0 multiplies the unit matrix, which is here suppressed.

Equation (34.23) then states $\langle\alpha_0\rangle = \langle\boldsymbol{\alpha}\rangle = 0$. Symmetry arguments immediately imply the following additional relations

$$\begin{aligned} \langle\alpha_0^2\rangle &= \langle\frac{1}{3}\boldsymbol{\alpha}^2\rangle = \langle\frac{1}{4}\alpha_\mu^2\rangle \\ \langle\alpha_0\boldsymbol{\alpha}\rangle &= 0 \\ \langle\alpha_i\alpha_j\rangle &= \langle\frac{1}{3}\boldsymbol{\alpha}^2\delta_{ij}\rangle \end{aligned} \tag{34.25}$$

The actual value of the expression $\langle\alpha_\mu^2\rangle$ is immediately evaluated

$$\langle\alpha_\mu^2\rangle = \frac{1}{\pi^2}\int d^4\alpha\,\delta(\alpha_\mu^2 - 1)\,\alpha_\mu^2 = 1 \tag{34.26}$$

The last of Eqs. (34.21), which provides the minimum requisite pairing of the link variables in the strong-coupling theory, is derived by explicitly evaluating the integral using these relations

$$\begin{aligned} &\int(dU_l)(U_l)^{\beta\alpha}(U_l^\dagger)^{\gamma\delta} \qquad\qquad (34.27) \\ &= \frac{1}{\pi^2}\int d^4\alpha\,\delta(\alpha_0^2 + \boldsymbol{\alpha}^2 - 1)(\alpha_0\delta^{\beta\alpha} + i\boldsymbol{\alpha}\cdot\boldsymbol{\tau}^{\beta\alpha})(\alpha_0\delta^{\gamma\delta} - i\boldsymbol{\alpha}\cdot\boldsymbol{\tau}^{\gamma\delta}) \\ &= \frac{1}{4}\left[\delta^{\beta\alpha}\delta^{\gamma\delta} + \boldsymbol{\tau}^{\beta\alpha}\cdot\boldsymbol{\tau}^{\gamma\delta}\right] \\ &= \frac{1}{4}\left[\delta^{\beta\alpha}\delta^{\gamma\delta} + \begin{pmatrix} 0 & 1 \\ 1 & 0 \end{pmatrix}^{\beta\alpha}\begin{pmatrix} 0 & 1 \\ 1 & 0 \end{pmatrix}^{\gamma\delta}\right. \\ &\qquad \left. + \begin{pmatrix} 0 & -i \\ i & 0 \end{pmatrix}^{\beta\alpha}\begin{pmatrix} 0 & -i \\ i & 0 \end{pmatrix}^{\gamma\delta} + \begin{pmatrix} 1 & 0 \\ 0 & -1 \end{pmatrix}^{\beta\alpha}\begin{pmatrix} 1 & 0 \\ 0 & -1 \end{pmatrix}^{\gamma\delta}\right] \end{aligned}$$

Indicate the possible sets of matrix indices by $(\beta\,\alpha, \gamma\,\delta)$; then the only nonvanishing values of the above are

$$\begin{aligned} (11, 11) &= (22, 22) = \frac{1}{2} \\ (12, 21) &= (21, 12) = \frac{1}{2} \end{aligned} \tag{34.28}$$

This establishes the last of Eqs. (34.21).

Strong-Coupling Limit ($\sigma \to 0$). Again expand the exponential of the action and keep the first nonvanishing terms. All of the link variables along the contour C in the Wilson loop (see below) must be paired, for any unpaired link variable integrates to zero. The links in all the plaquettes used for this pairing must themselves be paired for the same reason. Just as before, the minimum number of plaquettes required to achieve these pairings is the set of tiling plaquettes (Fig. 34.1).

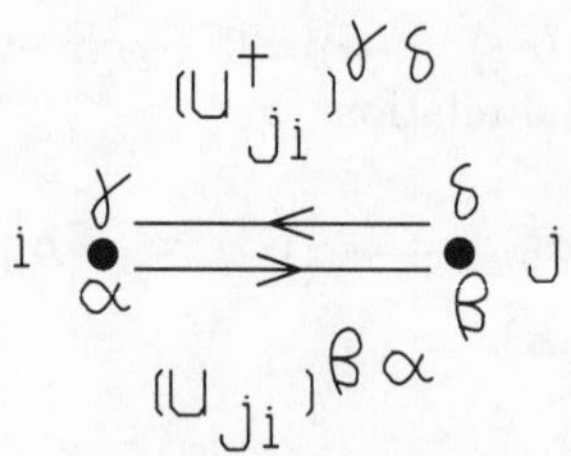

Figure 34.5: Pairing of link varibles in $SU(2)$ strong-coupling theory.

The use of Eqs. (34.20) and (34.17) leads to a pairing of link variables as indicated in Fig. 34.5. The Wilson loop is given by Eq. (33.25), and the integrals appearing in this expression must now be evaluated. As before, as $a \to 0$ the integral around the contour C can be broken up into the product of link variables around C (see Fig. 33.3).[60] Thus

$$\frac{1}{2}\mathrm{tr}\exp\left\{ig_0\frac{1}{2}\boldsymbol{\tau}\cdot\oint_C dx_\mu \mathbf{A}_\mu\right\} \to \frac{1}{2}\mathrm{tr}\prod\nolimits_{\text{links around } C} \mathrm{U}_l \tag{34.29}$$

The Wilson loop is then (it will turn out to be real)

$$W_J(m,n) = \frac{\prod_{\text{links}}\int(d\mathrm{U}_l)\exp\{-2\sigma\sum_\Box[1-(1/2)\mathrm{tr}(\mathrm{U}_\Box)_{\leftharpoonup}]\}(1/2)\mathrm{tr}\prod_{\text{links on } C}\mathrm{U}_l}{\prod_{\text{links}}\int(d\mathrm{U}_l)\exp\{-2\sigma\sum_\Box[1-(1/2)\mathrm{tr}(\mathrm{U}_\Box)_{\leftharpoonup}]\}} \tag{34.30}$$

Now expand the exponential and keep the first nonvanishing term. As before, the result is

$$\begin{aligned} W_J(m,n) &= \frac{\sigma^p}{2}\prod_{\text{all links}}\int(d\mathrm{U}_l)\mathrm{tr}(\mathrm{U}_{\Box 1})_{\leftharpoonup}\mathrm{tr}(\mathrm{U}_{\Box 2})_{\leftharpoonup}\cdots\mathrm{tr}(\mathrm{U}_{\Box p})_{\leftharpoonup} \\ &\quad\times\mathrm{tr}\left(\mathrm{U}_1\mathrm{U}_2\cdots\mathrm{U}_{2(m+n)}\right) \end{aligned} \tag{34.31}$$

All links are now paired. Use Eq. (34.21). It remains only to deal with the matrix indices. Again, repeated indices are summed. Since one is dealing with the tr in the plaquette contribution to the action and in the Wilson contour C, matrix indices close around the plaquettes and around C. Equation (34.21) and Fig. 34.5 indicate that the path integral over the link variables ties the matrix indices at the ends of each set of paired links together. Thus one has the three situations illustrated in Fig. 34.6. For the three cases one must evaluate, respectively

$$\begin{aligned} \delta_{\alpha\beta}\delta_{\beta\alpha} &= \delta_{\alpha\alpha} = 2 \\ \delta_{\beta\gamma}\delta_{\gamma\alpha}\delta_{\beta\alpha} &= \delta_{\alpha\alpha} = 2 \\ \delta_{\beta\gamma}\delta_{\gamma\delta}\delta_{\alpha\delta}\delta_{\beta\alpha} &= \delta_{\alpha\alpha} = 2 \end{aligned} \tag{34.32}$$

[60] There are a total of $2(m+n)$ links on the contour C and $p = mn$ plaquettes inside.

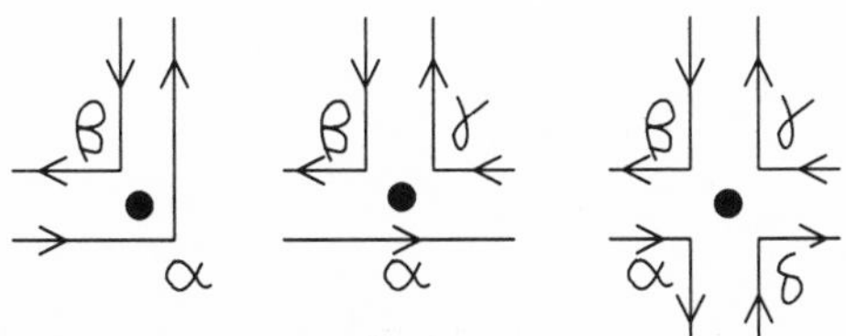

Figure 34.6: Matrix indices tied together in the three types of vertices required in the evaluation of the Wilson loop.

The result is 2 in all three cases.

One can now simply read off the answer: there is a factor of σ from each tiling plaquette, a factor of 1/2 from each set of paired link variables, and a factor of 2 from each vertex. Thus

$$W_J(m,n) = \frac{1}{2}(\sigma)^{\text{number plaquettes}}(\frac{1}{2})^{\text{number links}}(2)^{\text{number sites}} \tag{34.33}$$

The counting follows immediately from Fig. 33.3

$$\begin{aligned} \text{number plaquettes} &= mn \\ \text{number sites} &= (m+1)(n+1) \\ \text{number links} &= (m+1)n + (n+1)m \end{aligned} \tag{34.34}$$

Hence one again arives at a lovely, simple, analytical result for the Wilson loop in nonabelian lattice gauge theory based on $SU(2)$ in the strong-coupling limit

$$W_J(m,n) = \left(\frac{\sigma}{2}\right)^{mn} \tag{34.35}$$

Discussion. In this discussion we use the following notation

$$\begin{aligned} &\bar{\sigma} \equiv \text{string tension} \equiv \kappa \\ &g_0 \equiv g \qquad\qquad e_0 \equiv e \end{aligned} \tag{34.36}$$

Strong-Coupling $SU(2)$. The string tension that follows from the Wilson loop according to Eq. (33.27) is given in the strong-coupling limit by

$$a^2\kappa \;=\; -\frac{1}{mn}\ln W_J(m,n) \;=\; \ln\frac{2}{\sigma} \tag{34.37}$$

In four dimensions ($d = 4$) Eq. (33.32) implies

$$a^2\kappa \;=\; \ln\frac{2}{\sigma} \;=\; \ln g^2 \tag{34.38}$$

The strong-coupling limit corresponds to $\sigma \to 0$ or equivalently $g^2 \to \infty$. There are three important features of this result:

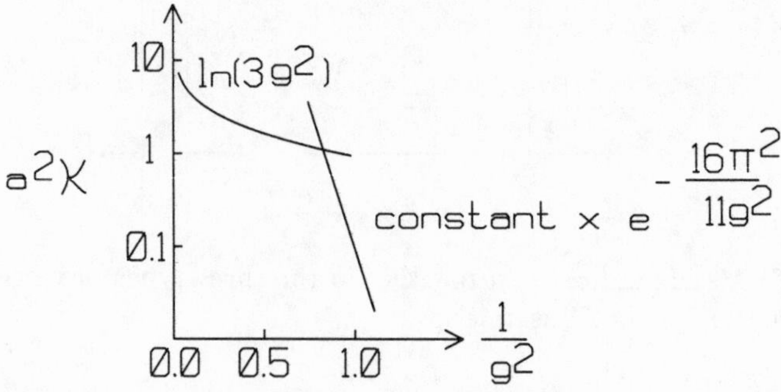

Figure 34.7: Sketch of strong and weak coupling limits for the string tension $a^2\kappa$ in $SU(3)$ in four dimensions ($d = 4$). Here $\kappa \equiv \bar{\sigma}$.

- It demonstrates confinement in the strong-coupling limit.
- It provides an analytical check on numerical calulations.
- It is not analytic in g^2, and hence is intrinsically nonperturbative.

Strong-Coupling SU(3). The discussion here is based on Refs. [Q39, Q37]. The corresponding result for $SU(3)$ and $d = 4$ is given by

$$a^2\kappa = \ln(3g^2) \tag{34.39}$$

In Section 33, in the *weak-coupling* limit where $g^2 \to 0$ it was shown that

$$\begin{aligned} a^2\kappa &= -\frac{1}{mn}\ln W_J(m,n) = \text{constant} \times f_0^2(g^2) \\ &= \text{constant} \times \exp\left\{\frac{-16\pi^2}{11g^2}\right\} \end{aligned} \tag{34.40}$$

The numerical calculations discussed in Section 33 interpolate between these two limiting cases as indicated in the log plot in Fig. 34.7. The fit to the numerical calculations yields the *constant* in Eq. (34.40). Now use

$$a^2 = \frac{1}{\lambda_1^2} f_0^2(g^2) \tag{34.41}$$

The string tension is then determined as $\kappa = \text{constant} \times \lambda_1^2$, leading to the result quoted in Section 33.

Strong - Coupling U(1). We have also evaluated the strong-coupling limit for the abelian $U(1)$ theory of lattice QED. Equation (34.13) and Prob. 29.3 give for $d = 4$

$$a^2\kappa = \ln\frac{1}{\sigma} = \ln(2e^2) \tag{34.42}$$

Several features of this result are also of interest:

- It provides an analytical check on numerical calculations.
- It is nonanalytic in the coupling constant.
- It implies that the abelian theory of QED on the lattice is *also* confining in the strong-coupling limit!

Here the reader is referred to the discussion at the end of Section 30. Clearly, one cannot simply pass to the continuum limit in this nonasymptotically free theory.

35

MONTE CARLO CALCULATIONS

The partition function or generating functional in field theory in the path integral approach involves the evaluation of multiple integrals over the local field variables where the dimension of the multiple integrals approaches infinity. In lattice gauge theory the dimension of the multiple integral over the link variables is finite, but very large. In many problems in statistical mechanics, for example, the Ising model, one is faced with the evaluation of multidimensional sums over dynamic variables where the dimension of the sum is again typically very large.[61]

The goal of this section is to describe the numerical methods commonly used to accurately evaluate many-dimensional multiple integrals (or sums). This section is based on Refs. [Q40, Q42, Q26]. Much of this material is taken from Ref. [Q26]. We start with a few preliminaries — some *statistics.*

Mean Values. Flip a coin N times. Assign $+1$ for heads and 0 for tails. Record the sequence of coin flips as indicated in Fig. 35.1. The set of all possible sequences of coin flips is said to form the *ensemble* of sequences. How many ways can one form a sequence that contains a total of m heads and $N-m$ tails in N tosses? The answer is a basic counting problem

$$\text{number with } m \text{ heads and } N-m \text{ tails} \;=\; \frac{N!}{m!(N-m)!} \tag{35.1}$$

The total number of sequences is obtained by summing this result over all m

$$\text{Total number sequences} = \sum_{m=0}^{N} \frac{N!}{m!(N-m)!} = (1+1)^N = 2^N \tag{35.2}$$

The second result follows immediately from the binomial theorem. This is clearly the correct total number of sequences since at each of the N steps in the sequence there are two possibilities — heads or tails.

[61] Consider some numbers: In an LGT calculation in d dimensions with N sites along one axis, the number of links is dN^d. For a lattice of size 16^4 in four dimensions the number of integrations over links in the multiple integrals $= 4 \times 16^4 = 262,144$; this must still be multiplied by the number of internal link variables. For the Ising model, the number of spin configurations is $2^{N_{\rm sites}} = 2^{N^d}$. For a 64×64 lattice in 2 dimensions the number of spin configurations, which is the number of terms in the partition function sum, is $2^{64\times 64} = 2^{4096} \approx 10^{1233}$!

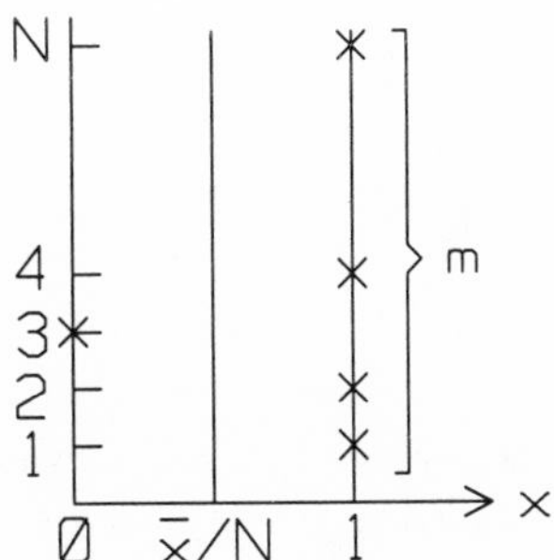

Figure 35.1: Record of N coin flips where the value +1 is assigned to heads and 0 to tails.

What is the *probability* $\mathcal{P}(m, N-m)$ that a sequence with m heads and $N-m$ tails will occur? This is just the probability that one would choose such a sequence at random from the ensemble of sequences; this, in turn, is just the number of such sequences divided by the total number of members of the ensemble of possible sequences

$$\mathcal{P}(m, N-m) = \frac{1}{2^N}\frac{N!}{m!(N-m)!} \tag{35.3}$$

Now let us use these probabilities to calculate the average value x that will occur if one repeats many sequences and for each flip in a sequence assigns an x as illustrated in Fig. 35.1. Denote the average value by $\bar{x}$. Since in a sequence with m heads and $N-m$ tails each head contributes $x = 1$ and each tail contributes $x = 0$ one has $x(m) = m \times 1 + (N-m) \times 0 = m$

$$\bar{x} = \sum_m \mathcal{P}(m, N-m)x(m) = \frac{1}{2^N}\sum_m \frac{N!}{m!(N-m)!}m \tag{35.4}$$

What is the mean value of x^2? Evidently

$$\overline{x^2} = \frac{1}{2^N}\sum_m \frac{N!}{m!(N-m)!}m^2 \tag{35.5}$$

The mean-square deviation is defined by

$$\begin{aligned}(\Delta x)^2 &\equiv \overline{(x-\bar{x})^2} = \overline{x^2} - 2\bar{x}\bar{x} + \bar{x}^2 \\ &= \overline{x^2} - \bar{x}^2\end{aligned} \tag{35.6}$$

The sums in Eqs. (35.4) and (35.5) can be explicitly evaluated with the aid of a generating function (Ref. [Q41])

$$(x_1 + x_2)^N = \sum_{m=0}^{N} \frac{N!}{m!(N-m)!}x_1^m x_2^{N-m} \tag{35.7}$$

Thus

$$\begin{aligned} 2^N \bar{x} &= \frac{\partial}{\partial x_1}(x_1 + x_2)^N \bigg|_{x_1=x_2=1} = N2^{N-1} \\ \bar{x} &= \frac{N}{2} \end{aligned} \tag{35.8}$$

On the average, each coin toss contributes 1/2, which is exactly what we would have guessed; however, any given sequence may yield a somewhat different value. How can one characterize the spread in values of x observed if this process is repeated many times, calculating x for each sequence? To this end, evaluate first

$$\begin{aligned} 2^N \overline{x^2} &= \frac{\partial}{\partial x_1} x_1 \frac{\partial}{\partial x_1}(x_1 + x_2)^N \bigg|_{x_1=x_2=1} = N2^{N-1} + N(N-1)2^{N-2} \\ \overline{x^2} &= \frac{N}{2} + \frac{N(N-1)}{4} \end{aligned} \tag{35.9}$$

Then determine the mean-square-deviation that follows as

$$(\Delta x)^2 = \frac{N}{4} \tag{35.10}$$

Now take the expression $\sqrt{(\Delta x)^2}/\bar{x}$ as a measure of the *deviation from the mean value*. Then

$$\frac{\sqrt{(\Delta x)^2}}{\bar{x}} = \frac{1}{\sqrt{N}} \tag{35.11}$$

The relative mean-square-deviation *decreases as* $1/\sqrt{N}$. As a consequence, as the length of the sequence N grows, it becomes more and more probable that a given sequence will yield a value of x very close to the mean value $\bar{x}$. For very large N, this becomes overwhelmingly probable (Ref. [Q41]).

It will be convenient in the following to define normalized values of these quantities by

$$\begin{aligned} \frac{\bar{x}}{N} &= \frac{1}{2} \\ \frac{\sqrt{(\Delta x)^2}}{N} &= \frac{1}{2\sqrt{N}} \end{aligned} \tag{35.12}$$

Let us now speak of $\mathcal{P}(x)$, the probability of observing a given x in any one sequence.[62] The above arguments indicate that this function has the shape sketched in Fig. 35.2. This probability is peaked at the value $\bar{x}/N = 1/2$ and its

[62] Really $\mathcal{P}[x(m)]$.

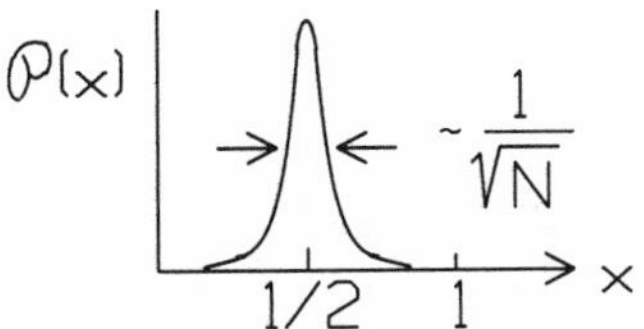

Figure 35.2: Sketch of probability that the value x is observed in any given sequence.

width goes as $1/\sqrt{N}$; the distribution grows sharper and sharper as the number of coin tosses in the sequence $N \to \infty$.

Monte Carlo Evaluation of a One-Dimensional Integral. Let us use these observations to find an alternate way to evaluate a one-dimensional integral. Consider

$$\frac{\int_0^1 dx\, f(x)}{\int_0^1 dx} \equiv \bar{f} \tag{35.13}$$

One can interpret this expression as the mean value of $f(x)$ on the interval $[0,1]$ as illustrated in Fig. 35.3. Now generate a random set of points along the x-axis between 0 and 1 (see Fig. 35.3 — note the x-axis is vertical in this figure). With very many points chosen at random, one will approach a uniform distribution along the axis. One can compute the *(normalized) mean value $\bar{f}$ by summing $f(x_p)$ at each of these points*

$$\bar{f} = \frac{1}{N}\sum_{p=1}^{N} f(x_p) \tag{35.14}$$

What does one expect the error to be in this *Monte Carlo* calculation? Denote any one such determination of the mean value by f and define $(\Delta f)^2 \equiv \overline{f^2} - \overline{f}^2$.

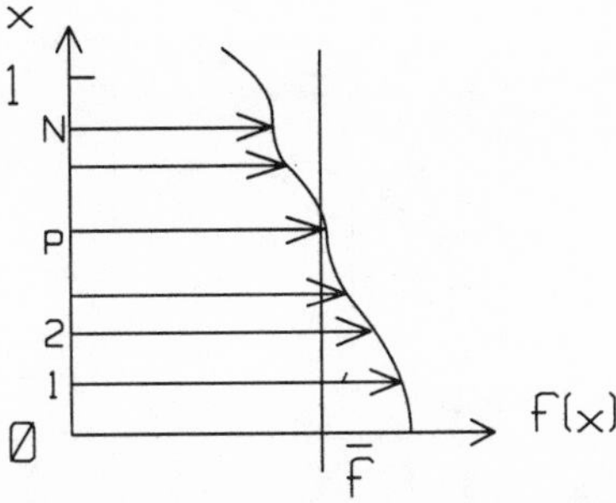

Figure 35.3: Mean value $\bar{f}$ of $f(x)$ on the interval $[0,1]$ and random set of points along the x-axis used to evaluate this quantity. Note the x-axis is vertical here.

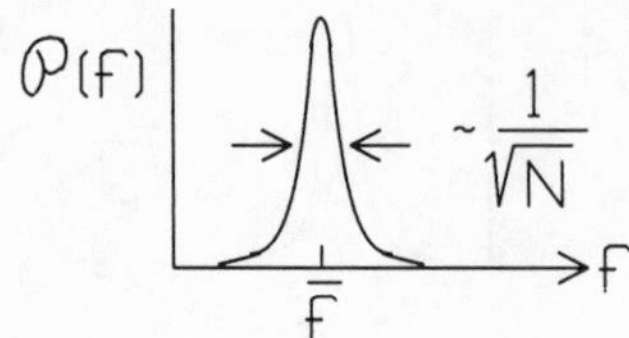

Figure 35.4: Probability of obtaining a given f in calculation of mean value $\bar{f}$ through the Monte Carlo procedure illustrated in Fig. 35.3.

Then from the previous discussion, in calculating this mean value one expects a *statistical error* of

$$\frac{\sqrt{(\Delta f)^2}}{\bar{f}} \approx \frac{1}{\sqrt{N}} \tag{35.15}$$

This is illustrated schematically in Fig. 35.4 (compare Fig. 35.2).

Standard Method for One-Dimensional Integrals. Let us compare the above with the standard method for evaluating a one-dimensional integral. Break the x-axis up into N intervals of length $h = 1/N$. Then add up the areas of each rectangle using, for example, the height at the right edge of each as illustrated in Fig. 35.5a. This estimate can be improved by using the trapezoid rule where the slope at the midpoint of each vertical element is fit with a straight line (Fig. 35.5b) or with Simpson's rule where one fits the slope and curvature at the midpoint (Fig. 35.5c), computing the area of the segment in each case, and then summing. What is the anticipated error when using these procedures? The error is expected to be $O(h = 1/N)$, $O(h^2 = 1/N^2)$, and $O(h^3 = 1/N^3)$, respectively, for these three methods. Now clearly this standard method is far superior to the Monte Carlo method for a one-dimensional integral since the anticipated error is $1/N \ll 1/\sqrt{N}$.

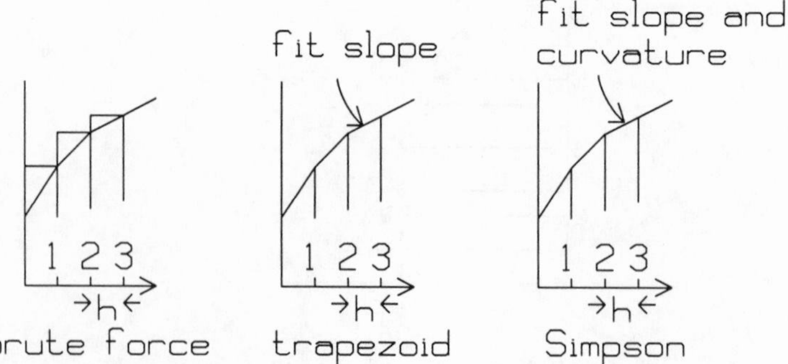

Figure 35.5: Standard method for evaluating a one-dimensional integral: (a) brute force; (b) trapezoid method; (c) Simpson's rule.

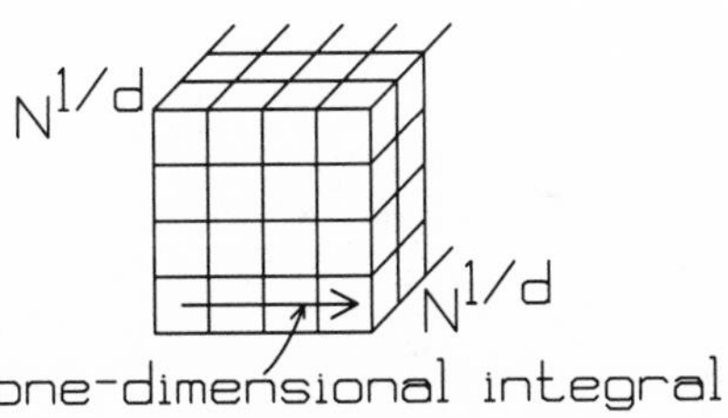

Figure 35.6: Evaluation of a multidimensional integral in d dimensions using N points. There are one-dimensional rows of $N^{1/d}$ points.

What about a *multidimensional* integral in d dimensions? The situation is sketched in Fig. 35.6. Suppose one is limited by computer power to a calculation of a total of N points. There are *rows* of $N^{1/d}$ points (Fig. 35.6). If one makes a series of brute force linear integrals, as above, one gets an accuracy $O(h = 1/N^{1/d})$. Refinements can reduce this to its square, or cube, but now *very quickly*

$$\frac{1}{\sqrt{N}} \ll \frac{1}{N^{1/d}} \tag{35.16}$$

Hence it is *much more accurate to use the Monte Carlo statistical method on multidimensional integrals.*

Importance Sampling. Suppose there is a particular region of $f(x)$ that is very important, as illustrated, for example, in Fig. 35.7. With just random sampling of the x-axis, one might miss this important region, or handle it only very crudely. Consider instead the expression

$$\frac{\int_0^1 m(x)dx\,[f(x)/m(x)]}{\int_0^1 m(x)dx} \equiv \bar{F} \tag{35.17}$$

Here $m(x)$ is a specified, positive, integrable function that emphasizes the important region in such a fashion that the new function $F(x) \equiv f(x)/m(x)$ is *flat*.

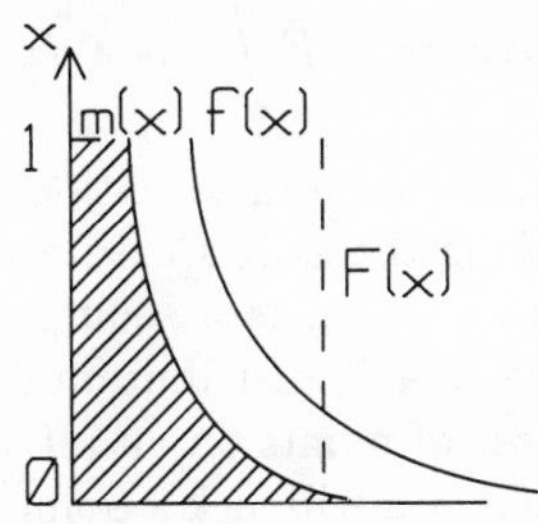

Figure 35.7: Importance sampling — region of $f(x)$ that is very important for the integral, here $x \to 0$. We define $F(x) \equiv f(x)/m(x)$.

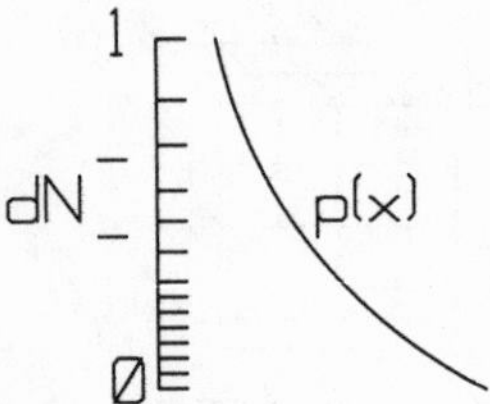

Figure 35.8: Points distributed along the x-axis with a probability density $dN/N = p(x)dx$.

One now faces exactly the same task as before, and the Monte Carlo method can be expected to work well. The only difference is that there is now a *measure* $p(x)$ in the integrals

$$p(x)dx \equiv \frac{m(x)dx}{\int_0^1 m(x)dx} \tag{35.18}$$

Suppose that one could generate and store a set of points $\{x_1, x_2, \ldots, x_N\}$ distributed along the x-axis with a *probability density* $p(x)$, that is, which satisfy (Fig. 35.8)

$$\text{fraction in } dx \equiv \frac{dN}{N} = p(x)dx \tag{35.19}$$

Then one could just evaluate $\bar{F}$ by summing $F(x_p)$ over these points

$$\bar{F} = \frac{1}{N}\sum_{p=1}^{N} F(x_p) \tag{35.20}$$

Now $F \approx \bar{F}$ everywhere, and the error in the Monte Carlo evaluation of the integral is expected to be small. At the end of the calculation, the desired integral is

$$\int_0^1 f(x)dx = \bar{F}\int_0^1 m(x)dx \tag{35.21}$$

Motivated by this discussion, we turn our attention to the problem of generating a set of points distributed along the x-axis with a probability density $p(x)dx$.

Markov Chains. A Markov chain is a sequence of distributions of points where the rule for getting the $(n+1)$ distribution depends only on the previous (n) distribution. Imagine a set of points on an interval, say $[0, 1]$. Pick a point. Let $p(x)$ be the *probability* that you will pick a point at the position x. Evidently

$$\int p(x)dx = 1 \tag{35.22}$$

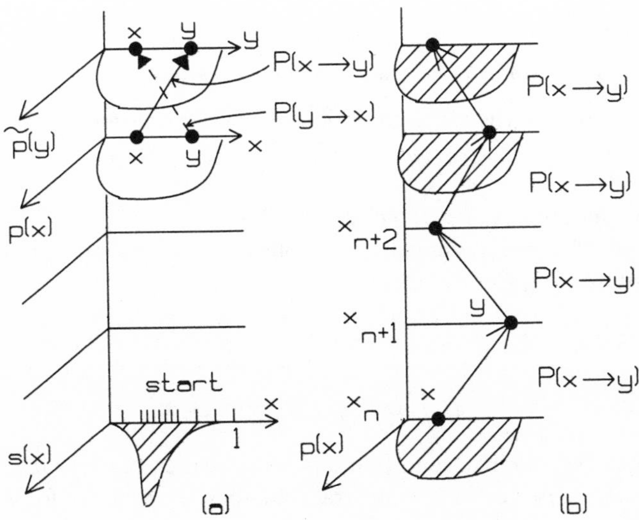

Figure 35.9: (a) Generation of new probability distribution of points $\tilde{p}(y)$ from probability distribution $p(x)$ according to probability rule $P(x \to y)$ in Markov chain; (b) Follow a given point through the steps of the Metropolis algorithm to generate a set of points $\{x_p\}$ distributed according to the probability $p(x)$.

Let $P(x \to y)$ be the ***probability*** that the point x is assigned to the new point y at the next step in the Markov chain (Fig. 35.9a). Since each point goes somewhere in the interval

$$\int dy P(x \to y) = 1 \tag{35.23}$$

What is the new distribution of points? Since one multiplies probabilities, the answer is

$$\tilde{p}(y) = \int dx P(x \to y) p(x) \tag{35.24}$$

In words, the new probability distribution at the point y is the intitial probability distribution at the point x multiplied by the probability of a point at x being taken into a point at y, summed over all x. One can verify that this rule maintains the normalization

$$\int dy\, \tilde{p}(y) = \int dx\, p(x) = 1 \tag{35.25}$$

What conditions must the rule $P(x \to y)$ satisfy so that the probability distribution is ***stable***, that is, eventually after each step one obtains the same probability distribution as the one before

$$\tilde{p}(y) = p(y) \; ; \qquad \text{stable distribution} \tag{35.26}$$

We claim that sufficient conditions for stability are the following:

- One must be able to reach all y with $P(x \to y)$.
- The rule must satisfy the condition of *microreversibility*

$$p(x)P(x \to y) = p(y)P(y \to x) \tag{35.27}$$

The second condition is illustrated in Fig. 35.9a.

The proof of these statements will follow in two steps. First show that microreversibility implies stability

$$\begin{aligned} \tilde{p}(y) &= \int dx P(x \to y)p(x) = \int dx P(y \to x)p(y) \\ &= p(y)\int dx P(y \to x) = p(y) \end{aligned} \tag{35.28}$$

Here the microreversibility condition has been used in the first line; this allows one to take the $p(y)$ out of the integral and the last equality follows from the normalization condition.

The second step in the proof is to show that the deviation from the equilibrium distribution is *nonincreasing* in the sequence of distributions in the Markov chain. Start with a distribution, say $s(x)$ (Fig. 35.9a). Define a measure of the deviation of this distribution from the equilibrium distribution by

$$D_{\text{old}} \equiv \int dx |s(x) - p(x)| \tag{35.29}$$

What is the new distribution generated from $s(x)$ and the new deviation from equilibrium?

$$\begin{aligned} \tilde{s}(y) &= \int dx P(x \to y)s(x) \\ D_{\text{new}} &= \int dy\, |\tilde{s}(y) - p(y)| \end{aligned} \tag{35.30}$$

The proof consists in demonstrating

$$D_{\text{new}} \le D_{\text{old}} \tag{35.31}$$

Here the equality holds only for the equilibrium distribution. The proof follows in a straightforward manner

$$\begin{aligned} D_{\text{new}} &= \int dy \left| \int s(x)P(x \to y)dx - p(y) \right| \\ &= \int dy \left| \int [s(x) - p(x)]P(x \to y)dx \right| \\ &\le \int dy \int dx P(x \to y)\, |s(x) - p(x)| \\ &= \int dx\, |s(x) - p(x)| = D_{\text{old}} \end{aligned} \tag{35.32}$$

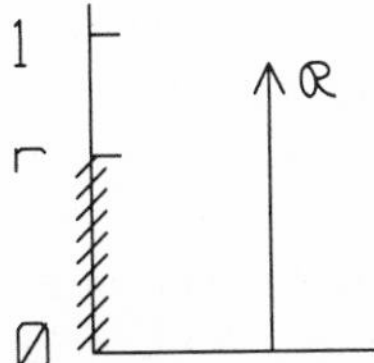

Figure 35.10: Criterion for accepting a move in $P(x \to y)$ if $r = p(y)/p(x) < 1$. Here $\mathcal{R}$ is a random number on the interval $[0,1]$. If $\mathcal{R} \leq r$ accept the move. If not, stay where you are.

The first and second line substitute definitions, the third follows from the positivity of the probability $P(x \to y)$, and the final equality comes from the normalization condition.

Can one find a probability rule $P(x \to y)$ that fulfills the two requirements and is readily adopted to computer calculations? The most widely used rule is the Metropolis algorithm given in Ref. [Q42].

The Metropolis Algorithm. Given the desired probability distribution $p(x)$, carry out the following sequence of steps:

1. Start with a point at position x.
2. Generate a new position y on the interval in a random manner.
3. Calculate

$$r = \frac{p(y)}{p(x)} \tag{35.33}$$

4. If $r > 1$, then *take the trial move*, that is, move the point to the region of higher probability density.

5. If $r < 1$, *accept the trial move with probability* r, that is, go to a region of lower probability *sometimes*. How does one do this? Generate a random number $\mathcal{R}$ on the interval $[0,1]$ (Fig. 35.10). If $\mathcal{R} \leq r$ accept the move. If not, stay where your are.

6) Repeat.

7. After equilibrium is achieved, store the points reached by following a given point through repeated steps of this process $\{x_n, x_{n+1}, \ldots, x_{n+N}\} \equiv \{x_p\}$ (Fig. 35.9b). *These points are now distributed according to* $p(x)$.[63]

8. Then

$$\int_0^1 F(x)p(x)dx = \frac{1}{N}\sum_{p=1}^{N} F(x_p) \tag{35.34}$$

[63] A little thought will convince the reader that following a single point through very many steps of the Metropolis algorithm is equivalent to taking the distribution of points from one step of the Markov chain to the next one. Since the probability distribution is *stable*, it is simply reproduced by this process.

Thermalization. If one starts with a set of points with some arbitrary distribution $s(x)$ (Fig. 35.9a), then repeating the steps in the Markov chain many times will take one to the equilibrium distribution $p(x)$. This process is known as thermalization.

The Metropolis algorithm is an efficient and powerful computational tool for doing multiple integrations or sums. We conclude this section with a proof of that algorithm.

Proof of the Metropolis Algorithm. One must show that the two conditions for stability stated below Eq. (35.26) are satisfied. Clearly the choice of y can reach all points on the interval. It remains to demonstrate microreversibility (see Fig. 35.9a).

If $r = p(y)/p(x) > 1$, then $P(x \to y) = 1$, and going the other way $y \to x$ would be accepted with probability $p(x)/p(y) = 1/r$. Thus in this case

$$P(y \to x) = \frac{p(x)}{p(y)} = \frac{1}{r}\,; \qquad r > 1 \tag{35.35}$$

Hence in this case

$$p(y)P(y \to x) = p(x) = p(x)P(x \to y) \tag{35.36}$$

If, on the other hand, $r = p(y)/p(x) < 1$, then $P(y \to x) = 1$ and going the other way $x \to y$ would be accepted with probability $p(y)/p(x) = r$. Therefore

$$P(x \to y) = \frac{p(y)}{p(x)} = r\,; \qquad r < 1 \tag{35.37}$$

Hence again

$$p(x)P(x \to y) = p(y) = p(y)P(y \to x) \tag{35.38}$$

Microreversibility thus holds for both cases and the Metropolis algorithm is established.

For clarity, the arguments in this section have all been formulated in terms of one-dimensional integrals. They are readily extended to multidimensional integrals (or sums). Suppose there are ν degrees of freedom. Replace x by the vector $\mathbf{q} \equiv \{q_1, q_2, \ldots, q_\nu\}$ and $p(x)dx$ by $p(\mathbf{q})d\mathbf{q} = p(q_1, q_2, \ldots, q_\nu)dq_1 dq_2 \ldots dq_\nu$. One step in the Markov chain then takes $p(\mathbf{q}) \to \tilde{p}(\mathbf{q}')$.

An appropriate choice of the probability density for importance sampling in the multidimensional integrals in LGT is $p = Ne^{-S(U)}$ (and in statistical mechanics $p = Ne^{-\beta H}$).

36

INCLUDE FERMIONS

So far the discussion of lattice gauge theory has been based on a gauge-invariant treatment of the nonlinear boson (gluon) couplings in a Yang-Mills nonabelian local gauge theory. In this section we discuss the extension of LGT to include fermions (quarks). This section is based on Refs. [Q17, Q43, Q44, Q45, Q46, Q34, Q47, Q48]— much of it is taken from Ref. [Q48].

Some Preliminaries. The free Dirac lagrangian density is given by

$$\begin{aligned} \mathcal{L}_{\mathrm{F}} &= -\bar{\psi}\left(\gamma_\mu \frac{\partial}{\partial x_\mu} + m\right)\psi \\ &\doteq -\frac{1}{2}\left[\bar{\psi}\left(\gamma_\mu \frac{\partial}{\partial x_\mu} + m\right)\psi + \bar{\psi}\left(-\gamma_\mu \frac{\overleftarrow{\partial}}{\partial x_\mu} + m\right)\psi\right] \end{aligned} \tag{36.1}$$

The second line is an equivalent form to be used in the action where a partial integration can be carried out.

In QED, a theory with $U(1)$ local gauge invariance, the lagrangian density takes the form

$$\begin{aligned} \mathcal{L}_{\mathrm{F}} &= -\frac{1}{2}\left\{\bar{\psi}\left[\gamma_\mu\left(\frac{\partial}{\partial x_\mu} - ieA_\mu\right) + m\right]\psi\right. \\ &\quad \left. + \bar{\psi}\left[-\gamma_\mu\left(\frac{\overleftarrow{\partial}}{\partial x_\mu} + ieA_\mu\right) + m\right]\psi\right\} \end{aligned} \tag{36.2}$$

In the *euclidian metric* of LGT one calculates the action

$$S = -\int_{\tau_1}^{\tau_2} d^3x\, d\tau \mathcal{L}_E \tag{36.3}$$

Here the four-vectors (x_μ, A_μ) are taken as

$$\begin{aligned} x_\mu &= (\mathbf{x}, it) \rightarrow (\mathbf{x}, \tau) \\ A_\mu &= (\mathbf{A}, i\Phi) \rightarrow (\mathbf{A}, A_0) \end{aligned} \tag{36.4}$$

In analogy, for the Dirac gamma matrices in the euclidian metric we will make the replacement

$$\gamma_\mu = (\boldsymbol{\gamma}, i\gamma_0) \rightarrow (\boldsymbol{\gamma}, \gamma_0) \tag{36.5}$$

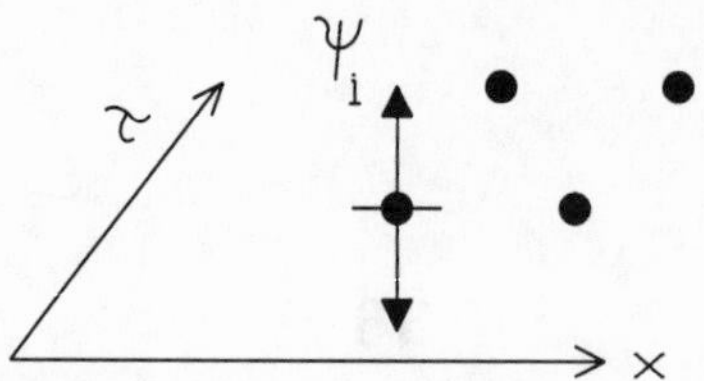

Figure 36.1: Associate fermion fields with each *site*.

Thus in the following we shall use

$$\gamma_4 \equiv i\gamma_0 \qquad\qquad \gamma_0 = -i\gamma_4 \tag{36.6}$$

Fermions in $U(1)$ *Lattice Gauge Theory.* Associate fermion fields with each *site* as indicated in Fig. 36.1. These fermion fields are taken to be *Grassmann variables*, that is, they are anticommuting c-numbers (Refs. [R1, R4] and Probs. 36.1-4).

Recall from Section 30 that

$$\sum_{\text{links}} = \sum_{\text{sites}} \sum_{\text{links/site}} \tag{36.7}$$

Here the second $\sum_{\text{links/site}}$ goes over the positive coordinate directions at each site (Fig. 36.2). This expression gives a complete enumeration of the terms in $\sum_{\text{links}}$. At the site i, as in Fig. 36.2, one can write the scalar product of two four vectors in terms of the components along the positive coordinate directions as

$$a_\mu b_\mu = a_1 b_1 + a_2 b_2 + \cdots \tag{36.8}$$

Hence the following expression has the character of a four-vector product at the site i

$$\sum_{\text{links/site}} (a_{ji} b_{ji}) = a_{ji} b_{ji} + a_{ki} b_{ki} + \cdots \tag{36.9}$$

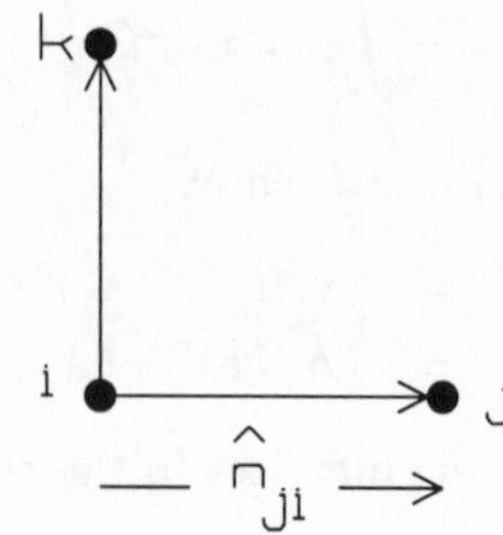

Figure 36.2: Positive coordinate directions at each site.

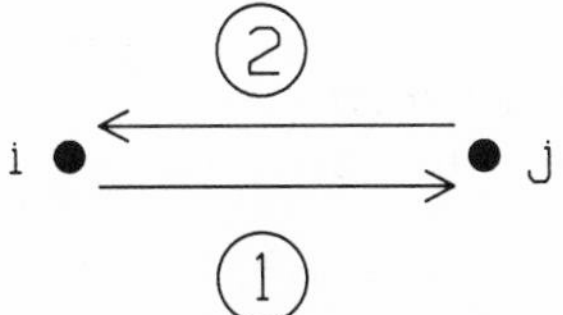

Figure 36.3: Two terms in fermion contribution to the action for a given link between two sites.

Associate a gamma matrix with the *link* according to

$$\gamma_{ji} \equiv \frac{(x_j - x_i)_\mu}{a}\gamma_\mu = (\hat{n}_{ji})_\mu \gamma_\mu \tag{36.10}$$

Here $\hat{n}_{ji}$ is a unit vector in the direction of the link. Associate an electromagnetic field variable with each *link* exactly as before

$$\begin{aligned} U_{ji} &\equiv \exp\left\{ie_0(x_j - x_i)_\mu A_\mu\left[\frac{1}{2}(x_j + x_i)\right]\right\} \\ &= \exp\left\{ie_0 a(\hat{n}_{ji})_\mu A_\mu\left[\frac{1}{2}(x_j + x_i)\right]\right\} \end{aligned} \tag{36.11}$$

We now state a model result for the *fermion contribution to the action in LGT* and then proceed to show that it has all the required properties

$$S_{\mathrm{F}} = -\sigma_{\mathrm{F}} \sum_{\text{links}} \left(\bar{\psi}_j \gamma_{ji} U_{ji} \psi_i + \bar{\psi}_i \gamma_{ij} U_{ij} \psi_j\right) \tag{36.12}$$

For simplicity the fermions are here assumed to be massless. The first and second terms run in the two directions along the link as illustrated in Fig. 36.3. Note the following

$$U_{ij} = U_{ji}^\dagger \qquad\qquad \gamma_{ji}^\dagger \gamma_4 = \gamma_4 \gamma_{ij} \tag{36.13}$$

Thus the second term in Eq. (36.12) is the adjoint of the first term

$$S_{\mathrm{F}} = -\sigma_{\mathrm{F}} \sum_{\text{links}} \left(\bar{\psi}_j \gamma_{ji} U_{ji} \psi_i + \text{h.c.}\right) \tag{36.14}$$

Hence S_{F} is real.

Gauge Invariance. Recall that a gauge transformation in this $U(1)$ theory takes the form

$$\begin{aligned} U_{ji} &\rightarrow e^{-i\Lambda(x_j)} U_{ji} e^{i\Lambda(x_i)} \\ \psi_i &\rightarrow e^{-i\Lambda(x_i)} \psi_i \\ \bar{\psi}_j &\rightarrow e^{i\Lambda(x_j)} \bar{\psi}_j \end{aligned} \tag{36.15}$$

The phases all *cancel at the sites*; this is one reason for putting the fermions at the sites. Hence S_F is gauge invariant.

Continuum Limit. Let $a \to 0$. Then

$$\begin{aligned} U_{ji} &\to 1 + ie_0(x_j - x_i)_\mu A_\mu = 1 + ie_0 a(\hat{n}_{ji})_\mu A_\mu \\ U_{ji} &\to 1 + ie_0 a A_{ji} \end{aligned} \tag{36.16}$$

The quantity in parenthesis P in Eq. (36.12) then takes the form

$$\begin{aligned} P &= \bar{\psi}_j \gamma_{ji}(1 + ie_0 a A_{ji} + \cdots)\psi_i - \bar{\psi}_i \gamma_{ji}(1 - ie_0 a A_{ji} + \cdots)\psi_j \\ &= \underbrace{\bar{\psi}_j \gamma_{ji} \psi_i - \bar{\psi}_i \gamma_{ji} \psi_j} + ie_0 a(\bar{\psi}_j \gamma_{ji} A_{ji} \psi_i + \bar{\psi}_i \gamma_{ji} A_{ji} \psi_j) \\ &\qquad = (\bar{\psi}_j - \bar{\psi}_i)\gamma_{ji}\psi_i - \bar{\psi}_i \gamma_{ji}(\psi_j - \psi_i) \\ &\qquad \approx a\left[\frac{\partial \bar{\psi}}{\partial x_{ji}} \gamma_{ji} \psi_i - \bar{\psi}_i \gamma_{ji} \frac{\partial \psi}{\partial x_{ji}}\right] \end{aligned} \tag{36.17}$$

Equations (36.9), (36.8), and (36.2) imply that when summed over links, this will yield the action of QED.

In two-dimensional ($d = 2$) space-tau define[64]

$$\sigma_F \equiv \frac{a^2}{2a} \tag{36.18}$$

Hence

$$\begin{aligned} S_F &= -\sigma_F \sum_{\text{sites}} \sum_{\text{links/site}} (\bar{\psi}_j \gamma_{ji} U_{ji} \psi_i + \bar{\psi}_i \gamma_{ij} U_{ij} \psi_j) \\ &\xrightarrow{a \to 0} -\int d^2x \, \mathcal{L}_F \end{aligned} \tag{36.19}$$

Where the lagrangian density is given by Eq. (36.2). Thus this model for including the fermions in $U(1)$ LGT has the correct continuum limit.

Path Integrals. With the addition of the fermions the partition function becomes

$$\begin{aligned} Z &= \int (d\bar{\psi})(d\psi)(dU) e^{-S(U,\bar{\psi},\psi)} \\ S &= S_G(U) + S_F(U, \bar{\psi}, \psi) \end{aligned} \tag{36.20}$$

Here $S_G(U)$ is the action arising purely from the gauge fields; it is the contribution we have been studying. $S_F(U, \bar{\psi}, \psi)$ is the fermion contribution to the action given by Eq. (36.12). The generating functional in the euclidian metric is obtained exactly as before. The problem is now well posed.

[64] The generalization to d dimensions is $\sigma_F = 1/2a^{1-d}$ (Prob. 36.7).

One can go further, however, since the integral over the fermion fields can be *explicitly evaluated.* This permits a reduction of the problem to path integrals over the gauge fields of the type we have been studying, only with a more complicated *effective action.*

To explicitly carry out the integration over the fermion fields, use the basic result for integration over Grassmann variables (Refs. [R1, R4]). If $\sum_i \sum_j$ go over all *sites*, then (see Prob. 36.1)[65]

$$\int (d\bar{\psi})(d\psi) \exp\left\{-\sum_i \sum_j \bar{\psi}_i \Delta_{ij}(U) \psi_j\right\} = \det \underline{\Delta}(U) \tag{36.21}$$

One can then also use the general relation for (positive) matrices (Refs. [R4, R1] and Prob. 36.5)

$$\ln \det \underline{\Delta} = \operatorname{Tr} \ln \underline{\Delta} \tag{36.22}$$

Here Tr indicates the trace of the (N_{sites}-dimensional) matrix. These two steps reduce Eq. (36.20) to the form

$$\begin{aligned} Z &= \int (dU) e^{-S_{\text{eff}}(U)} \\ S_{\text{eff}}(U) &= S_{\text{G}}(U) - \operatorname{Tr} \ln \underline{\Delta}(U) \end{aligned} \tag{36.23}$$

The problem has been reduced to ***integrals over the gauge fields with an effective action.*** One can bring down fermion fields for mean values in Eq. (36.21) by taking derivatives with respect Δ_{ij}.[66]

The additional term in S_{eff} adds significant complication to numerical evaluation of the path integrals. As we have seen, the confinement aspects of Yang-Mills nonabelian gauge theories in LGT arise from the nonlinear couplings of the gauge fields. Since virtual fermion pairs are not expected to qualitatively alter the results, one often employs the ***quenched*** approximation where in the action in the final calculation

$$S_{\text{eff}}(U) \approx S_{\text{G}}(U)\,; \qquad \text{quenched approximation} \tag{36.24}$$

This approximation neglects fermion loops in the partition function Z.

Problem — Fermion Doubling. There is a problem with this development, and it is referred to as fermion doubling (Refs. [Q17, Q43, Q44, Q45, Q46, Q34, Q47, Q48]). Consider free fermions on a lattice in 1+1 dimensions in Minkowski space. The Dirac equation for the stationary states is

$$\left(\frac{1}{i}\alpha\frac{d}{dx} + \beta m\right)\psi = E\psi \tag{36.25}$$

[65] One must rearrange the fermion action so that it is in this form; this is not hard to do since each site is connected to a finite number of links (Prob. 36.8).

[66] Use the general matrix relation $\partial(\det \underline{M})/\partial M_{ij} = (\underline{M}^{-1})_{ji}(\det \underline{M})$. The proof of this relation is not difficult if one uses the expansion of the determinant in terms of cofactors (Prob. 36.6).

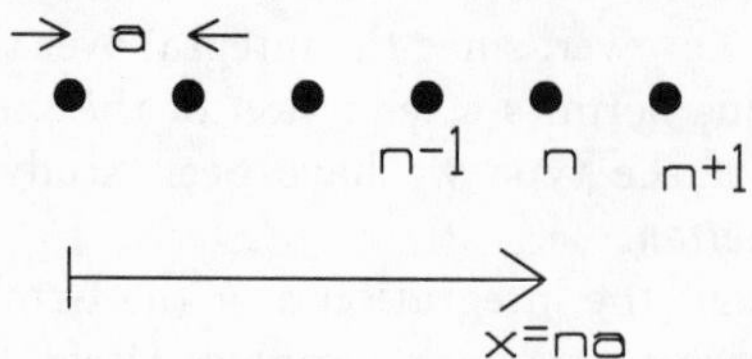

Figure 36.4: Dirac fermions on a lattice in one space and one time dimension.

Take[67]

$$\alpha = \sigma_x = \begin{pmatrix} 0 & 1 \\ 1 & 0 \end{pmatrix} \qquad \beta = \sigma_z = \begin{pmatrix} 1 & 0 \\ 0 & -1 \end{pmatrix}$$
$$\{\sigma_x, \sigma_z\} = 0 \qquad \sigma_x^2 = \sigma_z^2 = 1 \tag{36.26}$$

Look for solutions to this equation on a lattice where (Fig. 36.4)

$$\psi(x) = e^{ikx}u(k) \qquad x = na \tag{36.27}$$

Equivalently

$$\psi(n) = e^{ikna}u(k) \tag{36.28}$$

The straightforward definition of the derivative on the lattice is

$$\frac{d}{dx} \equiv \frac{1}{a}[\psi(n+1) - \psi(n)] \tag{36.29}$$

Substitution of these relations into the Dirac equation leads to the eigenvalue equation

$$\left[\frac{1}{ia}\sigma_x(e^{ika} - 1) + \sigma_z m - E\right] u(k) = 0 \tag{36.30}$$

This is a set of two linear homogeneous algebraic equations; for a solution, the determinant of the coefficients must vanish

$$\det\begin{vmatrix} m - E & (2/a)e^{ika/2}\sin ka/2 \\ (2/a)e^{ika/2}\sin ka/2 & -m - E \end{vmatrix} = 0 \tag{36.31}$$

It is evident that the solution to this equation yields a *complex* E. It thus appears that the definition of the derivative in Eq. (36.29) is too naive. Let us redefine the derivative to make it more symmetric[68]

$$\frac{d}{dx} \equiv \frac{1}{2a}\left[\psi(n+1) - \psi(n-1)\right] \tag{36.32}$$

[67] One needs only two anticommuting matrices now; this can be satisfied with a two-dimensional representation.

[68] Note $\{[\psi(n+1) - \psi(n)] + [\psi(n) - \psi(n-1)]\}/2a = [\psi(n+1) - \psi(n-1)]/2a$.

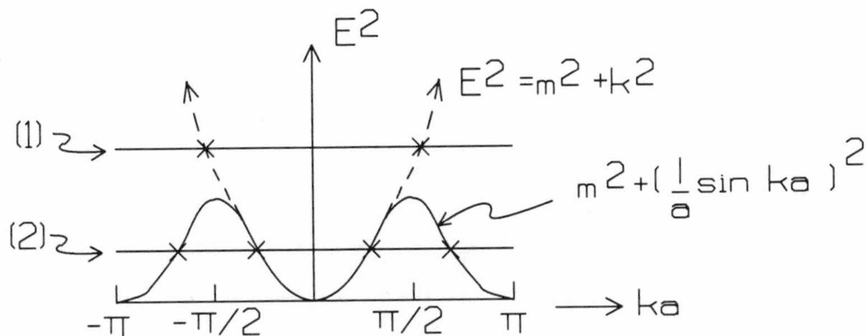

Figure 36.5: Solution to the eigenvalue equation for the stationary state solution to the Dirac equation in a one-dimensional lattice. (1) Two solutions corresponding to the two directions of motion in continuum limit $a \to 0$. (2) Four solutions on a lattice with finite a — problem of fermion doubling.

In this case the eigenvalue equation is changed to

$$\det\begin{vmatrix} m - E & (1/a)\sin ka \\ (1/a)\sin ka & -m - E \end{vmatrix} = 0 \tag{36.33}$$

This gives

$$E^2 = m^2 + \left(\frac{1}{a}\sin ka\right)^2 \tag{36.34}$$

Translation by $ka \to ka + 2\pi$ does not give a new solution in Eqs. (36.28) and (36.34). We can therefore limit ourselves to the basic interval

$$-\pi \leq ka < \pi \tag{36.35}$$

The eigenvalue equation can then be solved graphically as shown in Fig. 36.5. There are two interesting regions of solution indicated in the figure:

1. In the continuum limit as $a \to 0$, two solutions are obtained with $E^2 = m^2 + (\pm k)^2$. These correspond to a free particle going in the two different directions. This is the desired limit.[69]

2. In contrast, on the lattice with finite a, there are instead *four solutions* to the eigenvalue equation.[70] This situation arises because the function $(\sin ka/a)^2$ turns over and returns to zero at the endpoints of the interval in Eq. (36.35). This is the problem of *fermion doubling.*

Possible Solutions to the Problem of Fermion Doubling. Consider massless fermions with $m = 0$. One possible solution to the problem of fermion doubling is to use *Wilson fermions* (Refs. [Q17, Q43, Q46]). Here one modifies the Dirac equation with an additional term that raises the wings of the curve in Fig. 36.5,

[69] There are two values of k for given E^2.

[70] There are now four values of k for given E^2.

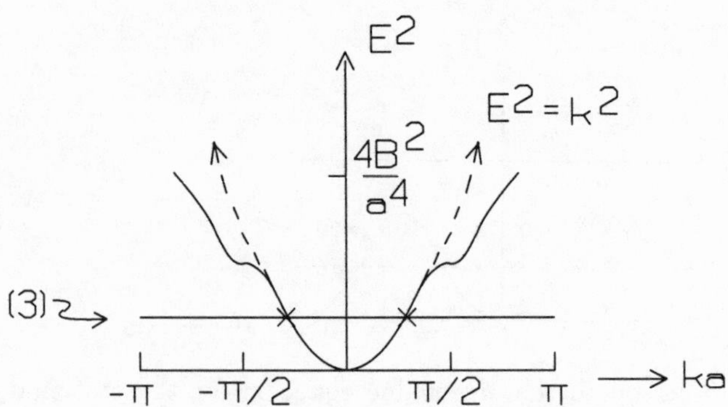

Figure 36.6: Eigenvalue equation with massless Wilson fermions. Here B is chosen large enough so that there are always only two solutions for k for given E^2.

and vanishes sufficiently fast as $a \to 0$ so that one recovers the proper continuum limit. Take

$$\left[\frac{1}{i}\sigma_x\frac{d}{dx} + \sigma_z\left(-\frac{B}{2}\frac{d^2}{dx^2}\right)\right]\psi = E\psi \tag{36.36}$$

Here the second derivative is defined as

$$\frac{d^2}{dx^2} \equiv \frac{1}{a}\left\{\frac{1}{a}[\psi(n+1)-\psi(n)] - \frac{1}{a}[\psi(n)-\psi(n-1)]\right\} \tag{36.37}$$

The eigenvalue equation then reads[71]

$$\det\begin{vmatrix} (2B/a^2)\sin^2 ka/2 - E & (1/a)\sin ka \\ (1/a)\sin ka & -(2B/a^2)\sin^2 ka/2 - E \end{vmatrix} = 0 \tag{36.38}$$

This leads to

$$E^2 = \left(\frac{1}{a}\sin ka\right)^2 + 4B^2\left(\frac{1}{a^2}\sin^2\frac{ka}{2}\right)^2 \tag{36.39}$$

The graphic solution to this eigenvalue equation in indicated in Fig. 36.6. The situation is now modified from the above:

3. For large enough B there are only two solutions for k for given E^2 and the problem of fermion doubling has been eliminated. One recovers the correct continuum limit provided $B \to 0$ sufficiently fast as $a \to 0$.[72]

Wilson fermions are often used in numerical calculations.

One additional problem is the retention of *chiral symmetry* that holds with massless fermions. We have seen that chiral symmetry is important in low

[71]Note $(-B/2a^2)[(e^{ika}-1)-(1-e^{-ika})] = (2B/a^2)\sin^2 ka/2$.

[72]For example, $B = \gamma a$ where γ is a constant.

energy hadron physics. Wilson fermions do not retain chiral symmetry. A possible solution here is the use of ***staggered fermions*** on the lattice (Refs. [Q44, Q45, Q34]). We refer the reader to Ref. [Q34] for further pursuit of this topic.

Observables. The calculation of observables in LGT is discussed in Section 33. One can obtain fermion observables by calculating the statistical average of appropriate fermion operators and, for example, by again looking at the exponential decay with τ for large τ in euclidian space.

37

QCD-INSPIRED MODELS

Solution to QCD presents formidable problems. It is often useful to make models that emphasize one or another aspect of QCD and that provide physical insight and guidance for further work. In this section we discuss QCD-inspired models of the internal structure of hadrons. Reference [Q49] provides good background reading and a more extended development on many of these topics.

Bag Models. We build on three features of QCD.

- Baryons have the *quantum numbers* of (qqq) systems and mesons of $(\bar{q}q)$ where the flavor quantum numbers of the quarks q are given in Table 37.1.
- Color and the strong color forces are *confined* to the interior of the hadrons. Quarks come in three colors (R, G, B). Lattice gauge theory calculations indicate that confinement arises from the strong nonlinear couplings of the gauge fields at large distances.
- QCD is *asymptotically free*; at short distances the renormalized coupling constant goes to zero. One can do perturbation theory at short distances.

M.I.T. Bag Model. The M.I.T. bag model (Refs. [Q50, Q51, Q52, Q53]) provides an extreme picture of each of the three items listed above. For baryons, consider three noninteracting quarks (correct quantum numbers) and treat the one-gluon-exchange interaction as a perturbation (asymptotic freedom). Put the quarks inside a vacuum *bubble* of radius R as illustrated in Fig. 37.1 (confinement). Assume it takes a positive amount of internal energy density to create this bubble in the vacuum

$$\left(\frac{E}{V}\right)_{\text{vac}} = +b \tag{37.1}$$

Some Simple Phenomenology. Consider some simple phenomenology based on this picture. Take the (u, d) quarks to be essentially massless in QCD. For simplicity, suppose to start with that they were scalar particles. Then for massless scalar particles in a cavity the Klein-Gordon equation for stationary states is

$$(\nabla^2 + k^2)\phi = 0\ ; \qquad\qquad r \leq R \tag{37.2}$$

Table 37.1: Flavor quantum numbers of the lightest quarks: isospin, third component of isospin, baryon number, strangeness, charm, and electric charge, respectively.

Quark/field	T	T_3	B	S	C	$Q = T_3 + (B + S + C)/2$
u	1/2	1/2	1/3	0	0	2/3
d	1/2	−1/2	1/3	0	0	−1/3
s	0	0	1/3	−1	0	−1/3
c	0	0	1/3	0	1	2/3

The s-wave solution to this equation, which is nonsingular at the origin and vanishes at the wall, is given by

$$\begin{aligned} \phi &= j_0(kr)\ ; && \text{nonsingular at r} = 0 \\ j_0(kR) &= 0\ ; && \text{vanishes at wall} \end{aligned} \tag{37.3}$$

The last is an eigenvalue equation. The roots are the zeros of j_0. The first zero of j_0, corresponding to the lowest lying state, is given by

$$\begin{aligned} k_{10}R &\equiv X_{10} = \pi \\ k_{10} &= E_{10} = \frac{\pi}{R} \end{aligned} \tag{37.4}$$

These results can now be used to estimate the internal energy of a hadron. Write

$$E = n_q\frac{\mathcal{K}}{R} + \left(\frac{4}{3}\pi R^3\right) b \tag{37.5}$$

The first term, where n_q is the number of quarks and $\mathcal{K}$ is a constant, is the kinetic energy of the quarks (in the above discussion, which motivates this form of the kinetic energy, $\mathcal{K} = \pi$); the second is the energy of the vacuum bubble. This expression gets large for $R \to 0$ because of the kinetic energy and also for $R \to \infty$ because of the bubble energy (Fig. 37.2). Minimization with respect to R requires $\partial E/\partial R = 0$ which yields

$$-\frac{n_q\mathcal{K}}{R^2} + 4\pi b R^2 = 0$$

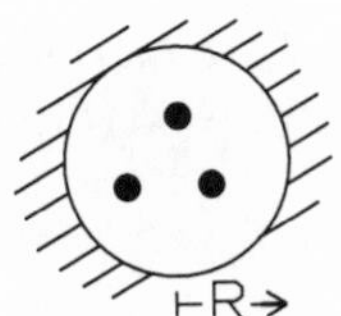

Figure 37.1: M.I.T. bag model—three quarks inside a bubble in the vacuum.

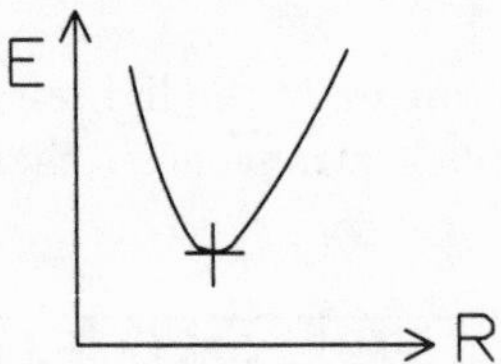

Figure 37.2: Energy of hadron in M.I.T. bag model.

$$R_{\rm min} \;=\; \left(\frac{n_q\mathcal{K}}{4\pi b}\right)^{1/4} \tag{37.6}$$

Substitution of this value leads to the energy at the minimum

$$E_{\rm min} \;=\; \frac{4}{3}n_q\mathcal{K}\left(\frac{4\pi b}{n_q\mathcal{K}}\right)^{1/4} \;=\; \left(\frac{4}{3}n_q\mathcal{K}\right)\frac{1}{R_{\rm min}} \tag{37.7}$$

Put in some numbers: for baryons $n_q = 3$; for massless Dirac particles $\mathcal{K} = 2.04$ (see below); $E_{\rm min} = M = 938.3\,{\rm MeV}$ is the nucleon mass; $\hbar c = 197.3\,{\rm MeV\,fm}$ restores the units. This gives

$$R_{\rm min} = 1.72\,{\rm fm} \tag{37.8}$$

This is not bad; it is certainly in the right ballpark, although it is too large. Recall that the experimental value of the root-mean-square charge radius of the proton is (Section 2)

$$\langle r^2\rangle_p^{1/2} \approx 0.8\,{\rm fm} \tag{37.9}$$

For comparison, for a uniform charge distribution[73] the mean square radius is $\langle r^2\rangle = 3R^2/5$ (Section 2); hence the experimental value of the equivalent uniform radius is approximately 1/2 as big as the bag value

$$R_{\rm uniform} = 1.03\,{\rm fm} \tag{37.10}$$

Dirac Equation in a Cavity. Consider now the Dirac equation for a particle of mass M in a spherical cavity. Start with the Dirac equation in spherically symmetric scalar and vector potentials $[U_0(r), V_0(r)]$, respectively. From Prob. 15.1 the solution to the Dirac equation in this case can be written

$$\psi_{n\kappa m}(\mathbf{r}) = \frac{1}{r}\begin{pmatrix} iG_{n\kappa}(r)\phi_{\kappa m} \\ -F_{n\kappa}(r)\phi_{-\kappa m}\end{pmatrix} \tag{37.11}$$

Here

$$\phi_{\kappa m} \;\equiv\; \sum_{m_l m_s}\langle l m_l \tfrac{1}{2} m_s | l \tfrac{1}{2} j m\rangle Y_{l m_l}(\theta,\phi)\chi_{m_s} \;\equiv\; \mathcal{Y}^m_{l\frac{1}{2}j} \tag{37.12}$$

Table 37.2: Low-lying states for massless Dirac particle in a scalar cavity. From Ref. [Q49].

κ	j	l_{upper}	l_{lower}	state	$\mathcal{K}$
-1	1/2	0	1	$1s_{1/2}$, $2s_{1/2}$	2.04, 5.40
1	1/2	1	0	$1p_{1/2}$	3.81
-2	3/2	1	2	$1p_{3/2}$	3.20
2	3/2	2	1	$1d_{3/2}$	5.12
-3	5/2	2	3	$1d_{5/2}$	4.33

In this expression $j = |\kappa| - 1/2$ and $l = \kappa$ if $\kappa > 0$ while $l = -(\kappa + 1)$ if $\kappa < 0$. The first few states are given in Table 37.2.

The coupled radial differential equations take the form (Prob. 15.2)

$$\frac{d}{dr}G_\alpha(r) + \frac{\kappa}{r}G_\alpha(r) - [E_\alpha - V_0(r) + M + U_0(r)]\,F_\alpha = 0$$
$$\frac{d}{dr}F_\alpha(r) - \frac{\kappa}{r}F_\alpha(r) + [E_\alpha - V_0(r) - M - U_0(r)]\,G_\alpha = 0 \qquad (37.13)$$

Note the signs in front of the potentials: the vector potential has the opposite sign from the energy; the scalar potential has the sign of the mass.

Let us find the lowest positive energy state ($E > 0$) for massless particles in a pure scalar potential ($M = V_0 = 0$). In this case the coupled radial equations reduce to

$$G' - \frac{G}{r} = (E + U_0)F$$
$$F' + \frac{F}{r} = -(E - U_0)G \qquad (37.14)$$

Apply $(d/dr + 1/r)$ to the first equation and substitute the second; the result is

$$G'' = -(E^2 - U_0^2)G \qquad (37.15)$$

Define $E \equiv k$, and assume a square-well potential as shown in Fig. 37.3. The differential Eq. (37.15) is readily solved in the two indicated regions:

Region I: Here $G'' = -k^2 G$ with solution

$$G_{\text{I}} = A \sin kr + B \cos kr \; ; \qquad E \equiv k \qquad (37.16)$$

For a solution that is nonsingular at the origin one must have $G(r) \to 0$ as $r \to 0$. This implies that the coefficient $B = 0$.

Region II: Assume here that $U_0 > k$ as indicated in the figure. Then

$$G'' = (U_0^2 - k^2)G \equiv \gamma^2 G \qquad (37.17)$$

[73] Which one does *not* have in this model.

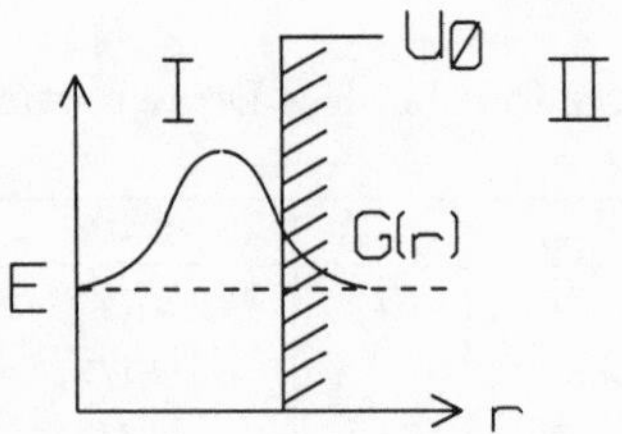

Figure 37.3: Ground state (lowest positive energy) of massless Dirac particle in spherically symmetric scalar potential.

The solution is

$$G_{\rm II} = Ce^{-\gamma(r-R)} + De^{\gamma(r-R)} \tag{37.18}$$

For a solution that is normalizable one must have $G(r) \to 0$ as $r \to \infty$. This implies that the coefficient $D = 0$.

Now from the first of Eqs. (37.14)

$$F = \frac{G' - G/r}{E + U_0} \tag{37.19}$$

Thus

$$\begin{aligned} F_{\rm I} &= A\left(\cos kr - \frac{\sin kr}{kr}\right) \\ F_{\rm II} &= \frac{C}{U_0+k}\left(-\gamma - \frac{1}{r}\right)e^{-(\gamma-R)} = -C\left(\sqrt{\frac{U_0-k}{U_0+k}} + \frac{1}{U_0+k}\frac{1}{r}\right)e^{-\gamma(r-R)} \end{aligned} \tag{37.20}$$

Now consider the case of a *scalar wall* where $U_0 \to \infty$. In this limit Eqs. (37.18) and (37.20) imply that for the outside solution in Region II

$$F_{\rm II}(r) = -G_{\rm II}(r) \tag{37.21}$$

If one demands that both (G, F) are continuous at the wall, then (see Fig. 37.4)

$$F_{\rm I}(R) = -G_{\rm I}(R) \tag{37.22}$$

Equations (37.16) and (37.20) then yield the *eigenvalue equation*[74]

$$j_0(kR) = j_1(kR) \tag{37.23}$$

The first few eigenvalues are shown in Table 37.2 (from Ref. [Q49]). They are displayed in Fig. 37.5. A combination of these results provides the Dirac wave

[74]Note $j_0(\rho) = \sin\rho/\rho$ and $j_1(\rho) = \sin\rho/\rho^2 - \cos\rho/\rho$.

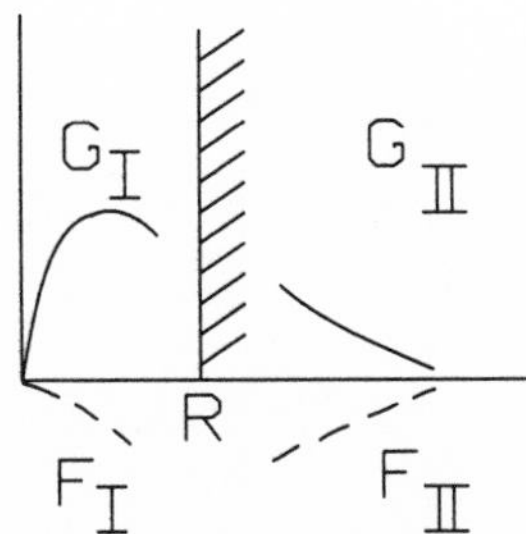

Figure 37.4: Match boundary conditions at a scalar wall.

function for the ground state (lowest positive energy)

$$\psi_{1s_{1/2}m} = N \begin{pmatrix} ij_0(kr)\mathcal{Y}^m_{0\frac{1}{2}\frac{1}{2}} \\ j_1(kr)\mathcal{Y}^m_{1\frac{1}{2}\frac{1}{2}} \end{pmatrix} \tag{37.24}$$

Here the eigenvalue is $k_{10}R \equiv x = 2.04$, which satisfies the boundary condition at the wall $j_0(x) = j_1(x)$.

Some Consequences. Let us determine the normalization constant N. First write the above solution in more detail

$$\psi_{1s_{1/2}m} = N \begin{pmatrix} ij_0(kr)\chi_m/\sqrt{4\pi} \\ j_1(kr)\sum_{m_l m_s}\langle 1m_l\frac{1}{2}m_s|1\frac{1}{2}\frac{1}{2}m\rangle Y_{1m_l}\chi_{m_s} \end{pmatrix} \tag{37.25}$$

It then follows that[75]

$$\begin{aligned} \psi^\dagger\psi &= \frac{N^2}{4\pi}[j_0^2(kr) + j_1^2(kr)] \\ \bar{\psi}\psi &= \frac{N^2}{4\pi}[j_0^2(kr) - j_1^2(kr)] \end{aligned} \tag{37.26}$$

Note that these equations imply that at the scalar wall

$$\begin{aligned} \psi^\dagger\psi|_{r=R} \neq 0 \\ \bar{\psi}\psi|_{r=R} = 0 \end{aligned} \tag{37.27}$$

[75]For $m = 1/2$ one has (the result is the same for $m = -1/2$)

$$\left(\sum_{m_l m_s}\langle 1m_l\frac{1}{2}m_s|1\frac{1}{2}\frac{1}{2}m\rangle Y_{1m_l}\chi_{m_s}\right)^\dagger\left(\sum_{m_l' m_s'}\langle 1m_l'\frac{1}{2}m_s'|1\frac{1}{2}\frac{1}{2}m\rangle Y_{1m_l'}\chi_{m_s'}\right)$$

$$= \sum_{m_l m_s}\langle 1m_l\frac{1}{2}m_s|1\frac{1}{2}\frac{1}{2}m\rangle^2 Y^*_{1m_l}Y_{1m_l} = \langle 10\frac{1}{2}\frac{1}{2}|1\frac{1}{2}\frac{1}{2}\frac{1}{2}\rangle^2|Y_{10}|^2 + \langle 11\frac{1}{2}-\frac{1}{2}|1\frac{1}{2}\frac{1}{2}\frac{1}{2}\rangle^2|Y_{11}|^2$$

$$= (1/3)(3/4\pi)\cos^2\theta + (2/3)(3/8\pi)\sin^2\theta = 1/4\pi$$

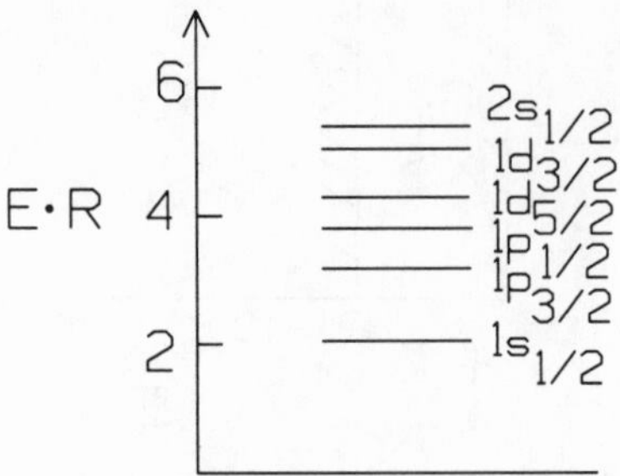

Figure 37.5: Low-lying spectrum of massless Dirac particle in a spherical scalar cavity. From Ref. [Q49].

Thus by explicitly solving the Dirac equation in a finite scalar well, demanding continuity of the Dirac wave function, and then letting the height of the scalar potential go to infinity, we have derived the boundary condition that the scalar density *vanishes* at a scalar wall. In contrast, the baryon density is *nonzero* at a scalar wall.

Now evaluate the remaining radial integrals using (see Prob. 37.1)

$$\int_0^x [j_0^2(\rho) + j_1^2(\rho)]\rho^2 d\rho = 2(x-1)\sin^2 x \tag{37.28}$$

Hence the normalization constant is determined to be

$$N^2 = \frac{k^3}{2(x-1)\sin^2 x} = \frac{x^3}{2R^3(x-1)\sin^2 x} \tag{37.29}$$

Here the eigenvalue is $x = 2.04$.

Magnetic Moment. Assume this massless Dirac particle in the $1s_{1/2}$ state carries a charge e_q. Let us compute the magnetic moment

$$\begin{aligned} \boldsymbol{\mu} &= \frac{1}{2}\int \mathbf{r}\times\mathbf{j}\, d^3r \\ \mathbf{j} &= e_q\psi^\dagger\boldsymbol{\alpha}\psi \qquad\qquad \boldsymbol{\alpha} = \begin{pmatrix} 0 & \boldsymbol{\sigma} \\ \boldsymbol{\sigma} & 0 \end{pmatrix} \end{aligned} \tag{37.30}$$

Substitute the Dirac wave function in Eq. (37.24)

$$\begin{aligned} \mu &\equiv \mu_z = \frac{e_q}{2}N^2\int d^3r\left\{-ij_0(kr)j_1(kr)\mathcal{Y}^{\frac{1}{2}\dagger}_{0\frac{1}{2}\frac{1}{2}}[\mathbf{r}\times\boldsymbol{\sigma}]_z\mathcal{Y}^{\frac{1}{2}}_{1\frac{1}{2}\frac{1}{2}}\right. \\ &\left.+ij_0(kr)j_1(kr)\mathcal{Y}^{\frac{1}{2}\dagger}_{1\frac{1}{2}\frac{1}{2}}[\mathbf{r}\times\boldsymbol{\sigma}]_z\mathcal{Y}^{\frac{1}{2}}_{0\frac{1}{2}\frac{1}{2}}\right\} \end{aligned} \tag{37.31}$$

Now express the required operator in spherical components[76]

$$
\begin{aligned}
[\mathbf{r}\times\boldsymbol{\sigma}]_z &= -i\sqrt{\frac{4\pi}{3}}\,r[Y_{11}\sigma_{-1} - Y_{1-1}\sigma_1] \\
\sigma_{\pm 1} &= \mp\frac{1}{\sqrt{2}}(\sigma_x \pm i\sigma_y) = \mp\sqrt{2}\sigma_\pm
\end{aligned}
\tag{37.32}
$$

The quantities $\sigma_\pm$ are the raising and lowering matrices for spin 1/2; their matrix elements follow directly. Thus

$$
\begin{aligned}
\mu &= \frac{e_q}{2}N^2\int d^3r\sqrt{\frac{2}{3}}rj_0(kr)j_1(kr)[-Y_{11}Y_{1-1} + Y_{11}^*Y_{11}]\langle 11\tfrac{1}{2} - \tfrac{1}{2}|1\tfrac{1}{2}\tfrac{1}{2}\tfrac{1}{2}\rangle \\
&= \frac{2e_q}{3}N^2\int r^3dr j_0(kr)j_1(kr) \\
&= \frac{2e_q}{3}N^2\frac{1}{k^4}\int_0^x \rho^3 j_0(\rho)j_1(\rho)d\rho
\end{aligned}
\tag{37.33}
$$

The second equality follows from the relation $Y_{1-1} = -Y_{11}^*$ and the normalization of the spherical harmonics; the C-G coefficient is $\sqrt{2/3}$. The final radial integral is evaluated in Prob. 37.1

$$
\int_0^x \rho^3 j_0(\rho)j_1(\rho)d\rho = \left(x - \frac{3}{4}\right)\sin^2 x
\tag{37.34}
$$

Substitution of the normalization constant in Eq. (37.29) then leads to the final expression for the magnetic moment

$$
\begin{aligned}
\mu &= \frac{e_qR}{12}\frac{4x-3}{x(x-1)}\,; \qquad\qquad x \equiv k_{10}R = 2.04 \\
&= 0.203\, e_qR
\end{aligned}
\tag{37.35}
$$

The Dirac particle is massless — there is no quark magneton in this model. The dimension for the magnetic moment arises here from the radius R of the cavity.[77] Multiplication and division by the nucleon mass M and by e_p yields

$$
\mu = 0.203\left(\frac{e_q}{e_p}\right)\left(\frac{e_p}{2M}\right)2MR
\tag{37.36}
$$

For a cavity of radius $R = 1\,\text{fm}$, one obtains in nuclear magnetons (n.m.)

$$
\mu = 1.93\left(\frac{e_q}{e_p}\right)\text{n.m.}\,; \qquad\qquad R = 1\,\text{fm}
\tag{37.37}
$$

[76] Start with $[\mathbf{a}\times\mathbf{b}]_z = a_xb_y - a_yb_x$ and the spherical components of a vector $v_{\pm 1} = \mp(v_x \pm iv_y)/\sqrt{2}$. Then $v_x = (v_{-1} - v_{+1})/\sqrt{2}$ and $v_y = i(v_{-1} + v_{+1})/\sqrt{2}$. Thus $[\mathbf{a}\times\mathbf{b}]_z = i(a_{-1}b_{+1} - a_{+1}b_{-1})$. Now use $r_{1m} = (4\pi/3)^{1/2}rY_{1m}$; this produces Eq. (37.32).

[77] In the same way one can calculate $\sqrt{\langle r^2\rangle} = 0.73R$ (Ref. [Q49] and Prob. 37.3).

This is certainly in the ballpark of the observed magnetic moment of the nucleon; however, to apply the result in that case, one must have the wave function for three quarks in the nucleon.

State Vectors. We have evaluated a few single-quark matrix elements in this bag model. To do a real calculation one needs the (qqq) [and $(\bar{q}q)$] wave functions, including all the quantum numbers. We make an independent-quark shell model of hadrons and work in the nuclear domain where only the lightest (u, d) quarks and their antiquarks are retained; thus the quark field is approximated by

$$\psi \doteq \begin{pmatrix} u \\ d \end{pmatrix} \; ; \qquad \text{nuclear domain} \tag{37.38}$$

Let us start with the simpler case of nonrelativistic quarks in a potential (where the spin and spatial wave functions decouple).[78] In this case one can write the one-quark wave function as

$$\psi = \underbrace{\psi_{nlm_l}(\mathbf{r})}_{\text{space}} \; \underbrace{\chi_{m_s}}_{\text{spin}} \; \underbrace{\eta_{m_t}}_{\text{isospin}} \; \underbrace{\rho_\alpha}_{\text{color}} \; ; \qquad \begin{aligned} m_s &= \pm\tfrac{1}{2} \\ m_t &= \pm\tfrac{1}{2} \\ \alpha &= (R, G, B) \end{aligned} \tag{37.39}$$

Consider the *color wave function* for the (qqq) system. The observed hadrons are color singlets. Hence the color wave function in this case is just the completely antisymmetric combination (a Slater determinant with respect to color)

$$\Psi_{\text{color}}(1,2,3) = \frac{1}{\sqrt{6}} \begin{vmatrix} \rho_R(1) & \rho_G(1) & \rho_B(1) \\ \rho_R(2) & \rho_G(2) & \rho_B(2) \\ \rho_R(3) & \rho_G(3) & \rho_B(3) \end{vmatrix} \; ; \qquad \text{antisymmetric} \tag{37.40}$$

If $G_\alpha^{\text{color}}; \alpha = 1, \ldots, 8$ are the generators of the color transformation among the quarks, then all of the generators annihilate this wave function[79]

$$G_\alpha^{\text{color}} \Psi_{\text{color}} = 0 \; ; \qquad \alpha = 1, \ldots, 8 \tag{37.41}$$

Since the total wave function must be antisymmetric in the interchange of any two fermions, the remaining space-spin-isospin wave function must be *symmetric*.

For the ground state in this shell model, the spatial wave functions $\psi_{n00}(\mathbf{r})$ will all be the same, all $1s$, and hence the spatial part of the wave function is totally symmetric

$$\Psi_{\text{space}}(1,2,3) = \psi_{1s}(\mathbf{r}_1)\psi_{1s}(\mathbf{r}_2)\psi_{1s}(\mathbf{r}_3) \; ; \qquad \text{symmetric} \tag{37.42}$$

[78] This is, in fact, the case for the very successful *constituent quark model* of nucleons and mesons (see Ref. [Q95]). Here one starts from an independent quark basis with nonrelativistic quarks of mass $m_q \approx m/3$. With the quark wave functions developed here, and the analysis of electroweak interactions with nuclei in Parts I and IV, one can understand many of the results of this model.

[79] Just as the fully occupied Slater determinant of spins has $S = 0$, or of j-shells has $J = 0$.

The spin-isospin wave function must thus be *totally symmetric*. Start with isospin. One is faced with the problem of coupling three angular momenta; however, the procedure follows immediately from the discussion of $6-j$ symbols in quantum mechanics (Ref. [N44]). An eigenstate of total angular momentum can be formed as follows

$$|(j_1j_2)j_{12}j_3jm\rangle = \sum_{m_1m_2m_3m_{12}} \langle j_1m_1j_2m_2|j_1j_2j_{12}m_{12}\rangle \times\langle j_{12}m_{12}j_3m_3|j_{12}j_3jm\rangle|j_1m_1\rangle|j_2m_2\rangle|j_3m_3\rangle \tag{37.43}$$

These states form a complete orthonormal basis for given (j_1, j_2, j_3). The states formed by coupling in the other order $|j_1(j_2j_3)j_{23}jm\rangle$ are linear combinations of these with $6-j$ symbols as coefficients.

For *isospin* in the nuclear domain all the $t_i = 1/2$, thus there are a total of $2 \times 2 \times 2 = 8$ basis states. Consider first the states with total $T = 3/2$. Here the only possible intermediate value is $t_{12} = 1$. The state with $T_3 = 3/2$ is readily constructed from the above as $\alpha(1)\alpha(2)\alpha(3)$. Now apply the total lowering operator $T_- = t(1)_- + t(2)_- + t(3)_-$ and use $t_-\alpha = \beta$, $t_-\beta = 0$. The set of states with $T = 3/2$ follows immediately

$$\begin{aligned}
\Phi[(\frac{1}{2}\frac{1}{2})1\frac{1}{2}\frac{3}{2}\frac{3}{2}] &= \alpha(1)\alpha(2)\alpha(3) \\
\Phi[(\frac{1}{2}\frac{1}{2})1\frac{1}{2}\frac{3}{2}\frac{1}{2}] &= \frac{1}{\sqrt{3}}[\beta(1)\alpha(2)\alpha(3) + \alpha(1)\beta(2)\alpha(3) + \alpha(1)\alpha(2)\beta(3)] \\
\Phi[(\frac{1}{2}\frac{1}{2})1\frac{1}{2}\frac{3}{2} -\frac{1}{2}] &= \frac{1}{\sqrt{3}}[\beta(1)\beta(2)\alpha(3) + \beta(1)\alpha(2)\beta(3) + \alpha(1)\beta(2)\beta(3)] \\
\Phi[(\frac{1}{2}\frac{1}{2})1\frac{1}{2}\frac{3}{2} -\frac{3}{2}] &= \beta(1)\beta(2)\beta(3) \; ; \qquad \text{4 symmetric states}
\end{aligned} \tag{37.44}$$

There are four symmetric states with $T = 3/2$.

Consider next the states with total $T = 1/2$. Here there are two possible intermediate values in the above, $t_{12} = 0, 1$. For the first of these values one finds

$$\begin{aligned}
\Phi^{\rho}[(\frac{1}{2}\frac{1}{2})0\frac{1}{2}\frac{1}{2}\frac{1}{2}] &= \frac{1}{\sqrt{2}}[\alpha(1)\beta(2) - \alpha(2)\beta(1)]\alpha(3) \\
\Phi^{\rho}[(\frac{1}{2}\frac{1}{2})0\frac{1}{2}\frac{1}{2} -\frac{1}{2}] &= \frac{1}{\sqrt{2}}[\alpha(1)\beta(2) - \alpha(2)\beta(1)]\beta(3) \; ; \qquad \text{2 states}
\end{aligned} \tag{37.45}$$

These two states have *mixed symmetry*; they are antisymmetric in the interchange of particles $(1 \leftrightarrow 2)$.

The second value $t_{12} = 1$ yields

$$\Phi^{\lambda}[(\frac{1}{2}\frac{1}{2})1\frac{1}{2}\frac{1}{2}\frac{1}{2}] = \frac{1}{\sqrt{6}}[2\alpha(1)\alpha(2)\beta(3) - \alpha(1)\beta(2)\alpha(3) - \beta(1)\alpha(2)\alpha(3)]$$

$$\Phi^{\lambda}[(\frac{1}{2}\frac{1}{2})1\frac{1}{2}\frac{1}{2}-\frac{1}{2}] = -\frac{1}{\sqrt{6}}[2\beta(1)\beta(2)\alpha(3)-\beta(1)\alpha(2)\beta(3)-\alpha(1)\beta(2)\beta(3)]\ ; \quad \text{2 states} \tag{37.46}$$

These two states also have mixed symmetry; they are symmetric in the interchange of particles $(1 \leftrightarrow 2)$.

Now look at the ***spin*** wave functions. The analysis is exactly the same! We have a set of spin states Ξ identical to those above.

For the overall spin-isospin wave function, we must take a product of these wave functions and make the result totally symmetric. Recall first from quantum mechanics how one makes a wave function totally antisymmetric. Introduce the antisymmetrizing operator

$$\mathcal{A} = N\sum_{(P)}(-1)^{p}P \tag{37.47}$$

Here the sum goes over all permutations, produced by the operator P, of a complete set of coordinates for each particle. The signature of the permutation is $(-1)^p$, and $N = 1/\sqrt{N_P}$ where N_P is the total number of permutations.

Similarly, to make a wave function totally symmetric introduce the (unnormalized) *symmetrizing operator*

$$\mathcal{S} = N\sum_{(P)}P \tag{37.48}$$

Note that if a wave function is antisymmetric under the interchange of any two particles, the application of $\mathcal{S}$ will give zero. This result is established as follows. Use

$$P_{12}\mathcal{S} = \mathcal{S}P_{12} \tag{37.49}$$

This follows since as P goes over all permutations, so does $P_{12}P$ or PP_{12}

$$\sum_{(P)}P_{12}P = \sum_{(P)}P = \sum_{(P)}PP_{12} \tag{37.50}$$

It follows that

$$P_{12}\mathcal{S}\psi = \mathcal{S}\psi = \mathcal{S}P_{12}\psi = -\mathcal{S}\psi = 0 \tag{37.51}$$

This is the stated result.

Note further that if the operator $\mathcal{S}$ is applied to the product of the totally symmetric 3/2 state and either of the 1/2 states with mixed symmetry, the result will vanish. The proof is as follows. Since $\mathcal{S}\Phi_{3/2} = \Phi_{3/2}\mathcal{S}$, one just needs to show that

$$\mathcal{S}[A\Phi^{\rho} + B\Phi^{\lambda}] = 0 \tag{37.52}$$

Table 37.3: Totally symmetric spin-isospin states for three nonrelativistic quarks.

T	S	Number of states
3/2	3/2	16
1/2	1/2	4
		20

The first term gives zero since Φ^ρ is antisymmetric in the interchange of the first pair of particles. The second vanishes because of the nature of the sums in Eq. (37.46) and the fact that $\mathcal{S}$ produces an identical result when applied to each term in the sum

$$\mathcal{S}(\alpha\alpha\beta) = \mathcal{S}(\alpha\beta\alpha) = \mathcal{S}(\beta\alpha\alpha) \tag{37.53}$$

It is a consequence of these two observations that *the only nonzero totally symmetric wave function will be obtained by combining the spin and isospin wave functions of the same symmetry.* Thus one must combine the two totally symmetric spin and isospin states and the other two pair of states with the same mixed symmetry; in the latter case there is only one totally symmetric linear combination (this is proven in Appendix J). This leads to the set of totally symmetric spin-isospin states shown in Table 37.3 and given by

$$\begin{aligned}&\Phi_{\frac{3}{2}m_t}\Xi_{\frac{3}{2}m_s}\\ &\frac{1}{\sqrt{2}}\left(\Phi^\lambda_{\frac{1}{2}m_t}\Xi^\lambda_{\frac{1}{2}m_s} + \Phi^\rho_{\frac{1}{2}m_t}\Xi^\rho_{\frac{1}{2}m_s}\right)\end{aligned} \tag{37.54}$$

These are all the baryons one can make in this model. Since all these states are degenerate in the model as presently formulated, one has a *supermultiplet* of baryons. The present calculation predicts the spins and isospins of the members of this supermultiplet.[80]

Masssless Dirac Particles in Their Ground State. Let us extend the arguments to the situation in the M.I.T. bag model where, in contrast to massive, nonrelativistic constituents, one has massless relativistic quarks. The problem is more complicated since the space-spin parts of the wave functions are now coupled; however, if the quarks occupy a common lowest positive energy $\psi_{1s_{1/2}m_j}(\mathbf{r})$ ground state , the problem is greatly simplified. Make the following replacement in the space-spin wave functions discussed above

$$\psi_{1s}(\mathbf{r})\chi_{m_s} \rightarrow \psi_{1s_{1/2}m_j}(\mathbf{r}) \tag{37.55}$$

[80] Define $\zeta_i \equiv \chi_{m_s}\eta_{m_t}$ with $(m_s, m_t) = (\pm 1/2, \pm 1/2)$. Then in a nonrelativistic quark model with spin-independent interactions one has an internal global $SU(4)$ (flavor) symmetry — this is just Wigner's supermultiplet theory. Here the baryons belong to the totally symmetric irreducible representation one gets from $4 \bigotimes 4 \bigotimes 4$; this is the [20] dimensional representation with spin-isospin content worked out in the text and shown in Table 37.3.

Instead of the spin $\mathbf{S}$, now talk about the total angular momentum $\mathbf{J}$; the angular momentum and symmetry arguments are then *exactly the same as before.*

Some Applications. Consider the nucleon (N) ground-state expectation value of the following operator

$$O = \sum_{i=1}^{3} O_i(\mathbf{r}_i, \boldsymbol{\sigma}_i) I_i(\boldsymbol{\tau}_i) \tag{37.56}$$

Assume that the isospin factor is diagonal $I_i = (1, \tau_3)_i$. Since the wave function is totally symmetric, it follows that one need evaluate the matrix element only for the third particle.[81]

$$\langle \Psi_N | \sum_{i=1}^{3} O_i I_i | \Psi_N \rangle = 3 \langle \Psi_N | O_3 I_3 | \Psi_N \rangle \tag{37.57}$$

Substitution of Eq. (37.54) then yields[82] for the state of total $m_j = 1/2$

$$\begin{aligned} 3\langle \Psi_N | O_3 I_3 | \Psi_N \rangle &= \frac{3}{2} \langle \Phi^\rho | I_3 | \Phi^\rho \rangle \langle \frac{1}{2}(3) | O_3 | \frac{1}{2}(3) \rangle \\ &+ \frac{3}{2} \langle \Phi^\lambda | I_3 | \Phi^\lambda \rangle \frac{1}{6} \left\{ 4 \langle -\frac{1}{2}(3) | O_3 | -\frac{1}{2}(3) \rangle + 2 \langle \frac{1}{2}(3) | O_3 | \frac{1}{2}(3) \rangle \right\} \end{aligned} \tag{37.58}$$

Here the remaining labels on the single-particle matrix elements of O_3 are $|m_j, (\text{particle number})\rangle$. The result is

$$\begin{aligned} \langle \Psi^N_{m_t \frac{1}{2}} | \sum_{i=1}^{3} O_i I_i | \Psi^N_{m_t \frac{1}{2}} \rangle &= \langle \frac{1}{2} | O | \frac{1}{2} \rangle \left[\frac{3}{2} \langle \Phi^\rho | I_3 | \Phi^\rho \rangle + \frac{1}{2} \langle \Phi^\lambda | I_3 | \Phi^\lambda \rangle \right] \\ &+ \langle -\frac{1}{2} | O | -\frac{1}{2} \rangle \left[\langle \Phi^\lambda | I_3 | \Phi^\lambda \rangle \right] \end{aligned} \tag{37.59}$$

This result is for total $m_j = 1/2$; the remaining isospin operator I_3 acts only on the third particle. For an *isoscalar* operator with $I_3 = 1$ this expression reduces to

$$\langle \Psi^N_{m_t \frac{1}{2}} | \sum_{i=1}^{3} O_i | \Psi^N_{m_t \frac{1}{2}} \rangle = 2 \langle \frac{1}{2} | O | \frac{1}{2} \rangle + \langle -\frac{1}{2} | O | -\frac{1}{2} \rangle \tag{37.60}$$

This is now just a sum of single-particle matrix elements. For an *isovector* operator with $I_3 = \tau_3$, the required isospin matrix elements for the proton with

[81] Assume the operators form the identity with respect to color; the color wave function then goes right through the matrix element, and it is normalized.

[82] Use $\langle \Phi^\rho | I_3 | \Phi^\lambda \rangle = 0$ if I_3 is diagonal; this follows immediately from the form of Eq. (37.45) and the orthogonality of the mixed-symmetry wave functions.

$m_t = 1/2$ follow from Eqs. (37.45) and (37.46)

$$\begin{aligned} \langle\Phi^\rho|\tau_3(3)|\Phi^\rho\rangle &= 1 \\ \langle\Phi^\lambda|\tau_3(3)|\Phi^\lambda\rangle &= \frac{1}{6}(-4+1+1) = -\frac{1}{3}\,; \qquad \text{proton } m_t = \frac{1}{2} \end{aligned} \tag{37.61}$$

For a neutron with $m_t = -1/2$, these isovector matrix elements simply change sign. It follows that

$$\begin{aligned} \langle\Psi^N_{\frac{1}{2}\frac{1}{2}}|\sum_{i=1}^{3} O_i\tau_3(i)|\Psi^N_{\frac{1}{2}\frac{1}{2}}\rangle &= \frac{4}{3}\langle\frac{1}{2}|O|\frac{1}{2}\rangle - \frac{1}{3}\langle-\frac{1}{2}|O|-\frac{1}{2}\rangle \\ \langle\Psi^N_{-\frac{1}{2}\frac{1}{2}}|\sum_{i=1}^{3} O_i\tau_3(i)|\Psi^N_{-\frac{1}{2}\frac{1}{2}}\rangle &= -\frac{4}{3}\langle\frac{1}{2}|O|\frac{1}{2}\rangle + \frac{1}{3}\langle-\frac{1}{2}|O|-\frac{1}{2}\rangle \end{aligned} \tag{37.62}$$

The notation here is $\Psi^N_{m_t,m_j}$.

In the nuclear domain with only (u,d) quarks the electric charge is given by

$$e_i = \left[\frac{1}{6} + \frac{1}{2}\tau_3(i)\right] e_p \tag{37.63}$$

Hence the expectation value of an operator proportional to the charge in the composite three-quark proton and neutron ground state is given by

$$\begin{aligned} \langle p|\sum_{i=1}^{3} O_i e_i|p\rangle &= e_p\left[\frac{1}{6}(2O_{1/2}+O_{-1/2}) + \frac{1}{2}(\frac{4}{3}O_{1/2} - \frac{1}{3}O_{-1/2})\right] \\ &= e_p\langle\frac{1}{2}|O|\frac{1}{2}\rangle \\ \langle n|\sum_{i=1}^{3} O_i e_i|n\rangle &= e_p\left[\frac{1}{6}(2O_{1/2}+O_{-1/2}) + \frac{1}{2}(-\frac{4}{3}O_{1/2} + \frac{1}{3}O_{-1/2})\right] \\ &= -\frac{e_p}{3}\langle\frac{1}{2}|O|\frac{1}{2}\rangle + \frac{e_p}{3}\langle-\frac{1}{2}|O|-\frac{1}{2}\rangle \end{aligned} \tag{37.64}$$

Let us apply this result to compute the magnetic moment of the ground state of the nucleon in the bag model using Eq. (37.35) for the expectation value of the single (massless) quark matrix element $\langle\frac{1}{2}|O|\frac{1}{2}\rangle = 0.203R$; since the magnetic moment is a vector operator, its expectation value in the state $m_j = -1/2$ must simply change sign $\langle-\frac{1}{2}|O|-\frac{1}{2}\rangle = -0.203R$. This yields

$$\mu_p = 0.203 e_p R \qquad\qquad \mu_n = -\frac{2\mu_p}{3} \tag{37.65}$$

The experimental results are

$$\mu_p = +2.79\,\text{n.m.} \qquad\qquad \mu_n = -1.91\,\text{n.m.} \tag{37.66}$$

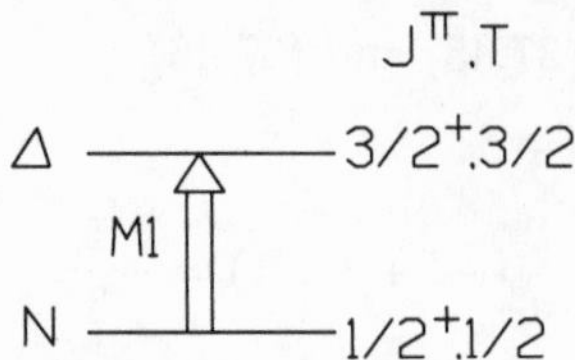

Figure 37.6: Transition magnetic dipole moment in the M.I.T. bag model.

The calculated ratio is quite impressive, and the absolute value is fit in the first relation with a radius $R = 1.44\,\text{fm}$, which, although too large, is certainly in the right ballpark.

Transition Magnetic Moment. Consider the transition magnetic dipole moment between the ground state (N) and the excited state (Δ) formed from the product of the totally symmetric isospin state and totally symmetric space-spin state. Since only different m_j states are involved in the latter, we are in a position to calculate this matrix element. The situation is illustrated in Fig. 37.6. The wave functions are given by

$$\begin{aligned}\Psi^N_{\frac{1}{2}\frac{1}{2}} &= \frac{1}{\sqrt{2}}\left[\Phi^\lambda_{\frac{1}{2}\frac{1}{2}}\Xi^\lambda_{\frac{1}{2}\frac{1}{2}} + \Phi^\rho_{\frac{1}{2}\frac{1}{2}}\Xi^\rho_{\frac{1}{2}\frac{1}{2}}\right] \\ \Psi^\Delta_{\frac{1}{2}\frac{1}{2}} &= \Phi_{\frac{3}{2}\frac{1}{2}}\Xi_{\frac{3}{2}\frac{1}{2}}\end{aligned} \tag{37.67}$$

The subscripts on the left are (m_t, m_j) and those of the right (Tm_t, Jm_j); in detail, these wave functions are

$$\Phi_{\frac{3}{2}\frac{1}{2}} = \frac{1}{\sqrt{3}}\left[\phi_{-\frac{1}{2}}(1)\phi_{\frac{1}{2}}(2)\phi_{\frac{1}{2}}(3) + \phi_{\frac{1}{2}}(1)\phi_{-\frac{1}{2}}(2)\phi_{\frac{1}{2}}(3) + \phi_{\frac{1}{2}}(1)\phi_{\frac{1}{2}}(2)\phi_{-\frac{1}{2}}(3)\right] \tag{37.68}$$

A similar expression holds for $\Xi_{\frac{3}{2}\frac{1}{2}}$. The transition magnetic dipole moment is now given by

$$\mu^* = \langle\Psi^\Delta_{\frac{1}{2}\frac{1}{2}}|\sum_{i=1}^{3}\mu(i)\frac{1}{2}\tau_3(i)e_p|\Psi^N_{\frac{1}{2}\frac{1}{2}}\rangle = \frac{3}{2}e_p\langle\Psi^\Delta_{\frac{1}{2}\frac{1}{2}}|\mu(3)\tau_3(3)|\Psi^N_{\frac{1}{2}\frac{1}{2}}\rangle \tag{37.69}$$

Here it has been observed that only the isovector part of the magnetic dipole operator can contribute to the transition and the total symmetry of the states has been used. It now follows from Eqs. (37.68), (37.45), and (37.46) that

$$\begin{aligned}\langle\Phi_{\frac{3}{2}\frac{1}{2}}|\tau_3(3)|\Phi^\rho_{\frac{1}{2}\frac{1}{2}}\rangle &= 0 \\ \langle\Phi_{\frac{3}{2}\frac{1}{2}}|\tau_3(3)|\Phi^\lambda_{\frac{1}{2}\frac{1}{2}}\rangle &= \frac{1}{\sqrt{18}}\left[2\langle-\frac{1}{2}|\tau_3|-\frac{1}{2}\rangle - 2\langle\frac{1}{2}|\tau_3|\frac{1}{2}\rangle\right] = -\frac{4}{\sqrt{18}} \\ \langle\Xi_{\frac{3}{2}\frac{1}{2}}|\mu(3)|\Xi^\lambda_{\frac{1}{2}\frac{1}{2}}\rangle &= \frac{1}{\sqrt{18}}\left[2\langle-\frac{1}{2}|\mu|-\frac{1}{2}\rangle - 2\langle\frac{1}{2}|\mu|\frac{1}{2}\rangle\right] = -\frac{4}{\sqrt{18}}\langle\frac{1}{2}|\mu|\frac{1}{2}\rangle\end{aligned} \tag{37.70}$$

Use of Eqs. (37.64) and (37.65) allows the final result for μ^* to be expressed in terms of the ground-state magnetic moment of the proton

$$\mu^* = \frac{3}{2}\frac{1}{\sqrt{2}}\frac{16}{18}\mu_p = \frac{4}{3\sqrt{2}}\mu_p \tag{37.71}$$

This is the matrix element for $(m_j, m_t) = (\frac{1}{2}\frac{1}{2}) \to (\frac{1}{2}\frac{1}{2})$; other components follow from the Wigner-Eckart theorem. This result agrees to about 30% with experimental observations of the transition magnetic dipole matrix element obtained from electroproduction of the first nucleon resonance (Ref. [Q54]).

Single-Particle Matrix Element of $i\gamma\gamma_5$. This matrix $i\gamma\gamma_5$ governs the spatial part of the axial-vector current in the weak interactions (Part IV); it is given by

$$i\gamma\gamma_5 = \begin{pmatrix} 0 & \boldsymbol{\sigma} \\ -\boldsymbol{\sigma} & 0 \end{pmatrix}\begin{pmatrix} 0 & -1 \\ -1 & 0 \end{pmatrix} = \begin{pmatrix} -\boldsymbol{\sigma} & 0 \\ 0 & \boldsymbol{\sigma} \end{pmatrix} \tag{37.72}$$

Taken between fields this matrix gives

$$\begin{aligned} \bar{\psi}i\gamma\gamma_5\psi &= \psi^\dagger \begin{pmatrix} 1 & 0 \\ 0 & -1 \end{pmatrix}\begin{pmatrix} -\boldsymbol{\sigma} & 0 \\ 0 & \boldsymbol{\sigma} \end{pmatrix}\psi = -\psi^\dagger \begin{pmatrix} \boldsymbol{\sigma} & 0 \\ 0 & \boldsymbol{\sigma} \end{pmatrix}\psi \\ &= -\psi^\dagger\boldsymbol{\sigma}\psi \end{aligned} \tag{37.73}$$

This is just the *spin* operator, whose nonrelativistic limit is

$$-\chi^\dagger_{m_s}\boldsymbol{\sigma}\chi_{m_s} \equiv -\langle\boldsymbol{\sigma}\rangle \tag{37.74}$$

Substitution of the ground-state (lowest positive energy) wave function $\psi_{1s_{1/2}m_j}$ for a massless Dirac particle in a spherical cavity in Eqs. (37.25) and (37.29) gives the following expression for this matrix element

$$\begin{aligned} &\bar{\psi}_{\frac{1}{2}} i\gamma_z\gamma_5\psi_{\frac{1}{2}} = \\ &-\frac{1}{2(x-1)\sin^2 x}\int_0^x \rho^2 d\rho \left[j_0^2(\rho)\chi^\dagger_{\frac{1}{2}}\sigma_z\chi_{\frac{1}{2}} + j_1^2(\rho) \right. \\ &\left. \times \underbrace{\int d\Omega \sum_{m_l m_s}\langle 1 m_l \tfrac{1}{2} m_s | 1\tfrac{1}{2}\tfrac{1}{2}\tfrac{1}{2}\rangle Y^*_{1m_l}\chi^\dagger_{m_s}\sigma_z \sum_{m_l' m_s'}\langle 1 m_l' \tfrac{1}{2} m_s' | 1\tfrac{1}{2}\tfrac{1}{2}\tfrac{1}{2}\rangle Y_{1m_l'}\chi_{m_s'}}\right] \\ &= \langle 10\tfrac{1}{2}\tfrac{1}{2}|1\tfrac{1}{2}\tfrac{1}{2}\tfrac{1}{2}\rangle^2 - \langle 11\tfrac{1}{2}-\tfrac{1}{2}|1\tfrac{1}{2}\tfrac{1}{2}\tfrac{1}{2}\rangle^2 = \frac{1}{3} - \frac{2}{3} = -\frac{1}{3} \end{aligned} \tag{37.75}$$

Hence the matrix element is

$$\bar{\psi}_{\frac{1}{2}} i\gamma_z\gamma_5\psi_{\frac{1}{2}} = -\frac{1}{2(x-1)\sin^2 x}\int_0^x \rho^2 \left[j_0^2(\rho) - \frac{1}{3}j_1^2(\rho) \right] d\rho \tag{37.76}$$

The remaining integral is $\tilde{I} = (2x/3)\sin^2 x$ (Prob. 37.1). Hence

$$\begin{aligned} \bar{\psi}_{\frac{1}{2}} i\gamma_z\gamma_5\psi_{\frac{1}{2}} &= -\frac{x}{3(x-1)} = -0.65 \\ \langle\sigma_z\rangle &= 0.65 \end{aligned} \tag{37.77}$$

This result exhibits the reduction of the expectation value of the spin of a single massless Dirac particle in the ground state in a scalar bubble. Of course for a composite nucleon, one must still take the expectation value using the three quark wave functions developed above. These results play an important role in the theory of the weak interactions, to which we will return in Part IV of this book.

38

MORE MODELS

Summary of Some Properties of QCD. Let us again summarize some of the general properties of QCD: Color is *confined* to the interior of hadrons. Lattice gauge theory calculations indicate that this is a dynamic property of QCD arising from the strong, nonlinear gluon interactions. The theory is *asymptotically free*; one can do perturbation theory at high momenta or short distances. In addition, the theory exhibits *spontaneously broken chiral symmetry*, a subject to which we now turn our attention.

Chiral Symmetry. Consider the nuclear domain where the quark field reduces to

$$\psi \doteq \begin{pmatrix} u \\ d \end{pmatrix} \; ; \qquad \text{nuclear domain} \tag{38.1}$$

Assume these quarks are massless.[83] The QCD lagrangian $\mathcal{L}_{\text{QCD}}$ is then invariant under the chiral transformation

$$\psi \to \exp\{\frac{i}{2}\gamma_5 \boldsymbol{\tau} \cdot \boldsymbol{\omega}\}\psi \tag{38.2}$$

As we have seen, this chiral invariance shows up in nature through

1. The partially conserved axial vector current (PCAC);
2. The decoupling of pions as $q_\lambda \to 0$ (soft-pion theorems).

Chiral symmetry appears to be exact in the limit as the pion mass goes to zero ($\mu = m_\pi \to 0$). This symmetry is the underlying basis for the σ-model discussed in Section 22. The generation of the mass of the physical hadrons appears to arise from the spontaneous breaking of chiral symmetry, as illustrated in the σ-model.

Large N_C Limit of QCD. Another interesting property of QCD is its behavior in the limit of very many colors. Although the physical world is evidently described by the Yang-Mills nonabelian gauge theory QCD with three colors ($N_C = 3$), the properties of QCD simplify in the limit $N_C \to \infty$. We proceed to discuss some features of this limit of the theory. The arguments are due to

[83]The mass terms in the QCD lagrangian for the (u, d) quarks are very small (a few MeV). The inclusion of these mass terms forms the basis of *chiral perturbation theory* (Ref. [Q71]). Here one uses a nonrenormalizable effective lagrangian (see Section 23 and Ref. [Q96] — also subsequent discussion) in a systematic low-energy approximation scheme. Recent nuclear physics results in chiral perturbation theory are summarized in Ref. [Q97].

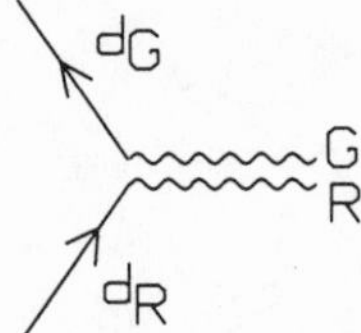

Figure 38.1: Basic Yukawa coupling of quarks to gluons in QCD.

't Hooft (Refs. [Q55, Q56]; see also Witten Ref. [Q57]).[84] The basic Yukawa coupling of quarks to gluons (Sections 19 and 27) is illustrated in Fig. 38.1. Gluons have the color properties of the $(q\bar{q})$ system (excluding the singlet). Color is then conserved along the connected lines in Fig. 38.1; it flows from the quark lines through the gluon lines, which effectively carry two colors.[85]

The following assumed dependence of the coupling constant on the number of colors $N_{\rm C}$ leads to a theory which is finite in the limit $N_{\rm C} \to \infty$

$$g \propto \frac{1}{\sqrt{N_{\rm C}}} \tag{38.3}$$

Loops in QCD can now be analyzed according to the number of free color indices which they contain, that is, over which one independently sums in the loop. Each free sum gives rise to one power of $N_{\rm C}$. A few elementary processes, the free color indices, and the inferred dependence on $N_{\rm C}$ are illustrated in Fig. 38.2. The results from this figure are

1. The quark loop is proportional to $g^2 = O(1/N_{\rm C})$.
2. The gluon loop is proportional to $g^2 N_{\rm C} = O(1)$.
3. The planar gluon loop is proportional to $g^4 N_{\rm C}^2 = O(1)$.

We leave it to the dedicated reader to extend the analysis. It is claimed in Refs. [Q55, Q56, Q49], for example, that the nonplanar gluon loops are of $O(1/N_{\rm C}^2)$. One can quite readily see the following:

1. The baryon mass is of $O(N_{\rm C})$; this follows since one requires $N_{\rm C}$ quarks to make a color-singlet state.[86]
2. Mesons are formed from $(q\bar{q})$ singlets; the meson mass is thus of $O(1)$.
3. Baryon-baryon interactions can take place, for example, through quark and gluon exchange as illustrated in Fig. 38.3a. Since one has $N_{\rm C}$ quarks to choose from in each baryon, the baryon-baryon interaction illustrated here is of $O(g^2 N_{\rm C}^2)$, which is $O(N_{\rm C})$.

[84] There is a nice discussion of this topic in Ref. [Q49].

[85] Gluons are "dichromatic."

[86] Since all the $N_{\rm C}$ quarks have the same spatial wave function in the color-singlet ground state, the mean square radius of the baryon will be $\langle r^2 \rangle = (1/N_{\rm C}) \sum_{i=1}^{N_{\rm C}} \langle r^2 \rangle_i = \langle r^2 \rangle_i$. Thus the baryon *size* is of $O(1)$.

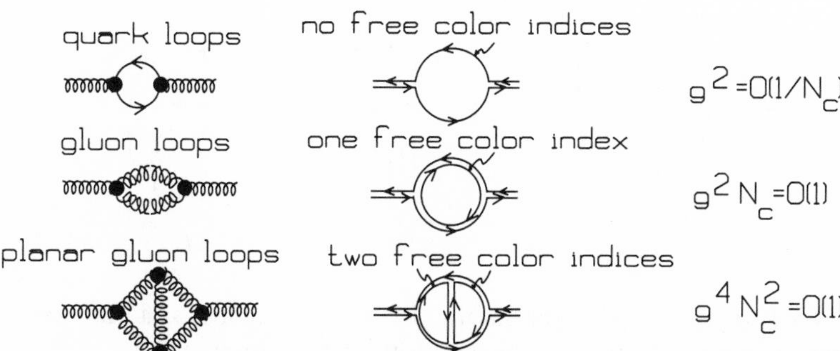

Figure 38.2: A few elementary processes, the free color indices, and the inferred dependence on N_C.

4) In classifying meson interactions we first note that the normalized color-singlet state vector for mesons is of the form

$$|q\bar{q}\rangle_C = \frac{1}{\sqrt{N_C}} \left[|\bar{R}R\rangle + |\bar{G}G\rangle + |\bar{B}B\rangle + \cdots\right] \tag{38.4}$$

There are N_C terms in the sum. The process of meson creation is illustrated in Fig. 38.3b. Since there are N_C quarks to choose from in the baryon, and since one picks up a factor $N_C/\sqrt{N_C}$ from the color part of the meson wave function, the meson-baryon interaction is of $O(g^2 N_C \sqrt{N_C})$, which is of $O(\sqrt{N_C})$.

5. Meson-meson interactions can take place through formation of an intermediate gluon as illustrated in Fig. 38.3c. The amplitude involves the initial and final wave function for each meson; there are then N_C colors to choose from for each meson. Hence the amplitude for the meson-meson interaction is of $O(g^2(1/N_C^2)N_C^2)$, which is of $O(1/N_C)$.

Summary. In summary the classification of various quantitites in the limit $N_C \to \infty$ from 't Hooft (Refs. [Q55, Q56, Q49]) is as follows:

1. Only planar gluon loops remain.
2. Mesons are stable and noninteracting.
3. Baryons are infinitely heavy and interact strongly with mesons and with each other.

And as extended by Witten (Refs. [Q57, Q49]), in this limit:

4. QCD becomes the theory of weakly coupled mesons in the meson sector.
5. Baryons are *soliton* solutions to the nonlinear meson field equations[87] with the mass of the baryon $M_B \propto 1/g^2$.

These results of the large color limit provide a theoretical basis for saying that hadrons are the effective low-energy degrees of freedom for QCD. We proceed

[87] We have already seen a good example of one type of soliton formation in the discussion of finite nuclei as the solutions to the set of nonlinear, Hartree field equations for given baryon number in Section 15. Recall the calculated charge densities of finite nuclei shown in Figs.15.2-4.

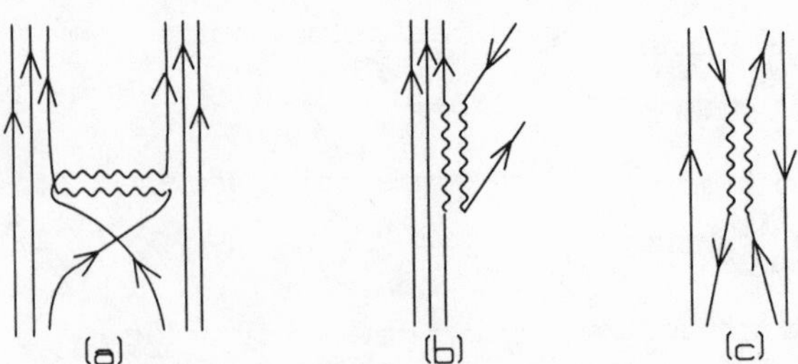

Figure 38.3: (a) Baryon-baryon interaction through quark and gluon exchange; (b) baryon-meson interaction; (c) meson-meson interaction.

to a discussion of an effective meson field theory that incorporates the chiral symmetry of QCD and manifests these $N_C \to \infty$ limits of the theory. It is the model developed some time ago by Skyrme (Refs. [Q58, Q59])[88] that provides a basis for describing low-energy meson interactions and gives a framework for building a soliton model of the baryon (Ref. [Q60, Q49]).

Skyrme Model. Given a hermitian isovector field $\boldsymbol{\phi}$: (ϕ_1, ϕ_2, ϕ_3), construct the following quantity

$$\begin{aligned} \underset{\sim}{U} &= \exp\{i\boldsymbol{\tau}\cdot\boldsymbol{\phi}\} = \exp\{i\mathbf{n}\cdot\boldsymbol{\tau}\phi\} \\ &= \cos\phi + i\mathbf{n}\cdot\boldsymbol{\tau}\sin\phi \\ &\equiv \sigma + i\boldsymbol{\pi}\cdot\tau \end{aligned} \tag{38.5}$$

$\underset{\sim}{U}$ evidently forms a 2×2 unitary, unimodular $SU(2)$ matrix (again denoted with a bar under the symbol), and by definition

$$\begin{aligned} \sigma &\equiv \cos\phi \qquad\qquad \boldsymbol{\pi} \equiv \mathbf{n}\sin\phi \\ \sigma^2 + \boldsymbol{\pi}^2 &= 1 \end{aligned} \tag{38.6}$$

The last relation is satisfied for any choice of $\boldsymbol{\phi}$. If the $(\sigma, \boldsymbol{\pi})$ fields are identified with those in Sections 20-22, the last relation is just that of chiral invariance.

Compute

$$\begin{aligned} \frac{\partial \underset{\sim}{U}}{\partial x_\mu} &= \frac{\partial\sigma}{\partial x_\mu} + i\boldsymbol{\tau}\cdot\frac{\partial\boldsymbol{\pi}}{\partial x_\mu} \\ \frac{\partial \underset{\sim}{U}}{\partial x_\mu}\underset{\sim}{U}^\dagger &= \left(\frac{\partial\sigma}{\partial x_\mu} + i\boldsymbol{\tau}\cdot\frac{\partial\boldsymbol{\pi}}{\partial x_\mu}\right)(\sigma - i\boldsymbol{\pi}\cdot\tau) \end{aligned} \tag{38.7}$$

Use $(\boldsymbol{\tau}\cdot\mathbf{a})(\boldsymbol{\tau}\cdot\mathbf{b}) = \mathbf{a}\cdot\mathbf{b} + i\boldsymbol{\tau}\cdot(\mathbf{a}\times\mathbf{b})$

$$\frac{\partial \underset{\sim}{U}}{\partial x_\mu}\underset{\sim}{U}^\dagger = \left(\sigma\frac{\partial\sigma}{\partial x_\mu} + \boldsymbol{\pi}\cdot\frac{\partial\boldsymbol{\pi}}{\partial x_\mu}\right) + i\boldsymbol{\tau}\cdot\left[\frac{\partial\boldsymbol{\pi}}{\partial x_\mu}\times\boldsymbol{\pi} + \sigma\frac{\partial\boldsymbol{\pi}}{\partial x_\mu} - \boldsymbol{\pi}\frac{\partial\sigma}{\partial x_\mu}\right] \tag{38.8}$$

[88] See also Ref. [Q61].

The first term on the right vanishes by the last of Eqs. (38.6). Thus one can define the matrix[89] [90]

$$\underset{\sim}{L}_\mu \equiv \frac{1}{2}\tau^a L_\mu^a = \frac{1}{2}\boldsymbol{\tau}\cdot\left[\frac{\partial\boldsymbol{\pi}}{\partial x_\mu}\times\boldsymbol{\pi}+\sigma\frac{\partial\boldsymbol{\pi}}{\partial x_\mu}-\boldsymbol{\pi}\frac{\partial\sigma}{\partial x_\mu}\right] = \frac{1}{2i}\frac{\partial\underset{\sim}{U}}{\partial x_\mu}\underset{\sim}{U}^\dagger \tag{38.9}$$

The fields (σ, π) are hermitian, hence so is the matrix $\underset{\sim}{L}_\mu$ [91]

$$\underset{\sim}{L}_\mu = \underset{\sim}{L}_\mu^\star \tag{38.10}$$

Skyrme used the vector field $\underset{\sim}{L}_\mu$ to construct a model lagrangian in the following manner. In analogy to Yang-Mills, define an equivalent *field tensor* [92]

$$\mathcal{F}_{\mu\nu}^a \equiv \frac{\partial L_\nu^a}{\partial x_\mu}-\frac{\partial L_\mu^a}{\partial x_\nu}+\varepsilon_{abc}L_\mu^b L_\nu^c = \frac{\partial L_\nu^a}{\partial x_\mu}-\frac{\partial L_\mu^a}{\partial x_\nu}+[\mathbf{L}_\mu\times\mathbf{L}_\nu]^a \tag{38.11}$$

Since the tau matrices obey the $SU(2)$ commutation rules $[\frac{1}{2}\tau^a, \frac{1}{2}\tau^b] = i\varepsilon_{abc}\frac{1}{2}\tau^c$, this result can be rewritten as

$$\underset{\sim}{\mathcal{F}}_{\mu\nu} \equiv \frac{1}{2}\tau^a\mathcal{F}_{\mu\nu}^a = \frac{\partial\underset{\sim}{L}_\nu}{\partial x_\mu}-\frac{\partial\underset{\sim}{L}_\mu}{\partial x_\nu}+\frac{1}{i}[\underset{\sim}{L}_\mu,\underset{\sim}{L}_\nu] \tag{38.12}$$

In this model, the following *Skyrme identity* holds

$$\frac{\partial\underset{\sim}{L}_\nu}{\partial x_\mu}-\frac{\partial\underset{\sim}{L}_\mu}{\partial x_\nu} = 2i[\underset{\sim}{L}_\mu,\underset{\sim}{L}_\nu] \tag{38.13}$$

This follows as an algebraic identity from Eq. (38.9); it is proven in Appendix K. Thus

$$\underset{\sim}{\mathcal{F}}_{\mu\nu} = i[\underset{\sim}{L}_\mu,\underset{\sim}{L}_\nu] \tag{38.14}$$

As a model lagrangian, Skyrme took

$$\mathcal{L}_S = -c_1 L_\mu^a L_\mu^a - c_2\mathcal{F}_{\mu\nu}^a\mathcal{F}_{\mu\nu}^a \tag{38.15}$$

The first term represents a "mass" contribution for the field and the second a "kinetic energy." Now use [see Eqs. (38.9) and (38.10)]

$$\begin{aligned} 2\mathrm{tr}\,\underset{\sim}{L}_\mu\underset{\sim}{L}_\mu &= 2\mathrm{tr}\,\underset{\sim}{L}_\mu\underset{\sim}{L}_\mu^\star = L_\mu^a L_\mu^a \\ &= \frac{1}{2}\mathrm{tr}\left(\frac{\partial\underset{\sim}{U}}{\partial x_\mu}\underset{\sim}{U}^\dagger\right)\left(\underset{\sim}{U}\frac{\partial\underset{\sim}{U}^\dagger}{\partial x_\mu}\right) = \frac{1}{2}\mathrm{tr}\left(\frac{\partial\underset{\sim}{U}}{\partial x_\mu}\frac{\partial\underset{\sim}{U}^\dagger}{\partial x_\mu}\right) \end{aligned} \tag{38.16}$$

[89]Repeated Latin indices are summed from 1 to 3.

[90]If the field $\vec{\phi}$ is a pseudoscalar, then it follows from the definitions in Eqs. (38.5) that $\vec{\pi}$ is a pseudoscalar and σ a scalar. The first term in brackets in Eq. (38.9) is thus a Lorentz vector, and the last two form an axial vector.

[91]Recall $v_\mu^\star \equiv (\vec{v}^\dagger, +iv_0^\dagger)$.

[92]Here the coupling constant has been scaled out of the field.

Also use

$$
\begin{aligned}
2\mathrm{tr}\,\mathcal{F}_{\mu\nu}\mathcal{F}_{\mu\nu} &= \mathcal{F}^a_{\mu\nu}\mathcal{F}^a_{\mu\nu} = -2\mathrm{tr}[\mathrm{L}_\mu,\mathrm{L}_\nu][\mathrm{L}_\mu,\mathrm{L}_\nu] \\
&= -\frac{1}{8}\mathrm{tr}\left[\frac{\partial \underset{\sim}{\mathrm{U}}}{\partial x_\mu}\underset{\sim}{\mathrm{U}}^\dagger, \frac{\partial \underset{\sim}{\mathrm{U}}}{\partial x_\nu}\underset{\sim}{\mathrm{U}}^\dagger\right]\left[\frac{\partial \underset{\sim}{\mathrm{U}}}{\partial x_\mu}\underset{\sim}{\mathrm{U}}^\dagger, \frac{\partial \underset{\sim}{\mathrm{U}}}{\partial x_\nu}\underset{\sim}{\mathrm{U}}^\dagger\right] \\
&= -\frac{1}{8}\mathrm{tr}\left[\underset{\sim}{\mathrm{U}}^\dagger\frac{\partial \underset{\sim}{\mathrm{U}}}{\partial x_\mu}, \underset{\sim}{\mathrm{U}}^\dagger\frac{\partial \underset{\sim}{\mathrm{U}}}{\partial x_\nu}\right]\left[\underset{\sim}{\mathrm{U}}^\dagger\frac{\partial \underset{\sim}{\mathrm{U}}}{\partial x_\mu}, \underset{\sim}{\mathrm{U}}^\dagger\frac{\partial \underset{\sim}{\mathrm{U}}}{\partial x_\nu}\right] \qquad (38.17)
\end{aligned}
$$

Here the cyclic property of the trace has been used to get the final $\underset{\sim}{\mathrm{U}}^\dagger$ to the left in the last line. A combination of these results yields

$$
\mathcal{L}_S = -\frac{c_1}{2}\mathrm{tr}\left(\frac{\partial \underset{\sim}{\mathrm{U}}}{\partial x_\mu}\frac{\partial \underset{\sim}{\mathrm{U}}^\dagger}{\partial x_\mu}\right) + \frac{c_2}{8}\mathrm{tr}\left[\underset{\sim}{\mathrm{U}}^\dagger\frac{\partial \underset{\sim}{\mathrm{U}}}{\partial x_\mu}, \underset{\sim}{\mathrm{U}}^\dagger\frac{\partial \underset{\sim}{\mathrm{U}}}{\partial x_\nu}\right]^2 \qquad (38.18)
$$

This is the *Skyrme lagrangian.* It has the following properties:

1. It is built out of the field variable $\underset{\sim}{\mathrm{U}}$, an $SU(2)$ matrix, and linear derivatives of $\underset{\sim}{\mathrm{U}}$.

2. The first term is quadratic in the field variable and the second term is quartic.

3. One *could* also have included an additional quartic term of the form

$$
c_3\left[\mathrm{tr}\left(\frac{\partial \underset{\sim}{\mathrm{U}}}{\partial x_\mu}\frac{\partial \underset{\sim}{\mathrm{U}}^\dagger}{\partial x_\mu}\right)\right]^2 \qquad (38.19)
$$

4. One could have included additional terms with higher powers of $\underset{\sim}{\mathrm{U}}$ and $\partial\underset{\sim}{\mathrm{U}}/\partial x_\mu$.

5. The theory is nonrenormalizable due to the derivative couplings.

6. Massless pions have been assumed.

7. The model is invariant under chiral transformations satisfying Eq.(38.6).[93]

8. Given the lagrangian density, one can proceed to the equations of motion and hamiltonian in the canonical fashion.

9. One can use this lagrangian as a chiral-invariant model for evaluating and correlating various processes in the soft-pion limit (Sections 21-23), a subject to which we now turn our attention.

Soft-Pion Limit ($q_\lambda \to 0$). Each derivative in the lagrangian brings in a power of the pion's momentum q_λ in the S-matrix for pion processes. The last term in Eq. (38.18) is thus of $O(q^4)$ and negligible in the soft-pion limit ($q_\lambda \to 0$). The scattering lengths for pion scattering thus come from the first term in Eq. (38.18). Insertion of Eq. (38.5) then gives

$$
-\mathcal{L}_S \doteq c_1\left[\left(\frac{\partial\sigma}{\partial x_\mu}\right)^2 + \left(\frac{\partial\boldsymbol{\pi}}{\partial x_\mu}\right)^2\right] \quad ; \qquad \sigma^2 + \boldsymbol{\pi}^2 = 1 \qquad (38.20)
$$

[93] See Prob. 38.2.

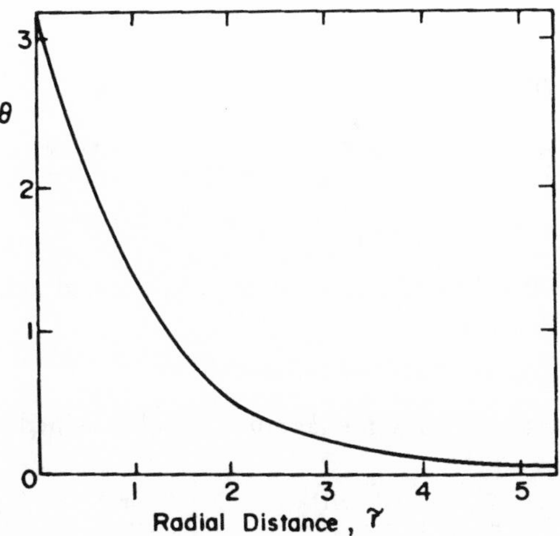

Figure 38.4: The chiral angle $\theta(\tilde{r})$ in the Skyrmion as a function of $\tilde{r} \equiv \mathcal{G}_\rho \mathcal{F}_\pi r$ that gives baryon number unity. From Ref. [Q49].

The second (chiral) constraint implies the problem is nonlinear. Solution of the second equation yields

$$\begin{aligned} \sigma &= (1 - \boldsymbol{\pi}^2)^{1/2} \\ \frac{\partial \sigma}{\partial x_\mu} &= -\boldsymbol{\pi} \cdot \frac{\partial \boldsymbol{\pi}}{\partial x_\mu} + \cdots (\text{higher powers of } \boldsymbol{\pi}^2) \end{aligned} \tag{38.21}$$

Insertion of this result then yields

$$-\mathcal{L}_{\rm S} \doteq c_1 \left[\left(\frac{\partial \boldsymbol{\pi}}{\partial x_\mu} \right)^2 + \left(\boldsymbol{\pi} \cdot \frac{\partial \boldsymbol{\pi}}{\partial x_\mu} \right)^2 + \cdots \right] \tag{38.22}$$

The first term is the pion kinetic energy, and the second gives rise to the soft-pion theorems;[94] it is evident from the form of the second term that the pions decouple in the soft-pion limit $q_\mu \to 0$.

It is convenient to rescale the pion field at this point to get the pion kinetic energy into the canonical form. Define

$$\begin{aligned} \tilde{\mathrm{U}} &\equiv \exp\{\frac{i}{\mathcal{F}_\pi} \boldsymbol{\tau} \cdot \boldsymbol{\phi}\} \equiv \frac{1}{\mathcal{F}_\pi}(\sigma + i\boldsymbol{\pi} \cdot \boldsymbol{\tau}) \\ &= \cos \frac{\phi}{\mathcal{F}_\pi} + i\boldsymbol{\tau} \cdot \mathbf{n} \sin \frac{\phi}{\mathcal{F}_\pi} \\ \sigma^2 + \boldsymbol{\pi}^2 &= \mathcal{F}_\pi^2 \end{aligned} \tag{38.23}$$

[94] Recall the pions are massless here in the chiral limit.

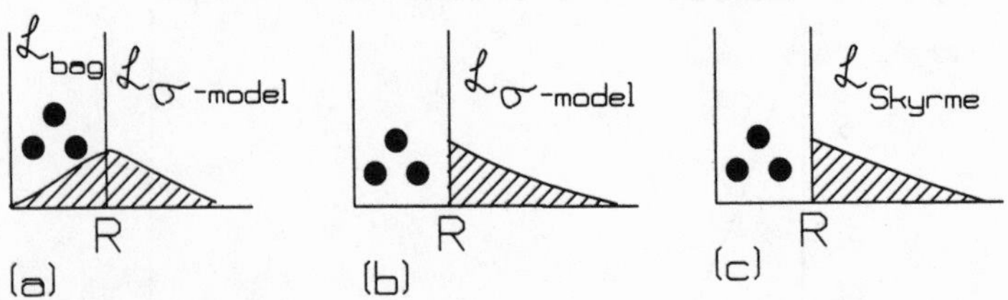

Figure 38.5: Extended bag models of hadrons: (a) cloudy bag model; (b) little bag model; (c) chiral bag.

Now take the lowest-order Skyrme lagrangian to be defined by

$$\begin{aligned} -\mathcal{L}_{\rm S}^{0} &\equiv \frac{\mathcal{F}_\pi^2}{4}{\rm tr}\left(\frac{\partial\tilde{\rm U}}{\partial x_\mu}\frac{\partial\tilde{\rm U}^\dagger}{\partial x_\mu}\right) = \frac{1}{2}\left[\left(\frac{\partial\boldsymbol{\pi}}{\partial x_\mu}\right)^2 + \left(\frac{\partial\sigma}{\partial x_\mu}\right)^2\right] \\ &= \frac{1}{2}\left[\left(\frac{\partial\boldsymbol{\pi}}{\partial x_\mu}\right)^2 + \frac{1}{\mathcal{F}_\pi^2}\left(\boldsymbol{\pi}\cdot\frac{\partial\boldsymbol{\pi}}{\partial x_\mu}\right)^2 + \cdots\right] \end{aligned} \tag{38.24}$$

This lagrangian can be used to compute the s-wave $\pi - \pi$ scattering lengths in tree approximation; the result quoted in Ref. [Q49] with an additional chiral symmetry breaking term $\delta\mathcal{L}_{\rm csb} \equiv \mathcal{F}_\pi m_\pi^2 \sigma$ (Section 22) and finite mass pions is (Prob. 38.1)

$$\begin{aligned} \frac{a_0^{(0)}}{m_\pi} &= \frac{7}{32\pi\mathcal{F}_\pi^2} \\ a_0^{(0)}|_{\rm expt} &= 0.28 \pm 0.05\,{\rm fm} \end{aligned} \tag{38.25}$$

Equations (38.25) allow a determination of the magnitude of the constant $\mathcal{F}_\pi$

$$\mathcal{F}_\pi \approx 83\,{\rm MeV} \tag{38.26}$$

Solitons. Consider the classical field equations generated from the lagrangian in Eq. (38.18). One can look for a classical *hedgehog* solution to these nonlinear

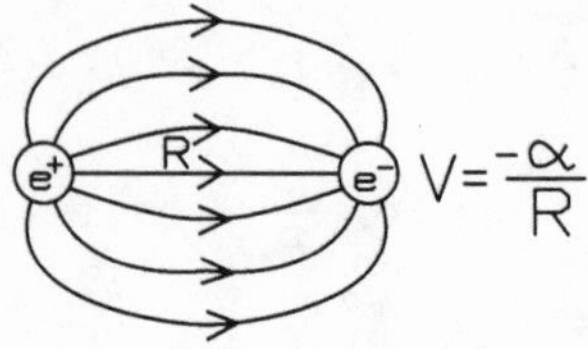

Figure 38.6: Electric field configuration and potential for two free charges in QED.

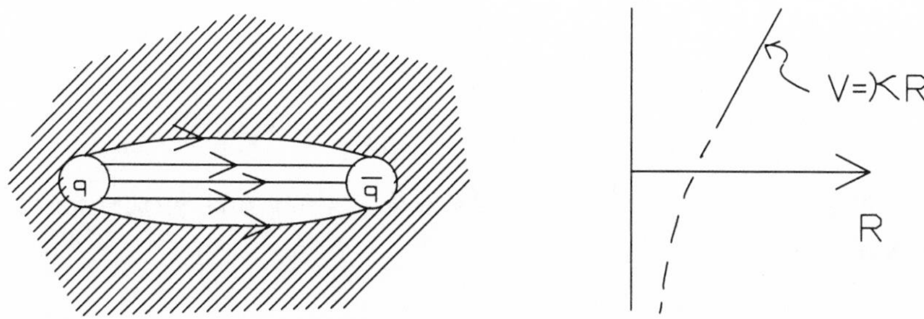

Figure 38.7: Color electric field configuration and corresponding potential at large separation for a $(q\bar{q})$ pair in lattice QCD.

field equations of the form

$$\underset{\sim}{U} = \exp\left\{ i\boldsymbol{\tau} \cdot \left(\frac{\mathbf{r}}{|\mathbf{r}|}\right) \theta(r) \right\} \tag{38.27}$$

The resulting *soliton* is referred to as a *Skyrmion*. The differential equation for $\theta(r)$ is written, and the solution is plotted in Ref. [Q49]; it is shown in Fig. 38.4.[95] Here $\theta(r)$ is a radial function satisfying the boundary conditions

$$\theta(0) = \pi \qquad\qquad \theta(\infty) = 0 \tag{38.28}$$

This problem has a topological quantum number that is strictly conserved; it can be identified with baryon number (Refs. [Q57, Q49]). The above boundary conditions on $\theta(r)$ correspond to baryon number unity.

As a more extensive introduction to this topic, a very simple soliton model of the nucleon, useful historically in estimating inelastic form factors for the excitation of nucleon resonances, is developed in Ref. [R90].

Extended Bag Models. At short distances the theory of QCD is asymptotically free. This is the basis for describing a hadron as a collection of free, or at most weakly interacting, quarks. The M.I.T. bag model emphasizes this aspect of hadron structure. At large distances, QCD is manifested in terms of mesons, and the large-distance part of the hadron consists of a meson cloud. The Skyrme model emphasizes this aspect of hadron structure in a manner that respects chiral symmetry.

There are several combinations and extensions of the QCD-inspired models discussed here that attempt to combine these two aspects of the internal structure of hadrons. We very briefly discuss three of the more successful *extended bag models* of hadrons illustrated in Fig. 38.5.

1. *Cloudy Bag.* The cloudy bag model is developed in Refs. [Q62, Q63, Q64] and illustrated in Fig. 38.5a. One assumes a bag of quarks out to a radius R. Beyond R one has meson fields described by the chiral-invariant σ-model

[95] Note $c_1 \equiv \mathcal{F}_\pi^2/2$ and $c_2 \equiv 1/4\mathcal{G}_\rho^2$. See Prob. 38.3.

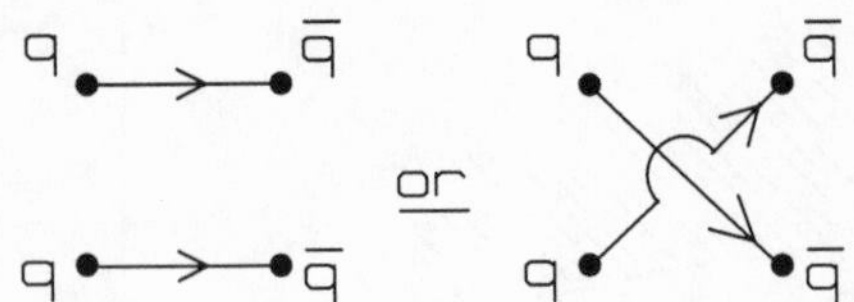

Figure 38.8: Potential model with links based on flux-tube picture for case of $(qq\bar{q}\bar{q})$ system.

(Section 22). In the cloudy bag, the meson field is extended *into the interior* of the hadron through interactions with the quarks and chiral-invariant boundary conditions imposed on the meson fields at the radius R. Thus there are meson fields both outside and inside the bag, and the axial current is conserved.

2. *Little Bag.* This model is developed in Refs. [Q65, Q66] and is illustrated in Fig. 38.5b. Here the meson fields, again described by the chiral-invariant σ-model, are restricted to the exterior of the hadron beyond R. Inside one has just quarks. Thus there are mesons outside, and quarks with only color interactions inside. The boundary condition at R is again conservation of the axial current.

3. *Chiral Bag.* This extension is developed in Refs. [Q67, Q68] and is illustrated in Fig. 38.5c. The physical picture is the same as in Fig. 38.5b; however, now the chiral-invariant Skyrme model, with its topological soliton solution, is used to describe the meson fields outside of R. Again the axial current is conserved across the boundary at R.

Flux Tube Models. The electric field configuration for two free charges in QED, and corresponding potential, are shown in Fig. 38.6. From the lattice gauge theory calculations in QCD (Sections 33 and 34), the corresponding color electric field configuration at large separation for a static $(q\bar{q})$ pair, and corresponding potential, take the form illustrated in Fig. 38.7. The physical picture here is that the vacuum acts to *exclude* the color electric flux, just as a superconductor acts to exclude magnetic flux in the Meissner effect. In QCD, the flux forms a tube between the color charges, and the potential $V = \kappa R$ then grows linearly with the length of the tube; here the string tension $\bar{\sigma} \equiv \kappa \approx 1\,\text{GeV/fm}$.

Two very useful models based on this flux tube picture exist:

1. *Potential Model with Links.* This model is developed in Ref. [Q69] and is illustrated in Fig. 38.8 for the case of a $(qq\bar{q}\bar{q})$ system. There is a potential V_l

Figure 38.9: Flux tube configuration for meson and baryon in the flux tube model.

associated with each link, and for every spatial configuration, one assigns that set of links that minimizes the total potential energy $V = \sum_l V_l$. One then solves the corresponding many-particle Schrödinger equation.

2. *Flux Tube Model.* This model is from Ref. [Q70] and is illustrated in Fig. 38.9. Here one chooses the flux tube configuration to minimize the energy and then does *dynamics* with the resulting system.

The reader is referred to Ref. [Q49] for a more extensive discussion of models of the nucleon; hopefully, with this background, that study will now prove more meaningful.

39

DEEP-INELASTIC SCATTERING

In the next two sections we discuss deep-inelastic electron scattering, where both the four-momentum transfer q^2 and energy transfer $\nu = q \cdot p/m$ become very large. It is through these experiments, initially carried out at the Stanford Linear Accelerator Center (SLAC), that the first dynamic evidence for a point-like substructure of hadrons was obtained (Refs. [Q72, Q73]). The structure functions exhibit this point-like substructure through Bjorken *scaling*, which implies $F_i(q^2, \nu) \to F_i(q^2/\nu)$ as $q^2 \to \infty$ and $\nu \to \infty$ at fixed q^2/ν. In this section we present some general considerations on electron scattering (Refs. [N14], [Q11]), summarize the deep-inelastic results (Refs. [Q11, Q72, Q73]), and introduce the quark-parton model through which the deep inelastic scaling can be understood (Refs. [Q74, Q75, Q76, Q77, Q78]). QCD then allows a calculation of the *corrections* to scaling and the evolution equations for doing this are developed in Section 40. Finally, the change of the structure functions in nuclei gives direct evidence for the modification of quark properties in the nuclear medium (EMC effect).

General Analysis. The kinematics for electron scattering employed in these two sections are shown in Fig. 39.1. Here the four-momentum transfer is defined by[96]

$$
\begin{aligned}
q &= k_2 - k_1 = p - p' \\
q^2 &= 4\varepsilon_1\varepsilon_2 \sin^2\frac{\theta}{2}
\end{aligned}
\tag{39.1}
$$

We further define

$$
\begin{aligned}
\nu &\equiv \frac{q \cdot p}{m} = \varepsilon_1 - \varepsilon_2 \\
x &\equiv \frac{q^2}{2m\nu}
\end{aligned}
\tag{39.2}
$$

These are the energy loss in the lab frame and Bjorken scaling variable, respectively.

The S-matrix for the process in Fig. 39.1 is given by

$$
S_{\mathrm{fi}} = -\frac{(2\pi)^4}{\Omega}\delta^{(4)}(k_1 + p - k_2 - p')ee_p\bar{u}(k_2)\gamma_\mu u(k_1)\frac{1}{q^2}\langle p'|J_\mu(0)|p\rangle \tag{39.3}
$$

[96] Massless electrons are assumed throughout this discussion.

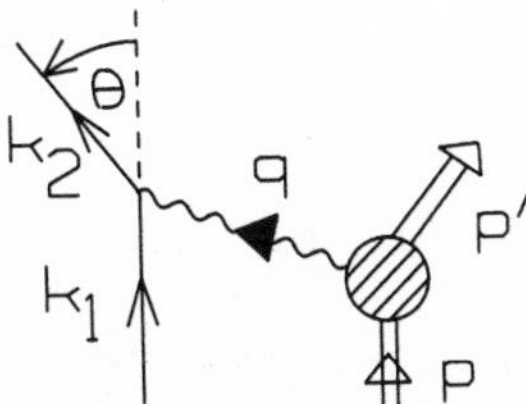

Figure 39.1: Kinematics in electron scattering; momenta are four-vectors.

Here $J_\mu(x)$ is the local electromagnetic current operator for the target system. With box normalization,[97] momentum conservation is actually expressed through the relation

$$\frac{(2\pi)^3}{\Omega}\delta^{(3)}(\mathbf{k}_1 + \mathbf{p} - \mathbf{k}_2 - \mathbf{p}') \doteq \delta_{\mathbf{k}_1+\mathbf{p},\mathbf{k}_2+\mathbf{p}'} \tag{39.4}$$

The incident flux in any frame where $\mathbf{k_1}||\mathbf{p}$ is given by

$$I_0 = \frac{1}{\Omega}\frac{\sqrt{(k_1 \cdot p)^2}}{\varepsilon_1 E_p} \tag{39.5}$$

Then for a one-body nuclear final state

$$\begin{aligned} S_{\mathrm{fi}} &\equiv -2\pi i\,\delta(\varepsilon_1 + E_p - \varepsilon_2 - E_{p'})\delta_{\mathbf{k}_1+\mathbf{p},\mathbf{k}_2+\mathbf{p}'}\bar{T}_{\mathrm{fi}} \\ d\sigma_{\mathrm{fi}} &= 2\pi|\bar{T}_{\mathrm{fi}}|^2\delta(W_f - W_i)\frac{\Omega d^3k_2}{(2\pi)^3}\left[\frac{1}{\Omega}\frac{\sqrt{(k_1\cdot p)^2}}{\varepsilon_1 E_p}\right]^{-1} \end{aligned} \tag{39.6}$$

Here $W_f = \varepsilon_2 + E_{p'}$ and $W_i = \varepsilon_1 + E_p$ are the total initial and final energies, respectively. It follows that the differential cross section in any frame where $\mathbf{k}_1||\mathbf{p}$ is given in Lorentz invariant form by (Prob. 39.1)

$$d\sigma = \frac{4\alpha^2}{q^4}\frac{d^3k_2}{2\varepsilon_2}\frac{1}{\sqrt{(k_1\cdot p)^2}}\eta_{\mu\nu}W_{\mu\nu} \tag{39.7}$$

In this expression the lepton and hadron tensors for unpolarized electrons and targets, generalized to include arbitrary nuclear final states, are defined by

$$\begin{aligned} \eta_{\mu\nu} &= -2\varepsilon_1\varepsilon_2\frac{1}{2}\sum_{s_1}\sum_{s_2}\bar{u}(k_1)\gamma_\nu u(k_2)\bar{u}(k_2)\gamma_\mu u(k_1) \\ W_{\mu\nu} &= (2\pi)^3\Omega\overline{\sum_i}\sum_f \delta^{(4)}(q + p' - p)\langle p|J_\nu(0)|p'\rangle\langle p'|J_\mu(0)|p\rangle E_p \end{aligned} \tag{39.8}$$

[97]That is, periodic boundary conditions in a big box of volume Ω.

The lepton tensor can be evaluated directly (recall the mass of the electron is neglected)

$$\begin{aligned}\eta_{\mu\nu} &= -2\varepsilon_1\varepsilon_2\frac{1}{2}\text{tr}\frac{(-ik_{1\lambda}\gamma_\lambda)}{2\varepsilon_1}\gamma_\nu\frac{(-ik_{2\rho}\gamma_\rho)}{2\varepsilon_2}\gamma_\mu \\ &= k_{1\mu}k_{2\nu} + k_{1\nu}k_{2\mu} - (k_1\cdot k_2)\delta_{\mu\nu}\end{aligned} \tag{39.9}$$

It follows from the definition in Eq. (39.8) that the lepton current is conserved

$$q_\mu\eta_{\mu\nu} = \eta_{\mu\nu}q_\nu = 0 \tag{39.10}$$

The hadron tensor depends on just the two four-vectors (q,p) and is also conserved; its general form is (Prob. 39.2)

$$\begin{aligned}W_{\mu\nu} &= W_1(q^2, q\cdot p)\left(\delta_{\mu\nu} - \frac{q_\mu q_\nu}{q^2}\right) \\ &\quad + W_2(q^2, q\cdot p)\frac{1}{m^2}\left(p_\mu - \frac{q\cdot p}{q^2}q_\mu\right)\left(p_\nu - \frac{q\cdot p}{q^2}q_\nu\right)\end{aligned} \tag{39.11}$$

The Heisenberg equations of motion are as follows:

$$\hat{O}(x) = e^{-i\hat{P}\cdot x}\hat{O}(0)e^{i\hat{P}\cdot x} \tag{39.12}$$

They can be used to exhibit the space-time dependence of a matrix element taken between eigenstates of four-momentum

$$\begin{aligned}W_{\mu\nu} &= \frac{1}{2\pi}(\Omega E)\overline{\sum_i}\sum_f\int e^{iq\cdot z}d^4z\langle p|J_\nu(z)|p'\rangle\langle p'|J_\mu(0)|p\rangle \\ &= \frac{1}{2\pi}(\Omega E)\overline{\sum_i}\int e^{iq\cdot z}d^4z\langle p|J_\nu(z)J_\mu(0)|p\rangle\end{aligned} \tag{39.13}$$

Completeness of the final set of hadronic states has been used to obtain the second line. Consider the matrix elements of the operators in the opposite order

$$\int e^{iq\cdot z}d^4z\langle p|J_\mu(0)J_\nu(z)|p\rangle \propto \sum_f(2\pi)^4\delta^{(4)}(p+q-p')\langle p|J_\mu(0)|p'\rangle\langle p'|J_\nu(0)|p\rangle \tag{39.14}$$

Here the kinematics are illustrated in Fig. 39.2

$$\begin{aligned}p+q &= p' \\ q_0 &= \varepsilon_2 - \varepsilon_1 < 0\end{aligned} \tag{39.15}$$

One cannot reach a physical state under these kinematic conditions since the nucleon is *stable*; thus the expression in Eq. (39.14) vanishes. One can subtract

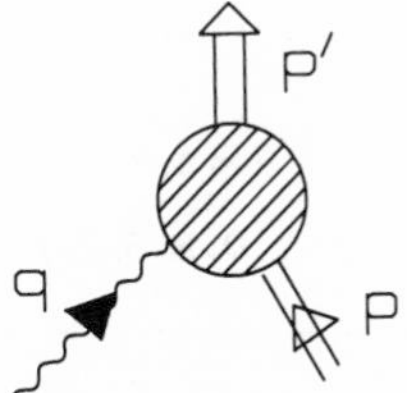

Figure 39.2: Kinematics for crossed term.

this vanishing term in Eq. (39.13) and write $W_{\mu\nu}$ as the Fourier transform of the commutator of the current density at two displaced space-time points

$$W_{\mu\nu} = \frac{1}{2\pi}(\Omega E)\overline{\sum_i}\int e^{iq\cdot z}d^4z\langle p|[J_\nu(z), J_\mu(0)]|p\rangle \tag{39.16}$$

Introduce states with *covariant norm*[98]

$$|p) \equiv \sqrt{2E\Omega}|p\rangle \tag{39.17}$$

Equation (39.13) can then be rewritten

$$-\pi W_{\mu\nu} \equiv t_{\mu\nu} = -\frac{1}{4}\overline{\sum_i}\int e^{iq\cdot z}d^4z\,(p|[J_\nu(z), J_\mu(0)]|p) \tag{39.18}$$

This expression is evidently covariant; it forms the absorptive part of the amplitude for forward, virtual Compton scattering.

A combination of Eqs. (39.7), (39.9), and (39.11) yields the general form of the laboratory cross section for the scattering of unpolarized (massless) electrons from an arbitrary, unpolarized hadronic target (Prob. 39.3)

$$\begin{aligned}\frac{d^2\sigma}{d\Omega_2 d\varepsilon_2} &= \sigma_M\frac{1}{m}\left[W_2(\nu,q^2) + 2W_1(\nu,q^2)\tan^2\frac{\theta}{2}\right] \\ \sigma_{\rm M} &= \frac{\alpha^2\cos^2\theta/2}{4\varepsilon_1^2\sin^4\theta/2}\end{aligned} \tag{39.19}$$

Here $\sigma_{\rm M}$ is the Mott cross section.

Bjorken Scaling. A qualitative overview of the SLAC data on deep-inelastic electron scattering from the proton is shown in Fig. 39.3 from Ref. [Q72]. On the basis of his analysis of various sum rules, Bjorken *predicted*, before the experiments, the following behavior of the structure functions in the deep inelastic regime (Ref. [Q73])

$$\begin{aligned}\frac{\nu}{m}W_2(\nu,q^2) &\to F_2(x)\,; \qquad q^2\to\infty, \quad \nu\to\infty \\ 2W_1(\nu,q^2) &\to F_1(x)\end{aligned} \tag{39.20}$$

[98]The norm of these states is $(\vec{p}|\vec{p}\,') = 2E(2\pi)^3\delta^{(3)}(\vec{p}-\vec{p}\,')$; this is Lorentz invariant.

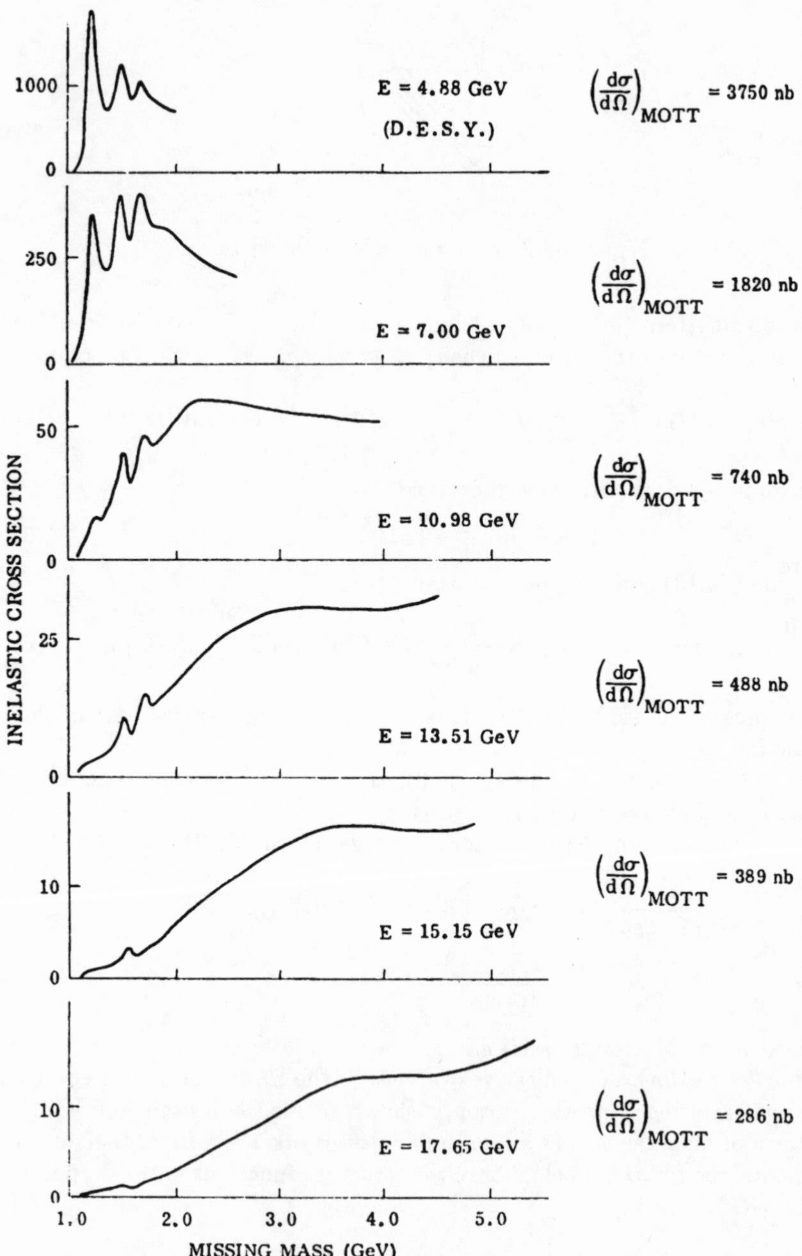

Figure 39.3: Visual fits to spectra showing the scattering of electrons from hydrogen at $\theta = 10^o$ for primary energies 4.88 to 17.65 GeV. The elastic peaks have been subtracted and radiative corrections applied. The cross sections are expressed in nanobarns/GeV/steradian. From Ref. [Q72].

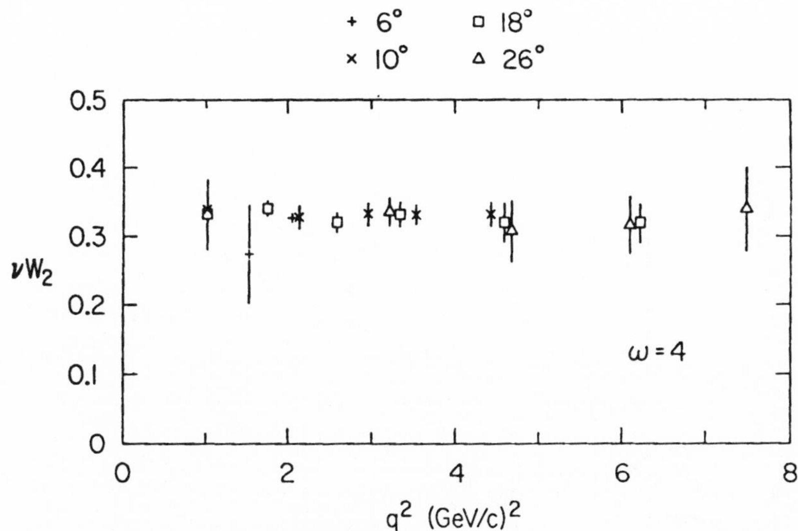

Figure 39.4: νW_2 for the proton as a function of q^2 and total C-M energy of the proton and virtual photon $W = [-(p-q)^2]^{1/2} > 2\,\mathrm{GeV}$ at $\omega = 1/x = 4$. From Ref. [Q72].

Here the scaling variable is defined by

$$x \equiv \frac{q^2}{2m\nu} \equiv \frac{1}{\omega} \tag{39.21}$$

These relations imply that the structure functions do not depend individually on (ν, q^2) but only on their *ratio*. The scaling behavior of the SLAC data is shown in Figs. 39.4 and 39.5. The first of these figures illustrates the independence from q^2 at fixed $\omega = 1/x$; the second shows the extracted structure functions $F_{1,2}(x)$.[99] [100]

Quark-Parton Model. The quark-parton model was developed by Feynman and Bjorken and Paschos to provide a framework for understanding the deep-inelastic scattering results (Refs. [Q74, Q75]). The basic idea is as follows:

1. The calculation of the structure functions is Lorentz invariant. Go to the C-M frame of the proton and incident electron with $\mathbf{p} = -\mathbf{k}_1$. Now let the proton move very fast with $|\mathbf{p}| \to \infty$. This forms the *infinite-momentum frame*; it is illustrated in Fig. 39.6.

2. The proper motion of the *parton* constituents of the hadron (proton) is slowed down by time dilation in this frame.

3. The partons are effectively *frozen* during the scattering process.

[99] These authors use $W_{1,2} \equiv (1/m)W_{1,2}^{\text{text}}$ where $W_{1,2}^{\text{text}}$ are the structure functions used here.

[100] From the SLAC data the ratio of longitudinal to transverse cross section is given by $R \equiv \sigma_l/\sigma_t = 0.18 \pm 0.10$ where $W_1/W_2 \equiv (1 + \nu^2/q^2)\sigma_t/(\sigma_t + \sigma_l)$.

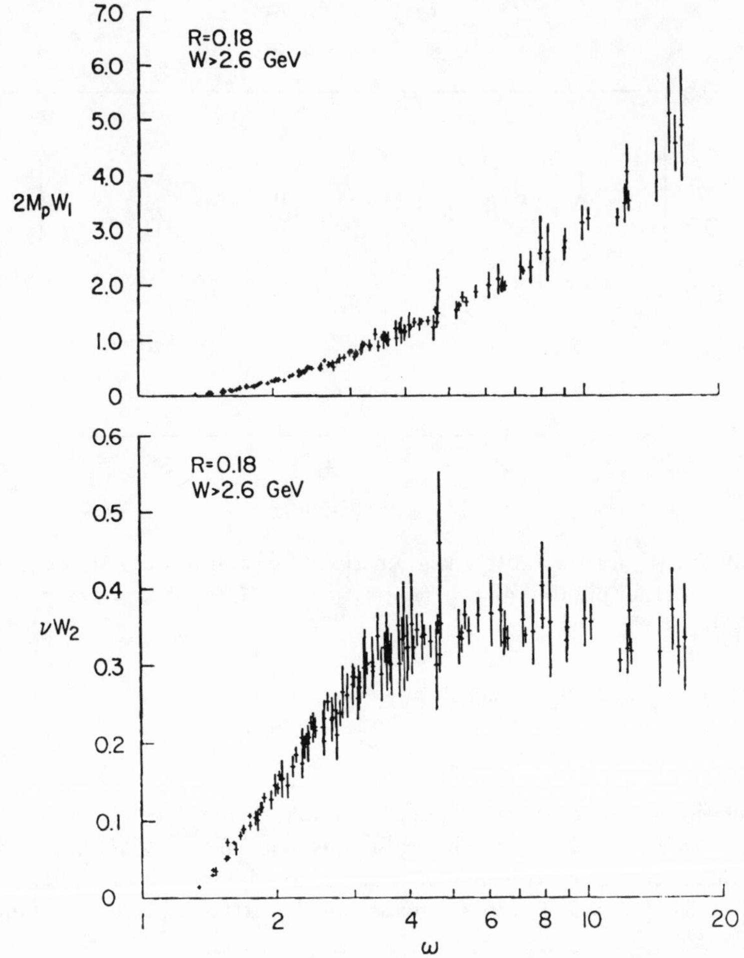

Figure 39.5: Structure functions $2mW_1$ and νW_2 for the proton vs ω for C-M energy $W > 2.6\,\text{GeV}$ and $q^2 > 1(\text{GeV/c})^2$, and using R = 0.18. From Ref. [Q72].

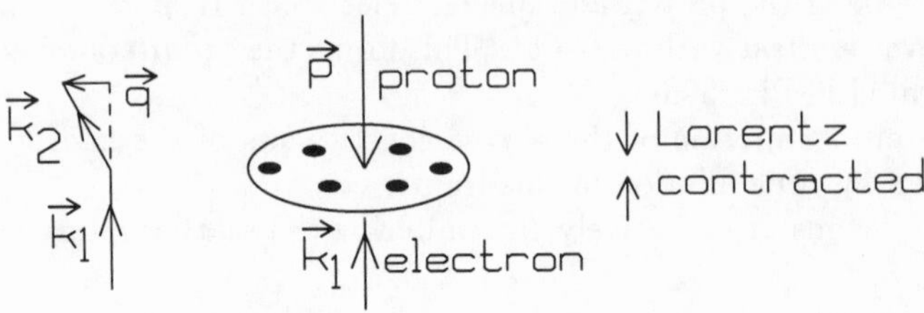

Figure 39.6: Situation in frame where the proton is moving very rapidly with momentum $\mathbf{p} = -\mathbf{k}_1$ (the infinite momentum frame).

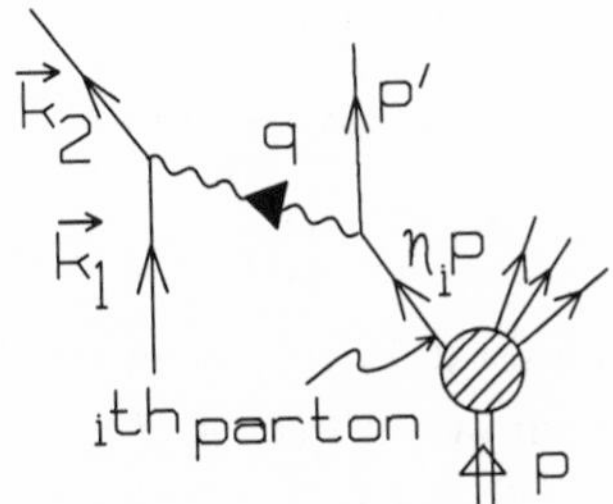

Figure 39.7: Scattering in impulse approximation in quark-parton model in infinite-momentum frame.

4. The interaction *between* the partons is then not important.
5. The electrons scatter *incoherently* from the constituents.
6. The electrons scatter from the constituents as if they are *pointlike*.
7. The parton constituents are quarks (charged) and gluons (neutral).
8. In the limit $q^2 \to \infty, \nu \to \infty$, the masses of the constituents can be neglected.[101]

Impulse Approximation. The remainder of this section is based on Ref. [Q76] (see also Refs. [Q77, Q78]). The calculation of the cross section is Lorentz invariant, and can be performed in any Lorentz frame, in particular in any frame where $\mathbf{p}||\mathbf{k}_1$. Go to the infinite-momentum frame. The scattering situation is then illustrated in Fig. 39.7. In this frame the ith parton carries the incident four-momentum

$$p_{\text{inc}} \quad = \quad \eta_i p \tag{39.22}$$

Here η_i is the fraction of the four-momentum p of the proton carried by the ith parton. Evidently

$$0 \leq \eta_i \leq 1 \tag{39.23}$$

The incident hadron is now just a collection of independent partons. The electron proceeds to scatter from one of the pointlike charged partons. We do not worry here about how the parton eventually gets converted into hadrons in the final state (*hadronization*). Only the quarks are charged with charges

$$q_i \quad \equiv \quad Q_i e_p \tag{39.24}$$

Now

> *Let $f_i(\eta_i)d\eta_i$ be the number of quarks of type i with four-momentum between $\eta_i p$ and $(\eta_i + d\eta_i)p$.*

[101] It is assumed also that the transverse momentum of the parton before the collision can be neglected in comparison with $\sqrt{q^2}$, the transverse momentum imparted as $|\vec{p}| \to \infty$.

The total four-momentum of the proton is then evidently given by

$$\begin{aligned} p &= p_{\text{gluons}} + p_{\text{quarks}} \\ p &= \zeta_g p + \sum_{i=1}^{N} \int_0^1 (\eta_i p) f_i(\eta_i) d\eta_i \end{aligned} \tag{39.25}$$

Here ζ_g is the fraction of the total four-momentum of the proton carried by all the gluons, and $\sum_{i=1}^{N}$ is a sum over all types of quarks.

Cancellation of an overall factor of the four-momentum p from the last of Eqs. (39.25) gives

$$1 = \zeta_g + \sum_{i=1}^{N} \int_0^1 \eta_i f_i(\eta_i) d\eta_i \tag{39.26}$$

Introduce a dummy variable x; this *momentum sum rule* can then be written

$$\begin{aligned} 1 &= \zeta_g + \sum_{i=1}^{N} \zeta_i \\ \zeta_i &\equiv \int_0^1 x f_i(x) dx \end{aligned} \tag{39.27}$$

Now calculate the process in Fig. 39.7 using the analysis of inelastic electron scattering presented at the beginning of this section. With the assumption of scattering from point-like Dirac quarks, the S-matrix is given by

$$\begin{aligned} S_{\mathrm{fi}}^{(i)} &= \frac{-i(2\pi)^4 e e_p Q_i}{\Omega^2 q^2} \delta^{(4)}(p' + q - \eta_i p) \bar{u}(k_2) \gamma_\mu u(k_1) \bar{u}(p') \gamma_\mu u(\eta_i p) \\ &\equiv -\frac{(2\pi)^4 i}{\Omega} \delta^{(4)}(p' + q - \eta_i p) \bar{T}_{\mathrm{fi}}^{(i)} \end{aligned} \tag{39.28}$$

The incident flux is given by

$$I_0 = \frac{1}{\Omega} \frac{\sqrt{[k_1 \cdot (\eta_i p)]^2}}{\varepsilon_1(\eta_i E_p)} = \frac{1}{\Omega} \frac{\sqrt{(k_1 \cdot p)^2}}{\varepsilon_1 E_p} \tag{39.29}$$

The cross section for inelastic electron scattering from the ith point-like quark, in the $|\mathbf{p}| \to \infty$ frame, in the impulse approximation follows as

$$\begin{aligned} d\sigma^{(i)} &= 2\pi |\bar{T}_{\mathrm{fi}}^{(i)}|^2 \delta(W_f - W_i) \frac{\Omega d^3 k_2}{(2\pi)^3} \frac{1}{I_0} \\ &= \frac{4\alpha^2}{q^4} \frac{d^3 k_2}{2\varepsilon_2} \frac{1}{\sqrt{(k_1 \cdot p)^2}} \eta_{\mu\nu} W_{\mu\nu}^{(i)} \end{aligned} \tag{39.30}$$

Here the response tensor for scattering from the ith quark is defined by

$$\begin{aligned} W_{\mu\nu}^{(i)} &= -Q_i^2 E_p \sum_{\mathbf{p}'} \frac{1}{2} \sum_{s_1} \sum_{s_2} \bar{u}(p') \gamma_\mu u(\eta_i p) \bar{u}(\eta_i p) \gamma_\nu u(p') \\ &\quad \times \delta_{\mathbf{p}', \eta_i \mathbf{p} - \mathbf{q}} \delta(p_0' - \eta_i p_0 + q_0) \end{aligned} \tag{39.31}$$

With the use of momentum conservation and the neglect of the masses of the participants, the energy-conserving delta function can be manipulated in the following manner

$$\begin{aligned}\delta(p'_0 - \eta_i p_0 + q_0) &= 2p'_0\,\delta[p'^2_0 - (\eta_i p_0 - q_0)^2] \\ &= 2p'_0\,\delta[p'^2 - (\eta_i p - q)^2] \\ &\approx 2p'_0\,\delta(2\eta_i p\cdot q - q^2) \\ &= \frac{2p'_0}{2p\cdot q}\,\delta(\eta_i - x)\end{aligned} \tag{39.32}$$

Here $x \equiv q^2/2m\nu$ is the scaling variable introduced in Eq. (39.21). Hence

$$\delta(p'_0 - \eta_i p_0 + q_0) = \frac{2E_{p'}}{2m\nu}\delta(\eta_i - x) \tag{39.33}$$

The required traces are the same as those evaluated in $\eta_{\mu\nu}$ at the beginning of this section, except that the initial momentum is $\eta_i p$. Thus

$$\begin{aligned}W^{(i)}_{\mu\nu} &= Q_i^2 E_p \frac{2E_{p'}}{2m\nu}\delta(\eta_i - x)\frac{4}{2E_{p'}2(\eta_i E_p)}\frac{1}{2}\times \\ &\quad \{p'_\mu(\eta_i p_\nu) + (\eta_i p_\mu)p'_\nu - (\eta_i p\cdot p')\delta_{\mu\nu}\} \\ &= \frac{Q_i^2}{2m\nu}\delta(\eta_i - x)\{p'_\mu p_\nu + p'_\nu p_\mu - (p\cdot p')\delta_{\mu\nu}\}\end{aligned} \tag{39.34}$$

Now use

$$\begin{aligned}p' &= \eta_i p - q \\ q_\mu \eta_{\mu\nu} &= \eta_{\mu\nu} q_\nu = 0\end{aligned} \tag{39.35}$$

Hence, again with the neglect of masses,

$$W^{(i)}_{\mu\nu} \doteq \frac{Q_i^2}{2}\delta(\eta_i - x)\left[\delta_{\mu\nu} + \frac{2\eta_i}{m\nu}p_\mu p_\nu\right] \tag{39.36}$$

The symbol $\doteq$ here indicates that the terms in q_μ and q_ν have been dropped because of Eq. (39.35).

An incoherent sum over all types of quarks now gives the response tensor for the composite nucleon

$$W_{\mu\nu} = \sum_{i=1}^{N}\int_0^1 d\eta_i f_i(\eta_i) W^{(i)}_{\mu\nu} \tag{39.37}$$

Substitution of Eq. (39.36) into Eq. (39.37) demonstrates that the response functions now explicitly *exhibit Bjorken scaling* and allows one to identify [see Eqs. (39.11), (39.20), and (39.21)]

$$F_1(x) = \sum_{i=1}^{N} Q_i^2 f_i(x) \qquad\qquad F_2(x) = \sum_{i=1}^{N} Q_i^2 x f_i(x) \tag{39.38}$$

Table 39.1: Quark sector used in discussion of deep-inelastic electron scattering from the nucleon.

	u	d	s
Q_i	2/3	−1/3	−1/3

Not only do these expressions explicitly exhibit scaling, but they also allow one to calculate the structure functions in terms of the charges of the various types of quarks and their momentum distributions as defined just below Eq. (39.24).

The Nucleon. Consider the nucleon to be made up of (u, d, s) quarks, with charges listed in Table 39.1, and their antiparticles. It then follows from Eq. (39.38) that

$$\begin{aligned}
\frac{F_2^p(x)}{x} &= \left(\frac{2}{3}\right)^2 [u^p(x) + \bar{u}^p(x)] + \left(\frac{1}{3}\right)^2 [d^p(x) + \bar{d}^p(x)] \\
&\quad + \left(\frac{1}{3}\right)^2 [s^p(x) + \bar{s}^p(x)] \\
\frac{F_2^n(x)}{x} &= \left(\frac{2}{3}\right)^2 [u^n(x) + \bar{u}^n(x)] + \left(\frac{1}{3}\right)^2 [d^n(x) + \bar{d}^n(x)] \\
&\quad + \left(\frac{1}{3}\right)^2 [s^n(x) + \bar{s}^n(x)]
\end{aligned} \tag{39.39}$$

Here an obvious notation has been introduced for the momentum distributions $f_i(x)$ of the various quark types in the proton and neutron.

Strong isospin symmetry implies that the quark distributions should be invariant under the interchange $(d \rightleftharpoons u)$ and hence $(p \rightleftharpoons n)$. Thus one defines

$$\begin{aligned}
u^p(x) = d^n(x) &\equiv u(x) \\
d^p(x) = u^n(x) &\equiv d(x) \\
s^p(x) = s^n(x) &\equiv s(x)
\end{aligned} \tag{39.40}$$

The quark contributions can be divided into two types: those from *valence* quarks, from whom the quantum numbers of the nucleon are constructed; and those from *sea* quarks, present, for example, from $(q\bar{q})$ pairs arising from strong vacuum polarization or mesons in the nucleon.

$$\begin{aligned}
u(x) &= u_V(x) + u_S(x) \\
d(x) &= d_V(x) + d_S(x) \\
s(x) &= s_V(x) + s_S(x)
\end{aligned} \tag{39.41}$$

Strong vacuum polarization should not distinguish greatly between the types of sea quarks; hence it shall be assumed for the purposes of the present arguments

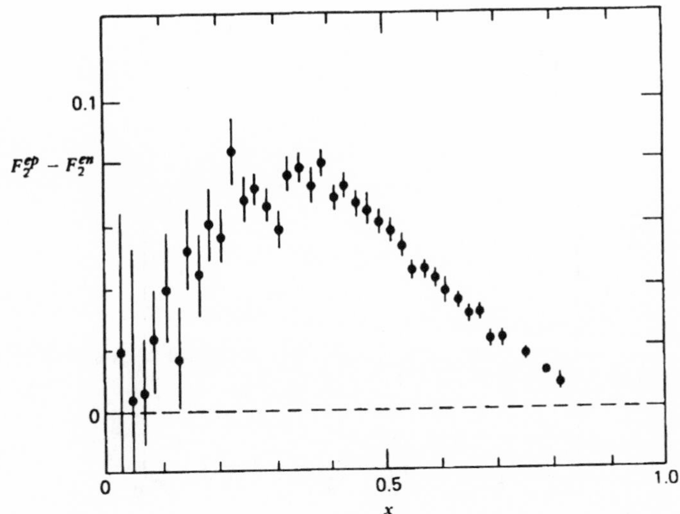

Figure 39.8: The difference $F_2^{\rm ep} - F_2^{\rm en}$ as a function of x, as measured in deep-inelastic scattering at the Stanford Linear Accelerator. From Ref. [Q77].

that the sea quark distributions are identical

$$S(x) \equiv u_{\rm S} = \bar{u}_{\rm S} = d_{\rm S} = \bar{d}_{\rm S} = s_{\rm S} = \bar{s}_{\rm S} \tag{39.42}$$

It follows that

$$\begin{aligned} \frac{F_2^p}{x} &= \frac{4}{9}u_{\rm V}(x) + \frac{1}{9}d_{\rm V}(x) + \frac{4}{3}S(x) \\ \frac{F_2^n}{x} &= \frac{1}{9}u_{\rm V}(x) + \frac{4}{9}d_{\rm V}(x) + \frac{4}{3}S(x) \end{aligned} \tag{39.43}$$

The SLAC data comparing the distributions functions $F_2^{p,n}$ is shown in Figs. 39.8 and 39.9 (taken from Refs. [Q77, Q76]). Evidently at small x the ratio $F_2^p/F_2^n \approx 1$ and the sea quark distribution $S(x)$ dominates the structure function; at large x the ratio $F_2^p/F_2^n \approx 4$ and it is the valence u quark distribution $u_{\rm V}(x)$ which dominates.

Consider the *momentum sum rule.* For simplicity, work in the nuclear domain where the nucleon is composed of (u, d) quarks and their antiquarks. The contribution of these quarks to the momentum sum rule in Eq. (39.27) takes the form

$$\begin{aligned} \zeta_u &\equiv \int_0^1 x\,dx(u+\bar{u}) \\ \zeta_d &\equiv \int_0^1 x\,dx(d+\bar{d}) \end{aligned} \tag{39.44}$$

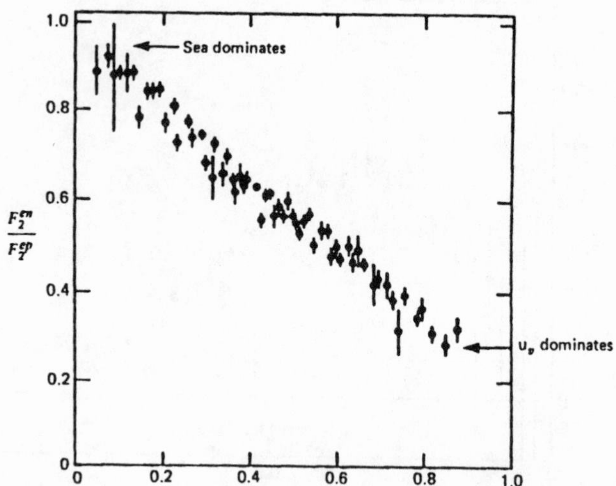

Figure 39.9: The ratio $F_2^{\text{en}}/F_2^{\text{ep}}$ as a function of x, as measured in deep-inelastic scattering. Data are from the Stanford Linear Accelerator. From Ref. [Q77].

From the SLAC results (Refs. [Q77, Q76]) one finds the sum rules

$$\begin{aligned}\int_0^1 dx F_2^p(x) &= \frac{4}{9}\zeta_u + \frac{1}{9}\zeta_d = 0.18 \\ \int_0^1 dx F_2^n(x) &= \frac{1}{9}\zeta_u + \frac{4}{9}\zeta_d = 0.12\end{aligned} \tag{39.45}$$

These results, together with Eq. (39.27), then imply

$$\begin{aligned}\zeta_u &= 0.36 \qquad\qquad \zeta_d = 0.18 \\ \zeta_g &= 0.46\end{aligned} \tag{39.46}$$

Hence one observes that the gluons carry approximately one-half of the momentum of the proton.

The EMC Effect. This material is from Refs. [Q79, Q80, Q81, Q82]. The most naive picture of the nucleus is that of a collection of free, noninteracting nucleons. In this picture the structure function one would observe from deep-inelastic electron scattering from a nucleus would be just N times the neutron structure function plus Z times that of the proton. It is an experimental fact, first established by the European Muon Collaboration (EMC), that the quark structure functions are *modified* inside the nucleus (Ref. [Q79]).

It is known that nucleons in the nucleus have a momentum distribution. The most elementary nuclear effect on the structure functions for the nucleus

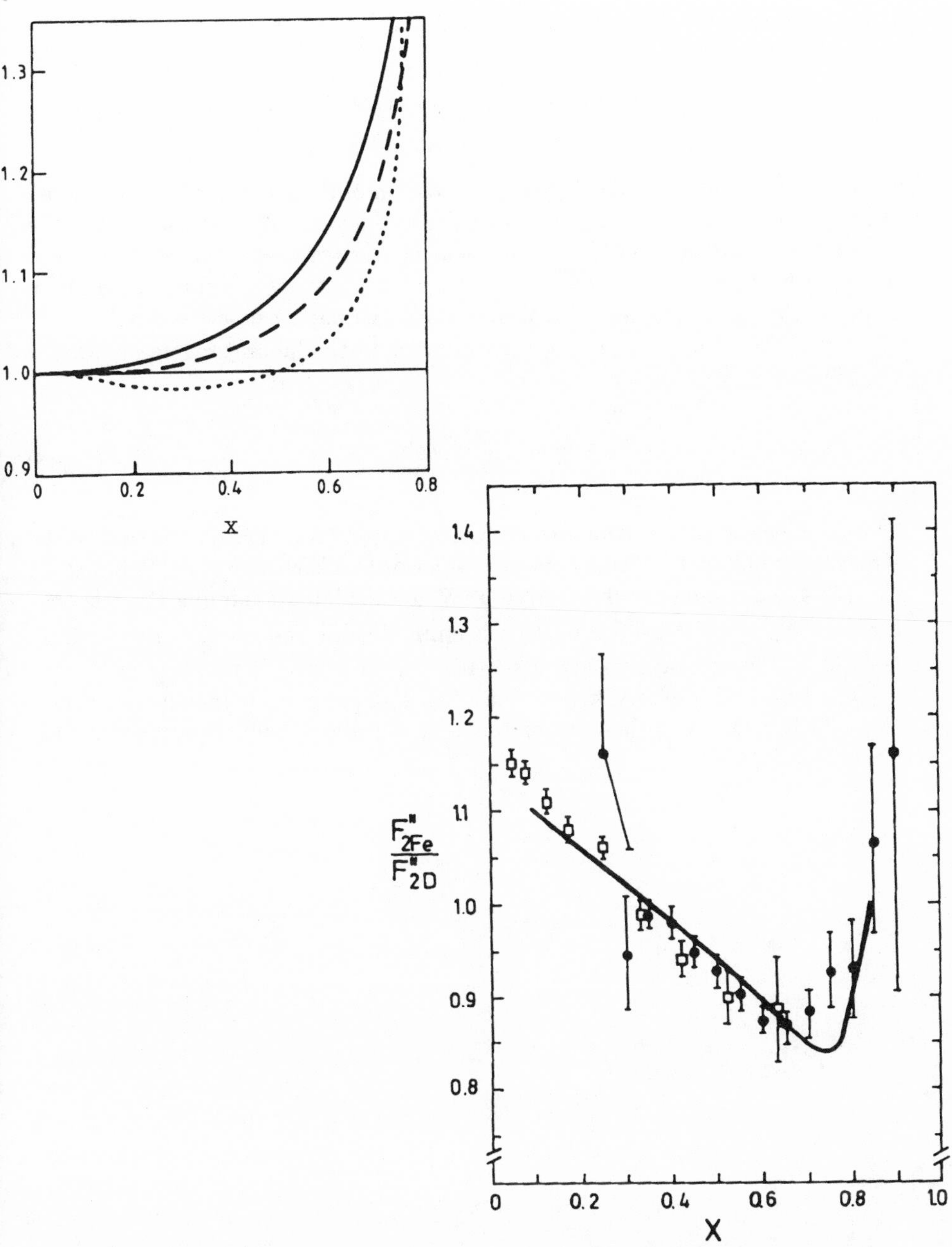

Figure 39.10: (a) A comparison of calculations of the effect of Fermi smearing on the ratio $\mathcal{R}$. From Ref. [Q80]; (b) The ratio $\mathcal{R}$ in a relativistic version of this single-particle model compared with some early experimental data. From Ref. [Q82].

A involves a simple average over the single-nucleon momentum distribution

$$W_{\mu\nu}^{(A)}(P,q) = \sum_{i=1}^{A} \int d^3p |\phi_i(\mathbf{p})|^2 W_{\mu\nu}^{(1)}(p,q) \tag{39.47}$$

We note an immediate difficulty in the extension of the theoretical analysis to an A-body nucleus; this expression is clearly model dependent in the sense that the integration is *not covariant.* It is only with a covariant description of the nuclear many-body system that one can freely transform between Lorentz frames, and, in particular, go to the $|\mathbf{p}| \to \infty$ frame where the parton model is developed.

It shall be assumed that Eq. (39.47) holds in the laboratory frame. Define the following ratio

$$\mathcal{R} \equiv \frac{F_2^{\mathrm{Fe}}(x)/A}{F_2^{\mathrm{D}}(x)/2} \tag{39.48}$$

This is the ratio of the structure function for iron (per nucleon) to that of the structure function for deuterium (per nucleon). Calculations of $\mathcal{R}$ based on Eq. (39.47) are shown in Fig. 39.10a. $\mathcal{R}$ is calculated assuming the response function $W_{\mu\nu}^{(1)}(p,q)$ for a free nucleon is unmodified in the nuclear interior (from Ref. [Q80]). Note that this Fermi smearing effect is sizable for large x.

The result of a relativistic version of this single-particle model is shown in Fig. 39.10b, along with some of the representative early experimental data (from Ref. [Q82]).

40

EVOLUTION EQUATIONS

The quark-parton model gives the structure functions in the scaling region in terms of the quark distribution functions[102]

$$F_1(x) = \frac{F_2(x)}{x} = \sum_i Q_i^2 q_i(x) \tag{40.1}$$

One cannot yet calculate the quark distribution functions from first principles. They result from the strong color interactions in the hadrons and are a consequence of strong-coupling QCD. Lattice gauge theories can, in principle, get at these distribution functions.[103]

One can, however, calculate the *evolution* of the structure functions with q^2 at high q^2 from perturbative QCD. The momentum transfer and spatial distance scale λ with which one examines the system bear an inverse relation to each other. From consideration of the Fourier transform, the relation is $|\mathbf{q}| = 2\pi/\lambda$. Suppose one examines the nucleon with higher and higher *resolution*, where the resolution will now be defined mathematically by (here q_0^2 is some initial value)

$$\tau \equiv \ln\left(\frac{q^2}{q_0^2}\right) \tag{40.2}$$

Then one expects to see finer and finer details of the substructure of the nucleon.

Consider first some kinematics in the quark-parton model. The situation in deep-inelastic electron scattering in the impulse approximation is shown in Figs. 39.6 and 39.7. The kinematics for the electron are shown again in Fig. 40.1. In the $|\mathbf{p}| \to \infty$ frame, and in the deep-inelastic region where $q^2 \to \infty$ and $\nu \to \infty$ at fixed q^2/ν, the magnitude of the three-momentum transfer is $|\mathbf{q}| \approx \sqrt{q^2}$, and as a vector it is perpendicular to $\mathbf{p}$ at small scattering angle θ (Prob. 40.1). Thus the situation at higher and higher resolution in this model is as illustrated in Fig. 40.2.

Now the evolution of the structure functions in perturbative QCD can be calculated with the *renormalization group equations* that sum the leading logarithms at large values of the space-like momentum transfer (here large τ).[104]

[102] Here $q_i(\eta_i)d\eta_i$ is the number of quarks of the ith type carrying momentum between $\eta_i p$ and $(\eta_i + d\eta_i)p$ in the proton in the infinite-momentum frame; it was called $f_i(\eta_i)d\eta_i$ in the previous section.

[103] And QCD sum rules can provide constraints on them.

[104] A discussion of this topic is contained, for example, in Refs. [Q14], [R45, R4].

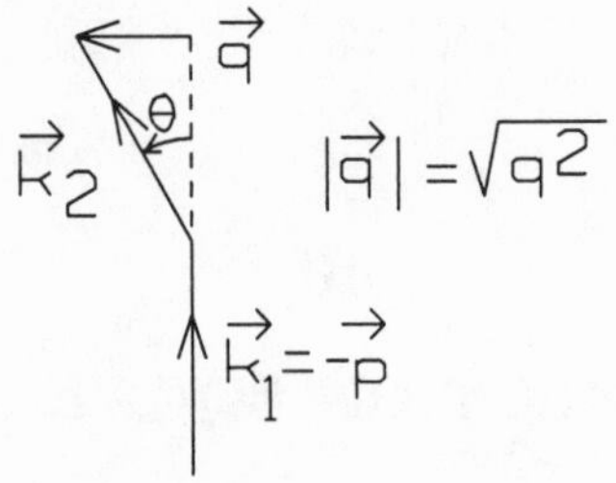

Figure 40.1: Electron variables in impulse approximation in quark-parton model for deep-inelastic scattering in $|\mathbf{p}| \to \infty$ frame.

This calculation is complicated, and involves a detailed examination of operator product expansions.

We here present, instead, a discussion of the *Altarelli-Parisi evolution equations* (Refs. [Q11, Q83]). They reproduce the renormalization group results and provide a simple, physical way of looking at renormalization group improved perturbation theory.

Evolution Equations in QED. To illustrate the basic ideas, we formulate the evolution equations within the abelian theory of QED; we then refer to the literature for the corresponding equations and results for QCD. This material is based on Refs. [Q11, Q83]; much of it is taken from Ref. [Q84].

Consider electron scattering as shown in Fig. 40.3. The first diagram is just electron scattering from a target, and the second is electron scattering with the emission of a photon, or bremsstrahlung. Now part of the time the incident electron is actually an electron plus a photon, and the photon will share some of the momentum of the incident electron. How can this be described? One can examine the effects of this compositeness by studying the top part of the diagrams in Fig. 40.3.

First introduce some definitions:

1. Let

$$\begin{aligned} z &\equiv \frac{\text{measured momentum of electron beam}}{\text{prepared momentum of electron beam}} \\ &= \text{momentum fraction carried by electron} \end{aligned} \tag{40.3}$$

2. Suppose one starts with a monochromatic electron beam. Then

$$\frac{dN}{dz} = N\delta(z-1) \tag{40.4}$$

This is the *prepared* electron momentum distribution with a monochromatic beam.[105]

[105] In this section $\int^1 dz\delta(1-z) \equiv \int^{1+} dz\delta(1-z) = 1$.

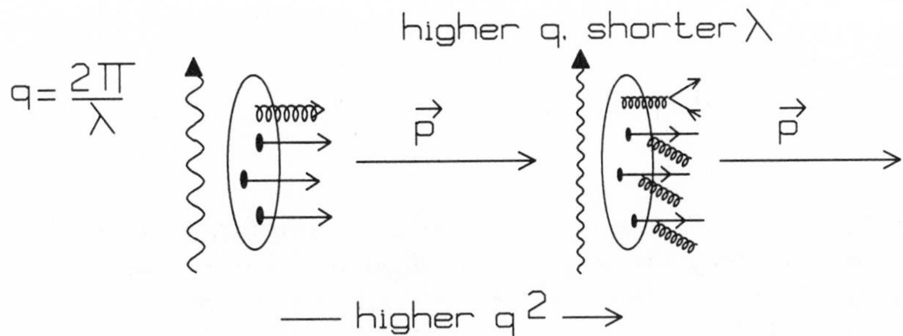

Figure 40.2: Examination of nucleon at higher and higher resolution in quark-parton model for deep-inelastic scattering in $|\mathbf{p}| \to \infty$ frame.

3. Let τ be the resolving power of the probe as defined in Eq. (40.2).

4. Let $e(z,\tau)dz$ be the number of electrons [(total number)×(probability)] with momentum fraction between z and $z+dz$ at this resolving power.

5. Let $(\alpha/2\pi)P_{e\leftarrow e}(z)d\tau dz$ be the *differential probability* of observing an electron carrying a fraction of momentum between z and $z+dz$ of the parent electron with a change in resolution $d\tau$.

We now write the *evolution* equation. Consider Fig. 40.4. With the above definitions, one can write[106]

$$de(x,\tau) \;=\; \int_0^1 [e(y,\tau)dy] \int_0^1 \left[\frac{\alpha(\tau)}{2\pi} P_{e\leftarrow e}(z) d\tau dz\right] [\delta(zy-x)] \quad (40.5)$$

The l.h.s. represents the change in probability density for electrons with momentum fraction x at a given resolution τ. The first factor on the r.h.s. represents the probability for electrons with momentum fraction y. The second factor on the r.h.s. is the differential probability that the process in Fig. 40.4 will actually yield an electron carrying momentum fraction z; recall that one *multiplies probabilities* (Section 35). The final factor guarantees that for the process in Fig. 40.4 feeding electrons from the interval at y into the interval at x on the l.h.s., one has (fraction x) = (fraction y) × (fraction z).

This result can be verified when one simply starts with an initial monochromatic beam; then from Eq. (40.4)

$$e(y,\tau) \;=\; \frac{dN}{dy} = N\delta(y-1) \quad (40.6)$$

The integrals over (y,z) can then be performed in Eq. (40.5) with the result

$$\frac{de(x,\tau)}{d\tau} \;=\; N\left[\frac{\alpha(\tau)}{2\pi} P_{e\leftarrow e}(x)\right] \quad (40.7)$$

[106]Note $de(x,\tau+d\tau) \approx de(x,\tau)$.

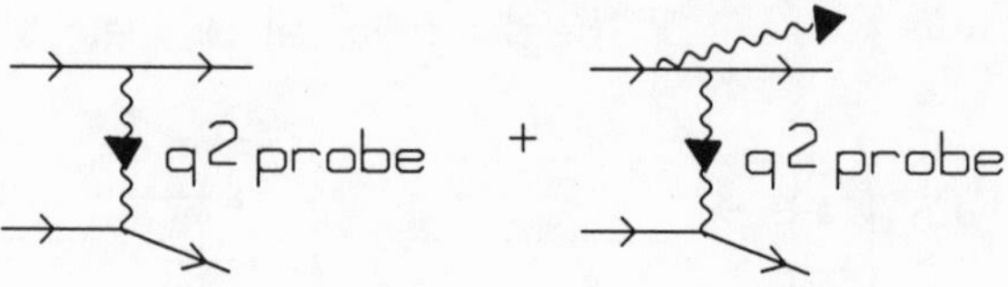

Figure 40.3: Electron scattering, and electron scattering with emission of a photon; we now study the top part of the diagrams. (Later an analogous study allows one to determine the behavior of a quark in a hadron.)

In words, this equation says that the rate of change (with respect to resolution) of the number of electrons carrying momentum fraction x is the initial number (N) times the "rate" [d(probability for fraction $x)/d\tau$] for producing electrons carrying that momentum fraction.

The presence of the δ-function on the r.h.s. of Eq. (40.5) allows one to perform the integral over $\int_0^1 dz\delta(zy-x)$ under arbitrary conditions; hence[107]

$$\frac{de(x,\tau)}{d\tau} = \int_x^1 \frac{dy}{y}\frac{\alpha(\tau)}{2\pi} P_{e\leftarrow e}\left(\frac{x}{y}\right) e(y,\tau) \tag{40.8}$$

This analysis is now *extended* in a similar fashion (Fig. 40.5). Introduce the following additional definitions:

1. Let $(\alpha/2\pi)P_{\gamma\leftarrow e}(z)d\tau dz$ be the differential probability of finding a *photon* carring a fraction z of the parent electron's momentum with an increase in resolution $d\tau$.

2. Let $\gamma(z,\tau)dz$ be the number of photons with momentum fraction between z and $z+dz$ at resolving power τ (Fig. 40.5a).

3. Let $(\alpha/2\pi)P_{e\leftarrow\gamma}d\tau dz$ be the differential probability of finding an e^- (or e^+) with momentum fraction z of the parent photon's momentum (Fig. 40.5b).

4. Let $\bar{e}(z,\tau)dz$ be the number of antiparticles (e^+) with momentum fraction between z and $z+dz$ at resolving power τ.

Master Equations. On the basis of the above example and the extended definitions, we are in a position to write the set of master equations for QED

$$\frac{de(x,\tau)}{d\tau} = \frac{\alpha(\tau)}{2\pi}\int_x^1 \frac{dy}{y}\left\{e(y,\tau)P_{e\leftarrow e}\left(\frac{x}{y}\right) + \gamma(y,\tau)P_{e\leftarrow\gamma}\left(\frac{x}{y}\right)\right\} \tag{40.9}$$

$$\frac{d\bar{e}(x,\tau)}{d\tau} = \frac{\alpha(\tau)}{2\pi}\int_x^1 \frac{dy}{y}\left\{\bar{e}(y,\tau)P_{e\leftarrow e}\left(\frac{x}{y}\right) + \gamma(y,\tau)P_{e\leftarrow\gamma}\left(\frac{x}{y}\right)\right\}$$

$$\frac{d\gamma(x,\tau)}{d\tau} = \frac{\alpha(\tau)}{2\pi}\int_x^1 \frac{dy}{y}\left\{[e(y,\tau)+\bar{e}(y,\tau)]P_{\gamma\leftarrow e}\left(\frac{x}{y}\right) + \gamma(y,\tau)P_{\gamma\leftarrow\gamma}\left(\frac{x}{y}\right)\right\}$$

[107] The lower limit on y follows from the condition $z \leq 1$.

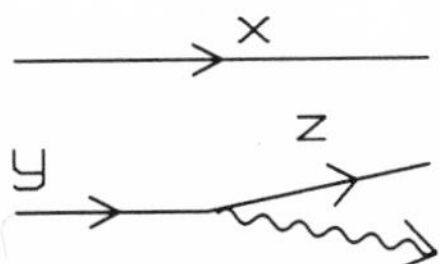

Figure 40.4: Primary electron, and electron with emitted photon. Here (x, y, z) denote the momentum fractions with $yz = x$.

Here the equality of particle and antiparticle processes feeding the various channels has been incorporated. There is no dynamic $P_{\gamma\leftarrow\gamma}$ to this order in QED (see Appendix L); there is a $P_{g\leftarrow g}$ in QCD.

These are the *Altarelli-Parisi evolution equations* for QED (Refs.[Q11, Q83]). The *splitting functions* $P_{b\leftarrow a}$ are assumed known; they are calculated from the structure of the theory, in this case QED — we demonstrate how one calculates these quantities below. The evolution equations for the distribution functions then form a set of coupled, linear, integrodifferential equations in the variables (x, τ);[108] their solution may be obtained by taking moments of the distribution functions (Prob. 40.6 and Refs. [Q11, Q83, Q84]).

The splitting functions obey various sum rules derivable from the master equations; we give examples in Appendix L.

Splitting Functions. The goal is now to calculate the splitting functions, which form the kernels in the master equations. For illustration we here concentrate on $(\alpha/2\pi)P_{\gamma\leftarrow e}(z)d\tau dz$. To do this, we relate the electron scattering process from a test target as shown in Fig. 40.6 to the corresponding *real photon* process. This will allow us to identify the *probability of finding a photon* in the field of the electron. The relationship between the processes illustrated in Fig. 40.6 is well known; it is just the *Weizsäcker-Williams approximation*, which becomes exact in the limit $\kappa^2 \to 0$. The classical basis for this approximation is described, for example, in Ref. [Q86]. The Coulomb field of a relativistic electron Lorentz contracts and becomes predominantly transverse; the electron current produces a transverse magnetic field of comparable magnitude (Fig. 40.7). This transverse field configuration is equivalent to a collection of *real* photons with a certain, specified momentum distribution.

The QED analysis here will follow Refs. [Q85], [N14]. Recall the structure of the hadronic response tensor (Fig. 40.6)

$$\begin{aligned} W_{\mu\nu} &= (2\pi)^3(\Omega E)\overline{\sum_i}\sum_f \delta^{(4)}(\kappa + p' - p)\langle p|J_\nu(0)|p'\rangle\langle p'|J_\mu(0)|p\rangle \\ &= W_1(\kappa^2, \kappa\cdot p)\left(\delta_{\mu\nu} - \frac{\kappa_\mu\kappa_\nu}{\kappa^2}\right) \\ &\quad + W_2(\kappa^2, \kappa\cdot p)\frac{1}{m^2}\left(p_\mu - \frac{p\cdot\kappa}{\kappa^2}\kappa_\mu\right)\left(p_\nu - \frac{p\cdot\kappa}{\kappa^2}\kappa_\nu\right) \end{aligned} \tag{40.10}$$

[108] Note that while the derivative is in the resolving power $\tau = \ln(q^2/q_0^2)$, the integral is only over the momentum fraction x; this is why the solution can be obtained relatively easily.

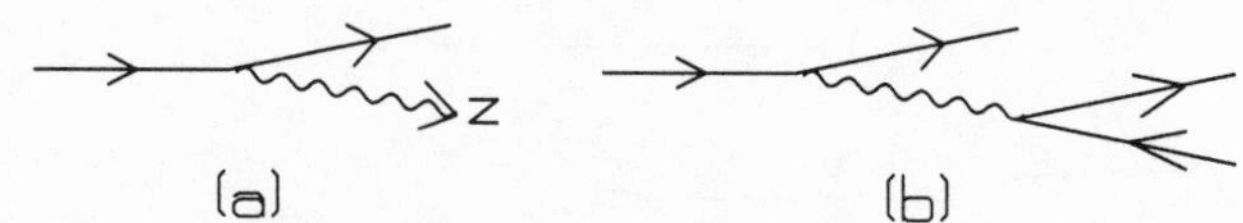

Figure 40.5: (a) Photon distribution; (b) another process feeding the electron distribution.

The (unpolarized) cross section for real photon processes follows directly from this response tensor. The relationship is derived in Refs. [Q85], [N14]; it is here left as an exercise for the reader (Prob. 40.2). The photoabsorption cross section (Fig. 40.6) is given by

$$\begin{aligned}\sigma_\gamma &= \frac{(2\pi)^2\alpha}{\sqrt{(k\cdot p)^2}}\frac{1}{2}W_{\mu\mu} \\ &= \frac{(2\pi)^2\alpha}{\sqrt{(k\cdot p)^2}}W_1(0,-k\cdot p)\end{aligned} \tag{40.11}$$

The first line follows from the covariant polarization sum, and the second from a change to incoming photon momentum (Fig. 40.6). Note that the real photon limit ($\kappa^2 \to 0$) of Eq. (40.10) is perfectly *finite*; there are no singularitites of the r.h.s. in this limit. Hence one establishes the following relations as $\kappa^2 \to 0$ (Refs. [Q85], [N14])

$$\begin{aligned}W_2(\kappa^2,\kappa\cdot p) &= O(\kappa^2)\,; \qquad\qquad \kappa^2 \to 0 \\ W_1(\kappa^2,\kappa\cdot p) &= \frac{(p\cdot\kappa)^2}{m^2\kappa^2}W_2(\kappa^2,\kappa\cdot p)\end{aligned} \tag{40.12}$$

These equations can be inverted to give for $\kappa^2 \to 0$

$$\begin{aligned}W_1 &\doteq \frac{\sqrt{(\kappa\cdot p)^2}}{(2\pi)^2\alpha}\sigma_\gamma\left(\frac{\kappa\cdot p}{m}\right) \\ W_2 &\doteq \frac{m^2\kappa^2}{(p\cdot\kappa)^2}W_1\end{aligned} \tag{40.13}$$

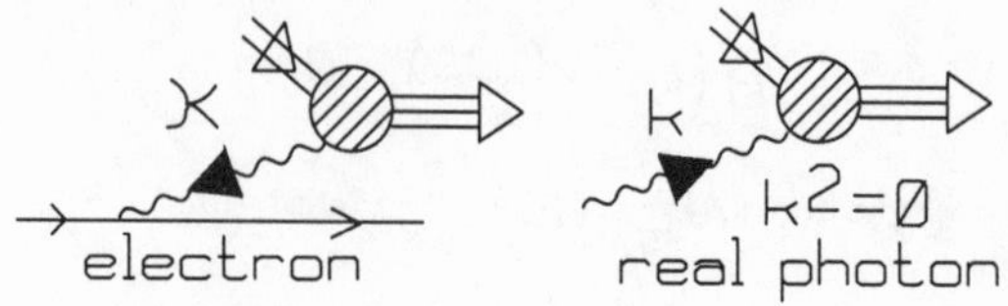

Figure 40.6: Relation of electron scattering process to real photon process.

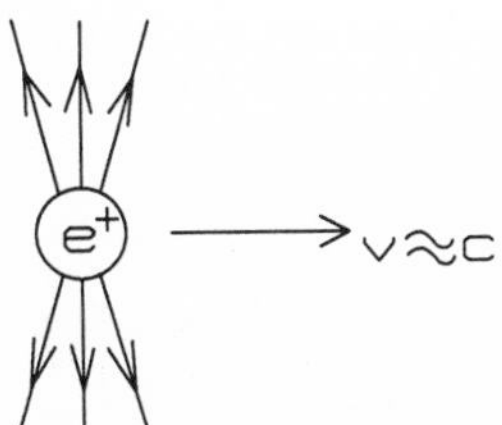

Figure 40.7: Lorentz contracted electric field of relativistic electron; basis for Weizsäcker-Williams approximation.

Electron Scattering Cross Section. The electron scattering cross section of Eqs. (39.7), (39.9), and (39.11) can be written in terms of the variables in Fig. 40.6 as

$$d\sigma_e = \frac{4\alpha^2}{\kappa^4}\frac{d\mathbf{k}_2}{2\varepsilon_2}\frac{1}{\sqrt{(k_1\cdot p)^2}}\left\{\kappa^2 W_1 + \left[\frac{2(k_1\cdot p)(k_2\cdot p)}{m^2} - \frac{1}{2}\kappa^2\right]W_2\right\} \tag{40.14}$$

The overall dependence of $1/\kappa^4$ coming from the square of the virtual photon propagator implies that in the integrated cross section, most of the contribution arises from the region where $\kappa^2 \to 0$. In this case, one can replace the structure functions by their limiting forms in Eqs. (40.13)[109]

$$\begin{aligned} d\sigma_e &\doteq \frac{4\alpha^2}{\kappa^4}\frac{d\mathbf{k}_2}{2\varepsilon_2}\frac{\sqrt{(\kappa\cdot p)^2}}{\sqrt{(k_1\cdot p)^2}}\frac{1}{(2\pi)^2\alpha}\sigma_\gamma\left(\frac{\kappa\cdot p}{m}\right) \\ &\quad\times\left\{\kappa^2 + \frac{m^2\kappa^2}{(p\cdot\kappa)^2}\left[\frac{2(k_1\cdot p)(k_2\cdot p)}{m^2} - \frac{1}{2}\kappa^2\right]\right\} \end{aligned} \tag{40.15}$$

This expression is Lorentz invariant. It is exact in the limit $\kappa^2 \to 0$ (Ref. [Q85]); at finite, but small κ^2, it forms the Weizsäcker-Williams approximation.

Now use some kinematics. From Fig. 40.6 one has in the lab frame

$$\begin{aligned} \kappa\cdot p &= m(\varepsilon_1 - \varepsilon_2) \equiv m\omega \\ k_1\cdot p &= -m\varepsilon_1 \end{aligned} \tag{40.16}$$

Also, the expression in brackets in Eq. (40.15) can be rewritten as

$$\{\cdots\} = \kappa^2 + \frac{4\varepsilon_1\varepsilon_2\sin^2\theta/2}{\omega^2}\left[2\varepsilon_1\varepsilon_2 - 2\varepsilon_1\varepsilon_2\sin^2\frac{\theta}{2}\right]$$

[109] Here the symbol $\doteq$ implies an approximate relation that is exact in the limit $\kappa^2 \to 0$. We will identify the splitting function from the coefficient of $d\kappa^2/\kappa^2$ as $\kappa^2 \to 0$. It is then used to generate the asymptotic $\ln q^2$ behavior as $q^2 \to \infty$. One can do this since $\int_{q_0^2}^{q^2} d\kappa^2/\kappa^2 = \ln(q^2/q_0^2)$ gives both the $q_0^2 \to 0$ and the $q^2 \to \infty$ behavior! In the $|\vec{p}| \to \infty$ frame with $\theta \to 0$, $\kappa^2/\vec{p}^2$ indeed provides a small parameter and one can first take $|\vec{p}|$ much larger than any other momentum of interest in the problem (Ref. [Q11]).

$$= \kappa^2 + \frac{8\varepsilon_1^2\varepsilon_2^2 \sin^2\theta/2\cos^2\theta/2}{\omega^2} = \kappa^2 + \frac{2\varepsilon_1^2\varepsilon_2^2\sin^2\theta}{\omega^2} \tag{40.17}$$

Hence the result in Eq. (40.15) becomes

$$d\sigma_e \doteq \frac{8\alpha^2}{\kappa^4}\frac{d\mathbf{k}_2}{2\varepsilon_2}\frac{\omega}{\varepsilon_1}\frac{1}{(2\pi)^2\alpha}\sigma_\gamma(\omega)\left[\frac{\varepsilon_1^2\varepsilon_2^2\sin^2\theta}{\omega^2} + \frac{1}{2}\kappa^2\right] \tag{40.18}$$

Now change variables using

$$\begin{aligned} \omega &= \varepsilon_1 - \varepsilon_2 \\ \kappa^2 &= 2\varepsilon_1\varepsilon_2(1-\cos\theta) \end{aligned} \tag{40.19}$$

Hence (after an immediate integration over $d\phi$)

$$\frac{d^3k_2}{2\varepsilon_2} = \frac{\varepsilon_2\varepsilon_2 d\omega}{2\varepsilon_2}2\pi\frac{d\kappa^2}{2\varepsilon_1\varepsilon_2} = \frac{\pi}{2\varepsilon_1}d\omega d\kappa^2 \tag{40.20}$$

The limit $\kappa^2 \to 0$ is achieved at finite ε_2 by going to small angles where $\theta \to 0$. In this case one has

$$\varepsilon_1^2\varepsilon_2^2\sin^2\theta \doteq \varepsilon_1^2\varepsilon_2^2\theta^2 \doteq \kappa^2\varepsilon_1\varepsilon_2 \tag{40.21}$$

Hence

$$d\sigma_e \doteq \frac{4\pi\alpha}{\kappa^2}\frac{d\omega}{\varepsilon_1}\frac{\omega}{\varepsilon_1}\frac{1}{(2\pi)^2}\left[\frac{\varepsilon_1\varepsilon_2}{\omega^2} + \frac{1}{2}\right]d\kappa^2\sigma_\gamma(\omega) \tag{40.22}$$

Now introduce the *momentum fraction*

$$\frac{\omega}{\varepsilon_1} \equiv z \qquad\qquad \frac{\varepsilon_2}{\varepsilon_1} = 1 - z \tag{40.23}$$

Also introduce the differential of the *resolution*

$$d\tau = d\ln\left(\frac{\kappa^2}{\kappa_0^2}\right) = \frac{d\kappa^2}{\kappa^2} \tag{40.24}$$

The electron scattering cross section in Eq. (40.22) can then be rewritten as

$$d\sigma_e \doteq \frac{\alpha}{2\pi}d\tau\, z\, dz\left[\frac{2(1-z)}{z^2} + 1\right]\sigma_\gamma(z) \tag{40.25}$$

We are now in a position to interpret this result in terms of previously introduced quantities. The contribution to the electron scattering cross section for a beam of N electrons from the accompanying photon field can be written as the following product: [number of photons $d\gamma(z,\tau)dz$ viewed with resolution between τ and $\tau + d\tau$ carrying a momentum fraction between z and $z + dz$ of

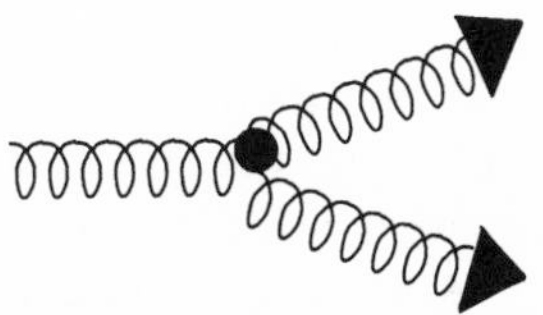

Figure 40.8: Additional splitting function for a gluon going into two gluons also present to first order in the coupling constant in QCD.

the beam]× (photoabsorption cross section at that z). The first factor can in turn be related to the probability that at that τ, a photon carrying momentum fraction z will be produced by an electron through the analog of Eq. (40.7). Hence

$$\begin{aligned} N d\sigma_e &\equiv [d\gamma(z,\tau)]\sigma_\gamma(z)dz \\ &\equiv \left[N\frac{\alpha}{2\pi}P_{\gamma\leftarrow e}(z)d\tau\right]\sigma_\gamma(z)dz \end{aligned} \tag{40.26}$$

One is now in a position to identify the *splitting function* through a comparison of Eqs. (40.25) and (40.26)

$$P_{\gamma\leftarrow e}(z) = z\left[\frac{2(1-z)}{z^2}+1\right] = \frac{1}{z}[(z-1)^2+1] \tag{40.27}$$

Note that the splitting function as calculated here is independent of τ. For the other splitting functions in QED, see Probs. 40.3-5.

QCD—Altarelli-Parisi Equations. The application to QCD follows the same basic ideas discussed here. In QCD, to lowest order in the coupling constant, one must also include an additional dynamic splitting function $P_{g\leftarrow g}(z)$ for a gluon going into two gluons as illustrated in Fig. 40.8 (Prob. 40.7). The master equations for QCD, and the required splitting functions are written down, for example, in Ref. [Q11], pp. 229-242, and their solution is discussed there. For solution, one takes moments of the evolution equations (Prob. 40.6). This procedure sums leading logarithms and leads to results of renormalization group improved perturbation theory in the QCD description of the approach to scaling in deep-inelastic electron scattering.

To give the reader a feeling for some of the applications, we are here content to simply quote two results:

1. From the asymptotic solution for the $n = 2$ moments one can compute the *momentum fraction of the gluons*

$$\zeta_g = \frac{16}{16+3N_f} \tag{40.28}$$

Here N_f is the number of flavors of quarks. for $N_f = 6$ this gives

$$\begin{aligned} \zeta_g &= 0.47\,; && \text{theory} \\ \zeta_g &= 0.46\,; && \text{experiment} \end{aligned} \tag{40.29}$$

The experimental result is from Eq. (39.46).

2. One can analyze the *evolution with increasing resolution* τ *of differences of quark distributions* $q_i(x)$. In particular, the "nonsinglet distribution" receives no contribution from the gluons (here nonsinglet refers to flavor). The nth moment of the nonsinglet (NS) quark distribution is predicted to evolve as

$$\begin{aligned} q_{\rm NS}^n(\tau) &= q_{\rm NS}^n(\tau_0)\left[\ln\left(\frac{q^2}{\Lambda^2}\right)\right]^{2A_n/\beta_0} \\ \beta_0 &\equiv 11-\frac{2N_f}{3} \end{aligned} \tag{40.30}$$

The numerical coefficients A_n for $n = (1,2,3,\ldots)$ are predicted to be $A_n = (0,-16/9,-25/9,\cdots)$.

There are very clear presentations of this subject in Refs. [Q83, Q11, Q84]; the reader is referred to the literature for further detailed developments.

41

CEBAF'S ROLE

This part of the book has been concerned with strong-coupling QCD. In this last section of Part III we return to the questions raised, and the projected role CEBAF can play, in understanding nuclear physics from this perspective.

Electron scattering from discrete states measures (Section 7 and 39; Ref. [N14])

$$w_1(q^2) = \frac{1}{2}\sum_{\lambda=\pm 1}\overline{\sum_i}\sum_f \left(\frac{EE'}{M_T^2}\right)\left|\langle f|\int e^{-i\mathbf{q}\cdot\mathbf{x}}\mathbf{e}_{\mathbf{q},\lambda}^{\dagger}\cdot\hat{\mathbf{J}}(\mathbf{x})\,d^3x|i\rangle\right|^2$$

$$w_2(q^2) = \frac{q^2}{\mathbf{q}^2}w_1(q^2) + \frac{q^4}{\mathbf{q}^4}\overline{\sum_i}\sum_f\left(\frac{EE'}{M_T^2}\right)\left|\langle f|\int e^{-i\mathbf{q}\cdot\mathbf{x}}\hat{\rho}(\mathbf{x})\,d^3x|i\rangle\right|^2 \quad (41.1)$$

Evidently one determines the Fourier transform of the transition matrix elements of the charge and current densities.

With coincidence experiments (Fig. 26.1), a wider range of phenomena and information becomes available (Section 26, Ref. [N14]).[110]

Static Electromagnetic Properties of Baryons. Baryons and mesons are composite objects with quantum numbers determined from the properties of the valence quarks and held together by the strong, nonlinear forces of QCD. A primary goal of lattice gauge theory is to calculate the static properties of these hadrons. *Experimental measurement of the static electromagnetic properties of the baryon will continue to provide benchmark tests of QCD.*[111]

On a more elementary level, recall, for example, that the mean-square charge radius of the nucleon in the QCD-inspired M.I.T. bag model is (Prob. 37.3)

$$e_p\langle r^2\rangle_{\text{ch}} = \sum_{i=1}^{3} e_i\langle r_i^2\rangle$$

$$(\langle r^2\rangle_p)^{1/2} = 0.73\,R \qquad\qquad (\langle r^2\rangle_n)^{1/2} = 0 \quad (41.2)$$

The experimental values of the latter two quantities are $(\langle r^2\rangle_p)^{1/2} \approx 0.8\,\text{fm}$ and $\langle r^2\rangle_n \approx -(0.119\,\text{fm})^2$, respectively. Clearly such comparisons teach us a great deal about the internal structure of the nucleon.

[110] Note the electron scattering convention used so far in this book: in (e,e') the momentum transfer is denoted by q, while in $(e,e'X)$ it is denoted by k and q is used for the momentum of the additional detected particle X in the C-M system.

[111] See, for example, Ref. [Q87].

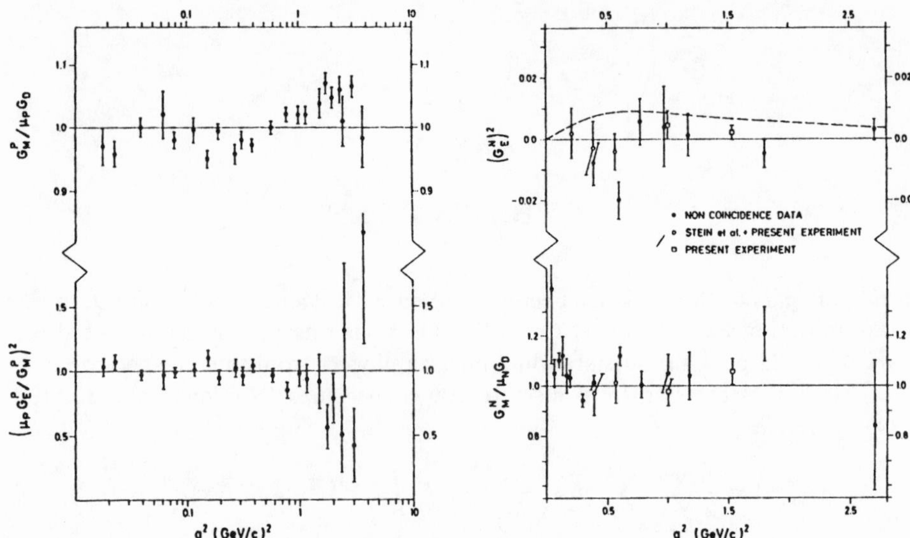

Figure 41.1: Compilation of the form factors of the proton. $(\mu_p G_E^p / G_M^p)^2$ and $G_M^p/\mu_p G_D$ are plotted versus q^2 on a logarithmic scale; also $(G_E^n)^2$ [the dashed curve is $G_E^n(q^2) = -\tau G_M^n(q^2)$] and $G_M^n/\mu_n G_D$. From Ref. [Q88].

One goal is to improve our knowledge of the electromagnetic form factors of the nucleon at all q^2. These determine the entire spatial distribution of charge and magnetization in the ground state; the above constitutes just one particular moment of the charge distribution. The current state of our knowledge of the static nucleon form factors (Section 8) is shown in Fig. 41.1. Here $\tau \equiv q^2/4m^2$ and

$$\begin{aligned} G_M(q^2) &\equiv F_1 + 2mF_2 \\ G_E(q^2) &\equiv F_1 - \frac{q^2 F_2}{2m} \end{aligned} \tag{41.3}$$

The dipole form factor, to which the measured values are compared, is defined by

$$G_D \equiv \frac{1}{(1 + q^2/0.71\,\mathrm{GeV}^2)^2} \tag{41.4}$$

One wants to significantly improve the knowledge of these form factors, particularly at higher q^2 where the separated values are only poorly known. It is the high q^2 behavior that determines the properties of the charge and magnetization distributions on a short distance scale.[112] Note especially that

[112]Reference [Q89] contains more recent high q^2 data.

the electric form factor of the neutron G_E^n, apart from its slope at $q^2 = 0$ determined from the scattering of slow neutrons by atomic electrons, is really not known.

Dynamic Electromagnetic Properties of Baryons. The internal dynamic motion of the baryons teaches us a great deal more about the implications of QCD in the strong-coupling nuclear domain. *Precision measurement of electromagnetic transition matrix elements for the nucleon will continue to provide benchmark tests for quark models and QCD.*

Consider, for example, transitions to the first excited state of the nucleon, the $\Delta(1232)$, as illustrated in Fig. 37.6 (Sections 26 and 37). We have seen that in the QCD-inspired M.I.T. bag model the transition magnetic dipole moment to this state is given by (Section 37)

$$\mu^* = \frac{4\mu_p}{3\sqrt{2}} \tag{41.5}$$

This expression describes the experimental value obtained from photoproduction to the order of 30% (Ref. [Q54]). One wants to determine the transition dipole moment $\mu^*(q^2)$ as a function of q^2 to again map out the detailed spatial distribution of the transition currents in the nucleon.

There are also transition charge and electric quadrupole moments for this process (Fig. 37.6). The quadrupole transition moment $Q^*(q^2)$ is particularly interesting because there are theoretical reasons for expecting the nucleon bag to be *deformed* by the interaction between the valence quarks, just as the presence of the tensor force from one pion exchange produces a deformation of the deuteron, or the presence of valence nucleons can deform the nucleus. A $J^\pi = 1/2^+$ object can have no static quadrupole moment, and, therefore, the first place a deformed bag would show up directly is in this transition quadrupole amplitude.

One also wants to examine the properties of other excited states of the baryon; there is a very rich structure here as indicated by the overview of some of the observed low-lying states in Fig. 41.2. The excitation dynamics of the nucleon, itself truly a complex many-body system, provides a fertile testing ground for the implications of strong-coupling QCD and various QCD-inspired quark models (Ref. [Q90]). Electron excitation of these resonances through $N(e,e')N^*$, and coincidence studies of the distribution of the various decay products of the resonances through $N(e,e'X)$ is the most powerful method we have for examining the internal dynamics of the nucleon.

There are complications in this study that will just have to be dealt with: these are mostly very broad, overlapping resonances (just as are the giant resonances in nuclei), and there are direct reaction backrounds present in addition to resonance formation (just as there are direct reactions, in addition to resonance or compound nucleus formation, in nuclei).

Particularly interesting is the strange baryon sector consisting of the hyperons Y and their resonances in Fig. 41.2; these states can be accessed in coin-

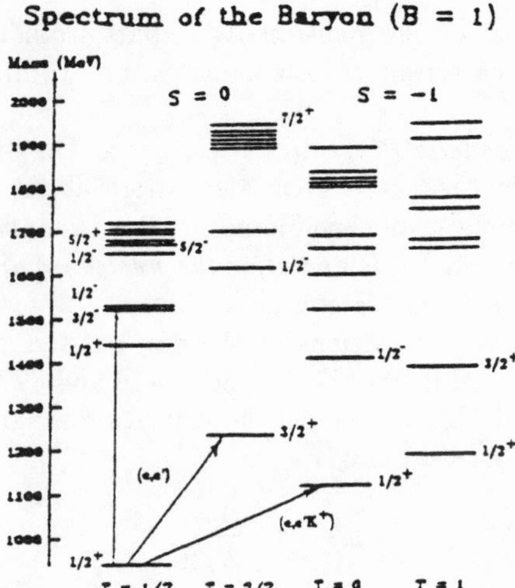

Figure 41.2: Some of the observed low-lying excitations of the baryon. From Ref. [Q92] .

cident electron scattering, and their properties studied, through the reaction $N(e,e'K^+)Y^*$.

Modification of Baryon Properties in the Nuclear Medium. Two extreme possible configurations of the quarks in the nucleon are illustrated in Fig. 41.3. At very small separations (Fig. 41.3a) one has weakly interacting quarks (asymptotic freedom) surrounded by nonlinear gluon interactions, with longer range interactions arising from meson fields. This is the basis for the various bag models that emphasize these aspects of nucleon structure. At large interquark separation (Fig. 41.3b), as indicated by lattice gauge theory calculations, one has tubes of color flux extending out to the quarks and keeping them in the nucleon (confinement).

Clearly, if a nucleon is immersed in the nuclear medium, its properties will be modified to some extent by the surrounding hadronic matter. One of the goals is to study this modification, and through it, the strong-interaction, confinement aspects of QCD. It is most likely that it will be a combination of the

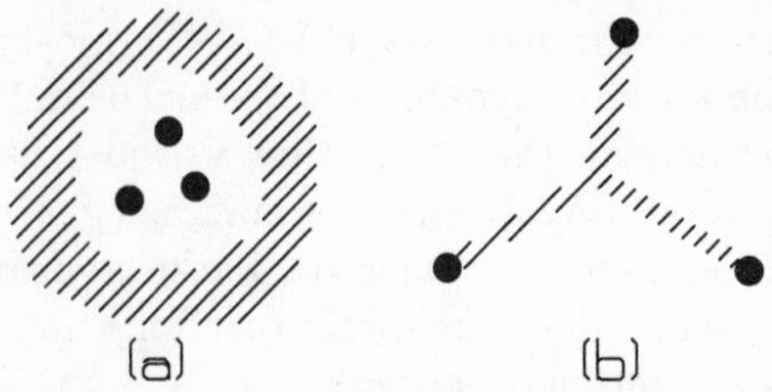

Figure 41.3: Two possible extreme configurations of quarks in the nucleon: (a) small interquark separation; (b) large interquark separation.

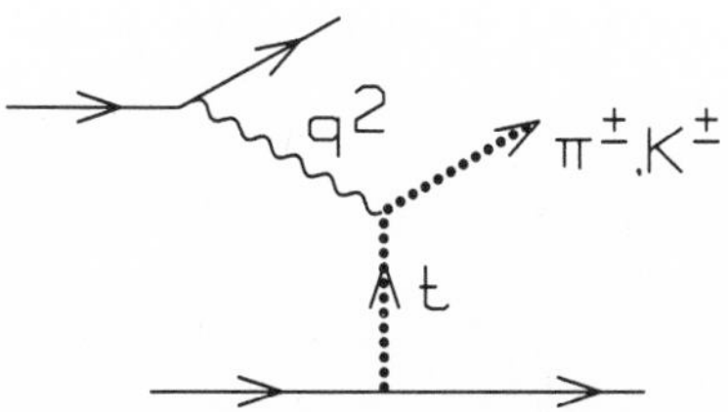

Figure 41.4: Meson exchange graph that dominates the production amplitude at the meson pole where $t = -m_\pi^2$ (or $-m_K^2$). At the pole, it is a real meson that is exchanged, and one can measure the meson form factor at the given q^2.

systematics of many precision electromagnetic measurements that will reveal these modifications of the properties of the nucleon in the nuclear interior.

Pion (and Kaon) Electromagnetic Form Factors. Mesons are themselves composite objects in QCD, with their own internal electromagnetic properties. One can determine the internal electromagnetic structure of the pion (as well as, in principle, the kaon) by extrapolating the meson production amplitude to the meson pole where the meson exchange graph illustrated in Fig. 41.4 dominates the process. After extrapolation to the pole where $t = -m_\pi^2$, it is a real meson that then dominates the amplitude; on the pole one can thus measure the form factor $F_\pi(q^2)$ of a real pion. Of course one cannot actually make a measurement *at* the pole, and it is always necessary to make some extrapolation of the physical measurements, or to extract $F_\pi(q^2)$ from measurements where pion exchange merely dominates the amplitude. At small q^2 one can determine $F_\pi(q^2)$ from the scattering of real, very high energy pions from atomic electrons. In the time-like region, one can measure $F_\pi(q^2)$ directly through the reaction $e^+ + e^- \rightarrow \pi^+ + \pi^-$. The present knowledge of the form factor of the pion from such electromagnetic measurements is shown in Fig. 41.5.

Electroproduction of Mesons in Nuclei. Production of mesons through the reaction $A(e, e'M)A^*$ will allow one to study both the internal structure of these QCD composite objects as well as their modification in the nuclear medium.

Short-Range Correlations in the Nuclear Medium. It should be possible to find an appropriate region in the phase space of the $(e, e'2N)$ triple coincidence reaction where one is sensitive to the high-momentum components induced by short-range nucleon-nucleon correlations in the nuclear medium. Clearly, one will have very interesting QCD effects arising from very short-range correlations where the nucleons actually overlap; this is indicated schematically in Fig. 41.6a.

Strange Quark in the Nuclear Medium – Flavor Physics. One can insert a strange s quark into the nuclear medium through the $(e, e'K^+)$ reaction, which carries off the corresponding $\bar{s}$-quark in associated production of strangeness. Through this reaction, one can access the new dimension of nuclear physics provided by nonzero strangeness, here $S = -1$ (Fig. 41.6b). More exotic reactions at higher energy, allow one, in principle, to access other sectors of flavor physics.

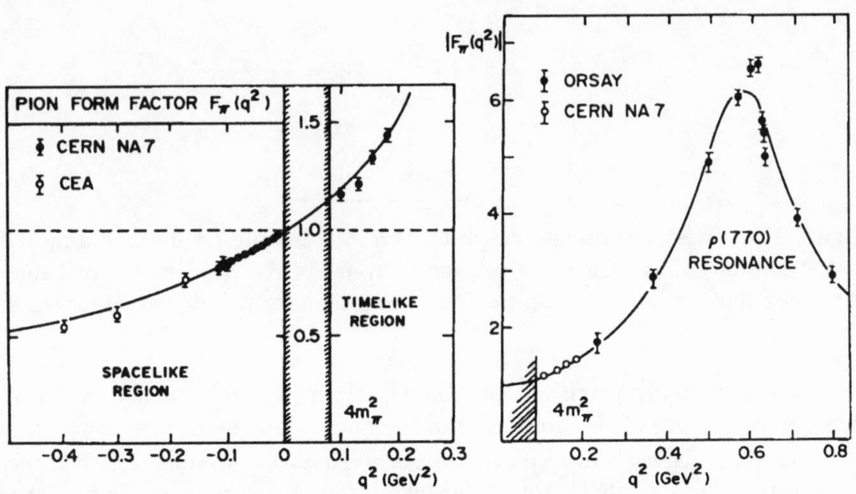

Figure 41.5: The pion form factor $F_\pi(q^2)$ for space- and time-like q^2 (in metric of Refs. [R2, R3]). From Refs. [Q49, Q93].

Possibility of New Phenomena in Nuclear Many-Body System. Any major venture into unknown regions, such as that made possible by CEBAF, offers the possibility of the discovery of unexpected phenomena. As two exotic possibilities consider the following (Fig. 41.7): Can one have color currents in the nucleus flowing between nucleons analogous to the electrical currents flowing in conductors? Can one find macroscopic color flux configurations extending over many nucleon dimensions in the nuclear interior? If such phenomena exist, how can one access them through electromagnetic means? Perhaps by observing very narrow (long-lived) states at relatively high energies?

All our experience has taught us that in any condensed-matter many-body system (here the nucleus), described in terms of new underlying degrees of

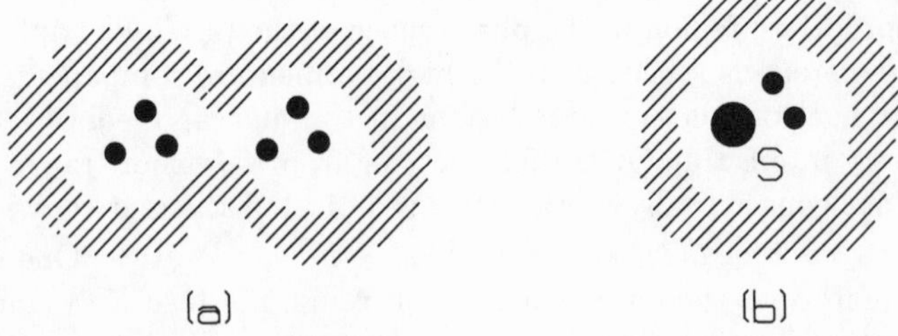

Figure 41.6: (a) Schematic illustration of overlapping nucleons in the nuclear medium; (b) schematic illustration of insertion of strange quark into the nuclear system.

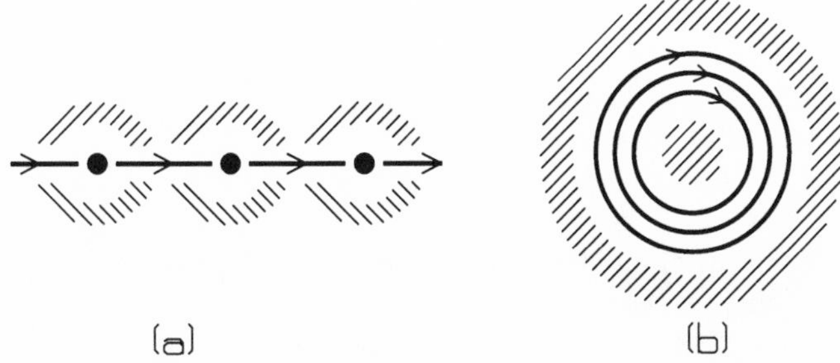

Figure 41.7: Two possible exotic configurations in the nuclear many-body system: (a) color currents flowing through nucleons; (b) macroscopic color flux configuration extending over many nucleon dimensions.

freedom (quarks and gluons) and new forces (QCD), there is the possibility of remarkable new quantum cooperative phenomena.

CEBAF's Role. CEBAF's role here is thus to

- Study the strong-coupling aspects of QCD under controlled conditions.
- Study the transition between the meson/baryon description of the nucleus, and the quark/gluon description. Note that mesons and baryons *are* the effective degrees of freedom at low energies and large distances in nuclear physics, while quarks and gluons *are* the appropriate degrees of freedom in the deep-inelastic regime.
- Look for new phenomena in nuclear systems.
- Lattice gauge theory allows one to compute some hadron properties directly from QCD. The electromagnetic interaction then allows one to put these predicted properties to the experimental test. Conversely, experimental measurements and their phenomenological interpretation provide insight into the immensely difficult problem of attempting to calculate nuclear properties from first principles in QCD.

Issues and Problems. We close this part of the book with a brief discussion of some issues and problems that will be resolved only in the next era of experimental and theoretical developments in nuclear physics:

1. With respect to lattice gauge theory, how does one get to the true continuum limit? What are the properties of *real* hadrons? How do these hadrons *move*? One can, for example, study the adiabatic motion of configurations obtained from lattice gauge theory (Fig. 41.8). (See Ref. [Q90] for a discussion of this approach).

2. How do hadrons *interact*? Baryons can interact through the exchange of mesons; this is the basis of QHD. In addition, they can interact through quark

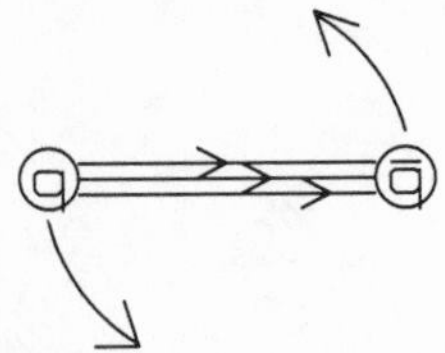

Figure 41.8: Example of adiabatic motion of meson configuration obtained from lattice gauge theory.

exchange. What is the relative importance of these contributions at various distance scales?

In the M.I.T. bag model the interaction of baryons at large distance arises from the tunneling of pairs (mesons), a situation reminiscent of the Josephson effect in superconductors.

3. Quarks and gluons are the immediate degrees of freedom at short distances, where asymptotic freedom indicates that QCD interactions become weak. Baryons and mesons are the effective degrees of freedom most directly describing nuclear interactions at large distances where the nonlinear, confining QCD forces become very strong. How can these descriptions be *combined*?

a. One approach is in coordinate space through bag models; these have been discussed in the text.

b. Another approach is in momentum space. Recall the discussion in Section 17 of the form factor, whose analytic properties in the complex q^2 plane are illustrated in Fig. 17.6. This leads to the spectral representation (for the isovector form factor) in Eq. (17.4). The calculation of the spectral weight function as the absorptive part of the amplitude is illustrated in Fig. 17.7. The contribution of the nearby singularities can be obtained explicitly from appropriate hadron amplitudes; the contribution of the far-away singularities, corresponding to very large momentum transfers, can be obtained from perturbative QCD (asymptotic freedom). This is one of the basic concepts of QCD sum rules (Refs. [Q98, Q99, Q100, Q91], [R46]).[113]

c. Still a third approach to combining explicit quark/gluon and meson/baryon degrees of freedom involves the analysis of the phase diagram for nuclear matter as discussed in Section 19.

4. Because of the phenomenological success of the meson/baryon description of nuclear physics, one goal is to calculate as well as possible within this picture and see where it fails to describe experimental data, and where one is forced to *explicitly* invoke quark/gluon degrees of freedom. In the deep-inelastic regime, this is already the case.

The situation in the nuclear domain is reminiscent of the history of the observation of exchange currents in nuclei (see Section 26). The consequences

[113] QCD sum rules also invoke quark and gluon condensates to phenomenologically describe the strong-coupling, confinement aspects of QCD.

of the traditional picture of nonrelativistic nucleons interacting through potentials can be exactly calculated in the three-body system through the solution of the Faddeev equations. For two-nucleon potentials fit to two-body scattering data, and electromagnetic interactions with free nucleons, one obtains for $^3\text{He}(e, e')_{\text{mag.}}$ the dashed line in Fig. 26.7. While the theoretical calculations agree quite well with the experimental data in the low-q^2 regime where one only samples the gross structure of the system, theory and experiment begin to diverge dramatically as the momentum transfer q^2 is increased and where one samples the system at shorter distance scales. There are exchange currents present in the nuclear system arising from the flow of charged mesons between the nucleons. The longest-range exchange current comes from the lightest meson, the pion; this longest-range contribution can be calculated exactly with the aid of low-energy theorems. Contributing processes are illustrated in Fig. 26.6. It is evident that the inclusion of the pion-exchange current remarkably improves the agreement between theory and experiment in Fig. 26.7 (solid line). At still larger q^2 and shorter distances, one sees the contribution of additional subnucleon degrees of freedom. The current theoretical uncertainty in the calculation of these contributions is indicated by the difference between the two solid lines in Fig. 26.7. Note, however, that one is able to describe $F_{\text{M}}^2(q^2)$ for ^3He down to the level of $\approx 10^{-7}$ out to $q^2 \approx 30\,\text{fm}^{-2}$ using a hadronic description in the nuclear domain.

Similarly, one will eventually have to invoke *explicit* quark/gluon degrees of freedom for this system at very large q^2. Just where this will occur is one of the central topics for the field.

5. It is not easy to *isolate* quark/gluon effects in the nuclear domain. For example, although the spin-flip transition to the $\Delta(1232)$ is one of the most immediate consequences of the quark model of the nucleon (Section 37), a hadronic description of the process $N(e, e')\Delta$, on which a great deal of work has been done over the years by a great many physicists (Section 26), is able to quantitatively describe the transition out to high q^2 as shown in Fig. 26.11 (Refs. [R90], [Q92]). It is only through systematic, precision studies of *all* of the baryon excitations that the true power of the quark model, and the full implications of QCD, will be established.

6. We have presented two distinct approaches to the nuclear many-body problem; one starts in the deep-inelastic regime where the quark/gluon degrees of freedom are manifest, and the other starts from the nuclear domain where hadrons are the appropriate effective degree of freedom for strong-coupling QCD. These two approaches consist of the following:

- The parton description in the $|\mathbf{p}| \to \infty$ frame; model building in this frame; the Altarelli-Parisi evolution equations for studying the QCD $\ln q^2$ modifications of scaling; physical insight in this regime.

- Nucleon and hadron descriptions of the nucleus in the laboratory frame; model building in this frame; physical insight in this regime.

These two descriptions have to be tied together. To do this, it is essential to have realistic, relativistic models and calculations that can be taken back and forth between the different frames. This is an *essential problem for CEBAF whose goal is to study the transition between these descriptions of the nucleus.* It is also a tough problem — witness the difficulty in just dealing correctly with the center-of-mass motion in bag models of the nucleon, or in relativistic Hartree descriptions of nuclei.

These issues and problems are just a subset of those that form an exciting challenge for the next generation of nuclear physicists.

References

[Q1] C. N. Yang and R. L. Mills, *Phys. Rev.* **96**, 191 (1954)

[Q2] D. J. Gross and F. Wilczek, *Phys. Rev. Lett.* **30**, 1343 (1973)

[Q3] D. J. Gross and F. Wilczek, *Phys. Rev.* **D8**, 3633 (1973)

[Q4] H. D. Politzer, *Phys. Rev. Lett.* **30**, 1346 (1973)

[Q5] H. D. Politzer, *Phys. Rep.* **14**, 129 (1974)

[Q6] H. Fritzsch and M. Gell-Mann in *Proc. XVI Int. Conf. on High Energy Physics*, eds. J. D. Jackson and A. Roberts, Vol II, FNAL, Batavia, Illinois (1972), p. 135

[Q7] H. Fritzsch, M. Gell-Mann, and H. Leutwyler, *Phys. Lett.* **47B**, 365 (1973)

[Q8] W. Marciano and H. Pagels, *Phys. Rep.* **36**, 137 (1978)

[Q9] E. Reya, *Phys. Rep.* **69**, 195 (1981)

[Q10] F.Wilczek, *Annu. Rev. Nucl. Part. Sci.* **32**, 177 (1982)

[Q11] C. Quigg, *Gauge Theories of the Strong, Weak, and Electromagnetic Interactions*, Benjamin/Cummings, Reading, Massachusetts (1983)

[Q12] T.-P. Cheng and L.-F. Li, *Gauge Theory of Elementary Particle Physics*, Clarendon Press, Oxford (1984)

[Q13] I. J. R. Aitchison and A. J. G. Hey, *Gauge Theories in Particle Physics*, 2nd edition, Adam Hilger, Bristol and Philadelphia (1989)

[Q14] M. Gell-Mann and F. E. Low, *Phys. Rev.* **95**, 1300 (1954)

[Q15] R. P. Feynman and A. R. Hibbs, *Quantum Mechanics and Path Integrals*, McGraw-Hill Book Co., Inc., New York (1965)

[Q16] E. S. Abers and B. W. Lee, *Phys. Rep.* **9**, 1 (1973)

[Q17] K. G. Wilson, *Phys. Rev.* **D10**, 2445 (1974)

[Q18] M. Creutz, *Quarks, Gluons, and Lattices*, Cambridge University Press, Cambridge, England (1983)

[Q19] C. Rebbi, *Lattice Gauge Theories and Monte Carlo Simulations*, World Scientific, Singapore (1983)

[Q20] F. Khan, Theory Study Group, CEBAF, Newport News, Virginia (1989) unpublished

[Q21] J. Schwinger, *Phys. Rev.* **82**, 664 (1951)

[Q22] K. Huang, *Statistical Mechanics*, John Wiley and Sons, Inc., New York (1963)

[Q23] R. Schiavilla, CEBAF Theory Study Group, CEBAF, Newport News, Virginia (1989) unpublished

[Q24] M. Creutz, L. Jacobs, and C. Rebbi, *Phys. Rev.* **D20**, 1915 (1979)

[Q25] B. Lautrup and M. Nauenberg, *Phys. Lett.* **95B**, 63 (1980)

[Q26] J. Dubach, CEBAF Theory Study Group, CEBAF, Newport News, Virginia (1989) unpublished

[Q27] R. Balian, J. M. Drouffe, and C. Itzykson, *Phys. Rev.* **D10**, 3376 (1974)

[Q28] R. Balian, J. M. Drouffe, and C. Itzykson, *Phys. Rev.* **D11**, 2098 (1975)

[Q29] R. Balian, J. M. Drouffe, and C. Itzykson, *Phys. Rev.* **D11**, 2104 (1975)

[Q30] K. Maung, CEBAF Theory Study Group, CEBAF, Newport News, Virginia (1989) unpublished

[Q31] M. Creutz and K. J. M. Moriarty, *Phys. Rev.* **D26**, 2166 (1982)

[Q32] S. W. Otto and J. D. Stack, *Phys. Rev. Lett.* **52**, 2328 (1984)

[Q33] M. Creutz, L. Jacobs, and C. Rebbi, *Phys. Rep.* **95**, 201 (1983)

[Q34] J. Kogut, *Rev. Mod. Phys.* **55**, 775 (1983)

[Q35] B. Berg and A. Billoire, *Nucl. Phys.* **B221**, 109 (1983)

[Q36] B. Berg and A. Billoire, *Nucl. Phys.* **B226**, 405 (1983)

[Q37] C. Carlson, CEBAF Theory Study Group, CEBAF, Newport News (1989) unpublished

[Q38] J. Schwinger, *Phys. Rev.* **125**, 397 (1962); *Phys. Rev.* **128**, 2425 (1962)

[Q39] H. Ito, CEBAF Theory Study Group, CEBAF, Newport News (1989) unpublished

[Q40] J.W. Negele and H. Orland, *Quantum Many-Particle Systems*, Addison-Wesley Publishing Co., Reading, Massachusetts (1988)

[Q41] *Fundamentals of Statistical Mechanics*, Manuscript and Notes of Felix Bloch, prepared by J. D. Walecka, Stanford University Press, Stanford, California (1989)

[Q42] N. Metropolis, A. Rosenbluth, M. Rosenbluth, A. Teller, and E. Teller, *J. Chem. Phys.* **21**, 1087 (1953)

[Q43] J. Kogut and L. Susskind, *Phys. Rev.* **D11**, 395 (1975)

[Q44] T. Banks, J. Kogut, and L. Susskind, *Phys. Rev.* **D13**, 1043 (1976)

[Q45] L. Susskind, *Phys. Rev.* **D16**, 3031 (1977)

[Q46] K. Wilson, in *New Phenomena in Subnuclear Physics,* Proc. Erice School, 1975, ed. A. Zichichi, Plenum, New York (1977), p. 13

[Q47] S. D. Drell, M. Weinstein, and S. Yankielowicz, *Phys. Rev.* **D14**, 487 (1976); **D14**, 1627 (1976)

[Q48] V. Dmitrašinović, CEBAF Theory Study Group, CEBAF, Newport News (1989) unpublished

[Q49] R. K. Bhaduri, *Models of the Nucleon*, Addison-Wesley Publishing Co., Reading, Massachusetts (1988)

[Q50] A. Chodos, R. L. Jaffe, K. Johnson, C. B. Thorn, and V. Weisskopf, *Phys. Rev.* **D9**, 3471 (1974)

[Q51] A. Chodos, R. L. Jaffe, K. Johnson, and C. B. Thorn, *Phys. Rev.* **D10**, 2599 (1974)

[Q52] T. DeGrand, R. L. Jaffe, K. Johnson, and J. Kiskis, *Phys. Rev.* **D12**, 2060 (1975)

[Q53] R. Jaffe and K. Johnson, *Phys. Lett.* **60B**, 201 (1976)

[Q54] G Kälbermann and J. M. Eisenberg, *Phys. Rev.* **D28**, 71 (1983)

[Q55] G. 't Hooft, *Nucl. Phys.* **B72**, 461 (1974)

[Q56] G. 't Hooft, *Nucl. Phys.* **B75**, 461 (1975)

[Q57] E. Witten, *Nucl. Phys.* **B160**, 57 (1979)

[Q58] T. H. R. Skyrme, *Proc. Roy. Soc. London* **A260**, 127 (1961)

[Q59] T. H. R. Skyrme, *Nucl. Phys.* **31**, 556 (1962)

[Q60] A. Chodos, E. Hadjimichael, and C. Tze, eds., *Solitons in Nuclear and Elementary Particle Physics*, Proc. of the Lewes Workshop, June, 1984, World Scientific, Singapore (1984)

[Q61] W. Pauli, *Meson Theory of Nuclear Forces*, Interscience, New York (1946)

[Q62] A. Chodos and C. B. Thorn, *Phys. Rev.* **D12**, 2733 (1975)

[Q63] S. Théberge, A. W. Thomas, and G. A. Miller, *Phys. Rev.* **D22**, 2838 (1980); **D23**, 2106 (1981); **D24**, 216 (1981)

[Q64] A. W. Thomas, *Advances in Nuclear Physics* **13**, eds. J. Negele and E. Vogt, Plenum Press, New York (1984), p. 1

[Q65] G. E. Brown and M. Rho, *Phys. Lett.* **82B**, 177 (1979)

[Q66] G. E. Brown, M. Rho, and V. Vento, *Phys. Lett.* **84B**, 383 (1979)

[Q67] A. D. Jackson, D. E. Kahana, L. Vepstas, H. Vershelde, and E. Wüst, *Nucl. Phys.* **A462**, 661 (1987)

[Q68] D. E. Kahana and J. Milana, *Nucl. Phys.* **A468**, 493 (1987)

[Q69] F. Lenz, J. T. Londergan, E. J. Moniz, R. Rosenfelder, M. Stingl, and K. Yazaki, *Ann. Phys.* **170**, 65 (1986)

[Q70] J. Carlson, J. Kogut, and V. R. Pandharipande, *Phys. Rev.* **D27**, 233 (1983)

[Q71] J. L. Goity, *Introduction to Chiral Perturbation Theory*, CEBAF Lecture Series, CEBAF, Newport News, Virginia (1992) (to be published)

[Q72] J. I. Friedman and H. W. Kendall, *Annu. Rev. Nucl. Sci.* **22**, 203 (1972)

[Q73] J. D. Bjorken, *Phys. Rev.* **179**, 1547 (1969)

[Q74] R. P. Feynman, (quoted in Ref. [Q75])

[Q75] J. D. Bjorken and E. A. Paschos, *Phys. Rev.* **185**, 1975 (1969)

[Q76] K. Maung, CEBAF Theory Study Group, CEBAF, Newport News, Virginia (1990) unpublished

[Q77] F. Halzen and A. D. Martin, *Quarks and Leptons*, John Wiley and Sons, Inc., New York (1984)

[Q78] I. J. R. Aitchison and A. J. G. Hey, *Gauge Theories in Particle Physics*, Adam Hilger, Bristol, England (1989)

[Q79] J. J. Aubert *et al.*, *Phys. Lett.* **123B**, 275 (1983)

[Q80] R. P. Bickerstaff and A. W. Thomas, *J. Phys.* **G15**, 1523 (1989)

[Q81] V. Dmitrašinović, CEBAF Theory Study Group, CEBAF, Newport News, Virginia (1990) unpublished

[Q82] P. Morley and I. Schmidt, *Phys. Rev.* **D34**, 1305 (1986)

[Q83] G. Altarelli and G. Parisi, *Nucl. Phys.* **B126**, 298 (1977)

[Q84] C. Carlson, CEBAF Theory Study Group, CEBAF, Newport News, Virginia (1990) unpublished

[Q85] S. D. Drell and J. D. Walecka, *Ann. Phys.* **28**, 18 (1964)

[Q86] J. D. Jackson, *Classical Electrodynamics*, John Wiley and Sons, Inc., New York (1962)

[Q87] D. B. Leinweber, R. M. Woloshyn, and T. Draper, *Phys. Rev.* **D43**, 1659 (1991); **D46**, 3067 (1992)

[Q88] W. Bartel, F.-W. Büsser, W.-R. Dix, R. Felst, D. Harmes, H. Krehbiel, P. E. Kuhlmann, J. McElroy, J. Meyer, and G. Weber, *Nucl. Phys.* **B58**, 429 (1973)

[Q89] R. G. Arnold, P. E. Bosted, C. C. Chang, J. Gomez, A. T. Katramatou, C. J. Martoff, G. G. Petratos, A. A. Rahbar, S. E. Rock, A. F. Sill, Z. M. Szalata, D. J. Sherdon, J. M. Lambert, and R. M. Lombard-Nelsen, *Phys. Rev. Lett.* **57**, 174 (1986)

[Q90] N. Isgur, in *Excited Baryons 1988*, eds. G. Adams, N. C. Mukhopadhyay, and P. Stoler, World Scientific, Singapore (1989), p. 1

[Q91] H. Ito, CEBAF Theory Study Group, CEBAF, Newport News, Virginia (1990) unpublished

[Q92] J. D. Walecka, in *Excited Baryons 1988*, eds. G. Adams, N. C. Mukhopadhyay, and P. Stoler, World Scientific, Singapore (1989), p. 495

[Q93] G. E. Brown, *Nucl. Phys.* **A446**, 3c (1985)

[Q94] S. Elitzer, *Phys. Rev.* **D12**, 3978 (1975)

[Q95] N. Isgur, in *The New Aspects of Subnuclear Physics*, ed. A. Zichichi, Plenum Press, New York (1980), p. 107; *Act. Phys. Aust. Suppl. XXVII*, Springer-Verlag, Vienna (1985), p. 177

[Q96] H. Georgi, *Weak Interactions and Modern Particle Theory*, Benjamin Cummings, Menlo Park, California (1984); *Effective Field Theories*, in *Annu. Rev. Nucl. Part. Sci.* **43**, 209 (1993)

[Q97] U. G. Meissner, *Rep. Prog. Phys.* **56**, 903 (1993)

[Q98] M. A. Shifman, A. I. Vainshtein, and V. I. Zakharov, *Nucl. Phys.* **B147**, 385 (1979); **B147**, 448 (1979)

[Q99] L. J. Reinders, H. Rubenstein, and S. Yazaki, *Phys. Rep.* **127**, 1 (1985)

[Q100] T. D. Cohen, R. J. Furnstahl, and D. K. Griegel, *Prog. in Part. and Nucl. Phys.*, ed. A. Faessler, (1994) to be published

PROBLEMS: PART III

27.1. Start from the definition of the field tensor $\vec{F}_{\mu\nu}$ in Eq. (27.21) and introduce the infinitesimal local gauge transformation in Eq. (27.18). Prove $\delta\vec{F}_{\mu\nu} = \vec{\theta} \times \vec{F}_{\mu\nu}$.

27.2. (a) Derive the Euler-Lagrange Eqs. (27.47) for QCD.
(b) Derive Eqs. (27.48) from Noether's theorem. What are the appropriate symmetries?

27.3. (a) Show the canonical momentum conjugate to the gluon field in QCD is $\Pi_j^a \equiv \Pi_{A_j^a} = i\mathcal{F}_{4j}^a$.
(b) Derive Gauss' law for the color fields

$$\vec{\nabla} \cdot \vec{\Pi}^a = -\frac{g}{2}\underline{\psi}^\dagger \underline{\lambda}^a \underline{\psi} + g f^{abc} \vec{\Pi}^b \cdot \vec{A}^c \equiv -g(\rho_{\text{quark}}^a + \rho_{\text{gluon}}^a)$$

27.4. (a) Show (with the aid of partial integration) that the hamiltonian density in QCD can be written $\mathcal{H}_{\text{QCD}} \doteq \frac{1}{2}\vec{\Pi}^a \cdot \vec{\Pi}^a + \frac{1}{4}\mathcal{F}_{ij}^a \mathcal{F}_{ij}^a + \underline{\psi}^\dagger\{\vec{\alpha} \cdot [\frac{1}{i}\vec{\nabla} - \frac{g}{2}\underline{\lambda}^a \vec{A}^a(x)] + \beta \underline{\text{M}}\}\underline{\psi}$.
(b) Prove that any vector field can be separated into $\vec{\Pi} = \vec{\Pi}_T + \vec{\Pi}_L$ where $\vec{\nabla} \cdot \vec{\Pi}_T = 0$ and $\vec{\nabla} \times \vec{\Pi}_L = 0$ with $\int_{\text{box}} \vec{\Pi}_L \cdot \vec{\Pi}_T \, d^3x = 0$. (Ref. [R5]).
(c) Hence show the hamiltonian density can be written

$$\begin{aligned}\mathcal{H}_{\text{QCD}} \doteq \;& \underline{\psi}^\dagger\{\vec{\alpha} \cdot [\frac{1}{i}\vec{\nabla} - \frac{g}{2}\underline{\lambda}^a \vec{A}^a(x)] + \beta \underline{\text{M}}\}\underline{\psi} + \frac{1}{2}\vec{\Pi}_T^a \cdot \vec{\Pi}_T^a + \frac{1}{4}\mathcal{F}_{ij}^a \mathcal{F}_{ij}^a \\ & + \frac{g^2}{8\pi}\int\int \rho^a(\vec{x})\frac{1}{|\vec{x} - \vec{x}'|}\rho^a(\vec{x}')d^3x\, d^3x'\end{aligned}$$

Here $\vec{\nabla} \cdot \vec{\Pi}^a = \vec{\nabla} \cdot \vec{\Pi}_L^a$ satisfies the constraint equation in Prob. 27.3(b) whose solution in terms of the color charge $\rho^a = \rho_{\text{quark}}^a + \rho_{\text{gluon}}^a$ is

$$\vec{\Pi}_L^a = \frac{g}{4\pi}\vec{\nabla}\int \frac{1}{|\vec{x} - \vec{x}'|}\rho^a(\vec{x}')d^3x'$$

Since ρ_{gluon}^a depends on $\vec{\Pi}_L$ this is an integral equation (or power series) for $\vec{\Pi}_L^a$.

27.5. Use the results of Prob. 27.4 and the analogy to QED to discuss the quantization of QCD in the Coulomb gauge where $\vec{\nabla} \cdot \vec{A}^a = 0$ (Ref. [R4]).

27.6. In QED charge renormalization comes entirely from vacuum polarization (Ward's indentity). Consider the Møller scattering of two electrons through one photon exchange at a momentum transfer k^2.
(a) Show that if $k^2 \gg M^2$ then to $O(\alpha)$ the modification of the photon propagator due to the vacuum polarization process in Fig. 27.4b is $e^2/k^2 \to e^2[1 + (\alpha/3\pi)(\ln k^2/M^2 - 5/3)]/k^2$ (Refs. [R2,R3,R4]).
(b) Use this result to define a new renormalized charge $e_2^2(k^2, e^2)/k^2$. Hence derive the first term in the power series expansion of Eq. (27.52). Note the sign.

27.7. In QCD color charge renormalization comes from vertex and self-energy parts as well as vacuum polarization. The following perturbation theory results in the Landau gauge are given for $k^2 \gg \lambda^2$ in Ref. [Q9]:
(a) From the graphs in Prob. 19.3, the quark propagator is $-\delta_{ij}\delta_{lm}/(i\gamma_\mu p_\mu + m) + O(g^4)$.
(b) From the graphs in Prob. 19.4, the gluon propagator is $(\delta^{ab}/k^2)(\delta_{\mu\nu} - k_\mu k_\nu/k^2)[1 - (13/2 - 2N_f/3)(g^2/16\pi^2)\ln k^2/\lambda^2]$. Here N_f is the number of quark flavors.
(c) From the graphs in Prob. 19.5, the quark vertex is $(\lambda^a/2)\gamma_\mu[1 - (9/4)(g^2/16\pi^2)\ln k^2/\lambda^2]$.

Consider the scattering of two quarks through one gluon exchange to $O(g^4)$. As in Prob. 27.6, use the above results to define a new renormalized color charge $g_2(k^2)/k^2$. Hence derive the first term in the power series expansion of Eq. (27.53). Note the sign.

27.8. Derive the result in Prob. 27.7(a). Define the finite renormalized functions by subtracting at the (euclidian) mass λ^2.

27.9. Repeat Prob. 27.8 for the result in Prob. 27(b).

27.10. Repeat Prob. 27.8 for the result in Prob. 27(c).

27.11. Represent gluons with double lines with a color assigned to each line. Then, as indicated in the text, color can be viewed as running continuously through Feynman diagrams built from the Yukawa quark-gluon coupling in Eq. (27.44). Can you extend this concept to the cubic gluon self-couplings? To the quartic couplings?

28.1. Use the analog of the derivation of the partition function in the text to derive the path integral expression for the quantum mechanical transition amplitude $\langle q_f t_f | q_i t_i \rangle = \langle q_f | \exp\{-\frac{i}{\hbar}\hat{h}(t_f - t_i)\} | q_i \rangle$ given in Eqs. (28.1)-(28.5).

28.2. A useful result for path integrals is provided by gaussian integration from ordinary analysis. Suppose $\underset{\sim}{q}^T \underset{\sim}{N} \underset{\sim}{q} = \sum_{i=1}^{n-1}\sum_{j=1}^{n-1} q_i N_{ij} q_j$ is a quadratic form and $N_{ij} = N_{ji}$ is a real, symmetric matix. Let $\underset{\sim}{J}^T \underset{\sim}{q} = \sum_{i=1}^{n-1} J_i q_i$ and (a, b) be real numbers. Prove the following for $a = \pm|a|$

$$\int \cdots \int dq_1 \cdots dq_{n-1} \exp\left\{\frac{i}{\hbar}[a\underset{\sim}{q}^T \underset{\sim}{N} \underset{\sim}{q} + b\underset{\sim}{J}^T \underset{\sim}{q}]\right\} = \frac{1}{(\det \underset{\sim}{N})^{1/2}} \left(\frac{\pi\hbar e^{\pm i\pi/2}}{|a|}\right)^{(n-1)/2}$$
$$\times \exp\left\{-\frac{i}{\hbar}\frac{b^2}{4a}\underset{\sim}{J}^T \underset{\sim}{N}^{-1} \underset{\sim}{J}\right\}$$

28.3. The variational derivative of a functional $W(f)$ can be defined by $\text{Lim}_{\lambda \to 0}$ $\{W[f(x) + \lambda\delta(x-y)] - W[f(x)]\}/\lambda \equiv \delta W(f)/\delta f(y)$.
(a) Let $W_x(f) \equiv \int K(x,y) f(y) dy$. Show $\delta W_x(f)/\delta f(z) = K(x,z)$.
(b) Let $K_n(x_1, \cdots, x_n)$ be a totally symmetric function and $W(f) \equiv \int \cdots \int K_n(x_1, \cdots)$ $f(x_1) \cdots f(x_n) dx_1 \cdots dx_n$. Show $\delta^n W(f)/\delta f(x_1) \cdots \delta f(x_n) = n! K_n(x_1, \cdots x_n)$.

28.4. Show with the aid of the result in Prob. 28.2 that the required path integral can be evaluated exactly for the one-dimensional s.h.o. to give (Ref. [Q15])

$$\langle q_f t | q_i 0 \rangle = \left(\frac{m\omega e^{-i\pi/2}}{2\pi\hbar \sin\omega t}\right)^{1/2} \exp\left\{\frac{i}{\hbar}\frac{m\omega}{2}\left[(q_i^2 + q_f^2)\cot\omega t - \frac{2q_f q_i}{\sin\omega t}\right]\right\}$$

Note this result is periodic in ωt and for $\omega t \to 0$ it reduces to that for a free particle.

28.5. Show from its definition that the result in Prob. 28.4 is the Green's function for the Schrödinger equation for the one-dimensional s.h.o.

28.6. Start from the definition in Prob. 28.1. Evaluate the trace through $\int dq\langle qt|q0\rangle$ and then make the replacement $t \to -i\beta\hbar$ in Prob. 28.4 to rederive the partition function $z = \exp\{\beta\hbar\omega/2\}(\exp\{\beta\hbar\omega\} - 1)^{-1}$ for the one-dimensional s.h.o.

28.7. Consider a free scalar field with source $\mathcal{L}_J = J\phi$ and set $\hbar = c = 1$.
(a) Assume the disturbance is confined to a finite region of space-time and carry out a partial integration to give $S \doteq (1/2)\int d^4x[-\phi\partial^2\phi/\partial t^2 + \phi\nabla^2\phi - (m^2 - i\eta)\phi^2 + 2J\phi]$.
(b) Divide space-time into finite cells of volume ε^4. Label each cell with α; in the continuum limit where $\varepsilon \to 0$ one has $\alpha \to x$, $\beta \to y$. Replace integrals by finite sums. In the action one can write $S \doteq \sum_\alpha \varepsilon^4 \sum_\beta \varepsilon^4 \phi_\alpha K_{\alpha\beta}\phi_\beta/2 + \sum_\alpha \varepsilon^4\phi_\alpha J_\alpha$. Here $\sum_\beta \varepsilon^4 \phi_\alpha K_{\alpha\beta}\phi_\beta \overset{\varepsilon\to 0}{\longrightarrow} \int d^4y\phi(x)K(x-y)\phi(y)$ and $\text{Lim}_{\varepsilon\to 0}K_{\alpha\beta} = [\partial^2/\partial x_\mu^2 - (m^2 - i\eta)]\delta^{(4)}(x-y)$. Use the result for gaussian integration in Prob. 28.2 to evaluate the generating functional in Eq. (28.49) in Minkowski space in terms of $\underline{K}^{-1}$ (Ref. [Q16]).
(c) Define the inverse matrix by $\sum_\gamma K_{\alpha\gamma}(K^{-1})_{\gamma\delta} = \delta_{\alpha\delta}$ with $\text{Lim}_{\varepsilon\to 0}(K^{-1})_{\alpha\beta}/\varepsilon^8 \equiv -\Delta_F(x-y)$. Show the generating functional is

$$\tilde{W}_0(J) = \exp\left\{\frac{i}{2}\int d^4x\int d^4y J(x)\Delta_F(x-y)J(y)\right\}$$

$$\Delta_F(x-y) = \int\frac{d^4k}{(2\pi)^4}\frac{1}{k^2 + (m^2 - i\eta)}e^{ik\cdot(x-y)}$$

Here Δ_F is the Feynman propagator.
(d) Show $(1/i)^2\delta^2\tilde{W}_0(0)/\delta J(x_1)\delta J(x_2) = (1/i)\Delta_F(x_1 - x_2) = \langle 0|T[\hat{\phi}(x_1), \hat{\phi}(x_2)]|0\rangle$.

29.1. The mathematical manipulations in Sections 29 and 30 involve analytically continuing the generating functional in complex time to produce the partition function and then rotating the path integrals over the fields to go from Minkowski to euclidian metric. As a very simple example illustrating some of these procedures (see Ref. [R5] for background), consider the following integral:

$$I(\lambda) \equiv \frac{1}{2}\int_{-\infty}^{\infty} e^{-\lambda x^2}dx = \int_0^\infty e^{-\lambda x^2}dx$$

(a) Show $I(\lambda)$ is an analytic function of λ for $\text{Re}\,\lambda > 0$.
(b) Evaluate $I(\lambda) = \sqrt{\pi/4\lambda} \equiv F(\lambda)$ on the positive real λ axis. Use this expression to analytically continue $I(\lambda)$ to the entire cut λ plane.
(c) Hence establish the integral representation $F(\lambda) = \int_C e^{-\lambda z^2}dz$ for $\text{Re}\,\lambda > 0$ where the contour C runs from the origin to ∞ along the positive real z-axis.
(d) Use the analyticity of the integrand and Cauchy's theorem to show that if λ lies on the positive real axis, one can rotate the contour C in the complex z plane to another contour C_1 so $F(\lambda) = \int_C e^{-\lambda z^2}dz = \int_{C_1} e^{-\lambda z^2}dz$ where C_1 is any ray running from the origin to ∞ with $\text{Re}\,z^2 > 0$.
(e) Let λ approach the negative imaginary axis from the right $\lambda \to |\lambda|e^{-i\pi/2} = -i|\lambda|$. Show the integral in (c) becomes $F(|\lambda|e^{-i\pi/2}) = \int_C \exp\{-e^{-i\pi/2}|\lambda|z^2\}dz$.

(f) Show the contour C in (e) can be rotated to any direction such that $\mathrm{Re}(e^{-i\pi/2}z^2) > 0$. For example, it can be rotated to

$$F(|\lambda|e^{-i\pi/2}) = \int_{C_2} \exp\{-e^{-i\pi/2}|\lambda|z^2\}dz$$

where C_2 runs along the positive imaginary axis.
(g) Make a simple change of variables in (f) to show $F(-i|\lambda|) = i\int_0^\infty \exp\{-i|\lambda|\zeta^2\}d\zeta$.

The net result is that the value of the function analytically continued from real to imaginary argument is obtained by rotating the contour in the integral representation of that function from the real to imaginary axis. Analytic continuation implies that the resulting function is unique, and all steps are mathematically justified as long as the integrals under consideration converge.

29.2. Extend the formulation of $U(1)$ lattice gauge theory in the text to two space and one time (2+1) dimensions. Demonstrate gauge invariance and the proper continuum limit.

29.3. Extend Prob. 29.2 to 3+1 dimensions. Show $\sigma \equiv 1/2e_0^2$ for the proper continuum limit in this case.

30.1. Extend the Ising model by letting the spin on each site take the values $\{s_i\} = \{s, s-1, \ldots, -s\}$. Work in mean field theory (MFT).
(a) Show the partition function is $z = [\sinh{(s+\frac{1}{2})x}]/\sinh{\frac{1}{2}x}$ where $x \equiv \beta J\gamma_l m$.
(b) Show the magnetization is $m = \partial \ln z/\partial x = (s+\frac{1}{2})\coth{(s+\frac{1}{2})x} - \frac{1}{2}\coth{\frac{1}{2}x}$.
(c) Find T_C; plot m vs. T/T_C.

30.2. Extend the Ising model by letting the spin on each site take the values $\{s_i\} = \{s\cos\theta_i\}$ with a measure $\int d\Omega_i/4\pi = \int\int \sin\theta_i\, d\theta_i\, d\phi_i/4\pi$. Work in MFT.
(a) Show the partition function is $z = [\sinh\zeta]/\zeta$ where $\zeta \equiv \beta J\gamma_l m s$.
(b) Show the magnetization is $m/s = \partial \ln z/\partial\zeta = \coth\zeta - 1/\zeta$.
(c) Find T_C; plot m/s vs. T/T_C.
(d) Let $J\gamma_l m \to \mathcal{H}_{\text{ext}}$ be an external magnetic field in the z-direction; write $m = \chi_{\text{mag}}\mathcal{H}_{\text{ext}}$ as $\mathcal{H}_{\text{ext}} \to 0$; and define $s \equiv \mu_0$. Use the above results to derive the Langevin expression for the paramagnetic susceptibility of a classical ensemble of magnetic moments $\chi_{\text{mag}} = \mu_0^2/3k_BT$.

30.3. Verify the MFT numerical results for lattice QED quoted in Eq. (30.28) and Fig. 30.6.

30.4. Consider lattice QED in $1+1$ dimensions on a 2×2 lattice with periodic boundary conditions in both directions.
(a) Introduce relative angles and show the full problem can be reduced to the following exact expression ($r_5 \equiv r_1$)

$$Z(\sigma) = \prod_{j=1}^4 \int_{-2\pi}^{2\pi} dr_j(2\pi - |r_j|)\exp\left\{-2\sigma\left(4 - \sum_{i=1}^4 \cos{(r_i - r_{i+1})}\right)\right\}/[\cdots]_{\sigma=0}$$

(b) Show $4\langle\langle S_\Box\rangle\rangle = -\sigma\, d\ln Z(\sigma)/d\sigma$.
(c) Extend these results to an $n\times n$ lattice.

31.1. The Baker-Haussdorf identity $e^Ae^B = e^{A+B+\frac{1}{2}[A,B]}$ holds for operators and matrices as long as $[A,[A,B]] = [B,[A,B]] = 0$. Prove this relation to third order in the operators by expanding the exponentials on both sides.

31.2. Construct a proof of the Baker-Haussdorf identity in Prob. 31.1 to all orders in the following fashion (Ref. [R3]):
(a) First show $e^B A e^{-B} = A + [B, A]$.
(b) Define $F(\lambda) = e^{\lambda(A+B)}e^{-\lambda B}e^{-\lambda A}$. Show $dF/d\lambda = e^{\lambda(A+B)}[A, e^{-\lambda B}]e^{-\lambda A}$.
(c) Hence show $[A, e^{-\lambda B}] = -\lambda[A, B]e^{-\lambda B}$ and $dF/d\lambda = -\lambda[A, B]F(\lambda)$ with $F(0) = 1$.
(d) Use (c) to conclude $F(\lambda) = e^{-\frac{\lambda^2}{2}[A,B]}$ and take $\lambda = 1$ to establish the identity.

31.3. (a) Prove that $\underline{U}_\square$ is again a 2×2 $SU(2)$ matrix and thus can be written $\underline{U}_\square = \exp\{\frac{i}{2}\vec{\tau} \cdot \vec{\phi}\}$ for some real $\vec{\phi} : (\phi_1, \phi_2, \phi_3)$.
(b) Show that as $\vec{\phi} \to 0$ one has $\frac{1}{2}\mathrm{tr}\,\underline{U}_\square = 1 - \frac{1}{8}\vec{\phi}^2 + \cdots$.

31.4. Generalize the discussion of the continuum limit in the text to the case of 2+1 and 3+1 dimensions (d=3,4). Show $\sigma = 2/g_0^2 a^{4-d}$.

31.5. Search the literature to find the appropriate gauge-invariant measure for $SU(3)$ nonabelian lattice gauge theory.

32.1. Plot the "magnetization" M^4 as a function of σ_C/σ for $SU(2)$ lattice gauge theory in $d = 4$ dimensions in MFT obtained from Eqs. (32.24), (32.29), and (32.35). Compare with Fig. 30.6.

32.2. (a) Show that to within an overall multiplicative constant the MFT free energy/site, and thus also the minimization condition for H, depends only on the combination $\sigma(d-1)$.
(b) Hence conclude that the critical value of σ for the development of spontaneous magnetization M in MFT in d dimensions scales as $\sigma_C(d-1) = 4.239$.
(c) Thus use the results of Prob. 32.1 to make a universal plot of M^4_{MFT} vs. σ_C/σ in d dimensions.

32.3. One of the original Monte Carlo calculations in nonabelian lattice gauge theory is due to Creutz,[114] who found clear evidence for a phase transition in the pure $SU(2)$ gauge theory in $d = 5$ dimensions.
(a) Use the scaling relation in Prob. 32.2 to show that for $SU(2)$ with $d = 5$ one has $\sigma_C = 1.060$ in MFT.
(b) Establish that $\beta_{\mathrm{Creutz}} \equiv 2\sigma$. (You may have to trace back in the literature here.)
(c) Hence establish the comparison with the Monte Carlo result that $\sigma_C = 0.821 \pm 0.008$ in this case.

33.1. At the end of Section 28 it was argued that the vacuum expectation value of a time-ordered product of Heisenberg operators, in the euclidian metric, can be obtained from variational derivatives with respect to the sources of the generating functional; these simply serve to bring down the field operators. Thus one can write

$$\langle Q(T)Q(0)\rangle = \frac{\int (dU)e^{-S(U)}Q(T)Q(0)}{\int (dU)e^{-S(U)}}$$

Here $S \equiv \bar{S}$ of Section 28, with $\int_{-\infty}^{\infty} d\tau$, and Q is constructed from the link variables. The correlation function of Eq. (33.42) is then given by $G(T) = \langle Q(T)Q(0)\rangle - \langle Q(T)\rangle\langle Q(0)\rangle$.

[114] M. Creutz, *Phys. Rev. Lett.* **43**, 553 (1979).

Let $Q = S_\square$ be the contribution to the action from an elementary plaquette. Formulate the problem of calculating the plaquette-plaquette correlation function in lattice gauge theory.

33.2. Discuss the following features of the calculation formulated in Prob. 33.1: gauge invariance, continuum limit, large $\mathcal{T}$ limit, and nature of the intermediate states contributing to the correlation function.

34.1. Derive the following strong-coupling $\sigma \to 0$ limits for $\langle S_\square \rangle$:
(a) For $U(1)$ $\langle S_\square \rangle / 2\sigma = 1 - \sigma + O(\sigma^3)$
(a) For $SU(2)$ $\langle S_\square \rangle / 2\sigma = 1 - \sigma/2 + O(\sigma^3)$

34.2. Plot the asymptotic strong-coupling result in Fig. 30.6.

34.3. Draw Fig. 34.7 for $SU(2)$ in $d = 4$ dimensions.

34.4. Evaluate the term of $O(\sigma^3)$ in Prob. 34.1(a).

35.1. Use the Monte Carlo method in Eq. (35.14) to evaluate the following integrals: $\int_0^1 x^2 dx$, $\int_0^1 \sin(\pi x/2)\, dx$, and $\int_0^1 e^{-x} dx$. Compare with the brute force method in Fig. 35.5 for the same number of points. Discuss convergence.

35.2. The Dirichlet integral $I_n = \int \cdots \int dx_1 \cdots dx_n$ with $x_1^2 + \cdots + x_n^2 \le 1$ is the volume of the unit sphere in n-dimensional euclidian space. Generate a random point in the unit cube, keep the point if the inequality is satisfied, and hence compute the ratio of the volume of the unit sphere to unit cube $I_n = (\sqrt{\pi})^n / \Gamma(1 + n/2)$ (see Prob. 16.6). Verify for several n, and discuss convergence.

35.3. Consider the following integral $\bar{f} = \int_0^1 f(x) m(x) dx / \int_0^1 m(x) dx$ with $p(x) \equiv m(x) / \int_0^1 m(x) dx$. Use the Metropolis algorithm to generate a distribution satisfying $dN/N = p(x)dx$ and evaluate $\bar{f}$ in the following cases: 1) $f(x) = x^2(1-x)^2$ with $m(x) = x(1-x)$; and 2) $f(x) = x^n$ with $m(x) = e^{-x}$. Discuss convergence.

35.4. Pick one of the integrals in Prob. 35.1. By repeated calculation for fixed N, and then for larger N, verify the main features of Fig. 35.4.

36.1. Fermion fields in path integrals are described through Grassmann algebras of anticommuting c-numbers c_i with $i = 1, \ldots, n$ satisfying $\{c_i, c_j\} = 0$. The concept of integration, here justified a posteriori (Prob. 36.4), is then defined by $\int dc_i = 0$ and $\int c_i dc_i = 1$ where it is assumed that all elements anitcommute $\{c_i, c_j\} = \{dc_i, c_j\} = \{dc_i, dc_j\} = 0$. Suppose one has two distinct Grassmann algebras $\bar{c}_i$ and c_i with all elements anticommuting. The basic integral relation corresponding to that in Prob. 28.2 is then (as usual, repeated indices are summed)

$$\int \cdots \int d\bar{c}_n \cdots d\bar{c}_1 dc_1 \cdots dc_n \exp\{-\bar{c}_i N_{ij} c_j\} = \det \underset{\sim}{N}$$

(a) Prove this result explicitly for $n = 2$.
(b) Generalize the proof to arbitrary n.

36.2. Let $\bar{\xi}_i$ and ξ_i be two additional Grassmann algebras of sources, everything again anticommuting. Prove

$$\int \cdots \int d\bar{c}_n \cdots d\bar{c}_1 dc_1 \cdots dc_n \exp\{-\bar{c}_i N_{ij} c_j + \bar{c}_i \xi_i + \bar{\xi}_j c_j\} = (\det \underset{\sim}{N}) \exp\{\bar{\xi}_i (N^{-1})_{ij} \xi_j\}$$

(*Hint*: change variables $\underset{\sim}{\eta} \equiv \underset{\sim}{c} - \underset{\sim}{N}^{-1}\underset{\sim}{\xi}$ with $d\eta_i = dc_i$.)

36.3. Let $\mathcal{L}_0 = -\bar{\psi}(\gamma_\mu \partial/\partial x_\mu + M)\psi$ be the lagrangian density of a free fermion field. Discretize space time as in Prob. 28.7, and repeat that problem dealing now with Grassmann variables $\bar{\psi}_\alpha$ and ψ_β and a measure $\mathcal{D}(\bar{\psi})\mathcal{D}(\psi)$. If $\bar{\zeta}$ and ζ are Grassmann sources, show the generating functional is (Ref. [R1, R4])

$$\begin{aligned}
\tilde{W}_0[\bar{\zeta},\zeta] &\equiv \int \mathcal{D}(\bar{\psi})\mathcal{D}(\psi)\exp\{i\int d^4x[\mathcal{L}_0 + \bar{\zeta}\psi + \bar{\psi}\zeta]\}\,/\,[\cdots]_{\zeta=\bar{\zeta}=0} \\
&= \exp\left\{-i\int d^4x\int d^4y\bar{\zeta}(x)S_{\mathrm{F}}(x-y)\zeta(y)\right\} \\
S_{\mathrm{F}}(x-y) &= -\int \frac{d^4k}{(2\pi)^4}\frac{1}{ik_\mu\gamma_\mu + (M - i\eta)}\exp\{ik\cdot(x-y)\}
\end{aligned}$$

36.4. Denote the continuum limit of the Grassmann algebra by $c_i \to c_x \to c(x)$, and define a functional and its variational derivatives by

$$\begin{aligned}
P[\bar{c},c] = \int & dx_1\cdots dx_n \int dy_1\cdots dy_n \\
& \times\bar{c}(x_1)\cdots\bar{c}(x_n)\frac{1}{n!}\frac{1}{n!}K_n(x_1,\ldots,x_n;y_1,\ldots,y_n)c(y_1)\cdots c(y_n)
\end{aligned}$$

$$\frac{\delta^n}{\delta\bar{c}(x_n)\cdots\delta\bar{c}(x_1)}P[\bar{c},c]\frac{\delta^n}{\delta c(y_n)\cdots\delta c(y_1)} \equiv K_n(x_1,\ldots,x_n;y_1,\ldots,y_n)$$

Here the kernal $K_n(x_1,\ldots,x_n;y_1,\ldots y_n)$ is assumed antisymmetric in $x_1,\ldots,x_n$ and in $y_1,\ldots y_n$; note the order is important on both sides of these relations.

Derive the following relation from the result in Prob. 36.3

$$\left[\frac{1}{i^2}\frac{\delta}{\delta\bar{\zeta}(x)}\tilde{W}_0[\bar{\zeta},\zeta]\frac{\delta}{\delta\zeta(y)}\right]_{\zeta=\bar{\zeta}=0} = iS_{\mathrm{F}}(x-y) = \langle 0|P[\psi(x),\bar{\psi}(y)]|0\rangle$$

36.5. (a) Assume $\underset{\sim}{A}_{\mathrm{D}}$ is a diagonal matrix. Show $\det\underset{\sim}{A}_{\mathrm{D}} = \exp\{\mathrm{Tr}\ln\underset{\sim}{A}_{\mathrm{D}}\}$.
(b) Assume the matrix $\underset{\sim}{A}$ can be diagonalized with a similarity transformation $\underset{\sim}{U}\underset{\sim}{A}\underset{\sim}{U}^{-1} = \underset{\sim}{A}_{\mathrm{D}}$. Use the result in (a) to prove $\det\underset{\sim}{A} = \exp\{\mathrm{Tr}\ln\underset{\sim}{A}\}$.

36.6. Use the expansion of the determinant in terms of cofactors to prove the general matrix relation $\partial(\det\underset{\sim}{M})/\partial M_{ij} = (\underset{\sim}{M}^{-1})_{ji}(\det\underset{\sim}{M})$.

36.7. Extend the demonstration of the correct continuum limit of the LGT fermion action to $d = 3$ and $d = 4$ dimensions. Show $\sigma_{\mathrm{F}} = 1/2a^{1-d}$.

36.8. Use Eq. (36.12) to find the matrix $\Delta_{ij}(U)$ in Eq. (36.21).

37.1. Assume x statisfies the eigenvalue equation $j_0(x) = j_1(x)$. Use the formulae for spherical Bessel functions in Ref. [N42] to show the following:
(a) $\int_0^x \rho^2 d\rho[j_0^2(\rho) + j_1^2(\rho)] = 2(x-1)\sin^2 x$
(b) $\int_0^x \rho^3 d\rho j_0(\rho)j_1(\rho) = \left(x - \frac{3}{4}\right)\sin^2 x$
(c) $\int_0^x \rho^2 d\rho[j_0^2(\rho) - \frac{1}{3}j_1^2(\rho)] = \left(\frac{2}{3}x\right)\sin^2 x$

37.2. (a) Use the arguments in Appendix A to derive the effective quark-quark potential arising from one-gluon exchange. Show that with neglect of retardation in the

gluon propagator one has $\nu_{q-q} = (g^2/4\pi r_{12})(\gamma_\mu)^{(1)}(\gamma_\mu)^{(2)}(\lambda^a/2)^{(1)}(\lambda^a/2)^{(2)}$. When added to the bag model, this gives rise to additional splitting of the $(q)^3$ levels.
(b) Use asymptotic freedom to discuss the expected behavior of $g^2(r_{12})$ as $r_{12} \to 0$.
(c) Discuss the $q - \bar{q}$ potential and the relevance to Fig. 33.5.

37.3. (a) Show that $(\langle r^2 \rangle)^{1/2} = 0.73\,R$ for a single massless quark in the $1s_{1/2}$ state in the bag model (Ref. [Q49]).
(b) Find the mean square charge radius of the proton and neutron in the bag model.

37.4. Construct the bag model $(q\bar{q})$ wave functions for the ground-state, color singlet, $(0^-, 1^-)$ mesons in the nuclear domain with massless quarks. (*Note*: Be careful with the antiquark contributions.)

38.1. Derive the scattering length in Eq. (38.25) from the Skyrme lagrangian in Eq. (38.24); include the term $\delta\mathcal{L}_{\rm csb} = \mathcal{F}_\pi m_\pi^2 \sigma$ and use finite mass pions (see Ref. [Q49]).

38.2. A global chiral transformation in the Skyrme model is represented by $\underset{\sim}{U} \to \underset{\sim}{U}' = \underset{\sim}{g}\underset{\sim}{U}\underset{\sim}{g}^{-1}$ where $\underset{\sim}{g}$ is another $SU(2)$ matrix, for then $\underset{\sim}{U}' = \exp\{i\vec{\tau}\cdot\vec{\phi}'\} = \sigma' + i\vec{\tau}\cdot\vec{\pi}'$ with $\sigma'^2 + \vec{\pi}'^2 = 1$. Prove the Skyrme lagrangian is chiral invariant.

38.3. Reference [Q49] gives an expression for the energy $E_{\rm S} = -\int \mathcal{L}_{\rm S}(r)d^3r$ of the static Skyrmion in Eq. (38.27)

$$E_{\rm S} = 4\pi \int_0^\infty dr \left\{ \frac{1}{2}\mathcal{F}_\pi^2 \left[r^2 \left(\frac{d\theta}{dr}\right)^2 + 2\sin^2\theta \right] + \frac{1}{2\mathcal{G}_\rho^2}\sin^2\theta \left[2\left(\frac{d\theta}{dr}\right)^2 + \frac{\sin^2\theta}{r^2} \right] \right\}$$

(a) Derive this result.
(b) Minimize the functional $E_{\rm S}(\theta)$ and derive the Euler-Lagrange equation for $\theta(r)$.
(c) Reproduce Fig. 38.4.[115]

39.1. Verify the expression for $d\sigma$ in Eq. (39.7).

39.2. Derive the general expression for the response tensor $W_{\mu\nu}$ in Eq. (39.11).

39.3. Verify the expression for $d^2\sigma/d\Omega_2 d\varepsilon_2$ in Eq. (39.19).

40.1. Show that in the $|\vec{p}| \to \infty$ frame in the deep inelastic region with small θ in the parton model one has $|\vec{q}| \approx \sqrt{q^2}$ and $\vec{q}$ is perpendicular to $\vec{k}_1 = -\vec{p}$ (Fig. 40.1).

40.2. Derive Eqs. (40.11) for the photoabsorption cross section σ_γ.

40.3. Derive the splitting function $P_{e\leftarrow\gamma} = [z^2 + (1-z)^2]/2$ in QED.

40.4. Derive the splitting function $P_{e\leftarrow e} = (1+z^2)/(1-z)_+ + (3/2)\delta(1-z)$ in QED. Here the singularity at $z = 1$ has been removed through $f(z)/(1-z)_+ \equiv [f(z) - f(1)]/(1-z)$; and the second term is added to properly account for depletion in this channel through satisfaction of the sum rules in Eqs. (L.4) and (L.7) (see Ref. [Q11]).

40.5. Show that to take into account depletion and to satisfy the sum rule in Eq. (L.7) $P_{\gamma\leftarrow\gamma} = -(1/3)\delta(1-z)$ in QED.

[115] One way to understand the boundary condition is to note that since the unit vector $\vec{r}/r$ is undefined at $r = 0$, $\sin\theta$ must vanish there.

40.6. Define moments of the distribution functions $e^n(\tau) \equiv \int_0^1 dx\, x^{n-1} e(x,\tau)$ with similar relations for $\bar{e}^n(\tau)$ and $\gamma^n(\tau)$. Also define $P^n_{b\leftarrow a} \equiv \int_0^1 dz\, z^{n-1} P_{b\leftarrow a}(z)$; note that these are now known quantities.
(a) Take moments of the master Eqs. (L.1) and show

$$\begin{aligned}
\frac{de^n(\tau)}{d\tau} &= \frac{\alpha(\tau)}{2\pi}[e^n(\tau)P^n_{e\leftarrow e} + \gamma^n(\tau)P^n_{e\leftarrow\gamma}] \\
\frac{d\bar{e}^n(\tau)}{d\tau} &= \frac{\alpha(\tau)}{2\pi}[\bar{e}^n(\tau)P^n_{e\leftarrow e} + \gamma^n(\tau)P^n_{e\leftarrow\gamma}] \\
\frac{d\gamma^n(\tau)}{d\tau} &= \frac{\alpha(\tau)}{2\pi}[\{e^n(\tau) + \bar{e}(\tau)\}P^n_{\gamma\leftarrow e} + \gamma^n(\tau)P^n_{\gamma\leftarrow\gamma}]
\end{aligned}$$

(b) Discuss the solution of these coupled, linear, first-order differential equations.

40.7. Show that in QCD (Refs. [Q83, Q84, Q11])

$$P_{g\leftarrow g}(z) = 2N_C\left[\frac{1-z}{z} + \frac{z}{(1-z)_+} + z(1-z) + \left(\frac{11}{12} - \frac{N_f}{6N_C}\right)\delta(1-z)\right]$$

APPENDICES: PART III

I Peierls' Inequality

This appendix is taken from Ref. [Q30]. The claim is that

$$\langle e^{f(x)} \rangle \geq e^{\langle f(x) \rangle} \tag{I.1}$$

Here the mean value is computed with any positive weighting function (measure).

For a proof, define first a convex function $\phi(x)$ that has $d^2\phi/dx^2 \geq 0$ in the interval $[a, b]$ as illustrated in Fig. I.1. Let x_0 be a point in the interval. The tangent to the curve at that point is given by

$$y = m(x - x_0) + \phi(x_0) \tag{I.2}$$

The tangent clearly stays below the curve everywhere in the interval

$$\phi(x) \geq m(x - x_0) + \phi(x_0) \tag{I.3}$$

Now suppose one has a one-to-one parameterization of x in the interval

$$x = f(t) \tag{I.4}$$

Compute the mean value of x with some positive weighting function $p(t)$ so that the mean value lies in the interval $[a, b]$; identify this mean value with the x_0 above. Thus

$$x_0 \equiv \langle x \rangle = \frac{\int p(t)f(t)dt}{\int p(t)dt} \tag{I.5}$$

Now take a similar mean value of the inequality in Eq. (I.3)

$$\begin{aligned} \frac{\int p(t)\phi[f(t)]dt}{\int p(t)dt} &\geq m\left[\frac{\int p(t)f(t)dt}{\int p(t)dt} - x_0\right] + \phi(x_0) \\ &\geq m(x_0 - x_0) + \phi(x_0) \\ &\geq \phi(x_0) \end{aligned} \tag{I.6}$$

Hence

$$\frac{\int p(t)\phi[f(t)]dt}{\int p(t)dt} \geq \phi\left(\frac{\int p(t)f(t)dt}{\int p(t)dt}\right) \tag{I.7}$$

This is Jensen's inequality. More concisely it states

$$\langle \phi[f(t)] \rangle \geq \phi[\langle f(t) \rangle] \tag{I.8}$$

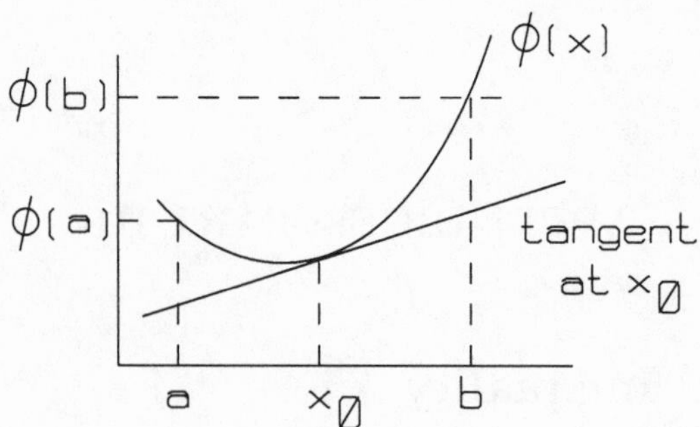

Figure I.1: Illustration of convex function used in proof of Peierls' theorem.

Now take as a specific convex function

$$\phi(x) = e^x \tag{I.9}$$

This function is convex on the entire interval $[-\infty, +\infty]$ as illustrated in Fig. I.2. Hence

$$\langle e^{f(t)} \rangle \geq e^{\langle f(t) \rangle} \tag{I.10}$$

Here $x = f(t)$ is any one-to-one mapping. This is the result that was to be proven.

J Symmetric $(T, S) = (\frac{1}{2}, \frac{1}{2})$ State

For $(T, S) = (\frac{1}{2}, \frac{1}{2})$ there are two states (λ, ρ) of each type. It is shown in the text that one must combine states of the same symmetry. The interchange of a pair of particles will mix the resulting states and only a linear combination can be totally symmetric. We demonstrate here that the following combination leads to a totally symmetric state

$$\Psi^N_{m_t,m_s} \equiv \frac{1}{\sqrt{2}} \left(\Phi^\lambda_{\frac{1}{2}m_t} \Xi^\lambda_{\frac{1}{2}m_s} + \Phi^\rho_{\frac{1}{2}m_t} \Xi^\rho_{\frac{1}{2}m_s} \right) \tag{J.1}$$

Start with $m_s = m_t = 1/2$, and multiply out the above wave function. Call $(\alpha, \beta) \equiv (\phi_{1/2}, \phi_{-1/2})$ etc. Then

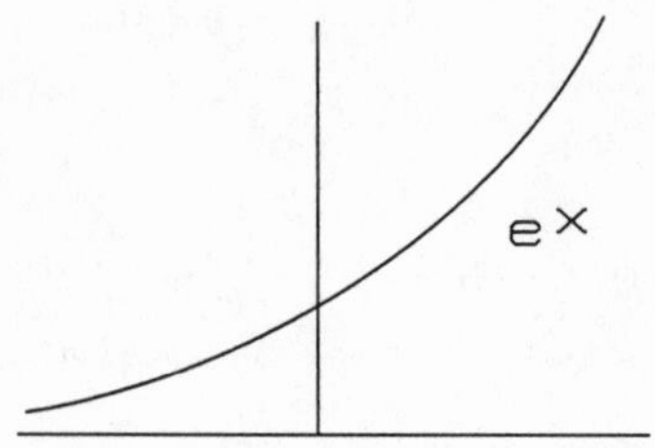

Figure I.2: e^x is convex for all x.

Table J.1: Coefficients of $2^{1/2}\Psi^N_{1/2,1/2}$ in direct product basis.

m_{t_1}	m_{t_2}	m_{t_3}	m_{s_1}	m_{s_2}	m_{s_3}	Coefficients	$P_{23}\Psi^N_{m_t m_s}$
1/2	1/2	−1/2	1/2	1/2	−1/2	2/3	2/3
			1/2	−1/2	1/2	−1/3	−1/3
			−1/2	1/2	1/2	−1/3	−1/3
1/2	−1/2	1/2	1/2	1/2	−1/2	−1/3	−1/3
			1/2	−1/2	1/2	2/3	2/3
			−1/2	1/2	1/2	−1/3	−1/3
−1/2	1/2	1/2	1/2	1/2	−1/2	−1/3	−1/3
			1/2	−1/2	1/2	−1/3	−1/3
			−1/2	1/2	1/2	2/3	2/3

$$\begin{aligned}
\sqrt{2}\,\Psi \;=\; & \frac{1}{\sqrt{2}}\left[\phi_{1/2}(1)\phi_{-1/2}(2)-\phi_{-1/2}(1)\phi_{1/2}(2)\right]\phi_{1/2}(3) \\
& \times\frac{1}{\sqrt{2}}\left[\chi_{1/2}(1)\chi_{-1/2}(2)-\chi_{-1/2}(1)\chi_{1/2}(2)\right]\chi_{1/2}(3) \\
& +\frac{1}{\sqrt{6}}\left[2\phi_{1/2}(1)\phi_{1/2}(2)\phi_{-1/2}(3)-\phi_{1/2}(1)\phi_{-1/2}(2)\phi_{1/2}(3)\right. \\
& \left.-\phi_{-1/2}(1)\phi_{1/2}(2)\phi_{1/2}(3)\right] \\
& \times\frac{1}{\sqrt{6}}\left[2\chi_{1/2}(1)\chi_{1/2}(2)\chi_{-1/2}(3)-\chi_{1/2}(1)\chi_{-1/2}(2)\chi_{1/2}(3)\right. \\
& \left.-\chi_{-1/2}(1)\phi_{1/2}(2)\chi_{1/2}(3)\right]
\end{aligned} \tag{J.2}$$

Read off the coefficients in the product basis $|m_{t_1}m_{t_2}m_{t_3}m_{s_1}m_{s_2}m_{s_3}\rangle$; they are shown in Table J.1. The application of P_{12} to this wave function gives $+1$ since all terms have a definite symmetry under $(1 \leftrightarrow 2)$. Apply P_{23} and read off the new coefficients in the direct product basis (last column); they are clearly identical to the starting coefficients (second last column). Application of P_{13} gives an identical result.The terms in the symmetrizing operator $\mathcal{S} = N\sum_{(P)} P$ are linear combinations of products of these particle interchange operators. Hence

$$\mathcal{S}\Psi^N_{1/2,1/2} = \Psi^N_{1/2,1/2} \tag{J.3}$$

Since $[\mathbf{T},\mathcal{S}] = [\mathbf{S},\mathcal{S}] = 0$ the indices (m_s, m_t) may be lowered with identical results.

K Skyrme Identity

Define

$$\underline{\mathrm{L}}_\mu = \frac{1}{2i}\frac{\partial \underline{\mathrm{U}}}{\partial x_\mu}\underline{\mathrm{U}}^\dagger = \frac{1}{2}\vec{\tau}\cdot\left(\frac{\partial \vec{\pi}}{\partial x_\mu}\times\vec{\pi}+\sigma\frac{\partial \vec{\pi}}{\partial x_\mu}-\vec{\pi}\frac{\partial \sigma}{\partial x_\mu}\right) \tag{K.1}$$

In this appendix we establish the Syrme identity

$$\frac{\partial \underline{\mathrm{L}}_\nu}{\partial x_\mu}-\frac{\partial \underline{\mathrm{L}}_\mu}{\partial x_\nu} = 2i[\underline{\mathrm{L}}_\mu,\underline{\mathrm{L}}_\nu] \tag{K.2}$$

Use will be made of the following relations

$$\sigma^2 + \vec{\pi}^2 = 1 \qquad\qquad \sigma\frac{\partial\sigma}{\partial x_\mu} = -\vec{\pi}\cdot\frac{\partial\vec{\pi}}{\partial x_\mu} \tag{K.3}$$

The l.h.s. of Eq. (K.2) takes the form

$$\begin{aligned}
\text{l.h.s.} &= \frac{1}{2}\vec{\tau}\cdot\left\{\frac{\partial}{\partial x_\mu}\left(\frac{\partial\vec{\pi}}{\partial x_\nu}\times\vec{\pi}+\sigma\frac{\partial\vec{\pi}}{\partial x_\nu}-\vec{\pi}\frac{\partial\sigma}{\partial x_\nu}\right)\right.\\
&\left.\quad-\frac{\partial}{\partial x_\nu}\left(\frac{\partial\vec{\pi}}{\partial x_\mu}\times\vec{\pi}+\sigma\frac{\partial\vec{\pi}}{\partial x_\mu}-\vec{\pi}\frac{\partial\sigma}{\partial x_\mu}\right)\right\}\\
&= \frac{1}{2}\vec{\tau}\cdot\left\{2\frac{\partial\vec{\pi}}{\partial x_\nu}\times\frac{\partial\vec{\pi}}{\partial x_\mu}+2\frac{\partial\sigma}{\partial x_\mu}\frac{\partial\vec{\pi}}{\partial x_\nu}-2\frac{\partial\sigma}{\partial x_\nu}\frac{\partial\vec{\pi}}{\partial x_\mu}\right\}
\end{aligned} \tag{K.4}$$

Use of the $SU(2)$ commutation relations for the tau matrices gives for the r.h.s. of Eq. (K.2)

$$\text{r.h.s.} = -\vec{\tau}\cdot\left(\frac{\partial\vec{\pi}}{\partial x_\mu}\times\vec{\pi}+\sigma\frac{\partial\vec{\pi}}{\partial x_\mu}-\vec{\pi}\frac{\partial\sigma}{\partial x_\mu}\right)\times\left(\frac{\partial\vec{\pi}}{\partial x_\nu}\times\vec{\pi}+\sigma\frac{\partial\vec{\pi}}{\partial x_\nu}-\vec{\pi}\frac{\partial\sigma}{\partial x_\nu}\right) \tag{K.5}$$

The cross product of the two terms in parentheses has nine terms

$$\begin{aligned}
(1) &= \frac{\partial\sigma}{\partial x_\mu}\frac{\partial\sigma}{\partial x_\nu}\vec{\pi}\times\vec{\pi}=0\\
(2) &= \sigma^2\frac{\partial\vec{\pi}}{\partial x_\mu}\times\frac{\partial\vec{\pi}}{\partial x_\nu}\\
(3+4) &= -\sigma\frac{\partial\sigma}{\partial x_\nu}\frac{\partial\vec{\pi}}{\partial x_\mu}\times\vec{\pi}-\sigma\frac{\partial\sigma}{\partial x_\mu}\vec{\pi}\times\frac{\partial\vec{\pi}}{\partial x_\nu}\\
(5+6) &= \sigma\left[\left(\frac{\partial\vec{\pi}}{\partial x_\mu}\times\vec{\pi}\right)\times\frac{\partial\vec{\pi}}{\partial x_\nu}+\frac{\partial\vec{\pi}}{\partial x_\mu}\times\left(\frac{\partial\vec{\pi}}{\partial x_\nu}\times\vec{\pi}\right)\right]\\
&= \sigma\left[-\frac{\partial\vec{\pi}}{\partial x_\mu}\left(\vec{\pi}\cdot\frac{\partial\vec{\pi}}{\partial x_\nu}\right)+\frac{\partial\vec{\pi}}{\partial x_\nu}\left(\vec{\pi}\cdot\frac{\partial\vec{\pi}}{\partial x_\mu}\right)\right]=\sigma^2\left[\frac{\partial\sigma}{\partial x_\nu}\frac{\partial\vec{\pi}}{\partial x_\mu}-\frac{\partial\sigma}{\partial x_\mu}\frac{\partial\vec{\pi}}{\partial x_\nu}\right]\\
(7+8) &= -\frac{\partial\sigma}{\partial x_\mu}\vec{\pi}\times\left(\frac{\partial\vec{\pi}}{\partial x_\nu}\times\vec{\pi}\right)-\frac{\partial\sigma}{\partial x_\nu}\left(\frac{\partial\vec{\pi}}{\partial x_\mu}\times\vec{\pi}\right)\times\vec{\pi}\\
&= -\frac{\partial\sigma}{\partial x_\mu}\left[\vec{\pi}^2\frac{\partial\vec{\pi}}{\partial x_\nu}-\vec{\pi}\left(\vec{\pi}\cdot\frac{\partial\vec{\pi}}{\partial x_\nu}\right)\right]+\frac{\partial\sigma}{\partial x_\nu}\left[\vec{\pi}^2\frac{\partial\vec{\pi}}{\partial x_\mu}-\vec{\pi}\left(\vec{\pi}\cdot\frac{\partial\vec{\pi}}{\partial x_\mu}\right)\right]\\
&= \vec{\pi}^2\left(\frac{\partial\sigma}{\partial x_\nu}\frac{\partial\vec{\pi}}{\partial x_\mu}-\frac{\partial\sigma}{\partial x_\mu}\frac{\partial\vec{\pi}}{\partial x_\nu}\right)\\
(9) &= \left(\frac{\partial\vec{\pi}}{\partial x_\mu}\times\vec{\pi}\right)\times\left(\frac{\partial\vec{\pi}}{\partial x_\nu}\times\vec{\pi}\right)\\
&= \frac{\partial\vec{\pi}}{\partial x_\nu}\left[\vec{\pi}\cdot\left(\frac{\partial\vec{\pi}}{\partial x_\mu}\times\vec{\pi}\right)\right]-\vec{\pi}\left[\frac{\partial\vec{\pi}}{\partial x_\nu}\cdot\left(\frac{\partial\vec{\pi}}{\partial x_\mu}\times\vec{\pi}\right)\right]\\
&= -\vec{\pi}\left[\vec{\pi}\cdot\left(\frac{\partial\vec{\pi}}{\partial x_\nu}\times\frac{\partial\vec{\pi}}{\partial x_\mu}\right)\right]
\end{aligned} \tag{K.6}$$

Addition of these nine terms and the use of Eq. (K.3) again gives

$$
\begin{aligned}
(1+\cdots+9) = & \left(\frac{\partial\sigma}{\partial x_\nu}\frac{\partial\vec{\pi}}{\partial x_\mu} - \frac{\partial\sigma}{\partial x_\mu}\frac{\partial\vec{\pi}}{\partial x_\nu}\right) + \sigma^2\frac{\partial\vec{\pi}}{\partial x_\mu}\times\frac{\partial\vec{\pi}}{\partial x_\nu} \\
& \underbrace{-\sigma\frac{\partial\sigma}{\partial x_\nu}\frac{\partial\vec{\pi}}{\partial x_\mu}\times\vec{\pi} - \sigma\frac{\partial\sigma}{\partial x_\mu}\vec{\pi}\times\frac{\partial\vec{\pi}}{\partial x_\nu} - \vec{\pi}\left[\vec{\pi}\cdot\left(\frac{\partial\vec{\pi}}{\partial x_\nu}\times\frac{\partial\vec{\pi}}{\partial x_\mu}\right)\right]} \\
= & \left(\vec{\pi}\cdot\frac{\partial\vec{\pi}}{\partial x_\nu}\right)\left(\frac{\partial\vec{\pi}}{\partial x_\mu}\times\vec{\pi}\right) + \left(\vec{\pi}\cdot\frac{\partial\vec{\pi}}{\partial x_\mu}\right)\left(\vec{\pi}\times\frac{\partial\vec{\pi}}{\partial x_\nu}\right) \\
& -\vec{\pi}\left[\vec{\pi}\cdot\left(\frac{\partial\vec{\pi}}{\partial x_\nu}\times\frac{\partial\vec{\pi}}{\partial x_\mu}\right)\right] \\
= & \vec{\pi}^2\left(\frac{\partial\vec{\pi}}{\partial x_\mu}\times\frac{\partial\vec{\pi}}{\partial x_\nu}\right)
\end{aligned} \qquad \text{(K.7)}
$$

The last relation follows from the vector identity

$$
(\vec{a}\cdot\vec{b})(\vec{c}\times\vec{a}) + (\vec{a}\cdot\vec{c})(\vec{a}\times\vec{b}) - \vec{a}[\vec{a}\cdot(\vec{b}\times\vec{c})] = \vec{a}^2(\vec{c}\times\vec{b}) \qquad \text{(K.8)}
$$

The proof of this relation follows by dotting $(\vec{a},\vec{b},\vec{c})$, assumed linearly independent, respectively, into both sides.[116]

Equations (K.5)-(K.7) then imply

$$
\begin{aligned}
\text{r.h.s.} &= \vec{\tau}\cdot\left\{\frac{\partial\vec{\pi}}{\partial x_\nu}\times\frac{\partial\vec{\pi}}{\partial x_\mu} + \frac{\partial\sigma}{\partial x_\mu}\frac{\partial\vec{\pi}}{\partial x_\nu} - \frac{\partial\sigma}{\partial x_\nu}\frac{\partial\vec{\pi}}{\partial x_\mu}\right\} \\
&= \text{l.h.s.}
\end{aligned} \qquad \text{(K.9)}
$$

This was the result to be proved.

L Sum Rules

In this appendix we derive two sum rules from the master equations for QED that we rewrite as

$$
\begin{aligned}
\frac{de(x,\tau)}{d\tau} &= \frac{\alpha(\tau)}{2\pi}\int_0^1 dz\int_0^1 dy\delta(zy-x)\left[e(y,\tau)P_{e\leftarrow e}(z) + \gamma(y,\tau)P_{e\leftarrow\gamma}(z)\right] \qquad \text{(L.1)} \\
\frac{d\bar{e}(x,\tau)}{d\tau} &= \frac{\alpha(\tau)}{2\pi}\int_0^1 dz\int_0^1 dy\delta(zy-x)\left[\bar{e}(y,\tau)P_{e\leftarrow e}(z) + \gamma(y,\tau)P_{e\leftarrow\gamma}(z)\right] \\
\frac{d\gamma(x,\tau)}{d\tau} &= \frac{\alpha(\tau)}{2\pi}\int_0^1 dz\int_0^1 dy\delta(zy-x)\left[\{e(y,\tau)+\bar{e}(y,\tau)\}P_{\gamma\leftarrow e}(z) + \gamma(y,\tau)P_{\gamma\leftarrow\gamma}(z)\right]
\end{aligned}
$$

The number of fermions is conserved. Hence

$$
\int_0^1 dx[e(x,\tau) - \bar{e}(x,\tau)] = N \qquad \text{(L.2)}
$$

[116] The case of linear dependence is left as an exercise for the reader.

Differentiation with respect to τ gives

$$\int_0^1 dx \left[\frac{de(x,\tau)}{d\tau} - \frac{d\bar{e}(x,\tau)}{d\tau} \right] = 0 \tag{L.3}$$

This result must hold for all distributions $[e(y,\tau), \bar{e}(y,\tau)]$. It then follows from the integrated difference of the first two of Eqs. (L.1) that

$$\int_0^1 dz P_{e\leftarrow e}(z) = 0 \tag{L.4}$$

Momentum is conserved. Hence

$$\int_0^1 x[e(x,\tau) + \bar{e}(x,\tau) + \gamma(x,\tau)]dx = 1 \tag{L.5}$$

Differentiate with respect to τ

$$\int_0^1 x dx \left[\frac{de(x,\tau)}{d\tau} + \frac{d\bar{e}(x,\tau)}{d\tau} + \frac{d\gamma(x,\tau)}{d\tau} \right] = 0 \tag{L.6}$$

This result must hold for all distributions $[e(x,\tau), \bar{e}(x,\tau), \gamma(x,\tau)]$. It follows from the integrated sum of Eqs. (L.1), each multiplied by x, that

$$\begin{aligned} \int_0^1 z dz [P_{e\leftarrow e}(z) + P_{\gamma\leftarrow e}(z)] &= 0 \\ \int_0^1 z dz [2P_{e\leftarrow\gamma}(z) + P_{\gamma\leftarrow\gamma}(z)] &= 0 \end{aligned} \tag{L.7}$$

Part IV

ELECTROWEAK INTERACTIONS WITH NUCLEI

42

WEAK INTERACTION PHENOMENOLOGY

In this section we review some basic phenomenology of the weak interactions. This material is the result of almost a half-century of beautiful experimental and theoretical work, and we cannot really do justice to this material in a superficial overview. Nonetheless, a familiarity with the basic weak interaction phenomenology is essential to the further development in this part of the book. The reader is referred to Refs. [W1, W2] for a much more extensive discussion of these topics and list of references to the original material.

Lepton Fields. Most leptons (l, ν_l) are light, or massless, and they can be created and destroyed in weak interactions. This indicates that they must be described with relativistic quantum fields. In the interaction representation, fermion fields take the following form

$$\psi(x) = \frac{1}{\sqrt{\Omega}} \sum_{\mathbf{k}\lambda} \left[a_{\mathbf{k}\lambda} u(\mathbf{k}\lambda) e^{ik\cdot x} + b^{\dagger}_{\mathbf{k}\lambda} v(-\mathbf{k}\lambda) e^{-ik\cdot x} \right] \tag{42.1}$$

In this expression a destroys a lepton, $b^{\dagger}$ creates an antilepton, and λ denotes the helicity with respect to the accompanying momentum variable.[1]

$V - A$ Theory. In the weak interactions, massless leptons are observed to couple through the following two-component fields

$$\phi \equiv \frac{1}{2}(1 + \gamma_5)\psi \tag{42.2}$$

Note the following properties of $\gamma_5 \equiv \gamma_1\gamma_2\gamma_3\gamma_4$

$$\begin{aligned}
\{\gamma_5, \gamma_\mu\} &= 0 \qquad\qquad \gamma_5^2 = 1 \\
\frac{1}{2}(1+\gamma_5)\frac{1}{2}(1+\gamma_5) &= \frac{1}{2}(1+\gamma_5) \\
\frac{1}{2}(1+\gamma_5)\frac{1}{2}(1-\gamma_5) &= 0 \\
\frac{1}{2}(1-\gamma_5)\frac{1}{2}(1-\gamma_5) &= \frac{1}{2}(1-\gamma_5)
\end{aligned} \tag{42.3}$$

[1]Hole theory implies that $v(-\mathbf{k}\lambda)$ is a negative-energy wave function with helicity λ with respect to $-\mathbf{k}$.

Equation (42.2) implies that the lepton coupling terms in the weak hamiltonian take the following form

$$\bar{\phi}_a O_i \phi_b = \frac{1}{4}\bar{\psi}_a(1-\gamma_5)O_i(1+\gamma_5)\psi_b \tag{42.4}$$

This expression *vanishes* for scalar (S), pseudoscalar (P), and tensor (T) couplings of the form $O_i = 1, \gamma_5, \sigma_{\mu\nu}$; and in the case of vector (V) and axial vector ($-A$) interactions $O_i = \gamma_\mu, \gamma_\mu\gamma_5$ the coupling is *unique*

$$\bar{\phi}_a O_i \phi_b = \frac{1}{4}\bar{\psi}_a(1-\gamma_5)\gamma_\mu(1+\gamma_5)\psi_b = \frac{1}{2}\bar{\psi}_a\gamma_\mu(1+\gamma_5)\psi_b \tag{42.5}$$

The early days of weak interactions are filled with studies to determine the nature of the couplings (S, P, T, V, A). The empirical evidence now is that it is entirely of this $V - A$ form (Ref. [W1]).

In the standard representation of the Dirac matrices

$$\begin{aligned} \gamma_5 &= \begin{pmatrix} 0 & -1 \\ -1 & 0 \end{pmatrix} \qquad\qquad \boldsymbol{\alpha} = \begin{pmatrix} 0 & \boldsymbol{\sigma} \\ \boldsymbol{\sigma} & 0 \end{pmatrix} \\ \gamma_5\boldsymbol{\alpha} = \boldsymbol{\alpha}\gamma_5 &= \begin{pmatrix} -\boldsymbol{\sigma} & 0 \\ 0 & -\boldsymbol{\sigma} \end{pmatrix} = -\boldsymbol{\sigma} \end{aligned} \tag{42.6}$$

The Dirac equation for the energy eigenstates in the case of massless particles is

$$\boldsymbol{\alpha}\cdot\mathbf{p}\psi = \pm E_p\psi \tag{42.7}$$

Multiplication by $(1+\gamma_5)/2$ and γ_5 in turn then leads to

$$\begin{aligned} \boldsymbol{\alpha}\cdot\mathbf{p}\phi &= \pm E_p\phi \\ -\boldsymbol{\sigma}\cdot\mathbf{p}\phi &= \pm E_p\phi \\ \boldsymbol{\sigma}\cdot(\mathbf{p}/|\mathbf{p}|)\phi &= \mp\phi \end{aligned} \tag{42.8}$$

The last equation exhibits the empirical *helicity* of the massless leptons; *particles are left-handed and antiparticles are right-handed* in the coupling in the weak interaction.

β-Decay Interaction. The empirical hamiltonian describing the weak interaction was first written by Fermi who, as legend has it, was teaching himself field theory. Written in terms of the ϕ with the corresponding unique $V - A$ coupling, the simplest form of the hamiltonian describing the β-decay process in Fig. 42.1 is

$$\begin{aligned} -\mathcal{L}_{\mathrm{W}} = \mathcal{H}_{\mathrm{W}} &= \frac{4G}{\sqrt{2}}(\bar{\phi}_p\gamma_\mu\phi_n)(\bar{\phi}_e\gamma_\mu\phi_{\nu_e}) + \text{h.c.} \\ &= \frac{G}{\sqrt{2}}[\bar{\psi}_p\gamma_\mu(1+\gamma_5)\psi_n][\bar{\psi}_e\gamma_\mu(1+\gamma_5)\psi_{\nu_e}] + \text{h.c.} \end{aligned} \tag{42.9}$$

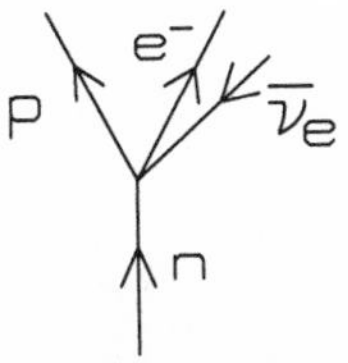

Figure 42.1: Basic β-decay process.

Here G is the Fermi coupling constant. The present convention is that $e \equiv e^-$ represents the particle in the charged lepton field, and e^+ the antiparticle.

Leptons. It is an empirical fact that if leptons are assigned a lepton number l then, as with the baryon number, the lepton number is a *conserved* quantity. Furthermore, suppose one groups the observed leptons into pairs in the following fashion

$$\begin{pmatrix} e^- \\ \nu_e \end{pmatrix} \qquad \begin{pmatrix} \mu^- \\ \nu_\mu \end{pmatrix} \qquad \begin{pmatrix} \tau^- \\ \nu_\tau \end{pmatrix} \tag{42.10}$$

Then all present experimental evidence indicates that the lepton numbers l_l of the individual pairs are *separately* conserved (Ref. [W1]).

Current-Current Theory. How do the other leptons couple in the weak interactions? The evidence is that there is a universal coupling to the charge-changing lepton current illustrated in Fig. 42.2a and given by

$$\begin{aligned} j_\lambda^{(-)}(\text{leptonic}) &= 2i\left[\bar{\phi}_e\gamma_\lambda\phi_{\nu_e} + \bar{\phi}_\mu\gamma_\lambda\phi_{\nu_\mu} + \bar{\phi}_\tau\gamma_\lambda\phi_{\nu_\tau} + \cdots\right] \qquad (42.11) \\ &= i\left[\bar{\psi}_e\gamma_\lambda(1+\gamma_5)\psi_{\nu_e} + \bar{\psi}_\mu\gamma_\lambda(1+\gamma_5)\psi_{\nu_\mu} + \bar{\psi}_\tau\gamma_\lambda(1+\gamma_5)\psi_{\nu_\tau} + \cdots\right] \end{aligned}$$

As is evident from β-decay, the total current in the charge-changing weak interactions has both a hadronic and leptonic part

$$\mathcal{J}_\lambda^{(-)} = \mathcal{J}_\lambda^{(-)}(\text{hadronic}) + j_\lambda^{(-)}(\text{leptonic}) \tag{42.12}$$

For nucleons with point Dirac couplings the weak charge-changing hadronic current describing, for example, the β-decay process $p \rightarrow n + e^+ + \nu_e$ as illustrated

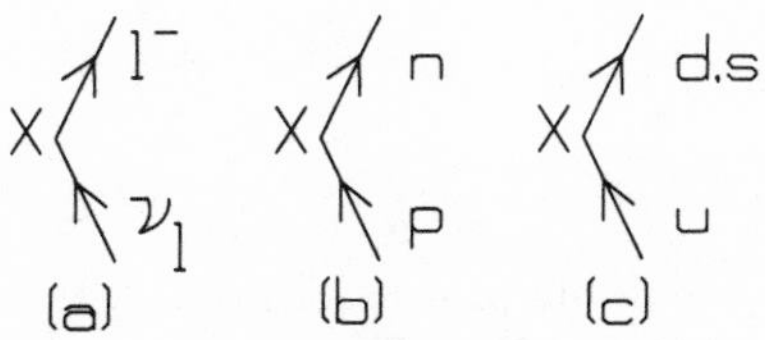

Figure 42.2: Charge-changing fermion currents in the weak interactions: (a) leptons; (b) point nucleons; (c) quarks.

in Fig. 42.2b takes the form

$$\begin{aligned} \mathcal{J}_\lambda^{(-)}(\text{hadronic}) &= 2i\bar{\phi}_n\gamma_\lambda\phi_p \\ &= i\bar{\psi}_n\gamma_\lambda(1+\gamma_5)\psi_p \end{aligned} \tag{42.13}$$

On a more basic level, the charge-changing weak interactions of the light quarks proceed through the reactions $u \to d + l^+ + \nu_l$ and $u \to s + l^+ + \nu_l$ as illustrated in Fig. 42.2c. The empirical form of the appropriate quark current is given by (Ref. [W1])

$$\begin{aligned} \mathcal{J}_\lambda^{(-)}(\text{hadronic}) &= 2i\left[\bar{\phi}_d\gamma_\lambda\phi_u\cos\theta_C + \bar{\phi}_s\gamma_\lambda\phi_u\sin\theta_C\right] \qquad (42.14) \\ &= i\left[\bar{\psi}_d\gamma_\lambda(1+\gamma_5)\psi_u\cos\theta_C + \bar{\psi}_s\gamma_\lambda(1+\gamma_5)\psi_u\sin\theta_C\right] \end{aligned}$$

Here $\sin\theta_C \approx \sin 13^o$ represents the Cabbibo angle; it is a *rotated quark field* $(\psi_d\cos\theta_C + \psi_s\sin\theta_C)$ that couples into the charge-changing weak interactions.

The adjoint of the above current describes weak processes where the charge is raised

$$\mathcal{J}_\lambda^{(+)} \equiv (\mathcal{J}_1^{(-)\dagger}, \mathcal{J}_2^{(-)\dagger}, \mathcal{J}_3^{(-)\dagger}, +i\mathcal{J}_0^{(-)\dagger}) \tag{42.15}$$

It is then an empirical fact that all the charge-changing weak interactions can be described through a *universal current-current interaction* of the currents in Eqs. (42.12) and (42.15)

$$\mathcal{H}_W(x) = -\frac{G}{\sqrt{2}}\mathcal{J}_\lambda^{(+)}\mathcal{J}_\lambda^{(-)} \tag{42.16}$$

Written in terms of the lepton and quark currents in Eqs. (42.11) and (42.14), this point four-fermion hamiltonian appears to provide the correct description of weak interactions in the nuclear domain (Ref. [W1]). One must realize, however, that the evaluation of the matrix elements of the quark fields in the nuclear domain involves all the complexities of strong-coupling QCD.

In fact, a remarkably good starting description of semileptonic weak interactions in the nuclear domain is given by the Fermi hamiltonian and the hadronic current of Eq. (42.13) with point Dirac fields

$$\begin{aligned} \mathcal{H}_W(x) &= \frac{G}{\sqrt{2}}\left[\bar{\psi}_{\nu_e}\gamma_\lambda(1+\gamma_5)\psi_e + \cdots + \bar{\psi}_p\gamma_\lambda(1+\gamma_5)\psi_n + \cdots\right] \\ &\quad \times \left[\bar{\psi}_e\gamma_\lambda(1+\gamma_5)\psi_{\nu_e} + \cdots + \bar{\psi}_n\gamma_\lambda(1+\gamma_5)\psi_p + \cdots\right] \end{aligned} \tag{42.17}$$

μ-Decay. The purely leptonic process of μ-decay is described by the hamiltonian in Eqs. (42.16) and (42.17) as illustrated in Fig. 42.3. Since no strong interactions are involved, the calculation of the decay amplitude and rate is a straightforward exercise in perturbation theory (Prob. 42.6). The experimental

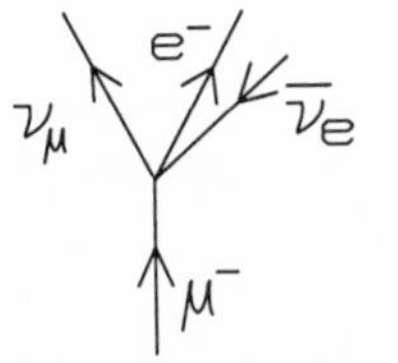

Figure 42.3: μ-decay as described by the point four-fermion interaction.

value of the decay rate can then be used to determine the value of the Fermi constant (Ref. [W1])[2]

$$G_\mu = \frac{1.0267 \times 10^{-5}}{m_p^2} \tag{42.18}$$

Note that the Fermi constant is not dimensionless; it has dimensions $1/m_p^2$. Thus the hamiltonian in Eq. (42.16) would appear to be an effective low-energy representation of a deeper underlying theory.

Conserved Vector Current Theory (CVC). The conserved vector current theory of Feynman and Gell-Mann is one of the loveliest and most powerful weak-interaction results (Ref. [W1]). To *motivate* it, start with point nucleons. The charge-changing weak current from above is then

$$\mathcal{J}_\lambda^{(+)} = i\bar{\psi}_p\gamma_\lambda(1+\gamma_5)\psi_n \tag{42.19}$$

Introduce the isodoublet nucleon field and Pauli matrices

$$\psi = \begin{pmatrix} p \\ n \end{pmatrix} \qquad\qquad \frac{1}{2}(1+\tau_3) = \begin{pmatrix} 1 & 0 \\ 0 & 0 \end{pmatrix}$$
$$\tau_+ = \frac{1}{2}(\tau_1 + i\tau_2) = \begin{pmatrix} 0 & 1 \\ 0 & 0 \end{pmatrix} \qquad \tau_- = \frac{1}{2}(\tau_1 - i\tau_2) = \begin{pmatrix} 0 & 0 \\ 1 & 0 \end{pmatrix} \tag{42.20}$$

The expression in Eq. (42.19) can then be rewritten as

$$\begin{aligned}\mathcal{J}_\lambda^{(+)} &= i\bar{\psi}\gamma_\lambda(1+\gamma_5)\tau_+\psi = i\bar{\psi}\gamma_\lambda(1+\gamma_5)\frac{1}{2}(\tau_1+i\tau_2)\psi \\ &\equiv J_\lambda^{(+)} + J_{\lambda 5}^{(+)}\end{aligned} \tag{42.21}$$

The electromagnetic current can similarly be rewritten as

$$J_\lambda^\gamma = i\bar{\psi}_p\gamma_\lambda\psi_p = i\bar{\psi}\gamma_\lambda\frac{1}{2}(1+\tau_3)\psi \tag{42.22}$$

[2]This value includes the lowest-order electromagnetic correction to the rate; the uncorrected value is $G_\mu^{\text{uncorr.}} = 1.024 \times 10^{-5}/m_p^2$ (Ref. [W1]).

Now the actual hadronic current depends on the details of hadronic structure and the strong interactions; even within a purely hadronic picture there will be additional mesonic currents in the weak interaction, and without some guiding principle they are unconstrained.[3] One can, however, attempt to *abstract the general symmetry properties of the currents* as provided by the above very simplistic model. If we use a subscript to denote the properties under Lorentz transformations, and a superscript to denote the transformation properties under isospin, then it is evident by inspection that the above currents have the following general characteristics

$$\begin{aligned}
\mathcal{J}_\lambda &= J_\lambda + J_{\lambda 5} &;\quad& \text{V} - \text{A} \\
\mathcal{J}_\lambda^{(\pm)} &= \mathcal{J}_\lambda^{V_1} \pm i\mathcal{J}_\lambda^{V_2} &;\quad& \text{Isovector} \\
J_\lambda^\gamma &= J_\lambda^S + J_\lambda^{V_3} &;\quad& \text{EM current} \\
J_\lambda^{(\pm)} &= J_\lambda^{V_1} \pm iJ_\lambda^{V_2} &;\quad& \text{CVC}
\end{aligned} \tag{42.23}$$

The first equation indicates that the weak current is the sum of a Lorentz vector and axial-vector, the second that the charge-changing weak current is an isovector, and the third that the electromagnetic current is the sum of an isoscalar and third component of an isovector; the last equation is the statement of CVC. The last relation states that the Lorentz vector part of the weak charge-changing current is simply obtained from the other spherical isospin components of the *same isovector operator that appears in the electromagnetic current.* As a consequence, one can relate matrix elements of the Lorentz vector part of the charge-changing weak currents to those of the isovector part of the electromagnetic current by use of the Wigner-Eckart theorem applied to isospin. The resulting relations are then *independent of the details of hadronic structure*; they depend only on the existence of the isospin symmetry of the strong interactions. CVC is a powerful, deep, and far-reaching result for it established the first direct relation between the electromagnetic and weak interactions which *a priori have nothing to do with each other!*

In fact, CVC goes further than this. The electromagnetic current J_λ^γ is conserved, and one expects the dynamically independent isoscalar and isovector contributions in Eq. (42.23) to be separately conserved. In CVC one identifies $J_\lambda^{\mathbf{V}}$ with the conserved isovector current arising from strong isospin symmetry (Section 21). The integral over the fourth component then yields the strong isospin operator,[4] and the full weak Lorentz vector, charge-changing current is then conserved (CVC). All known applications of CVC are consistent with experiment. A stronger motivation for CVC in terms of quarks will be presented later.

[3]CVC was developed before the discovery of the quark substructure of hadrons; indeed, one of the major triumphs of the quark picture with point electroweak couplings is the simple form and symmetry properties of the predicted electroweak currents (see later).

[4]See, for example, Prob. 42.2.

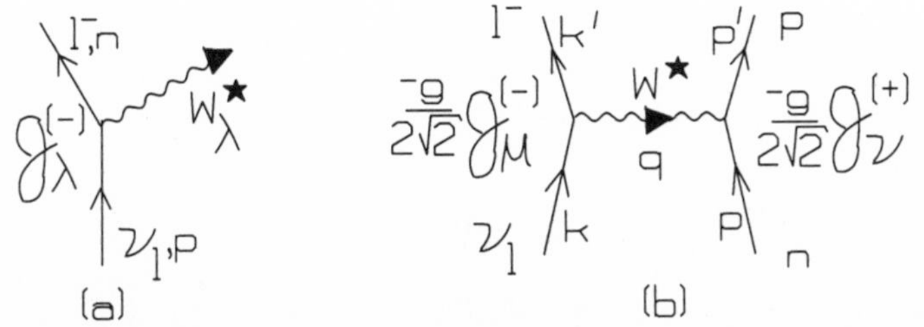

Figure 42.4: Interaction with charged weak vector boson: (a) basic vertex; (b) second-order Feynman diagram for $n + \nu_l \to p + l^-$ through boson exchange.

Intermediate Vector Bosons. The interaction in QED, the most successful physical theory we have, is mediated by the exchange of a vector boson, the photon. The Lorentz vector nature of the weak charge-changing current, the form of the effective low-energy current-current hamiltonian in Eq. (42.16), and the dimensional form of the Fermi constant all strongly suggest that the weak interactions are *also mediated by the exchange of a vector boson.* In contrast to the massless, neutral photon of QED, this weak vector boson must be *massive and charged.* The coupling of such a boson is illustrated in Fig. 42.4a and can be described with the following hermitian interaction lagrangian

$$\mathcal{L}(x) \quad = \quad \frac{g}{2\sqrt{2}}[\mathcal{J}_\mu^{(-)}W_\mu^\star + \mathcal{J}_\mu^{(+)}W_\mu] \tag{42.24}$$

Here $W_\mu = (\mathbf{W}, iW_0)$ is the weak vector boson field, which destroys a W^+ and creates a W^-, while $W_\mu^\star \equiv (\mathbf{W}^*, iW_0^*)$ creates a W^+ and destroys a W^-. g is a dimensionless coupling constant, and the charge-changing weak currents are those discussed above. The interaction hamiltonian follows as

$$\mathcal{H}_\text{I} \quad = \quad -\mathcal{L}_\text{I} = -\frac{g}{2\sqrt{2}}[\mathcal{J}_\mu^{(-)}W_\mu^\star + \mathcal{J}_\mu^{(+)}W_\mu] \tag{42.25}$$

Now use this expression to compute, for example, the S-matrix to order g^2 for the process $n+\nu_l \to p+l^-$ through W exchange as illustrated in Fig. 42.4b. The Feynman rules imply

1. There are two cross terms in $S^{(2)}$ arising from this hamiltonian; this cancels the 1/2! in the S-operator.
2. There is a factor of $(2\pi)^4\delta^{(4)}(\sum p_i)$ at each vertex.
3. There is an overall factor of $(-i)^2(-g/2\sqrt{2})^2$ for second order.
4. The vector meson propagator is[5]

$$\frac{1}{(2\pi)^4 i}\left(\delta_{\mu\nu} + \frac{q_\mu q_\nu}{M_\text{W}^2}\right)\frac{1}{q^2 + M_\text{W}^2} \tag{42.26}$$

[5]Note the factor of $q_\mu q_\nu/M_\text{W}^2$ in the numerator, which appears to preclude the possibility of making this a *renormalizable* theory.

5. One must take $\int d^4q$ over the internal four-momenta.

6. For the external wave functions, leave the following general expression, which takes the indicated form in the above Feynman diagram

$$\langle f|\mathcal{J}_\mu^{(-)}(0)|i\rangle \xrightarrow{\text{above}} \frac{i}{\Omega}\bar{u}_l(k')\gamma_\mu(1+\gamma_5)u_{\nu_l}(k)\ ; \qquad \text{etc.} \tag{42.27}$$

A combination of these results then gives the S-matrix

$$\begin{aligned} S_{\text{fi}}^{(2)} &= (2\pi)^8\delta^{(4)}(p+k-p'-k')(-i)^2\left(\frac{-g}{2\sqrt{2}}\right)^2\left[\frac{1}{(2\pi)^4 i}\right] \\ &\quad \times\left(\delta_{\mu\nu}+\frac{q_\mu q_\nu}{M_{\text{W}}^2}\right)\frac{1}{q^2+M_{\text{W}}^2}\langle f|\mathcal{J}_\mu^{(-)}(0)|i\rangle\langle f|\mathcal{J}_\nu^{(+)}(0)|i\rangle \end{aligned} \tag{42.28}$$

The momentum transfer $q = p' - p = k - k'$ is here controlled by the external variables. Suppose that the weak boson is very *heavy* so that $|q|/M_{\text{W}} \ll 1$. In this case the S-matrix simplifies to

$$S_{\text{fi}}^{(2)} \approx (2\pi)^4 i\delta^{(4)}(p+k-p'-k')\frac{g^2}{8M_{\text{W}}^2}\langle f|\mathcal{J}_\mu^{(-)}(0)|i\rangle\langle f|\mathcal{J}_\mu^{(+)}(0)|i\rangle \tag{42.29}$$

One can now define an *effective* lagrangian that, when treated in lowest order, gives exactly the same result as in Eq. (42.29)

$$\begin{aligned} -\mathcal{L}_{\text{eff}}(x) &= \mathcal{H}_{\text{eff}}(x) = -\frac{G}{\sqrt{2}}\mathcal{J}_\mu^{(-)}(x)\mathcal{J}_\mu^{(+)}(x) \\ \frac{G}{\sqrt{2}} &\equiv \frac{g^2}{8M_{\text{W}}^2} \end{aligned} \tag{42.30}$$

This is precisely the expression in Eq. (42.16).

These arguments are truly compelling; searches for this massive weak vector boson extended over several decades, resulting finally in its discovery at CERN in 1983 (Ref. [Q78]).

Applications of the semileptonic part of this lagrangian, describing semileptonic weak nuclear processes, will be discussed in detail later in this part of the book (Ref. [W2]).

Neutral Currents. The question immediately arises, if there are charged weak vector mesons, why not neutral ones also? The observation of the effects of weak neutral currents at low energies is difficult because many of their effects are masked by electromagnetic interactions.

The following hermitian lagrangian describes the interaction with a weak neutral meson as illustrated in Fig. 42.5a

$$\mathcal{L}_{\text{I}}(x) = \frac{f}{2}\mathcal{J}_\mu^{(0)}Z_\mu^{(0)} \tag{42.31}$$

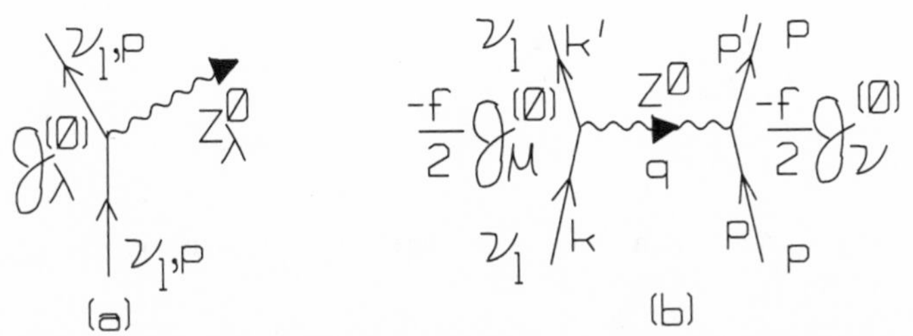

Figure 42.5: Interaction with neutral weak vector boson: (a) vertex; (b) second-order Feynman diagram for for process of $p+\nu_l \to p+\nu_l$ through boson exchange.

Here $Z_\mu^{(0)}$ is the neutral boson field and $\mathcal{J}_\mu^{(0)}$ a hermitian weak neutral current, as yet unspecified.

Consider the second-order Feynman diagram for the process $p + \nu_l \to p + \nu_l$ through boson exchange as illustrated in Fig. 42.5b. Proceed through the same steps as above. If the $Z_\mu^{(0)}$ is heavy, one arrives at an effective lagrangian of the form

$$-\mathcal{L}_{\text{eff}} = \mathcal{H}_{\text{eff}} = -\frac{f^2}{4M_Z^2} j_\lambda^{(0)}(\text{lepton})\mathcal{J}_\lambda^{(0)}(\text{nucleon}) \tag{42.32}$$

Here it is assumed, as before, that the weak neutral current is the sum of a leptonic and a hadronic part

$$\mathcal{J}_\lambda^{(0)} = \mathcal{J}_\lambda^{(0)}(\text{hadron}) + j_\lambda^{(0)}(\text{lepton}) \tag{42.33}$$

There will now be *two cross terms* in the current-current interaction of the form in Eq. (42.32), and hence the total current-current interaction is

$$\begin{aligned} -\mathcal{L}_{\text{eff}}(x) &= \mathcal{H}_{\text{eff}}(x) = -\frac{G'}{\sqrt{2}}\mathcal{J}_\lambda^{(0)}(x)\mathcal{J}_\lambda^{(0)}(x) \\ \frac{G'}{\sqrt{2}} &\equiv \frac{f^2}{8M_Z^2} \end{aligned} \tag{42.34}$$

The standard model of electroweak interactions of Weinberg and Salam (Section 43) relates the charged and neutral coupling constants and masses.[6] The heavy neutral weak vector meson Z_μ^0 was also discovered at CERN in 1983 (Ref. [Q78]).

Single-Nucleon Matrix Elements of the Currents. Consider the full single-nucleon matrix element of the charge-changing weak current (Fig. 42.6). The strong interactions determine the exact value of the matrix elements of the currents at any momentum transfer; however, the use of Lorentz invariance,

[6] We will derive the following relations between the coupling constants and masses in the standard model: $G'/\sqrt{2} = f^2/8M_Z^2 = g^2/8M_Z^2\cos^2\theta_W = g^2/8M_W^2 = G/\sqrt{2}$.

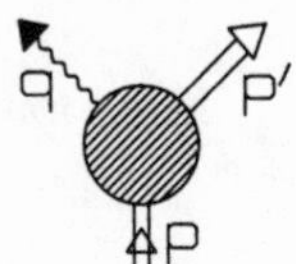

Figure 42.6: Kinematics in single-nucleon weak vertex.

parity invariance, isospin invariance, and the Dirac equation permits one to write the *general form* of these matrix elements (Ref. [R4], Probs. 42.3-5)

$$\langle p'|J_\mu^{(-)}(0)|p\rangle = \frac{i}{\Omega}\bar{u}(p')[F_1\gamma_\mu + F_2\sigma_{\mu\nu}q_\nu + iF_S q_\mu]\tau_- u(p)$$
$$\langle p'|J_{\mu 5}^{(-)}(0)|p\rangle = \frac{i}{\Omega}\bar{u}(p')[F_A\gamma_5\gamma_\mu - iF_P\gamma_5 q_\mu - F_T\gamma_5\sigma_{\mu\nu}q_\nu]\tau_- u(p) \quad (42.35)$$

Here $q \equiv p - p'$ with $p^2 = p'^2 = -m^2$, and all the form factors $F_i(q^2)$ are functions of q^2. We suppress the isospin wave functions.

It is a theorem due to Weinberg that if the currents have the same transformation properties under time reversal $\hat{T}$, parity $\hat{P}$, charge conjugation $\hat{C}$, and strong isospin $\hat{\mathbf{T}}$ as the point fermion currents in Eqs. (42.21) and (42.22), then (Refs. [W1, W2], Probs. 42.3-5)

$$F_S = F_T = 0 \quad (42.36)$$

These terms are labeled *second class currents* by Weinberg; they are absent in the standard model.

The general structure of the single-nucleon matrix element of the *electromagnetic* current follows in the same fashion; it has already been employed in Section 8.

$$\langle p'|J_\mu^\gamma(0)|p\rangle = \frac{i}{\Omega}\bar{u}(p')[F_1^\gamma\gamma_\mu + F_2^\gamma\sigma_{\mu\nu}q_\nu]u(p) \quad (42.37)$$

The isospin structure is given by

$$\begin{aligned} F_i^\gamma &= \frac{1}{2}(F_i^S + \tau_3 F_i^V)\,; \qquad i = 1,2 \\ F_1^S(0) &= F_1^V(0) = 1 \\ 2mF_2^S(0) &= \lambda_p' + \lambda_n = -0.120 \\ 2mF_2^V(0) &= \lambda_p' - \lambda_n = 3.706 \end{aligned} \quad (42.38)$$

The conserved vector current theory in Eq. (42.23) now provides nontrivial relations between the weak and electromagnetic form factors of the nucleon. The observation that $(\tau_1 \pm i\tau_2)/2 = \tau_\pm$ immediately leads to the relations

$$F_i = F_i^V\,; \qquad i = 1,2 \quad (42.39)$$

Hence CVC implies that

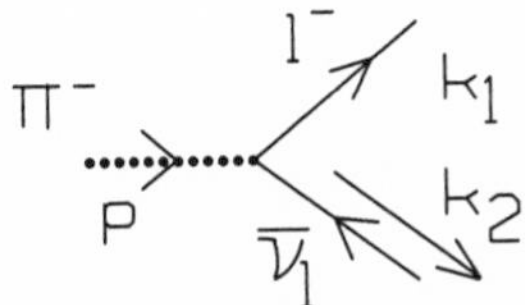

Figure 42.7: Weak decay of the pion.

the entire Lorentz vector part of the single-nucleon matrix element of the charge-changing weak current, whatever the detailed dynamic structure of the nucleon, can be obtained from elastic electron scattering through the electromagnetic interaction!

Pion Decay. The weak decay of the pion (Fig. 42.7) plays a central role in the phenomenology of the weak interactions. By Lorentz covariance, the hadronic part of the decay amplitude must take the following form

$$\langle \text{vac}|\mathcal{J}_\lambda^{(+)}(0)|\pi^-;p\rangle = \frac{1}{\sqrt{2\omega_p\Omega}} iF_\pi(p^2)p_\lambda \tag{42.40}$$

Here $p^2 = -\mu^2 = -m_\pi^2$. Go to the rest frame of the pion. It is then evident that since the pion has $J^\pi = 0^-$, only the axial vector current can contribute to this matrix element.

It follows from the weak hamiltonian density in Eq. (42.16), that the S-matrix for this process is given by[7]

$$S_{\text{fi}} = -\frac{(2\pi)^4 i}{\sqrt{\Omega^3}}\delta^{(4)}(p-k_1-k_2)\frac{G}{\sqrt{2}}\frac{1}{\sqrt{2\omega_p}}F_\pi \bar{u}(k_1)\gamma_\lambda p_\lambda(1+\gamma_5)v(-k_2) \tag{42.41}$$

The pion decay rate is then given by (Prob. 42.1)

$$\omega(\pi \to l+\nu_l) = \frac{G^2F_\pi^2}{8\pi}m_\pi^2\left(\frac{m_l}{m_\pi}\right)^2\left(1-\frac{m_l^2}{m_\pi^2}\right)^2 m_\pi \tag{42.42}$$

Comparison with experiment then yields the result (Prob. 42.1)

$$F_\pi \equiv m_\pi \bar{f} \approx 0.92\, m_\pi \tag{42.43}$$

Note only the magnitude of F_π is determined here.

Pion-Pole Dominance of the Induced Pseudoscalar Coupling. There is a one-pion exchange process that contributes to the weak charge-changing lepton-nucleon scattering amplitude $l^- + p \to \nu_l + n$ (Fig. 42.8). This process produces a *pole* in the scattering amplitude at a four-momentum transfer $q^2 = -m_\pi^2$. The

[7]Note that the field $\pi = (\pi_1 - i\pi_2)/\sqrt{2}$ creates π^-, $\pi^* = (\pi_1 + i\pi_2)/\sqrt{2}$ creates π^+, and $\vec{\tau}\cdot\vec{\pi} \equiv \sqrt{2}(\pi\tau_+ + \pi^*\tau_-) + \pi_3\tau_3$.

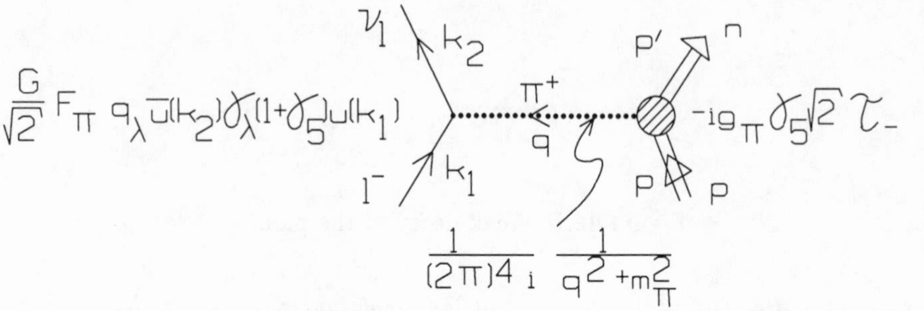

Figure 42.8: Pion pole contribution to the weak charge-changing lepton-nucleon scattering amplitude.

contribution to the S-matrix from the Feynman diagram in Fig. 42.8 follows from the preceding analysis as

$$
\begin{aligned}
S_{\mathrm{f\,i}} &= \frac{(2\pi)^4}{\Omega^2}\delta^{(4)}(p+k_1-p'-k_2) \\
&\quad \times \frac{GF_\pi}{\sqrt{2}} q_\lambda \bar{u}(k_2)\gamma_\lambda(1+\gamma_5)u(k_1)\frac{1}{q^2+m_\pi^2}\bar{u}(p')g_\pi\gamma_5\sqrt{2}\tau_- u(p)
\end{aligned} \tag{42.44}
$$

This expression is exact on the pole, where the vertices are to be evaluated for real particles on the mass shell.

On the other hand, the general form of the S-matrix for the process $p+l^- \to n+\nu_l$ follows from the current-current interaction and the weak nucleon vertex Γ_λ defined in Eqs. (42.35)

$$
S_{\mathrm{f\,i}} = \frac{-iG}{\sqrt{2}}\frac{(2\pi)^4}{\Omega^2}\delta^{(4)}(p+k_1-p'-k_2)\bar{u}(k_2)\gamma_\lambda(1+\gamma_5)u(k_1)\bar{u}(p')\Gamma_\lambda u(p) \tag{42.45}
$$

Comparison of these results allows an identification of the pion-pole contribution to the induced pseudoscalar form factor F_P

$$
-iF_P\gamma_5 q_\lambda \tau_- = iF_\pi\gamma_5 q_\lambda\sqrt{2}g_\pi\tau_-\frac{1}{q^2+m_\pi^2} \tag{42.46}
$$

The result is

$$
F_P = \frac{-\sqrt{2}g_\pi F_\pi}{q^2+m_\pi^2} \tag{42.47}
$$

If one now makes the bold assumption that the pion-pole contribution dominates in the experimentally accessible region of small momentum transfer, then Eq. (42.47) provides an explicit expression for the induced pseudoscalar coupling.

Goldberger-Treiman Relation. We previously introduced the notion of the partially conserved axial vector current (PCAC) in Sections 21 and 22

$$
\frac{\partial J_{\lambda 5}(x)}{\partial x_\lambda} = O(m_\pi^2)
$$

$$\stackrel{m_\pi \to 0}{\longrightarrow} 0 \tag{42.48}$$

While historically the argument ran in the opposite direction, if one *assumes PCAC*, then for the single-nucleon matrix element of the divergence of the axial vector current[8]

$$\begin{aligned} \langle p'|\frac{\partial J_{\lambda 5}(0)}{\partial x_\lambda}|p\rangle &= iq_\lambda \langle p'|J_{\lambda 5}(0)|p\rangle \\ &= \frac{i}{\Omega}\bar{u}(p')[F_A\gamma_5 i\gamma_\lambda q_\lambda + q^2 F_P \gamma_5]\tau_- u(p) \end{aligned} \tag{42.49}$$

Here the second equality follows from Eqs. (42.35) and (42.36). Now use the Dirac equation on the first term with $q = p - p'$

$$\begin{aligned} \bar{u}(p')\gamma_5(i\gamma_\lambda p_\lambda - i\gamma_\lambda p'_\lambda)u(p) &= \bar{u}(p')(i\gamma_\lambda p'_\lambda \gamma_5 + \gamma_5 i\gamma_\lambda p_\lambda)u(p) \\ &= -2m\bar{u}(p')\gamma_5 u(p) \end{aligned} \tag{42.50}$$

Hence

$$\langle p'|\frac{\partial J_{\lambda 5}(0)}{\partial x_\lambda}|p\rangle = \frac{i}{\Omega}\bar{u}(p')[-2mF_A + q^2 F_P]\gamma_5\tau_- u(p) \tag{42.51}$$

Now use the pion-pole result in Eq. (42.47); the expression in square brackets becomes

$$\begin{aligned} -2mF_A + q^2 F_P &= -2mF_A + \frac{q^2}{q^2+m_\pi^2}(-\sqrt{2}g_\pi F_\pi) \\ &\equiv -2mF_A - \sqrt{2}g_\pi F_\pi - m_\pi^2\frac{(-\sqrt{2}g_\pi F_\pi)}{q^2+m_\pi^2} \end{aligned} \tag{42.52}$$

For this result to satisfy Eq. (42.48), the first two terms must cancel to $O(m_\pi^2)$

$$-2mF_A = \sqrt{2}g_\pi\, F_\pi \tag{42.53}$$

This is the Goldberger-Treiman relation (Ref. [W1]); PCAC implies that this relation is exact in the limit $m_\pi \to 0$.

Use the following numerical values

$$\begin{aligned} F_A(0) &= -1.23 \pm 0.01 & m_{\pi\pm} &= 139.6\,\text{MeV} \\ \frac{g_\pi^2}{4\pi} &= 14.4 & m_p &= 938.3\,\text{MeV} \end{aligned} \tag{42.54}$$

These lead to the following prediction for the pion decay constant from the Goldberger-Treiman relation

$$F_\pi \equiv \bar{f}m_\pi = 0.87\,m_\pi \tag{42.55}$$

It agrees with the result obtained from pion decay in Eq. (42.43) to better than 10%.[9]

[8] From the Heisenberg equations of motion $\langle p'|J_{\mu5}(x)|p\rangle = e^{i(p-p')\cdot x}\langle p'|J_{\mu5}(0)|p\rangle$.

[9] Note that in this section, as in Appendix A, we have explicitly included a factor of $(-i)^n$ where n is the order in the Feynman rules for the S-matrix.

43

INTRODUCTION TO THE STANDARD MODEL

The development of a unified theory of the electroweak interactions surely must be regarded as one of the great intellectual achievements of our era. After the brief introduction to the phenomenology of the weak interactions in the previous section, we are now in a position to discuss the so-called *standard model* of the electroweak interactions originally presented in Refs. [W3, W4, W5, W6] (see also Ref. [W1].)

Spinor Fields. A spinor field can always be decomposed as follows

$$\begin{aligned}
\psi &= \frac{1}{2}(1+\gamma_5)\psi + \frac{1}{2}(1-\gamma_5)\psi \equiv \psi_L + \psi_R \\
\bar{\psi}\gamma_\mu \frac{\partial}{\partial x_\mu}\psi &= \bar{\psi}_L\gamma_\mu \frac{\partial}{\partial x_\mu}\psi_L + \bar{\psi}_R\gamma_\mu \frac{\partial}{\partial x_\mu}\psi_R \\
\bar{\psi}\psi &= \bar{\psi}_L\psi_R + \bar{\psi}_R\psi_L
\end{aligned} \tag{43.1}$$

Leptons. The lepton fields for the electron and electron neutrino[10] will be combined in the following fashion:

$$\psi_l = \begin{pmatrix} \psi_{\nu_e} \\ \psi_e \end{pmatrix} \equiv \begin{pmatrix} \nu \\ e \end{pmatrix} \tag{43.2}$$

The fields (L, R) are defined by

$$\begin{aligned}
L &\equiv \begin{pmatrix} \nu_L \\ e_L \end{pmatrix} = \frac{1}{2}(1+\gamma_5)\psi_l \\
R &\equiv e_R = \frac{1}{2}(1-\gamma_5)\psi_e
\end{aligned} \tag{43.3}$$

The *kinetic energy* of the leptons is then given by[11]

$$\begin{aligned}
\mathcal{L}^0_{\text{lepton}} &= -\left[\bar{\psi}_e\gamma_\mu \frac{\partial}{\partial x_\mu}\psi_e + \bar{\nu}_L\gamma_\mu \frac{\partial}{\partial x_\mu}\nu_L\right] \\
&= -\left[\bar{L}\gamma_\mu \frac{\partial}{\partial x_\mu}L + \bar{R}\gamma_\mu \frac{\partial}{\partial x_\mu}R\right]
\end{aligned} \tag{43.4}$$

[10] And similarly for the other leptons.

[11] There is only one neutrino field in the standard model $\nu_L \equiv \frac{1}{2}(1+\gamma_5)\psi_\nu$; it describes left-handed neutrinos and right-handed antineutrinos. This is put in by hand, as is the fact that this neutrino is massless $m_\nu = 0$.

This lagrangian is invariant under a global $SU(2)_W$ symmetry — a weak (left-handed) isospin — which treats the field L as a weak isodoublet and R as a weak isosinglet. From our previous discussions, the generators for this $SU(2)_W$ symmetry can be immediately written in terms of the above fields as

$$\begin{aligned} \hat{T}_W^i &= \int L^\dagger(\mathbf{x})\frac{1}{2}\tau_i L(\mathbf{x})d^3x \\ &= \int \psi_l^\dagger(\mathbf{x})\frac{1}{2}\tau_i\frac{1}{2}(1+\gamma_5)\psi_l(\mathbf{x})d^3x \end{aligned} \tag{43.5}$$

It follows immediately from the canonical (anti)commutation relations that these generators satisfy an $SU(2)$ algebra

$$[\hat{T}_W^i, \hat{T}_W^j] = i\varepsilon_{ijk}\hat{T}_W^k \tag{43.6}$$

The finite symmetry transformations are given by

$$\begin{aligned} \exp\{i\boldsymbol{\theta}\cdot\hat{\mathbf{T}}_W\}\,L\,\exp\{-i\boldsymbol{\theta}\cdot\hat{\mathbf{T}}_W\} &= [e^{-\frac{i}{2}\boldsymbol{\theta}\cdot\boldsymbol{\tau}}]\,L\,; \qquad \text{doublet} \\ \exp\{i\boldsymbol{\theta}\cdot\hat{\mathbf{T}}_W\}\,R\,\exp\{-i\boldsymbol{\theta}\cdot\hat{\mathbf{T}}_W\} &= [1]\,R \qquad ; \qquad \text{singlet} \end{aligned} \tag{43.7}$$

These equations follow from the projection properties of $(1\pm\gamma_5)/2$, which imply

$$\begin{aligned} [\hat{T}_W^i, L] &= -\frac{1}{2}\tau_i\left[\frac{1}{2}(1+\gamma_5)\right]^2\psi_l = -\frac{1}{2}\tau_i L \\ [\,\hat{T}_W^i, R] &= (0,1)\left[-\frac{1}{2}\tau_i\right]\frac{1}{2}(1-\gamma_5)\frac{1}{2}(1+\gamma_5)\psi_l = 0 \end{aligned} \tag{43.8}$$

The *mass term* for the electron has the following form

$$-m_e\bar{\psi}_e\psi_e = -m_e[\bar{e}_L e_R + \bar{e}_R e_L] \tag{43.9}$$

This expression is *not* invariant under $SU(2)_W$. Hence if one wants to build on this symmetry, it is necessary to start with *massless fermions.*

Point Nucleons. As in the previous section, the corresponding lagrangian for point Dirac nucleon fields illustrates the general structure of the theory (Ref. [W4]), and for clarity of concept, we start with this simple description of the hadronic sector. The extension to matrix elements for *physical nucleons* then follows from general symmetry considerations, leading to Eqs. (42.35) and (42.36). The deeper formulation of the standard model in terms of quarks is discussed in the next section.

Thus we here include proton and neutron fields in a manner analogous to the above

$$\begin{aligned} N_L &= \begin{pmatrix} p_L \\ n_L \end{pmatrix} = \frac{1}{2}(1+\gamma_5)\psi_N\,; \qquad \text{doublet} \\ & \quad p_R, n_R \qquad\qquad\qquad ; \qquad \text{singlets} \\ \mathcal{L}^0_{\text{nucleon}} &= -\left[\bar{N}_L\gamma_\mu\frac{\partial}{\partial x_\mu}N_L + \bar{p}_R\gamma_\mu\frac{\partial}{\partial x_\mu}p_R + \bar{n}_R\gamma_\mu\frac{\partial}{\partial x_\mu}n_R\right] \end{aligned} \tag{43.10}$$

Table 43.1: Weak symmetry quantum numbers in the standard model.

Particle/field	T_W	T_{3W}	Y_W	Q
$(\nu_e)_L$	1/2	1/2	−1	0
e_L	1/2	−1/2	−1	−1
e_R	0	0	−2	−1
p_L	1/2	1/2	1	1
n_L	1/2	−1/2	1	0
p_R	0	0	2	1
n_R	0	0	0	0
ϕ^+	1/2	1/2	1	1
ϕ^0	1/2	−1/2	1	0
$\tilde{\phi}^0$	1/2	1/2	−1	0
$\tilde{\phi}^-$	1/2	−1/2	−1	−1

This lagrangian is now also invariant under $SU(2)_W$; again this is true only if one starts with massless fermions.

Weak Hypercharge. Introduce an additional global $U(1)_W$ symmetry — weak hypercharge — defined so that the fields transform according to

$$\exp\{i\alpha\hat{Y}_W\}\,\phi\,\exp\{-i\alpha\hat{Y}_W\} \quad = \quad e^{-i\alpha Y_W}\phi \tag{43.11}$$

Assign quantum numbers to the fields (and corresponding particles) so that the lagrangian is invariant and the *electric charge* is still given by the Gell-Mann - Nishijima relation

$$Q \quad = \quad (T_3 + \frac{1}{2}Y)_W \tag{43.12}$$

Conservation of electric charge will always be imposed as an exact symmetry of the theory. Assignments of the weak quantum numbers for the fields introduced so far are shown in Table 43.1.

The *generator* for the weak hypercharge symmetry for the fermions is readily constructed in second quantization, as are those for the electric charge operator and third component of weak isospin, by (Table 43.1)

$$\begin{aligned}
\hat{Y}_W &= \int [L^\dagger(-1)L + R^\dagger(-2)R + N_L^\dagger N_L + 2p_R^\dagger p_R]d^3x \\
\hat{Q} &= \int [L^\dagger\frac{1}{2}(\tau_3 - 1)L - R^\dagger R + N_L^\dagger\frac{1}{2}(1+\tau_3)N_L + p_R^\dagger p_R]d^3x \\
&= \int [-\psi_e^\dagger\psi_e + \psi_p^\dagger\psi_p]d^3x \\
\hat{T}_{3W} &= \int [L^\dagger\frac{1}{2}\tau_3 L + N_L^\dagger\frac{1}{2}\tau_3 N_L]d^3x
\end{aligned} \tag{43.13}$$

Hence

$$
\begin{aligned}
\hat{Q} &= \hat{T}_{3W} + \frac{1}{2}\hat{Y}_W \\
[\,\hat{T}^i_W, \hat{Y}_W] &= 0
\end{aligned}
\tag{43.14}
$$

Local Gauge Symmetry. Now make this a Yang-Mills local gauge theory based on the symmetry group $SU(2)_W \bigotimes U(1)_W$. The technique for doing this has been previously discussed in detail. The only slight new complexity is that now one has the direct product of two symmetry groups with commuting generators [Eq. (43.14)]; however, an examination of the basic concept shows that this is an inessential complication. The steps of the Yang-Mills construction are as follows:

1. Add *gauge bosons*, one for each of the generators $(\hat{T}^i_W, \hat{Y}_W)$

$$
\begin{aligned}
&A^i_\mu(x)\,; \qquad\qquad i = 1, 2, 3 \\
&B_\mu(x)
\end{aligned}
\tag{43.15}
$$

2. Use the *covariant derivative* in the lagrangian

$$
\begin{aligned}
\frac{\partial}{\partial x_\mu} \to \frac{D}{D x_\mu} &\equiv \left(\frac{\partial}{\partial x_\mu} - \frac{i}{2} g' Y_W B_\mu - \frac{i}{2} g \boldsymbol{\tau}\cdot\mathbf{A}_\mu\right)\,; \qquad \text{on doublets} \\
&\equiv \left(\frac{\partial}{\partial x_\mu} - \frac{i}{2} g' Y_W B_\mu\right)\,; \qquad \text{on singlets}
\end{aligned}
\tag{43.16}
$$

3. Include a kinetic energy term for the gauge bosons

$$
\mathcal{L}_{\text{gauge}} = -\frac{1}{4}\left(\frac{\partial B_\nu}{\partial x_\mu} - \frac{\partial B_\mu}{\partial x_\nu}\right)^2 - \frac{1}{4}\left(\frac{\partial \mathbf{A}_\nu}{\partial x_\mu} - \frac{\partial \mathbf{A}_\mu}{\partial x_\nu} + g\mathbf{A}_\mu \times \mathbf{A}_\nu\right)^2 \tag{43.17}
$$

4. Mass terms of the form $m_B^2 B_\mu B_\mu$ or $m_A^2 \mathbf{A}_\mu \cdot \mathbf{A}_\mu$ break the local gauge invariance; hence the gauge bosons must be *massless*.

The Yang-Mills lagrangian thus takes the form

$$
\begin{aligned}
\mathcal{L}_{\text{lepton}} &= -\left[\bar{L}\gamma_\mu\left(\frac{\partial}{\partial x_\mu} - \frac{i}{2}(-1)g' B_\mu - \frac{i}{2} g\boldsymbol{\tau}\cdot\mathbf{A}_\mu\right) L\right. \\
&\qquad \left. + \bar{R}\gamma_\mu\left(\frac{\partial}{\partial x_\mu} - \frac{i}{2}(-2)g' B_\mu\right) R\right] \\
\mathcal{L}_{\text{nucleon}} &= -\left[\bar{N}_L\gamma_\mu\left(\frac{\partial}{\partial x_\mu} - \frac{i}{2}(1)g' B_\mu - \frac{i}{2} g\boldsymbol{\tau}\cdot\mathbf{A}_\mu\right) N_L\right. \\
&\qquad \left. + \bar{p}_R\gamma_\mu\left(\frac{\partial}{\partial x_\mu} - \frac{i}{2}(2)g' B_\mu\right) p_R + \bar{n}_R\gamma_\mu\left(\frac{\partial}{\partial x_\mu} - \frac{i}{2}(0)g' B_\mu\right) n_R\right] \\
\mathcal{L}_{\text{gauge}} &= -\frac{1}{4} B_{\mu\nu} B_{\mu\nu} - \frac{1}{4}\mathcal{F}^i_{\mu\nu}\mathcal{F}^i_{\mu\nu}
\end{aligned}
\tag{43.18}
$$

Vector Meson Masses. As in our discussion of the σ-model in Section 22, the masses for the gauge bosons will be generated by *spontaneous symmetry breaking.* One proceeds to

1. Introduce a weak isodoublet of *complex scalar mesons*

$$\underline{\phi} \equiv \begin{pmatrix} \phi^+ \\ \phi^0 \end{pmatrix} \tag{43.19}$$

2. Assign weak quantum numbers as indicated in Table 43.1.
3. Use the covariant derivative of Eq. (43.16).
4. Add a term to the lagrangian for this scalar field that is invariant under local $SU(2)_W \bigotimes U(1)_W$

$$\mathcal{L}_{\text{scalar}} = -\left(\frac{D\underline{\phi}}{Dx_\mu}\right)^\star \left(\frac{D\underline{\phi}}{Dx_\mu}\right) - V(\underline{\phi}^\dagger\underline{\phi}) \tag{43.20}$$

Here $(D\underline{\phi}/Dx_\mu)^\star \equiv [(D\underline{\phi}/D\mathbf{x})^\dagger, +(1/i)(D\underline{\phi}/Dx_0)^\dagger]$.[12] Thus

$$\left(\frac{D\underline{\phi}}{Dx_\mu}\right)^\star \left(\frac{D\underline{\phi}}{Dx_\mu}\right) =$$

$$\underline{\phi}^\dagger \left(\frac{\overleftarrow{\partial}}{\partial x_\mu} + \frac{i}{2}g'B_\mu + \frac{i}{2}g\tau\cdot\mathbf{A}_\mu\right)\left(\frac{\partial}{\partial x_\mu} - \frac{i}{2}g'B_\mu - \frac{i}{2}g\tau\cdot\mathbf{A}_\mu\right)\underline{\phi} \tag{43.21}$$

5. Assume the most general form of the scalar self-interaction potential V for a *renormalizable theory*

$$V = \mu^2\underline{\phi}^\dagger\underline{\phi} + \lambda(\underline{\phi}^\dagger\underline{\phi})^2 \tag{43.22}$$

Spontaneous Symmetry Breaking. For the generation of mass for the gauge bosons, while maintaining the local gauge symmetry, one now employs essentially the same argument that was used for the generation of nucleon mass in the chiral-invariant σ-model in Section 22.

Assume that $\mu^2 < 0$ and $\lambda > 0$ so that the potential V has the shape shown in Fig. 43.1. The minimum of the potential no longer occurs at the origin with $\underline{\phi} = 0$, but now at a finite value of $\underline{\phi}$. Hence the scalar field acquires a *vacuum expectation value.* Only the neutral component of the field can be allowed to develop a vacuum expectation value in order to preserve electric charge conservation. Furthermore, the (constant) phase of the field can always be redefined so that this vacuum expectation value is real. Thus we write

$$\langle\phi^0\rangle = \langle\phi^{0*}\rangle \equiv \frac{v}{\sqrt{2}} \tag{43.23}$$

[12] The metric is not complex conjugated in $v_\mu^\star \equiv (\mathbf{v}^\dagger, +iv_0^\dagger)$.

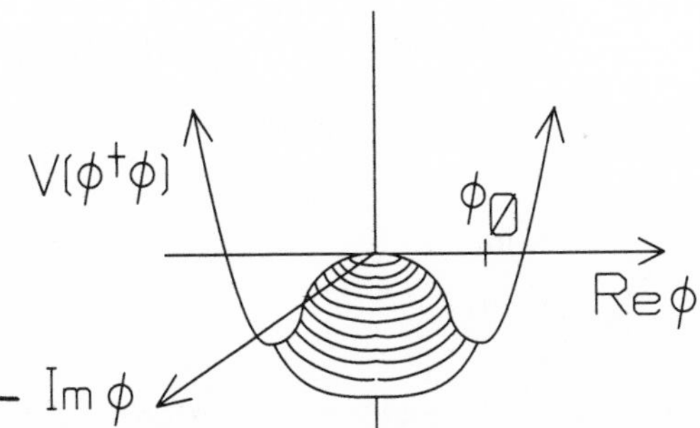

Figure 43.1: Form of the scalar self-interaction potential to generate mass for the gauge bosons by spontaneous symmetry breaking. The illustration is for a single, neutral, complex ϕ.

At the minimum of the vacuum expectation value of the potential one finds

$$v^2 \quad = \quad -\frac{\mu^2}{\lambda} \tag{43.24}$$

Without loss of generality, one can now *parameterize* the complex scalar field $\underline{\phi}$ in terms of four real parameters $\{\boldsymbol{\xi}(x), \eta(x)\}$ describing the fluctuations around the vacuum expectation value in the following fashion (Ref. [W7]):

$$\underline{\phi} \quad \equiv \quad \exp\{\frac{-i}{2v}\boldsymbol{\xi}\cdot\boldsymbol{\tau}\}\begin{pmatrix} 0 \\ \frac{1}{\sqrt{2}}(v+\eta) \end{pmatrix} \tag{43.25}$$

The theory has been constructed to be locally gauge invariant. Make use of this fact to simplify matters. *Make a gauge transformation to eliminate the first factor in this equation.* Define

$$\underline{\phi}\,' \equiv \exp\{\frac{+i}{2v}\boldsymbol{\xi}\cdot\boldsymbol{\tau}\}\underline{\phi} = \underline{\mathrm{U}}(\boldsymbol{\xi})\underline{\phi} = \frac{1}{\sqrt{2}}\begin{pmatrix} 0 \\ v+\eta \end{pmatrix} \tag{43.26}$$

Written in terms of the new field $\underline{\phi}\,'$, the three scalar field variables $\{\boldsymbol{\xi}(x)\}$ now no longer appear in the lagrangian; and, as we proceed to demonstrate, the free lagrangian has instead a simple interpretation in terms of massive vector and scalar particles. The lagrangian in this form is said to be written in the *unitary gauge* where the particle content of the theory is manifest. This procedure for generating the mass of the gauge bosons in this fashion is known as the *Higgs mechanism* (see Refs. [W1, W7]).

Substitution of the expression in Eq. (43.26) in the scalar lagrangian in Eqs. (43.20)-(43.22) leads to

$$\mathcal{L}_{\text{scalar}} = -V[\frac{1}{2}(v+\eta)^2] - \frac{1}{2}\chi_\downarrow^\dagger \left[\frac{\partial \eta}{\partial x_\mu} + \frac{ig'}{2}(v+\eta)B_\mu + \frac{ig}{2}(v+\eta)\boldsymbol{\tau}\cdot\mathbf{A}_\mu\right]$$
$$\times \left[\frac{\partial \eta}{\partial x_\mu} - \frac{ig'}{2}(v+\eta)B_\mu - \frac{ig}{2}(v+\eta)\boldsymbol{\tau}\cdot\mathbf{A}_\mu\right]\chi_\downarrow \qquad (43.27)$$

Here $\chi_\downarrow \equiv \begin{pmatrix} 0 \\ 1 \end{pmatrix}$. An evaluation of the potential term, utilizing the minimization condition in Eq. (43.24), gives

$$\begin{aligned} V[\frac{1}{2}(v+\eta)^2] &= \frac{\mu^2}{2}(v+\eta)^2 + \frac{\lambda}{4}(v+\eta)^4 \\ &= v^2\left(\frac{\mu^2}{4}\right) + \eta^2(-\mu^2) + \eta^3(\lambda v) + \eta^4\left(\frac{\lambda}{4}\right) \end{aligned} \qquad (43.28)$$

Note that there is no term linear in η when one expands about the true minimum in V. The coefficient of the term linear in $\partial\eta/\partial x_\mu$ similarly vanishes in Eq. (43.27).

The remaining boson interactions in $\mathcal{L}_{\text{scalar}}$ are proportional to

$$\begin{aligned} &\chi_\downarrow^\dagger(g'B_\mu + g\boldsymbol{\tau}\cdot\mathbf{A}_\mu)(g'B_\mu + g\boldsymbol{\tau}\cdot\mathbf{A}_\mu)\chi_\downarrow \\ &\quad = \chi_\downarrow^\dagger(g'^2B_\mu^2 + g^2\mathbf{A}_\mu^2 + 2gg'B_\mu\boldsymbol{\tau}\cdot\mathbf{A}_\mu)\chi_\downarrow \\ &\quad = (g'^2B_\mu^2 + g^2\mathbf{A}_\mu^2 - 2gg'B_\mu A_\mu^{(3)}) \end{aligned} \qquad (43.29)$$

Hence the scalar lagrangian in the unitary gauge is given by

$$\begin{aligned} \mathcal{L}_{\text{scalar}} &= -\frac{1}{2}\left[\left(\frac{\partial\eta}{\partial x_\mu}\right)^2 + (-2\mu^2)\eta^2\right] - \frac{\lambda}{4}(4v\eta^3 + \eta^4) - \frac{1}{4}\mu^2v^2 \\ &\quad -\frac{1}{8}(v+\eta)^2(g'^2B_\mu^2 + g^2\mathbf{A}_\mu^2 - 2gg'B_\mu A_\mu^{(3)}) \end{aligned} \qquad (43.30)$$

The term in v^2 in the second line now provides the sought-after mass for the gauge bosons. The coefficient of this term is a quadratic form in the gauge fields, which can be put on principal axes with the introduction of the following linear combinations of fields:

$$\begin{aligned} W_\mu^{(+)} &\equiv W_\mu^\star \equiv \frac{1}{\sqrt{2}}(A_\mu^{(1)} + iA_\mu^{(2)}) \\ W_\mu^{(-)} &\equiv W_\mu \equiv \frac{1}{\sqrt{2}}(A_\mu^{(1)} - iA_\mu^{(2)}) \\ Z_\mu &\equiv \frac{-gA_\mu^{(3)} + g'B_\mu}{(g^2+g'^2)^{1/2}} \\ A_\mu &\equiv \frac{g'A_\mu^{(3)} + gB_\mu}{(g^2+g'^2)^{1/2}} \end{aligned} \qquad (43.31)$$

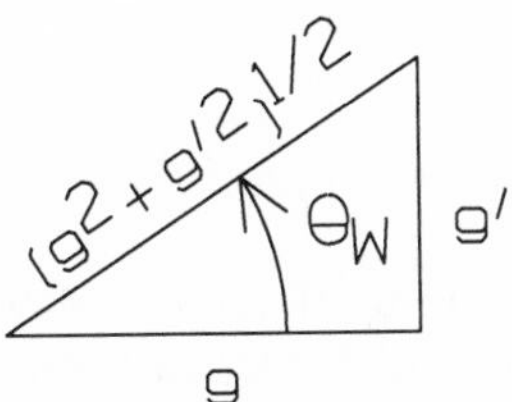

Figure 43.2: Coefficients in the orthogonal transformation that diagonalizes the vector meson mass matrix in the standard model.

The fields $(W_\mu^\star, W_\mu)$ will create particles (W_μ^+, W_μ^-), respectively, the third field describes a neutral Z_μ^0 vector boson, and the fourth is the photon field. The relation between $(B_\mu, A_\mu^{(3)})$ and (Z_μ, A_μ) is an *orthogonal transformation*, which is illustrated in Fig. 43.2. Note in particular that the weak angle is defined by

$$\sin\theta_W \equiv \frac{g'}{(g^2+g'^2)^{1/2}} \tag{43.32}$$

The scalar lagrangian can thus finally be written in the unitary gauge as

$$\begin{aligned}\mathcal{L}_{\text{scalar}} &= -\frac{1}{2}\left[\left(\frac{\partial\eta}{\partial x_\mu}\right)^2 + (-2\mu^2)\eta^2\right] - \frac{\lambda}{4}(4v\eta^3+\eta^4) - \frac{1}{4}\mu^2v^2 \\ &\quad -\frac{1}{4}v^2(g^2+g'^2)\frac{1}{2}Z_\mu^2 - \frac{1}{4}v^2g^2W_\mu W_\mu^\star \\ &\quad -\frac{1}{8}\eta(2v+\eta)[(g^2+g'^2)Z_\mu^2 + 2g^2W_\mu W_\mu^\star]\end{aligned} \tag{43.33}$$

The first term in the first line is the lagrangian for a free, neutral scalar field of mass $-2\mu^2$ — the *Higgs field*; this is the only remaining physical degree of freedom from the complex doublet of scalar fields introduced previously, in this unitary gauge. The second term describes cubic and quartic self-couplings of the Higgs field; the third term in the first line is simply an additive constant.

The terms in the second line proportional to the constant v^2 represent the *quadratic mass terms for the gauge bosons.* Note, in particular, that *no mass term has been generated for the photon field*, which thus remains massless, as it must.

Finally, the terms in the last line proportional to $(2v\eta + \eta^2)$ represent cubic and quartic couplings of the Higgs to the massive gauge bosons.

Since the transformation in Eq. (43.31) is orthogonal, the quadratic part of the kinetic energy of the gauge bosons remains on principal axes and Eq. (43.17) can be rewritten as

$$\begin{aligned}\mathcal{L}_{\text{gauge}} &= -\frac{1}{2}W^{\star}_{\mu\nu}W_{\mu\nu} - \frac{1}{4}Z_{\mu\nu}Z_{\mu\nu} - \frac{1}{4}F_{\mu\nu}F_{\mu\nu} \\ &\quad -\frac{g}{2}\mathbf{F}_{\mu\nu}\cdot(\mathbf{A}_\mu \times \mathbf{A}_\nu) - \frac{g^2}{4}(\mathbf{A}_\mu \times \mathbf{A}_\nu)^2 \end{aligned} \tag{43.34}$$

Here the field tensors are defined by the linear Maxwell form $V_{\mu\nu} \equiv \partial V_\nu/\partial x_\mu - \partial V_\mu/\partial x_\nu$ and the original gauge field $\mathbf{A}_\mu$ in the nonlinear terms must still be expressed in terms of the physical fields defined through Eqs. (43.31). The second line in the above result represents cubic and quartic couplings of the physical gauge fields.

Particle Content. The particle content of the theory is now made manifest in this unitary gauge, since the free lagrangian has the required quadratic form in the kinetic energy and masses. In addition to the original (still massless!) fermions, the theory evidently now contains

1. A massive neutral weak vector meson Z^0_μ with mass given by

$$M_Z^2 = \frac{v^2(g^2 + g'^2)}{4} \tag{43.35}$$

2. Massive charged weak vector mesons $W_\mu^{(\pm)}$ with masses

$$M_W^2 = \frac{v^2 g^2}{4} = M_Z^2 \cos^2\theta_W \tag{43.36}$$

3. A massless photon

$$M_\gamma^2 = 0 \tag{43.37}$$

The lagrangian retains the exact local $U(1)$ gauge invariance generated by the electric charge $\hat{Q}$, corresponding to QED.

Lagrangian. The total lagrangian for the standard model as presented so far is the sum of the individual contributions discussed above

$$\mathcal{L} = \mathcal{L}_{\text{lepton}} + \mathcal{L}_{\text{nucleon}} + \mathcal{L}_{\text{gauge}} + \mathcal{L}_{\text{scalar}} \tag{43.38}$$

This lagrangian now contains all the electroweak interactions; in particular, it yields *all the weak interaction phenomenology of Section 42.* It is still necessary to put in the fermion mass, while preserving the underlying local gauge symmetry and accompanying renormalizability; this will be done by again appealing to spontaneous symmetry breaking, employing the already-introduced complex scalar field.

First, however, let us continue to investigate some of the consequences of the lagrangian in Eq. (43.38). The coupling of the leptons to the gauge bosons follows immediately from Eqs. (43.18) and (43.31) (the details of this algebra, central to applications of the standard model, are provided in Appendix M)

$$
\begin{aligned}
\mathcal{L}^{(\pm)}_{\text{lepton}} &= \frac{g}{2\sqrt{2}}[j^{(+)}_\mu W_\mu + j^{(-)}_\mu W^\star_\mu] \\
\mathcal{L}^{(0)}_{\text{lepton}} &= -\frac{g}{2\cos\theta_W} j^{(0)}_\mu Z_\mu \\
\mathcal{L}^{\gamma}_{\text{lepton}} &= e_p j^\gamma_\mu A_\mu
\end{aligned}
\tag{43.39}
$$

Here the electric charge e_p is defined by[13]

$$
e_p \equiv \frac{gg'}{(g^2+g'^2)^{1/2}} \tag{43.40}
$$

The lepton currents are given by the following expressions

$$
\begin{aligned}
j^{(\pm)}_\mu &= i\bar{\psi}_l\gamma_\mu(1+\gamma_5)\tau_\pm\psi_l \\
j^\gamma_\mu &= i\bar{\psi}_l\gamma_\mu\left[-\frac{1}{2}(1-\tau_3)\right]\psi_l \\
j^{(0)}_\mu &= i\bar{\psi}_l\gamma_\mu(1+\gamma_5)\frac{1}{2}\tau_3\psi_l - 2\sin^2\theta_W j^\gamma_\mu
\end{aligned}
\tag{43.41}
$$

The interaction of the point nucleons with the gauge fields takes exactly the same form as in Eqs. (43.39), with hadronic currents given by

$$
\begin{aligned}
\mathcal{J}^{(\pm)}_\mu &= i\bar{\psi}\gamma_\mu(1+\gamma_5)\tau_\pm\psi \\
J^\gamma_\mu &= i\bar{\psi}\gamma_\mu\left[\frac{1}{2}(1+\tau_3)\right]\psi \\
\mathcal{J}^{(0)}_\mu &= i\bar{\psi}\gamma_\mu(1+\gamma_5)\frac{1}{2}\tau_3\psi - 2\sin^2\theta_W J^\gamma_\mu
\end{aligned}
\tag{43.42}
$$

The lepton and nucleon doublets appearing in these currents are defined by

$$
\psi_l = \begin{pmatrix}\psi_{\nu_e}\\ \psi_e\end{pmatrix} \qquad \psi = \begin{pmatrix}\psi_p\\ \psi_n\end{pmatrix} \tag{43.43}
$$

Effective Low-Energy Lagrangian. The analysis in Section 42 shows how interactions with the gauge bosons of the form in Eqs. (43.39) lead to an effective current-current lagrangian in the low-energy, nuclear domain where $\mathbf{q}^2 \ll M_W^2, M_Z^2$. In particular, comparison with that analysis immediately establishes the following relationships between the gauge couplings and masses of the standard model and the coupling constants introduced in that section

$$
\frac{G}{\sqrt{2}} = \frac{g^2}{8M_W^2} = \frac{g^2}{8M_Z^2\cos^2\theta_W} = \frac{f^2}{8M_Z^2} = \frac{G'}{\sqrt{2}} \tag{43.44}
$$

[13]Note that both (g, g') must be nonzero for nonzero e_p.

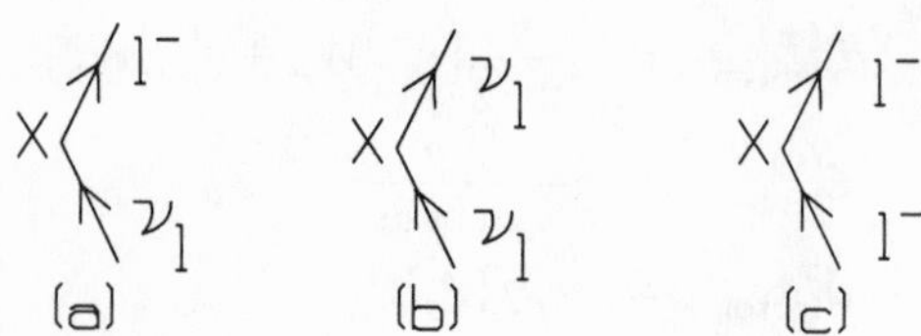

Figure 43.3: Semileptonic processes described by the effective low-energy semileptonic lagrangians in the text: (a) charge-changing [±]; (b) weak neutral current neutrino scattering [ν]; and (c) weak neutral current, charged-lepton interaction [l].

It is also evident that the total weak currents here receive additive contributions from the leptons and hadrons

$$\begin{aligned}\mathcal{J}_\lambda^{(\pm)} &= \mathcal{J}_\lambda^{(\pm)}(\text{hadrons}) + j_\lambda^{(\pm)}(\text{leptons}) \\ \mathcal{J}_\lambda^{(0)} &= \mathcal{J}_\lambda^{(0)}(\text{hadrons}) + j_\lambda^{(0)}(\text{leptons}) \end{aligned} \tag{43.45}$$

Effective Low-Energy Semileptonic Lagrangian. The semileptonic parts of this effective low-energy lagrangian will form the basis of most of the subsequent discussion of nuclear applications; they describe the semileptonic processes illustrated in Fig. 43.3.[14] The corresponding lagrangians are

$$\begin{aligned}\mathcal{L}_{\text{eff}}^{(\pm)} &= \frac{iG}{\sqrt{2}}\Big\{[\bar{\psi}_e\gamma_\lambda(1+\gamma_5)\psi_{\nu_e} + (e \leftrightarrow \mu)]\mathcal{J}_\lambda^{(+)}(\text{hadrons}) \\ &\quad +[\bar{\psi}_{\nu_e}\gamma_\lambda(1+\gamma_5)\psi_e + (e \leftrightarrow \mu)]\mathcal{J}_\lambda^{(-)}(\text{hadrons})\Big\} \\ \mathcal{L}_{\text{eff}}^{(\nu)} &= \frac{iG}{\sqrt{2}}\left[\bar{\psi}_{\nu_e}\gamma_\lambda(1+\gamma_5)\psi_{\nu_e} + (e \leftrightarrow \mu)\right]\mathcal{J}_\lambda^{(0)}(\text{hadrons}) \\ \mathcal{L}_{\text{eff}}^{(l)} &= -\frac{iG}{\sqrt{2}}\left[\bar{\psi}_e\gamma_\lambda(1+\gamma_5)\psi_e - 4\sin^2\theta_W\bar{\psi}_e\gamma_\lambda\psi_e + (e \leftrightarrow \mu)\right]\mathcal{J}_\lambda^{(0)}(\text{hadrons})\end{aligned} \tag{43.46}$$

In both of the last two lagrangians there is a suppressed multiplicative factor 2(cross terms)/2(from lepton current) = 1.

These lagrangians give all the phenomenology of Section 42. We discuss their application to the calculation of semileptonic weak nuclear processes in some detail in the subsequent sections (Ref. [W2]).

Fermion Mass. The theory as formulated assumes massless fermions. The fermion mass will now be *put in by hand.* One adds Yukawa couplings of the fermions to the previously introduced complex scalar field that preserve the local $SU(2)_W \bigotimes U(1)_W$ local gauge symmetry. One such coupling is introduced for each fermion field. The fermions then acquire mass when the scalar field

[14] Formulation in terms of quarks simply changes the underlying structure of $\mathcal{J}_\lambda$(hadrons). See next section.

develops its vacuum expectation value. As a consequence of this procedure, each fermion also has a prescibed Yukawa coupling to the *fluctuation* of the scalar field about its vacuum expectation value — the real scalar Higgs.

We illustrate the procedure in the case of leptons and point nucleons. In the latter case, in addition to the complex scalar field $\underline{\phi}$ in Eq. (43.19), one also needs the field $\tilde{\underline{\phi}}$ derived from it as follows:[15]

$$\tilde{\underline{\phi}} \equiv i\tau_2\underline{\phi}^* = \begin{pmatrix} 0 & 1 \\ -1 & 0 \end{pmatrix}\begin{pmatrix} \phi^{+*} \\ \phi^{0*} \end{pmatrix} = \begin{pmatrix} \phi^{0*} \\ -\phi^- \end{pmatrix} \equiv \begin{pmatrix} \tilde{\phi}^0 \\ \tilde{\phi}^- \end{pmatrix} \tag{43.47}$$

Under the weak hypercharge transformation one simply transforms ϕ^*, and hence this field has *opposite weak hypercharge* $Y_W = -1$; however, *it still transforms as a weak isodoublet.* This is readily established since

$$\exp\{i\boldsymbol{\theta}\cdot\hat{\mathbf{T}}_W\}\,\tilde{\underline{\phi}}\,\exp\{-i\boldsymbol{\theta}\cdot\hat{\mathbf{T}}_W\} = i\tau_2[e^{-\frac{i}{2}\boldsymbol{\theta}\cdot\boldsymbol{\tau}}\underline{\phi}\,]^* = [e^{-\frac{i}{2}\boldsymbol{\theta}\cdot\boldsymbol{\tau}}]\tilde{\underline{\phi}} \tag{43.48}$$

Hence one now has an additional field with which to build invariant Yukawa couplings (Table 43.1).

Start with the following lagrangian with Yukawa couplings of the fermions to the complex scalar field and invariant under local $SU(2)_W \otimes U(1)_W$

$$\mathcal{L}_{\text{int}} = -G_e\bar{R}(\underline{\phi}^{\,\dagger}\underline{\text{L}}) - G_1(\bar{\underline{N}}_L\tilde{\underline{\phi}})p_R - G_2(\bar{\underline{N}}_L\underline{\phi})n_R + \text{h.c.} \tag{43.49}$$

Each term is a weak isoscalar, and each term is neutral in weak hypercharge (Table 43.1).

Now with the previously discussed spontaneous symmetry breaking, and in the unitary gauge

$$\underline{\phi} = \begin{pmatrix} 0 \\ \frac{1}{\sqrt{2}}(v+\eta) \end{pmatrix} \qquad\qquad \tilde{\underline{\phi}} = \begin{pmatrix} \frac{1}{\sqrt{2}}(v+\eta) \\ 0 \end{pmatrix} \tag{43.50}$$

Substitution into Eq. (43.49) gives

$$\mathcal{L}_{\text{int}} = -G_e\bar{e}_R\left[(0, \frac{1}{\sqrt{2}}(v+\eta))\begin{pmatrix} \nu_L \\ e_L \end{pmatrix}\right] \tag{43.51}$$
$$-G_1\left[(\bar{p}_L, \bar{n}_L)\begin{pmatrix} \frac{1}{\sqrt{2}}(v+\eta) \\ 0 \end{pmatrix}\right]p_R - G_2\left[(\bar{p}_L, \bar{n}_L)\begin{pmatrix} 0 \\ \frac{1}{\sqrt{2}}(v+\eta) \end{pmatrix}\right]n_R + \text{h.c.}$$

Hence

$$\begin{aligned}\mathcal{L}_{\text{int}} &= -\frac{1}{\sqrt{2}}(v+\eta)[G_e(\bar{e}_Le_R + \bar{e}_Re_L) \\ &\quad + G_1(\bar{p}_Lp_R + \bar{p}_Rp_L) + G_2(\bar{n}_Ln_R + \bar{n}_Rn_L)] \\ &= -\frac{1}{\sqrt{2}}(v+\eta)[G_e\bar{e}e + G_1\bar{p}p + G_2\bar{n}n]\end{aligned} \tag{43.52}$$

[15]Here $\underline{\phi}^{\,*}$ indicates hermitian adjoint in the Hilbert space, while $\underline{\phi}^{\,\dagger}$ includes a matrix transpose.

For *strong isospin symmetry*, one must evidently impose the condition

$$G_1 = G_2 \tag{43.53}$$

The final result is

$$\mathcal{L}_{\text{int}} = -\frac{v}{\sqrt{2}}[G_e\bar{e}e + G_1\bar{\psi}\psi] - \frac{\eta}{\sqrt{2}}[G_e\bar{e}e + G_1\bar{\psi}\psi] \; ; \qquad \psi = \begin{pmatrix} p \\ n \end{pmatrix} \tag{43.54}$$

The first term is the sought-after fermion mass, with one adjustable coupling constant for each fermion mass in the theory. The second term is the remaining Yukawa interaction with the real scalar Higgs particle, with a prescribed coupling determined by the mass of the fermion.

For clarity of concept, this introduction to the standard model has been presented in terms of a hadronic sector consisting of point nucleons. The extension to matrix elements of the hadronic current for *physical nucleons* can be obtained using general symmetry properties of the hadronic current, preserved in the extension to an underlying quark structure. Just as in Sections 7 and 8, the resulting Eqs. (42.35) and (42.36) can then be used to obtain a description of semileptonic processes in nuclei; this is discussed in detail in later sections.

The deeper formulation of the standard model is in terms of quarks, and it is to this topic that we now turn our attention in the following section.

44

QUARKS IN THE STANDARD MODEL

With this introduction to the methodology of the standard model, we turn our attention to the formulation of the theory of electroweak interactions with the underlying quark fields.

Weak Multiplets. At first glance, one might expect that the first quark weak isodoublet would just be that constructed from (u, d) quarks. The actual quark weak isospin doublets that couple in the electroweak interaction have a more complicated form (Refs. [W6, W8]). They are

$$\begin{aligned} q_L &= \begin{pmatrix} u_L \\ d_L \cos\theta_C + s_L \sin\theta_C \end{pmatrix} \equiv \begin{pmatrix} u_L \\ d_{cL} \end{pmatrix} \\ Q_L &= \begin{pmatrix} c_L \\ -d_L \sin\theta_C + s_L \cos\theta_C \end{pmatrix} \equiv \begin{pmatrix} c_L \\ D_{cL} \end{pmatrix} \; ; \quad \text{weak doublets} \end{aligned} \tag{44.1}$$

The fact that it is a slightly rotated combination of fields in the charge-changing current, which includes a small strangeness-changing component, was first noted by Cabbibo (Ref. [W8]). The discovery that one requires a second doublet with an additional c quark and the orthogonal rotated combination is due to Glashow, Iliopolous, and Maiani (GIM) (Ref. [W6]) who in fact *predicted* the existence of the c quark on the basis of the arguments given below.[16]

As before, the right-handed quark fields form weak isosinglets

$$u_R, d_R, s_R, c_R \; ; \qquad \text{weak singlets} \tag{44.2}$$

The quarks are assigned the weak quantum numbers in Table 44.1. The assignments are again made so that the electric charge operator is given by

$$\hat{Q} = (\hat{T}_3 + \frac{1}{2}\hat{Y})_W \tag{44.3}$$

GIM Identity. Because one has two orthogonal linear combinations, the following (GIM) identity holds

$$\begin{aligned} \bar{d}_c d_c + \bar{D}_c D_c &= (\bar{d}\cos\theta_C + \bar{s}\sin\theta_C)(d\cos\theta_C + s\sin\theta_C) \\ &\quad + (-\bar{d}\sin\theta_C + \bar{s}\cos\theta_C)(-d\sin\theta_C + s\cos\theta_C) \\ &= \bar{d}d + \bar{s}s \end{aligned} \tag{44.4}$$

[16] The extension to include still another (heavy) quark family is discussed in Section 49.

Table 44.1: Weak isospin and weak hypercharge assignments for the quarks.

Field /particle	q_L	Q_L	u_R	d_R	s_R	c_R
T_W	1/2	1/2	0	0	0	0
Y_W	1/3	1/3	4/3	−2/3	−2/3	4/3

No off-diagonal, strangeness-changing terms appear in this expression; as a consequence, the neutral currents generated in the standard model have no lowest-order strangeness-changing components — an empirical observation that was the primary motivation for the introduction of the c quark in Ref. [W6].

The GIM identity can be used to rewrite the noninteracting quark kinetic energy as

$$\mathcal{L}^0_{\text{quark}} = -\left[\bar{q}_L\gamma_\mu\frac{\partial}{\partial x_\mu}q_L + \bar{Q}_L\gamma_\mu\frac{\partial}{\partial x_\mu}Q_L \right.$$
$$\left. +\bar{u}_R\gamma_\mu\frac{\partial}{\partial x_\mu}u_R + \bar{d}_R\gamma_\mu\frac{\partial}{\partial x_\mu}d_R + \bar{s}_R\gamma_\mu\frac{\partial}{\partial x_\mu}s_R + \bar{c}_R\gamma_\mu\frac{\partial}{\partial x_\mu}c_R\right] \tag{44.5}$$

Covariant Derivative. The standard model is a Yang-Mills theory based on local $SU(2)_W \bigotimes U(1)_W$ gauge invariance. It starts from massless fermions and massless gauge bosons and introduces mass by coupling to a complex scalar weak isodoublet and spontaneous symmetry breaking (Section 43). The covariant derivatives acting on the quark fields are as before (see Table 44.1)

$$\begin{aligned}
&(\frac{\partial}{\partial x_\mu} - \frac{i}{2}g'Y_W B_\mu - \frac{i}{2}g\boldsymbol{\tau}\cdot\mathbf{A}_\mu)\ ; && \text{on isodoublets} \\
&(\frac{\partial}{\partial x_\mu} - \frac{i}{2}g'Y_W B_\mu)\ ; && \text{on isosinglets}
\end{aligned} \tag{44.6}$$

The gauge boson and Higgs sectors of the theory are exactly the same as discussed in the previous section.

Electroweak Quark Currents. The electroweak currents representing the interaction with the physical gauge bosons can now be identified exactly as before (Appendix M).

The charge-changing weak current is given by

$$\begin{aligned}
\mathcal{J}_\mu^{(\pm)} &= i\bar{q}\gamma_\mu(1+\gamma_5)\tau_\pm q + i\bar{Q}\gamma_\mu(1+\gamma_5)\tau_\pm Q \\
\mathcal{J}_\mu^{(+)} &= i\bar{u}\gamma_\mu(1+\gamma_5)(d\cos\theta_C + s\sin\theta_C) \\
&\quad +i\bar{c}\gamma_\mu(1+\gamma_5)(-d\sin\theta_C + s\cos\theta_C)
\end{aligned} \tag{44.7}$$

Note it is the Cabbibo-rotated combination that enters into these charge-changing currents.

The electromagnetic current of QED is just the point Dirac current multiplied by the correct charge

$$J_\mu^\gamma = i\left[\frac{2}{3}(\bar{u}\gamma_\mu u + \bar{c}\gamma_\mu c) - \frac{1}{3}(\bar{d}\gamma_\mu d + \bar{s}\gamma_\mu s)\right] \tag{44.8}$$

The weak neutral current is

$$\begin{aligned} \mathcal{J}_\mu^{(0)} &= i\bar{q}\gamma_\mu(1+\gamma_5)\frac{1}{2}\tau_3 q + i\bar{Q}\gamma_\mu(1+\gamma_5)\frac{1}{2}\tau_3 Q - 2\sin^2\theta_W J_\mu^\gamma \\ \mathcal{J}_\mu^{(0)} &= \frac{i}{2}[\bar{u}\gamma_\mu(1+\gamma_5)u + \bar{c}\gamma_\mu(1+\gamma_5)c - \bar{d}\gamma_\mu(1+\gamma_5)d - \bar{s}\gamma_\mu(1+\gamma_5)s] \\ &\quad -2\sin^2\theta_W J_\mu^\gamma \end{aligned} \tag{44.9}$$

The second equality follows with the aid of the GIM identity. Terms of the form $(\bar{s}d)$ or $(\bar{d}s)$ have been eliminated; hence there are no strangeness-changing weak neutral currents in this quark-based standard model, as advertised.

The quarks can be given mass in the same fashion as were the nucleons in the previous section. Here we are content to refer the reader to Prob. 44.1 and the literature for the details (Refs. [W1, W7]).

QCD. How does the standard model of electroweak interactions get combined with QCD, the theory of the *strong* forces binding quarks into hadrons? Consider for simplicity the nuclear domain of (u, d) quarks. Quarks now carry an additional color index that takes three values (R, G, B), and the quark field gets extended to

$$\psi = \begin{pmatrix} u \\ d \end{pmatrix} \to \begin{pmatrix} u_R & u_G & u_B \\ d_R & d_G & d_B \end{pmatrix} \equiv (\psi_R, \psi_G, \psi_B) \tag{44.10}$$

These get combined into a three component (actually multicomponent) field $\underline{\psi}$ defined as

$$\underline{\psi} \equiv \begin{pmatrix} \psi_R \\ \psi_G \\ \psi_B \end{pmatrix} \tag{44.11}$$

Electroweak Currents. Let $\mathbb{O}$ be a matrix that is the *identity* with respect to color, but an *arbitrary* matrix O with respect to flavor so that

$$\mathbb{O} \equiv \begin{pmatrix} O & & \\ & O & \\ & & O \end{pmatrix} \tag{44.12}$$

Then under the extension of the quark fields to include color, all electroweak currents are defined to be correspondingly extended to

$$\begin{aligned} \bar{\psi}\gamma_\mu O\psi &\to \bar{\psi}_R\gamma_\mu O\psi_R + \bar{\psi}_G\gamma_\mu O\psi_G + \bar{\psi}_B\gamma_\mu O\psi_B \\ &\equiv \bar{\underline{\psi}}\gamma_\mu \mathbb{O}\underline{\psi} \end{aligned} \tag{44.13}$$

Such currents have the following important properties:

- They are invariant under strong $SU(3)_C$.
- The vector currents are conserved in QCD (Section 27).

Symmetry Group. The full lagrangian of the strong and electroweak interactions thus takes the form (see Ref. [W37] for an extended discussion)

$$\mathcal{L} = \mathcal{L}^0 + \mathcal{L}_{\text{QCD}}^{\text{int}} + \mathcal{L}_{\text{EW}}^{\text{int}} \tag{44.14}$$

This lagrangian is locally gauge invariant under the full symmetry group

$$SU(3)_C \bigotimes SU(2)_W \bigotimes U(1)_W \tag{44.15}$$

This full theory is renormalizable. It has the following characteristic properties:

- The electroweak interactions are colorblind — they are the same, independent of the color of the quarks.
- The gluons are absolutely *neutral* to the electroweak interactions — the electroweak interactions couple to the quarks.

Nuclear Currents. Let us examine the implications of this development for nuclear physics. To summarize the weak and electromagnetic quark currents in the standard model, we have

$$\begin{aligned}
\mathcal{J}_\mu^{(+)} &= i\bar{u}\gamma_\mu(1+\gamma_5)[d\cos\theta_C + s\sin\theta_C] \\
&\quad + i\bar{c}\gamma_\mu(1+\gamma_5)[-d\sin\theta_C + s\cos\theta_C] \\
\mathcal{J}_\mu^{(0)} &= \frac{i}{2}[\bar{u}\gamma_\mu(1+\gamma_5)u + \bar{c}\gamma_\mu(1+\gamma_5)c \\
&\quad - \bar{d}\gamma_\mu(1+\gamma_5)d - \bar{s}\gamma_\mu(1+\gamma_5)s] - 2\sin^2\theta_W J_\mu^\gamma \\
J_\mu^\gamma &= i\left[\frac{2}{3}(\bar{u}\gamma_\mu u + \bar{c}\gamma_\mu c) - \frac{1}{3}(\bar{d}\gamma_\mu d + \bar{s}\gamma_\mu s)\right]
\end{aligned} \tag{44.16}$$

Each current is actually a sum over three colors $\sum_{\text{colors}}(\cdots)$ leading to an operator which is an $SU(3)_C$ - singlet as discussed above.

Nuclear Domain. To a good approximation, the hadrons that make up the nucleus are composed of (u,d) quarks. As a starting point for nuclear physics, consider that subspace of the full Hilbert space consisting of any number of (u,d) quarks and their antiquarks $(\bar{u},\bar{d})$. The quark field in this sector takes the form

$$\psi \doteq \begin{pmatrix} u \\ d \end{pmatrix}; \qquad \text{nuclear domain} \tag{44.17}$$

Assume that the (u,d) quarks have the *same mass* in the lagrangian; they are in fact both nearly massless. In this case, the lagrangian of the strong interactions, with the full complexity of QCD, has an *exact symmetry* — the $SU(2)$ of strong isospin. This is the familiar isotopic spin symmetry of nuclear physics. It is important to note that one still has the full complexity of strong-coupling QCD with colored quarks and gluons in this truncated flavor sector of the nuclear domain; nevertheless, one can draw conclusions that are exact to all orders in the strong interactions using this strong isospin symmetry.

The quark field ψ in Eq. (44.17) forms an isodoublet under this strong isospin. The quark currents in Eq. (44.16) can then be written in terms of this isospinor in the nuclear domain as follows

$$\begin{aligned} J_\mu^\gamma &= i\bar{\psi}\gamma_\mu\left(\frac{1}{6}+\frac{1}{2}\tau_3\right)\psi \\ \mathcal{J}_\mu^{(\pm)} &= i\bar{\psi}\gamma_\mu(1+\gamma_5)\tau_\pm\psi \\ \mathcal{J}_\mu^{(0)} &= i\bar{\psi}\gamma_\mu(1+\gamma_5)\frac{1}{2}\tau_3\psi - 2\sin^2\theta_W J_\mu^\gamma \end{aligned} \tag{44.18}$$

As in the previous section, the properties of these currents under general symmetry properties of the theory now follow by inspection

$$\begin{aligned} \mathcal{J}_\mu &= J_\mu + J_{\mu 5} &&; \quad V-A \\ \mathcal{J}_\mu^{(\pm)} &= \mathcal{J}_\mu^{V_1} \pm i\mathcal{J}_\mu^{V_2} &&; \quad \text{isovector} \\ J_\mu^\gamma &= J_\mu^S + J_\mu^{V_3} &&; \quad \text{EM current} \\ J_\mu^{(\pm)} &= J_\mu^{V_1} \pm iJ_\mu^{V_2} &&; \quad \text{CVC} \\ \mathcal{J}_\mu^{(0)} &= \mathcal{J}_\mu^{V_3} - 2\sin^2\theta_W J_\mu^\gamma &&; \quad \text{standard model} \end{aligned} \tag{44.19}$$

Here the Cabbibo angle has been absorbed into the definition of the hadronic weak charge-changing Fermi coupling constant

$$G^{(\pm)} \equiv G\cos\theta_C \qquad\qquad \cos\theta_C = 0.974 \tag{44.20}$$

Note that the numerical value of $\cos\theta_C$ is, in fact, very close to 1 (Ref. [W1]).[17]

Consequences of the Lorentz structure of various nuclear matrix elements, and relations of various matrix elements through the use of the Wigner-Eckart theorem applied to strong isospin, now follow immediately from Eqs. (44.19).

If the discussion is extended to that sector of the full theory with no *net* strangeness or charm, and the electroweak interactions are treated in lowest order, then the first four of Eqs. (44.19) still hold; however, the weak neutral current is modified by the *addition of an isoscalar contribution*

$$\delta\mathcal{J}_\mu^{(0)} = \frac{i}{2}[\bar{c}\gamma_\mu(1+\gamma_5)c - \bar{s}\gamma_\mu(1+\gamma_5)s] \tag{44.21}$$

[17]If a rate calculation uses $G^{(\pm)} = G_\mu$, as in Eqs. (42.53)-(42.55) and Prob. 42.1, then one must reinterpret $F_i \to F_i\cos\theta_C$.

In this sector of the theory, (s, c) quarks and and their antiparticles $(\bar{s}, \bar{c})$ enter through loop processes.

We shall return to discussions of the extension to include an additional heavy quark family and to radiative corrections in the standard model later in this part of the book. First, however, we explore in some detail the consequences of the theory developed so far for electroweak interactions with nuclei.

45

WEAK INTERACTIONS WITH NUCLEI

Introduction and Multipole Analysis. The topic of semileptonic weak interactions with nuclei includes the rich variety of processes illustrated in Fig. 45.1. The kinematic variables used to describe the processes are shown in Fig. 45.2. The theoretical framework presented here will closely parallel that in Sections 7 and 8 on *electromagnetic* interactions with nuclei.[18]

We start from the semileptonic weak hamiltonian of the standard model, which in the Schrödinger picture takes the form

$$\hat{H}_{\mathrm{W}} = -\frac{G}{\sqrt{2}} \int d^3x\, j_\mu^{\mathrm{lept}}(\mathbf{x})\hat{\mathcal{J}}_\mu(\mathbf{x}) \tag{45.1}$$

The appropriate lepton and hadron currents have been discussed in the previous two sections.

We work to first order in the weak coupling constant G; thus the leptons will be treated in lowest-order perturbation theory. In contrast, the strong interactions will be treated to all orders, and one is still required to evaluate the exact transition matrix elements of the hadronic weak currents. With this in mind, the matrix element of this weak hamiltonian required to describe the semileptonic processes in Fig. 45.2 takes the form

$$\begin{aligned} \langle f|\hat{H}_{\mathrm{W}}|i\rangle &= -\frac{G}{\sqrt{2}} l_\mu \int d^3x\, e^{-i\mathbf{q}\cdot\mathbf{x}} \langle f|\hat{\mathcal{J}}_\mu(\mathbf{x})|i\rangle \\ &= -\frac{G}{\sqrt{2}} \int d^3x\, e^{-i\mathbf{q}\cdot\mathbf{x}} [\mathbf{l}\cdot\boldsymbol{\mathcal{J}}(\mathbf{x})_{\mathrm{fi}} - l_0\mathcal{J}_0(\mathbf{x})_{\mathrm{fi}}] \end{aligned} \tag{45.2}$$

Here the matrix element of the leptonic current is written

$$\langle f|j_\mu^{\mathrm{lept}}(\mathbf{x})|i\rangle = l_\mu e^{-i\mathbf{q}\cdot\mathbf{x}} \tag{45.3}$$

Define a complete othonormal set of spatial unit vectors with z-axis $\mathbf{q}/|\mathbf{q}|$ as illustrated in Fig. 45.3. Now any vector can be expanded in this set as follows

$$\mathbf{l} = \sum_{\lambda=0,\pm1} l_\lambda \mathbf{e}_\lambda^\dagger \tag{45.4}$$

[18] See also the detailed analysis of electron scattering in Ref. [N14].

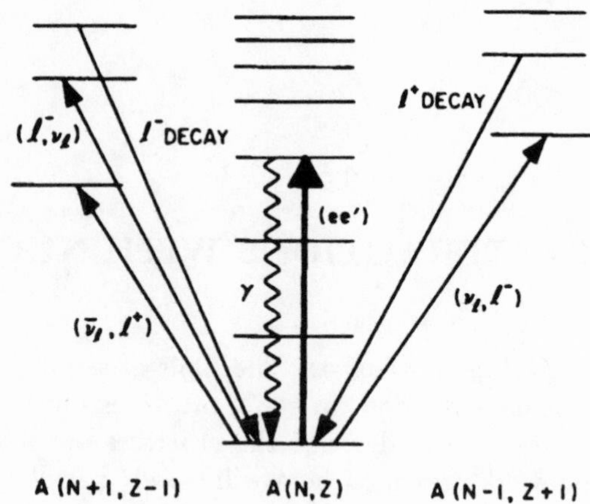

Figure 45.1: Charge-changing processes in semileptonic weak interactions with nuclei.

Here

$$\begin{aligned} \mathbf{e}_{\pm 1} &\equiv \mp\frac{1}{\sqrt{2}}(\mathbf{e}_{\mathbf{q}1} \pm i\mathbf{e}_{\mathbf{q}2}) \\ \mathbf{e}_0 &\equiv \mathbf{q}/|\mathbf{q}| \equiv \mathbf{e}_{\mathbf{q}0} \\ \mathbf{e}_\lambda^\dagger \cdot \mathbf{e}_{\lambda'} &= \delta_{\lambda,\lambda'} \qquad\qquad \mathbf{e}_\lambda^\dagger = (-1)^\lambda \mathbf{e}_{-\lambda} \end{aligned} \tag{45.5}$$

Hence $l_\lambda = \mathbf{e}_\lambda \cdot \mathbf{l}$, and to avoid confusion with the time component, we write $\mathbf{e}_0 \cdot \mathbf{l} \equiv l_3$

$$l_{\pm 1} = \mp\frac{1}{\sqrt{2}}(l_1 \pm i l_2) \qquad\qquad l_{\lambda=0} \equiv l_3 \tag{45.6}$$

Now make a multipole expansion of the hadronic current to project irreducible tensor operators (ITO), which permits the use of the entire theory of angular momentum on the nuclear matrix elements. The required multipole expansion was derived in Section 7

$$\mathbf{e}_{\mathbf{q}\lambda} e^{i\mathbf{q}\cdot\mathbf{x}} = -\sum_{J\geq 1}^{\infty} \sqrt{2\pi(2J+1)}\, i^J \left\{ \lambda j_J(\kappa x) \boldsymbol{\mathcal{Y}}_{JJ1}^\lambda + \frac{1}{\kappa}\nabla \times \left[j_J(\kappa x) \boldsymbol{\mathcal{Y}}_{JJ1}^\lambda \right] \right\}; \quad \text{for } \lambda = \pm 1 \tag{45.7}$$

To avoid confusion with the four-momentum transfer, we henceforth define

$$\kappa \equiv |\mathbf{q}| \tag{45.8}$$

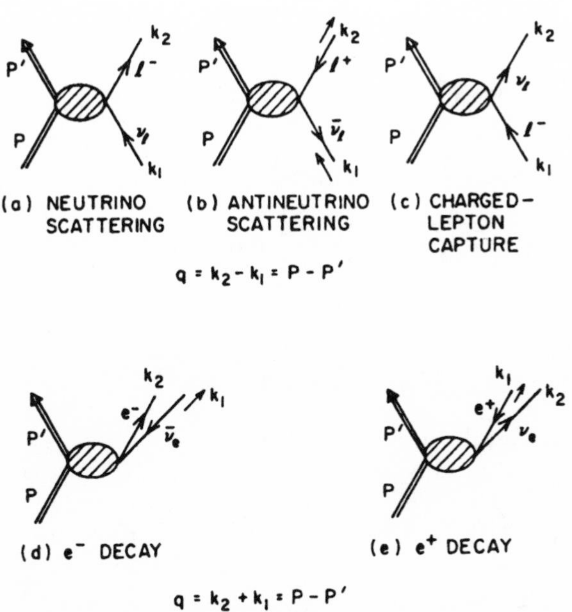

Figure 45.2: Kinematic variables for semileptonic processes in Fig. 45.1.

For weak interactions, because the axial-vector current is not conserved as is the electromagnetic current, one is also required to deal explicitly with its longitudinal matrix elements. The derivation of the longitudinal multipoles exactly parallels that in Section 7 for the transverse multipoles. One needs the following identity (Ref. [N44])

$$\begin{aligned}\nabla_\rho[j_J(\rho)Y_{JM}] &= \left(\frac{J+1}{2J+1}\right)^{1/2} j_{J+1}(\rho)\boldsymbol{\mathcal{Y}}^M_{J,J+1,1} \\ &\quad + \left(\frac{J}{2J+1}\right)^{1/2} j_{J-1}(\rho)\boldsymbol{\mathcal{Y}}^M_{J,J-1,1} \qquad (45.9)\end{aligned}$$

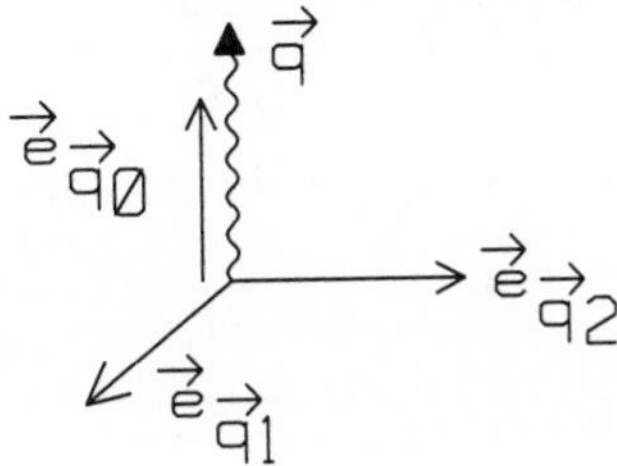

Figure 45.3: Complete orthonormal set of spatial unit vectors.

The result is (Prob. 45.1, Ref. [W2])[19]

$$\mathbf{e}_{\mathbf{q}0}e^{i\mathbf{q}\cdot\mathbf{x}} = \frac{-i}{\kappa}\sum_{J\geq 0}^{\infty}\sqrt{4\pi(2J+1)}\, i^J\nabla[j_J(\kappa x)Y_{J0}] \tag{45.10}$$

The adjoint relations follow as in Section 7

$$\mathbf{e}^{\dagger}_{\mathbf{q}\lambda}e^{-i\mathbf{q}\cdot\mathbf{x}} = -\sum_{J\geq 1}^{\infty}\sqrt{2\pi(2J+1)}(-i)^J\left\{\lambda j_J(\kappa x)\boldsymbol{\mathcal{Y}}^{-\lambda}_{JJ1}+\frac{1}{\kappa}\nabla\times\left[j_J(\kappa x)\boldsymbol{\mathcal{Y}}^{-\lambda}_{JJ1}\right]\right\};$$

$$\text{for}\,\lambda=\pm 1$$

$$\mathbf{e}^{\dagger}_{\mathbf{q}0}e^{-i\mathbf{q}\cdot\mathbf{x}} = \frac{i}{\kappa}\sum_{J\geq 0}^{\infty}\sqrt{4\pi(2J+1)}(-i)^J\nabla[j_J(\kappa x)Y_{J0}] \tag{45.11}$$

Insertion of these results in Eq. (45.2) leads to

$$\langle f|\hat{H}_{\mathrm{W}}|i\rangle = \frac{-G}{\sqrt{2}}\langle f|\left\{-\sum_{J\geq 1}\sqrt{2\pi(2J+1)}(-i)^J\sum_{\lambda=\pm 1}l_\lambda\left[\lambda\hat{\mathcal{T}}^{\mathrm{mag}}_{J-\lambda}(\kappa)+\hat{\mathcal{T}}^{\mathrm{el}}_{J-\lambda}(\kappa)\right]\right.$$
$$\left.+\sum_{J\geq 0}\sqrt{4\pi(2J+1)}(-i)^J\left[l_3\hat{\mathcal{L}}_{J0}(\kappa)-l_0\hat{\mathcal{M}}_{J0}(\kappa)\right]\right\}|i\rangle \tag{45.12}$$

Here the multipole operators are defined by

$$\begin{aligned}
\hat{\mathcal{M}}_{JM}(\kappa) &\equiv \hat{M}_{JM}+\hat{M}^5_{JM} = \int d^3x[j_J(\kappa x)Y_{JM}(\Omega_x)]\hat{\mathcal{J}}_0(\mathbf{x})\\
\hat{\mathcal{L}}_{JM}(\kappa) &\equiv \hat{L}_{JM}+\hat{L}^5_{JM} = \frac{i}{\kappa}\int d^3x\{\nabla[j_J(\kappa x)Y_{JM}(\Omega_x)]\}\cdot\hat{\boldsymbol{\mathcal{J}}}(\mathbf{x})\\
\hat{\mathcal{T}}^{\mathrm{el}}_{JM}(\kappa) &\equiv \hat{T}^{\mathrm{el}}_{JM}+\hat{T}^{\mathrm{el5}}_{JM} = \frac{1}{\kappa}\int d^3x[\nabla\times j_J(\kappa x)\boldsymbol{\mathcal{Y}}^M_{JJ1}(\Omega_x)]\cdot\hat{\boldsymbol{\mathcal{J}}}(\mathbf{x})\\
\hat{\mathcal{T}}^{\mathrm{mag}}_{JM}(\kappa) &\equiv \hat{T}^{\mathrm{mag}}_{JM}+\hat{T}^{\mathrm{mag5}}_{JM} = \int d^3x[j_J(\kappa x)\boldsymbol{\mathcal{Y}}^M_{JJ1}(\Omega_x)]\cdot\hat{\boldsymbol{\mathcal{J}}}(\mathbf{x})
\end{aligned} \tag{45.13}$$

As in Section 7, these operators are now irreducible tensor operators (ITO) in the nuclear Hilbert space. In contrast to the previous discussion, each operator contains contributions of *both parities,* as explicitly indicated above, since the weak hadronic currents have a $V-A$ structure

$$\hat{\mathcal{J}}_\mu = \hat{J}_\mu+\hat{J}_{\mu 5} \tag{45.14}$$

The Wigner-Eckart theorem can now be employed on the ITO

$$\langle J_fM_f|\hat{\mathcal{T}}_{JM}|J_iM_i\rangle = (-1)^{J_f-M_f}\begin{pmatrix} J_f & J & J_i\\ -M_f & M & M_i\end{pmatrix}\langle J_f||\hat{\mathcal{T}}_J||J_i\rangle \tag{45.15}$$

[19]Note again that this is just an algebraic identity, following from the definition of the vector spherical harmonics.

Equations (45.12) and (45.15) are now completely general in that they hold for any nuclear wave functions and any local nuclear weak current. One can proceed to calculate *any* semileptonic weak nuclear process — with polarized leptons, polarized targets, recoil polarizations, etc.

With *unoriented and unobserved* targets, one sums over final target states, and averages over initial states. The orthonormality of the $3-j$ coefficients then yields

$$\frac{1}{2J_i+1}\sum_{M_f}\sum_{M_i}\begin{pmatrix} J_f & J & J_i \\ -M_f & M & M_i \end{pmatrix}\begin{pmatrix} J_f & J' & J_i \\ -M_f & M' & M_i \end{pmatrix} = \frac{1}{(2J+1)(2J_i+1)}\delta_{JJ'}\delta_{MM'} \qquad (45.16)$$

Hence from Eq. (45.12)

$$\frac{1}{(2J_i+1)}\sum_{M_i}\sum_{M_f}|\langle f|\hat{H}_{\rm W}|i\rangle|^2 = \frac{G^2}{2}\frac{1}{(2J_i+1)}\left\{\sum_{\lambda=\pm1} l_\lambda l_\lambda^* \sum_{J\geq1} 2\pi \right.$$
$$\times|\langle J_f||\lambda\hat{\mathcal{T}}_J^{\rm mag}+\hat{\mathcal{T}}_J^{\rm el}||J_i\rangle|^2 + \sum_{J\geq0} 4\pi\left[l_3l_3^*|\langle J_f||\hat{\mathcal{L}}_J||J_i\rangle|^2\right.$$
$$\left.\left.+l_0l_0^*|\langle J_f||\hat{\mathcal{M}}_J||J_i\rangle|^2 - 2\,{\rm Re}\left(l_3l_0^*\langle J_f||\hat{\mathcal{L}}_J||J_i\rangle\langle J_f||\hat{\mathcal{M}}_J||J_i\rangle^*\right)\right]\right\} \qquad (45.17)$$

Now use

$$\sum_{\lambda=\pm1} l_\lambda l_\lambda^*|a+\lambda b|^2 = |a+b|^2\frac{1}{2}(l_1l_1^*+l_2l_2^*+il_2l_1^*-il_1l_2^*)$$
$$+|a-b|^2\frac{1}{2}(l_1l_1^*+l_2l_2^*-il_2l_1^*+il_1l_2^*)$$
$$= (|a|^2+|b|^2)(\mathbf{l}\cdot\mathbf{l}^*-l_3l_3^*) - i(\mathbf{l}\times\mathbf{l}^*)_3\,2\,{\rm Re}\,(ab^*) \qquad (45.18)$$

Thus

$$\frac{1}{(2J_i+1)}\sum_{M_i}\sum_{M_f}|\langle f|\hat{H}_{\rm W}|i\rangle|^2 = \frac{G^2}{2}\frac{4\pi}{(2J_i+1)}\times$$
$$\left\{\sum_{J\geq1}\left[\frac{1}{2}(\mathbf{l}\cdot\mathbf{l}^*-l_3l_3^*)\left(|\langle J_f||\hat{\mathcal{T}}_J^{\rm mag}||J_i\rangle|^2+|\langle J_f||\hat{\mathcal{T}}_J^{\rm el}||J_i\rangle|^2\right)\right.\right.$$
$$\left.-\frac{i}{2}(\mathbf{l}\times\mathbf{l}^*)_3\left(2\,{\rm Re}\langle J_f||\hat{\mathcal{T}}_J^{\rm mag}||J_i\rangle\langle J_f||\hat{\mathcal{T}}_J^{\rm el}||J_i\rangle^*\right)\right]$$
$$+\sum_{J\geq0}\left[l_3l_3^*|\langle J_f||\hat{\mathcal{L}}_J||J_i\rangle|^2+l_0l_0^*|\langle J_f||\hat{\mathcal{M}}_J||J_i\rangle|^2\right.$$
$$\left.\left.-2\,{\rm Re}\left(l_3l_0^*\langle J_f||\hat{\mathcal{L}}_J||J_i\rangle\langle J_f||\hat{\mathcal{M}}_J||J_i\rangle^*\right)\right]\right\} \qquad (45.19)$$

This is a *general result*; it holds for any semileptonic nuclear process. In addition to lowest-order perturbation theory in the weak coupling constant G, it assumes only

- The existence of a local weak nuclear current operator;
- That the initial and final nuclear states, whatever they may be, are eigenstates of angular momentum.

Nuclear Current Operator. The next step is to construct the nuclear current operator. In the traditional nuclear physics picture, the electroweak current is constructed from the properties of free nucleons, and we start with this approach. The full matrix element of the hadronic weak current operator for a free nucleon in the standard model follows from general symmetry considerations; it was constructed in Section 42. The kinematic situation is shown in Fig. 42.6. Here $q \equiv p - p'$ and all the form factors $F_i(q^2)$ are functions of q^2.

For the *vector* current one has

$$\langle \mathbf{p}'\sigma'\rho'|\hat{J}_\mu^{(\pm)}(0)|\mathbf{p}\sigma\rho\rangle = \frac{i}{\Omega}\bar{u}(\mathbf{p}'\sigma')\eta_{\rho'}^\dagger[F_1\gamma_\mu + F_2\sigma_{\mu\nu}q_\nu]\tau_\pm\eta_\rho u(\mathbf{p}\sigma) \quad (45.20)$$

Note the following features of this result:

1. It is assumed here that, as in the standard model, there are no second class currents.

2. From CVC

$$F_i = F_i^V \; ; \qquad i = 1, 2 \quad (45.21)$$

Here F_i^V is the isovector form factor in the *electromagnetic* interaction measured in (e, e') (Section 42).

3. Consider the quark description in the *nuclear domain* where the quark Hilbert space is restricted to contain only (u, d) quarks and their antiquarks and the quark field reduces to

$$\psi \doteq \begin{pmatrix} u \\ d \end{pmatrix} \quad (45.22)$$

The near equality of the mass of the (u, d) quarks (they are both almost zero) implies that the QCD lagrangian possesses an $SU(2)$ symmetry with respect to flavor mixtures of these quarks; this symmetry is *strong isospin* and the field in Eq. (45.22) then forms a *strong isodoublet*. In this case, all the preceding arguments on the general symmetry properties of the hadronic weak current follow from the discussion in Section 44; the sole exception is that the charge-changing weak coupling constant must be modified to take into account the presence of the Cabbibo angle

$$G^\pm \equiv G\cos\theta_C \quad (45.23)$$

The single-nucleon matrix element of the *axial vector current* takes the form

$$\langle \mathbf{p}'\sigma'\rho'|\hat{J}^{(\pm)}_{\mu 5}(0)|\mathbf{p}\sigma\rho\rangle = \frac{i}{\Omega}\bar{u}(\mathbf{p}'\sigma')\eta^{\dagger}_{\rho'}[F_A\gamma_5\gamma_\mu - iF_P\gamma_5 q_\mu]\tau_{\pm}\eta_\rho u(\mathbf{p}\sigma) \quad (45.24)$$

This result has the following features:

1. From pion-pole dominance of the induced pseudoscalar coupling and the Goldberger-Treiman relation one has (Section 42)

$$F_P \;=\; \frac{2mF_A}{q^2+m_\pi^2} \quad (45.25)$$

This implies the PCAC relation

$$\langle f|\frac{\partial}{\partial x_\mu}J^{(\pm)}_{\mu 5}(0)|i\rangle \;=\; O(m_\pi^2) \quad (45.26)$$

2. The form factor $F_A(q^2)$ describing the internal axial vector structure of the nucleon must be measured through some weak process.

As in Sections 7 and 8, these results can now be used to construct the *nuclear* current operator in the traditional nuclear physics picture. Assume the nuclear current density operator at the origin is given in second quantization by

$$\hat{\mathcal{J}}_\mu(0) \;=\; \sum_{\mathbf{p}'\sigma'\rho'}\sum_{\mathbf{p}\sigma\rho} c^{\dagger}_{\mathbf{p}'\sigma'\rho'}\langle \mathbf{p}'\sigma'\rho'|\mathcal{J}_\mu(0)|\mathbf{p}\sigma\rho\rangle c_{\mathbf{p}\sigma\rho} \quad (45.27)$$

Here the matrix element is taken to be that for free nucleons, as discussed above.

Write the current density operator in first quantization as

$$\hat{\mathcal{J}}_\mu(\mathbf{x}) \;=\; \sum_{i=1}^{A}\left[\hat{\mathcal{J}}^{(1)}_\mu(i)\delta^{(3)}(\mathbf{x}-\mathbf{x}_i)\right] \quad (45.28)$$

The prescription for the transition from first to second quantization then identifies the single-particle matrix element of the current density appearing in Eq. (45.27) as (Ref. [N2])

$$\langle \mathbf{p}'\sigma'\rho'|\mathcal{J}_\mu(\mathbf{x})|\mathbf{p}\sigma\rho\rangle = \int d^3y\,\phi^{\dagger}_{\mathbf{p}'\sigma'\rho'}(\mathbf{y})\left[\mathcal{J}^{(1)}_\mu(\mathbf{y})\delta^{(3)}(\mathbf{x}-\mathbf{y})\right]\phi_{\mathbf{p}\sigma\rho}(\mathbf{y}) \quad (45.29)$$

By comparison with the single-particle matrix element for a free nucleon, one can now identify the appropriate single-particle densities.

As in Section 7, we anticipate the form of the results and write the nuclear current densities in this approach as follows

$$\begin{aligned}
\hat{\mathbf{J}}^{(\pm)}(\mathbf{x}) &= \hat{\mathbf{J}}^{(\pm)}_{\mathrm{C}}(\mathbf{x}) + \nabla\times\hat{\boldsymbol{\mu}}^{(\pm)}(\mathbf{x}) \\
\hat{\mathbf{J}}^{(\pm)}_5(\mathbf{x}) &= \hat{\mathbf{A}}^{(\pm)}(\mathbf{x}) + \nabla\hat{\phi}^{(\pm)}_{PS}(\mathbf{x}) \\
\hat{J}^{(\pm)}_0(\mathbf{x}) &= \hat{\rho}^{(\pm)}(\mathbf{x}) \\
\hat{J}^{(\pm)}_{05}(\mathbf{x}) &= \hat{\rho}^{(\pm)}_5(\mathbf{x}) + \frac{1}{i}\left[\hat{H},\hat{\phi}^{(\pm)}_{PS}(\mathbf{x})\right]
\end{aligned} \quad (45.30)$$

Now substitute the explicit form of the solutions to the Dirac equation in the single-nucleon matrix element, use the standard representation of the gamma matrices, and make an expansion in powers of $1/m$; a calculation exactly paralleling that in Section 8 then leads to (Prob. 45.2)[20]

$$\begin{aligned}
\langle \mathbf{p}'\sigma'\rho'|\mathcal{J}_\mu^{(-)}(0)|\mathbf{p}\sigma\rho\rangle &= \frac{1}{\Omega}\chi_{\sigma'}^\dagger\eta_{\rho'}^\dagger\left[M_\mu - q_\mu\left(F_P\frac{\boldsymbol{\sigma}\cdot\mathbf{q}}{2m}\right)\right]\tau_-\eta_\rho\chi_\sigma + O(1/m^2) \\
M_\mu &\equiv (\mathbf{M}, iM_0) \\
\mathbf{M} &= F_A\boldsymbol{\sigma} - (F_1+2mF_2)\frac{i\boldsymbol{\sigma}\times\mathbf{q}}{2m} + F_1\left(\frac{2\mathbf{p}-\mathbf{q}}{2m}\right) \\
M_0 &= F_1 + \boldsymbol{\sigma}\cdot\left[\left(F_A\frac{2\mathbf{p}-\mathbf{q}}{2m}\right)\right] \qquad (45.31)
\end{aligned}$$

As in Section 8, use the definition of the kinematics in Eq. (45.2) and Fig. 45.2, and assume that in this discussion the nuclear target is *localized* so that partial integrations are permitted with vanishing surface terms; this allows the following identification in Eq. (45.30)

$$\nabla \leftrightarrow i\mathbf{q} \qquad (45.32)$$

A comparison of these results then yields the first-quantized nuclear density operators

$$\begin{aligned}
\hat{\rho}^{(\pm)}(\mathbf{x}) &= F_1\sum_{j=1}^A \tau_\pm(j)\delta^{(3)}(\mathbf{x}-\mathbf{x}_j) \\
\hat{\mathbf{J}}_C^{(\pm)}(\mathbf{x}) &= F_1\sum_{j=1}^A \tau_\pm(j)\left[\frac{\mathbf{p}(j)}{m}, \delta^{(3)}(\mathbf{x}-\mathbf{x}_j)\right]_{\text{sym}} \\
\hat{\mathbf{A}}^{(\pm)}(\mathbf{x}) &= F_A\sum_{j=1}^A \boldsymbol{\sigma}(j)\tau_\pm(j)\delta^{(3)}(\mathbf{x}-\mathbf{x}_j) \\
\hat{\rho}_5^{(\pm)}(\mathbf{x}) &= F_A\sum_{j=1}^A \tau_\pm(j)\boldsymbol{\sigma}(j)\cdot\left[\frac{\mathbf{p}(j)}{m}, \delta^{(3)}(\mathbf{x}-\mathbf{x}_j)\right]_{\text{sym}} \qquad (45.33)
\end{aligned}$$

In addition

$$\begin{aligned}
\hat{\boldsymbol{\mu}}^{(\pm)}(\mathbf{x}) &= \frac{F_1+2mF_2}{2mF_A}\hat{\mathbf{A}}^{(\pm)}(\mathbf{x}) \\
\hat{\phi}_{PS}^{(\pm)}(\mathbf{x}) &= \frac{F_P}{2mF_A}\nabla\cdot\hat{\mathbf{A}}^{(\pm)}(\mathbf{x}) \qquad (45.34)
\end{aligned}$$

[20] Recall $i\gamma_4\gamma_5\vec{\gamma} = \begin{pmatrix} \vec{\sigma} & 0 \\ 0 & \vec{\sigma} \end{pmatrix}$ and $\gamma_4\gamma_5 = \begin{pmatrix} 0 & -1 \\ 1 & 0 \end{pmatrix}$.

Matrix elements of this weak nuclear current operator are tabulated in Ref. [W9] for three-dimensional harmonic oscillator wave functions, and also in Ref. [W10] for arbitrary radial wave functions; either set consititutes a complete basis of single-particle wave functions for the nuclear many-body problem.

Long-Wavelength Reduction. The long-wavelength reduction of the multipoles is useful both to obtain insight into their character and as a starting point in analyzing low momentum transfer processes. This reduction is obtained with exactly the *same* analysis carried out in detail for the electromagnetic interaction in Section 7. It assumes only a localized nuclear transition current density.

The results are as follows (Ref. [W2], Prob. 45.3)

$$\hat{\mathcal{M}}_{JM}(\kappa) \rightarrow \frac{\kappa^J}{(2J+1)!!}\int d^3x\, x^J Y_{JM}\hat{\mathcal{J}}_0(\mathbf{x})\ ; \qquad \kappa \equiv |\mathbf{q}| \rightarrow 0$$
$$\hat{\mathcal{L}}_{JM}(\kappa) \rightarrow \frac{-i\kappa^{J-1}}{(2J+1)!!}\int d^3x\, x^J Y_{JM}\nabla\cdot\hat{\mathcal{J}}(\mathbf{x}) \tag{45.35}$$

The one *exception* occurs with the monopole longitudinal multipole where the first nonvanishing contribution is

$$\hat{\mathcal{L}}_{00}(\kappa) \rightarrow \frac{i\kappa}{6}\int d^3x\, x^2 Y_{00}\nabla\cdot\hat{\mathcal{J}}(\mathbf{x}) \tag{45.36}$$

The long-wavelength reduction of the transverse multipoles is exactly the same as before (here $\mathbf{x} \equiv \mathbf{r}$)

$$\hat{T}^{\text{el}}_{JM} \rightarrow \frac{1}{i}\frac{\kappa^{J-1}}{(2J+1)!!}\left(\frac{J+1}{J}\right)^{1/2}\int d^3x\, x^J Y_{JM}\nabla\cdot\hat{\mathcal{J}}(\mathbf{x})\ ; \qquad \kappa \equiv |\mathbf{q}| \rightarrow 0$$
$$\hat{T}^{\text{mag}}_{JM} \rightarrow -\frac{1}{i}\frac{\kappa^{J}}{(2J+1)!!}\left(\frac{J+1}{J}\right)^{1/2}\int d^3x \left[\frac{1}{J+1}\mathbf{r}\times\hat{\mathcal{J}}(\mathbf{x})\right]\cdot\nabla x^J Y_{JM} \tag{45.37}$$

The conserved vector current theory (CVC) allows a rewriting of the integrands. For the vector current

$$\nabla\cdot\hat{\mathbf{J}}(\mathbf{x}) \quad = \quad -\frac{\partial\hat{J}_0}{\partial t} = -i[\hat{H},\hat{J}_0(\mathbf{x})] \tag{45.38}$$

Hence the transverse electric multipoles and the charge multipoles for the vector current are related exactly as in the electromagnetic case

$$\langle f|\hat{T}^{\text{el}}_{JM}(\kappa)|i\rangle \rightarrow -\frac{(E_f-E_i)}{\kappa}\left(\frac{J+1}{J}\right)^{1/2}\langle f|\hat{M}_{JM}(\kappa)|i\rangle\ ; \qquad \kappa \equiv |\mathbf{q}| \rightarrow 0$$
$$\hat{M}_{JM}(\kappa) \rightarrow \frac{\kappa^J}{(2J+1)!!}\int d^3x\, x^J Y_{JM}\hat{J}_0(\mathbf{x}) \tag{45.39}$$

Example – "Allowed" Processes. Consider semileptonic weak nuclear processes in the long-wavelength limit where

$$\kappa \equiv |\mathbf{q}| \rightarrow 0 \tag{45.40}$$

From the above analysis, *the only surviving multipoles in this limit are*

$$\begin{aligned}\hat{\mathcal{T}}^{\text{el}}_{1M} &= \sqrt{2}\hat{\mathcal{L}}_{1M} = \frac{i\sqrt{2}}{3}\sqrt{\frac{3}{4\pi}}\int \hat{\mathcal{J}}_{1M}(\mathbf{x})d^3x \\ \hat{\mathcal{M}}_{00} &= \sqrt{\frac{1}{4\pi}}\int \hat{\mathcal{J}}_0(\mathbf{x})d^3x\end{aligned} \tag{45.41}$$

This is a general result.

Now assume further that one is dealing with *slow nucleons* so that

$$\frac{\mathbf{p}}{m} \sim \left(\frac{v}{c}\right)_{\text{nucleon}} \rightarrow 0 \tag{45.42}$$

In this limit, the only surviving charge-changing nuclear densities are

$$\begin{aligned}\hat{\rho}^{(\pm)}(\mathbf{x}) &= F_1\sum_{i=1}^{A}\tau_\pm(i)\delta^{(3)}(\mathbf{x}-\mathbf{x}_i) \\ \hat{\mathbf{A}}^{(\pm)}(\mathbf{x}) &= F_A\sum_{i=1}^{A}\tau_\pm(i)\boldsymbol{\sigma}(i)\delta^{(3)}(\mathbf{x}-\mathbf{x}_i)\end{aligned} \tag{45.43}$$

A combination of Eqs. (45.41) and (45.43) then leads to

$$\begin{aligned}\hat{\mathcal{T}}^{\text{el}}_{1M} = \sqrt{2}\hat{\mathcal{L}}_{1M} &= \frac{i}{\sqrt{6\pi}}F_A\sum_{i=1}^{A}\tau_\pm(i)\sigma_{1M}(i)\ ; \qquad \text{Gamow-Teller} \\ \hat{\mathcal{M}}_{00} &= \frac{1}{\sqrt{4\pi}}F_1\sum_{i=1}^{A}\tau_\pm(i) \qquad\qquad ; \qquad \text{Fermi}\end{aligned} \tag{45.44}$$

Several features of these results are of interest:

1. These operators give rise to the *allowed* weak transitions in the traditional picture of the nucleus.

2. The operators and transitions they give rise to are known as Gamow-Teller and Fermi, respectively.[21]

3. This is the form of the operators in the limit $(v/c)_{\text{nucleon}} \rightarrow 0$. The general results for the nuclear densities, keeping relativistic corrections up through $O(1/m)$, have been presented previously in this section.

[21] The nuclear selection rules for Fermi and Gamow-Teller transitions follow immediately from the form of the operators (Prob. 45.4).

4. It is evident that at long wavelengths[22]

$$\hat{\mathcal{M}}_{00} = \frac{1}{\sqrt{4\pi}}\hat{T}_{\pm} \tag{45.45}$$

Here $\hat{T}_{\pm}$ is the isospin raising and lowering operator in this traditional picture. A great beauty of CVC is that this relation is predicted to continue to hold in *any hadronic picture* of the strong interactions, for example, in QHD. The standard model predicts that this relation continues to hold on the *quark-gluon level* with the strong interactions described by QCD.

5. In the standard model, the operators governing neutral current weak interactions in the nuclear domain are obtained immediately from those for charge-changing processes (discussed here) and the electromagnetic current (Section 44).

The Relativistic Nuclear Many-Body Problem. In Part II of this book we discussed the relativistic nuclear many-body problem in terms of quantum field theories based on hadronic degrees of freedom. QHD-I, a simple model with neutral scalar and vector meson fields (σ, ω) was shown to enjoy some phenomenological success at the mean-field level. There an effective electromagnetic current, which attempts to take into account the internal charged-meson structure of the nucleon, was introduced as follows:

$$\begin{aligned} J_{\mu}^{\gamma}(x) &= i\bar{\psi}\gamma_{\mu}\underset{\sim}{Q}\psi + \frac{1}{2m}\frac{\partial}{\partial x_{\nu}}\left(\bar{\psi}\sigma_{\mu\nu}\underset{\sim}{\lambda}'\psi\right) \; ; \qquad \psi = \begin{pmatrix} p \\ n \end{pmatrix} \\ \underset{\sim}{Q} &= \frac{1}{2}(1+\tau_3) \\ \underset{\sim}{\lambda}' &= \lambda_p'\frac{1}{2}(1+\tau_3) + \lambda_n\frac{1}{2}(1-\tau_3) \end{aligned} \tag{45.46}$$

Recall that this effective electromagnetic current has the following properties:

1. It is local.
2. It is covariant.
3. It is conserved in QHD-I.
4. When used with the effective Møller potential $f_{\rm SN}(q^2)/q^2$, it correctly describes electron scattering from a free nucleon.

Now perform *the same analysis with the weak current.* The free nucleon matrix elements are given in Eqs. (45.20) and (45.24). In an exactly analogous fashion one then has

$$\begin{aligned} J_{\mu}^{(\pm)} &\equiv i\bar{\psi}\gamma_{\mu}\tau_{\pm}\psi + \frac{(\lambda_p' - \lambda_n)}{2m}\frac{\partial}{\partial x_{\nu}}\left(\bar{\psi}\sigma_{\mu\nu}\tau_{\pm}\psi\right) \\ J_{\mu 5}^{(\pm)} &\equiv \left(\delta_{\mu\nu} + \frac{1}{m_{\pi}^2 - \Box}\frac{\partial}{\partial x_{\mu}}\frac{\partial}{\partial x_{\nu}}\right) F_A(0)\, i\bar{\psi}\gamma_5\gamma_{\nu}\tau_{\pm}\psi \end{aligned}$$

[22]Recall $F_1(0) = F_1^V(0) = 1$.

$$\begin{aligned}
\mathcal{J}_\mu^{(\pm)} &\equiv J_\mu^{(\pm)} + J_{\mu5}^{(\pm)} \\
\mathcal{J}_\mu^{(0)} &\equiv \mathcal{J}_\mu^{V_3} - 2\sin^2\theta_{\rm W} J_\mu^\gamma
\end{aligned} \tag{45.47}$$

Here for simplicity a common single nucleon form factor has again been assumed

$$\frac{F_A(q^2)}{F_A(0)} \approx f_{\rm SN}(q^2) \tag{45.48}$$

The resulting weak nuclear current has the following features to recommend it:

1. It is covariant.
2. It satisfies PCAC

$$\frac{\partial J_{\mu5}}{\partial x_\mu} = O(m_\pi^2) \tag{45.49}$$

3. It satisfies all the *general symmetry properties of the standard model* in the nuclear domain [Eqs. (44.19)].

4. It gives the correct result for semileptonic weak interactions on a free nucleon.

5. In conjunction with QHD-I, it provides a model for summing relativistic effects in nuclei to all orders.[23]

Summary. In summary, we now have a general multipole analysis of the transition matrix element of the weak current for any semileptonic weak nuclear process; the cross section or rate for that process follows immediately from Fermi's Golden Rule. The weak charge-changing nuclear current operators have been constructed in the traditional nuclear physics picture where the electroweak currents are obtained from the properties of free nucleons, retaining relativistic corrections through $O(1/m)$. The nuclear weak neutral current operator in the standard model in the nuclear domain is simply a linear combination of a rotated isospin component of the charge-changing current and the electromagnetic current (Sections 43 and 44). We have extended this analysis to the relativistic nuclear many-body problem through the construction of an effective covariant

[23]The electromagnetic current in Eq. (45.46) assumes that $F_1(q^2)/F_1(0) \approx F_2(q^2)/F_2(0) \approx f_{\rm SN}(q^2)$. This relation breaks down at large q^2 where the data indicate that it is the Sachs form factors which scale (see Fig. 41.1)

$$G_{\rm M} \equiv F_1 + 2mF_2 \qquad\qquad G_{\rm E} = F_1 - q^2F_2/2m$$

To incorporate this observation, make the following replacements:

$$\begin{aligned}
\frac{f_{\rm SN}(q^2)}{q^2} &\to \frac{f_{\rm SN}(q^2)}{q^2}\frac{1}{1+q^2/4m^2}\,; \qquad \text{effective Møller potential} \\
J_\mu^\gamma &\to J_\mu^\gamma - \frac{i}{4m^2}\frac{\partial}{\partial x_\nu}\frac{\partial}{\partial x_\nu}\left(\bar\psi\underline{\mu}\,\gamma_\mu\psi\right)
\end{aligned}$$

Here $\underline{\mu}$ is the full magnetic moment. We leave the demonstration of this result as a problem (Prob. 45.5). The assumption in Eq. (45.48) can similarly be relaxed.

weak current operator, possessing all the symmetry properties of the full theory, to be used in conjunction with QHD-I. Given nuclear wave functions, one can now calculate any semileptonic process. We proceed to discuss applications of these results.

46

SEMILEPTONIC WEAK PROCESSES

In this section we consider in detail the basic semileptonic weak nuclear processes: neutrino (antineutrino) reactions, charged lepton (muon) capture, and β-decay. General expressions for cross sections and rates are derived in terms of relevant multipoles of the nuclear weak currents, and some useful limiting forms of these results are discussed.

Neutrino Reactions. Consider neutrino reactions with nuclear targets. In addition to the charge-changing processes illustrated in Figs. 45.1 and 45.2 one also has the processes of neutrino scattering through the weak neutral current as shown in Fig. 46.1.[24] Such reactions are of central importance in astrophysics where the transport of neutrinos determines the rate of cooling of many stellar objects, and their detection provides a unique way of looking at such fascinating astrophysical phenomena as the interior workings of our sun (Ref. [W11]) and supernovae explosions (Ref. [W12]). Furthermore, transitions between discrete nuclear states are now being measured in the laboratory using neutrinos from the decays of accelerator-produced particles (Refs. [W13, W14, W15]).

Lepton Matrix Elements. The lepton matrix elements relevant to the neutrino reactions, as well as sign factors used in the presentation of various results, are given in Table 46.1. They are calculated with the aid of Eq. (42.1).

Cross Sections from the Golden Rule. To obtain the cross section, it is necessary to perform the appropriate sums and averages over the lepton spins; here it is essential to keep two facts in mind: (1) *Do not average over initial neutrino helicities* since there is only one kind of neutrino (and antineutrino) in nature that is produced and absorbed in the electroweak interactions; (2) instead, just *sum over initial and final neutrino helcities* since the $(1+\gamma_5)$ in the interaction acts as a projection operator, selecting the appropriate helicity for the particle (and antiparticle). The cross section then follows from the Golden Rule as

$$d\sigma = 2\pi \sum_{\text{helicities}} |\langle f|\hat{H}_{\mathrm{W}}|i\rangle|^2 \delta(W_f - W_i)\frac{\Omega d^3k}{(2\pi)^3}\frac{1}{1/\Omega} \tag{46.1}$$

[24] Unless specifically stated otherwise, the phrase *neutrino reaction* used in this section will now refer generically to the production of charged leptons by neutrinos and antineutrinos, as well as to the elastic and inelastic scattering of neutrinos and antineutrinos, by nuclei.

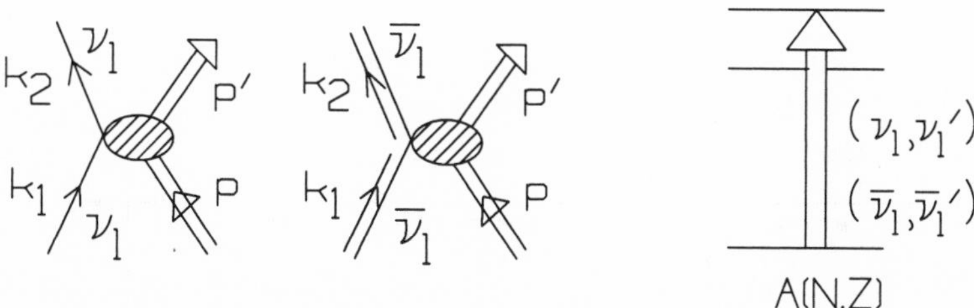

Figure 46.1: Neutrino scattering through the weak neutral current. Here $q = k_2 - k_1 = p - p'$.

The last three factors are the energy-conserving delta function, the number of final states (in a big box with periodic boundary conditions), and the initial flux, respectively.

For unpolarized and unobserved targets, one can now use the general result in Eq. (45.19) for the square of the matrix element summed and averaged over nuclear orientations. Note that the quantization volume Ω *cancels* in the cross section, as it must.

To perform the integration over the energy-conserving delta function, use

$$\begin{aligned} kdk &= \varepsilon d\varepsilon \\ \int \delta(W_f - W_i)d\varepsilon &= \int \delta(M_T^* + \varepsilon - M_T - \nu)d\varepsilon = 1 \end{aligned} \tag{46.2}$$

Thus the cross section is given by[25]

$$\frac{d\sigma}{d\Omega} = \frac{2k\varepsilon}{(2\pi)^2}\left\{\frac{\Omega^2}{2}\sum_{\text{lepton spins}}\frac{1}{2J_i+1}\sum_{M_i}\sum_{M_f}|\langle f|\hat{H}_{\text{W}}|i\rangle|^2\right\} \tag{46.3}$$

Lepton Traces. The required lepton traces produced from the lepton matrix elements in Table 46.1 by $(\Omega^2/2)\sum_{\text{lepton spins}}$ are now summarized in Table 46.2. Note that for neutrino scattering, only the extreme relativistic limit (ERL) with $|\boldsymbol{\beta}| \to 1$ is relevant.

[25] If one includes target recoil in the density of states, then $\int \delta(W_f - W_i)d\varepsilon = (\partial\varepsilon/\partial W_f) \equiv r$. To compute r use $W_f = \sqrt{M_T^{*2} + (\vec{k} - \vec{\nu})^2} + \varepsilon$. Then

$$\begin{aligned} \frac{\partial W_f}{\partial \varepsilon} &= \frac{\partial W_f}{\partial k}\frac{\partial k}{\partial \varepsilon} = \frac{\varepsilon}{k}\left[\frac{k}{\varepsilon} + \frac{k - \nu\cos\theta}{E_f}\right] = \frac{1}{E_f}\left[M_T + \nu - \frac{\varepsilon}{k}\nu\cos\theta\right] \\ r^{-1} &= \frac{M_T}{E_f}\left[1 + \frac{\nu}{M_T}\left(1 - \frac{\varepsilon}{k}\cos\theta\right)\right] \approx \left[1 + \frac{\nu}{M_T}\left(1 - \frac{\varepsilon}{k}\cos\theta\right)\right] \end{aligned}$$

Table 46.1: Lepton matrix elements and sign factors for semileptonic nuclear weak interactions: the sign factors S_n are used in Table 46.2.

Process	q	$(\Omega/i)l_\lambda$	S_1	S_2	S_3
Neutrino reaction $(\nu_l, l^-), (\nu_l, \nu_l')$	$k - \nu$	$\bar{u}(\mathbf{k})\gamma_\lambda(1+\gamma_5)u(\boldsymbol{\nu})$	-1	$+1$	$+1$
Antineutrino reaction[a] $(\bar{\nu}_l, l^+), (\bar{\nu}_l, \bar{\nu}_l')$	$k - \nu$	$\bar{v}(-\boldsymbol{\nu})\gamma_\lambda(1+\gamma_5)v(-\mathbf{k})$	$+1$	$+1$	$+1$
Lepton capture (l^-, ν_l)	$\nu - k$	$\bar{u}(\boldsymbol{\nu})\gamma_\lambda(1+\gamma_5)u(\mathbf{k})$ $(\mathbf{k} \to 0)$	$+1$	-1	-1
β^--decay $(e^-\bar{\nu}_e)$	$k + \nu$	$\bar{u}(\mathbf{k})\gamma_\lambda(1+\gamma_5)v(-\boldsymbol{\nu})$	-1	-1	sgn $\varepsilon - \nu$
β^+-decay $(e^+\nu_e)$	$k + \nu$	$\bar{u}(\boldsymbol{\nu})\gamma_\lambda(1+\gamma_5)v(-\mathbf{k})$	$+1$	-1	

[a] The antilepton requires $-l_\lambda$; this is immaterial for the subsequent bilinear forms.

As for the derivation of the results in Table 46.2, we will do one and let the reader verify the rest (Prob. 46.1). Consider

$$\begin{aligned}
\frac{\Omega^2}{2}\sum_{\text{lepton spins}} l_0 l_0^* &= \frac{1}{2}\text{tr } \gamma_4(1+\gamma_5)\left(\frac{-i\gamma_\lambda \nu_\lambda}{2\nu}\right)\gamma_4(1+\gamma_5)\left(\frac{\pm m_l - i\gamma_\sigma k_\sigma}{2\varepsilon}\right) \\
&= -\frac{1}{8\varepsilon\nu}2\,\text{tr } \gamma_4\gamma_\lambda\nu_\lambda\gamma_4(1+\gamma_5)\gamma_\sigma k_\sigma = -\frac{1}{\varepsilon\nu}(2\nu_4 k_4 - \nu\cdot k) \\
&= \frac{1}{\varepsilon\nu}(\boldsymbol{\nu}\cdot\mathbf{k} + \varepsilon\nu) = 1 + \frac{\boldsymbol{\nu}\cdot\mathbf{k}}{\nu\varepsilon} \\
\frac{\Omega^2}{2}\sum_{\text{lepton spins}} l_0 l_0^* &= 1 + \hat{\boldsymbol{\nu}}\cdot\boldsymbol{\beta}
\end{aligned} \tag{46.4}$$

Note the term in the lepton mass m_l in the first line goes out since it is multiplied by $(1+\gamma_5)(1-\gamma_5)$.[26]

At threshold, $\boldsymbol{\beta} \to 0$ and this expression goes to one. In the ERL

$$1 + \hat{\boldsymbol{\nu}}\cdot\boldsymbol{\beta} = 1 + \cos\theta = 2\cos^2\frac{\theta}{2} \tag{46.5}$$

Here θ is the scattering angle of the outgoing lepton with respect to the incident direction.

The general neutrino reaction cross section now follows from Eqs. (46.3) and (45.19) and the lepton traces in Table 46.2.[27] Two limiting cases are particularly simple.

[26] Use the cyclic property of the tr.

[27] For charge-changing semileptonic processes in nuclei, one must use the standard model result from Eq. (45.23) that $G^{(\pm)} = G\cos\theta_C \approx 0.974G$.

Table 46.2: Lepton traces: $(\Omega^2/2)\sum_{\text{lepton spins}}$ produced from lepton matrix elements l_λ in Table 46.1 as required for semileptonic weak nuclear processes.

Summand	General result[a]	Threshold $\lvert\boldsymbol{\beta}\rvert \to 0$	ERL $\lvert\boldsymbol{\beta}\rvert \to 1$
$\frac{1}{2}(\mathbf{l}\cdot\mathbf{l}^* - l_3 l_3^*)$	$1-(\hat{\boldsymbol{\nu}}\cdot\hat{\mathbf{q}})(\boldsymbol{\beta}\cdot\hat{\mathbf{q}})$	1	$(q_\mu^2/\mathbf{q}^2)\cos^2\theta/2 +$ $2\sin^2\theta/2$
$l_0 l_0^*$	$1+\hat{\boldsymbol{\nu}}\cdot\boldsymbol{\beta}$	1	$2\cos^2\theta/2$
$l_3 l_3^*$	$1-\hat{\boldsymbol{\nu}}\cdot\boldsymbol{\beta}+$ $2(\hat{\boldsymbol{\nu}}\cdot\hat{\mathbf{q}})(\boldsymbol{\beta}\cdot\hat{\mathbf{q}})$	1	$(q_0^2/\mathbf{q}^2)2\cos^2\theta/2$
$-l_3 l_0^*$	$-\hat{\mathbf{q}}\cdot(\hat{\boldsymbol{\nu}}+\boldsymbol{\beta})$	S_2	$-(q_0/\lvert\mathbf{q}\rvert)2\cos^2\theta/2$
$-\frac{i}{2}(\mathbf{l}\times\mathbf{l}^*)_3$	$-S_1\hat{\mathbf{q}}\cdot(\hat{\boldsymbol{\nu}}-\boldsymbol{\beta})$	$S_1 S_2$	$(S_1 S_3/\lvert\mathbf{q}\rvert)\,2\sin\theta/2\times$ $\sqrt{q_\mu^2\cos^2\theta/2+\mathbf{q}^2\sin^2\theta/2}$

[a] Here the hats indicate unit vectors and the massive lepton velocity is $\vec{\beta} = \vec{k}/\varepsilon$.

Extreme Relativistic Limit (ERL). The ERL is defined by $\boldsymbol{\beta} = \mathbf{k}/\varepsilon \to 1$. This is the case for relativistic final massive leptons; and it is *always* the case for neutrino scattering. In the ERL, the result for neutrino reactions with incident ν_l or $\bar{\nu}_l$ becomes

$$\left(\frac{d\sigma}{d\Omega}\right)^{\text{ERL}}_{\substack{\nu\\ \bar{\nu}}} = \frac{G^2\varepsilon^2}{2\pi^2}\frac{4\pi}{2J_i+1}\left\{\cos^2\frac{\theta}{2}\sum_{J=0}^{\infty}\lvert\langle J_f\lvert\lvert\hat{\mathcal{M}}_J - \frac{q_0}{\lvert\mathbf{q}\rvert}\hat{\mathcal{L}}_J\rvert\rvert J_i\rangle\rvert^2\right. \tag{46.6}$$

$$+\left[\frac{q_\mu^2}{2\mathbf{q}^2}\cos^2\frac{\theta}{2}+\sin^2\frac{\theta}{2}\right]\sum_{J=1}^{\infty}\left[\lvert\langle J_f\lvert\lvert\hat{\mathcal{T}}_J^{\text{mag}}\rvert\rvert J_i\rangle\rvert^2+\lvert\langle J_f\lvert\lvert\hat{\mathcal{T}}_J^{\text{el}}\rvert\rvert J_i\rangle\rvert^2\right]$$

$$\left.\mp\frac{\sin\theta/2}{\lvert\mathbf{q}\rvert}\sqrt{q_\mu^2\cos^2\frac{\theta}{2}+\mathbf{q}^2\sin^2\frac{\theta}{2}}\sum_{J=1}^{\infty}2\text{Re}\left[\langle J_f\lvert\lvert\hat{\mathcal{T}}_J^{\text{mag}}\rvert\rvert J_i\rangle\langle J_f\lvert\lvert\hat{\mathcal{T}}_J^{\text{el}}\rvert\rvert J_i\rangle^*\right]\right\}$$

Note that the charge and longitudinal multipoles enter only in the combination $\hat{\mathcal{M}}_J - (q_0/|\mathbf{q}|)\hat{\mathcal{L}}_J$ in the ERL limit. Suppose the current and charge density have the following form (as they do in the case of the induced pseudoscalar interaction – see Section 45)

$$\begin{aligned} \hat{\boldsymbol{\mathcal{J}}}^{\phi} &= \nabla\hat{\phi} \\ \hat{\mathcal{J}}_0^{\phi} &= \frac{1}{i}[\hat{H},\hat{\phi}] \end{aligned} \tag{46.7}$$

The charge and longitudinal multipoles of this current then have the following form (Section 45)

$$\begin{aligned} \langle f|\hat{\mathcal{M}}_{JM}^{\phi}|i\rangle &= \frac{(E_f - E_i)}{i}\langle f|\int d^3x j_J(\kappa x) Y_{JM}(\Omega_x)\hat{\phi}(\mathbf{x})|i\rangle \\ &= iq_0\langle f|\int d^3x j_J(\kappa x) Y_{JM}(\Omega_x)\hat{\phi}(\mathbf{x})|i\rangle \\ \langle f|\hat{\mathcal{L}}_{JM}^{\phi}|i\rangle &= \frac{i\kappa^2}{\kappa}\langle f|\int d^3x j_J(\kappa x) Y_{JM}(\Omega_x)\hat{\phi}(\mathbf{x})|i\rangle \end{aligned} \tag{46.8}$$

Here energy conservation has been used in the second equality, and a partial integration performed in arriving at the third. As a consequence of these relations one finds

$$\langle f|\hat{\mathcal{M}}_{JM}^{\phi}|i\rangle = \frac{q_0}{|\mathbf{q}|}\langle f|\hat{\mathcal{L}}_{JM}^{\phi}|i\rangle \tag{46.9}$$

Hence we establish the result that *the induced pseudoscalar interaction does not contribute to the neutrino cross section in the ERL.*[28]

Threshold Cross Section. The other simple limiting case is at threshold in charged lepton production, where one has just enough energy to produce the massive final lepton and

$$\boldsymbol{\beta} = \frac{\mathbf{k}}{\varepsilon} = \frac{\mathbf{k}}{\sqrt{\mathbf{k}^2 + m_l^2}} \to 0 \tag{46.10}$$

In this case, the kinematics are as follows[29]

$$\begin{aligned} \mathbf{q} &= -\boldsymbol{\nu} \\ q_0 &= \varepsilon - \nu \to m_l - \nu\,; \qquad \text{threshold} \end{aligned} \tag{46.11}$$

[28] This can be seen directly from the S-matrix for the neutrino reaction on a nucleon. The induced pseudoscalar coupling is proportional to q_λ and on the lepton current, this gives a contribution proportional to the lepton mass $q_\lambda l_\lambda \propto m_l$. Hence this term does not contribute to the scattering in the ERL where the lepton mass is negligible.

[29] This assumes a sufficiently heavy target.

The cross section for (ν_l, l^-) and $(\bar{\nu}_l, l^+)$ then takes the form

$$\left(\frac{d\sigma}{d\Omega}\right)^{\text{Th}}_{\substack{\nu \\ \bar{\nu}}} = \beta\frac{G^2 m_l^2}{4\pi^2}\frac{4\pi}{2J_i+1}\left\{\sum_{J=0}^{\infty}|\langle J_f||\hat{\mathcal{L}}_J+\hat{\mathcal{M}}_J||J_i\rangle|^2 + \sum_{J=1}^{\infty}|\langle J_f||\hat{\mathcal{T}}_J^{\text{mag}} \mp \hat{\mathcal{T}}_J^{\text{el}}||J_i\rangle|^2\right\} \tag{46.12}$$

Here, for the current $\hat{\mathcal{J}}_\lambda^\phi$ in Eq. (46.7) one has

$$\begin{aligned}\langle f|\hat{\mathcal{M}}_{JM}^\phi + \hat{\mathcal{L}}_{JM}^\phi|i\rangle &= i(q_0+|\mathbf{q}|)\langle f|\int d^3x j_J(\kappa x)Y_{JM}(\Omega_x)\hat{\phi}(\mathbf{x})|i\rangle \\ &= im_l\langle f|\int d^3x j_J(\kappa x)Y_{JM}(\Omega_x)\hat{\phi}(\mathbf{x})|i\rangle\end{aligned} \tag{46.13}$$

Thus, in the threshold neutrino cross section, the entire effect of the induced pseudoscalar coupling can be taken into account by the following simple replacement in the axial charge density

$$\hat{J}_{05} \rightarrow \hat{J}_{05} + im_l\hat{\phi}_{\text{PS}}\ ; \qquad \text{threshold} \tag{46.14}$$

With the expression for the cross section obtained from Eqs. (46.3) and (45.19) and the lepton traces in Table 46.2, which has the two limiting forms discussed above, one is now in a position to calculate *any neutrino reaction on a nuclear target, under any kinematic conditions.* It simply requires calculating the appropriate multipoles of the weak current operator (Section 45). We discuss some specific examples in Section 47.

Charged Lepton (Muon) Capture. We next discuss the process of charged lepton capture where the basic nucleon process is

$$l^- + p \rightarrow n + \nu_l \tag{46.15}$$

With a nuclear target, the relevant processes are

$$\begin{aligned}&\mu^- + A(N,Z) \rightarrow A^*(N+1, Z-1) + \nu_\mu \\ &e^- + A(N,Z) \rightarrow A^*(N+1, Z-1) + \nu_e\end{aligned} \tag{46.16}$$

This is illustrated in Fig. 45.1. Although the formula to be derived for the capture rate is applicable to both processes, we shall focus the discussion in this section on *muon capture.*

The kinematics for this process are shown in Fig. 46.2. The relevant lepton matrix element and sign factors are given in Table 46.1.

We are primarily concerned here with processes where the negatively charged lepton is captured from the lowest $1s_{1/2}$ atomic orbital; for the case of muon

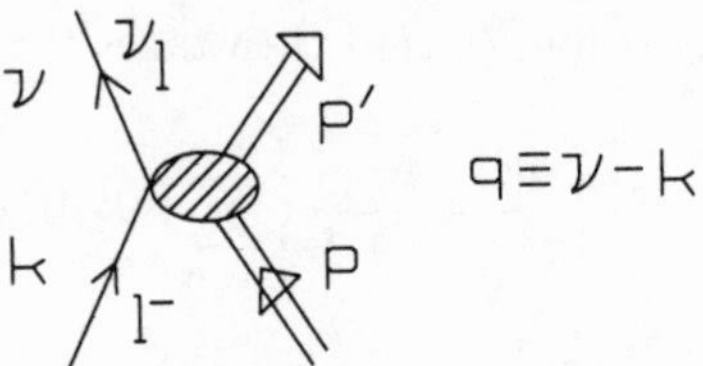

Figure 46.2: Charged lepton capture by nuclei.

capture, this necessitates a brief survey of some of the essential features of muonic atoms.

Muonic Atoms. When a muon (produced, for example, from cosmic rays or particle decay) passes through matter, it is finally captured into high-lying atomic orbits. It then *quickly* cascades down to the $1s$ Bohr orbit through Auger processes with atomic electrons and the emission of some X-rays. Here it can do one of two things: it can decay with its characteristic free lifetime

$$\begin{aligned} \mu^- &\rightarrow e^- + \nu_\mu + \bar{\nu}_e \\ \tau_\mu &= 2.197 \times 10^{-6}\,\text{sec} \end{aligned} \tag{46.17}$$

Or it can be captured by the nucleus, where the basic nucleon process is

$$\mu^- + p \rightarrow n + \nu_\mu \tag{46.18}$$

The $1s$ orbit for a muon has a Bohr radius given by

$$\frac{a_0^\mu}{4\pi} = \frac{1}{Z}\frac{\hbar^2}{m_\mu e^2} \tag{46.19}$$

Hence the ratio of the muon Bohr radius to that of an atomic electron is

$$\frac{a_0^\mu}{a_0^e} = \frac{m_e}{m_\mu} = \frac{1}{206.8} \tag{46.20}$$

The situation is illustrated in Fig. 46.3. The muon evidently sits well *outside* the nucleus, for light nuclei, and well *inside* all of the atomic electrons. In this case, the muon is accurately described by the simple, one-particle Bohr atom![30]

From quantum mechanics, the square of the $1s$ wave function at the origin for a Bohr atom is given by (Ref. [N42])

$$|\phi_{1s}^0(0)|^2 = \frac{(Z\alpha m_\mu)^3}{\pi}\left(\frac{1}{1+m_\mu/M_T}\right)^3 \tag{46.21}$$

[30] In a closed-shell atom, the electron distribution is spherically symmetric; the electrostatic potential inside such a spherically symmetric charge distribution is constant, and thus has no effect on the motion of the muon. In addition, if the muon is outside of a nucleus with a spherically symmetric charge distribution, the nucleus acts as a point charge.

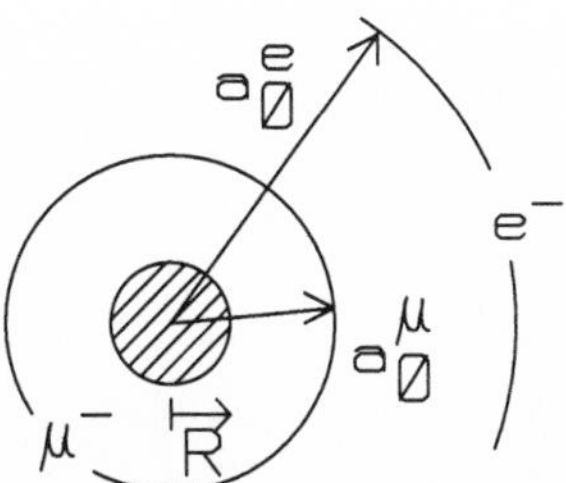

Figure 46.3: Ratio of Bohr radii characterizing size of $1s$ Bohr orbits for muon and atomic electrons (not to scale).

The final factor takes into account the reduced mass of the system.

The nucleus actually has a charge density of finite extent. As Z gets larger, the muon wave function is pulled into the nuclear Coulomb potential, where it no longer feels the full strength of a point charge; thus the magnitude of the $1s$ atomic wave function in the region of the nucleus will be reduced from the point-charge value quoted above. The situation is illustrated in Fig. 46.4. We therefore write

$$|\phi_{1s}|^2_{\text{av}} \equiv \mathcal{R}|\phi^0_{1s}(0)|^2 \tag{46.22}$$

Here $\mathcal{R}$ is a reduction factor obtained by averaging the actual $1s$ wave function over the nuclear volume.[31] Values of $\mathcal{R}$ obtained from numerical integration of the Dirac equation for the electron in the finite nuclear charge distribution are shown in Table 46.3 (from Ref. [W2]).

There is another limiting case of the muonic atom that has a very simple quantum mechanical interpretation. Take the other extreme of a very heavy, large nucleus where the muon lies completely inside the nucleus. If the nucleus is modeled with a spherically symmetric uniform charge distribution (Fig. 46.5), then the electrostatic potential is that of a three-dimensional simple harmonic oscillator and the atomic energy levels and wave functions will be those for that simple system (Prob. 46.6). Intermediate situations between the two simple limiting cases can be characterized by interpolation.

Z^4 Law for μ-Capture Rate. From Eq. (46.18), the nuclear muon capture rate is proportional to the number of protons in the nucleus. Since the semileptonic weak nuclear interaction is effectively a contact interaction (Section 45), the capture rate is also proportional to the probability of finding the muon at the nucleus. Hence

$$\begin{aligned} \omega_{\mu-\text{capt}} &\propto \text{(probability at nucleus)} \times \text{(number of protons)} \\ &\propto |\phi_{1s}|^2_{\text{av}}\, Z \end{aligned} \tag{46.23}$$

[31]It is sensible to characterize this effect with a single gross factor $\mathcal{R}$ because the $1s$ wave function continues to be a slowly varying function over the nuclear volume up to medium-weight nuclei.

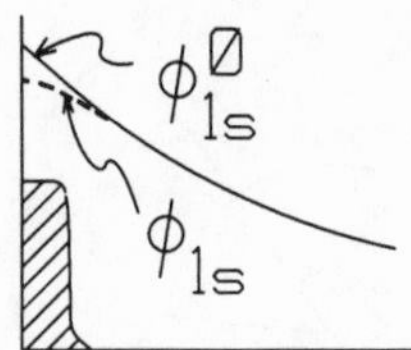

Figure 46.4: Reduction of $1s$ atomic wave function in the vicinity of the nucleus from that of the point-charge Bohr atom caused by the finite extent of the nuclear charge distribution.

For light nuclei $\mathcal{R} \approx 1$, and hence one finds from Eq. (46.21)

$$\omega_{\mu-\text{capt}} \propto Z^4 \tag{46.24}$$

This is the celebrated Z^4 law for nuclear muon capture. It is not until about Mg that the nuclear capture rate is equal to the free decay rate.

Calculation of μ-Capture Rate. The kinematics are illustrated in Fig. 46.2. To a good approximation, in all cases one can treat the initial charged lepton in the $1s$ atomic orbit as a nonrelativistic particle. Thus one can write the Dirac wave function for the initial charged lepton as

$$u_{1s\frac{1}{2}m} \approx \phi_{1s}(x)\begin{pmatrix}\chi_m \\ 0\end{pmatrix} \equiv \frac{\phi_{1s}(x)}{1/\sqrt{\Omega}}\frac{u_m(\mathbf{0})}{\sqrt{\Omega}} \tag{46.25}$$

The matrix element of the semi-leptonic weak interaction in Section 45 thus takes the form

$$\langle f|\hat{H}_{\text{W}}|i\rangle = -\frac{G}{\sqrt{2}}l_\mu \int d^3x \frac{\phi_{1s}(x)}{1/\sqrt{\Omega}} e^{-i\boldsymbol{\nu}\cdot\mathbf{x}}\langle f|\hat{\mathcal{J}}_\mu(\mathbf{x})|i\rangle \tag{46.26}$$

Here we have defined (Table 46.1)

$$-i\Omega l_\lambda \equiv \bar{u}(\boldsymbol{\nu})\gamma_\lambda(1+\gamma_5)u(\mathbf{k}) \; ; \qquad \mathbf{k} \to 0 \tag{46.27}$$

The capture rate follows from the Golden Rule

$$\omega_{\text{fi}} = 2\pi\frac{\Omega 4\pi\nu^2}{(2\pi)^3}\frac{1}{2}\sum_{\text{lepton spins}}\frac{1}{2J_i+1}\sum_{M_i}\sum_{M_f}|\langle f|\hat{H}_{\text{W}}|i\rangle|^2 \tag{46.28}$$

Table 46.3: Reduction factor of square of $1s$ Bohr wave function obtained by averaging solution of Dirac equation in finite nuclear charge distribution over the nuclear volume (from Ref. [W2]).

Element	^{4}He	^{12}C	^{16}O	^{28}Si	^{40}Ca	^{56}Ni
$\mathcal{R}$	0.98	0.86	0.79	0.60	0.44	0.30

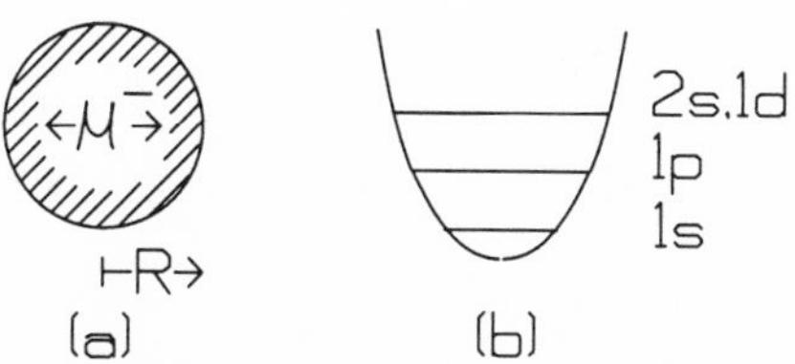

Figure 46.5: (a) Muon inside a spherically symmetric uniform charge distribution; (b) resulting spectrum.

Here one, indeed, must *average* over the initial lepton spins for spinless nuclei.[32]

Now assume that, because of its slow variation over the nuclear volume, one can remove a factor of $\langle\phi_{1s}\rangle_{\text{av}}$ from the nuclear matrix element.[33] In this case, the nuclear multipole analysis and sum over nuclear states *is precisely the same* as that performed in Section 45 and leads to Eq. (45.19). Note the factors of Ω again cancel in the rate, as they must. It remains to perform the lepton traces.

Lepton Traces. Because of the form into which the amplitude has been cast, the result for the lepton traces is just that in Table 46.2, using the sign factors from Table 46.1. It is the *threshold* value that must be used since here $\boldsymbol{\beta} = \mathbf{k}/\varepsilon \to 0$. To confirm these results, we will again verify one line in detail, and let the reader check the others. Consider

$$
\begin{aligned}
\frac{\Omega^2}{2}\sum_{\text{lepton spins}} l_0 l_0^* &= \frac{1}{2}\frac{1}{4\nu m_l}\text{tr}\ \gamma_4(1+\gamma_5)(m_l - ik_\lambda\gamma_\lambda)\gamma_4(1+\gamma_5)(-i\gamma_\sigma\nu_\sigma) \\
&= -\frac{1}{4\nu m_l}\text{tr}\ \gamma_4\gamma_\lambda k_\lambda\gamma_4(1+\gamma_5)\gamma_\sigma\nu_\sigma \\
&= -\frac{1}{\nu m_l}(2k_4\nu_4 - k\cdot\nu) = -\frac{1}{\nu m_l}(-m_l\nu - \mathbf{k}\cdot\boldsymbol{\nu}) \\
\frac{\Omega^2}{2}\sum_{\text{lepton spins}} l_0 l_0^* &= +1\ ; \qquad\qquad \mathbf{k}\to 0
\end{aligned}
\tag{46.29}
$$

Note that the term in m_l in the first line again goes out.

Result. The final result for the charged-lepton (muon) capture rate is thus given by

$$
\begin{aligned}
\omega_{\text{fi}} &= \frac{G^2\nu^2}{2\pi}\frac{4\pi}{2J_i+1}\left\{\sum_{J=0}^{\infty}|\langle J_f||\hat{\mathcal{M}}_J(\nu) - \hat{\mathcal{L}}_J(\nu)||J_i\rangle|^2\right. \\
&\left. + \sum_{J=1}^{\infty}|\langle J_f||\hat{\mathcal{T}}_J^{\text{el}}(\nu) - \hat{\mathcal{T}}_J^{\text{mag}}(\nu)||J_i\rangle|^2\right\}|\phi_{1s}|^2_{\text{av}}
\end{aligned}
\tag{46.30}
$$

[32]If nuclear recoil is included in the density of states, the result is to multiply the expression for the rate by a recoil correction factor $r = (1+\nu/M_T)^{-1}$ (Prob. 46.3).

[33]Improved approximations here are (1) average ϕ_{1s} over the *actual* nuclear transition density; (2) leave the spherically symmetric factor ϕ_{1s} in the nuclear matrix element.

Several features of this result are of interest:

1. Here the neutrino energy is determined by energy conservation from the relation

$$m_l - \varepsilon_b + E_i = E_f + \nu \tag{46.31}$$

In this expression ε_b is the binding energy of the muonic atom.

2. Note that the momentum transfer in the nuclear multipoles is $\kappa = \nu$. The nucleus must absorb the momentum of the final neutrino for the reaction to go, since the initial charged lepton in the atomic orbit furnishes its rest mass, but negligible momentum.

3. One could have left the muon wave function $\phi_{1s}(r)$ inside the nuclear multipoles, since it is just a spherically symmetric factor that does not affect the angular momentum analysis.

4. This result should be compared with the threshold antineutrino cross section in Eq. (46.12) (the threshold antineutrino process is obtained from the charged lepton capture process through *crossing*); it has the same form except for signs obtained from the lepton traces.

β-Decay. Finally, we discuss nuclear β-decay. This is the process by which nuclei of the same baryon number transform into one another until stable isobars are obtained (Section 2). The basic nucleon processes are[34]

$$\begin{aligned} n &\rightarrow p + e^- + \bar{\nu}_e \\ p &\rightarrow n + e^+ + \nu_e \end{aligned} \tag{46.32}$$

With a nuclear target, the relevant processes are

$$\begin{aligned} A(N,Z) &\rightarrow A^*(N-1, Z+1) + e^- + \bar{\nu}_e \\ A(N,Z) &\rightarrow A^*(N+1, Z-1) + e^+ + \nu_e \end{aligned} \tag{46.33}$$

The kinematics for this process are shown in Fig. 46.6. The relevant lepton matrix element and sign factors are given in Table 46.1.

The analysis proceeds as in the previous sections. The only new feature is that one now has *two* particles in the final state, and hence a factor for the number of final states must be included for each of them. The decay rate is thus

$$d\omega = 2\pi \sum_{\text{lepton spins}} \frac{1}{2J_i+1} \sum_{M_i} \sum_{M_f} |\langle f|\hat{H}_W|i\rangle|^2 \delta(W_f - W_i) \frac{\Omega d^3k}{(2\pi)^3} \frac{\Omega d^3\nu}{(2\pi)^3} \tag{46.34}$$

[34] The first gives rise to the β-decay of the free neutron; since the proton is stable, the second can take place only inside a nucleus where nuclear binding effects allow it to proceed.

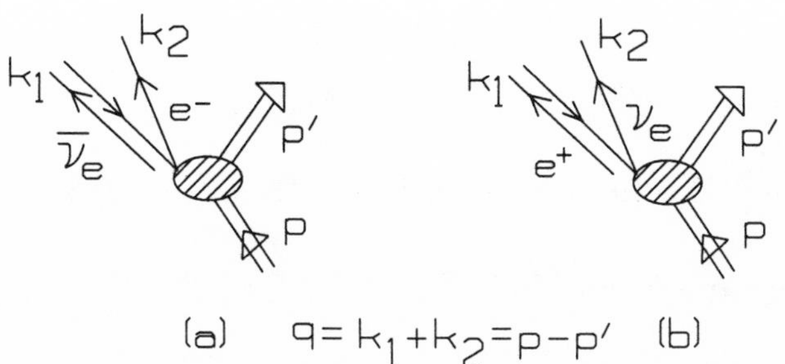

Figure 46.6: Nuclear β-decay: (a) $(e^-\bar{\nu}_e)$ decay; (b) $(e^+\nu_e)$ decay

The volume factors again cancel in the rate, as they must. The integral over the energy conserving delta-function can be performed as follows[35]

$$\begin{aligned}\int \delta(W_f - W_i)d\nu &= \int \delta(\varepsilon + \nu + E_f - E_i)d\nu = 1 \\ E_i - E_f &\equiv W_0\ ; \qquad \text{energy release} \\ \varepsilon + \nu &= W_0\ ; \qquad \text{max. electron energy}\end{aligned} \tag{46.35}$$

The final state lepton kinematics thus become

$$\int_{\text{neutrino energy}} \delta(W_f - W_i)d^3k d^3\nu = k\varepsilon d\varepsilon d\Omega_k (W_0 - \varepsilon)^2 d\Omega_\nu \tag{46.36}$$

Hence the decay rate is

$$d\omega = \frac{\Omega^2}{(2\pi)^5}(W_0 - \varepsilon)^2 k\varepsilon d\varepsilon d\Omega_k d\Omega_\nu \sum_{\text{lepton spins}} \frac{1}{2J_i + 1} \sum_{M_i} \sum_{M_f} |\langle f|\hat{H}_W|i\rangle|^2 \tag{46.37}$$

General Expression. The general expression in terms of nuclear multipoles for the sum and average over nuclear orientations in Eq. (45.19) can now again be employed. One then has to sum over final lepton spins. The required lepton traces obtained from the expression $(\Omega^2/2)\sum_{\text{lepton spins}}$ are given in Table 46.2 for general final lepton kinematics; here the sign factors in Table 46.1 are to be employed. A combination of these results gives the following β-decay rate for $(e^-\bar{\nu}_e)$ and $(e^+\nu_e)$

$$\begin{aligned} d\omega_{\beta\mp} &= \frac{G^2}{2\pi^3} k\varepsilon (W_0 - \varepsilon)^2 d\varepsilon \frac{d\Omega_k}{4\pi}\frac{d\Omega_\nu}{4\pi}\frac{4\pi}{2J_i + 1} \\ &\times \left\{ \sum_{J=0}^{\infty} \left[(1 + \hat{\boldsymbol{\nu}} \cdot \boldsymbol{\beta}) |\langle J_f || \hat{\mathcal{M}}_J || J_i \rangle|^2 + \right. \right. \end{aligned}$$

[35] We leave the calculation of the target recoil correction as a problem (Prob. 46.3); it is usually unimportant since the energy release in β-decay is small.

$$
\begin{aligned}
&+\{1-\hat{\boldsymbol{\nu}}\cdot\boldsymbol{\beta}+2(\hat{\boldsymbol{\nu}}\cdot\hat{\mathbf{q}})(\hat{\mathbf{q}}\cdot\boldsymbol{\beta})\}|\langle J_f||\hat{\mathcal{L}}_J||J_i\rangle|^2 \\
&-\hat{\mathbf{q}}\cdot(\hat{\boldsymbol{\nu}}+\boldsymbol{\beta})2\mathrm{Re}\,\langle J_f||\hat{\mathcal{L}}_J||J_i\rangle\langle J_f||\hat{\mathcal{M}}_J||J_i\rangle^*\Big] \\
&+\sum_{J=1}^{\infty}\Big[\{1-(\hat{\boldsymbol{\nu}}\cdot\hat{\mathbf{q}})(\hat{\mathbf{q}}\cdot\boldsymbol{\beta})\}\left(|\langle J_f||\hat{\mathcal{T}}_J^{\mathrm{mag}}||J_i\rangle|^2+|\langle J_f||\hat{\mathcal{T}}_J^{\mathrm{el}}||J_i\rangle|^2\right) \\
&\pm\hat{\mathbf{q}}\cdot(\hat{\boldsymbol{\nu}}-\boldsymbol{\beta})2\,\mathrm{Re}\,\langle J_f||\hat{\mathcal{T}}_J^{\mathrm{mag}}||J_i\rangle\langle J_f||\hat{\mathcal{T}}_J^{\mathrm{el}}||J_i\rangle^*\Big]\Big\}
\end{aligned}
\tag{46.38}
$$

Here $\hat{\boldsymbol{\nu}}\equiv\boldsymbol{\nu}/\nu$, $\hat{\mathbf{q}}\equiv\mathbf{q}/|\mathbf{q}|$, and $\boldsymbol{\beta}\equiv\mathbf{k}/\varepsilon$.

All the mutipoles in this expression are to be evaluated at a momentum transfer

$$
\kappa\equiv|\mathbf{q}|=|\mathbf{k}+\boldsymbol{\nu}|
\tag{46.39}
$$

Thus there is still a complicated *angle dependence on $\theta_{\mathbf{k}\boldsymbol{\nu}}$ contained in the multipoles.* The long-wavelength expansion of these multipoles in Section 45 makes this angle dependence explicit.

Equation (46.38) provides a general expression for the β-decay rate between any two nuclear states.

Allowed β-Decay. Consider the simplification of the above expression in the long-wavelength *allowed* limit where from Section 45 the only remaining, nonzero nuclear multipoles are

$$
\begin{aligned}
\hat{\mathcal{T}}_{1M}^{\mathrm{el}} &= \sqrt{2}\hat{\mathcal{L}}_{1M}\,; \qquad \text{Gamow-Teller} \\
\hat{\mathcal{M}}_{00} & \qquad\qquad ; \qquad \text{Fermi}
\end{aligned}
\tag{46.40}
$$

One now has the significant simplification that these multipoles are *independent of κ.* The rate then becomes

$$
\begin{aligned}
d\omega_{\mathrm{fi}}^{\beta} &= \frac{2G^2}{\pi^2}k\varepsilon(W_0-\varepsilon)^2d\varepsilon\frac{d\Omega_k}{4\pi}\frac{d\Omega_\nu}{4\pi}\left\{(1+\hat{\boldsymbol{\nu}}\cdot\boldsymbol{\beta})\frac{1}{2J_i+1}|\langle J_f||\hat{\mathcal{M}}_0||J_i\rangle|^2\right. \\
&\left.+3\left(1-\frac{1}{3}\hat{\boldsymbol{\nu}}\cdot\boldsymbol{\beta}\right)\frac{1}{2J_i+1}|\langle J_f||\hat{\mathcal{L}}_1||J_i\rangle|^2\right\}
\end{aligned}
\tag{46.41}
$$

We comment on a few features of this result:

1. This is the general expression for the *allowed* $\beta^{\mp}$-decay rate.

2. Note the characteristic allowed energy spectrum for the electron arising from phase space arguments.

3. Note the characteristic $(\hat{\boldsymbol{\nu}}\cdot\boldsymbol{\beta})$ angular correlations between the momenta of the emitted leptons in the allowed rate.

4. The neutrino momentum $\boldsymbol{\nu}$ can be determined from measurements of the electron and nuclear recoil momenta and the use of momentum conservation; these are very lovely, but difficult, measurements.

5. To close the loop, a direct derivation from start to finish of this result for the allowed β-decay rate is given in Prob. 46.5.

Final-State Coulomb Interaction. The electron (positron) emitted in β-decay is charged, and has a Coulomb interaction with the final charged nucleus. This interaction will modify the wave function of the final charged lepton from the plane-wave value used in the calculation of the rate above; by changing the electron wave function over the nuclear volume, the final-state Coulomb interaction modifies the matrix element of the contact four-fermion interaction, and hence changes the rate. A full treatment of this final-state interaction is available through numerical integration of the Schrödinger (or Dirac) equation of the electron in the Coulomb field of the final nucleus and atom.

An *approximate* treatment of this final-state Coulomb interaction can be obtained by using the ratio at the origin of the wave function for scattering at energy ε from a point nuclear charge Ze_p to the noninteracting plane wave. This ratio can be calculated analytically for the Schrödinger equation with the result (Ref. [N42])

$$\begin{aligned} F(Z,\varepsilon) &\equiv \left|\frac{\phi_{\mathbf{k}}(0)_{\text{Coul}}}{\phi_{\mathbf{k}}(0)}\right|^2 = \frac{2\pi\eta}{e^{2\pi\eta}-1} \\ \eta &= \frac{zZ\alpha}{|\beta|}\;; \qquad z \equiv \text{lepton charge} \end{aligned} \tag{46.42}$$

Here α is the fine-structure constant. This factor $F(Z,\varepsilon)$ then multiplies the above expressions for the β-decay rates.

Slow Nucleons and Allowed β-Decay . Up to this point, the general expressions for the multipole operators appearing in the expression for the allowed β-decay rate are given by

$$\begin{aligned} \hat{\mathcal{M}}_{00} &= \frac{1}{\sqrt{4\pi}}\int \hat{\mathcal{J}}_0^{(\pm)}(\mathbf{x})d^3x \;; \qquad \text{Fermi} \\ \hat{\mathcal{L}}_{1M} = \frac{1}{\sqrt{2}}\hat{\mathcal{T}}^{\text{el}}_{1M} &= \frac{i}{\sqrt{12\pi}}\int \hat{\mathcal{J}}_{1M}^{(\pm)}(\mathbf{x})d^3x\,; \qquad \text{Gamow-Teller} \end{aligned} \tag{46.43}$$

With the further approximation that the nucleons are moving *slowly* so that $(v/c)_{\text{nucleon}} \ll 1$, these expressions simplify to the result given in Eq. (45.44)

$$\begin{aligned} \hat{\mathcal{M}}_{00} &= \frac{1}{\sqrt{4\pi}}F_1\sum_{j=1}^{A}\tau_{\pm}(j) \qquad ; \qquad \text{Fermi} \\ \hat{\mathcal{L}}_{1M} = \frac{1}{\sqrt{2}}\hat{\mathcal{T}}^{\text{el}}_{1M} &= \frac{i}{\sqrt{12\pi}}F_A\sum_{j=1}^{A}\tau_{\pm}(j)\sigma_{1M}(j)\,; \qquad \text{Gamow-Teller} \end{aligned} \tag{46.44}$$

47

SOME APPLICATIONS

There have been many, many theoretical and experimental studies of weak interactions with nuclei (see, e.g., Refs. [W1, W2] and the Conference Proceedings in Ref. [W12]). It is impossible, and even inappropriate, to attempt a comprehensive, up-to-date summary here. Rather, we present a few selected examples that illustrate the study of weak interactions with nuclei, and that demonstrate the accuracy with which one can do nuclear physics in selected cases by employing a unified analysis of electromagnetic and weak interactions with nuclear systems.

One-Body Operators. Recall the discussion of nuclear structure in Part I. An arbitrary one-body multipole operator can be expanded in second quantization as

$$\hat{T}_{JM_J;TM_T}(q) = \sum_{\alpha}\sum_{\beta} c_{\alpha}^{\dagger}\langle\alpha|T_{JM_J;TM_T}(q)|\beta\rangle c_{\beta} \tag{47.1}$$

Here $\alpha = \{nljm_j;\frac{1}{2}m_t\} \equiv \{a;m_j,m_t\}$ is a complete set of single-particle quantum numbers.

The matrix element of this operator between an arbitrary pair of nuclear states thus takes the form[36]

$$\begin{aligned}\langle\Psi_f|\hat{T}_{JM_J;TM_T}(q)|\Psi_i\rangle &= \sum_{\alpha}\sum_{\beta}\langle\alpha|T_{JM_J;TM_T}(q)|\beta\rangle\psi^{fi}_{\alpha\beta} \\ \psi^{fi}_{\alpha\beta} &= \langle\Psi_f|c_{\alpha}^{\dagger}c_{\beta}|\Psi_i\rangle \end{aligned} \tag{47.2}$$

In words, this result says that the *nuclear matrix element* of this multipole operator between any two states can be written as a *sum of single-particle matrix elements multiplied by numerical coefficients*; the numerical coefficients are simply the matrix elements of the creation and destruction operators as defined in the second line.

Unified Analysis of Electromagnetic and Weak Interactions with Nuclei. Consider the traditional nuclear physics picture as presented in Parts I and IV of this book: nucleons interacting through static potentials, with quantum dynamics governed by the nonrelativistic many-body Schrödinger equation, and with

[36] See Ref. [N14] for the derivation of some general properties of the matrix elements of these multipole operators and their evaluation in several specific cases.

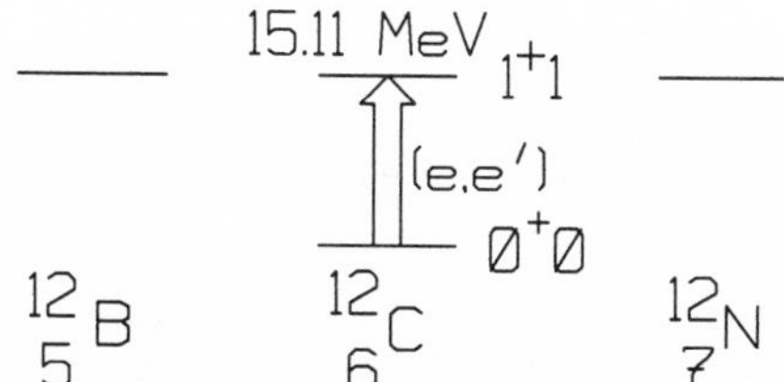

Figure 47.1: System of nuclear levels in first application.

electroweak currents constructed from the properties of free nucleons. Within this picture, the above result can be utilized in the following way:

1. Within this traditional nuclear physics picture, Eq. (47.2) is *exact*. For example, any shell-model calculation, no matter how complicated, must give a result in this form.

2. Now suppose that one truncates Eq. (47.2) to a finite sum of single-particle transitions. Within this truncation one still has the flexibility of describing very complicated nuclear states. For example, in any shell model calculation within the *s-d* shell, no matter how many particles and basis states are included, any final nuclear transition matrix element must be a simple sum over the few possible transition matrix elements between the single-particle *s-d* basis states; furthermore, any TDA or RPA calculation of nuclear excitations as in Sections 9-11 must give a result of this form.

3. The single-particle matrix elements (s.p.m.e.) appearing in Eq. (47.2) can be constructed from the radial wave functions of the single-particle basis states (Sections 6) and the one-body currents (Sections 8 and 45). This is carried out in Refs. [W9, W10].

4. These s.p.m.e. have different $\mathbf{q}^2$ dependence.

5. One can make use of this additional, invaluable, $\mathbf{q}^2$ dependence to *synthesize* the experimental electromagnetic (e, e') form factors (Section 7), and hence determine the set of numerical coefficients $\psi^{fi}_{\alpha\beta}$.

6. These same one-body densities can then be used to compute the semi-leptonic weak processes (Section 46).

We proceed to discuss a few selected applications of this approach.

Applications. These applications are all taken from Refs. [W16, W17, W18, W19, W20]. They will be presented in the following format:

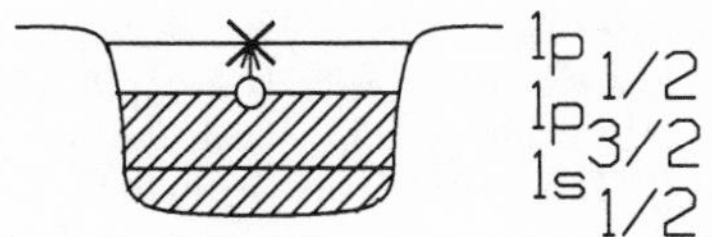

Figure 47.2: Simplest model $|0\rangle \rightarrow |1p_{1/2}(1p_{3/2})^{-1}\rangle$ for first application.

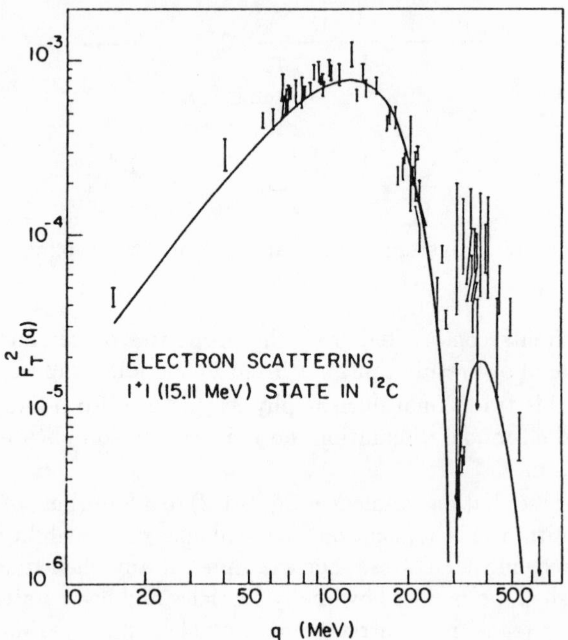

Figure 47.3: $F_T^2(q)$ for the 1^+1 state in ^{12}C (15.11 MeV). Here $q = |\mathbf{q}|$. The curve is a best fit with $b = 1.77$ fm and $\xi = 2.25$. From Ref. [W16].

1. The nuclear transition is identified and the (e, e') data shown;[37]

2. The truncation scheme and parameterization used in the determination of the one-body densities is discussed;[38]

3. A comparison is then made with all existing weak rates. These applications all assume the weak nucleon couplings from Section 45; there are no second-class currents and $F_P = 2mF_A/(q^2 + m_\pi^2)$. The value of $F_A(0)$ is taken from the β-decay of the neutron (Ref. [W1])[39]

$$F_A(0)\cos\theta_C = -1.23 \qquad\qquad \cos\theta_C = 0.974 \tag{47.3}$$

The *first* set of nuclear levels is shown in Fig. 47.1. It involves the $0^+0 \to 1^+1$ transition in the $B = 12$ system.

The simplest model for this transition is that ^{12}C forms a closed $1p_{3/2}$ shell and the excited state is $1p_{1/2}(1p_{3/2})^{-1}$ as illustrated in Fig. 47.2. In this case,

[37] We here use (Section 7) $d\sigma/d\Omega = 4\pi\sigma_M F^2[1 + (2\varepsilon_1 \sin^2\theta/2)/M_T]^{-1}$ where $F^2 \equiv (q_\mu^4/\vec{q}^{\,4})F_L^2 + (q_\mu^2/2\vec{q}^{\,2} + \tan^2\theta/2)F_T^2$. Also $f_{SN} = [1 + q_\mu^2/(855\,\text{MeV})^2]^{-2}$ throughout.

[38] An overall factor $f_{CM} = e^{y/A}$ is included in all calculations to take into account the C-M motion — see Refs. [N14, N47]. This factor is also included in the calculations in Section 11.

[39] It is assumed for historical reasons in these applications that $G^{(\pm)} = G\cos\theta_C \approx G$ in the vector couplings, introducing a ≈ 2.6 % error in these terms. The Particle Data Booklet gives $g_A/g_V = F_A(0)/F_1(0) = -1.2573 \pm 0.0028$.

Table 47.1: Partial weak rates with ^{12}C(g.s.) (from Ref. [W16]).

Process	Experiment	Theory
β^--decay rate	$32.98 \pm 0.10\,\text{sec}^{-1}$	$33.8\,\text{sec}^{-1}$
β^+-decay rate	$59.55 \pm 0.22\,\text{sec}^{-1}$	$66.9\,\text{sec}^{-1}$
μ^--capture rate	$6.75^{+0.30}_{-0.75} \times 10^3\,\text{sec}^{-1}$	$6.64 \times 10^3\,\text{sec}^{-1}$

only a single s.p.m.e. $\langle(1p_{1/2})1/2 \vdots\vdots T_{1,1}(q) \vdots\vdots (1p_{3/2})1/2\rangle$ is required (Sections 9 and 10). The resulting nuclear matrix element of the transverse magnetic dipole operator is then given by (Refs. [N14, N47], Prob. 47.1)

$$
\begin{aligned}
i\sqrt{4\pi}\langle 1^+;10||\hat{T}_1^{\text{mag}}||0^+,0\rangle &= \frac{2}{3}\frac{|\mathbf{q}|}{2m}\left[1 - 2(\lambda_p - \lambda_n)\left(1 - \frac{1}{2}y\right)\right]e^{-y} \\
y &\equiv \left(\frac{|\mathbf{q}|b_{\text{osc}}}{2}\right)^2
\end{aligned}
\tag{47.4}
$$

Here harmonic oscillator wave functions have been assumed. This is the TDA result; the RPA result (Section 9) gives the same answer reduced by a factor ξ_{RPA}; calculation in the open-shell RPA gives the same answer reduced by a still larger factor ξ_{OSRPA}. We choose to compare the electron scattering data for this transition with Eq. (47.4) multiplied by an empirical overall reduction factor ξ and with an empirical oscillator parameter. The result is shown in Fig. 47.3.

The predicted charge-changing semileptonic weak transition rates for this system using a one-body density parametrized in terms of these values of (ξ, b_{osc}) are shown in Table 47.1.

The *second* example consists of the $0^+0 \to J^\pi 1$ transitions to the first $(0^-, 1^-, 2^-, 3^-)$ states in the $B = 16$ system as illustrated in Fig. 47.4; only

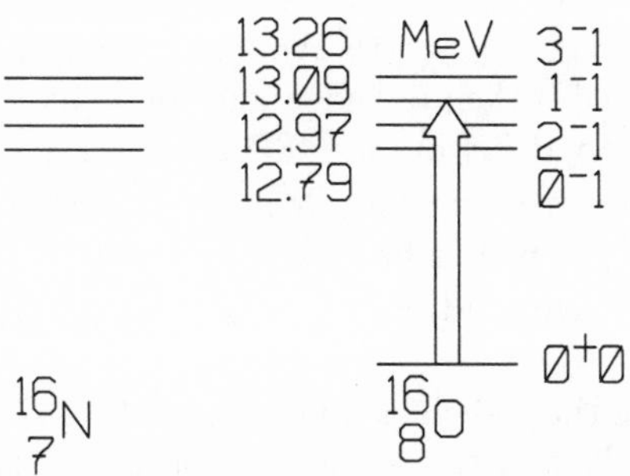

Figure 47.4: The $0^+0 \to J^\pi 1$ transitions to the first $(0^-, 1^-, 2^-, 3^-)$ states in the $B = 16$ system; only the latter three are excited in electron scattering with one-photon exchange.

Table 47.2: Partial weak rates with ^{16}O(g.s.) (from Ref. [W16]).

Process	Experiment	Theory		
		Set 1	Set 2	Set 3
μ^--capture ($10^3\,\text{sec}^{-1}$)				
0^-	1.1 ± 0.2	0.86	0.86	0.70
	1.6 ± 0.2			
	$0.85^{+0.145}_{-0.060}$			
1^-	1.88 ± 0.10	1.42	1.28	1.16
	1.4 ± 0.2			
	$1.85^{+0.355}_{-0.170}$			
2^-	6.17 ± 0.71	7.54	6.65	7.44
	7.9 ± 0.8			
3^-	≤ 0.08	0.060	0.054	0.077
β^--decay ($10^{-2}\,\text{sec}^{-1}$)				
2^-	2.53 ± 0.20	2.18	1.92	2.29
0^-	43 ± 10	42	46	-

the latter three are excited in electron scattering with one-photon exchange.

The simplest model of these transitions is to assume that ^{16}O forms a closed p-shell and the excitations are linear combinations of the $2s(1p)^{-1}, 1d(1p)^{-1}$ p-h states shown in Fig. 11.1. The resulting excitation spectrum calculated in TDA with a simple two-nucleon potential fit to scattering data is presented in Section 11; the calculation is discussed there. An improvement is an RPA calculation as discussed in Sections 9 and 10.

The existing (e, e') data are shown in Fig. 47.5. The curves are three fits to the data: (1) TDA with harmonic oscillator wave functions and $b_{\text{osc}} = 1.77\,\text{fm}$ as determined from a fit to *elastic* (e, e) scattering; (2) RPA, but otherwise the same as (1); (3) same as set (1) with Woods-Saxon radial wave functions. An overall reduction factor ξ is then included and the individual p-h amplitudes are allowed to vary by up to 10% from the TDA and RPA values.[40]

The partial weak rates with ^{16}O calculated using the resulting parameterizations of the one-body densities are shown in Table 47.2.

The *third* example involves the $1^+0 \to 0^+1$ transition in the $B = 6$ system as shown in Fig. 47.6.

Here we truncate to the p-shell as illustrated in Fig. 47.7. The core is assumed to form a closed s-shell, and the wave function of the two valence nucleons is

[40] The actual values used in the fit are given in Ref. [W18]; the 0^- is assumed to behave similarly to the other three states.

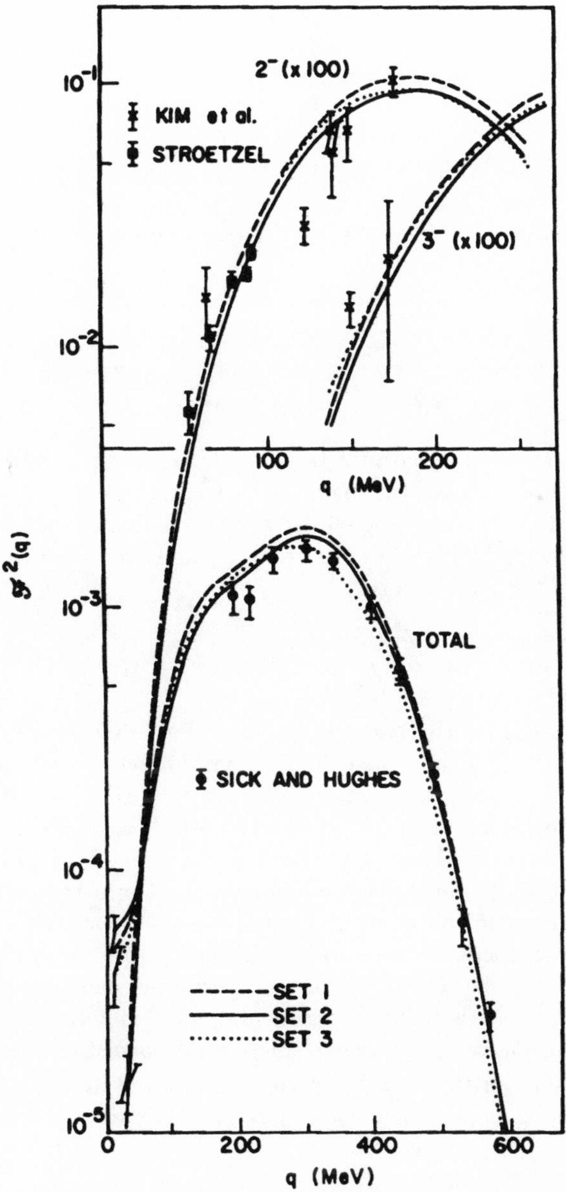

Figure 47.5: Inelastic form factors $\mathcal{F}^2 \equiv F^2/(1/2 + \tan^2\theta/2)$ at large θ for the first $(1^-, 2^-, 3^-)$ states in ^{16}O. Here $q = |\mathbf{q}|$. The three fits are described in the text. From Ref. [W16].

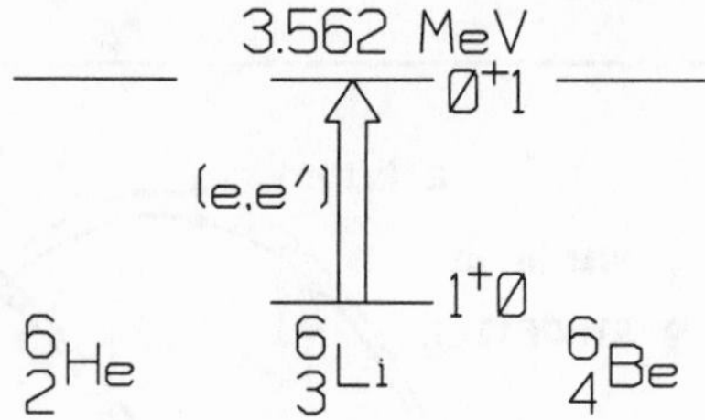

Figure 47.6: Third example of $1^+0 \to 0^+1$ transition in the $B = 6$ system.

written quite generally in this space as

$$\begin{aligned} |1^+0\rangle &= A|(1p_{3/2})^2 1^+0\rangle + B|(1p_{3/2}1p_{1/2})1^+0\rangle + C|(1p_{1/2})^2 1^+0\rangle \\ |0^+1\rangle &= D|(1p_{3/2})^2 0^+1\rangle + E|(1p_{1/2})^2 0^+1\rangle \end{aligned} \tag{47.5}$$

All existing electromagnetic data for these levels are fit with the parameter set $\{A, \ldots, E; b_{\text{osc}}\}$; this includes the magnetic dipole and electric quadrupole moments of the ground state and the elastic and inelastic magnetic electron scattering form factors. When the full elastic magnetic form factor is calculated in this basis it takes the form (Prob. 47.2)

$$\frac{\langle 1^+0||\hat{T}^{\text{mag}}_{1,0}||1^+0\rangle}{\sqrt{3}(i|\mathbf{q}|/m)e^{-y}f_{\text{SN}}f_{\text{CM}}} \equiv p(y) = \alpha_e + \beta_e y \tag{47.6}$$

A similar result holds for the inelastic magnetic form factor. Hence this truncation predicts a straight line when the combination on the left side is plotted against y. The experimental result is shown in Fig. 47.8. In contrast to the previous logarithmic plots, this comparison is a much more accurate one made on a *linear* scale. The resulting parameter set determined by a fit to the data is shown in Table 47.3. Note that in this case one has determined the entire *nuclear wave function* for both levels through the aid of electron scattering.

The resulting semileptonic weak rates calculated from these wave functions are shown in Table 47.4. Note this is now an absolute calculation of the ^{6}He β-decay rate, which agrees with the experimental value to better that 5%.

Finally, the *fourth* example is an extremely simple one. Consider the ground-state $\frac{1}{2}^+\frac{1}{2}$ isodoublet in the $B = 3$ system as illustrated in Fig. 47.9. Here the lifetime of ^{3}H is long enough that one can carry out (e, e) experiments on both nuclei.

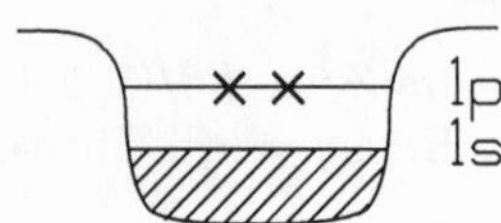

Figure 47.7: Truncation to p-shell in third example.

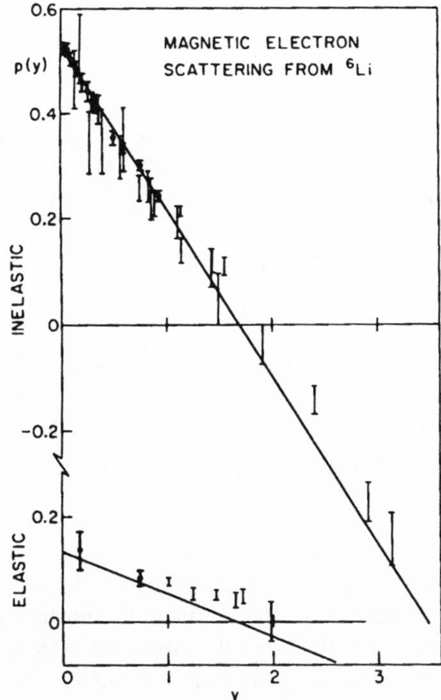

Figure 47.8: Magnetic electron scattering from ^{6}Li in terms of $p(y)$ (see text). The straight line is a minimum χ^2 fit to the accurate data at $|\mathbf{q}| < 200\,\text{MeV}$ (heavy error bars). From Ref. [W19].

The simplest model for this is $(1s_{1/2})^{-1}$ as illustrated in Fig. 26.5. If harmonic oscillator wave functions are used, there is only one parameter left in the one-body density, and Fig. 47.10 shows a fit to some of the early, low-q^2 data for these nuclei with an oscillator parameter chosen as $b_{\text{osc}} = 1.59\,\text{fm}$. Although much more precise experimental and theoretical results now exist to much higher q^2, this is an acceptable one-parameter fit in the low-q^2 regime for the present purposes.

A	B	C	D	E	b_{osc}(fm)
0.810	−0.581	0.084	0.80	0.60	2.03
±0.001	±0.001	±0.002	±0.03	±0.04	±0.02

Table 47.3: Parameter set determined by fit to electromagnetic data involving two valence nucleons in ^{6}Li (from Ref. [W16]).

Table 47.4: Semileptonic weak rates for $0^+1 \leftrightarrow 1^+0$ transition in $B = 6$ system (from Ref. [W16]).

Process	Experiment	Theory
β^--decay (sec^{-1})	0.864 ± 0.003	0.877 ± 0.023
μ^--capture (10^3sec^{-1})	$1.6^{+0.33}_{-0.13}$	1.39 ± 0.04^a

^{a} Statistical average over hyperfine states ($\bar{\omega}_\mu$).

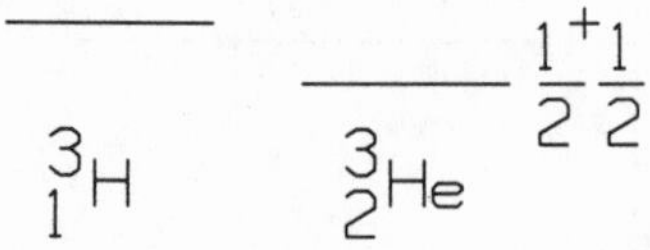

^3_1H $\quad$ ^3_2He $\quad$ $\frac{1}{2}^+\frac{1}{2}$

Figure 47.9: Example of the ground-state $\frac{1}{2}^+\frac{1}{2}$ isodoublet in the $B = 3$ system.

The calculated magnetic moments in this model are independent of the choice of radial wave functions; the fit is well known as is shown in Table 47.5. The discrepancy with the isovector moment was one of the first pieces of evidence for the role of exchange currents in nuclei.

The resulting semileptonic weak rates for $^3\text{He} - {}^3\text{H}$ using this one-body density are shown in Table 47.5. These are now absolute calculations of the weak rates since the required one-body nuclear densities have been determined from (e, e); the weak rates agree with the data in this case to better than 5%.[41]

Some Predictions for New Processes. This analysis uses electron scattering (e, e') through the electromagnetic interaction to determine specific nuclear transition densities; these densities are then used to compute semileptonic weak processes within the traditional nuclear physics framework. Comparison with experiment indicates that this procedure can be carried out to quite high accuracy in selected cases. The analysis is "model independent" in the sense that it eliminates any theoretical calculation of the underlying nuclear structure.

On the basis of this analysis, one can make reliable predictions for some as-yet-unmeasured semileptonic weak processes. For example, Fig. 47.11 shows predicted charge-changing neutrino cross sections for the two nuclear systems shown in Figs. 47.6 and 47.4.

Predictions can also be made for the inelastic scattering of neutrinos (antineutrinos) through the weak neutral current in the standard model using the analysis in Sections 45 and 46. Figure 47.12 presents these cross sections for the

[41] The requisite calculations with this state are carried out in Refs. [N14, N47] (Probs.47.3-4).

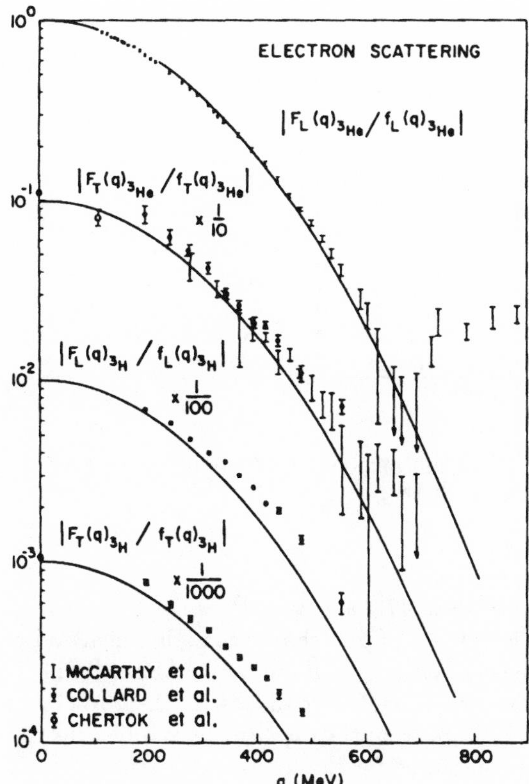

Figure 47.10: Longitudinal and transverse elastic (e, e) form factors for $B = 3$ in terms of $f_{\text{L}}(q)_{M_T} \equiv Zf_{\text{SN}}/(4\pi)^{1/2}$ and $f_{\text{T}}(q)_{M_T} \equiv (q/m)(\lambda_p \delta_{M_T,-1/2} + \lambda_n \delta_{M_T,+1/2})f_{\text{SN}}/(8\pi)^{1/2}$ where f_{SN} is the single-nucleon form factor of Sections 8 and 45. Here $q = |\mathbf{q}|$. Theoretical curves use $b_{\text{osc}} = 1.59\,\text{fm}$. From Ref. [W20].

Table 47.5: Magnetic moments and semileptonic weak rates for $^3\text{He} - {}^3\text{H}$ (from Ref. [W20]). (Here $\mathbf{F} = \mathbf{J} + \mathbf{S}$.)

Quantity	Experiment	Theory
mag. moment		
$\mu_{^3\text{H}} + \mu_{^3\text{He}}$	0.8513	0.8795
$\mu_{^3\text{H}} - \mu_{^3\text{He}}$	5.106	4.706
β^--decay		
$\omega_\beta(\text{sec}^{-1})$	$(1.7906 \pm 0.0067) \times 10^{-9}$	1.84×10^{-9}
μ^--capture		
$\bar{\omega}_\mu(\text{sec}^{-1})$	1505 ± 46	1531
$(F = 0)$		5740

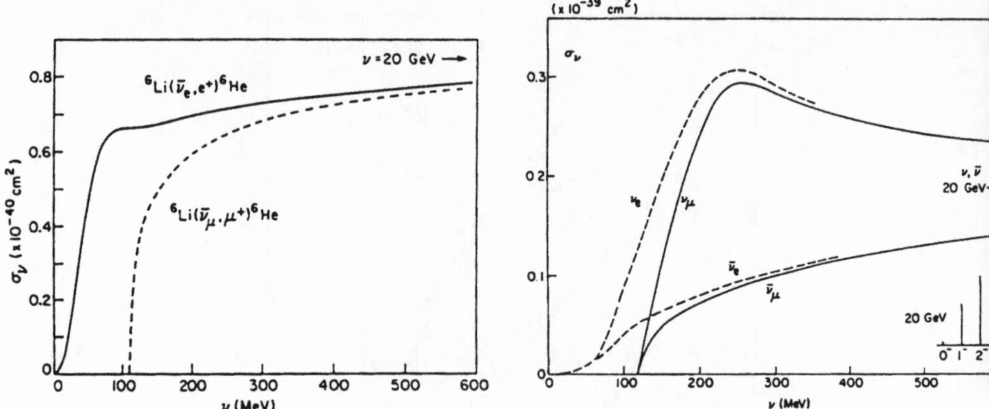

Figure 47.11: Predicted charge-changing neutrino cross sections for transitions analyzed in the text with targets: (a) ^{6}Li; (b) ^{16}O. From Refs. [W17, W18, W19].

two systems shown in Figs. 47.1 and 47.6.[42]

The direct measurement of one charge-changing neutrino cross section between discrete nuclear levels has been reported in Refs. [W14, W15, W13]. The first two refer to an experiment carried out at the LAMPF neutrino facility viewing the LAMPF beam stop; the experiment involves the nuclei in Fig. 47.1. The cross section is measured for the following process

$$\nu_e + {}^{12}\text{C} \rightarrow e^- + {}^{12}\text{N(g.s.)} \tag{47.7}$$

The neutrinos are produced through the decay chain $\pi^+ \rightarrow \mu^+ + \nu_\mu \rightarrow e^+ + \nu_e + \bar{\nu}_\mu + \nu_\mu$. The number of stopped π^+ is known and the subsequent decay spectra can be accurately calculated; hence the incident neutrino flux is *known*. Since the incident proton beam is pulsed, one has a signal for initializing timing measurements; the low duty factor serves to reduce cosmic-ray backgrounds. The appearance of an e^- indicates that a neutrino event has taken place.

The nucleus ^{12}N has only one bound state; its presence is therefore a clear indication of direct production, as all excited states decay by particle emission. The presence of ^{12}N(g.s.) is detected by the time-delayed β^+-decay back to the ground state of ^{12}C.

$$\tau[{}^{12}\text{N(g.s.)} \rightarrow {}^{12}\text{C(g.s.)} + e^+ + \nu_e] \;=\; 15.9\,\text{msec} \tag{47.8}$$

The previous analysis clearly allows a theoretical prediction for the cross section in Eq. (47.7). Table 47.1 gives an indication of the kind of accuracy

[42]The cross sections are shown for two values of the weak mixing angle $\theta_W = 0^o$ and 35°; one can easily interpolate to $\theta_W = 28.7^o$ corresponding to $\sin^2\theta_W = 0.23$. They also assume $F_A(0) = -1.23$.

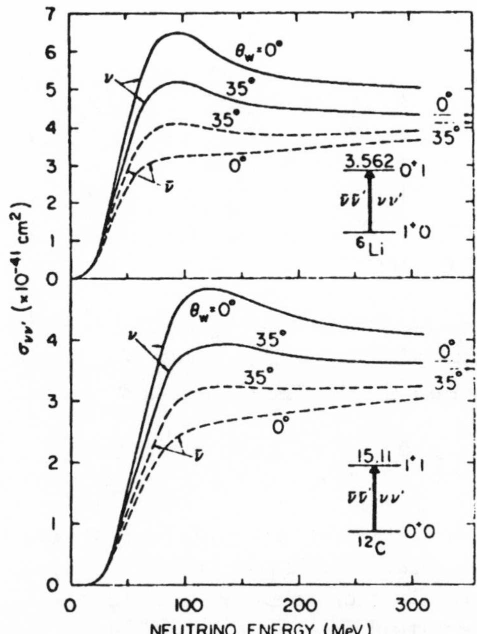

Figure 47.12: Predicted cross sections for inelastic neutrino (antineutrino) scattering through the weak neutral current for transitions analyzed in text with targets: (a) ^{6}Li; (b) ^{12}C. From Refs. [W17, W16].

one can expect. The average over the beamstop neutrino spectrum has been performed by Donnelly (Refs. [W14, W15]).[43] The results are (Ref. [W15])

$$\begin{aligned}\sigma[\nu_e + {}^{12}\text{C} \to e^- + {}^{12}\text{N(g.s.)}] &= (1.05 \pm 0.14) \times 10^{-41}\,\text{cm}^2\;; \quad \text{experiment} \\ &= 0.94 \times 10^{-41}\,\text{cm}^2\;; \quad \text{theory}\end{aligned} \tag{47.9}$$

Variation with Weak Coupling Constants. We now let the weak coupling constants vary to see how well they are determined by the above comparison between theory and experiment. Define

$$C_P \equiv \frac{m_\mu F_P}{F_A} \qquad\qquad \mu \equiv F_1 + 2mF_2 \tag{47.10}$$

The singlet μ-capture rates for $^1\text{H}(\mu^-, \nu_\mu)n$ and statistically averaged rate for $^3\text{He}(\mu^-, \nu_\mu)^3\text{H}$ are shown as functions of their induced-pseudoscalar $C_P(0)$ and weak magnetism $\mu^{(1)}(0) \equiv \mu^V(0)$ contributions in Fig. 47.13. The pion-pole value of $C_P(0)$ and CVC value of $\mu^{(1)}(0)$ are indicated by the vertical lines in

[43] Donnelly has also extended the theoretical analysis to include all possible p-shell s.p.m.e.; the effect on the cross section for this transition is not large.

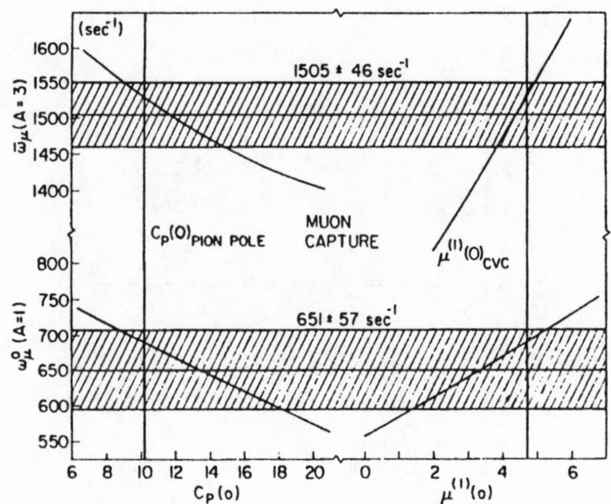

Figure 47.13: The $B = 1$ singlet and statistically weighted $B = 3$ μ-capture rates as functions of their induced-pseudoscalar $C_P(0)$ and weak magnetism $\mu^{(1)}(0)$ contributions. The pion-pole value of $C_P(0)$ and CVC value of $\mu^{(1)}(0)$ are indicated by the vertical lines. From Ref. [W20].

the figure.[44] If one asssumes the latter quantity to be given by CVC, then the rate for $^3\text{He}(\mu^-, \nu_\mu)^3\text{H}$ gives $C_P(0)/C_P(0)_{\text{pion pole}} \approx 1 \pm 30\%$ as the best determination; it is not determined very well.

The μ-capture rate for $^{12}\text{C}(\mu^-, \nu_\mu)^{12}\text{B(g.s.)}$ is shown as a function of these same two quantities in Fig. 47.14. The upper curve shows the dependence on the deviation of $C_P(0)$ from the pion-pole value; evidently this rate is *completely insensitive to* $C_P(0)$ since the theoretical rate lies within the experimental error bars for all plotted values. One can make use of this insensitivity to $C_P(0)$ and use the comparison as a measure of the role of *weak magnetism* in μ-capture. One has $\mu^V(0) \approx \mu^V(0)_{\text{CVC}} \pm 1.5 = 4.71 \pm 1.5$, a confirmation of the role of the magnetic part of the weak vector coupling predicted by CVC.

The Relativistic Nuclear Many-Body Problem. We have discussed in Part II the motivation for, and development of a relativistic quantum field theory description of nuclear structure based on hadronic degrees of freedom. In particular, the relativistic Hartree description in QHD-I provides a minimal explanation of many essential features of nuclear structure such as charge densities, the shell model, and the spin dependence of nucleon-nucleus scattering. Although the complete calculation of the hadronic contribution to the electroweak currents in QHD is a formidable problem, an effective local electroweak current to be used in lowest order with QHD-I was constructed in Section 45; it is covariant, the

[44] The pion-pole value from Section 45 is $F_P = 2mF_A/(q^2 + m_\pi^2)$.

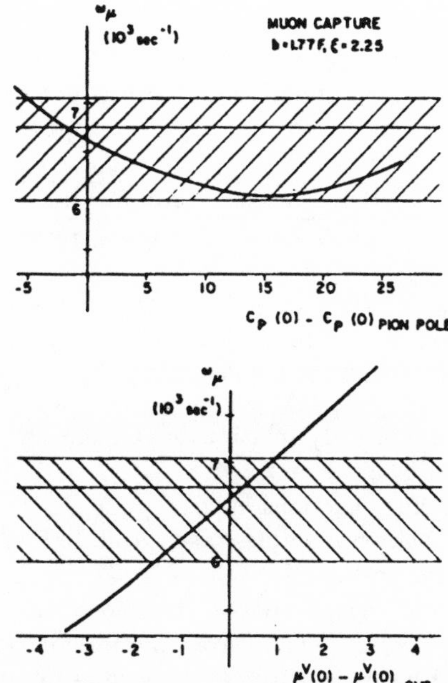

Figure 47.14: The μ-capture rate for $^{12}\text{C}(\mu^-, \nu_\mu)^{12}\text{B}(\text{g.s.})$ as a function of the deviation of the induced-pseudoscalar $C_P(0)$ and weak magnetism $\mu^{(1)}(0) \equiv F_1^V(0) + 2mF_2^V(0)$ contributions from the pion-pole and CVC values, respectively. From Refs. [W17, W16] .

vector current is conserved, the axial vector current satisfies PCAC, and the full current has all the symmetry properties of the standard model.

The traditional approach to nuclear structure and a unified analysis of electroweak processes in nuclei can, in selected cases, provide a quantitative analysis as discussed above. In this approach, relativistic corrections are treated in perturbation theory. Furthermore, as with electron scattering, the approach clearly becomes inadequate when the momentum transfer becomes large compared to the mass of the nucleon.

It is therefore of interest to examine a few semileptonic processes within the context of QHD with at least two goals:

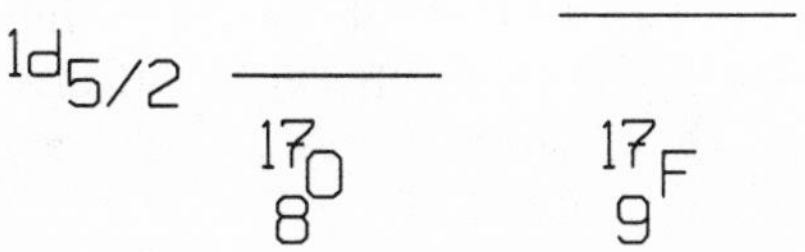

$1d_{5/2}$

$^{17}_{8}\text{O}$ $^{17}_{9}\text{F}$

Figure 47.15: Nuclear system considered in first example of relativistic calculation in QHD (Ref. [W21]).

Table 47.6: β-decay and μ-capture rates (sec^{-1}) for relativistic Hartree calculations; also shown is the nonrelativistic limit as described in the text (from Ref. [W21]).

Process	Rel. Hartree	Nonrel. lim.	Experiment
$^3\text{H} \to {}^3\text{He} + e^- + \bar{\nu}_e$	1.815×10^{-9}	1.818×10^{-9}	$(1.7906 \pm 0.0067) \times 10^{-9}$
$^{17}\text{F} \to {}^{17}\text{O} + e^+ + \nu_e$	1.223×10^{-2}	1.228×10^{-2}	1.075×10^{-2}
$\mu^- + {}^3\text{He} \to {}^3\text{H} + \nu_\mu$	1458	1378	1505 ± 46

1. To carry out a completely relativistic calculation where the relativistic corrections are included to all orders;

2. To calculate neutrino reactions at high q^2.

Examples. We present two examples taken from Ref. [W21]. Consider as the *first* example the nuclear system shown in Fig. 47.15. The configuration here is a $1d_{5/2}$ isodoublet, and the relativistic Hartree wave function of Section 15 is used for the valence nucleon. The wave function and elastic magnetic cross section for $^{17}\text{O}(e, e)^{17}\text{O}$ have already been presented in Section 17.

The effective relativistic electroweak current of Section 45 is now employed, appropriate reduced matrix elements evaluated, and the rates and cross sections are just those of Section 46.

The differential and integrated cross sections for the charge-changing and neutral current reactions $^{17}\text{O}(\nu_l, l^-)^{17}\text{F(g.s.)}$ and $^{17}\text{O}(\nu_l, \nu_l)^{17}\text{O}$ are shown in Fig. 47.16.[45]

The nonrelativistic limit is obtained by first writing the Dirac wave function in the following manner

$$\psi = \begin{pmatrix} \chi(r) \\ \frac{\boldsymbol{\sigma}\cdot\mathbf{p}}{2m}\chi(r) \end{pmatrix} \tag{47.11}$$

An expansion through $O(1/m)$ then reproduces the results of Section 45; numerical results with the Hartree $\chi(r)$ are indicated by dashed curves in Fig. 47.16.

The β-decay rate for this system calculated in the same manner is shown in Table 47.6.

As a *second* example consider the nuclear system in Fig. 47.9. The configuration is $(1s_{1/2})^{-1}$. Relativistic Hartreé calculations of integrated cross sections for $^3\text{He}(\nu_l, \nu_l)^3\text{He}$, $^3\text{He}(\bar{\nu}_l, \bar{\nu}_l)^3\text{He}$, and $^3\text{He}(\bar{\nu}_l, l^+)^3\text{H}$ are shown in Ref. [W21] together with the nonrelativistic result. The β-decay and μ-capture rates for this system are also given in Table 47.6.

[45] The calculation is now completely relativistic, except for the C-M correction where the previous expression f_{CM} is used.

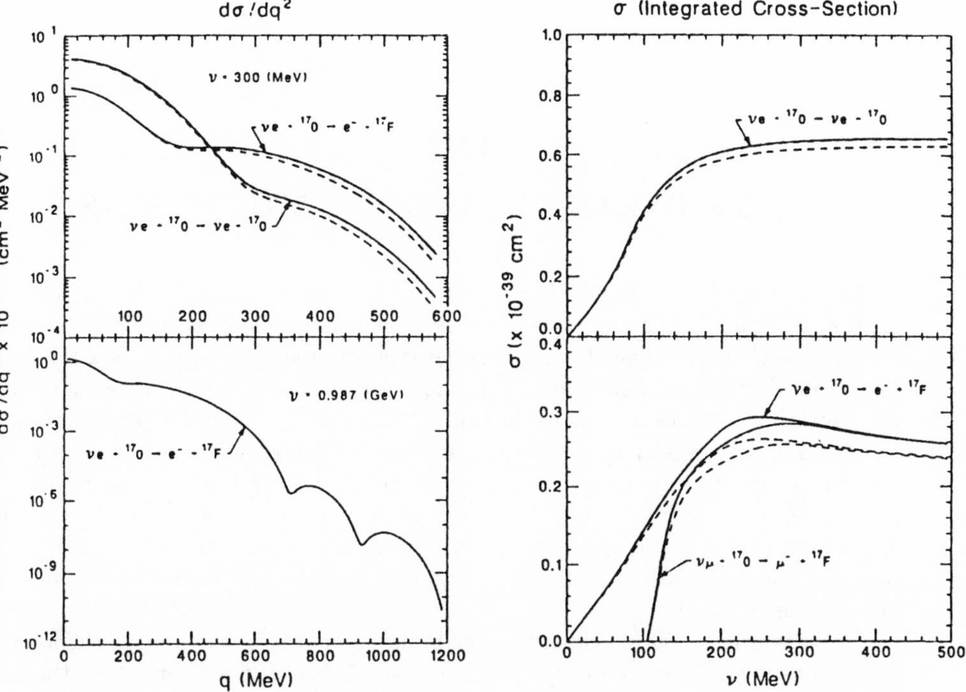

Figure 47.16: Differential $d\sigma/dq^2$ and integrated cross section for the charge-changing $^{17}\mathrm{O}(\nu_l, l^-)^{17}\mathrm{F}$(g.s.) and neutral current reaction $^{17}\mathrm{O}(\nu_l, \nu_l)^{17}\mathrm{O}$. The dashed curve (where calculated) is the nonrelativistic limit as described in the text. From Ref. [W21] .

The conclusions from this work are as follows:

- One has a closed-form summation of relativistic corrections in the semileptonic weak interactions with nuclei.
- There is agreement with the nonrelativistic results for the rates and integrated cross sections at the level of $\leq 9\%$.
- The deviation of the nonrelativistic result for the differential cross section can be larger at high q^2.
- Nothing now limits these calculations to low q^2 — one can make predictions for any new semileptonic process at any q^2.

48

ELECTROWEAK RADIATIVE CORRECTIONS

In this section we discuss loop corrections in the standard model of electroweak interactions based on the underlying local gauge symmetry $SU(2)_W \otimes U(1)_W$ (Section 43). The standard model was originally motivated as a renormalizable theory, and hence one of its most salient theoretical features is that the radiative corrections can be calculated in terms of renormalized coupling constants and masses in an unambiguous manner. This remains true in the presence of the strong interactions described by QCD and the full gauge symmetry $SU(3)_C \otimes SU(2)_W \otimes U(1)_W$. While the radiative corrections in the full theory are calculable in the region where the strong interactions are asymptotically free, one faces the full complexity of hadronic physics in the strong-coupling, nuclear domain.

The Feynman rules provide the basic building blocks for calculating the loop corrections, and for consistency with the rest of this text we here state those rules. There are a few caveats:

1. The stated rules are *not complete* — there are additional couplings of the gauge bosons. The rules do, however, provide the necessary ingredients for the lowest-order radiative corrections to the processes we have been discussing. An enlarged set of Feynman rules is given, for example, in Appendix B of Ref. [W22]. A full set of Feynman rules, which includes the Faddeev-Popov ghosts coupled to the vector bosons, is given in Ref. [W25].[46]

2. The Feynman rules in Ref. [W22], for example, are given in the metric of Bjorken and Drell. For consistency, the rules are stated here in the present metric. It is thus essential to have a metric (and convention) conversion table; this is given in Appendix N. Every effort has been made to establish the consistency of the rules as stated here and those in Refs. [W1, W22] and in Ref. [R1]. Where there is any discrepancy in these references, it will be explictly footnoted.[47]

3. The rules will be presented for leptons and point nucleons in the strong-interaction sector; the latter provides a simple model in which to examine the structure of the theory (Section 43). The extension to the underlying quark sector requires further discussion and this is postponed to the next section.

[46] The author thanks V. Dmitrašinovič for calling this reference to his attention.

[47] Anyone performing a calculation of radiative corrections will certainly derive their own consistent set of Feynman rules from scratch (Ref. [W22] or Refs. [R4, R1]). The rules as stated here provide insight into the structure of the theory, and provide a nice check.

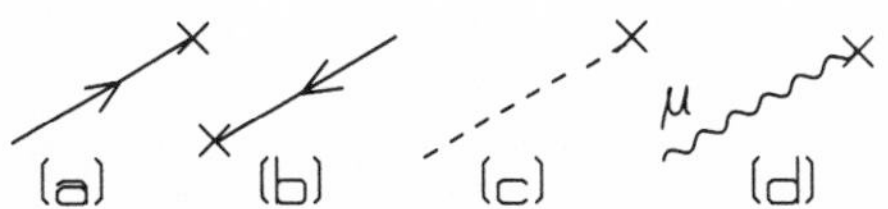

Figure 48.1: External lines: (a) fermion; (b) antifermion; (c) scalar; (d) vector.

Feynman Rules for the S-Matrix. The S-matrix for an electroweak process in the standard model is given by the following set of rules:

1. Draw all topologically distinct connected diagrams.[48]
2. Conserve four-momentum at each vertex with a factor $(2\pi)^4\delta^{(4)}(\sum_i p_i)$.
3. Integrate over internal four-momenta $\int d^4q/(2\pi)^4$.
4. Include a factor (-1) for each closed fermion loop.[49]
5. Include a factor for each external line (Fig. 48.1):
   Fermion : $u(\mathbf{p})/\sqrt{\Omega}$ or $v(-\mathbf{p})/\sqrt{\Omega}$ with $u^\dagger u = v^\dagger v = 1$
   Scalar: $1/\sqrt{2\omega\Omega}$
   Vector: $\varepsilon_\mu(\lambda)/\sqrt{2\omega\Omega}$
6. Include a factor for each propagator (Fig. 48.2)

$\surd$ *Fermion*[50]

$$\frac{1}{i}\frac{1}{ip_\mu\gamma_\mu + m} \tag{48.1}$$

Vector mesons

$$\begin{aligned}
W_\mu^{(\pm)}: &\quad \frac{1}{i}\frac{1}{k^2+m_W^2}\left[\delta_{\mu\nu} - \frac{k_\mu k_\nu}{k^2+m_W^2/\xi}\left(1-\frac{1}{\xi}\right)\right] \\
Z_\mu: &\quad \frac{1}{i}\frac{1}{k^2+m_Z^2}\left[\delta_{\mu\nu} - \frac{k_\mu k_\nu}{k^2+m_Z^2/\zeta}\left(1-\frac{1}{\zeta}\right)\right] \\
A_\mu: &\quad \frac{1}{i}\frac{1}{k^2}\left[\delta_{\mu\nu} - \frac{k_\mu k_\nu}{k^2}(1-\bar{\alpha})\right]
\end{aligned} \tag{48.2}$$

Here the choice of the parameters $(\xi, \zeta, \bar{\alpha})$ define various gauges:
$\surd$ In the unitary gauge $\xi \to 0$ the W-propagator is

$$W_\mu^{(\pm)}: \quad \frac{1}{i}\frac{1}{k^2+m_W^2}\left(\delta_{\mu\nu} + \frac{k_\mu k_\nu}{m_W^2}\right) \tag{48.3}$$

[48] Any ambiguity here can be resolved by an appeal to Wick's theorem.
[49] For any additional symmetry factor in a diagram, go back to Wick's theorem (see Ref. [W22] , Appendix B).
[50] Elements explicitly derived in this book are denoted with a $\surd$.

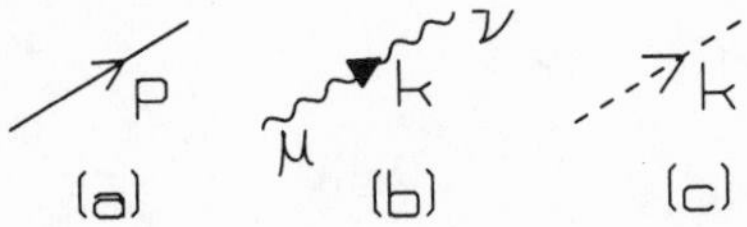

Figure 48.2: Propagators: (a) fermion; (b) vector meson; (c) scalar.

√ In the 't Hooft-Feynman gauge $\xi = 1$ the W-propagator is

$$W_\mu^{(\pm)}: \qquad \frac{1}{i}\frac{1}{k^2+m_W^2}\delta_{\mu\nu} \tag{48.4}$$

In the Landau gauge $\xi \to \infty$ the W-propagator is

$$W_\mu^{(\pm)}: \qquad \frac{1}{i}\frac{1}{k^2+m_W^2}\left(\delta_{\mu\nu} - \frac{k_\mu k_\nu}{k^2}\right) \tag{48.5}$$

The same relations hold for ζ.

Scalar mesons

$$\begin{aligned}
\surd\ \eta(\phi_1) - \text{Higgs}: &\qquad \frac{1}{i}\frac{1}{k^2+m_H^2} \\
\chi(\phi_2) - \text{unphysical scalar}: &\qquad \frac{1}{i}\frac{1}{k^2+m_Z^2/\zeta} \\
s^\pm(\phi^\pm) - \text{unphysical scalar}: &\qquad \frac{1}{i}\frac{1}{k^2+m_W^2/\xi}
\end{aligned} \tag{48.6}$$

Note that the unphysical scalars *decouple* in the unitary gauge with $(\xi, \zeta) = 0$.

7. Include the following factors for each vertex (Fig. 48.3)

Lepton sector[51]

$$\begin{aligned}
\surd\ (l^- l^- A_\mu): &\qquad +e\gamma_\mu \\
\surd\ (l^- \nu_l W_\mu^\star): &\qquad \frac{-g}{2\sqrt{2}}\gamma_\mu(1+\gamma_5) \\
\surd\ (\nu_l \nu_l Z_\mu): &\qquad \frac{g}{4\cos\theta_W}\gamma_\mu(1+\gamma_5) \\
\surd\ (l^- l^- Z_\mu): &\qquad \frac{-g\gamma_\mu}{4\cos\theta_W}\left[(1-4\sin^2\theta_W)+\gamma_5\right] \\
\surd\ (l^- l^- \eta): &\qquad \frac{-igm_l}{2m_W} \\
(l^- l^- \chi): &\qquad \frac{-gm_l}{2m_W}\gamma_5 \\
(l^- \nu_l s^\star): &\qquad \frac{-ig}{\sqrt{2}}\frac{m_l}{2m_W}(1+\gamma_5)
\end{aligned} \tag{48.7}$$

[51] Reference [W22] differs in the sign of the γ_5 in the $(l^- l^- Z_\mu)$ coupling; Reference [W1] differs in the overall sign of the $(\nu_l \nu_l Z_\mu)$, $(l^- l^- Z_\mu)$, and $(l^- \nu_l W_\mu^\star)$ couplings and has a coefficient of $(-e)$ in the second of Eqs. (48.8).

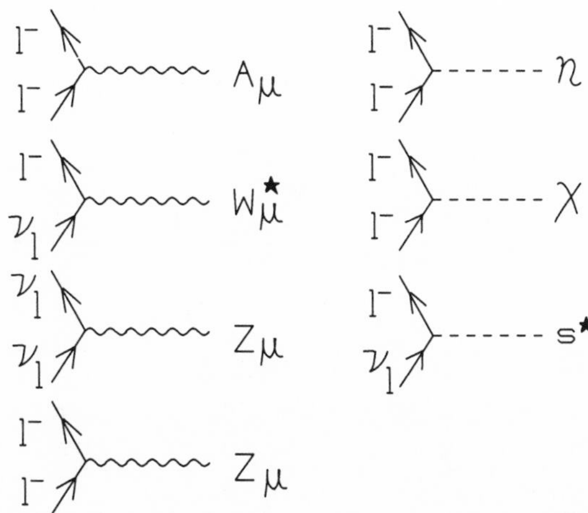

Figure 48.3: Vertices in lepton sector referred to in the text.

Note that here $e \equiv |e|$.

Boson sector. All momenta are incoming (see Fig. 48.4)[52] [53]

$$\begin{aligned}
[\, W^\star_\nu A_\mu s] &: \quad -iem_W\,\delta_{\mu\nu} \\
[\, W^\star_\lambda(k_3) W_\nu(k_2) A_\mu(k_1)] &: \quad +ie[(k_1-k_2)_\lambda \delta_{\mu\nu} + (k_2-k_3)_\mu \delta_{\nu\lambda} + (k_3-k_1)_\nu \delta_{\lambda\mu}] \\
[\, s^\star(p_-) s(p_+) A_\mu] &: \quad -ie(p_- - p_+)_\mu
\end{aligned} \tag{48.8}$$

Point Nucleon Sector (Fig. 48.5).

$$\begin{aligned}
\surd\,(ppA_\mu) &: \quad -e\gamma_\mu \\
\surd\,(npW^\star_\mu) &: \quad \frac{-g}{2\sqrt{2}}\gamma_\mu(1+\gamma_5) \\
\surd\,(ppZ_\mu) &: \quad \frac{g\gamma_\mu}{4\cos\theta_W}\left[(1-4\sin^2\theta_W)+\gamma_5\right] \\
\surd\,(nnZ_\mu) &: \quad \frac{-g}{4\cos\theta_W}\gamma_\mu(1+\gamma_5) \\
\surd\,(pp\eta) &: \quad \frac{-igm_N}{2m_W} \\
\surd\,(nn\eta) &: \quad \frac{-igm_N}{2m_W} \\
(pp\chi) &: \quad \frac{gm_N}{2m_W}\gamma_5
\end{aligned}$$

[52]The second vertex can be derived from the term $\mathcal{L}_{\rm gauge} \doteq -(g/2)\vec{F}_{\mu\nu}\cdot(\vec{A}_\mu \times \vec{A}_\nu)$ after expressing it in terms of the physical fields (Prob. 48.1).

[53]There are many more boson vertices — see Refs. [W22, W25] and Ref. [R1].

Table 48.1: Contribution to anomalous magnetic moment of the muon $a_\mu = (g_\mu - 2)/2$ in units of $Gm_\mu^2/8\pi^2(2)^{1/2}$ (from Ref. [W1]).

Quantity	Gauge		
	Unitary $(\xi,\zeta) \to 0$	't Hooft -Feynman $(\xi,\zeta) = 1$	Landau $(\xi,\zeta) \to \infty$
(a)	10/3	7/3	4/3
(b) + (c)	0	1	1
(d)	0	0	1
Subtotal	10/3	10/3	10/3
(e)	$a(\theta_W)^a$	$a(\theta_W)$	$a(\theta_W) + 1$
(f)	0	0	-1
Subtotal	$a(\theta_W)$	$a(\theta_W)$	$a(\theta_W)$

[a] Here $3a(\theta_W) \equiv (3 - 4\cos^2\theta_W)^2 - 5$.

$$
\begin{aligned}
(nn\chi) : &\quad \frac{-gm_N}{2m_W}\gamma_5 \\
(nps^\star) : &\quad \frac{-ig}{\sqrt{2}}\frac{m_N}{m_W}\gamma_5
\end{aligned}
\tag{48.9}
$$

An Application. We briefly discuss one application of electroweak radiative corrections. Consider the 1-loop electroweak correction to the magnetic moment of the muon. The second-order QED correction is worked out in detail in Refs. [R2, R4].

The Feynman diagrams contributing to the second-order correction to the magnetic moment of the muon are shown in Fig. 48.6. The results for graphs $\{a, b, \ldots, f\}$ are shown in Table 48.1 (taken from Ref. [W1]). Note the contribution of the different pieces in different gauges. The combination $(a) + (b) + (c) + (d)$ involving the charged bosons is *gauge invariant*, as is $(e) + (f)$ involving neutral boson exchange.

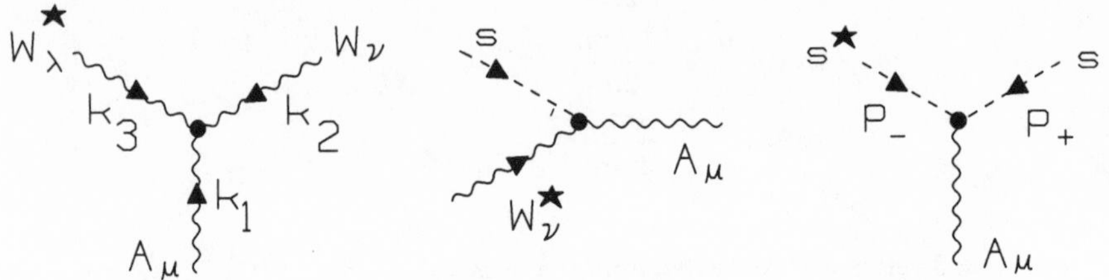

Figure 48.4: Vertices in boson sector referred to in the text.

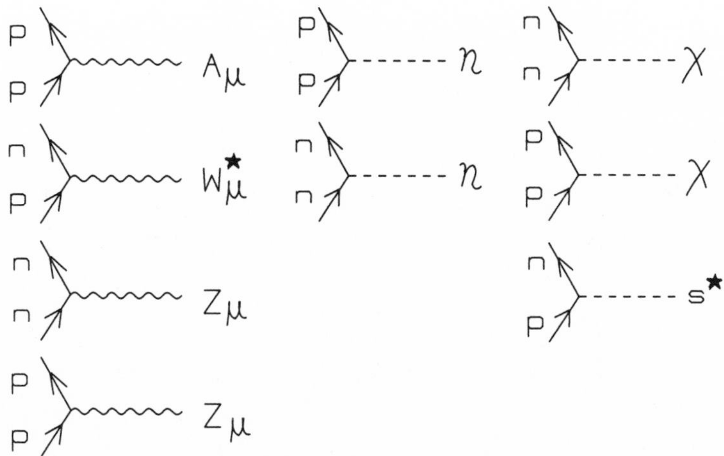

Figure 48.5: Vertices in point-nucleon sector referred to in the text.

The *Higgs exchange* is described by graph (g). In the same units as in Table 48.1, the contribution from this process is (Ref. [W23] and Prob. 48.2)

$$\begin{aligned} I &= 2\int_0^1 dx \frac{x^2(2-x)}{x^2+r(1-x)} \\ r &= \left(\frac{m_\eta}{m_\mu}\right)^2 \end{aligned} \tag{48.10}$$

This quantity is negligible if the mass of the Higgs m_η is large enough.

Other Contributions. If one is to detect these electroweak contributions to a_μ, it is necessary to compute the *electromagnetic* corrections to this accuracy. A systematic expansion in $\alpha \equiv e^2/4\pi$, though extremely tedious, will permit such an evaluation. There is one contribution, however, that of hadronic vacuum polarization as illustrated in Fig. 48.7, that involves the strong interaction and cannot as of now be fully calculated in the hadronic, nuclear domain. Fortunately this contribution can be determined experimentally through use of a dispersion relation in q^2 for the virtual photon propagator and measurement of the absorptive part in the reaction $e^+ + e^- \rightarrow$ hadrons.

Results. The current status of the anomalous magnetic moment of the muon is as follows (taken from Ref. [W23]). The individual contributions are

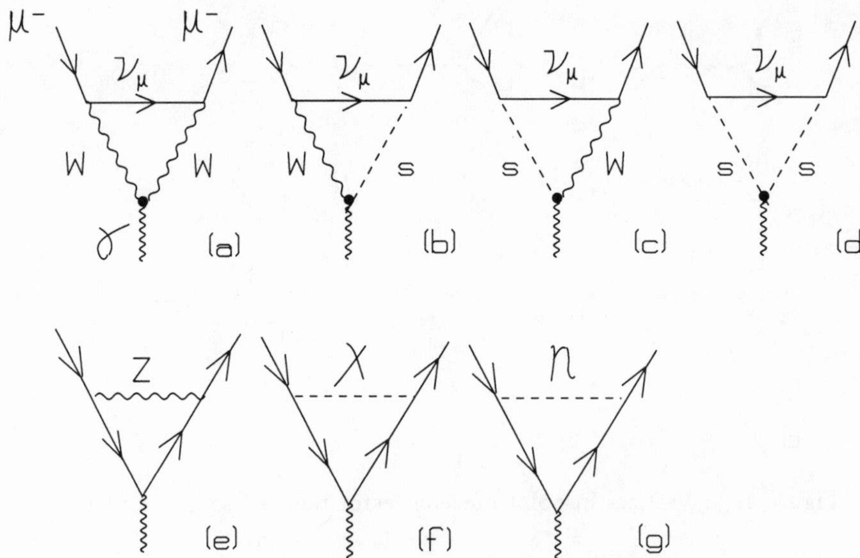

Figure 48.6: Feynman diagrams contributing to second-order electroweak correction to the magnetic moment of the muon.

$$
\begin{aligned}
a_\mu^{\rm QED} &= 1,165,848.1(1.2) \times 10^{-9} \; ; && \text{through 6th order} \\
a_\mu^{\rm hadron} &= 66.7(9.4) \times 10^{-9} \; ; && \text{vacuum polarization} \\
a_\mu^{\rm W-S} &= 2.1(0.2) \times 10^{-9} \; ; && \text{standard model (above)}
\end{aligned} \tag{48.11}
$$

The sum total is

$$
\begin{aligned}
a_\mu^{\rm theory} &= 1,165,921(13) \times 10^{-9} \\
a_\mu^{\rm expt} &= 1,165,922(9) \times 10^{-9}
\end{aligned} \tag{48.12}
$$

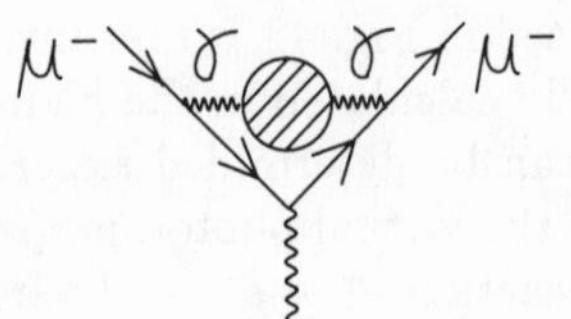

Figure 48.7: Hadronic vacuum polarization contribution to the anomalous magnetic moment of the muon.

These results use (Ref. [W25])[54] [55]

$$
\begin{aligned}
a &\equiv \frac{(\mu-\mu_0)}{\mu_0} = \frac{(g-2)}{2} \\
\alpha^{-1} &= 137.035,987(29)
\end{aligned}
\tag{48.13}
$$

An experiment is currently underway at Brookhaven National Laboratory to determine a_μ to the level where the contribution of the standard model $a_\mu^{\text{W-S}}$ can be discerned.[56]

[54]The theoretical result includes $a_\mu^{\text{QED}}(\text{8th order}) = (3.7 \pm 2.1) \times 10^{-9}$ (Ref. [W23]).
[55]The experimental result is from Ref. [W24].
[56]One also requires improved $e^+ + e^- \rightarrow$ hadron measurements to decrease the uncertainty in the contribution from hadronic vacuum polarization.

49

FULL QUARK SECTOR OF THE STANDARD MODEL

In this section we extend the discussion of the standard model of electroweak interactions to the full underlying quark sector. To do this, an extended discussion of quark mixing in the electroweak interactions is required. We first review the development in Section 44 involving quark mixing in the case of two families of quarks. Color indices will here be suppressed for clarity; their inclusion is discussed in Section 44.

Quark Mixing in the Electroweak Interactions: Two-Families — A Review. Start with the kinetic energy of massless (u, d, s, c) quarks

$$-\mathcal{L}_0 \quad = \quad \bar{u}\gamma_\mu \frac{\partial}{\partial x_\mu} u + \bar{d}\gamma_\mu \frac{\partial}{\partial x_\mu} d + \bar{s}\gamma_\mu \frac{\partial}{\partial x_\mu} s + \bar{c}\gamma_\mu \frac{\partial}{\partial x_\mu} c \tag{49.1}$$

Use the previously established decomposition into left- and right-handed fields

$$\begin{aligned}-\mathcal{L}_0 = \bar{u}_{\rm L}\gamma_\mu \frac{\partial}{\partial x_\mu} u_{\rm L} + \bar{d}_{\rm L}\gamma_\mu \frac{\partial}{\partial x_\mu} d_{\rm L} + \bar{s}_{\rm L}\gamma_\mu \frac{\partial}{\partial x_\mu} s_{\rm L} + \bar{c}_{\rm L}\gamma_\mu \frac{\partial}{\partial x_\mu} c_{\rm L} \\ + \bar{u}_{\rm R}\gamma_\mu \frac{\partial}{\partial x_\mu} u_{\rm R} + \bar{d}_{\rm R}\gamma_\mu \frac{\partial}{\partial x_\mu} d_{\rm R} + \bar{s}_{\rm R}\gamma_\mu \frac{\partial}{\partial x_\mu} s_{\rm R} + \bar{c}_{\rm R}\gamma_\mu \frac{\partial}{\partial x_\mu} c_{\rm R}\end{aligned} \tag{49.2}$$

Now define a rotated combination of (d, s) fields

$$\begin{pmatrix} d_{\rm CL} \\ s_{\rm CL} \end{pmatrix} = \begin{pmatrix} \cos\theta_{\rm C} & \sin\theta_{\rm C} \\ -\sin\theta_{\rm C} & \cos\theta_{\rm C} \end{pmatrix} \begin{pmatrix} d_L \\ s_L \end{pmatrix} \tag{49.3}$$

Here $\theta_{\rm C}$ is the Cabbibo angle.[57] The GIM identity allows the first line in Eq. (49.2) to be rewritten

$$-\mathcal{L}_{0{\rm L}} = \bar{u}_{\rm L}\gamma_\mu \frac{\partial}{\partial x_\mu} u_{\rm L} + \bar{c}_{\rm L}\gamma_\mu \frac{\partial}{\partial x_\mu} c_{\rm L} + \bar{d}_{\rm CL}\gamma_\mu \frac{\partial}{\partial x_\mu} d_{\rm CL} + \bar{s}_{\rm CL}\gamma_\mu \frac{\partial}{\partial x_\mu} s_{\rm CL} \tag{49.4}$$

Now introduce two weak (left-handed) isospin doublets

$$q_{\rm L} \equiv \begin{pmatrix} u_{\rm L} \\ d_{\rm CL} \end{pmatrix} \qquad\qquad Q_{\rm L} \equiv \begin{pmatrix} c_{\rm L} \\ s_{\rm CL} \end{pmatrix} \tag{49.5}$$

[57] Note the change of notation from Section 44. Now $D_{\rm CL} \equiv s_{\rm CL}$.

Equation (49.4) then takes the form

$$-\mathcal{L}_{0\mathrm{L}} = \bar{q}_{\mathrm{L}}\gamma_\mu \frac{\partial}{\partial x_\mu} q_{\mathrm{L}} + \bar{Q}_{\mathrm{L}}\gamma_\mu \frac{\partial}{\partial x_\mu} Q_{\mathrm{L}} \tag{49.6}$$

The lagrangian $\mathcal{L}_0 = \mathcal{L}_{0\mathrm{L}} + \mathcal{L}_{0\mathrm{R}}$ now possesses a *global* weak isospin symmetry with left-handed doublets and right-handed singlets. Convert this into a Yang-Mills theory with a *local* symmetry. The charge-changing weak interaction then takes the form (Appendix M)[58]

$$\begin{aligned} \mathcal{L}^{(\pm)} &= \frac{g}{2\sqrt{2}}(\mathcal{J}_\mu^{(+)}W_\mu + \mathcal{J}_\mu^{(-)}W_\mu^\star) \\ \mathcal{J}_\mu^{(\pm)} &\equiv 2i\bar{q}_{\mathrm{L}}\gamma_\mu\tau_\pm q_{\mathrm{L}} + 2i\bar{Q}_{\mathrm{L}}\gamma_\mu\tau_\pm Q_{\mathrm{L}} \end{aligned} \tag{49.7}$$

Let us recast this last result in a slightly different form. Evaluation of the indicated matrix products shows this result is identical to the following

$$\begin{aligned} \mathcal{J}_\mu^{(+)} &= i(\bar{u},\bar{c})\gamma_\mu(1+\gamma_5)\begin{pmatrix} d_{\mathrm{C}} \\ s_{\mathrm{C}} \end{pmatrix} \\ &= i(\bar{u},\bar{c})\gamma_\mu(1+\gamma_5)\begin{pmatrix} \cos\theta_{\mathrm{C}} & \sin\theta_{\mathrm{C}} \\ -\sin\theta_{\mathrm{C}} & \cos\theta_{\mathrm{C}} \end{pmatrix}\begin{pmatrix} d \\ s \end{pmatrix} \end{aligned} \tag{49.8}$$

The effect of quark mixing in the two-family sector is to modify the charge-changing weak quark currents leading from (d,s) to (u,c) by the inclusion of a *mixing matrix*

$$\begin{pmatrix} \cos\theta_{\mathrm{C}} & \sin\theta_{\mathrm{C}} \\ -\sin\theta_{\mathrm{C}} & \cos\theta_{\mathrm{C}} \end{pmatrix} \equiv \begin{pmatrix} U_{ud} & U_{us} \\ U_{cd} & U_{cs} \end{pmatrix} \tag{49.9}$$

This mixing matrix has the following properties:

1. It is a unitary 2×2 matrix;
2. As long as the mixing matrix in Eq. (49.9) is unitary, then the weak neutral currents will be diagonal in flavor by the GIM mechanism (Section 44);
3. Since the quarks (d,s) and (u,c) are interconverted only through the weak interactions, the relative phases of the quark fields can be chosen so that this unitary 2×2 matrix is, in fact, a real rotation;
4. The mixing matrix here is just that of Cabbibo (Ref. [W8]) discussed in Section 44;
5. The empirical value of the mixing angle is $\cos\theta_{\mathrm{C}} = 0.974$ (Section 47).[59]

One can now proceed to generate mass by spontaneous symmetry breaking, and form the mass eigenstates (u,d,s,c), as discussed previously.

[58] With the GIM mechanism, the corresponding weak neutral current is *diagonal in flavor* (Section 44).

[59] Although many explanations have been put forward, there is no simple way to understand why the weak interactions sample this particular combination of quark fields.

Extension to Three Families of Quarks. We have already observed that leptons come in three families

$$\begin{pmatrix} \nu_e \\ e^- \end{pmatrix} \qquad \begin{pmatrix} \nu_\mu \\ \mu^- \end{pmatrix} \qquad \begin{pmatrix} \nu_\tau \\ \tau^- \end{pmatrix} \tag{49.10}$$

Quarks evidently also come in three families[60]

$$\begin{pmatrix} u \\ d \end{pmatrix} \qquad \begin{pmatrix} c \\ s \end{pmatrix} \qquad \begin{pmatrix} t \\ b \end{pmatrix} \tag{49.11}$$

Now one can go through *exactly the same arguments as presented above*, but this time including all three families of quarks. Evidently the charge-changing weak current in Eqs. (49.8) and (49.9) is then generalized to

$$\mathcal{J}_\mu^{(+)} = i(\bar{u}, \bar{c}, \bar{t})\gamma_\mu(1+\gamma_5)\begin{pmatrix} U_{ud} & U_{us} & U_{ub} \\ U_{cd} & U_{cs} & U_{cb} \\ U_{td} & U_{ts} & U_{tb} \end{pmatrix}\begin{pmatrix} d \\ s \\ b \end{pmatrix} \tag{49.12}$$

In this case one can have a 3×3 unitary quark mixing matrix — the Cabbibo-Kobayashi-Moskawa (CKM) matrix (Ref. [W26]).

The existing experimental data can be summarized in a matrix of the form (Refs. [W27, W28])

$$\underset{\sim}{U} \approx \begin{pmatrix} 1-\lambda^2/2 & \lambda & A\lambda^3(\rho - i\eta) \\ -\lambda & 1-\lambda^2/2 & A\lambda^2 \\ A\lambda^3(1-\rho-i\eta) & -A\lambda^2 & 1 \end{pmatrix} \tag{49.13}$$

The experimental values of the parameters (λ, A) appearing in this expression are given by

$$\begin{aligned} \lambda &\approx 0.220 \qquad\qquad 1-\frac{\lambda^2}{2} \approx 0.974 \\ A &= 1.05 \pm 0.17 \end{aligned} \tag{49.14}$$

The parameter η admits the possibility of a small CP (or T)-violating phase; the two parameters (ρ, η) remain to be determined.

One can now again proceed to generate mass by spontanteously symmetry breaking, and form the mass eigenstates (u, d, s, c, b, t).[61]

Feynman Rules in the Quark Sector. We proceed to give the resulting Feynman rules for the standard model in the quark sector. Here the following notation will be employed for the quark fields q: p denotes a p-type quark (u, c, t); n

[60] Most physicists believe the top quark t exists, although at the time of this writing it remains to be discovered.

[61] If the neutrinos were to have a small mass, then by analogy, one might also expect mixing in the *lepton* sector; so far there is no experimental evidence to support this.

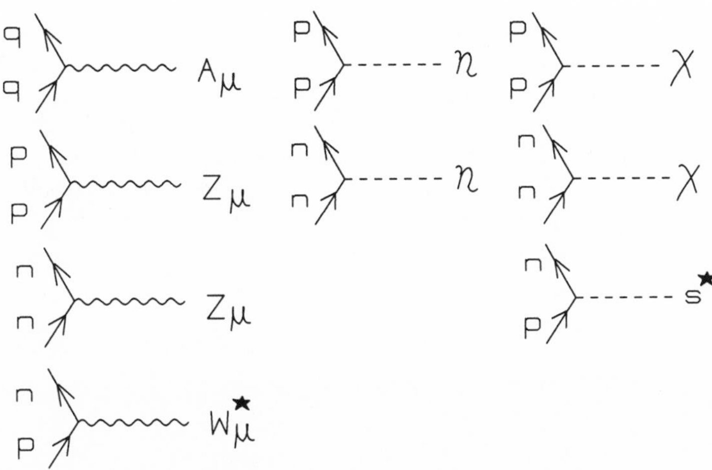

Figure 49.1: Vertices for p-type (u, c, t) and n-type (d, s, b) quarks in the standard model of electroweak interactions.

denotes an n-type quark (d, s, b); and U_{pn} denotes the appropriate CKM mixing matrix element. The overall factors and propagators are as discussed previously. The vertex components are shown in Fig. 49.1; the corresponding factors in the S-matrix are as follows:[62] [63]

$$\begin{aligned}
\surd(qqA_\mu): &\quad -eQ_q\gamma_\mu \\
\surd(ppZ_\mu): &\quad \frac{g}{4\cos\theta_W}\gamma_\mu[(1-\frac{8}{3}\sin^2\theta_W)+\gamma_5] \\
\surd(nnZ_\mu): &\quad \frac{-g}{4\cos\theta_W}\gamma_\mu[(1-\frac{4}{3}\sin^2\theta_W)+\gamma_5] \\
\surd(npW_\mu^\star): &\quad \frac{-g}{2\sqrt{2}}\gamma_\mu(1+\gamma_5)U_{pn}^* \\
\surd(pp\eta): &\quad \frac{-igm_p}{2m_W} \\
\surd(nn\eta): &\quad \frac{-igm_n}{2m_W} \\
(pp\chi): &\quad \frac{gm_p}{2m_W}\gamma_5 \\
(nn\chi): &\quad \frac{-gm_n}{2m_W}\gamma_5 \\
(nps^\star): &\quad \frac{-ig}{2\sqrt{2}}\frac{1}{m_W}[m_n(1+\gamma_5)-m_p(1-\gamma_5)]U_{pn}^*
\end{aligned} \tag{49.15}$$

[62]Reference [W22] differs in the sign of the γ_5 in the (ppZ_μ) and (nnZ_μ) couplings; in addition in that reference, the mixing matrix factor is U_{pn} in the $(npW_\mu^\star)$ and $(nps^\star)$ vertices.

[63]A result derived in this text is denoted with a $\surd$.

Experimental data from LEP at CERN are now regularly compared with radiatively corrected theoretical results; some of the Feynman rules for calculating these radiative corrections in the standard model have now been given in this and the preceding section. This beautiful particle physics work takes us well beyond the framework of the present text. As a final topic in this book, we proceed to discuss one nuclear physics application of these results — parity violation in $(\vec{e}, e')$.

50

PARITY VIOLATION IN $(\vec{e}, e')$

The measurement of parity violation in the scattering of longitudinally polarized electrons in deep inelastic electron scattering from deuterium at SLAC is a classic experiment which played a pivotal role in the establishment of the weak neutral current structure of the standard model (Ref. [W29]). The measurement of parity violation in $A(\vec{e}, e')$, where A includes the nucleon, promises to play a central role in future developments in nuclear physics (Ref. [W30]). In this section we use the previous results to develop a general description of this process.

To start the discussion, consider the scattering of a relativistic (massless) longitudinally polarized electron from a point proton. The contributing diagrams in the unitary gauge are shown in Fig. 50.1. From the Feynman rules in Section 48, the S-matrix is given by

$$
\begin{aligned}
S_{\mathrm{fi}} = {} & \frac{-(2\pi)^4 i}{\Omega^2}\delta^{(4)}(k_1+p_1-k_2-p_2)\Big\{\bar{u}(k_2)(e\gamma_\mu)u(k_1)\frac{\delta_{\mu\nu}}{q^2}\bar{u}(p_2)(-e\gamma_\nu)u(p_1) \\
& + \bar{u}(k_2)\left[\frac{-g\gamma_\mu}{4\cos\theta_W}[(1-4\sin^2\theta_W)+\gamma_5]\right]u(k_1)\frac{(\delta_{\mu\nu}+q_\mu q_\nu/m_Z^2)}{q^2+m_Z^2} \\
& \times\bar{u}(p_2)\left[\frac{g\gamma_\nu}{4\cos\theta_W}[(1-4\sin^2\theta_W)+\gamma_5]\right]u(p_1)\Big\}
\end{aligned} \tag{50.1}
$$

At low energy $|\mathbf{q}|/M_Z \ll 1$ and the momentum-dependent terms can be neglected in the Z-propagator. Take the standard model values

$$
\begin{aligned}
e^2 &= 4\pi\alpha & \frac{g^2}{8m_Z^2}\cos^2\theta_W &\equiv \frac{G}{\sqrt{2}} \\
a &\equiv -(1-4\sin^2\theta_W) & b &\equiv -1
\end{aligned} \tag{50.2}
$$

Then

$$
\begin{aligned}
S_{\mathrm{fi}} &= \frac{-(2\pi)^4 i}{\Omega^2}\delta^{(4)}(k_1+p_1-k_2-p_2)T_{\mathrm{fi}} \\
T_{\mathrm{fi}} &= -\frac{4\pi\alpha}{q^2}\Big\{\bar{u}(k_2)\gamma_\mu u(k_1)\bar{u}(p_2)\gamma_\mu u(p_1) - \frac{Gq^2}{4\pi\alpha\sqrt{2}}\bar{u}(k_2)\gamma_\mu[a+b\gamma_5]u(k_1) \\
& \quad \times \bar{u}(p_2)\gamma_\mu[\frac{1}{2}(1+\gamma_5)-2\sin^2\theta_W]u(p_1)\Big\}
\end{aligned} \tag{50.3}
$$

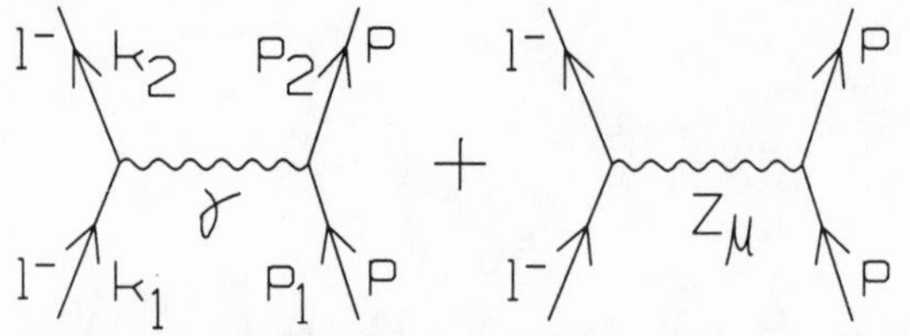

Figure 50.1: Contributing Feynman diagrams (unitary gauge) for parity-violating asymmetry in scattering of longitudinally polarized electrons from point protons. Here $q = k_2 - k_1$.

This result is easily extended to point neutrons using the Feynman rules of Section 48 through the replacement

$$\begin{aligned} T_{\mathrm{f\,i}} &= -\frac{4\pi\alpha}{q^2}\left\{\bar{u}(k_2)\gamma_\mu u(k_1)\bar{u}(p_2)\gamma_\mu \frac{1}{2}(1+\tau_3)u(p_1)\right. \\ &\quad -\frac{Gq^2}{4\pi\alpha\sqrt{2}}\bar{u}(k_2)\gamma_\mu[a+b\gamma_5]u(k_1) \\ &\quad \left.\times\bar{u}(p_2)\gamma_\mu[(1+\gamma_5)\frac{1}{2}\tau_3 - 2\sin^2\theta_W\frac{1}{2}(1+\tau_3)]u(p_1)\right\} \end{aligned} \tag{50.4}$$

At this juncture one can redefine things so that the result is more general than for just point nucleons

$$\begin{aligned} S_{\mathrm{f\,i}} &= \frac{-(2\pi)^4 i}{\Omega}\delta^{(4)}(k_1+p_1-k_2-p_2)T_{\mathrm{f\,i}} \\ T_{\mathrm{f\,i}} &= \frac{4\pi\alpha}{q^2}\left\{i\bar{u}(k_2)\gamma_\mu u(k_1)\langle p_2|J_\mu^\gamma(0)|p_1\rangle\right. \\ &\quad \left.-\frac{Gq^2}{4\pi\alpha\sqrt{2}}i\bar{u}(k_2)\gamma_\mu(a+b\gamma_5)u(k_1)\langle p_2|\mathcal{J}_\mu^{(0)}(0)|p_1\rangle\right\} \end{aligned} \tag{50.5}$$

Now these are single-nucleon matrix elements of the full electromagnetic and weak neutral current densities taken between exact Heisenberg states; for point nucleons, this expression reduces to Eq. (50.4).

The dimensionless ratio $Gq^2/4\pi\alpha\sqrt{2}$ forms the small parameter in these nuclear physics parity-violation calculations.

Cross Sections. The first term in Eq. (50.5) leads to the electron scattering cross section derived in Section 39 (Ref. [N14])[64]

$$d\sigma = \frac{4\alpha^2}{q^4}\frac{d^3k_2}{2\varepsilon_2}\frac{1}{\sqrt{(k_1\cdot p)^2}}\eta_{\mu\nu}W_{\mu\nu}$$

[64] Here $p_1 \equiv p$ and $p_2 \equiv p'$ in the previous notation, and we again generalize to include the possibility of inelastic processes.

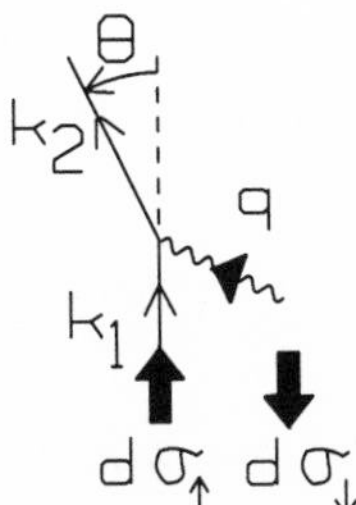

Figure 50.2: Cross section for right- and left-handed electrons.

$$
\begin{aligned}
\eta_{\mu\nu} &= -2\varepsilon_1\varepsilon_2\frac{1}{2}\sum_{s_1}\sum_{s_2}\bar{u}(k_1)\gamma_\nu u(k_2)\bar{u}(k_2)\gamma_\mu u(k_1) \\
&= k_{1\mu}k_{2\nu} + k_{1\nu}k_{2\mu} - (k_1\cdot k_2)\delta_{\mu\nu} \\
W_{\mu\nu} &= (2\pi)^3\overline{\sum_i}\sum_f \delta^{(4)}(q+p'-p)\langle p|J_\nu^\gamma(0)|p'\rangle\langle p'|J_\mu^\gamma(0)|p\rangle(\Omega E_p) \\
&= W_1^\gamma(q^2, q\cdot p)\left(\delta_{\mu\nu} - \frac{q_\mu q_\nu}{q^2}\right) \\
&\quad + W_2^\gamma(q^2, q\cdot p)\frac{1}{M_T^2}\left(p_\mu - \frac{p\cdot q}{q^2}q_\mu\right)\left(p_\nu - \frac{p\cdot q}{q^2}q_\nu\right)
\end{aligned} \tag{50.6}
$$

Polarized Cross Sections. From Section 42 we know that the following are projections for left- and right-handed (massless) Dirac electrons

$$
P_\uparrow = \frac{1}{2}(1-\gamma_5) \qquad\qquad P_\downarrow = \frac{1}{2}(1+\gamma_5) \tag{50.7}
$$

To calculate the cross sections for such particles (Fig. 50.2) one simply modifies $\eta_{\mu\nu}$ with the appropriate insertion of these projections and removes the average over the initial helicities[65]

$$
\begin{aligned}
\text{for } d\sigma_\uparrow: \quad \eta_{\mu\nu}^\uparrow &= \cdots\overbrace{(\tfrac{1}{2})}^{\text{omit}}\sum_{s_1}\sum_{s_2}\bar{u}(k_1)\ldots\frac{1}{2}(1-\gamma_5)u(k_1) \\
\text{for } d\sigma_\downarrow: \quad \eta_{\mu\nu}^\downarrow &= \cdots\sum_{s_1}\sum_{s_2}\bar{u}(k_1)\ldots\frac{1}{2}(1+\gamma_5)u(k_1) \\
\text{for } d\sigma_\uparrow - d\sigma_\downarrow: \quad \eta_{\mu\nu}^{(-)} &= \cdots\sum_{s_1}\sum_{s_2}\bar{u}(k_1)\ldots(-\gamma_5)u(k_1) \\
\text{for } d\sigma_\uparrow + d\sigma_\downarrow: \quad \eta_{\mu\nu}^{(+)} &= \cdots\sum_{s_1}\sum_{s_2}\bar{u}(k_1)\ldots(1)u(k_1)
\end{aligned} \tag{50.8}
$$

[65]Note $d\sigma^\uparrow + d\sigma^\downarrow = 2d\sigma_{\text{unpolarized}}$.

Thus one now has either $(-\gamma_5)$ or (1) in the lepton trace. Since *all common factors cancel in the ratio* the asymmetry is given by

$$\mathcal{A} \equiv \frac{d\sigma_\uparrow - d\sigma_\downarrow}{d\sigma_\uparrow + d\sigma_\downarrow} = -\frac{Gq^2}{4\pi\alpha\sqrt{2}} \frac{\eta_{\mu\nu}^{(1)}W_{\mu\nu}^{(1)} + \eta_{\mu\nu}^{(2)}W_{\mu\nu}^{(2)}}{2\eta_{\mu\nu}W_{\mu\nu}} \tag{50.9}$$

Here

$$\eta_{\mu\nu}^{(1)} = -2\varepsilon_1\varepsilon_2 \sum_{s_1}\sum_{s_2} \bar{u}(k_1)\gamma_\nu u(k_2)\bar{u}(k_2)\gamma_\mu(a+b\gamma_5)(-\gamma_5)u(k_1)$$

$$W_{\mu\nu}^{(1)} = (2\pi)^3 \overline{\sum_i}\sum_f \delta^{(4)}(q+p'-p)\langle p|J_\nu^\gamma(0)|p'\rangle\langle p'|\mathcal{J}_\mu^{(0)}(0)|p\rangle(\Omega E_p)$$

$$\eta_{\mu\nu}^{(2)} = -2\varepsilon_1\varepsilon_2 \sum_{s_1}\sum_{s_2} \bar{u}(k_1)\gamma_\nu(a+b\gamma_5)u(k_2)\bar{u}(k_2)\gamma_\mu(-\gamma_5)u(k_1)$$

$$W_{\mu\nu}^{(2)} = (2\pi)^3 \overline{\sum_i}\sum_f \delta^{(4)}(q+p'-p)\langle p|\mathcal{J}_\nu^{(0)}(0)|p'\rangle\langle p'|J_\mu^\gamma(0)|p\rangle(\Omega E_p) \tag{50.10}$$

The lepton traces are evaluated in Prob. 50.1. The result is[66]

$$\eta_{\mu\nu}^{(1)} = \eta_{\mu\nu}^{(2)} = -2(b\eta_{\mu\nu} + a\varepsilon_{\mu\nu\rho\sigma}k_{1\rho}k_{2\sigma}) \tag{50.11}$$

Thus in the numerator of Eq. (50.9) one needs $\eta_{\mu\nu}^{(1)}(W_{\mu\nu}^{(1)} + W_{\mu\nu}^{(2)})$ and

$$\begin{aligned} W_{\mu\nu}^{(1)} + W_{\mu\nu}^{(2)} &= (2\pi)^3 \overline{\sum_i}\sum_f \delta^{(4)}(q+p'-p)\Big[\langle p|J_\nu^\gamma(0)|p'\rangle\langle p'|\mathcal{J}_\mu^{(0)}(0)|p\rangle \\ &\quad + \langle p|\mathcal{J}_\nu^{(0)}(0)|p'\rangle\langle p'|J_\mu^\gamma(0)|p\rangle\Big](\Omega E_p) \end{aligned} \tag{50.12}$$

Now separate

$$\mathcal{J}_\mu^{(0)} = J_\mu^{(0)} + J_{\mu 5}^{(0)}\,; \qquad \mathrm{V-A} \tag{50.13}$$

Since the asymmetry is already explicitly of order $Gq^2/4\pi\alpha\sqrt{2}$, one can then use the good parity of the nuclear states to write

$$W_{\mu\nu}^{(1)} + W_{\mu\nu}^{(2)} = W_{\mu\nu}^{\mathrm{int}} + W_{\mu\nu}^{\mathrm{A-V}} \tag{50.14}$$

Here

1. The first term $W_{\mu\nu}^{\mathrm{int}}$ comes from $J_\mu^{(0)}$; it has the same general structure as $W_{\mu\nu}^\gamma$ in Eq. (50.6)[67]

$$\begin{aligned} W_{\mu\nu}^{\mathrm{int}} &= W_1^{\mathrm{int}}(q^2, q\cdot p)\left(\delta_{\mu\nu} - \frac{q_\mu q_\nu}{q^2}\right) \\ &\quad + W_2^{\mathrm{int}}(q^2, q\cdot p)\frac{1}{M_T^2}\left(p_\mu - \frac{p\cdot q}{q^2}q_\mu\right)\left(p_\nu - \frac{p\cdot q}{q^2}q_\nu\right) \end{aligned} \tag{50.15}$$

[66] Note that the first term is symmetric in $\mu \leftrightarrow \nu$ while the second term is antisymmetric.
[67] The proof of this result uses the fact that the current $J_\mu^{(0)}$ is conserved.

2. The second term, coming from $J^{(0)}_{\mu 5}$, is a *pseudotensor*; the only pseudotensor that can be constructed from the two four-vectors (p_μ, q_μ) is [68]

$$W^{A-V}_{\mu\nu} = W_8(q^2, q\cdot p)\frac{1}{M_T^2}\varepsilon_{\mu\nu\rho\sigma}p_\rho q_\sigma \tag{50.16}$$

Now combine these expressions with Eq. (50.11). The result follows from simple algebra and kinematics (Ref. [N14], Prob. 50.2). The only nonzero terms are

$$\begin{aligned} 2\eta_{\mu\nu}W_{\mu\nu} &= 4\varepsilon_1\varepsilon_2[W_2^\gamma \cos^2\frac{\theta}{2} + 2W_1^\gamma \sin^2\frac{\theta}{2}] \\ -2b\eta_{\mu\nu}W^{\text{int}}_{\mu\nu} &= (-b)4\varepsilon_1\varepsilon_2[W_2^{\text{int}} \cos^2\frac{\theta}{2} + 2W_1^{\text{int}} \sin^2\frac{\theta}{2}] \end{aligned} \tag{50.17}$$

and

$$\begin{aligned} &(-2a\varepsilon_{\mu\nu\rho\sigma}k_{1\rho}k_{2\sigma})\left[W_8(q^2, q\cdot p)\frac{1}{M_T^2}\varepsilon_{\mu\nu\alpha\beta}p_\alpha q_\beta\right] = \\ &\left(\frac{2a}{M_T}W_8\right)4\varepsilon_1\varepsilon_2 \sin\frac{\theta}{2}\left(q^2\cos^2\frac{\theta}{2} + \mathbf{q}^2\sin^2\frac{\theta}{2}\right)^{1/2} \end{aligned} \tag{50.18}$$

The final result is

$$\begin{aligned} &\left[\frac{d\sigma_\uparrow - d\sigma_\downarrow}{d\sigma_\uparrow + d\sigma_\downarrow}\right]\left[W_2^\gamma \cos^2\frac{\theta}{2} + 2W_1^\gamma \sin^2\frac{\theta}{2}\right] = \\ &\frac{Gq^2}{4\pi\alpha\sqrt{2}}\left\{b\left[W_2^{\text{int}}\cos^2\frac{\theta}{2} + 2W_1^{\text{int}}\sin^2\frac{\theta}{2}\right]\right. \\ &\left. -a\left(\frac{2W_8}{M_T}\right)\sin\frac{\theta}{2}\left(q^2\cos^2\frac{\theta}{2} + \mathbf{q}^2\sin^2\frac{\theta}{2}\right)^{1/2}\right\} \end{aligned} \tag{50.19}$$

Several features of this result are of interest:

1. This is the general expression for the parity-violating asymmetry in relativistic polarized electron scattering from a hadronic target arising from the interference of one-photon and one-Z exchange (Fig. 50.1).[69]

2. The left-hand side is the product of the asymmetry $\mathcal{A}$ [Eq. (50.9)] and the basic (e, e') response [Eqs. (50.6) and (50.17)].

3. The characteristic scale of parity violation in nuclear physics from the process $(\vec{e}, e')$ is set by the parameter $Gq^2/4\pi\alpha\sqrt{2}$ appearing on the right-hand side.

[68] Note that this expression is antisymmetric in $\mu \leftrightarrow \nu$.

[69] Additional contributions to the parity-violating asymmetry can arise from parity admixtures in the nuclear states coming from weak parity-violating nucleon-nucleon interactions. These contributions are generally negligible, except perhaps at very small q^2 (Refs. [W31, W32]).

4. The parameter b characterizes the lepton axial-vector weak neutral current [Eq. (50.2)]; its coefficient here arises from the interference of the vector part of the weak neutral and electromagnetic hadronic currents [Eqs. (50.12), (50.13), and (50.15)]

$$\begin{aligned} W_{\mu\nu}^{\text{int}} &= (2\pi)^3 \overline{\sum_i} \sum_f \delta^{(4)}(q+p'-p) \left[\langle p|J_\nu^\gamma(0)|p'\rangle\langle p'|J_\mu^{(0)}(0)|p\rangle \right. \\ &\quad \left. + \langle p|J_\nu^{(0)}(0)|p'\rangle\langle p'|J_\mu^\gamma(0)|p\rangle \right] (\Omega E_p) \\ &= W_1^{\text{int}}(q^2, q\cdot p)\left(\delta_{\mu\nu} - \frac{q_\mu q_\nu}{q^2}\right) \\ &\quad + W_2^{\text{int}}(q^2, q\cdot p)\frac{1}{M_T^2}\left(p_\mu - \frac{p\cdot q}{q^2}q_\mu\right)\left(p_\nu - \frac{p\cdot q}{q^2}q_\nu\right) \end{aligned} \tag{50.20}$$

5. The parameter a characterizes the lepton vector weak neutral current [Eq. (50.2)]; its coefficient here arises from the interference of the axial vector part of the weak neutral and electromagnetic hadronic currents [Eqs. (50.12)-(50.14) and (50.16)]

$$\begin{aligned} W_{\mu\nu}^{\text{A}-\text{V}} &= (2\pi)^3 \overline{\sum_i} \sum_f \delta^{(4)}(q+p'-p) \left[\langle p|J_\nu^\gamma(0)|p'\rangle\langle p'|J_{\mu 5}^{(0)}(0)|p\rangle \right. \\ &\quad \left. + \langle p|J_{\nu 5}^{(0)}(0)|p'\rangle\langle p'|J_\mu^\gamma(0)|p\rangle \right] (\Omega E_p) \\ &= W_8(q^2, q\cdot p)\frac{1}{M_T^2}\varepsilon_{\mu\nu\rho\sigma}p_\rho q_\sigma \end{aligned} \tag{50.21}$$

6. The three response functions on the right-hand side of Eq. (50.19) can be separated by varying the electron scattering angle θ at fixed $(q^2, q\cdot p)$.[70]

7. The parity violation arises from the interference of the transition matrix element of the electromagnetic and the weak neutral currents. If the electromagnetic matrix elements have been measured, then *parity violation in* $(\vec{e}, e')$ *provides a measurement of the matrix elements of the weak neutral current in nuclei at all* q^2.

An Example — $(\vec{e}, e)$ *from a* 0^+ *Target.* We give one example (Ref. [N14]). Consider elastic scattering from a 0^+ target (Fig. 50.3a). Then from Lorentz invariance and current conservation the transition matrix elements of the electromagnetic and weak neutral currents must have the form[71]

$$\begin{aligned} \langle p'|J_\mu^\gamma(0)|p\rangle &= \left(\frac{m^2}{EE'\Omega^2}\right)^{1/2} F_0^\gamma(q^2)\frac{1}{m}\left(p_\mu - \frac{p\cdot q}{q^2}q_\mu\right) \\ \langle p'|J_\mu^{(0)}(0)|p\rangle &= \left(\frac{m^2}{EE'\Omega^2}\right)^{1/2} F_0^{(0)}(q^2)\frac{1}{m}\left(p_\mu - \frac{p\cdot q}{q^2}q_\mu\right) \end{aligned}$$

[70] This is known as a *Rosenbluth* separation.

[71] Hermiticity of the current implies that the form factors, as defined here, are real.

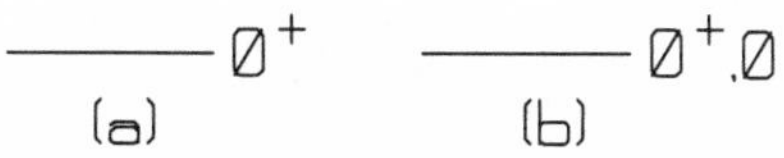

Figure 50.3: Example of parity-violating asymmetry in scattering from (a) $J^\pi = 0^+$, and (b) $J^\pi, T = 0^+, 0$ target.

$$\langle p'|J_{\mu 5}^{(0)}(0)|p\rangle \quad = \quad 0 \tag{50.22}$$

The last relation follows since it is impossible to construct an axial vector from only two four-vectors (p_μ, q_μ).

Insertion of these relations in the defining equations yields

$$\begin{aligned} W_1^{\text{int}} &= W^{\text{A-V}} = 0 \\ \mathcal{A} &= \frac{Gq^2}{4\pi\alpha\sqrt{2}} b \frac{2F_0^{(0)}(q^2)}{F_0^\gamma(q^2)} \end{aligned} \tag{50.23}$$

Hence

$$\mathcal{A} \quad = \quad -\frac{Gq^2}{2\pi\alpha\sqrt{2}} \frac{F_0^{(0)}(q^2)}{F_0^\gamma(q^2)} \tag{50.24}$$

This expression allows one to measure the ratio of the weak neutral current and electromagnetic form factors — the latter measures the distribution of electromagnetic charge in the 0^+ target, and the former the distribution of weak neutral charge.

Nuclear Domain. Now suppose that, in addition, the target has isospin $T = 0$ (Fig. 50.3b). Then only isoscalar operators can contribute to the matrix elements. In the nuclear domain of (u, d) quarks and antiquarks, the only isoscalar piece of the weak neutral current in the standard model arises from the electromagnetic current itself, and hence in this case (Section 44)

$$J_\mu^{(0)} \quad \doteq \quad -2\sin^2\theta_W J_\mu^\gamma \tag{50.25}$$

This implies

$$F_0^{(0)}(q^2) \quad = \quad -2\sin^2\theta_W F_0^\gamma(q^2) \tag{50.26}$$

The ratio of form factors is then the constant $-2\sin^2\theta_W$ at all q^2 — a truly remarkable prediction![72] Insertion of this equality in the expression for the asymmetry leads to (Ref. [W33])

$$\mathcal{A} \quad = \quad \frac{Gq^2}{\pi\alpha\sqrt{2}} \sin^2\theta_W \tag{50.27}$$

[72]This result depends on the assumption of isospin invariance that is broken to $O(\alpha)$ in nuclei.

Several comments are of interest:

1. It is important to note that this result holds *to all orders in the strong interactions (QCD).*

2. This expression is linear in q^2 with a coefficient that depends only on fundamental constants.

3. It can be used to measure $\sin^2\theta_W$ in the low-energy quark sector, complementing other measurements of this quantity.

4. It can be used to test the remarkable prediction in Eq. (50.26) that holds in the nuclear domain.

5. A measurement of this parity-violating asymmetry for elastic scattering from ^{12}C at $q = 150$ MeV has been carried out in a tour de force experiment at the Bates Laboratory (Ref. [W34]). Take

$$\begin{aligned} q &= 150 \text{ MeV} & \sin^2\theta_W &= 0.2325 \\ \alpha^{-1} &= 137.0 & G &= \frac{1.027\times 10^{-5}}{m_p^2} \\ \mathcal{A} &= 1.882\times 10^{-6} \end{aligned} \tag{50.28}$$

Then, with an electron beam polarization P_e, one has (Refs. [W34, W35])

$$\begin{aligned} \mathcal{A}P_e &= 0.696\times 10^{-6} & &; \text{ theory } (P_e = 0.37) \\ \mathcal{A}P_e &= 0.60 \pm 0.14 \pm 0.02 \times 10^{-6} & &; \text{ experiment} \end{aligned} \tag{50.29}$$

The first error is statistical.

This experiment provides the prototype for the next generation of electron scattering parity-violation $(\vec{e}, e')$ studies.

Extended Domain. Consider next the *extended domain* of (u, d, s, c) quarks and their antiquarks. The standard model then has an additional isoscalar term in the weak neutral current (Section 44)

$$\delta\mathcal{J}_\mu^{(0)} = \frac{i}{2}[\bar{c}\gamma_\mu(1+\gamma_5)c - \bar{s}\gamma_\mu(1+\gamma_5)s] \tag{50.30}$$

This leads to an additional contribution $\delta F_0^{(0)}$ in the form factor in Eq. (50.26); the asymmetry for elastic scattering of polarized electrons on a $0^+, 0$ nucleus such as ^{4}He then takes the form

$$\mathcal{A} = \frac{Gq^2}{\pi\alpha\sqrt{2}}\sin^2\theta_W\left[1 - \frac{\delta F_0^{(0)}(q^2)}{2\sin^2\theta_W F_0^\gamma(q^2)}\right] \tag{50.31}$$

The additional weak neutral current form factor comes from the vector current in Eq. (50.30) — expected to arise predominantly from the much lighter strange quarks. Hence one has a direct measure of the *strangeness current* in nuclei. The total strangeness of the nucleon must vanish in the strong and electromagnetic sector, and hence $\delta F_0^{(0)}(0) = 0$; however, just as with electromagnetic charge in the neutron, there can be a strangeness *distribution*, which is determined in this experiment.

The measurement of weak neutral currents in nuclei through $(\vec{e}, e')$ will be one of CEBAF's most important roles (Refs. [W36, W38]).

References

[W1] E. D. Cummins and P. H. Buchsbaum, *Weak Interactions of Leptons and Quarks*, Cambridge University Press, Cambridge, England (1983)

[W2] J. D. Walecka, *Semileptonic Weak Interactions in Nuclei*, in *Muon Physics* Vol. II, eds V. W. Hughes and C. S. Wu, Academic Press, New York (1975), pp. 113-218

[W3] S. Weinberg, *Phys. Rev. Lett.* **19**, 1264 (1967)

[W4] S. Weinberg, *Phys. Rev.* **D5**, 1412 (1972)

[W5] A. Salam and J. C. Ward, *Phys. Lett.* **13**, 168 (1964)

[W6] S. L. Glashow, J. Iliopoulos, and L. Maiani, *Phys. Rev.* **D2**, 1285 (1970)

[W7] E. S. Abers and B. W. Lee, *Phys. Rep.* **C9**, 1 (1973)

[W8] N. Cabbibo, *Phys. Rev. Lett.* **10**, 531 (1963)

[W9] T. W. Donnelly and W. C. Haxton, *Atomic Data and Nuclear Data Tables* **23**, 103 (1979)

[W10] T. W. Donnelly and W. C. Haxton, *Atomic Data and Nuclear Data Tables* **25**, 1 (1980)

[W11] J. N. Bahcall and M. H. Pinsonneault, *Rev. Mod. Phys.* **64**, 885 (1992)

[W12] A. Burrows, *Proc. Int. Conf. on Weak and Electromagnetic Interactions in Nuclei* (WIEN-89), ed. P. Depommier, Éditions Frontière, Gif-sur-Yvette, Cedex-France (1989), p. 373

[W13] KARMEN Collaboration, *Phys. Lett.* **B280**, 198 (1992)

[W14] R. C. Allen, H. H. Chen, P. J. Doe, R. Hausammann, W. P. Lee, X.-Q. Lu, H. J. Mahler, M. E. Potter, K. C. Wang, T. J. Bowles, R. L. Burman, R. D. Carlini, D. R. F. Cochran, J. S. Frank, E. Piasetzky, V. D. Sandberg, D. A. Krakauer, and R. L. Talaga, *Phys. Rev. Lett.* **64**, 1871 (1990)

[W15] D. A. Krakauer *et al.*, *Phys. Rev.* **C45**, 2450 (1992)

[W16] T. W. Donnelly and J. D. Walecka, *Annu. Rev. Nucl. Sci.* **25**, 329 (1975)

[W17] J. D. Walecka, *Weak Interactions - 1977*, ed. D. B. Lichtenberg, *A. I. P. Conf. Proc.* **37**, A. I. P., New York (1977), p. 125

[W18] T. W. Donnelly and J. D. Walecka, *Phys. Lett.* **41B**, 275 (1972)

[W19] T. W. Donnelly and J. D. Walecka, *Phys. Lett.* **44B**, 330 (1973)

[W20] T. W. Donnelly and J. D. Walecka, *Nucl. Phys.* **A274**, 368 (1976)

[W21] E. J. Kim, *Phys. Lett.* **B198**, 9 (1987)

[W22] T.-P. Cheng and L.-F. Li, *Gauge Theory of Elementary Particle Physics*, Clarendon Press, Oxford, England (1984)

[W23] J. Calmet, S. Narison, M. Perrottet, and E. de Rafael, *Rev. Mod. Phys.* **49**, 21 (1977)

[W24] J. Bailey, K. Borer, F. Combley, H. Drumm, F. J. M. Farley, J. H. Field, W. Flegel, P. M. Hattersley, F. Krienen, F. Lange, E. Picasso, and W. Von Rüden, *Phys. Lett.* **68B**, 191 (1977)

[W25] K. Aoki, Z. Hioki, R. Kawabe, M. Konuma, and T. Muta, *Suppl. Prog. Th. Phys.* **73**, 1-225 (1982)

[W26] M. Kobayashi and T. Maskawa, *Prog. Th. Phys. Jpn.*, **49**, 652 (1973)

[W27] E. M. Henley, *Proc. Int. Conf. on Weak and Electromagnetic Interactions in Nuclei (WEIN-89)*, ed. P. Depommier, Éditions Frontières, Gif-sur-Yvette, France (1989), p. 181

[W28] J. Ng, *Proc. Int. Conf. on Weak and Electromagnetic Interactions in Nuclei (WEIN-89)*, ed. P. Depommier, Éditions Frontières, Gif-sur-Yvette, France (1989), p. 167

[W29] C. Y. Prescott, W. B. Atwood, R. L. A. Cottrell, H. DeStaebler, E. L. Garwin, A. Gonidec, R. H. Miller, L. S. Rochester, T. Sato, D. J. Shereden, C. K. Sinclair, S. Stein, R. E. Taylor, J. E. Clendenin, V. W. Hughes, N. Sasao, K. P. Schüler, M. G. Borghini, L. Lübelsmeyer, and W. Jentschke, *Phys. Lett.* **77B**, 347 (1978); (with C. Young) **84B**, 524 (1979)

[W30] *Parity Violation in Electron Scattering*, Proc. Workshop at Cal. Inst. of Tech. Feb. 23-24, 1990, eds. E. J. Biese and R. D. McKeown, World Scientific, Singapore (1990)

[W31] B. D. Serot, *Nucl. Phys.* **A322**, 408 (1979)

[W32] V. Dmitrašinović, *Nucl. Phys.* **A537**, 551 (1992)

[W33] G. Feinberg, *Phys. Rev.* **D12**, 3575 (1975)

[W34] P. A. Souder, R. Holmes, D.-H. Kim, K. S. Kumar, M. E. Schulze, K. Isakovich, G. W. Dodson, K. A. Dow, M Farkhondeh, S. Kowalski, M. S. Lubell, J. Bellanca, M. Goodman, S. Patch, R. Wilson, G. D. Cates, S. Dhawan, T. J. Gay, V. W. Hughes, A. Magnon, R. Michaels, and H. R. Schaefer, *Phys. Rev. Lett.* **65**, 694 (1990)

[W35] E. Moniz, private communication

[W36] J. D. Walecka, *Overview of CEBAF Scientific Program*, CEBAF Summer Workshop, June 15, 1992, *A.I.P. Conf. Proc.* **269**, eds. F. Gross and R. Holt, A.I.P., New York (1993), pp. 87-136

[W37] J. F. Donoghue, E. Golowich, and B. R. Holstein, *Dynamics of the Standard Model*, Cambridge University Press, New York (1993)

[W38] M. J. Musolf, T. W. Donnelly, J. Dubach, S. J. Pollock, S. Kowalski, and E. J. Biese, *Phys. Rep.* **239**, 1 (1994)

PROBLEMS: PART IV

42.1. Equation (42.41) is the S-matrix for pion decay $\pi^- \to l^- + \bar{\nu}_l$ (Fig. 42.7).
(a) Verify the general form of the hadronic matrix element of the current in Eq. (42.40).
(b) Use the Dirac equation to show $\bar{u}(\vec{k}_1)\gamma_\lambda p_\lambda(1+\gamma_5)v(-\vec{k}_2) = im_l\bar{u}(\vec{k}_1)(1+\gamma_5)v(-\vec{k}_2)$.
(c) Asssume a big box with p.b.c. so that $[(2\pi)^3/\Omega]\delta^{(3)}(\vec{p}-\vec{k}_1-\vec{k}_2) \doteq \delta\vec{p},\vec{k}_1+\vec{k}_2$. Let $\vec{k}$ be the momentum of the l^- in the pion rest frame. Show the decay rate is

$$d\omega = 2\pi\delta(W_f - W_i)\frac{G^2F_\pi^2}{4\omega_p\Omega}\frac{\Omega d^3k}{(2\pi)^3}\frac{m_l^2}{4\varepsilon_1\varepsilon_2}\text{tr}[(1+\gamma_5)(-i\gamma_\lambda k_{2\lambda})(1-\gamma_5)(m_l - i\gamma_\rho k_{1\rho})]$$

(d) With $W_f = k + (k^2+m_l^2)^{1/2}$ and $W_i = m_\pi$, show $dW_f/dk = m_\pi/\varepsilon_1$.
(e) Show $\text{tr}\{\cdots\} = -8(k_1 \cdot k_2) = 8km_\pi$ in (c).
(f) Hence show the pion decay rate for this channel is given by Eq. (42.42).
(g) It is observed that $\omega(\pi^- \to \mu^- + \bar{\nu}_\mu) = 3.841 \times 10^7\ \text{sec}^{-1}$. Use $G_\mu = 1.024 \times 10^{-5}/m_p^2$, $m_\mu = 105.7\,\text{MeV}$, and the numbers in Section 20 to show $F_\pi \equiv m_\pi\bar{f} = 0.92m_\pi$.

42.2. Consider a nuclear β-decay transition $\{0^+, T, M_T\} \to \{0^+, T, M_T \pm 1\}$.
(a) Prove this transition can proceed only through the Lorentz vector part of the weak charge-changing hadronic current $\hat{J}_\lambda^{(\pm)}$.
(b) Take $p' = p - q$. Use the Heisenberg equations of motion to show as $q_\lambda \to 0$ that $\langle p'|\hat{J}_\lambda^{\vec{V}}(0)|p\rangle \to (1/\Omega)\langle p|\int d^3x\,\hat{J}_\lambda^{\vec{V}}(\vec{x})|p\rangle$.
(c) As in Section 21, identify $\int d^3x\,\hat{J}_0^{\vec{V}}(\vec{x}) = \hat{\vec{T}}$ as the strong isospin operator (CVC). Show that in the target rest frame $\Omega\langle p|\hat{J}_\lambda^{(\pm)}(0)|p\rangle = i\delta_{\lambda 4}\sqrt{(T \mp M_T)(T \pm M_T + 1)}$. Hence conclude that the hadronic matrix element is known exactly in this case.

42.3. Assume the matrix element of the electromagnetic current in Eq. (42.37) were to be augmented by a term of the form $(i/\Omega)\bar{u}(p')[iF_3(q^2)\gamma_5\sigma_{\mu\nu}q_\nu]u(p)$.
(a) Use hermiticity of the EM current to prove that the form factors F_i are all real.
(b) Use invariance under parity $\hat{P}$ (Refs. [R2-R4]) to prove $F_3 = 0$.
(c) Use invariance under time reversal $\hat{T}$ (Refs. [R2-4]) to seperately prove $F_3 = 0$.
(d) Make a nonrelativistic reduction of the current and give a physical interpretation of the term in F_3.

42.4. (a) Use Lorentz invariance, the Dirac equation, and strong isospin invariance to derive the general form of the single-nucleon matrix element of the Lorentz vector, isovector current in the first of Eqs. (42.35).
(b) Assume the symmetry properties of the current in Eq. (42.21) under strong isospin and the hermiticity properties of $\hat{J}_\lambda^{(\pm)}$ in Eq. (42.15); prove F_1, F_2, iF_S are real.
(c) Assume the properties of the current in Eq. (42.21) under time reversal; show $F_S = 0$.
(d) Show that current conservation also implies $F_S = 0$ if $q^2 \neq 0$.

42.5. (a) Repeat Prob. 42.4(a) for the Lorentz axial vector, isovector current.
(b) Repeat Prob. 42.4(b) in this case; prove F_A, F_P, iF_T are real.
(c) Repeat Prob. 42.4(c) in this case; prove $F_T = 0$ (see Prob. 42.3).

42.6. Use the interaction in Eq. (42.16) to compute the rate for μ decay $\mu^- \rightarrow e^-(\vec{p}) + \nu_\mu(\vec{k}) + \bar{\nu}_e(\vec{q})$.
(a) Show the differential decay rate is (here $E_{\max} \equiv W_0 \approx m_\mu/2$)

$$d\omega_{fi} = \frac{4G^2}{(2\pi)^5}\delta(W_f - m_\mu)p^2\,dp\,d\Omega_p k^2\,dk\,d\Omega_k\left(1 - \frac{p}{E}\cos\theta_{\vec{p},\vec{k}}\right)$$

(b) Show the electron spectrum is $d\omega_{fi} = (G^2/6\pi^3)m_\mu p\,dE[3E(W_0 - E) + p^2]$.
(c) Assume relativistic electrons; show the integrated rate is $\omega_{fi} = G^2 W_0^5/6\pi^3$.

Prob. 43.1. Consider the charge-changing weak reactions (ν_l, l^-) and $(\bar{\nu}_l, l^+)$ on a hadronic target [denoted $(\nu, l\mp)$]. Use the effective interaction in Eq. (42.30), and remember the initial neutrinos are polarized.
(a) Show the analogue of Eq. (39.7) is

$$d\sigma_{\nu,l\mp} = \frac{G^2}{2\pi^2}\frac{1}{\sqrt{(k_1\cdot p)^2}}\frac{d^3k_2}{2\varepsilon_2}\overline{\eta}_{\mu\nu}\overline{W}_{\mu\nu}$$

Here $\overline{\eta}_{\mu\nu}$ contains the additional term $\pm\varepsilon_{\mu\nu\rho\sigma}k_{1\rho}k_{2\sigma}$, and $\overline{W}_{\mu\nu}$ is calculated from the matrix elements of $\{\mathcal{J}_\nu^{(\mp)}(0), \mathcal{J}_\mu^{(\pm)}(0)\}$. [Note Eq. (44.20).]
(b) Assume the ERL ($m_l \rightarrow 0$); assume also that $m_\pi \rightarrow 0$ so that the weak current is conserved. Show the form of the target response tensor in Eq. (39.11) is now augmented by a term $\overline{W}_3(q^2, q\cdot p)(1/m^2)\varepsilon_{\mu\nu\rho\sigma}p_\rho q_\sigma$. Hence derive the analog of Eq. (39.19) (Ref. [W2])

$$\left(\frac{d^2\sigma}{d\Omega_2 d\varepsilon_2}\right)^{\text{ERL}}_{\nu,l\mp} = \frac{G^2\varepsilon_2^2}{2\pi^2}\frac{1}{m}\left[\overline{W}_2\cos^2\frac{\theta}{2} + 2\overline{W}_1\sin^2\frac{\theta}{2}\right.$$
$$\left.\mp\left(\frac{2\overline{W}_3}{m}\right)\left(q^2\cos^2\frac{\theta}{2} + \vec{q}^{\,2}\sin^2\frac{\theta}{2}\right)^{1/2}\sin\frac{\theta}{2}\right]$$

43.2. Assume a transition to a discrete state of mass M_T^* in (e, e') (Ref. [N14]).
(a) Show the cross section in Eq. (39.19) is $(d\sigma/d\Omega) = \sigma_M[w_2(q^2) + 2w_1(q^2)\tan^2\theta/2]r$ where $r^{-1} \equiv 1 + (2\varepsilon_1/M_T)\sin^2\theta/2$. Here the response tensor in Eq. (39.8) is now the covariant expression $w_{\mu\nu} = \overline{\sum_i}\sum_f'(EE'\Omega^2/M_T^2)\langle i|J_\nu(0)|f\rangle\langle f|J_\mu(0)|i\rangle$, which has the tensor structure of Eq. (39.11) with coefficients $w_i(q^2)$.
(b) Consider elastic scattering from a $J^\pi = 0^+$ target. Construct the general form of the matrix element of the current; show that $w_1 = 0$ and $w_2 = |F_0(q^2)|^2$.
(c) Consider elastic scattering from a $J^\pi = (1/2)^+$ target. Use the general form of the matrix element of the current in Eq. (42.37); show $w_1 = q^2(F_1 + 2mF_2)^2/4m^2$ and $w_2 = F_1^2 + q^2(2mF_2)^2/4m^2$. Hence derive the celebrated Rosenbluth cross section.
(d) Restore the spatial dependence to the matrix elements through the use of the Heisenberg equations of motion and derive Eqs. (41.1).

43.3. Consider the weak charge changing exclusive reactions $\nu_l + n \to l^- + p$ and $\bar{\nu}_l + p \to l^+ + n$. Use the general form of the matrix elements of the weak currents in Eq. (42.35); show the cross section is given in the ERL by (Ref. [W2])

$$\left(\frac{d\sigma}{d\Omega}\right)^{\rm ERL}_{\nu,l\mp} = \frac{G^2\varepsilon^2}{2\pi^2}\left\{\left[F_1^2 + F_A^2 + \frac{q^2}{4m^2}(2mF_2)^2 + \frac{q^2}{4m^2}(2mF_T)^2\right]\cos^2\frac{\theta}{2}\right.$$
$$+2\left[F_A^2\left(1+\frac{q^2}{4m^2}\right) + \frac{q^2}{4m^2}(F_1+2mF_2)^2\right]\sin^2\frac{\theta}{2}$$
$$\left.\mp\frac{2F_A}{m}(F_1+2mF_2)(q^2\cos^2\frac{\theta}{2}+\vec{q}^{\,2}\sin^2\frac{\theta}{2})^{1/2}\sin\frac{\theta}{2}\right\} r$$

The term in F_T (assumed real) is absent in the standard model. [Note Eq. (44.20).]

43.4. (a) Use Eq. (43.39) to write the lowest order S-matrix for $\nu_\mu + e^- \to \nu_e + \mu^-$.
(b) Repeat (a) for $\nu_e + e^- \to \nu_e + e^-$. (*Hint*: there are two Feynman diagrams here.)
(c) Calculate the C-M cross section for (a).

43.5. Use Eq. (43.39) to compute the rate for $Z^0 \to \nu_l + \bar{\nu}_l$ in the standard model. (*Note*: this reaction can be used to determine the number of ν_l with $2m_{\nu_l} < M_Z$.)

43.6. Rewrite the interaction in $\mathcal{L}_{\rm gauge}$ [Eq. (43.34)] in terms of the physical vector meson fields. Discuss.

44.1. (a) Consider the Yukawa coupling $\mathcal{L}_{\rm int} = G_u(\bar{\underline{q}}_L\tilde{\underline{\phi}})u_R + G_c(\bar{\underline{Q}}_L\tilde{\underline{\phi}})c_R + \text{h.c.}$ Show this is invariant under $SU(2)_W \bigotimes U(1)_W$.
(b) Show that with spontaneous symmetry breaking, in the unitary gauge, this term gives rise to masses and Higgs couplings for the (u, c) quarks.
(c) Construct all $SU(2)_W \bigotimes U(1)_W$ invariant quark Yukawa couplings from the fields in Tables 43.1 and 44.1.
(d) Discuss how the diagonal quark mass matrix arises from these couplings.

44.2. Include the possibility of production of heavy quark flavor in the reactions (ν_l, l^-) and $(\bar{\nu}_l, l^+)$ through the additional currents in Eq. (44.16), relax the condition of conservation of axial vector current, and stay in the ERL for the leptons. Show the results in Prob. 43.1 still hold.

44.3. It is an empirical observation that the strangeness-changing, charge-changing weak hadronic currents in nuclear physics satisfy $\Delta S = \Delta Q$ and $|\Delta\vec{T}| = 1/2$. Derive these selection rules from the currents in Eq. (44.16).

45.1. Derive the expansion of $e_{\vec{q}0}e^{i\vec{q}\cdot\vec{x}}$ in Eqs. (45.10) and (45.9).

45.2. Derive the nonrelativistic reduction of the single-nucleon matrix element of the weak current in Eq. (45.31). This is the analog of Eq. (8.22) and employs the same assumptions.

45.3. Derive the long-wavelength reduction of $\hat{\mathcal{L}}_{JM}$ in Eqs. (45.35) and (45.36).

45.4. Give the nuclear selection rules for the allowed Fermi and Gamow-Teller operators in Eq. (45.44).

45.5. (a) Derive the improved form of the effective electromagnetic currrent operator given in the footnote at the end of Section 45; this incorporates the scaling of the Sachs form factors, which holds to very high q^2.

(b) Discuss a corresponding extension of the effective weak current in Eq. (45.47).

46.1. One line in Table 46.2 is derived in the text in Eqs. (46.4), (46.5), and (46.29). Verify the remaining entries.

46.2. (a) As in Section 42, show that the effective interaction for electron scattering from a nuclear target with one photon exchange is given by the Møller potential

$$\langle f|\hat{H}^{\gamma}|i\rangle = \frac{-ee_p}{\Omega} i\bar{u}(k_2)\gamma_\lambda u(k_1)\frac{1}{q_\mu^2}\langle f|\int d^3x e^{-i\vec{q}\cdot\vec{x}}\hat{J}_\lambda^{\gamma}(\vec{x})|i\rangle$$

Here $q = p - p' = k_2 - k_1$.
(b) Use the analysis in Section 46 to now derive the (e, e') cross section in Eq. (7.76).

46.3. (a) Show that if nuclear recoil is allowed in the density of final states, the lepton capture rate in Eqs. (46.28) and (46.30) is multiplied by $r = (1 + \nu/M_T)^{-1}$.
(b) Repeat part (a) for β-decay; determine r.

46.4. The longitudinal polarization of the emitted $e^-(e^+)$ in β-decay is defined as $P_\uparrow = (N_\uparrow - N_\downarrow)/(N_\uparrow + N_\downarrow)$. Construct helicity projection operators from $\vec{\sigma}\cdot\vec{p}$ to use in the lepton traces, and show that in the ERL one finds $[P_\uparrow]_{\beta\mp}^{\text{ERL}} = \mp 1$.

46.5. Consider β^--decay with the kinematics in Fig. 46.6a; work in the allowed limit $\kappa \equiv |\vec{q}| \to 0$ where only the multipoles in Eq. (46.40) remain;
(a) Start from the Golden Rule and show explicitly

$$\frac{2\pi}{2J_i+1}\sum_{M_i}\sum_{M_f}|\langle f|\hat{H}_W|i\rangle|^2 = \frac{4\pi^2G^2}{2J_i+1}\left\{|\langle J_f||\hat{\mathcal{L}}_1||J_i\rangle|^2\vec{l}\cdot\vec{l}^* + |\langle J_f||\hat{\mathcal{M}}_0||J_i\rangle|^2 l_0 l_0^*\right\}$$

(b) Evaluate $(\Omega^2/2)\sum_{\text{lepton spins}} l_0 l_0^* = 1 + \hat{\nu}\cdot\vec{\beta}$ and $(\Omega^2/2)\sum_{\text{lepton spins}}\vec{l}\cdot\vec{l}^* = 3 - \hat{\nu}\cdot\vec{\beta}$.
(c) Show the density of final states is $\int_{\text{neutrino energy}}\delta(W_f - W_i)\Omega^2(2\pi)^{-6}d^3k\,d^3\nu = \Omega^2(2\pi)^{-6}k\varepsilon\,d\varepsilon\,d\Omega_k(W_0 - \varepsilon)^2 d\Omega_\nu$.
Hence independently derive the allowed β-decay rate in Eq. (46.41).

46.6. (a) Show that a μ^- moving entirely inside a uniform spherically symmetric charge distribution feels a three-dimensional simple harmonic oscillator potential; hence compute the spectrum of this muonic atom (Fig. 46.5).
(b) The case where the muon moves entirely outside the charge distribution reduces to the Bohr atom. Interpolate the spectrum of the muonic atom between these two limiting cases of nuclear size and discuss.

47.1 Use the analysis in Sections 6-8 to find $\langle(1p_{3/2})^{-1}1p_{1/2};1^+,10||\hat{T}_1^{\text{mag}}(q)||0^+,0\rangle$ in Eq. (47.4).

47.2. Prove that when calculated with valence particles in the p-shell and s.h.o. wave functions, the matrix element in Eq. (47.6) must yield a straight line in y.

47.3. Assume a configuration $(1s_{1/2})^{-1}$ for the three nucleon system and s.h.o. wave functions. Reproduce the EM results in Table 47.5 and Fig. 47.10.

47.4. Repeat Prob. 47.3 for the weak rates in Table 47.5: (a) for β-decay; (b) for μ-capture.

47.5. Consider the cross sections for $^4\text{He}(\nu_l,\nu_l)^4\text{He}$ and $^4\text{He}(\bar{\nu}_l,\bar{\nu}_l)^4\text{He}$ in the standard model; work to lowest order in G:

(a) Prove that in the nuclear domain with strong isospin invariance this cross section is determined by ${}^4\text{He}(e,e){}^4\text{He}$; derive the relation between the differential cross sections.
(b) How is this relation modified in the full standard model?

47.6. Calculate the rate for $\mu^- + p \rightarrow n + \nu_\mu$ using the single-nucleon matrix elements of the weak current in Eqs. (45.20) and (45.24) and full kinematics.
(a) Show that for statistical occupancy of the initial atomic hyperfine states one has [Note Eq. (44.20)]

$$\begin{aligned} \bar{\omega}_\mu &= \frac{G^2\nu^2}{2\pi}|\phi_{1s}(0)|^2 \left\{ F_1^2 + F_A^2\left(1 - \frac{\nu}{2m}\frac{2mm_\mu}{q^2+m_\pi^2}\right)^2 \right. \\ &\quad \left. +2\left[F_A - \frac{\nu}{2m}(F_1 + 2mF_2)\right]^2\right\}\left(1+\frac{\nu}{m}\right)^{-1} \end{aligned}$$

(b) Calculate the rate from the individual hyperfine states (see, e.g., Ref. [W20]).

47.7. Extend the analysis of the relativistic Hartree single-particle matrix elements of the EM current in Prob. 17.2 to include the contributions of the additional weak currents in Eq. (45.47).

47.8. Reduce the S-matrix $S_{fi}^{(1)}$ for $\nu_l + {}^2\text{H} \rightarrow n + p + \nu_l$ and $\nu_l + {}^2\text{H} \rightarrow p + p + l^-$ to the evaluation of a nuclear matrix element; here $S^{(1)}$ is exact to lowest order in G. Work in the standard model. Discuss the evaluation of the nuclear matrix element in: (a) traditional nuclear physics; (b) QHD; and (c) QCD. These reactions are the basis for the new Sudbury solar neutrino detector (SNO).

47.9. Extend the Fermi gas results in Prob. 17.3 to compute the quasielastic nuclear response to the neutrino reaction (ν_l, l^-). Assume $N > Z$ and introduce separate Fermi momenta for the neutrons and protons with $\rho_p + \rho_n = \rho =$ const. Take the Fermi energy to be $\varepsilon_F = k_{Fn}^2/2m = k_{Fp}^2/2m + \varepsilon_0$.
(a) Show the appropriate replacement in Prob. 17.3(b) is

$$\int e^{-i\vec{q}\cdot\vec{x}}\hat{\rho}_N^{(+)}(\vec{x})d^3x = \sum\nolimits_{\vec{k}\sigma} a^\dagger_{\vec{k}-\vec{q},\sigma,p}a_{\vec{k},\sigma,n}$$

Introduce dimensionless variables $\vec{\Delta} \equiv \vec{q}/k_{Fn}, \xi \equiv m\omega_{\text{eff}}/k_{Fn}^2$ with $\omega_{\text{eff}} \equiv \omega - \varepsilon_0$, and $\lambda \equiv k_{Fp}/k_{Fn}$. Show :
(b) If $\Delta > 1 + \lambda$ (proton and neutron spheres do not intersect) and $\Delta/2 + 1 > \xi/\Delta > \Delta/2 - 1$

$$\left(\frac{3N}{4\pi}\frac{m}{k_{Fn}^2}\right)^{-1} R^{(+)}(q,\omega) = \frac{\pi}{\Delta}\left[1 - \left(\frac{\xi}{\Delta} - \frac{\Delta}{2}\right)^2\right] \equiv \mathcal{R}_\text{I}$$

(c) If $1 + \lambda > \Delta > 1 - \lambda$ (proton and neutron spheres intersect)
(i) If $\Delta/2 + 1 > \xi/\Delta > \lambda - \Delta/2$, one has $\mathcal{R}_\text{I}$.
(ii) If $\lambda - \Delta/2 > \xi/\Delta > -(1-\lambda^2)/2\Delta$

$$\left(\frac{3N}{4\pi}\frac{m}{k_{Fn}^2}\right)^{-1} R^{(+)}(q,\omega) = \frac{\pi}{\Delta}[(1-\lambda^2) + 2\xi] \equiv \mathcal{R}_\text{II}$$

(d) If $1 - \lambda > \Delta > 0$ (proton sphere inside neutron sphere)

(i) If $\Delta/2+1>\xi/\Delta>\lambda-\Delta/2$, one has $\mathcal{R}_{\rm I}$.
(ii) If $\lambda-\Delta/2>\xi/\Delta>-(\lambda+\Delta/2)$, one has $\mathcal{R}_{\rm II}$.
(iii) If $-(\lambda+\Delta/2)>\xi/\Delta>-(1-\Delta/2)$, one has $\mathcal{R}_{\rm I}$.
(e) Recover the results in Prob. 17.3 as $\lambda\to 1$.
(f) Sketch and discuss (see Ref. [W2]).

47.10. Use Prob. 47.9 to extend the Coulomb sum rule in Prob. 17.4 to (ν_l, l^-). Define $C^{(+)}(q)\equiv(1/N)\int_0^\infty d\omega R^{(+)}(q,\omega)$. Show

$$\begin{aligned} C^{(+)}(q) &= 1 && ; \quad \Delta>1+\lambda \\ &= \frac{1}{2}(1-\lambda^2)+\frac{3}{8}(1+\lambda^2)\Delta-\frac{1}{16}\Delta^3+\frac{3(1-\lambda^2)^2}{16\Delta} && ; \quad 1+\lambda>\Delta>1-\lambda \\ &= 1-\lambda^2 && ; \quad 1-\lambda>\Delta>0 \end{aligned}$$

Recover the result in Prob. 17.4 as $\lambda\to 1$.

47.11. In coincidence reactions $(e,e'X)$ or (ν_l, l^-X) the final-state interaction of the emitted hadrons must be taken into account. The optical potential for doing this is analyzed in Probs. 15.6-10. The Glauber approximation[73] then provides an excellent high energy approximation for determining the scattering state wave function.
(a) Look for a solution $\psi=u/r$ to the radial Schrödinger equation in Prob. 1.3 of the form $u_l(r)=e^{\pm ik\phi(r)}$. Show that for large k, one can write the solution $\phi(r)=r-r_0+\int_{r_0}^r dr[(1-v_{\rm eff}(r)/k^2)^{1/2}-1]+\phi(r_0)$. Here $v_{\rm eff}(r)\equiv v(r)+l(l+1)/r^2$.
(b) Assume the radial solution vanishes at the classical turning point $v_{\rm eff}(r_0)=k^2$ and write $u_l\approx a\{e^{ik[\phi(r)-\phi(r_0)]}-e^{-ik[\phi(r)-\phi(r_0)]}\}$. Identify the phase shift through the asymptotic form in Probs. 1.3-4 to get $\delta_l=l\pi/2-kr_0+k\int_{r_0}^\infty dr[(1-v_{\rm eff}(r)/k^2)^{1/2}-1]$.
(c) To satisfy the condition $\delta_l=0$ when $v=0$, it is necessary to allow the wave function to slightly leak into the barrier; hence define

$$\begin{aligned} \delta_l^{\rm WKB} &\equiv (l+\frac{1}{2})\frac{\pi}{2}-kr_0+k\int_{r_0}^\infty dr\left[\left(1-\frac{v_{\rm eff}(r)}{k^2}\right)^{1/2}-1\right] \\ v_{\rm eff}(r) &= v(r)+\frac{(l+\frac{1}{2})^2}{r^2} \qquad\qquad v_{\rm eff}(r_0)=k^2 \end{aligned}$$

Show $\delta_l^{\rm WKB}=0$ when $v=0$.
(d) Define the impact parameter by $l+\frac{1}{2}=kb$. Let $v(r)/k^2\to 0$ and show

$$\delta_l^{\rm WKB}\ \overset{v/k^2\to 0}{\longrightarrow}\ \delta_l^{\rm Glauber}\equiv-\frac{1}{4k}\int_{-\infty}^\infty dz\, v(\sqrt{b^2+z^2})$$

Here the Glauber phase shift is calculated by integrating on a straight line eikonal trajectory through the potential at impact parameter b.

47.12. Write $2i\delta_l^{\rm Glauber}\equiv i\chi(b,k)=-(i/2k)\int_{-\infty}^\infty dz\, v(\sqrt{b^2+z^2})$.

[73] R. Glauber, *Lectures in Theoretical Physics Vol. I*, Interscience, New York (1959) (see also Ref. [N42]).

(a) Justify replacing $\sum_l \to k\int db$; use Heine's relation $\text{Lim}_{l\to\infty} P_l(1-z^2/2l^2) = J_0(z)$; recall $\vec{q}^{\,2} = 2k^2(1-\cos\theta)$; and hence show at high energy the scattering amplitude in Prob. 1.3 can be written

$$f(k,\theta) \to \frac{k}{i}\int_0^\infty b\,db\,J_0(qb)[e^{i\chi(b,k)} - 1] = \frac{k}{2\pi i}\int d^2b\, e^{-i\vec{q}\cdot\vec{b}}[e^{i\chi(b,k)} - 1]$$

The last expression is the (transverse) two-dimensional Fourier transform of the eikonal.
(b) Change variables to $d\cos\theta = -q\,dq/k^2$, and assume the scattering amplitude falls off fast enough so that one can perform $\int d^2q$ over the entire transverse plane. Show

$$\sigma_{\text{el}} = \int d^2b|e^{i\chi(k,b)} - 1|^2$$

(c) Use the optical theorem $\text{Im}\, f_{\text{el}}(0) = (k/4\pi)\sigma_{\text{tot}}$ (Ref. [N42]) to show

$$\sigma_{\text{tot}} = \int d^2b\, 2\,\text{Re}(1 - e^{i\chi(k,b)})$$

$$\sigma_{\text{r}} \equiv \sigma_{\text{tot}} - \sigma_{\text{el}} = \int d^2b[1 - |e^{i\chi(b,k)}|^2]$$

47.13. Construct the incoming wave $\psi_{\vec{k}}^{(-)}(\vec{x})$ that is to be used for calculating final-state interactions (Section 12) in Glauber's approximation.

47.14. The matrix elements of the nuclear weak currents in the traditional nuclear physics picture in Section 45 can be calculated through the use of the results in Probs. 8.3-4 and the following relation (Ref. [W2])

$$\begin{aligned}\langle n'(l'\tfrac{1}{2})j'||M_J\vec{\nabla}\cdot\vec{\sigma}||n(l\tfrac{1}{2})j\rangle &= \sum\nolimits_{J'}(-1)^{J'-J}[6(2j+1)(2j'+1)(2J'+1)]^{1/2} \\ &\quad\times\begin{Bmatrix} l' & l & J' \\ \frac{1}{2} & \frac{1}{2} & 1 \\ j' & j & J\end{Bmatrix}\langle n'l'||\vec{M}_{J'J}\cdot\vec{\nabla}||nl\rangle\end{aligned}$$

(a) Derive this result (see Ref. [N44]).
(b) Use the result in Prob. 8.4 to obtain an explicit expression in terms of radial integrals.

48.1. Express the term in $\mathcal{L}_{\text{gauge}} \doteq -(g/2)\vec{F}_{\mu\nu}\cdot(\vec{A}_\mu\times\vec{A}_\nu)$ in the physical fields (Prob. 43.6) and derive the covariant $[W_\lambda^\star(k_3)W_\nu(k_2)A_\mu(k_1)]$ vertex in Eq. (48.8).

48.2. Derive the contribution of the exchange of a scalar Higgs in Fig. 48.6g to the anomalous magnetic moment of the muon in Eq. (48.10) and Table 48.1.

49.1. Demonstrate the approximate unitarity of the parameterization of the mixing matrix $\underset{\sim}{U}$ in Eq. (49.13).

49.2. Introduce weak isodoublets whose lower components are the fully mixed (d,s,b) expressions in Eq. (49.12). Construct the weak neutral current, and show that it is diagonal in flavor.

50.1. Establish the following traces for massless (relativistic) electrons [see Eqs. (50.10) and (50.6)]:

(a) tr $\gamma_\mu\gamma_\nu\gamma_\rho\gamma_\sigma\gamma_5 = 4\varepsilon_{\mu\nu\rho\sigma}$
(b) $\eta^{(1)}_{\mu\nu} = \frac{-1}{2}\text{tr}\ \gamma_\nu(-i\gamma_\rho k_{2\rho})(a\gamma_\mu + b\gamma_\mu\gamma_5)(-\gamma_5)(-i\gamma_\sigma k_{1\sigma}) = -2(b\eta_{\mu\nu} + a\varepsilon_{\mu\nu\rho\sigma}k_{1\rho}k_{2\sigma})$
(c) $\eta^{(2)}_{\mu\nu} = \frac{-1}{2}\text{tr}\ (a\gamma_\nu + b\gamma_\nu\gamma_5)(-i\gamma_\rho k_{2\rho})\gamma_\mu(-\gamma_5)(-i\gamma_\sigma k_{1\sigma}) = -2(b\eta_{\mu\nu} + a\varepsilon_{\mu\nu\rho\sigma}k_{1\rho}k_{2\sigma})$
(d) Hence conclude

$$\eta^{(1)}_{\mu\nu} = \eta^{(2)}_{\mu\nu} = -2(b\eta_{\mu\nu} + a\varepsilon_{\mu\nu\rho\sigma}k_{1\rho}k_{2\sigma})$$
$$\eta_{\mu\nu} = k_{2\nu}k_{1\mu} + k_{2\mu}k_{1\nu} - k_1\cdot k_2\delta_{\mu\nu}$$

50.2. Derive Eqs. (50.17) and (50.18).

50.3. Assume the single-nucleon matrix element of the weak neutral current has the form $\langle p'|\mathcal{J}^{(0)}_\mu(0)|p\rangle = (i/\Omega)\bar{u}(p')[F_1^{(0)}\gamma_\mu + F_2^{(0)}\sigma_{\mu\nu}q_\nu + F_A^{(0)}\gamma_5\gamma_\mu - iF_P^{(0)}\gamma_5 q_\mu]u(p)$. Assume the matrix element of the electromagnetic current has the form in Eq. (42.37). Show that for relativistic electrons the parity-violating asymmetry for $N(\vec{e},e)N$ has the form (here $G_{\text{M}} \equiv F_1 + 2mF_2$)[74]

$$\mathcal{A}\left\{[(F_1^\gamma)^2 + q^2(F_2^\gamma)^2]\cos^2\frac{\theta}{2} + \frac{q^2}{2m^2}(G_{\text{M}}^\gamma)^2\sin^2\frac{\theta}{2}\right\} = -\frac{Gq^2}{2\pi\alpha\sqrt{2}}$$
$$\times\left\{[F_1^{(0)}F_1^\gamma + q^2F_2^{(0)}F_2^\gamma]\cos^2\frac{\theta}{2} + \frac{q^2}{2m^2}G_{\text{M}}^{(0)}G_{\text{M}}^\gamma\sin^2\frac{\theta}{2}\right.$$
$$\left. - \frac{\sin\theta/2}{m}\sqrt{q^2\cos^2\frac{\theta}{2} + \vec{q}^{\,2}\sin^2\frac{\theta}{2}}\ G_{\text{M}}^\gamma(1 - 4\sin^2\theta_W)F_A^{(0)}\right\}$$

50.4. Discuss the form factors $F_i^{(0)}$ in Prob. 50.3 within the standard model under the following assumptions about the strong interactions:
(a) Point nucleons;
(b) QCD in the nuclear domain of (u,d) quarks and strong isospin invariance;
(c) QCD in the extended domain of (u,d,s,c) quarks and strong isospin invariance.

50.5. Consider $\nu_l + {}^4\text{He} \to \nu_l + {}^4\text{He}$. Calculate the differential cross section under the following assumptions about the strong interactions:
(a) QCD in the nuclear domain of (u,d) quarks and strong isospin invariance;
(b) QCD in the extended domain of (u,d,s,c) quarks and strong isospin invariance.
(c) Discuss the relation to the parity violation measurement ${}^4\text{He}(\vec{e},e){}^4\text{He}$.

50.6. Carry out the following simplified calculation of ${}^2\text{H}(\vec{e},e')$ in the deep inelastic region (Ref. [N14]):
(a) Assume forward angles with $\theta_e \to 0$; assume also $\sin^2\theta_W \approx 1/4$. Show $\mathcal{A} = -(Gq^2/4\pi\alpha\sqrt{2})[\nu W_2(\nu,q^2)^{\text{int}}/\nu W_2(\nu,q^2)^\gamma]$.
(b) For a nucleon, assume just three valence quarks and identical quark distributions; use the quark parton model (Section 39) to reduce the required ratio of structure functions to a ratio of charges $[\sum_i 2Q_i^\gamma Q_i^{(0)}]/[\sum_i(Q_i^\gamma)^2]$.
(c) For the deuteron, take an incoherent sum of structure functions and show $\mathcal{A}_{2\text{H}} = -(Gq^2/4\pi\alpha\sqrt{2})\,2\,[(\sum_i Q_i^\gamma Q_i^{(0)})_p + (\sum_i Q_i^\gamma Q_i^{(0)})_n]\,/\,[(\sum_i Q_i^\gamma)_p^2 + (\sum_i Q_i^\gamma)_n^2]$.

[74] S. J. Pollock, Ph.D. Thesis, Stanford University (1987).

(d) Hence show that under these assumptions[75]

$$\mathcal{A}_{^2\mathrm{H}} = -\frac{Gq^2}{2\pi\alpha\sqrt{2}}\frac{2}{5}$$

(e) Compare with the experimental results in Refs. [W29].
The last two problems go beyond the standard model:

50.7. Consider just the first two lepton families in Eq. (49.10). Suppose the neutrinos (ν_e, ν_μ) have a small mass, and in analogy with the quarks, these leptons enter the weak interactions with some mixing angle $\cos\theta_{\mathrm{CW}}$. Discuss the consequences.

50.8. Suppose there were a very heavy charged vector boson $\tilde{W}_\mu$, right-handed neutrinos $(\nu_l)_{\mathrm{R}}$, and an interaction with coupling $(1-\gamma_5)$ in Eqs. (43.39), (43.41), and (43.42).
(a) Discuss the experimental consequences.
(b) Compute the allowed rate for $n \to p + e^- + (\nu_e)_{\mathrm{R}}$ in terms of $(\tilde{g}, M_{\tilde{W}})$.

[75] For a better treatment see R. N. Cahn and F. J. Gilman, *Phys. Rev.* **D17**, 1313 (1978).

APPENDICES: PART IV

M Standard Model Currents

This appendix details the algebra leading from the initial lagrangian to the interaction of the fermion currents with the physical gauge fields. Consider first the leptons.

Leptons. The lepton lagrangian is

$$\mathcal{L}_{\text{lepton}} = -\bar{L}\gamma_\mu \left(\frac{\partial}{\partial x_\mu} + \frac{ig'}{2}B_\mu - \frac{ig}{2}\vec{\tau}\cdot\vec{A}_\mu \right) L - \bar{R}\gamma_\mu \left(\frac{\partial}{\partial x_\mu} + ig'B_\mu \right) R \tag{M.1}$$

Insert the definitions of the initial gauge fields $(B_\mu, \mathbf{A}_\mu)$ in terms of the physical fields

$$\frac{1}{\sqrt{2}}(A_\mu^{(1)} + iA_\mu^{(2)}) = W_\mu^\star \qquad A_\mu^{(3)} = \frac{-gZ_\mu + g'A_\mu}{(g^2+g'^2)^{1/2}}$$
$$\frac{1}{\sqrt{2}}(A_\mu^{(1)} - iA_\mu^{(2)}) = W_\mu \qquad B\mu = \frac{g'Z_\mu + gA_\mu}{(g^2+g'^2)^{1/2}} \tag{M.2}$$

Recall $\tau_\pm \equiv (\tau_1 \pm i\tau_2)/2$ and use

$$\vec{\tau}\cdot\vec{A}_\mu \equiv \sqrt{2}(\tau_+ W_\mu + \tau_- W_\mu^\star) + \tau_3 A_\mu^{(3)} \tag{M.3}$$

One can thus immediately identify the charge-changing lepton current

$$\begin{aligned}\mathcal{L}_{\text{lepton}}^{(\pm)} &= \frac{ig}{\sqrt{2}}(\bar{L}\gamma_\mu\tau_+ L\, W_\mu + \bar{L}\gamma_\mu\tau_- L\, W_\mu^\star) \\ &= \frac{ig}{2\sqrt{2}}[\bar{\psi}_l\gamma_\mu(1+\gamma_5)\tau_+\psi_l W_\mu + \bar{\psi}_l\gamma_\mu(1+\gamma_5)\tau_-\psi_l W_\mu^\star]\end{aligned} \tag{M.4}$$

Hence

$$\begin{aligned}\mathcal{L}_{\text{lepton}}^{(\pm)} &= \frac{g}{2\sqrt{2}}(j_\mu^{(+)}W_\mu + j_\mu^{(-)}W_\mu^\star) \\ j_\mu^{(\pm)} &= i\bar{\psi}_l\gamma_\mu(1+\gamma_5)\tau_\pm\psi_l\ ; \qquad \psi_l = \begin{pmatrix} \nu_e \\ e \end{pmatrix}\end{aligned} \tag{M.5}$$

There are additive contributions to this current from the other lepton doublets.

Next collect coefficients of A_μ

$$\mathcal{L}_{\text{lepton}}^{\gamma} = \frac{igg'}{(g^2+g'^2)^{1/2}}A_\mu \left[\bar{L}\gamma_\mu(-\frac{1}{2}+\frac{1}{2}\tau_3)L - \bar{R}\gamma_\mu R\right] \tag{M.6}$$

Note the form of the expression in square brackets and recall the definition of the electric charge operator for the fermions as $\hat{Q} = (\hat{T}_3 + \hat{Y}/2)_W$. Use

$$-\bar{e}_{\text{L}}\gamma_\mu e_{\text{L}} - \bar{e}_{\text{R}}\gamma_\mu e_{\text{R}} = -\bar{e}\gamma_\mu e \tag{M.7}$$

Hence the electromagnetic interaction is

$$\begin{aligned} \mathcal{L}^{\gamma}_{\text{lepton}} &= e_p j^{\gamma}_{\mu} A_{\mu} \\ j^{\gamma}_{\mu} &= (-)i\bar{\psi}_e \gamma_{\mu} \psi_e \end{aligned} \tag{M.8}$$

This is just the lagrangian of QED! We have defined

$$e_p \equiv \frac{gg'}{(g^2+g'^2)^{1/2}} > 0 \tag{M.9}$$

Note that here $e_e = -e_p$, and there is again an additive term in the current for each charged lepton.

Finally, collect the coefficients of Z_{μ}

$$\begin{aligned} \mathcal{L}^{(0)}_{\text{lepton}} &= \frac{-iZ_{\mu}}{2(g^2+g'^2)^{1/2}} \underbrace{[\bar{L}\gamma_{\mu}(g'^2 + g^2\tau_3)L + \bar{R}\gamma_{\mu}(2g'^2)R]} \\ &= \bar{L}\gamma_{\mu}(g^2+g'^2)\tau_3 L + \underbrace{\bar{L}\gamma_{\mu}(g'^2)(1-\tau_3)L + \bar{R}\gamma_{\mu}(2g'^2)R} \\ &= 2g'^2(\bar{e}_{\text{L}}\gamma_{\mu}e_{\text{L}} + e_{\text{R}}\gamma_{\mu}e_{\text{R}}) \end{aligned} \tag{M.10}$$

Hence

$$\begin{aligned} \mathcal{L}^{(0)}_{\text{lepton}} &= -\frac{(g^2+g'^2)^{1/2}}{2} Z_{\mu} j^{(0)}_{\mu} \\ j^{(0)}_{\mu} &= i\left[\bar{L}\gamma_{\mu}\tau_3 L + \frac{2g'^2}{(g^2+g'^2)}\bar{\psi}_e\gamma_{\mu}\psi_e\right] \\ &= i\bar{\psi}_l\gamma_{\mu}(1+\gamma_5)\frac{1}{2}\tau_3\psi_l - \frac{2g'^2}{(g^2+g'^2)} j^{\gamma}_{\mu} \end{aligned} \tag{M.11}$$

The introduction of the weak mixing angle in Fig. 43.2 then gives the final form of the weak neutral current interaction of the leptons

$$\begin{aligned} \mathcal{L}^{(0)}_{\text{lepton}} &= -\frac{g}{2\cos\theta_W} Z_{\mu} j^{(0)}_{\mu} \\ j^{(0)}_{\mu} &= i\bar{\psi}_l\gamma_{\mu}(1+\gamma_5)\frac{1}{2}\tau_3\psi_l - 2\sin^2\theta_W j^{\gamma}_{\mu} \end{aligned} \tag{M.12}$$

Point Nucleons. Start from the lagrangian for point nucleons

$$\begin{aligned} \mathcal{L}_{\text{nucleon}} &= -\bar{N}_{\text{L}}\gamma_{\mu}\left(\frac{\partial}{\partial x_{\mu}} - \frac{ig'}{2}B_{\mu} - \frac{ig}{2}\vec{\tau}\cdot\vec{A}_{\mu}\right)N_{\text{L}} \\ &\quad -\bar{p}_{\text{R}}\gamma_{\mu}\left(\frac{\partial}{\partial x_{\mu}} - ig'B_{\mu}\right)p_{\text{R}} - \bar{n}_{\text{R}}\gamma_{\mu}\frac{\partial}{\partial x_{\mu}}n_{\text{R}} \end{aligned} \tag{M.13}$$

Now repeat the above calculation. The charge-changing interaction follows immediately as

$$\begin{aligned} \mathcal{L}^{(\pm)}_{\text{nucleon}} &= \frac{g}{2\sqrt{2}}(\mathcal{J}^{(+)}_{\mu}W_{\mu} + \mathcal{J}^{(-)}_{\mu}W^{\star}_{\mu}) \\ \mathcal{J}^{(\pm)}_{\mu}(\text{nucleon}) &= i\bar{\psi}\gamma_{\mu}(1+\gamma_5)\tau_{\pm}\psi\,; \qquad \psi = \begin{pmatrix} p \\ n \end{pmatrix} \end{aligned} \tag{M.14}$$

The electromagnetic current in this case is identified as[76]

$$
\begin{aligned}
J_\mu^\gamma &= i\left[\bar{N}_{\rm L}\gamma_\mu(\frac{1}{2}+\frac{1}{2}\tau_3)N_{\rm L}+\bar{p}_{\rm R}\gamma_\mu p_{\rm R}\right] \\
&= i[\bar{p}_{\rm L}\gamma_\mu p_{\rm L}+\bar{p}_{\rm R}\gamma_\mu p_{\rm R}] = i\bar{p}\gamma_\mu p \qquad \text{(M.15)}
\end{aligned}
$$

Hence

$$
\begin{aligned}
\mathcal{L}^\gamma_{\rm nucleon} &= e_p J_\mu^\gamma A_\mu \\
J_\mu^\gamma(\text{nucleon}) &= i\bar{p}\gamma_\mu p = i\bar{\psi}\gamma_\mu\frac{1}{2}(1+\tau_3)\psi \qquad \text{(M.15)}
\end{aligned}
$$

The weak neutral current is identified through

$$
\begin{aligned}
\mathcal{L}^{(0)}_{\rm nucleon} &= \frac{iZ_\mu}{2(g^2+g'^2)^{1/2}}\underbrace{[\bar{N}_{\rm L}\gamma_\mu(g'^2-g^2\tau_3)N_{\rm L}+\bar{p}_{\rm R}\gamma_\mu(2g'^2)p_{\rm R}]} \\
&= \bar{N}_{\rm L}\gamma_\mu[-(g^2+g'^2)\tau_3+g'^2(1+\tau_3)]N_{\rm L}+\bar{p}_{\rm R}\gamma_\mu(2g'^2)p_{\rm R} \\
&= -(g^2+g'^2)\bar{N}_{\rm L}\gamma_\mu\tau_3N_{\rm L}+2g'^2\bar{p}\gamma_\mu p \qquad \text{(M.16)}
\end{aligned}
$$

Hence

$$
\begin{aligned}
\mathcal{L}^{(0)}_{\rm nucleon} &= -\frac{(g^2+g'^2)^{1/2}}{2}\mathcal{J}_\mu^{(0)}Z_\mu \\
\mathcal{J}_\mu^{(0)}(\text{nucleon}) &= i\left[\bar{N}_{\rm L}\gamma_\mu\tau_3N_{\rm L}-\frac{2g'^2}{(g^2+g'^2)}\bar{p}\gamma_\mu p\right] \\
&= i\bar{\psi}\gamma_\mu(1+\gamma_5)\frac{1}{2}\tau_3\psi-\frac{2g'^2}{(g^2+g'^2)}J_\mu^\gamma(\text{nucleon}) \qquad \text{(M.17)}
\end{aligned}
$$

With the introduction of the weak mixing angle these expressions become

$$
\begin{aligned}
\mathcal{L}^{(0)}_{\rm nucleon} &= -\frac{g}{2\cos\theta_W}\mathcal{J}_\mu^{(0)}Z_\mu \\
\mathcal{J}_\mu^{(0)}(\text{nucleon}) &= i\bar{\psi}\gamma_\mu(1+\gamma_5)\frac{1}{2}\tau_3\psi-2\sin^2\theta_W J_\mu^\gamma(\text{nucleon}) \qquad \text{(M.18)}
\end{aligned}
$$

Quarks. When the hadronic structure is described in terms of quarks, the lagrangian of the standard model takes the form

$$
\begin{aligned}
\mathcal{L}_{\rm quark} = &-\bar{q}_{\rm L}\gamma_\mu\left(\frac{\partial}{\partial x_\mu}-\frac{ig'}{2}(\frac{1}{3})B_\mu-\frac{ig}{2}\vec{\tau}\cdot\vec{A}_\mu\right)q_{\rm L} \qquad \text{(M.19)} \\
&-\bar{Q}_{\rm L}\gamma_\mu\left(\frac{\partial}{\partial x_\mu}-\frac{ig'}{2}(\frac{1}{3})B_\mu-\frac{ig}{2}\vec{\tau}\cdot\vec{A}_\mu\right)Q_{\rm L} \\
&-\bar{u}_{\rm R}\gamma_\mu\left(\frac{\partial}{\partial x_\mu}-\frac{ig'}{2}(\frac{4}{3})B_\mu\right)u_{\rm R}-\bar{c}_{\rm R}\gamma_\mu\left(\frac{\partial}{\partial x_\mu}-\frac{ig'}{2}(\frac{4}{3})B_\mu\right)c_{\rm R} \\
&-\bar{d}_{\rm R}\gamma_\mu\left(\frac{\partial}{\partial x_\mu}-\frac{ig'}{2}(\frac{-2}{3})B_\mu\right)d_{\rm R}-\bar{s}_{\rm R}\gamma_\mu\left(\frac{\partial}{\partial x_\mu}-\frac{ig'}{2}(\frac{-2}{3})B_\mu\right)s_{\rm R}
\end{aligned}
$$

[76]Recall again that $\hat{Q} = (\hat{T}_3+\hat{Y}/2)_W$.

The electroweak interactions of the quarks then follow *exactly as above.* The charge-changing weak interaction is given by

$$\begin{aligned}
\mathcal{L}^{(\pm)}_{\text{quark}} &= \frac{g}{2\sqrt{2}}(\mathcal{J}^{(+)}_\mu W_\mu + \mathcal{J}^{(-)}_\mu W^\star_\mu) \qquad \text{(M.20)}\\
\mathcal{J}^{(\pm)}_\mu(\text{quark}) &= i\bar{q}\gamma_\mu(1+\gamma_5)\tau_\pm q + i\bar{Q}\gamma_\mu(1+\gamma_5)\tau_\pm Q\\
q &\equiv \begin{pmatrix} u \\ d\cos\theta_C + s\sin\theta_C\end{pmatrix} \qquad Q \equiv \begin{pmatrix} c \\ -d\sin\theta_C + s\cos\theta_C\end{pmatrix}
\end{aligned}$$

The electromagnetic interaction of the quarks follows as[77]

$$\begin{aligned}
\mathcal{L}^{\gamma}_{\text{quark}} &= \frac{igg'}{(g^2+g'^2)^{1/2}}A_\mu\Big[\bar{q}_L\gamma_\mu(\frac{1}{6}+\frac{1}{2}\tau_3)q_L + \bar{Q}_L\gamma_\mu(\frac{1}{6}+\frac{1}{2}\tau_3)Q_L\\
&\quad +\frac{4}{6}(\bar{u}_R\gamma_\mu u_R + \bar{c}_R\gamma_\mu c_R) - \frac{2}{6}(\bar{d}_R\gamma_\mu d_R + \bar{s}_R\gamma_\mu s_R)\Big] \qquad \text{(M.21)}
\end{aligned}$$

Hence

$$\begin{aligned}
\mathcal{L}^{\gamma}_{\text{quark}} &= e_p J^\gamma_\mu A_\mu\\
J^\gamma_\mu(\text{quark}) &= i\left[\frac{2}{3}(\bar{u}\gamma_\mu u + \bar{c}\gamma_\mu c) - \frac{1}{3}(\bar{d}\gamma_\mu d + \bar{s}\gamma_\mu s)\right] \qquad \text{(M.22)}
\end{aligned}$$

The weak neutral current interaction is given by[78]

$$\begin{aligned}
\mathcal{L}^{(0)}_{\text{quark}} &= \frac{iZ_\mu}{2(g^2+g'^2)^{1/2}}\Big[\bar{q}_L\gamma_\mu(\frac{g'^2}{3} - g^2\tau_3)q_L + \bar{Q}_L\gamma_\mu(\frac{g'^2}{3} - g^2\tau_3)Q_L\\
&\quad +\frac{4g'^2}{3}(\bar{u}_R\gamma_\mu u_R + \bar{c}_R\gamma_\mu c_R) - \frac{2g'^2}{3}(\bar{d}_R\gamma_\mu d_R + \bar{s}_R\gamma_\mu s_R)\Big]\\
&= -\frac{i(g^2+g'^2)^{1/2}}{2}Z_\mu\Big\{\bar{q}_L\gamma_\mu\tau_3 q_L + \bar{Q}_L\gamma_\mu\tau_3 Q_L\\
&\quad -\frac{2g'^2}{(g^2+g'^2)}\left[\frac{2}{3}(\bar{u}\gamma_\mu u + \bar{c}\gamma_\mu c) - \frac{1}{3}(\bar{d}\gamma_\mu d + \bar{s}\gamma_\mu s)\right]\Big\} \qquad \text{(M.23)}
\end{aligned}$$

It follows that

$$\begin{aligned}
\mathcal{L}^{(0)}_{\text{quark}} &= -\frac{(g^2+g'^2)^{1/2}}{2}\mathcal{J}^{(0)}_\mu Z_\mu \qquad \text{(M.24)}\\
\mathcal{J}^{(0)}_\mu(\text{quark}) &= i\left[\bar{q}\gamma_\mu(1+\gamma_5)\frac{1}{2}\tau_3 q + \bar{Q}\gamma_\mu(1+\gamma_5)\frac{1}{2}\tau_3 Q\right] - \frac{2g'^2}{(g^2+g'^2)}J^\gamma_\mu
\end{aligned}$$

With the introduction of the weak mixing angle these expressions become

$$\begin{aligned}
\mathcal{L}^{(0)}_{\text{quark}} &= -\frac{g}{2\cos\theta_W}\mathcal{J}^{(0)}_\mu Z_\mu \qquad \text{(M.25)}\\
\mathcal{J}^{(0)}_\mu(\text{quark}) &= i\left[\bar{q}\gamma_\mu(1+\gamma_5)\frac{1}{2}\tau_3 q + \bar{Q}\gamma_\mu(1+\gamma_5)\frac{1}{2}\tau_3 Q\right] - 2\sin^2\theta_W J^\gamma_\mu(\text{quark})
\end{aligned}$$

[77] Recall again that $\hat{Q} = (\hat{T}_3 + \hat{Y}/2)_W$.
[78] Here in the first line we write $g'^2/3 - g^2\tau_3 \equiv g'^2/3 + g'^2\tau_3 - (g^2+g'^2)\tau_3$.

Table N.1: Convention comparison table.

Bjorken and Drell		Present text
$g_{\mu\nu} = \begin{pmatrix} 1 & 0 & 0 & 0 \\ 0 & -1 & 0 & 0 \\ 0 & 0 & -1 & 0 \\ 0 & 0 & 0 & -1 \end{pmatrix}$	$\leftrightarrow^a$	$\delta_{\mu\nu}$
$a^\mu = (a^0, \vec{a})$	$\leftrightarrow$	$a_\mu = (a_1, a_2, a_3, a_4) = (\mathbf{a}, ia_0)$
$a_\mu b^\mu = g_{\mu\nu} a^\mu b^\nu = a^0 b^0 - \vec{a} \cdot \vec{b}$	$\leftrightarrow$	$a_\mu b_\mu = \vec{a} \cdot \vec{b} - a^0 b^0$
$x^\mu = (t, \vec{x})$	$\leftrightarrow$	$x_\mu = (\vec{x}, it)$
$x_\mu = g_{\mu\nu} x^\nu = (t, -\vec{x})$	$\leftrightarrow$	$x^\mu \equiv x_\mu$
$\partial_\mu = \partial/\partial x^\mu = (\partial/\partial t, \vec{\nabla})$	$\leftrightarrow$	$\partial/\partial x_\mu = (\vec{\nabla}, \partial/i\partial t)$
$\gamma^\mu = (\beta, \beta\vec{\alpha})$	$\leftrightarrow$	$\gamma_\mu = (i\vec{\alpha}\beta, \beta)$
$\gamma^\mu\gamma^\nu + \gamma^\nu\gamma^\mu = 2g^{\mu\nu}$	$\leftrightarrow$	$\gamma_\mu\gamma_\nu + \gamma_\nu\gamma_\mu = 2\delta_{\mu\nu}$
$\gamma^{\mu\dagger} = \gamma^0\gamma^\mu\gamma^0$	$\leftrightarrow$	$\gamma_\mu^\dagger = \gamma_\mu$
$(i\gamma^\mu\partial_\mu - M)\psi = 0$	$\leftrightarrow$	$(\gamma_\mu\partial/\partial x_\mu + M)\psi = 0$
$(k_\mu\gamma^\mu - M)u(k) = 0$	$\leftrightarrow$	$(i\gamma_\mu k_\mu + M)u(k) = 0$
$\gamma^5 = i\gamma^0\gamma^1\gamma^2\gamma^3 = \gamma_5$	$\leftrightarrow$	$\gamma_5 = \gamma_1\gamma_2\gamma_3\gamma_4$
$\sigma^{\mu\nu} = \frac{i}{2}[\gamma^\mu, \gamma^\nu]$	$\leftrightarrow$	$\sigma_{\mu\nu} = \frac{1}{2i}[\gamma_\mu, \gamma_\nu]$

[a] Note $g_{\mu\nu} = g^{\mu\nu}$.

N Metric and Convention Conversion Tables

In this appendix we give a set of convention and metric conversion tables between the present text and some other standard references.

Comparison with Conventions of Bjorken and Drell. A comparison of the conventions and metric used in Bjorken and Drell with those in the present text is shown in Table N.1.

Metric Conversion. It follows that the conversion of expressions presented in the metric of Bjorken and Drell (used in Refs. [R1], [W1], and [W22]) to results in the metric used in this text (and in Ref. [R4]) is obtained by the substitutions shown in Table N.2 (see Ref. [R1]).

Electroweak Convention Conversion. The conversion between the conventions used in Refs. [W22, W1] for the standard model, and those in the present text, is shown in Table N.3.

Table N.2: Metric conversion table (from Ref. [R1]).

Bjorken and Drell Refs. [R1], [W1, W22]		Present metric[a]
$a_\mu b^\mu$	$\rightarrow$	$-a_\mu b_\mu$
$g_{\mu\nu}$	$\rightarrow$	$-\delta_{\mu\nu}$
γ^μ	$\rightarrow$ [b]	$i\gamma_\mu$
∂^μ	$\rightarrow$	$-\partial/\partial x_\mu$
γ_5	$\rightarrow$	$-\gamma_5$
$\partial_\mu J^\mu$	$\rightarrow$	$\partial J_\mu/\partial x_\mu$
$\sigma^{\mu\nu}$	$\rightarrow$	$\sigma_{\mu\nu}$
$\varepsilon_{\mu\nu\rho\sigma}$	$\rightarrow$	$i\varepsilon_{\mu\nu\rho\sigma}$

[a]Some examples:

$$a_\mu b^\mu = a^\mu g_{\mu\nu} b^\nu \quad \overset{\text{conv}}{\longrightarrow} \quad a_\mu(-\delta_{\mu\nu})b_\nu = -a_\mu b_\mu$$

$$(i\gamma^\mu \partial_\mu - M) = (i\gamma^\mu g_{\mu\nu}\partial^\nu - M) \quad \overset{\text{conv}}{\longrightarrow} \quad -(\gamma_\mu \partial/x_\mu + M)$$

$$\partial_\mu \phi \partial^\mu \phi - m_s^2\phi^2 \quad \overset{\text{conv}}{\longrightarrow} \quad -[(\partial\phi/\partial x_\mu)^2 + m_s^2\phi^2]$$

$$F^{\mu\nu} = \partial^\mu V^\nu - \partial^\nu V^\mu \quad \overset{\text{conv}}{\longrightarrow} \quad -(\partial V_\nu/\partial x_\mu - \partial V_\mu/\partial x_\nu) = -F_{\mu\nu}$$

$$\partial_\nu F^{\nu\mu} = \partial^\lambda g_{\lambda\nu} F^{\nu\mu} \quad \overset{\text{conv}}{\longrightarrow} \quad -\partial F_{\nu\mu}/\partial x_\nu = +\partial F_{\mu\nu}/\partial x_\nu$$

$$\gamma_\mu p^\mu - M = \gamma^\nu g_{\nu\mu} p^\mu - M \quad \overset{\text{conv}}{\longrightarrow} \quad -(i\gamma_\mu p_\mu + M)$$

$$4(a\cdot b) = \text{tr}\ a_\mu \gamma^\mu b_\nu \gamma^\nu \quad \overset{\text{conv}}{\longrightarrow} \quad -\text{tr}\ a_\mu\gamma_\mu b_\nu\gamma_\nu = -4(a\cdot b)$$

$$4i\varepsilon_{\mu\nu\rho\sigma} = \text{tr}\ \gamma_\mu\gamma_\nu\gamma_\rho\gamma_\sigma\gamma_5 \quad \overset{\text{conv}}{\longrightarrow} \quad (-i^4)\text{tr}\ \gamma_\mu\gamma_\nu\gamma_\rho\gamma_\sigma\gamma_5 = -4\varepsilon_{\mu\nu\rho\sigma}$$

[b]The lowering and raising of the Lorentz index on the overall vertex Γ^μ itself is controlled by the $g_{\mu\nu}$ in the propagator; thus $\Gamma_1^\mu g_{\mu\nu}\Gamma_2^\nu \equiv \Gamma_{1\mu} g^{\mu\nu}\Gamma_{2\nu}$.

Table N.3: Electroweak convention conversion table.

Cheng and Li Ref. [W22]	Cummins and Buchsbaum Ref. [W1]	Present text[a]	Comments
W_μ^+	W_μ^+	W_μ	Creates W_μ^- Destroys W_μ^+
W_μ^-	W_μ^-	$W_\mu^\star$	Creates W_μ^+ Destroys W_μ^-
$-Z_\mu$	$-Z_\mu$	Z_μ	
ξ	$1/\xi$	$1/\xi$	
ζ	$1/\eta$	$1/\zeta$	
$e \equiv \lvert e\rvert$	$e \equiv \lvert e\rvert$	$e \equiv \lvert e\rvert$	
ϕ_1	σ	η	Higgs
ϕ_2	χ	χ	Unphysical scalar
$\phi^\pm$	$s^\pm$	$s, s^\star$	Unphysical scalar

[a] In this text $v_\mu^\star \equiv (\vec{v}^\dagger, +iv_0^\dagger)$.

Index

p denotes a problem